Lecture Notes in Networks and Systems

Volume 547

The series "Lecture Notes in Networks and Systems" publishes the latest developments in Networks and Systems—quickly, informally and with high quality. Original research reported in proceedings and post-proceedings represents the core of LNNS.

Volumes published in LNNS embrace all aspects and subfields of, as well as new challenges in, Networks and Systems.

The series contains proceedings and edited volumes in systems and networks, spanning the areas of Cyber-Physical Systems, Autonomous Systems, Sensor Networks, Control Systems, Energy Systems, Automotive Systems, Biological Systems, Vehicular Networking and Connected Vehicles, Aerospace Systems, Automation, Manufacturing, Smart Grids, Nonlinear Systems, Power Systems, Robotics, Social Systems, Economic Systems and other. Of particular value to both the contributors and the readership are the short publication timeframe and the world-wide distribution and exposure which enable both a wide and rapid dissemination of research output.

The series covers the theory, applications, and perspectives on the state of the art and future developments relevant to systems and networks, decision making, control, complex processes and related areas, as embedded in the fields of interdisciplinary and applied sciences, engineering, computer science, physics, economics, social, and life sciences, as well as the paradigms and methodologies behind them.

Indexed by SCOPUS, INSPEC, WTI Frankfurt eG, zbMATH, SCImago.

All books published in the series are submitted for consideration in Web of Science.

For proposals from Asia please contact Aninda Bose (aninda.bose@springer.com).

Manoj Thakur · Samar Agnihotri ·
Bharat Singh Rajpurohit · Millie Pant ·
Kusum Deep · Atulya K. Nagar
Editors

Soft Computing for Problem Solving

Proceedings of the SocProS 2022

Springer

Editors
Manoj Thakur
School of Mathematical and Statistical Sciences
Indian Institute of Technology Mandi
Mandi, Himachal Pradesh, India

Samar Agnihotri
School of Computing and Electrical Engineering
Indian Institute of Technology Mandi
Mandi, Himachal Pradesh, India

Bharat Singh Rajpurohit
School of Computing and Electrical Engineering
Indian Institute of Technology Mandi
Mandi, Himachal Pradesh, India

Millie Pant
Department of Applied Science and Engineering
Indian Institute of Technology Roorkee
Roorkee, Uttarakhand, India

Kusum Deep
Department of Mathematics
Indian Institute of Technology Roorkee
Roorkee, Uttarakhand, India

Atulya K. Nagar
School of Mathematics, Computer Science and Engineering
Liverpool Hope University
Liverpool, UK

ISSN 2367-3370 ISSN 2367-3389 (electronic)
Lecture Notes in Networks and Systems
ISBN 978-981-19-6524-1 ISBN 978-981-19-6525-8 (eBook)
https://doi.org/10.1007/978-981-19-6525-8

This Springer imprint is published by the registered company Springer Nature Singapore Pte Ltd.
The registered company address is: 152 Beach Road, #21-01/04 Gateway East, Singapore 189721, Singapore

Preface

We are delighted that the 11th International Conference on Soft Computing for Problem Solving, SocProS 2022, was hosted by the Indian Institute of Technology Mandi, India, during May 14–15, 2022.

SocProS is a yearly international conference that started in 2011. It is the signature event of the Soft Computing Research Society (SCRS), India. The earlier editions of the conference have been hosted at various prestigious institutions. SocProS aims to bring together the researchers, engineers, and practitioners to present the latest achievements and innovations in the interdisciplinary areas of soft computing, machine learning, and artificial intelligence and to discuss thought-provoking developments and challenges to select potential future directions. The primary objective of the conference is to encourage the participation of young researchers carrying out research in these areas internationally.

The 11th edition of this mega event has touched further heights in terms of quality research papers and fruitful research discussions. The theme of SocProS 2022 was "Unlocking the power of Soft Computing, Machine Learning, and Data Science". The proceeding of the conference consists of a collection of selected high-quality articles that cover recent developments in various topics related to the theme of the conference. Many research articles contribute to real-life applications arising in different domains.

We hope that this edited volume will serve as a comprehensive source of reference for students, researchers, and practitioners interested in the current advancements and applications of soft computing, machine learning, and data science.

We express our heartfelt gratitude to all the authors, reviewers, and Springer personnel for their motivation and patience.

Mandi, India — Manoj Thakur
Mandi, India — Samar Agnihotri
Mandi, India — Bharat Singh Rajpurohit
Roorkee, India — Millie Pant
Roorkee, India — Kusum Deep
Liverpool, UK — Atulya K. Nagar

Contents

Editors and Contributors

About the Editors

Dr. Manoj Thakur is Professor at the School of Mathematical and Statistical Sciences, Indian Institute of Technology Mandi, Mandi, India. His research interests include optimization, machine learning, and computational finance.

Dr. Samar Agnihotri received the M.Sc. (Engg.) and Ph.D. degrees in electrical sciences from IISc Bangalore. From 2010 to 2012, he was Postdoctoral Fellow with the Department of Information Engineering, the Chinese University of Hong Kong. He is currently Associate Professor with the School of Computing and Electrical Engineering, IIT Mandi. His research interests include communication and information theory.

Dr. Bharat Singh Rajpurohit received the M.Tech. degree in power apparatus and electric drives from the Indian Institute of Technology Roorkee, Roorkee, India, in 2005 and the Ph.D. degree in electrical engineering from the Indian Institute of Technology Kanpur, Kanpur, India, in 2010. He is currently Professor with the School of Computing and Electrical Engineering, Indian Institute of Technology Mandi, Mandi, India. His major research interests include electric drives, renewable energy integration, and intelligent and energy-efficient buildings. He is Member of the International Society for Technology in Education, the Institution of Engineers, India, and the Institution of Electronics and Telecommunication Engineers.

Dr. Millie Pant is Professor at the Department of Applied Mathematics and Scientific Computing, Indian Institute of Technology Roorkee (IIT Roorkee) in India. Her areas of interests include numerical optimization, operations research, decision-making techniques, and artificial intelligence.

Dr. Kusum Deep is Professor at the Department of Mathematics, Indian Institute of Technology Roorkee. Her research interests include numerical optimization, nature-inspired optimization, computational intelligence, genetic algorithms, parallel genetic algorithms, and parallel particle swarm optimization.

Prof. Atulya K. Nagar holds Foundation Chair as Professor of Mathematical Sciences and is Pro-Vice-Chancellor (Research) at Liverpool Hope University, UK. He is responsible for developing Sciences and Engineering and has been the Head of the School of Mathematics, Computer Science, and Engineering which he established at the university. He received a prestigious Commonwealth Fellowship for pursuing his doctorate (D.Phil.) in Applied Nonlinear Mathematics, which he earned from the University of York (UK) in 1996. He holds B.Sc. (Hons), M.Sc., and M.Phil. (with distinction) in Mathematical Physics from the MDS University of Ajmer, India. Prior to joining Liverpool Hope, he was with the Brunel University, London. He is an internationally respected scholar working at the cutting edge of nonlinear mathematics, theoretical computer science, and systems engineering. He has edited volumes on intelligent systems and applied mathematics. He is well published with over 450 publications in prestigious publishing outlets. He has an extensive background and experience of working in universities in the UK and India. He has been an expert reviewer for the Biotechnology and Biological Sciences Research Council (BBSRC) grants peer-review committees for Bioinformatics Panel; Engineering and Physical Sciences Research Council (EPSRC) for High Performance Computing Panel; and served on the Peer-Review College of the Arts and Humanities Research Council (AHRC) as Scientific Expert member. Prof. Nagar sits on the JISC Research Strategy group, and he is Fellow of the Institute of Mathematics and its applications (FIMA) and Fellow of the Higher Education Academy (FHEA).

Contributors

Abhigna Raptadu School of Electrical Engineering, Vellore Institute of Technology, Vellore, India

Abubakar Mustapha Yusuf Kano State Polytechnic, Kano, Nigeria

Agarwal Purab Department of Computer Science and Engineering, SRM University-AP, Amaravati, India

Agarwal Shivi Department of Mathematics, Birla Institute of Technology and Science Pilani, Pilani, India

Arora Rishu Department of Mathematics and Humanities, MM Engineering College, Maharishi Markandeshwar (Deemed to be University), Mullana, Ambala, Haryana, India

Assad Assif Islamic University of Science and Technology, Awantipora, J&K, India

Bagde Ashutosh D. Department of R & D, Jawaharlal Nehru Medical College, Datta Meghe Institute of Medical Sciences, Wardha, Maharashtra, India

Barbhuiya Tariq Arshad Machine Intelligence and Bio-motion Lab, Department of Computer Science and Engineering, National Institute of Technology, Rourkela, India

Barukula Snehitha Department of Computer Science and Engineering, SRM University-AP, Amaravati, India

Basha C. H. Hussaian NITTE Meenakshi Institute of Technology, Bangalore, Karnataka, India

Bhambu Aryan Department of Mathematics, Indian Institute of Technology Guwahati, Assam, India

Bhanu Prakash N. School of Maritime Management, Indian Maritime University, Visakhapatnam, India

Bharati Ambuj Department of Computer Application, National Institute of Technology Kurukshetra, Kurukshetra, Haryana, India

Bhowmick Bhaskar IIT Kharagpur, Kharagpur, West Bengal, India

Chakraborty Anikash Delhi Technological University, New Delhi, India

Chakraborty Saikat School of Computer Engineering, Kalinga Institute of Industrial Technology (KIIT), Bhubaneswar, Odisha, India;
Machine Intelligence and Bio-motion Research Lab, Department of Computer Science and Engineering, National Institute of Technology Rourkela, Rourkela, Odisha, India

Chiroma Haruna University of Hafr Al-Batin, Hafr Al-Batin, Saudi Arabia

Choubey Dilip Kumar Department of Computer Science and Engineering, Indian Institute of Information Technology Bhagalpur, Bhagalpur, Bihar, India

Choudhary Himanshu Indian Institute of Technology-Mandi, Mandi, Himachal Pradesh, India

Choudhuri Rudrajit St. Thomas College of Engineering and Technology, Kolkata, India

Dasari K. V. School of Computer and Information Sciences, University of Hyderabad, Hyderabad, Telangana, India

Deep Kusum Indian Institute of Technology Roorkee, Roorkee, Uttrakhand, India

Deshpande Varuun A. Indian Institute of Technology, Mandi, HP, India

Dhal Krishna Gopal Department of Computer Science and Application, Midnapore College (Autonomous), West Bengal, India

Dhankhar Chirag Department of Mathematics, Amity School of Applied Sciences, Amity University Haryana, Gurugram, Haryana, India

Dhingra Madhavi Amity University Madhya Pradesh, Gwalior, MP, India

Ejim Samson Abubakar Tafawa Balewa University, Bauchi, Nigeria

Garg Rahul Dev Geomatics Engineering Group, Department of Civil Engineering, Indian Institute of Technology Roorkee, Roorkee, Uttarakhand, India

Gital Abdulsalam Ya'u Abubakar Tafawa Balewa University, Bauchi, Nigeria

Goyal Rajeev Department of CSE, Vellore Institute of Technology, Bhopal, India

Goyal Samta Jain Department of CSE, Amity University Madhya Pradesh, Gwalior, India

Goyal Swati Department of Mathematics, BITS Pilani, Pilani Campus, India

Gunjan Amartya School of Electronics Engineering, Vellore Institute of Technology, Vellore, India

Gupta Dhruv IIT (BHU), Varanasi, India

Gupta Sandeep Dr. K. N. Modi University, Rajasthan, India

Haidari Moazzam Saharsa College of Engineering, Saharsa, India

Halder Amiya St. Thomas College of Engineering and Technology, Kolkata, India

Henry Alan Department of Computer Science and Engineering, Amrita School of Engineering, Amrita Vishwa Vidyapeetham, Amrita University, Coimbatore, India

Iswarya School of Advanced Sciences, Vellore Institute of Technology, Vellore, Tamil Nadu, India

Jadon Rakesh Singh Department of Computer Applications, MITS, Gwalior, India

Jain S. C. Amity University Madhya Pradesh, Gwalior, MP, India

Jena Pravat Kumar University of Petroleum and Energy Studies, Dehradun, India

Kamma Samhitha Department of Computer Science and Engineering, SRM University-AP, Amaravati, India

Kavya Kudumuu School of Electrical Engineering, Vellore Institute of Technology, Vellore, India

Khandelwal Hitesh Vidyashilp University, Bangalore, India

Khodabux Mohammad Kaleem Faculty of Sustainable Development and Engineering, Université Des Mascareignes, Beau Bassin-Rose Hill, Mauritius

Kubi Ganaka Musa The Federal Polytechnic Nasarawa, Nasarawa, Nigeria

Kumar Atul Department of Applied Mathematics and Scientific Computing, Indian Institute of Technology Roorkee, Roorkee, India

Kumar Hement Indian Institute of Technology Mandi, Himachal Pradesh, India

Kumar Indubhushan Saharsa College of Engineering, Saharsa, India

Kumar Jitendra School of Advanced Sciences, Vellore Institute of Technology, Vellore, Tamil Nadu, India

Kumar Kamal Department of Mathematics, Amity School of Applied Sciences, Amity University Haryana, Gurugram, Haryana, India

Kumar Komal Indian Institute of Technology Mandi, Himachal Pradesh, India

Kumar Mohit Department of IT, Dr. B.R. Ambedkar National Institute of Technology, Jalandhar, India

Kumar Rajeev Data to Knowledge (D2K) Lab, School of Computer and Systems Sciences, Jawaharlal Nehru University, New Delhi, India

Kumar Sandeep Department of Mathematics, Ch. Charan Singh University, Uttar Pradesh, India

Kumar Sanjay Delhi Technological University, New Delhi, India

Kumar Subham Birla Institute of Technology and Science, Pilani, India

Kumar Sumit Department of Applied Mathematics and Scientific Computing, Indian Institute of Technology Roorkee, Roorkee, Uttarakhand, India

Kumar Sushil Department of Computer Science and Engineering, National Institute of Technology Warangal, Telangana, India

Kumar Tarun Indian Institute of Technology Mandi, Mandi, Himachal Pradesh, India

Kumbhar Abhishek SCRC, Nanasaheb Mahadik College of Engineering, Walwa, India

Kundalakkaadan Junaciya Data to Knowledge (D2K) Lab, School of Computer and Systems Sciences, Jawaharlal Nehru University, New Delhi, India

Lawal Mustapha Abdulrahman Abubakar Tafawa Balewa University, Bauchi, Nigeria

Mahapatra Monalisha Machine Intelligence and Bio-motion Lab, Department of Computer Science and Engineering, National Institute of Technology, Rourkela, India

Mahla Deepak Department of Mathematics, Birla Institute of Technology and Science Pilani, Pilani, India

Mathur Trilok Department of Mathematics, Birla Institute of Technology and Science Pilani, Pilani, India

Mittal Raman Department of Computer Application, National Institute of Technology Kurukshetra, Kurukshetra, Haryana, India

Mogha Sandeep Kumar Department of Mathematics, Chandigarh University, Mohali, Punjab, India

Mohapatra Subrajeet Department of Computer Science Engineering, Birla Institute of Technology Mesra, Ranchi, Jharkhand, India

Mukherjee Sourish Department of Computer Science and Engineering, SRM University-AP, Amaravati, India

Mukherjee Trishita Data to Knowledge (D2K) Lab, School of Computer & Systems Sciences, Jawaharlal Nehru University, New Delhi, India

Nadaf S. M. SCRC, Nanasaheb Mahadik College of Engineering, Walwa, India

Nagar Atulya K. Liverpool Hope University, Liverpool, UK

Naidu Rani Chinnappa Faculty of Sustainable Development and Engineering, Université Des Mascareignes, Beau Bassin-Rose Hill, Mauritius

Nandy Anup Machine Intelligence and Bio-motion Research Lab, Department of Computer Science and Engineering, National Institute of Technology Rourkela, Rourkela, Odisha, India

Narule M. SCRC, Nanasaheb Mahadik College of Engineering, Walwa, India

Naskar Prabir Kumar Department of Computer Science and Engineering, Government College of Engineering and Textile Technology, Serampore, West Bengal, India

Nehra Tarun Rajasthan Institute of Engineering and Technology, Jaipur, India

Orra Arishi Indian Institute of Technology-Mandi, Mandi, Himachal Pradesh, India

Panchal Ashay Department of Information and Communication Technology, Pandit Deendayal Energy University (PDEU), Gandhinagar, India

Panda Prashansa Machine Intelligence and Bio-motion Research Lab, Department of Computer Science and Engineering, National Institute of Technology Rourkela, Rourkela, Odisha, India

Pannala R. K. Pavan Kumar Department of Mathematics, Sharda University, Greater Noida, India

Pant Millie Department of Applied Mathematics and Scientific Computing, Indian Institute of Technology Roorkee, Roorkee, Uttarakhand, India;

Mehta Family School of Data Science and Artificial Intelligence, Indian Institute of Technology Roorkee, Roorkee, Uttarakhand, India

Panwar Ankita Department of Applied Mathematics and Scientific Computing, Indian Institute of Technology Roorkee, Roorkee, India

Parouha Raghav Prasad Indira Gandhi National Tribal University, Amarkantak, Madhya Pradesh, India

Pasayat Ajit Kumar IIT Kharagpur, Kharagpur, West Bengal, India

Pathade Aniket Department of R & D, Jawaharlal Nehru Medical College, Datta Meghe Institute of Medical Sciences, Wardha, Maharashtra, India

Patil Nikita SCRC, Nanasaheb Mahadik College of Engineering, Walwa, India

Prasad Shitala Institute for Infocomm Research, A*Star, Singapore, Singapore

Preeti Indian Institute of Technology Roorkee, Roorkee, Uttrakhand, India

Prince Indian Institute of Technology Roorkee, Roorkee, Uttrakhand, India

Priya ITER College, SOA University, Bhubaneshwar, Odisha, India

Rajesh Kumar M. Faculty of Sustainable Development and Engineering, Université Des Mascareignes, Beau Bassin-Rose Hill, Mauritius

Rajesh Chilukamari Department of Computer Science and Engineering, National Institute of Technology Warangal, Telangana, India

Rajput Jahanvi Institute of Technical Education and Research, Siksha 'O' Anusandhan, Bhubaneswar, Odisha, India

Raman Sundaresan Birla Institute of Technology and Science, Pilani, India

Rani Anjana Banasthali Vidyapith, Tonk, India

Rathod Asmita Ajay School of Electrical Engineering, Vellore Institute of Technology, Vellore, India

Rawat Akhilesh Data to Knowledge (D2K) Lab, School of Computer and Systems Sciences, Jawaharlal Nehru University, New Delhi, India

Ray Swarnajit Department of Computer Science and Engineering, Maulana Abul Kalam Azad University of Technology, Nadia, West Bengal, India

Sahoo Debashis Indian Institute of Technology Mandi, Mandi, Himachal Pradesh, India

Sahoo Kartik Indian Institute of Technology Mandi, Mandi, Himachal Pradesh, India

Salkuti Surender Reddy Department of Railroad and Electrical Engineering, Woosong University, Daejeon, South Korea

Sambhavi Sruti Machine Intelligence and Bio-motion Research Lab, Department of Computer Science and Engineering, National Institute of Technology Rourkela, Rourkela, Odisha, India

Satapathy Santosh Kumar Department of Information and Communication Technology, Pandit Deendayal Energy University (PDEU), Gandhinagar, India

Satya Shekar Varma P. Department of Computer Science and Engineering, National Institute of Technology Warangal, Telangana, India

Saxena Deepika Department of Computer Application, National Institute of Technology Kurukshetra, Kurukshetra, Haryana, India

Saxena Monika Banasthali Vidyapith, Tonk, India

Shafi Sadaf Islamic University of Science and Technology, Awantipora, J&K, India

Shah Bhavya Department of Information and Communication Technology, Pandit Deendayal Energy University (PDEU), Gandhinagar, India

Shah Het Indian Institute of Technology, Roorkee, UK, India

Shah Khelan Department of Information and Communication Technology, Pandit Deendayal Energy University (PDEU), Gandhinagar, India

Shah Shrey Department of Information and Communication Technology, Pandit Deendayal Energy University (PDEU), Gandhinagar, India

Sharma Akshat School of Electrical Engineering, Vellore Institute of Technology, Vellore, India

Sharma Anish Data to Knowledge (D2K) Lab School of Computer and Systems Sciences, Jawaharlal Nehru University, New Delhi, India

Sharma Hemlata Sheffield Hallam University, Sheffield, England

Sharma Kumuda ITER College, SOA University, Bhubaneshwar, Odisha, India

Sharma Manika Department of Computer Application, National Institute of Technology Kurukshetra, Kurukshetra, Haryana, India

Sharma Neha Department of Mathematics, Chandigarh University, Mohali, Punjab, India

Sharma Pankaj School of Electrical Engineering, Vellore Institute of Technology, Vellore, India

Shingare Haresh Department of CSE, Dr. B.R. Ambedkar National Institute of Technology, Jalandhar, India

Shukla K. K. IIT (BHU), Varanasi, India

Singh A. School of Computer and Information Sciences, University of Hyderabad, Hyderabad, Telangana, India

Singh Aman Department of Computer Science Engineering, Birla Institute of Technology Mesra, Ranchi, Jharkhand, India

Singh Ashutosh Kumar Department of Computer Application, National Institute of Technology Kurukshetra, Kurukshetra, Haryana, India

Singh Pankaj Pratap Department of Computer Science and Engineering, Central Institute of Technology Kokrajhar, Kokrajhar, Assam, India

Singh Pooja Data to Knowledge (D2K) Lab School of Computer and Systems Sciences Jawaharlal Nehru University, New Delhi, India

Singh Priyanka Department of Computer Science and Engineering, SRM University-AP, Amaravati, India

Singh Satnam Indian Institute of Technology Mandi, Mandi, Himachal Pradesh, India

Singh Shivendra Department of R & D, Jawaharlal Nehru Medical College, Datta Meghe Institute of Medical Sciences, Wardha, Maharashtra, India

Sreekeessoon Bhamini Faculty of Sustainable Development and Engineering, Université Des Mascareignes, Beau Bassin-Rose Hill, Mauritius

Srivastava Shilpa CHRIST (Deemed to be University), NCR, Delhi, India

Suhakar Bait Yash Department of Computer Science and Engineering, SRM University-AP, Amaravati, India

Sujee R. Department of Computer Science and Engineering, Amrita School of Engineering, Amrita Vishwa Vidyapeetham, Amrita University, Coimbatore, India

Sybol Sumin Samuel CHRIST (Deemed to be University), NCR, Delhi, India

Talwar Manpreet Singh Department of Mathematics, BITS Pilani, Pilani Campus, India

Telrandhe Shital Department of R & D, Jawaharlal Nehru Medical College, Datta Meghe Institute of Medical Sciences, Wardha, Maharashtra, India

Terhuja Khriesavinyu Indian Institute of Technology Mandi, Mandi, Himachal Pradesh, India

Tyagi Reshu Department of Mathematics, Ch. Charan Singh University, Uttar Pradesh, India

Umate Roshan Department of R & D, Jawaharlal Nehru Medical College, Datta Meghe Institute of Medical Sciences, Wardha, Maharashtra, India

Valadi Jayaraman Vidyashilp University, Bangalore, India

Venkata Manish Reddy G. School of Advanced Sciences, Vellore Institute of Technology, Vellore, Tamil Nadu, India

Wadhwa Pratishtha Indian Institute of Technology Mandi, Himachal Pradesh, India

Wanjari Mayur Department of R & D, Jawaharlal Nehru Medical College, Datta Meghe Institute of Medical Sciences, Wardha, Maharashtra, India

Yadav A. K. Department of Mathematics, Amity School of Applied Sciences, Amity University Haryana, Gurugram, Haryana, India

Yadav Arti Department of CSE, Amity University Madhya Pradesh, Gwalior, India

Benchmarking State-of-the-Art Methodologies for Optic Disc Segmentation

Subham Kumar and Sundaresan Raman

Abstract Glaucoma and Diabetic Retinopathy are widely prevalent diseased eye conditions which gradually lead to blindness. Early and timely diagnosis requires the help of expert ophthalmologists which is not available everywhere. As a result, many attempts have been made to come up with fully automated and intelligent systems to address this issue. A very important component of this task is detecting the optic disc. This work aims to establish a clear and concise picture of the present state-of-the-art models on this problem and benchmark their robustness and versatility to adapt to a variety of scans and images. This paper aims to deploy and review various deep learning architectures on a uniform test bed to establish the best models in optic disc detection. GAN approaches (pOSAL and CFEA) give the best performance, giving Dice Coefficient of 0.96 and 0.94. This is followed by specialized CNN architectures such as U-Net, M-Net and P-Net, giving Dice Coefficients between 0.93 and 0.86.

Keywords Optic disc · Generative adversarial networks · Deep learning

1 Introduction

Glaucoma is one of the most widespread eye diseases in the world [17]. It affects the optic nerve, gradually causing blindness. Diabetic Retinopathy is another disease which affects the blood vessels in the eye, particularly near the retina. It is another leading cause of blindness in the world. However, these diseases can be prevented if appropriate steps are taken early, for which early detection and diagnosis is paramount [2]. However, this requires expert medical knowledge, and specialist ophthalmologists are not available for consultation in many parts of the world. This warrants the need of an intelligent, end-to-end CAD system which can help detect and diagnose these diseases in a timely manner.

S. Kumar (✉) · S. Raman
Birla Institute of Technology and Science, Pilani, India
e-mail: h20180123@pilani.bits-pilani.ac.in

S. Raman
e-mail: sundaresan.raman@pilani.bits-pilani.ac.in

M. Thakur et al. (eds.), *Soft Computing for Problem Solving*, Lecture Notes in Networks and Systems 547, https://doi.org/10.1007/978-981-19-6525-8_1

A key challenge in these domain-specific tasks is detecting the optic disc. Traditional approaches to this problem have utilized handcrafted features and operators to try and isolate this region. However, such solutions are not scalable and robust to be used in modern-day CAD systems. Edge detection algorithms such as the Canny edge detector have also been used, with mixed results. Some machine learning models have also been attempted, such as fuzzy c-means segmentation [8]. However, when it comes to the medical domain, these models performed below par of the accuracies which are expected.

Deep learning is one of the most exciting and fastest growing paradigms under the umbrella of Artificial Intelligence. Feature engineering is not explicitly carried out, and the model itself learns to extract features and create increasingly complex representations of the input data to provide intelligent insights [9]. Convolutional Neural Networks are a class of neural networks which revolutionized the task of computer vision, enabling computers to 'see' and 'understand' images. Since then, a plethora of architectures have been designed, many exclusively for the domain of medical imaging and optic disc segmentation such as U-Net [15]. Generative Adversarial Networks [5] are a new class of deep learning models which use unsupervised learning to improve the performance of the classic deep learning approaches such as CNNs. Section 2 contains details of the dataset, Sect. 3 details the methodology used, Sect. 4 details the results, and the final section brings up the conclusion.

The key contributions are as follows:

- Many approaches have tested and reported their results on a subset of the training data, which may not be indicative enough of the robustness of the method. Hence, all the state-of-the-art models have been trained and evaluated on two completely different sets of images, to rigorously evaluate their adaptability to a variety of scans and images.
- This paper aims to provide a comprehensive comparison across different algorithms which have been devised for this task and outline possible avenues for further research.

2 Dataset

The dataset used to measure the effectiveness of the models is the REFUGE dataset [14]. They released a large database of 1200 CFP (Colour Fundus Photography) with reliable and standardized annotations for segmenting the optic disc and other tasks.

The REFUGE dataset is composed of images taken from two different cameras, the Zeus Visucam 500 and the Canon CR-2 as shown in Fig. 1. Typically, the reporting of accuracies of deep learning models is on images which have a similar nature to the images on which the deep networks have been trained. However, this is not a very good indicator of robustness. The starkly different nature of the images in the dataset is shown in Fig. 1.

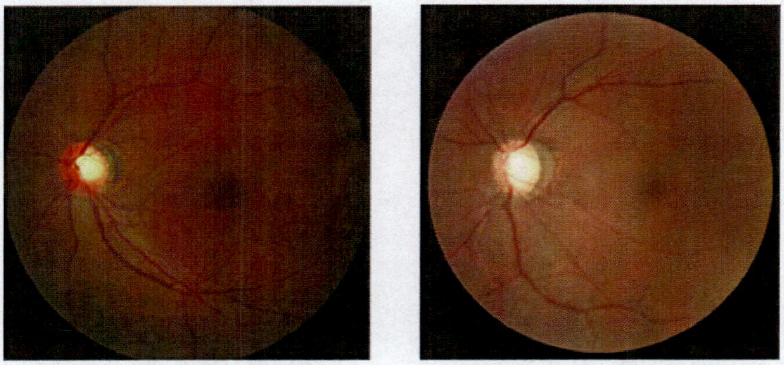

(a) Training Image (Zeiss Visucam 500) (b) Test Image (Canon CR-2)

Fig. 1 The REFUGE dataset. **a** Shows the nature of training images taken using the Zeus Visucam 500 camera, while **b** Shows the test dataset images, comprising images taken from the Canon CR-2 camera

There are 400 images of the Zeiss Visucam 500 type, which are provided as the training set. 400 images of the Canon CR-2 dataset are provided as the validation dataset and the test dataset each. The validation dataset does not contain the ground truths; it is up to the algorithm to make the best use of such images, typically in an unsupervised fashion. This dataset was used to train and evaluate all the networks.

3 Methodology

3.1 Deep Learning Techniques (CNN-Based)

These techniques involve the use of deep learning architectures for detecting the optic disc. Because of their automatic feature extraction, deep learning networks automatically discover the representations needed for detection, classification and segmentation tasks, reducing the need for supervision and speeding up the process of extracting tangible insights from datasets that have not been as extensively curated as normally required.

U-Net U-Net [15] is a CNN designed primarily for the medical domain. It uses fully convolutional architectures along with specialized loss functions, contracting/expanding paths and skip connections as shown in Fig. 2 to provide better performances compared to standard CNN architectures. The increasing receptive field is combined with the features corresponding to that layer in the expanding path so as to combine low-level and high-level information effectively. The skip connections running across the network allow it to propagate contextual information to the upper

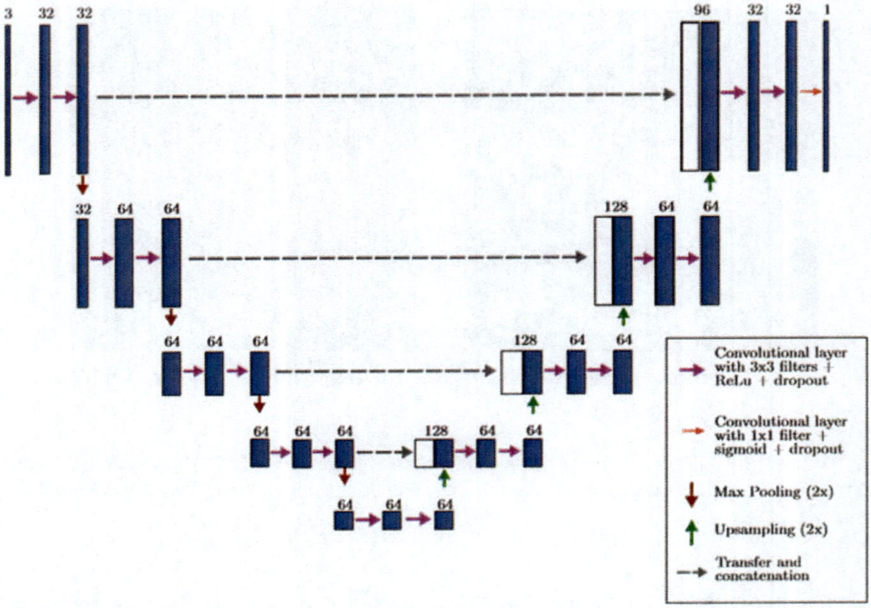

Fig. 2 Architecture of the fully convolutional U-Net [15]

layers. Images of arbitrarily large sizes can be segmented without a hitch using an 'overlap-tile' strategy.

M-Net M-Net [4] is a deep network based on U-Net. It is an end-to-end multi-label CNN, consisting of three components. The first is a multi-scale layer used to construct an image pyramid input and achieve multi-level receptive field fusion. The second is a U-Net architecture, which is the backbone network for the purpose of learning an effective representation. The third component is a side-output layer to produce output images at each of the levels.

The architecture of the network is shown in Fig. 3. The input and output are the polar transformed images. The multi-scale input improves the segmentation quality [10]. M-Net uses average pooling layers for downsampling. This helps in the integration of multi-scale inputs into the layers present in the contracting path. This results in a lesser growth of parameters and manageable network width.

Position-Encoded CNNs and P-Net Position-encoded CNNs [1] are neural network architectures which are based on DenseNet [6] and ResNet. A 57-layered densely connected semantic segmentation CNN is used as shown in Fig. 4. The network comprises of long and short skip connections so as to effectively reuse the features learned by the network. The architecture is an ensemble of 2 different deep networks, which differ in the number of channels provided as the input. P-Net (Prior Network) [13] is a CNN which is also based on DenseNet. They parameterized an efficient flow of information by connecting the first layer to all subsequent layers and passing

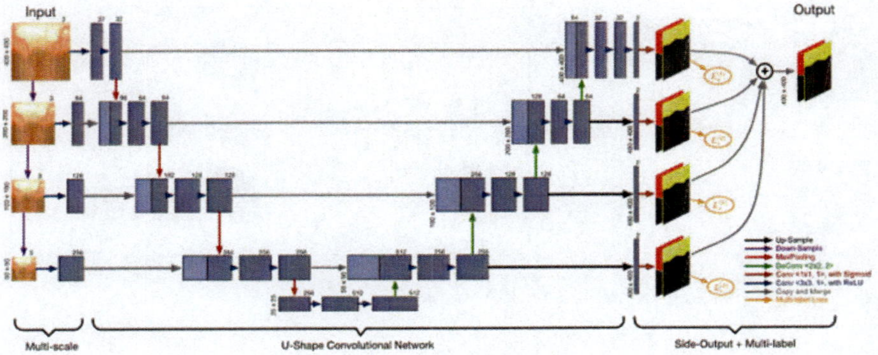

Fig. 3 The M-Net architecture showing the multi-scale inputs, side outputs and multi-labels in addition to the main architecture [4]

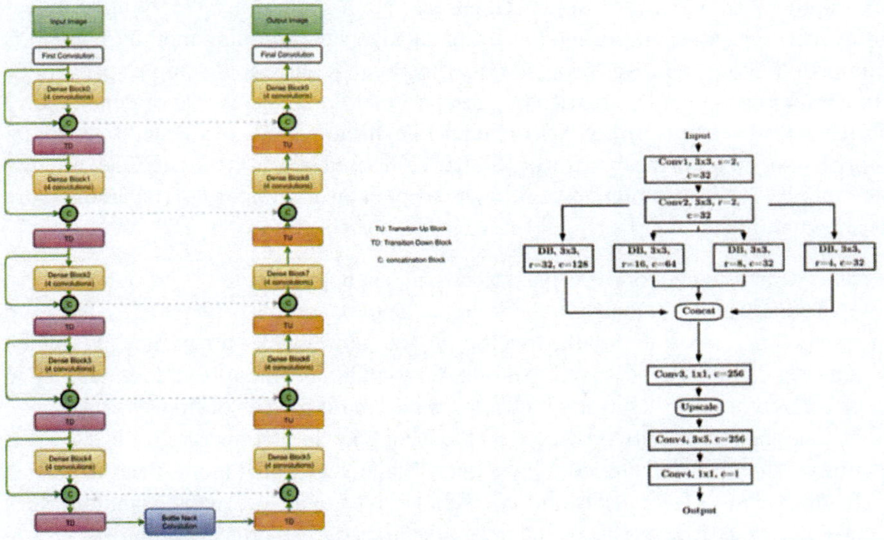

(a) Position encoded CNN (b) P-Net

Fig. 4 Position-encoded CNN and P-Net

the concatenated feature maps which leads to an increased variance, which allows narrower networks with a lesser number of filters to achieve a performance similar to other wider and bigger networks.

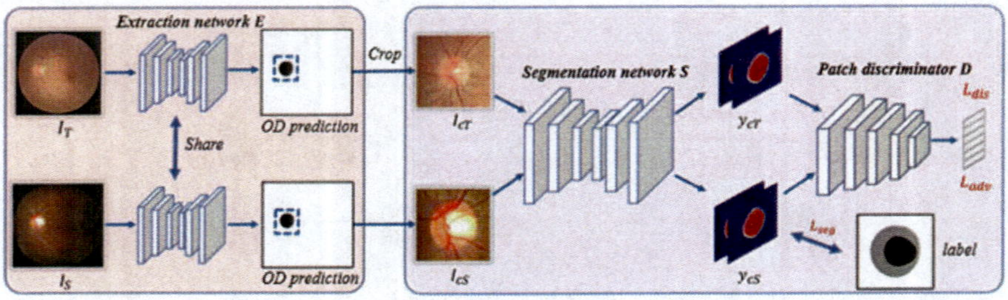

Fig. 5 End-to-end architecture of the pOSAL network [18]

3.2 Adversarial Deep Learning Techniques

Generative Adversarial Networks (GANs) [5] are a class of deep learning architectures which train generative models using an adversarial training methodology. It is comprised of a generative model G that tries to model the input sample space and a discriminative model D that tries to guess whether the image is real or generated.
Patch-based Output Space Adversarial Learning pOSAL is a generative framework for segmenting the optic disc [18]. It is composed of 3 modules, a ROI extraction network E, a segmentation network S and a patch discriminator D. The architecture is shown in Fig. 5.

- The extraction network E is used to provide an approximate segmentation of the optic disc and crop the ROI accordingly. The extraction network is U-Net, and the final layer is a sigmoid activation. It generalizes well on new images (which are of a different type, taken from the Canon CR-2 camera) due to its ability to effectively model the invariant characteristics in both types of images.
- The segmentation network based on the DeepLabv3+ architecture [3] is shown in Fig. 6. The first convolutional layer and MobileNetV2 [16] are utilized to extract features. This is followed by the concatenation of the feature maps. Semantic clues are aggregated across different levels through the generated combined feature maps.
- A patch discriminator D is attached to gauge the outputs of the segmentation network S, and then, adversarial learning is employed to train the entire framework. The segmentation architecture S generates similar images as expected in the source or target domains, while the discriminator tries to differentiate between the generated images and the images coming from the original data distribution.
- The discriminator network based on PatchGAN [19] is used to conduct adversarial training. The joint segmentation loss function is given as follows:

$$L_{\text{seg}} = \lambda_1 L_{\text{DL}}\left(p^d, y^d\right) + \lambda_2 L_{SL}\left(p^d, y^d\right) \quad (1)$$

λ_1 and λ_2 are empirically set weights. p^d and y^d are the predicted probability maps and binary ground truth masks of the optic disc. The loss function contains

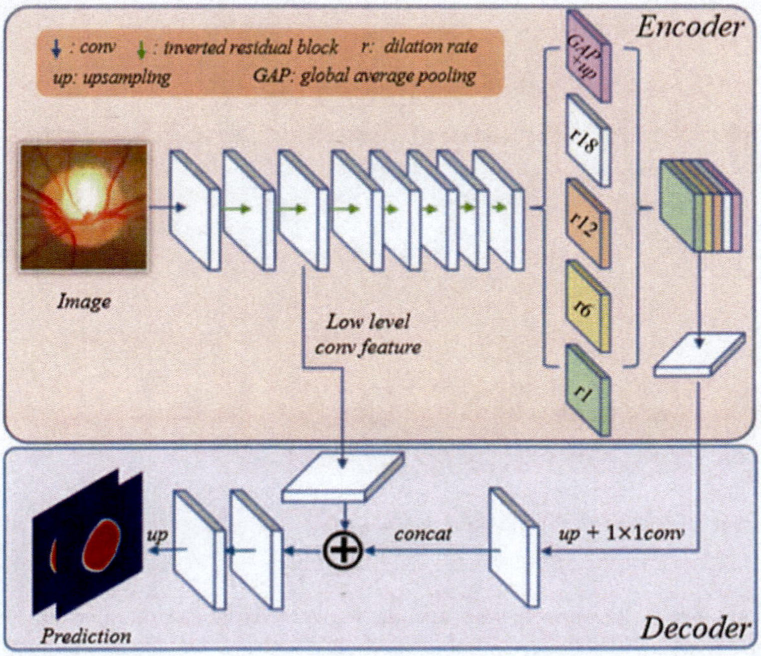

Fig. 6 The pOSAL segmentation network, based on the DeepLabv3+ architecture [3]

two more terms for the optic cup, but the values of their weights are set to very small values, as this is not our main objective. L_{DL} is the Dice Coefficient loss, and L_{SL} is the smoothness loss. The smoothness loss provides the network an incentive to produce homogeneous predictions within neighbouring regions.

Collaborative Feature Ensembling Adaptation Collaborative Feature Ensembling Adaptation (CFEA) [11] is an unsupervised domain adaptation framework, which uses adversarial learning and self-ensembling of weights. Multiple adversarial loss functions in the encoder and decoder components help in the modelling of domain-invariant features. The architecture is showed in Fig. 7. Some key characteristics of the network are as follows:

- The framework is composed of three networks, the Source Network (SN), the Target Student Network (TSN) and the Target Teacher Network (TTN). The Source Network focuses on supervised learning from the labelled samples of the Zeus camera type, while the Target-domain Student Network model is used for the unsupervised adversarial learning. The TTN architecture works on unlabeled target images. Different data augmentation techniques are applied to address the Vanishing Gradient problem.

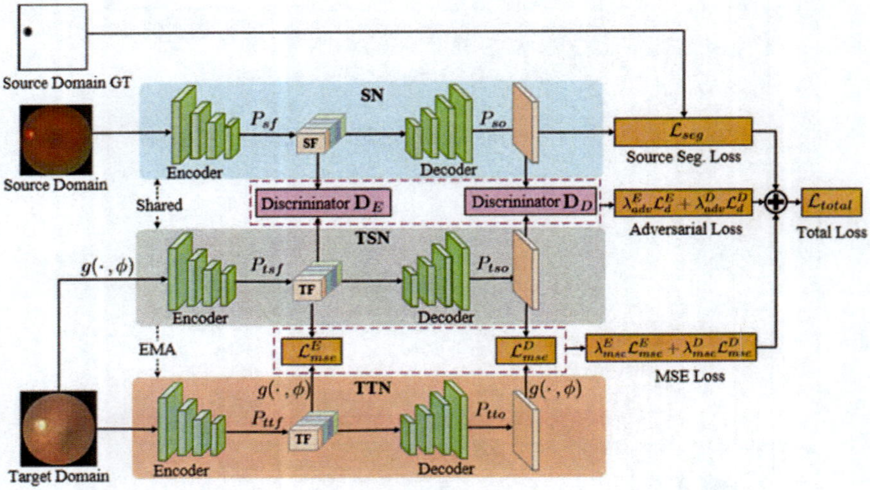

Fig. 7 Complete architecture of the CFEA model [11]

- U-Net is used as the base encoder-decoder network. Two discriminators are applied to the encoder and decoder components each. There are two adversarial loss functions calculated between the Source Network and the Target Student Network.
- The loss function is given as follows:

$$\mathcal{L}_{\text{total}}(X_s, X_t) = \mathcal{L}_{\text{seg}}(X_s) + \lambda_{\text{adv}}^{E}\mathcal{L}_{d}^{E}(X_s, X_t) + \lambda_{\text{adv}}^{D}\mathcal{L}_{d}^{D}(X_s, X_t) + \lambda_{\text{mse}}^{E}\mathcal{L}_{\text{mse}}^{E}(X_t) + \lambda_{\text{mse}}^{D}\mathcal{L}_{\text{mse}}^{D}(X_t) \tag{2}$$

where $\lambda_{\text{adv}}^{E}, \lambda_{\text{adv}}^{D}, \lambda_{\text{mse}}^{E}, \lambda_{\text{mse}}^{D}$ are regularization parameters. $L_{\text{seg}}(X_s)$ is the Dice segmentation loss. $\mathcal{L}_{d}^{E}$ and $\mathcal{L}_{d}^{D}$ are the discriminator losses for the encoder and decoder. $\mathcal{L}_{\text{mse}}^{E}$ and $\mathcal{L}_{\text{mse}}^{D}$ are the MSE losses between the encoders and decoders of TSN and TTN.

4 Results

The performance metric used to evaluate the models was the Dice Coefficient [7]. Its expression is given as follows:

$$\text{Dice\_Coefficient}(p, y) = \frac{2\sum_{i\in\Omega} p_i \cdot y_i}{\sum_{i\in\Omega} p_i^2 + \sum_{i\in\Omega} y_i^2} \tag{3}$$

where p is the predicted probability map and y the ground truth mask, respectively. Ω is all the pixels in the image.

Table 1 Results of the deep learning approaches compared to image processing techniques

Method	Dice coefficient	Jaccard coefficient
Bilateral median	0.59	0.53
Morphological operators	0.64	0.60
Fuzzy c-means	0.73	0.71
Canny edge detector	0.77	0.72
Sobel edge detector	0.62	0.61
Kirsch operator	0.81	0.78
U-Net	0.91	0.89
M-Net	0.93	0.92
Position-encoded CNN	0.88	0.85
P-Net	0.86	0.82
CFEA	0.94	0.93
pOSAL	0.96	0.94

Higher dice and Jaccard coefficient is better

All of the deep learning models were trained on a RTX 2080 Ti GPU. The models were implemented using PyTorch and Keras. Table 1 summarizes all the results. They clearly illustrate the superior performance of deep learning models, because of their implicit feature modelling.

The baseline U-Net architecture performed well, giving a Dice Coefficient of 0.91 even though it was unable to take advantage of the validation data of REFUGE. This shows the versatility of the model in various medical segmentation challenges. This is why it is used as the base architecture for most of the deep learning models. P-Net and position-encoded CNNs gave relatively modest performances, with Dice Coefficients of 0.86 and 0.88, respectively. These networks are designed for natural images. M-Net gives the best performance among Convolutional Neural Networks (CNNs) with a Dice Coefficient of 0.93. The use of multi-labels and polar transformations improves the performance over U-Net. A sample result is shown in Fig. 8f. The red disc is the optic disc.

Figure 8 shows the results for the top deep learning approaches, and this is contrasted with the best image processing techniques. CFEA and pOSAL give the best performance, taking full advantage of the unsupervised data of REFUGE's validation data. CFEA reported a Dice Coefficient of 0.94, while pOSAL reported a Dice Coefficient of 0.96.

CFEA uses the lesser-known concept of self-ensembling to great effect along with adversarial learning. The performance of the network without the adversarial part achieved a Dice Coefficient of 0.85, illustrating the advantage of incorporating the unsupervised approach. The result is shown in Fig. 8h. pOSAL is the present state of the art for optic disc segmentation. The use of multiple highly specialized architectures for extracting features, then segmentation with a carefully chosen loss function and patch-based adversarial learning all contribute to a state-of-the-art performance.

Fig. 8 Results of bilateral median, Kirsch operator, M-Net, CFEA and pOSAL

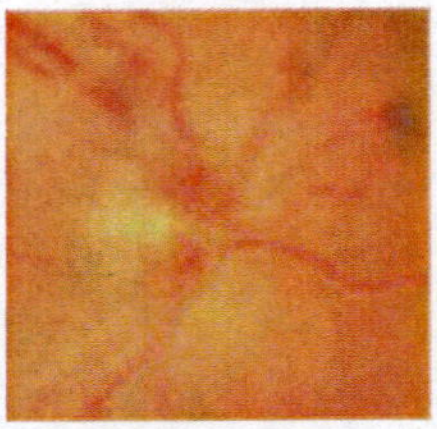

(a) B-median OD original

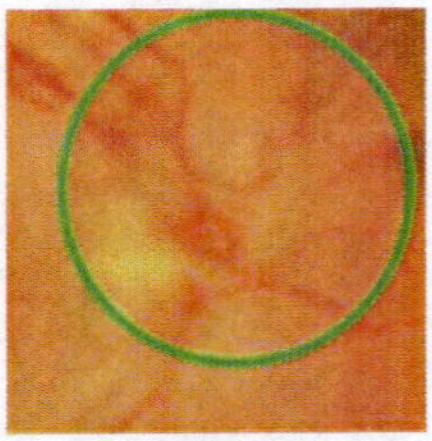

(b) B-median OD result

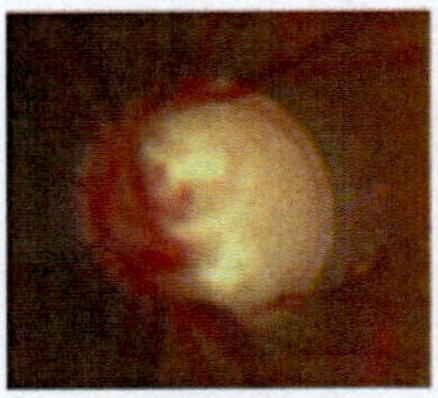

(c) Kirsch operator OD original

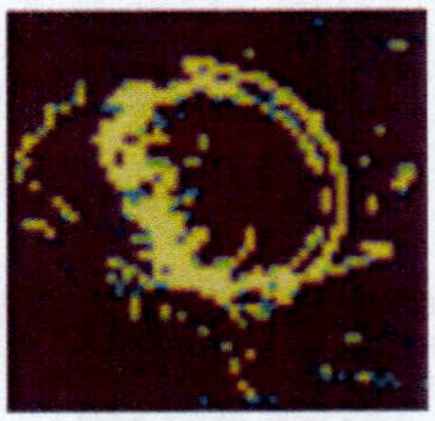

(d) Kirsch operator OD result

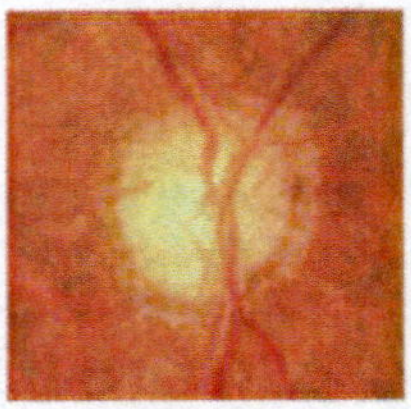

(e) M-Net OD original

(f) M-Net OD result

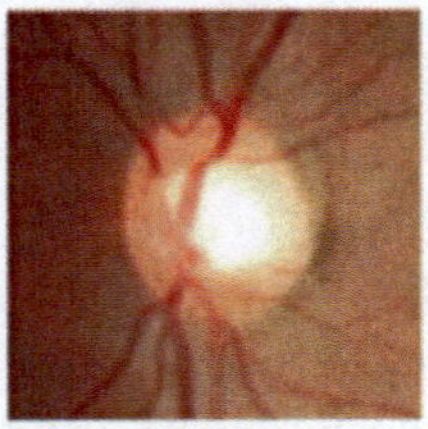

(g) CFEA OD original

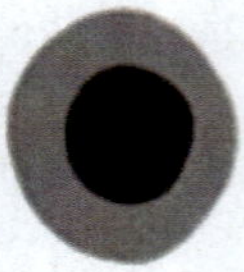

(h) CFEA OD result

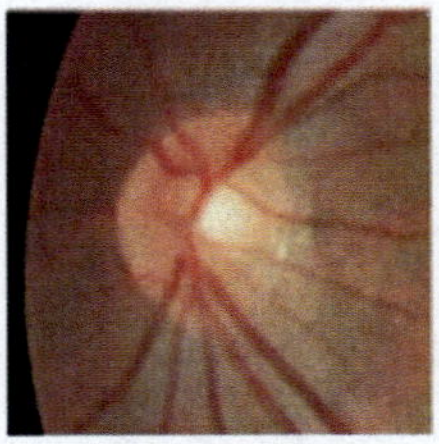

(i) pOSAL OD original

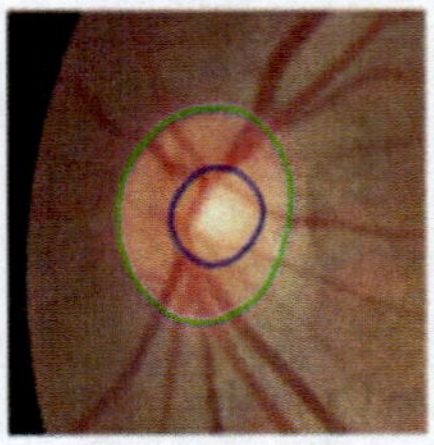

(j) pOSAL OD result

5 Conclusion

Various deep learning networks have been compared in this work. The deep learning models outperform image processing and edge detection techniques. This work aims to establish a clear benchmark for segmentation of the optic disc. However, these rankings have to be interpreted with care [12].

References

1. Agrawal V, Kori A, Alex V, Krishnamurthi G (2018) Enhanced optic disk and cup segmentation with glaucoma screening from fundus images using position encoded CNNS. ArXiv preprint arXiv:1809.05216
2. Bourne RR, Stevens GA, White RA, Smith JL, Flaxman SR, Price H, Jonas JB, Keeffe J, Leasher J, Naidoo K et al (2013) Causes of vision loss worldwide, 1990–2010: a systematic analysis. The Lancet Global Health 1(6):e339–e349
3. Chen LC, Zhu Y, Papandreou, G, Schroff F, Adam H (2018) Encoder-decoder with atrous separable convolution for semantic image segmentation. In: Proceedings of the European conference on computer vision (ECCV), pp 801–818
4. Fu H, Cheng J, Xu Y, Wong DWK, Liu J, Cao X (2018) Joint optic disc and cup segmentation based on multi-label deep network and polar transformation. IEEE Trans Med Imaging 37(7):1597–1605
5. Goodfellow I, Pouget-Abadie J, Mirza M, Xu B, Warde-Farley D, Ozair S, Courville A, Bengio Y (2014) Generative adversarial nets. In: Advances in neural information processing systems, pp 2672–2680
6. Huang G, Liu Z, Van Der Maaten L, Weinberger KQ (2017) Densely connected convolutional networks. In: Proceedings of the IEEE conference on computer vision and pattern recognition, pp 4700–4708
7. Jadon S (2020) A survey of loss functions for semantic segmentation. In: 2020 IEEE Conference on computational intelligence in bioinformatics and computational biology (CIBCB). IEEE, pp 1–7
8. Khalid NEA, Noor NM, Ariff NM (2014) Fuzzy c-means (FCM) for optic cup and disc segmentation with morphological operation. Proced Comput Sci 42(C):255–262
9. LeCun Y, Bengio Y, Hinton G (2015) Deep learning. Nature 521(7553):436–444
10. Li G, Yu Y (2016) Visual saliency detection based on multiscale deep CNN features. IEEE Trans Image Proc 25(11):5012–5024
11. Liu P, Kong B, Li Z, Zhang S, Fang R (2019) Cfea: collaborative feature ensembling adaptation for domain adaptation in unsupervised optic disc and cup segmentation. In: International conference on medical image computing and computer-assisted intervention. Springer, pp 521–529
12. Maier-Hein L, Eisenmann M, Reinke A, Onogur S, Stankovic M, Scholz P, Arbel T, Bogunovic H, Bradley AP, Carass A et al (2018) Why rankings of biomedical image analysis competitions should be interpreted with care. Nature Commun 9(1):1–13
13. Mohan D, Kumar JH, Seelamantula CS (2019) Optic disc segmentation using cascaded multiresolution convolutional neural networks. In: 2019 IEEE international conference on image processing (ICIP). IEEE, pp 834–838
14. Orlando JI, Fu H, Breda JB, van Keer K, Bathula DR, Diaz-Pinto A, Fang R, Heng PA, Kim J, Lee J et al (2020) Refuge challenge: a unified framework for evaluating automated methods for glaucoma assessment from fundus photographs. Med Image Analysis 59:101570
15. Ronneberger O, Fischer P, Brox T (2015) U-net: convolutional networks for biomedical image segmentation. In: International conference on medical image computing and computer-assisted intervention. Springer, pp 234–241

16. Sandler M, Howard A, Zhu M, Zhmoginov A, Chen LC (2018) Mobilenetv2: inverted residuals and linear bottlenecks. In: Proceedings of the IEEE conference on computer vision and pattern recognition, pp 4510–4520
17. Varma R, Lee PP, Goldberg I, Kotak S (2011) An assessment of the health and economic burdens of glaucoma. Am J Ophthalmol 152(4):515–522
18. Wang S, Yu L, Yang X, Fu CW, Heng PA (2019) Patch-based output space adversarial learning for joint optic disc and cup segmentation. IEEE Trans Med Imag 38(11):2485–2495
19. Zhu JY, Park T, Isola P, Efros AA (2017) Unpaired image-to-image translation using cycle-consistent adversarial networks. In: Proceedings of the IEEE international conference on computer vision, pp 2223–2232

Automated Student Emotion Analysis During Online Classes Using Convolutional Neural Network

Sourish Mukherjee, Bait Yash Suhakar, Samhitha Kamma, Snehitha Barukula, Purab Agarwal, and Priyanka Singh

Abstract In non-verbal communication, facial emotions play a very crucial role. Facial recognition can be useful in various ways, such as understanding people better and using the collected data in various fields. In an e-learning platform, students' facial expressions determine their comprehension levels. Students' facial emotions can have a favorable or unfavorable impact on their academic performance. As a result, instructors need to create a positive, emotionally secure classroom environment to optimize student learning. In this paper, a novel Facial Emotion Recognition for improving our understanding of students during e-learning is proposed. Suggested model detects different students' facial emotions such as anger, disgust, fear, happiness, sadness, surprise, and neutral and utilizing them for better teaching and learning during a lecture in an e-learning platform. Convolutional neural networks (CNNs) have been used for detecting facial emotions of students in e-learning platforms, and the proposed model shows an outcome of test accuracy of 67.5%.

Keywords Facial expression recognition · Artificial intelligence · Deep learning · Convolutions neural networks · Education · E-learning system

S. Mukherjee (✉) · B. Y. Suhakar · S. Kamma · S. Barukula · P. Agarwal · P. Singh
Department of Computer Science and Engineering, SRM University-AP, Amaravati 522502, India
e-mail: sourish_mukherjee@srmap.edu.in

B. Y. Suhakar
e-mail: bait_yash@srmap.edu.in

S. Kamma
e-mail: kamma_samhitha@srmap.edu.in

S. Barukula
e-mail: barukula_snehitha@srmap.edu.in

P. Agarwal
e-mail: purab_agarwal@srmap.edu.in

P. Singh
e-mail: priyanka.s@srmap.edu.in

M. Thakur et al. (eds.), *Soft Computing for Problem Solving*, Lecture Notes in Networks and Systems 547, https://doi.org/10.1007/978-981-19-6525-8_2

1 Introduction

One of the most efficient, natural, and expeditious ways for humans to communicate their emotions and intentions is through facial expression. Facial expressions can also be one of the most genuine reactions that can be given. Individuals may be prohibited from expressing their emotions in some situations, such as being sick or physically challenged where they cannot communicate verbally. In those cases, human emotion detection is one of the very functional methods that can be utilized [1]. Recent advances in computer vision and machine learning have made it possible to recognize emotions from images. In this paper, Facial Emotion Recognition is used to improve the analysis of students' emotions during e-learning. When positive emotions are mediated by self-regulated student motivation, they have a favorable effect on academic attainment [2]. Recent advances in neurology have revealed that emotions, cognitive and auditory functions are linked, implying that there is a link between learning and emotion. Students' emotions are crucial throughout the lecture.

In the proposed scheme, seven elementary emotions are detected using facial emotions, i.e., anger, disgust, fear, happiness, sadness, surprise, and neutral to achieve our target of achieving feedback from students. This will improvise teaching methods as the lecturer can get genuine feedback without any influence. The teacher can adapt the course content in such a way that every student is comfortable which makes learning very much effective. This intelligent learning technology has mechanisms for responding to a student's emotional states, such as encouraging students and modifying materials to suit them. This facial detection can be used for both online and in-person lectures. There has been some research on Facial Emotion Recognition [3], but there are not many cases where it has been used for the advantage of e-learning. Hence, the primary goal of this research is to provide a novel approach for analyzing emotions that could be used in e-learning systems.

Motivation and contributions of the proposed scheme: Traditional emotion recognition systems based on facial expressions have vital issues like being unable to recognize variations in viewing angles of the face. For example, head rotation is an important orientation to consider for e-learning models. To address these problems in the proposed scheme, a CNN model is proposed to recognize variations in viewing angles.

The rest of the paper is organized as follows: Sect. 2 reviews the state-of-the-art schemes, Sect. 3 elucidates the proposed scheme in detail, and experimental results and discussion are shown in Sect. 4. Finally, Sect. 5 concludes the paper.

2 Related Works

Researchers have proposed various schemes to recognize the emotional states of a person. Studies have shown that a student's likability toward a particular subject can be easily determined by the student's emotional response during classroom interaction. Students usually have a very positive response toward subjects that seem interesting, whereas students seem to show negative emotional responses toward subjects that don't seem too appealing. Thus, taking measures such as collecting data that projects a particular student's emotional response to a particular subject at a particular time in the classroom and analyzing the data will be proven to be of great help in determining the student's interests, likes, and dislikes. Drawing useful insights from this data can also be used to map them with the classroom interaction techniques employed. The usage of convolutional neural networks to identify facial emotions has so far been proven to be the most useful technique.

It has been observed that there are six, globally acknowledged facial expressions: anger, fear, happiness, sadness, fear, and disgust. Building automated systems that can distinguish one facial expression from another is known to have several real-world applications ranging in the fields such as medicine, psychology, advertisement, and warfare. The traditional FER schemes are based on 2-step machine learning approach, where firstly features are extracted from images and then a classifier such as SVM, neural network, or random forest is used to detect the emotions. It is observed that these approaches show higher performance on simpler datasets, but lag behind for complex datasets. Ch [4] have presented an efficient Facial Emotion Recognition (FER) system by utilizing a novel Deep Learning Neural Network-regression activation (DR) classifier. Hajarolasvadi et al. [5] have presented a system for Facial Emotion Recognition in video sequences. Then, the system is evaluated for person-dependent and person-independent cases. Depending on the purpose of the designed system, the importance of training a personalized model versus a non-personalized one differs. Mansouri-Benssassi and Ye [6] have addressed the challenges which arise from cross-dataset generalization. They have proposed the application of a spiking neural network (SNN) in predicting emotional states based on facial expression and speech data, then investigate, and compare their accuracy when facing data degradation or unseen new input. Saravanan et al. [7] created a model, which predicts the individual probability of each facial emotion. This leads to the gathering of valuable insights regarding how closely certain emotions are related to one another. Ninad Mehendale has implemented a facial emotion recognizer by calculating the Euclidean distances between the major components of the face like eyes, ears, nose, mouth, forehead, etc. [8]. This technique had introduced a novel and at the time, a somewhat efficient way of predicting facial expressions. In their work, the researchers gave the whole face as an input to the algorithm. Every single pixel in a face is treated like a feature instead of connecting different parts of a face to Facial Action Units proposed by Rzayeva and Alasgarov [9]. A somewhat more efficient method using the Gabor filter was suggested by Zadeh et al. in their work [10]. Gabor filters are generally used for texture analysis and edge detection. An EEG-based emotional feature learning and classification method using a deep convolutional neural network (CNN) was proposed by Chen et al. [11], based on temporal features, frequential features, and their combinations of EEG signals in the DEAP dataset. DEAP is a database for Emotion Analysis Using Physiological Signals. Pranav et al. [12] have suggested a CNN model using the Adam optimizer for the reduction of the loss function, thus leading to the model giving an accuracy of 78.04%. Thuseethan et al. [13] proposed a metric-based approach for defining the various intensity levels of primary emotions.

Several other attempts have been made to recognize the facial expressions of individuals. At the same time, not many attempts have been made to integrate facial expression classifiers with the classroom environment and with e-learning systems. Integrating facial expression classifiers with such environments will serve as an opening for several modern data collection techniques. The data collected using such systems will prove to be helpful for several educational institutions. The proposed hybrid deep learning architecture to learn and classify the various intensities of emotions during e-learning/physical classroom is aimed at integrating FER with learning for improved teaching and learning.

3 Proposed Scheme

In this paper, a Facial Emotion Recognition (FER) system is proposed using CNN. The FER-2013 dataset is used for the training of the CNN model. In the proposed method, images are preprocessed using the TensorFlow library to enhance the element of variation in the dataset. After preprocessing, features of the images were extracted by processing them through a series of convolutional filters, and then the output of those filters is passed through a fully connected neural network. The fully connected neural network classifies the emotion of the input image. The proposed method is explained in detail in the following subsections. The flow diagram of the proposed system can be seen in Fig. 1.

3.1 Dataset

The proposed model is trained on the traditional FER-2013 dataset. The dataset consists of about 35,000 low-resolution images of different facial expressions with size of each image restricted to 48×48 pixels. The main labels of expressions are of seven types, i.e., anger, disgust, fear, happiness, sadness, surprise, and neutral. The expression of disgust has 600 samples of it, while the other expressions have around 5000 samples each. The images in this dataset are of dissimilar age groups, and some of the images are taken in extreme circumstances (i.e., 'taken from a certain angle', 'taken from a random distance'). Since the images are of such stature, therefore the

Fig. 1 Flow diagram of the proposed system

Fig. 2 Emotions in the dataset

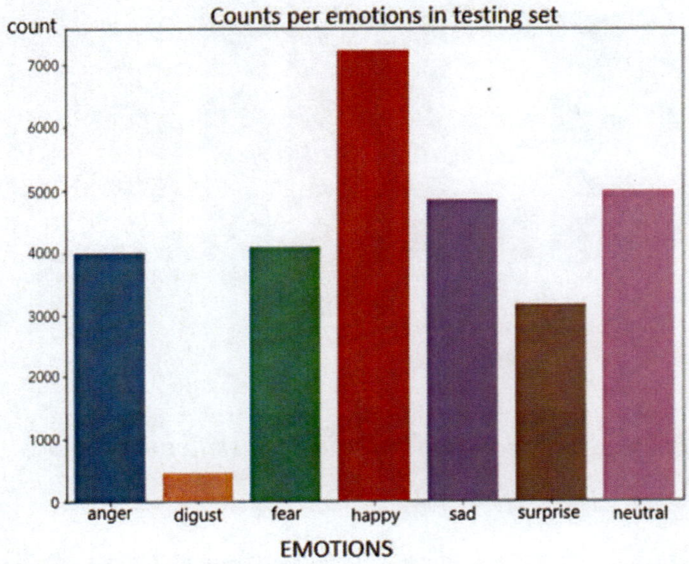

Fig. 3 Counts of emotions in testing dataset

proposed model is trained on FER-2013 dataset. Images of the different expressions from the dataset can be seen in Fig. 2. The general counts for emotion can seen in Figs. 3 and 4.

3.2 Data Preprocessing

Firstly, the images in the dataset were converted into grayscale images. Then the images were grouped based on their specific expression into their respective subgroups. The dataset was then divided into a training set and a validation set in a 3:1 ratio, i.e., 3/4th of the dataset for training and 1/4th for the validation set. To add a sense of variation to the training dataset, image augmentation techniques using the TensorFlow library have been used. In image augmentation, each image was rotated in a range of certain angles, horizontally flipped, zoomed in, and zoomed out in a specific range, and brightness was changed in a specific range. All the images resulting from such augmentations were added to the training dataset.

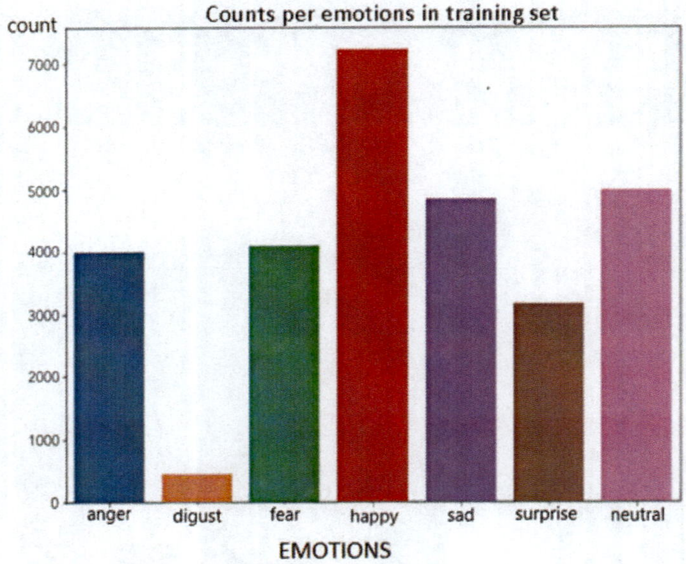

Fig. 4 Counts of emotions in training dataset

3.3 *Feature Extraction and Emotion Classification Using CNN*

The images from the preprocessing are then served as input to the CNN. The convolutional layers are then used to extract features such as corners, edges, and shapes. Convolutional filters can accept 2D images as input. To classify a facial expression, we have a total of 10 layers, i.e., 4 convolutional layers, 4 pooling layers, and 2 fully connected neural layers, shown in Fig. 5. Layers 0, 2, 4, and 6 are the convolutional layers; layers 1, 3, 5, and 7 are the pooling layers; layers 8 and 9 are the fully connected neural layers. The block diagram of the layers of CNN is provided in Fig. 5. The description for each layer is as follows:

Layer 0 (Conv1): The input image to this layer is of size 48×48. The kernel size is 3×3 pixels. A total of 64 filters are applied to the image. Batch normalization is carried out on the resulting pixels to reduce the parameters of the image, and an activation function 'ReLu' was used. To further reduce the parameters, the image is passed through a max pooling layer (layer 1) of size 2×2.

Layer 2 (Conv2): A total of 128 filters are applied to the image passing through this layer. The kernel size of the filter is 5×5. Batch normalization is used with the activation function being 'ReLu'. A max pooling layer (layer 3) of size 2×2 has been applied.

Layer 4 and Layer 6 (Conv3 and Conv4): In layers 4 and 6, 512 filters are applied to the passing images. The kernel size of the filter for both layers is 3×3. As usual

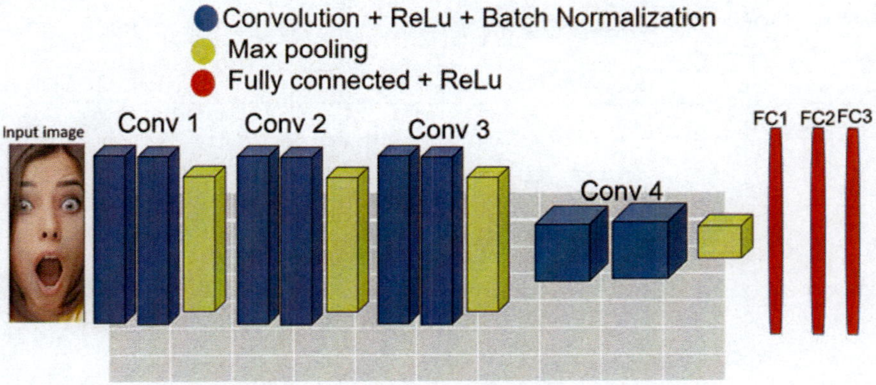

Fig. 5 Block diagram of the layers of CNN

batch normalization is applied with the function of activation being 'ReLu' for both layers. The max pooling layers (layers—5, 7) are of size 2 × 2.

Layer 8: Fully connected neural layer: The outputs from the filters are flattened to feed it into this layer as a 1-D vector. This neural layer is dense with 256 neurons. The output is batch normalized and passed through the activation function 'ReLu'.

Layer 9: Fully connected neural layer: This neural layer consists of 512 neurons. Batch normalization has been used. The output is passed through the activation function—'ReLu'. The output from the CNN is passed through a 'Softmax' activation function to obtain the classified facial expression.

4 Experimental Results and Analysis

The initially proposed experiment was carried out using the Adam optimizer and an adaptive learning rate method called 'Reduce LR on Plateau'. The activation function for the first two experiments was taken as Rectified Linear Unit (ReLU). The model was trained for a total of 100 epochs with an initial learning rate of 0.0001 and a batch size of 64. The test accuracy for this model came out to be 64%. The confusion matrix can be seen in Table 1.

The second experiment analysis was done with a few modifications to the model. The batch size was taken as 64, and the initial learning rate was increased to 0.0005. The SGD optimizer was used instead of Adam to see how it affects the accuracy and the training speed. The accuracy for this analysis was 6% less than when Adam was used, and the training speed was slower. The confusion matrix for this model is in Table 2.

The third experiment analysis of the model was done using the Nadam optimizer (Nesterov-accelerated Adaptive Moment Estimation) which is an extension of the

Table 1 Confusion matrix using ADAM optimizer

True	Predicted						
	Anger	Disgust	Fear	Happy	Sad	Surprise	Neutral
Anger	550	15	72	43	136	118	24
Disgust	34	55	1	2	5	10	4
Fear	116	6	373	38	159	217	115
Happy	21	0	21	1577	102	23	30
Sad	114	6	84	71	307	645	20
Surprise	18	2	63	43	29	21	655
Neutral	49	3	35	77	881	164	24

Table 2 Confusion matrix using SGD optimizer

True	Predicted						
	Anger	Disgust	Fear	Happy	Sad	Surprise	Neutral
Anger	97	0	31	106	47	40	111
Disgust	11	0	3	16	5	1	13
Fear	32	0	66	137	58	93	97
Happy	14	0	19	674	26	29	39
Sad	36	0	53	138	127	25	159
Surprise	10	0	26	54	6	245	36
Neutral	22	0	31	118	44	23	312

Table 3 Confusion matrix using Nadam optimizer

True	Predicted						
	Anger	Disgust	Fear	Happy	Sad	Surprise	Neutral
Anger	235	7	53	22	67	4	44
Disgust	16	13	8	1	9	1	1
Fear	68	4	175	31	100	47	58
Happy	19	1	20	697	19	18	27
Sad	50	0	66	38	268	7	109
Surprise	11	0	58	21	5	269	13
Neutral	32	1	30	34	91	14	348

Adam optimizer which adds Nesterov momentum. Alongside these, the batch size was decreased to 32 with the learning rate being 0.001 and the activation function being Exponential Linear Unit (ELU). HeNormal kernel initializer was also implemented into the structure. It was found that this model was producing an accuracy that was 3% lower on average than the previous two models.

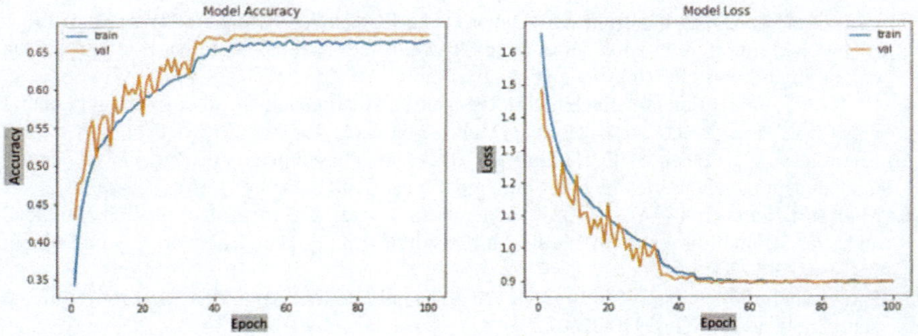

Fig. 6 Loss and accuracy analysis of the final model

The final analysis was done using the modifications mentioned in the first analysis. It was carried out using the Adam optimizer and an adaptive learning rate method called 'Reduce LR on Plateau'. The activation function for the first two experiments was taken as ReLU. The model was trained for a total of 100 epochs with a batch size of 72 and an initial learning rate of 0.0001. The test accuracy for this model was found to be the highest for our proposed method, 67.5%. The confusion matrix can be seen in Table 3. The graph in Fig. 6 shows the loss and accuracy analysis.

5 Conclusion

In this paper, a Facial Emotion Recognition (FER) system is proposed using the CNN for detecting different students' facial emotions and utilizing them for better teaching and learning during a lecture. The FER-2013 dataset was used for training the CNN model. Using the proposed method, we were able to achieve an accuracy of 67.5%. Further advancements will be made by trying to merge our model with pre-existing models to achieve better accuracy. The model also needs to be tested on real-time data by recording videos of students and checking the accuracy of the datasets generated from them. Our future works will include integrating it into surveillance cameras during in-class sessions for better analysis resulting in improved education quality. Implementing VGG and ResNets for analysis can also be seen as future work.

References

1. Hassouneh A, Mutawa AM, Murugappan M (2020) Development of a real-time emotion recognition system using facial expressions and EEG based on machine learning and deep neural network methods. Inform Med Unlocked 20:100372
2. Mega C, Ronconi L, De Beni R (2014) What makes a good student? How emotions, self-regulated learning, and motivation contribute to academic achievement. J Educ Psychol 106(1):121

3. El Hammoumi O, Benmarrakchi F, Ouherrou N, El Kafi J, El Hore A (2018, May) Emotion recognition in e-learning systems. In: 2018 6th International conference on multimedia computing and systems (ICMCS). IEEE, pp 1–6
4. Ch S (2021) An efficient facial emotion recognition system using novel deep learning neural network-regression activation classifier. Multimedia Tools Appl 80(12):17543–17568
5. Hajarolasvadi N, Bashirov E, Demirel H (2021) Video-based person-dependent and person-independent facial emotion recognition. Signal Image Video Process 15(5):1049–1056
6. Mansouri-Benssassi E, Ye J (2021) Generalisation and robustness investigation for facial and speech emotion recognition using bio-inspired spiking neural networks. Soft Comput 25(3):1717–1730
7. Saravanan A, Perichetla G, Gayathri DK (2019) Facial emotion recognition using convolutional neural networks. arXiv:1910.05602
8. Mehendale N (2020) Facial emotion recognition using convolutional neural networks (FERC). SN Appl Sci 2(3):1–8
9. Rzayeva Z, Alasgarov E (2019, October) Facial emotion recognition using convolutional neural networks. In 2019 IEEE 13th international conference on application of information and communication technologies (AICT). IEEE, pp 1–5
10. Zadeh MMT, Imani M, Majidi B (2019, Feb) Fast facial emotion recognition using convolutional neural networks and Gabor filters. In: 2019 5th Conference on knowledge based engineering and innovation (KBEI). IEEE, pp 577–581
11. Chen JX, Zhang PW, Mao ZJ, Huang YF, Jiang DM, Zhang YN (2019) Accurate EEG-based emotion recognition on combined features using deep convolutional neural networks. IEEE Access 7:44317–44328
12. Pranav E, Kamal S, Chandran CS, Supriya MH (2020, Mar) Facial emotion recognition using deep convolutional neural network. In: 2020 6th International conference on advanced computing and communication systems (ICACCS). IEEE, pp 317–320
13. Thuseethan S, Rajasegarar S, Yearwood J (2019, July) Emotion intensity estimation from video frames using deep hybrid convolutional neural networks. In: 2019 International joint conference on neural networks (IJCNN). IEEE, pp 1–10

Transfer Learning-Based Malware Classification

Anikash Chakraborty and Sanjay Kumar

Abstract Any software created with malignant intent to harm others in terms of monetary damage, reputation damage, privacy infringement, etc., is known as malware. Therefore, classifying malware into their families is crucial for developing anti-malware software. The work essentially offers a malware detection method based on transfer learning where we use the pre-trained deep convolutional-based AlexNet architecture having ImageNet weights for feature extraction. The extracted features are then used to categorize malware samples into their corresponding malware families using a dense neural network architecture. For our study, we use the benchmarked MalImg dataset. The performance of our suggested model is compared to that of various other contemporary ImageNet models. As indicated by the experimental findings, the families to which the malware sample belongs have effectively been found by our proposed method.

Keywords Malware analysis · Transfer learning · ImageNet · AlexNet · Deep neural network · Convolutional neural network · Visual malware

1 Introduction

An executable program or software that is designed with an intent to hamper computer operations and harm with an unwanted interruption is termed malware. Such programs primarily steal critical and sensitive information, cause loss of privacy, and compromise the security of the system. Quick financial gains are a major motivator for the authors of the malware. According to Kaspersky Labs, there has been a 5.7% growth in discovering new malicious files daily. Malware is being used to target government entities in the field of energy, the military, banks, financial institutions, and transport. The traditional antiviruses which are currently in the market are mostly

A. Chakraborty (✉) · S. Kumar
Delhi Technological University, New Delhi, India
e-mail: akh.chakraborty11@gmail.com

S. Kumar
e-mail: sanjay.kumar@dtu.ac.in

M. Thakur et al. (eds.), *Soft Computing for Problem Solving*, Lecture Notes in Networks and Systems 547, https://doi.org/10.1007/978-981-19-6525-8_3

effective against already known malware. The method used by such antiviruses is either signature based or heuristic based. The signature-based method uses an algorithm or a hash to uniquely identify a specific malware by generating a signature (sensitive to program variations) of the encountered malware and then mapping it to the existing signatures present in the database. Heuristic-based methods analyze the behavior of the software and categorize it as malware accordingly. However, the above methods do not work in cases when the antivirus encounters a new unseen malicious program as the prerequisite of the methods is to analyze the malware before and store it in their database and it is impractical to analyze and study every new sample.

Machine learning and deep learning-based approaches have been effective in solving many real-life applications like object detection [1, 2], medical imaging [3, 4], fake news detection [5], link predication [6, 7], influential node detection [8, 9], and many others [10–14]. Machine learning is being actively used to provide effective solutions for malware classification and detection as applications of image classification continue to rise at a tremendous rate. Statistical analysis of their characteristics including API calls is used to classify malware, but these require a wide domain knowledge for feature extraction. It gathers information regarding the similarities among the instruction set. The popularity of the usage of deep convolutional neural networks for the classification of malware, especially the VGG-16 architecture, has increased over the years. A 28.5% top-1 classification error was displayed in the ImageNet competition (2014) by the VGG-16 model which has a depth of 23. In the following year, a 24.1% classification error was displayed by the ResNet model having a depth of 168. Subsequently, in 2017, a 21.0% error was displayed by the Xception model having a depth of 126 layers.

Working upon advancing the existing malware detection techniques, we present this work. The work essentially offers a malware detection method based on transfer learning. We start by converting the malware files into their binary files which are stacked and post that grayscale images are generated by conversion. Subsequently, the grayscale images are fed to an AlexNet feature extractor which extracts all the relevant features for the task of malware detection. We use the pre-trained deep convolutional-based AlexNet architecture having ImageNet weights. As a result, the extracted features are used to categorize malware samples into their respective malware families using a dense neural network architecture. For our study, we use the benchmarked MalImg dataset. To counter the high-class imbalance of the MalImg dataset, we perform data augmentation using various data augmentation techniques. The performance of our suggested model is compared to that of various other contemporary ImageNet models.

As indicated by the experimental findings, the families to which the malware sample belongs have effectively been found by our proposed method. The remainder of the paper has been described in the given manner. Section 2 discusses some of the current malware detection research and the advancements which have been made till now. It briefly explains the methodology used in various works and the accuracy achieved by them. Section 3 describes AlexNet which is a pre-trained model trained on ImageNet dataset and about transfer learning and its idea. Section 4 describes the proposed work along with a flowchart for a better understanding of the proposed work. Section 5 lists the description of datasets used in this study. In Sect. 6, the performed experimental analysis is discussed. Finally, Sect. 7 concludes the work.

2 Related Work

Software or programs comprising of code are just a composition of binary files as a computational machine can only interpret binary code. Over time, GUI-based editors have blossomed like the text and binary editors which help in composing binary data through visualization. Malware in simple words is also a composition of binary data, and this property can be exploited for its visual representation. Nowadays, classification of malware into its families is being studied upon extensively, and machine learning techniques are being employed.

The authors in [15] identified substantial perceptible similarity in picture texture among malwares from the same family. This led to the proposal of a visualization approach for classifying malware into families. Texture features were computed using GIST which decomposes the image using wavelet, and classification was subsequently done by KNN classifier which led them to achieve an accuracy of 97.18%. In [16], the author proposed a technique which used SVM to extract the textural patterns from malware to classify malware. After converting the raw malware binary data from the file into grayscale, the image is resized and then sub-band filer is applied to generate bands which are used by Gabor wavelet to extract gradient information. Following this a feature vector was generated upon which SVM classification is done to segregate malware into 24 different families with accuracy touching 89.86%. Gibert [17] proposed a method which used CNN on MalIMG and the Microsoft Malware Classification Challenge dataset wherein they achieved an accuracy of 98.48% and 97.49%, respectively. In [18], the author proposed a method utilizing M-CNN which they built using VGG-16 architecture and converted the malware files into images achieving an accuracy of 98.52%. Cui et al. [19] also proposed a method for classification of malware using CNN by converting executable files of malware into grayscale images, but they argued that the MalImg dataset is imbalanced. To overcome this, they employed a genetic sorting algorithm. [20] proposed a mechanism to classify malware into families using CNN with attention mechanism. The malware sample is converted into a image using this approach, and then an attention map is generated that is the regions which have higher importance in classification using the attention mechanism. It then generates outputs of the regions which are used for classification so that they can be used for manual analysis in the future. Recently, a lot of techniques based on deep convolutional neural networks have been explored, and extensive studies are conducted for classification of malware [21, 22]. Another method for classification of malware into families by utilizing CNN in combination with the Xception model has been proposed by Lo et al. [23]. It employs the transfer learning strategy which performs an improvement on the current task through transferring the knowledge acquired from an already learnt-related task. Therefore, the pre-trained model is transferred onto the classification task. The results achieved show a validation accuracy of 99.04% on MalImg dataset and 99.17% on Microsoft Malware dataset. Ren et al. [24] proposed two methods based on visualization for classification of malware by utilizing byte sequences generated by n-gram features. First is the space filling curve mapping method which visualizes one-gram features of malware files. Second is the Markov dot plot method which visualized bi-gram features of malware files. The accuracy achieved by these two methods when applied on the Microsoft Malware sample was 99.21% and 98.74%, respectively. Turker et al. [25] proposed a novel malware recognition method that employed local neighborhood binary pattern (LNBP) for feature extraction. It extracts information using all neighborhood relations. It achieved an accuracy of 89.40%. Kolosnjaji et al. [26] worked on the classification of malware using a unique method wherein they utilized deep neural networks to analyze system call sequences by combining convolutional and recurrent layers using system call n-grams obtained from dynamic analysis. Rezende et al. [27] proposed a malware classification technique using a deep CNN model based on 50 layers ResNet architecture. Byteplot images were used for representing malware samples, and it employs transfer learning approach to transfer the pre-trained parameters of the ResNet-50 model to the classification problem. It achieved an accuracy of 98.62%. Another interesting work that compares the performance of CNN and extreme learning machines (ELM) for the classification of malware was given by Jain et al. [28]. Although CNN techniques have been used widely for the classification problem, the authors display how ELMs with fine parameter tuning can achieve accuracies comparable to CNNs and ELMs utilizing much less training cost.

3 Preliminary

This section explains AlexNet and transfer learning methodology.

3.1 AlexNet

AlexNet is a CNN architecture created by Alex Krizhevsky, Ilya Sutskever, and Geoffrey Hinton. The architecture competed in the ImageNet challenge in 2012 and became the winner with a classification error of just 15.3%.

The AlexNet architecture comprises eight layers and has the parameters that can be learned. There are five layers out of which some have a combination of max pooling layers with the convolutional layer. The subsequent three layers are fully connected. Each of these three layers, except the output layer, uses ReLU activation. An increase of six times in the training process speed was discovered when ReLU was used as an activation function. Overfitting of the model was prevented using dropout layers and the dataset used for trained using the ImageNet dataset. This dataset has close to 14 million images spread across thousands of classes. Figure 1 shows the architecture of AlexNet.

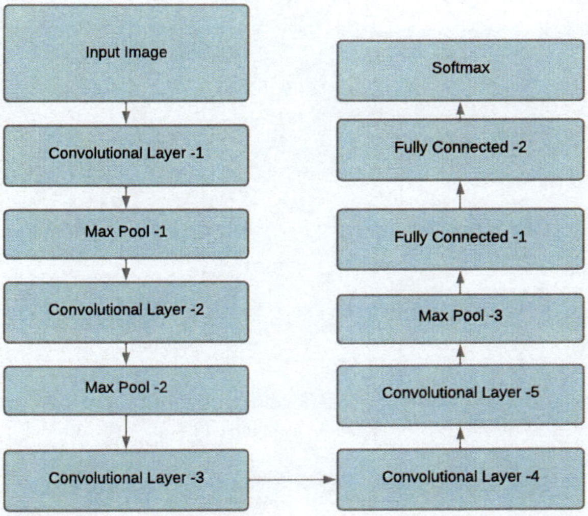

Fig. 1 Graphical representation of the AlexNet architecture

3.2 Transfer Learning

Transfer learning is essentially the transfer of knowledge gained by a previous task or issue and applying it to a new issue for solving it. When it is associated with machine learning, the computer uses the knowledge gained from a previous task to improve the results of the in-hand task. Figure 2 shows the general idea behind transfer learning. There are universal, low-level properties that are common between images, and because of this, we can employ a network trained on unrelated categories in a huge dataset and use the result to solve our problem.

There are several pre-trained models used for transfer learning. A brief about some of the pre-trained models which would be utilized to compare with our method are given below:

- **VGG-16**
  The VGG-16 model comprises thirteen convolutional layers along with three fully connected layers, and it achieves an accuracy of 92% on the ImageNet dataset.
- **ResNet-50**
  The ResNet-50 model comprises fifty convolutional layers along with one max pool layer and one average pool layer, and it is widely used for image classification.
- **InceptionV3**
  The InceptionV3 model consists of 42 layers, and although the number of layers is high, the complexity is the same as that of a VGG Net.
- **Xception**
  The Xception model consists of 71 layers and can classify images into 1000 categories. It is also trained on the ImageNet dataset.

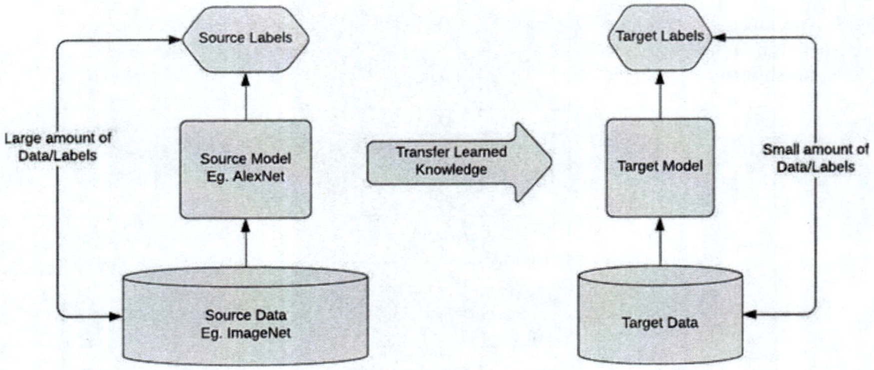

Fig. 2 Transfer learning methodology used in our proposed work. AlexNet model which is pre-trained on ImageNet dataset is transferred to our target model

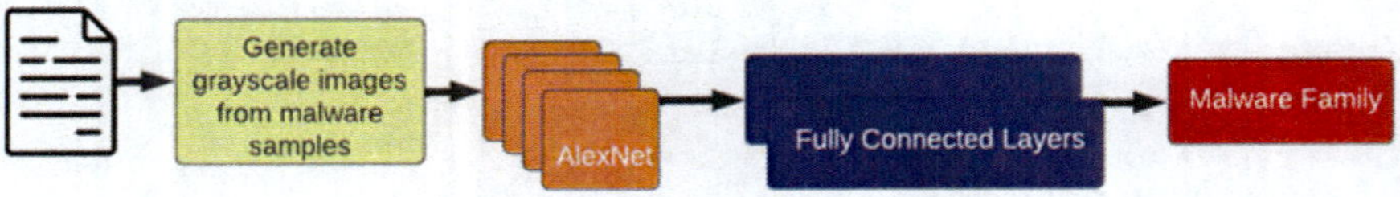

Fig. 3 Malware sample is fed and grayscale image is generated according to Fig. 4. The image is then passed through AlexNet for feature extraction and then through our deep convolutional neural network layers for classification

4 Proposed Work

This section elaborates on the proposed transfer learning technique utilized for the classification of malware samples into 25 families which act as our class labels. We start by converting the malware samples into grayscale images. The effects of dataset imbalance are countered by performing the technique of data augmentation to the images. The augmented grayscale images are then passed through AlexNet which is a pre-trained deep convolutional neural network architecture. The AlexNet architecture is further augmented by dense neural network architecture. Ultimately, the malware sample is classified into their families. The various steps of our proposed work are elaborated below along with Fig. 3 displaying the flowchart of the proposed work.

4.1 Generating Grayscale Images from Malware Samples

We start by converting the malware samples into grayscale, starting with obtaining the binary files of the malware samples. Figure 4 describes the process of generating grayscale image from malware samples. Firstly, binary data was transformed into 8-bit vector representations (or a grayscale pixel), converting them into an integer

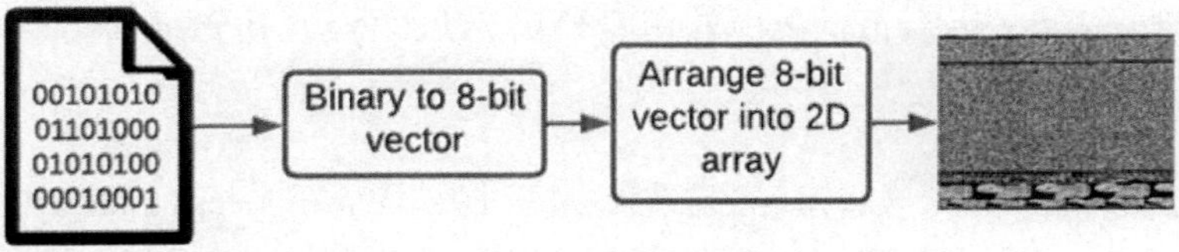

Fig. 4 Generating grayscale images from malware samples by transforming into 2D matrix

between 0 to 255, thus forming a 1D vector. Then a row width of 256 pixels was fixed to form a 2D matrix. The height of the matrix varies depending on the malware files' size. The value of each element of the matrix is the same as the grayscale pixel value thus forming the grayscale image. These grayscale malware images generate a well-labeled visual malware dataset with labels representing the classes of the malware samples.

4.2 Augmenting the Dataset

The dataset obtained in the previous step is usually highly imbalanced with class bias occurring due to a higher number of malware samples belonging to one class and a lesser number of samples belonging to the other class. Training our model on such an imbalanced dataset would lead to overfitting to a certain group of classes as the number of samples in them is more. This leads to decreased performance of the algorithm. To overcome this, we perform data augmentation by utilizing techniques like clockwise and anticlockwise image rotation, shifting, and horizontal and vertical flipping. This generates a balanced dataset for our algorithm. This balanced dataset is then used by our algorithm to derive features and make the malware classification.

4.3 Feature Extraction Using AlexNet

The well-balanced and well-labeled dataset generated in the previous step is passed through an AlexNet-based feature extractor to extract the most relevant features. The AlexNet is used in a pre-trained version with the ImageNet weights. The final classification layer of the AlexNet is removed as it is to be used only to extract the features and not classify them. The reason for using AlexNet is that it has been extensively used for making image classification in various domains like medical diagnosis, image segmentation, etc.

4.4 *Sorting the Malware Samples into the Appropriate Families*

The extracted features from the previous section are used in this step to make the final malware classification. For this purpose, we use a dense layer with five layers having 16 neurons, 64 neurons, 64 neurons, 16 neurons, and 25 neurons for classifying. A dense structure helps in appropriately processing the features extracted in the previous step. The output layer with softmax activation having 25 neurons finally generates a probability of the malware belonging to each class.

5 Datasets

The dataset that we utilized in our proposed work is discussed in this part. For performing experimental simulations, we have used the standardized MalImg dataset which was introduced by the authors of Nataraj et al. [15]. The dataset is constituted of several malware samples that were collected from various sources and compiled together. There are a total of 9348 malware samples and twenty-five families of malware present in the dataset. The imbalance is relatively high in MalImg dataset and is clearly visible in Table 1. Certain abbreviations, namely "TD" for "Trojan Downloader" and "WA" for "Worm:AutoIT", have been used. A total of 25,000 malware samples with 1000 samples from every family were resulted from the stabilization of dataset with image augmentation techniques.

6 Experimental Analysis

In this section, we present the experimental analysis performed by us for our study. We ran simulations on the dataset mentioned in Sect. 5. We also compared the performance of our model with several other contemporary ImageNet models. We performed the simulations by augmenting the images of the dataset as mentioned in Sect. 4. We split the entire dataset in an 80:20 ratio. Training data consisted of 80% of the entire dataset, and the testing data consisted of the remaining 20%. We also split the training dataset into 70:30 ratios keeping 70% for training and the remaining 30% for validation purposes. The performance of our proposed work was compared with VGG19, InceptionV3, Xception, DenseNet201, ResNet-50, MobileNetV2, and NASNetLarge. The various experimental results are as follows.

Table 2 shows the results obtained by the various algorithms on the training dataset and compares their accuracy. From Table 2, we see that our proposed algorithm performs the best. DenseNet201 is the second-best performer after our proposed algorithm which achieved an accuracy of 99.12%. All the other algorithms follow thereby. We also note that NasNetLarge is the worst performer on the training dataset. Table 3

Table 1 Malware sample distribution among 25 malware families

S.No	Class	Family	Specimens
1	Dialer	Adialer.C	122
2	Backdoor	Agent.FYI	116
3	Worm	Allaple.A	2949
4	Worm	Allaple.L	1591
5	Trojan	Alueron.gen!J	198
6	WA	Autorun.K	106
7	Trojan	C2Lop.P	146
8	Trojan	C2Lop.gen!G	200
9	Dialer	Dialplatform.B	177
10	TD	Dontovo.A	162
11	Rogue	Fakerean	381
12	Dialer	Instantaccess	431
13	PWS	Lolyda.AA1	213
14	PWS	Lolyda.AA2	184
15	PWS	Lolyda.AA3	123
16	PWS	Lolyda.AT	159
17	Trojan	Male.gen!J	136
18	TD	Obfuscator.AD	142
19	Backdoor	Rbot!gen	158
20	Trojan	Skintrim.N	80
21	TD	Swizzor.gen!E	128
22	TD	Swizzor.gen!I	132
23	Worm	VB.AT	408
24	TD	Wintrim.BX	97
25	Worm	Yuner.A	800

Table 2 Comparison of accuracies achieved on the training dataset

Method	Training accuracy (%)
Xception	93.87
InceptionV3	96.79
DenseNet201	98.68
ResNet-50	97.81
NasNetLarge	93.65
MobileNetV2	96.78
VGG19	95.61
Proposed work	**99.12**

Table 3 Comparison of accuracies achieved on the validation dataset

Method	Validation accuracy (%)
Xception	93.46
InceptionV3	91.78
DenseNet201	94.59
ResNet-50	93.28
NasNetLarge	87.10
MobileNetV2	90.09
VGG19	87.28
Proposed work	**95.71**

Table 4 Comparison of accuracies achieved on the testing dataset

Method	Testing accuracy (%)
Xception	92.25
InceptionV3	91.65
DenseNet201	94.50
ResNet-50	93.19
NasNetLarge	86.63
MobileNetV2	90.06
VGG19	86.93
Proposed work	**95.59**

shows the accuracy results obtained by our proposed algorithm and other contemporary ImageNet models on the validation dataset. Our suggested method outperforms all others with an accuracy of 95.71%. The DenseNet201 is still the second-best performer, while the NasNetLarge remains to be the worst performer. VGG19 is the second-worst performer. Now, moving on to the testing dataset. Table 4 shows the experimental results obtained by our proposed algorithm and other ImageNet models for the testing dataset. The obtained results show that our proposed algorithm is the best performer with an accuracy of 95.59%. The DenseNet201 is the second-best performer. The NasNetLarge and the VGG19 are the worst performer with VGG19 performing slightly better than the NasNetLarge. The above experimental analysis demonstrates the utility of our proposed work for malware detection as it outperforms all other methods. It also shows that our choice for using the AlexNet was optimal.

7 Conclusion

The work essentially offers a malware detection method based on transfer learning. The inexpensive method of converting binary files to grayscale images for input has been employed which has made the proposed work independent of file type. Image augmentation techniques used help us to get rid of class bias and generate a more generalized framework for malware classification. The proper hyperparameter tuning performed by us also helped us in obtaining better results. The performance of our suggested model is compared to that of various other contemporary ImageNet models, and our proposed work has outperformed all other techniques mentioned in the literature with an accuracy of 95.59 % on the testing dataset. As a future update to our work, several other sophisticated ImageNet models could be used by us, and instead of just using grayscale images, we could also utilize colored images.

References

1. Parihar AS, Singh K, Rohilla H, Asnani G (2021) Fusion-based simultaneous estimation of reflectance and illumination for low-light image enhancement. IET Image Process 15:1410–1423. https://doi.org/10.1049/ipr2.12114
2. Singh K, Parihar AS (2021) Variational optimization based single image dehazing. J Vis Commun Image Represent 79:103241. https://doi.org/10.1016/j.jvcir.2021.103241
3. Bhowmik A, Kumar S, Bhat N (2019) Eye disease prediction from optical coherence tomography images with transfer learning. In: Pädiatrie. Springer International Publishing, Cham, pp 104–114
4. Katyal S, Kumar S, Sakhuja R, Gupta S (2018) Object detection in foggy conditions by fusion of saliency map and YOLO. In: 2018 12th International Conference on Sensing Technology (ICST), IEEE
5. Raj C, Meel P (2022) ARCNN framework for multimodal infodemic detection. Neural Netw 146:36–68
6. Anand S, Mallik A, Kumar S (2012) Integrating node centralities, similarity measures, and machine learning classifiers for link prediction. In: Multimedia tools and applications, pp 1–29
7. Kumar S, Mallik A, Panda BS (2022) Link prediction in complex networks using node centrality and light gradient boosting machine. In: World wide web, pp 1–27
8. Kumar S, Panda A (2021) Identifying influential nodes in weighted complex networks using an improved WVoteRank approach. In: Applied intelligence, pp 1–15
9. Kumar S, Gupta A, Khatri I (2022) CSR: a community based spreaders ranking algorithm for influence maximization in social networks. In: World wide web, pp 1–20
10. Sharma G, Johri A, Goel A, Gupta A (2018) Enhancing RansomwareElite app for detection of ransomware in android applications. In: 2018 eleventh International Conference on Contemporary Computing (IC3). IEEE, pp 1–4
11. Dahiya S, Tyagi R, Gaba N (2020) Comparison of ML classifiers for image data. No 3815 EasyChair
12. Dahiya S, Gosain A, Mann S (2021) Experimental analysis of fuzzy clustering algorithms. In: Advances in intelligent systems and computing. Springer, Singapore, pp 311–320
13. Jain M, Beniwal R, Ghosh A, Grover T, Tyagi U (2019) Classifying question papers with bloom's taxonomy using machine learning techniques. In: Communications in computer and information science. Springer, Singapore, pp 399–408
14. Beniwal R, Gupta V, Rawat M, Aggarwal R (2018) Data mining with linked data: past, present, and future. In: 2018 second International Conference on Computing Methodologies and Communication (ICCMC), IEEE
15. Nataraj L, Karthikeyan S, Jacob G, Manjunath BS (2011) Malware images: visualization and automatic classification. In: Proceedings of the 8th international symposium on visualization for cyber security—VizSec '11. ACM Press, New York, USA

16. Makandar A, Patrot A (2015) Malware image analysis and classification using Support Vector Machine
17. Gibert D (2016) Convolutional neural networks for malware classification. University Rovira i Virgili, Tarragona, Spain
18. Kalash M, Rochan M, Mohammed N, Bruce NDB, Wang Y, Iqbal F (2018) Malware classification with deep convolutional neural networks. In: 2018 9th IFIP international conference on New Technologies, Mobility and Security (NTMS), IEEE
19. Cui Z, Du L, Wang P, Cai X, Zhang W (2019) Malicious code detection based on CNNs and multi-objective algorithm. J Parallel Distrib Comput 129:50–58. https://doi.org/10.1016/j.jpdc.2019.03.010
20. Yakura H, Shinozaki S, Nishimura R, Oyama Y, Sakuma J (2019) Neural malware analysis with attention mechanism. Comput Secur 87:101592. https://doi.org/10.1016/j.cose.2019.101592
21. Mallik A, Khetarpal A, Kumar S (2022) ConRec: malware classification using convolutional recurrence. J Comput Virol Hacking Tech. https://doi.org/10.1007/s11416-022-00416-3
22. Khetarpal A, Mallik A (2021) Visual malware classification using transfer learning. In: 2021 fourth International Conference on Electrical, Computer and Communication Technologies (ICECCT), IEEE
23. Lo WW, Yang X, Wang Y (2019) An xception convolutional neural network for malware classification with transfer learning. In: 2019 10th IFIP international conference on New Technologies, Mobility and Security (NTMS), IEEE
24. Ren Z, Chen G, Lu W (2020) Malware visualization methods based on deep convolution neural networks. Multimed Tools Appl 79:10975–10993
25. Tuncer T, Ertam F, Dogan S (2020) Automated malware recognition method based on local neighborhood binary pattern. Multimed Tools Appl 79:27815–27832. https://doi.org/10.1007/s11042-020-09376-6
26. Kolosnjaji B, Zarras A, Webster G, Eckert C (2016) Deep learning for classification of malware system call sequences. In: AI 2016: advances in artificial intelligence. Springer International Publishing, Cham, pp 137–149
27. Rezende E, Ruppert G, Carvalho T, Ramos F, de Geus P (2017) Malicious software classification using transfer learning of ResNet-50 deep neural network. In: 2017 16th IEEE International Conference on Machine Learning and Applications (ICMLA), IEEE
28. Jain M, Andreopoulos W, Stamp M (2020) Convolutional neural networks and extreme learning machines for malware classification. J Comput Virol Hacking Tech 16:229–244

A Study on Metric-Based and Initialization-Based Methods for Few-Shot Image Classification

Dhruv Gupta and K. K. Shukla

Abstract Few-shot learning (FSL), or learning to generalize using few training data samples, is a particularly challenging problem in machine learning. This paper discusses various state-of-the-art distance metric-based and initialization-based FSL methods. It also gives a background on the meta-learning framework employed by many of the discussed models to generalize to novel classification tasks after training on multiple training tasks. We also discuss other techniques, such as whole-class classification, that have produced better results than meta-learning for metric-based methods.

Keywords Few-shot learning · Meta-learning · Image classification · Metric-based learning · Initialization-based learning · Feature extractor

1 Introduction

Conventional machine learning algorithms have surpassed human performance on a variety of tasks, such as image classification on the ImageNet dataset [11]. However, it is worth noting that each training class in the ImageNet dataset comprises 1200 images on average for an ML model to learn from. Comparatively, a human can learn to identify a new class using only a few samples of data, in fact, a description of the class itself may suffice in many cases. Recently, a research has surged in the discovery of algorithms that can overcome this disparity between machines and humans. Algorithms that aim to classify using only a few samples of training data are called few-shot learning algorithms [29]. An algorithm that uses k training samples per class is termed as a k-shot learner. Conventionally, $k \in \{1,5\}$ for supervised learning.

An initial strategy for few-shot learning was to augment training data to create a large enough dataset on which conventional ML algorithms can be trained [8, 14, 23]. For image classification, such techniques are primarily based on standard

D. Gupta (✉) · K. K. Shukla
IIT (BHU), Varanasi, India
e-mail: dhruv.gupta.cse17@iitbhu.ac.in

M. Thakur et al. (eds.), *Soft Computing for Problem Solving*, Lecture Notes in Networks and Systems 547, https://doi.org/10.1007/978-981-19-6525-8_4

preprocessing methods such as rotation, cropping, and scaling [3, 28, 33]. Most image classification networks, such as convolutional networks, are not invariant to the orientation and size of objects in images, and their accuracy benefits significantly from the augmented dataset. However, using manual rules to augment data does not capture all forms of invariance in the images belonging to the same class; more is required to solve the FSL problem [29]. Recently, models based on generative adversarial networks [2, 4, 10, 18] have been used to augment few-shot datasets and have significantly enhanced the performance of FSL models. Data augmentation is still usually used in few-shot learning, but it is considered a preprocessing step to aid classification rather than an approach to FSL in itself.

Meta-learning, or *learning to learn* is commonly employed by algorithms aimed at few-shot datasets. The goal of meta-learning is to enable an algorithm to generalize performance to an unseen novel task after training on similar training tasks. In meta-learning, the dataset is grouped into M tasks such that each task has an n number of classes. Meta-learning algorithms consist of a *meta-learner* and a *base-learner*. The meta-learner consists of *meta-parameters* that store information about features that are common across the task distribution. On the other hand, the base-learner consists of parameters that hold task-specific information to distinguish between classes within the given task. Since the base parameters are unique for each task, the base-learner must be re-trained for every novel task. Consider the following example: a set of tasks $T_{1..M}$, where each task requires the algorithm to differentiate between two species of birds (Sparrows vs. Parrots, Crows vs. Pigeons, etc.). On training a meta-learning algorithm on this dataset, the meta-parameters hold information that helps distinguish two species of birds in general. At the same time, the base parameters are used to distinguish the birds within a specific task, conditioned on the meta-parameters. In meta-learning, a model that can distinguish among n classes in a given task is called an *n-way learner* (Fig. 1).

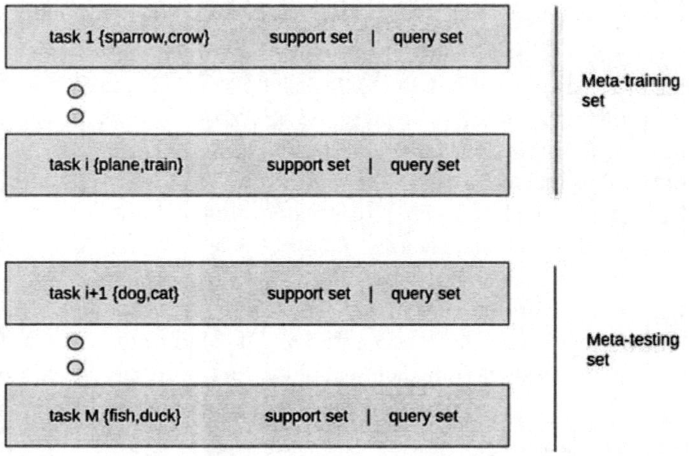

Fig. 1 Division of a dataset into tasks [34]

Metric-based FSL methods employ a feature extractor that embeds the input into an embedding space where a classifier can easily distinguish between classes in each task (see Fig. 2). The feature extractor remains the same across all tasks, while the classifier is re-trained for each task. Classification is performed by comparing the distance between feature embeddings of the support set and the query image. The exact distance metric used depends from model to model. The model predicts the class closest to the query embedding as the labeling class. Conventionally, metric-based methods were trained using meta-learning to enable them to generalize to unseen novel tasks. However, the recent papers [5, 27] show that it is possible to obtain a good accuracy on few-shot datasets without meta-learning using *whole-class classification* to train the feature extractor.

Initialization-based methods work on the principle that similar tasks have similar features, and instead of using a separate neural network for each task, we can use a single neural network whose parameters are re-adjusted for the current task at hand. Initialization-based methods are more efficient as only a small number of gradient descent steps can be used to retrain the network, provided that the variance across tasks in the dataset is not too large. The meta-parameters are used to set the default initial state of the network. When we wish to perform classification for a particular task, a few gradient descent steps are used to obtain the base parameters optimal for that task.

2 Background

Most algorithms discussed in this paper use meta-learning to train their models. In meta-learning, each task is said to be sampled from some task distribution, i.e., $T_m \sim p(T)$. The smaller the variance across tasks in the distribution, the easier it is for the meta-learner to infer details about the testing tasks using training tasks.

The training step in meta-learning optimizing both meta-parameters and base parameters as follows [12]:

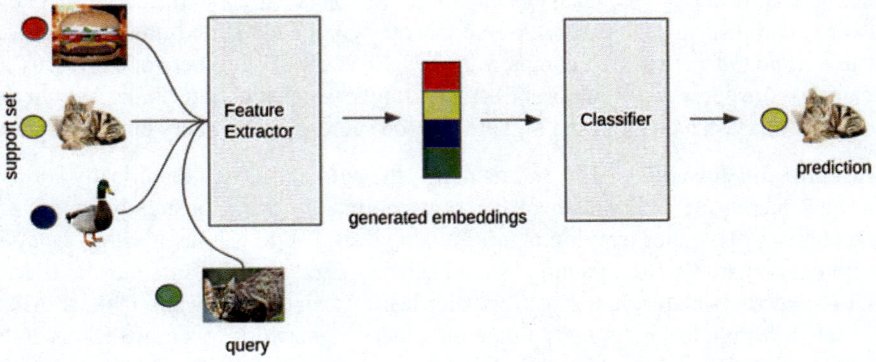

Fig. 2 Distance metric learning

$$\omega^* = \arg\min_{\omega} \sum_{i=1}^{M} \mathcal{L}^{\text{meta}} \left(\theta^{*(i)}(\omega), \omega, \mathcal{D}_{\text{source}}^{\text{query}\,(i)}\right)$$
$$\text{s.t. } \theta^{*(i)}(\omega) = \arg\min \mathcal{L}^{\text{task}} \left(\theta, \omega, \mathcal{D}_{\text{source}}^{\text{support}\,(i)}\right) \tag{1}$$

ω refers to the meta-learning parameters. $\theta^{*(i)}$ refers to the base learning parameters for the ith training task.

Loss functions: For an algorithm that uses meta-learning, two losses are computed, the task-loss and the meta-loss. In metric-based methods that use meta-learning, the task-loss is obtained by computing the distance (Euclidean, cosine, polynomial, among others) of support set embeddings to the current class embedding. The meta-loss is the sum of task losses for the query set portion of the data. It is used to update the parameters of the feature extractor, which holds the joint parameters used across all tasks. For initialization-based methods such as MAML, in the context of image classification, cross-entropy loss is used as the task-loss function. Similar to metric methods, meta-loss is calculated by summing up the regularized losses on the query sets for each task.

3 Comparison of Few-Shot Learning Papers

3.1 Distance Metric-Based Learning Methods

Prototypical Networks [24]: Prototypical networks compute a prototype to represent each class in the task. Each example in the support set is embedded, and the centroid of all examples is used as the prototype. The query embedding is then compared with the class prototypes using Euclidean distance. The network predicts the query label as the class whose prototype is closest to the query embedding in the embedding space. The prototypical network is a benchmark for metric-based meta-learning methods due to its simplicity and reasonably high accuracy on FSL datasets.

The **Relation Net** [26] takes the idea of prototypical networks further. Instead of using a Euclidean distance metric to compare query embeddings and class prototypes, it uses a neural network to compare the two. The classifying network gives us a relation score between the query and each prototype which is used to predict the query label as the class which has the highest relation score with the query embedding.

Baseline and Baseline++ [5]: In this paper, the authors start by empirically showing that increasing the number of layers in the feature extractor greatly improves the ability of few-shot learning algorithms to generalize to test tasks drawn from a different domain than the training tasks. Their proposed models, Baseline and Baseline++, perform better than most few-shot learning algorithms when the test tasks are taken from a different dataset than on which the network was trained. Baseline uses linear distance as its metric to compare feature maps, while the Baseline++ uses cosine distance for the same. The feature extractor is pre-trained using a *whole-class classification strategy* where instead of dividing the dataset into tasks and training on each individual task, the feature extractor network is trained to classify all $M * n$ classes available in the training dataset, where the dataset was originally divided into M tasks with n classes in each task. Generally, the logit layer is discarded after pre-training and the network up to the penultimate layer is used as the feature extractor. However, there is no evidence that discarding the logit layer after whole-class pre-training improves accuracy, in fact the converse may be true: the accuracy of the RFS algorithm (discussed later) on the MiniImageNet dataset was reported to be higher ($\leq 1\%$) when the logit layer was retained on the feature extractor.

RFS [27]: Yonglong Tian et al. showed in their paper that a powerful feature extractor could give high accuracies even for basic classifiers like linear models. They illustrated this using their proposed RFS model as well as some concurrent results [6, 17]. Notably, RFS only uses whole-class classification to train the feature extractor, it does not use fine-tuning on the network using meta-learning. They also showed that logistic regression seems superior to nearest-neighbor in the classifier stage (e.g., Euclidean distance used by prototypical networks). However, on feature normalization, both techniques work equally well for $k \leq 5$. The ablation study conducted by the authors also showed that self-distillation [1] can be used to improve the feature extractor network.

Feature Map Reconstruction Networks [31] use the output of the feature extractor to generate a feature map for both the support set as well the query image. Then, the feature map of each class in the support set is regressed to reconstruct the query feature map. The paper claims that empirically, the reconstruction of the query features is easier using the support set of the same class compared to other classes, and so the query label is assigned to the class that best reconstructs the query feature map. It uses a closed-form solution rather than a convergence method and is therefore highly efficient. It uses meta-learning to learn a parameter that controls the degree of regularization. It too uses whole-class classification to pre-train the feature extractor on the training dataset.

Algorithm 1 Distance Metric Method trained using meta-learning

Input: Distribution of tasks $p(T)$, distance based classifier with parameters θ, Feature Extractor network with meta-learned parameters ϕ

Randomly Initialize ϕ
while not done **do**
  $metaLoss \leftarrow 0$
  for $T_m \sim p(T)$ **do**
    $supportEmbedding \leftarrow featureExtractor(\phi, supportSet(T_m))$
    $queryEmbedding \leftarrow featureExtractor(\phi, querySet(T_m))$
    $\theta \leftarrow \arg\min_\theta \mathcal{L}(\theta, supportEmbedding)$ ▷ Train the classifier
    $metaLoss \mathrel{+}= \mathcal{L}(\theta, queryEmbedding)$
  end for
  Update ϕ using $\nabla_\phi metaLoss$ ▷ Update meta-learner parameters
end while

Algorithm 2 Initialization-based methods (Maml) [9]

Input: Distribution of tasks $p(T)$, hyperparameters α and β, classifier network f_θ where θ is the meta-learning parameter, each task has its own adapted parameter θ_m for classification

Randomly Initialize ϕ
while not done **do**
  $metaLoss \leftarrow 0$
  for $T_m \sim p(T)$ **do**
    for i in (1,numAdaptSteps) **do** ▷ numAdaptSteps is a small integer value
      $trainLoss \leftarrow 0$
      for $trainExample$ in RandomShuffle($supportSet(T_m)$) **do**
        $trainLoss \mathrel{+}= \mathcal{L}(f_\theta, trainExample)$
      end for
      $\theta_m \leftarrow \theta - \alpha \nabla_\theta trainLoss$ ▷ Update adapted parameter
    end for
    for $query$ in $querySet(T_m)$ **do**
      $metaLoss \mathrel{+}= \mathcal{L}(f_\theta, query)$
    end for
  end for
  $\theta \leftarrow \theta - \beta \nabla_\theta metaLoss$ ▷ Update meta-parameter θ
end while

3.2 *Initialization-Based Methods*

Initialization-based methods aim to find a set of model parameters such that only a few steps of gradient descent can be used to obtain optimal task parameters for any task in the task distribution.

Model-Agnostic Meta-learning [9] is an algorithm that can be applied over any ML model trained using gradient descent, irrespective of whether the model performs classification, regression, or even reinforcement learning. MAML set the foundation for initialization-based methods. Despite its broad applicability, the core algorithm is quite straightforward. At a given training epoch, optimal parameters are calculated using gradient descent on the initial parameters for each training task. The sum of regularized task losses is then used as the meta-loss to update the model parameters for the next epoch. Eventually, the model reaches a state where just a few gradient descent steps allow it to obtain optimal task parameters from its default parameters. MAML's performance relies on feature reuse among various tasks rather than its ability to infer highly general network parameters [20, 27], similar to conventional transfer learning methods [25, 35].

LEO [22] aims to decouple initialization-based learning from the high-dimensional model parameters. LEO uses an encoder-decoder network to directly generate the model parameters from the training set. The encoder network generates a lower-dimensional embedding z from the dataset, which is then decoded to obtain the model parameters. Unlike MAML, during base training on a specific task, instead of updating the model parameters, the algorithm updates z instead. z is updated till finally, decoding it gives the optimal model parameters for that task. The parameters of the encoder-decoder network are updated using meta-learning. Overall, LEO is more efficient and is less prone to overfitting than MAML as it performs learning in a smaller dimensional space.

4 Experimental Results

The MiniImageNet dataset [28] is currently the most widely used FSL dataset. It contains, in total, 100 different classes, where each class itself has 600 images. The images are in color. This dataset is, in fact, a subset of the ImageNet dataset. There are a variety of classes in the MiniImageNet dataset, such as bicycle, spider-web, arctic fox, jellyfish, and horse, among others. The *tiered*ImageNet [21] is a subset of the ILSVRC dataset. It contains 608 classes that are hierarchically grouped into nodes. Each leaf node is divided into disjoint sets of classes for training, validation, and testing. Previously, the Omniglot dataset [15] was used to evaluate few-shot learning algorithms. It contains a total of 1623 handwritten characters in 50 different language alphabets. However, as algorithms were able to get upwards of 99% on Omniglot, it has been discontinued in favor of the *tiered*ImageNet.

The hardware consists of an Intel®Core™i5-8250U Processor and Nvidia Geforce GTX 1050 Ti 4GB graphics card. The operating system used is Ubuntu 18.04 LTS (Bionic Beaver). Pytorch [19] has been used as the machine learning framework to carry out experiments. The code used to run the experiments have been obtained from Github repositories [7, 13, 16, 30, 32].

All the published accuracies (see Tables 1 and 2) are for the 5-way case: each training and test task has exactly 5 classes that are to be classified. In the case of 1-shot, only a single train example is given to a model from each class. For the 5-shot case, each model is given five training examples from each class.

Each algorithm can be more suitable in a particular few-shot scenario. All algorithms perform well on simple datasets, such as a dataset of characters. A simpler model such as a prototype network may be preferable in such a case. On challenging

Table 1 5-way accuracy in percentage on the MiniImageNet dataset

Model	Backbone	1-shot	5-shot
ProtoNet	Conv-4	48.34	66.13
RelationNet	Conv-4	49.67	64.83
Baseline	Conv-4	43.01	60.42
Baseline++	Conv-4	48.93	64.57
RFS	ResNet-12	61.73	76.38
RFS-distill	ResNet-12	63.26	80.37
FRN	ResNet-12	65.97	79.60
MAML	Conv-4	47.49	60.71
LEO	–	60.22	77.96

Table 2 5-way accuracy in percentage on the *tiered*ImageNet Dataset

Model	Backbone	1-shot	5-shot
ProtoNet	Conv-4	51.26	72.30
RelationNet	Conv-4	55.81	70.39
RFS	ResNet-12	69.58	83.95
RFS-distill	ResNet-12	71.35	85.02
FRN	ResNet-12	71.69	86.40
MAML	Conv-4	51.67	72.38
LEO	–	66.07	81.45

datasets such as the MiniImageNet or the *tiered*ImageNet, based on our experiments, FRN and RFS-distill have the highest accuracies. If performance is a priority, the FRN algorithm is the superior choice between the two as it uses a closed-form solution and does not rely on gradient descent, making it much faster. Initialization-based methods rely on the similarity of features of the various classes. Hence, they may be more useful in situations where the tasks are very similar, for instance, if each task requires the algorithm to distinguish between two species of cats.

5 Conclusion

In this paper, we have cherry-picked and discussed some important metric-based and initialization-based methods for few-shot image classification. We start by giving a background on few-shot classification and the meta-learning framework. We then elaborate on the underlying training algorithms for metric-based and initialization-based methods using pseudocode. Meta-learning used to be the favored approach to train the joint parameters of metric-based methods. However, we have found that recent models such as RFS were able to surpass previous results by pre-training their feature extractor using the *whole-class classification* approach. The experiments indicate that metric-based FSL models perform better than their optimization-based counterparts. Current research on metric-based learning focuses primarily on the training methodology or choosing the right classifier, while there is a lack of attention on improving the feature extractor. So far, researchers have just used conventional image recognition networks such as ConvNet or Resnet-12 for the feature extractor. However, ResNet-12 and most image classification networks work best in image classification scenarios where large amounts of data are present. Future work could improve the feature extractor of metric-based models and make them more suitable for few-shot learning. Furthermore, more studies need to be performed to measure the cross-domain accuracies across different datasets to understand how FSL algorithms perform under extreme task variance. In particular, more research must be done to make initialization-based methods robust to higher levels of variance in the task distribution.

References

1. Allen-Zhu Z, Li Y (2020) Towards understanding ensemble, knowledge distillation and self-distillation in deep learning. ArXiv preprint arXiv:2012.09816
2. Antoniou A, Storkey A, Edwards H (2017) Data augmentation generative adversarial networks. ArXiv preprint arXiv:1711.04340
3. Benaim S, Wolf L (2018) One-shot unsupervised cross domain translation. advances in neural information processing systems **31**
4. Bowles C, Chen L, Guerrero R, Bentley P, Gunn R, Hammers A, Dickie DA, Hernández MV, Wardlaw J, Rueckert D (2018) Gan augmentation: augmenting training data using generative adversarial networks. ArXiv preprint arXiv:1810.10863
5. Chen WY, Liu YC, Kira Z, Wang YCF, Huang JB (2019) A closer look at few-shot classification. ArXiv preprint arXiv:1904.04232
6. Chen Y, Liu Z, Xu H, Darrell T, Wang X (2021) Meta-baseline: exploring simple meta-learning for few-shot learning. In: Proceedings of the IEEE/CVF international conference on computer vision, pp 9062–9071
7. Davis Wertheimer LT, Frn. https://github.com/Tsingularity/FRN
8. Edwards H, Storkey A (2016) Towards a neural statistician. ArXiv preprint arXiv:1606.02185
9. Finn C, Abbeel P, Levine S (2017) Model-agnostic meta-learning for fast adaptation of deep networks
10. Goodfellow I, Pouget-Abadie J, Mirza M, Xu B, Warde-Farley D, Ozair S, Courville A, Bengio Y (2014) Generative adversarial nets. Adv Neural Inf Proc Syst 27
11. He K, Zhang X, Ren S, Sun J (2015) Delving deep into rectifiers: surpassing human-level performance on imagenet classification. In: Proceedings of the IEEE international conference on computer vision, pp 1026–1034
12. Hospedales T, Antoniou A, Micaelli P, Storkey A (2020) Meta-learning in neural networks: a survey. ArXiv preprint arXiv:2004.05439
13. Hung-Ting C (2020) Leo. https://github.com/timchen0618/pytorch-leo
14. Kozerawski J, Turk MA (2018) Clear: cumulative learning for one-shot one-class image recognition. In: 2018 IEEE/CVF conference on computer vision and pattern recognition, pp 3446–3455
15. Lake BM, Salakhutdinov R, Tenenbaum JB (2015) Human-level concept learning through probabilistic program induction 350(6266):1332–1338. https://doi.org/10.1126/science.aab3050
16. Li W, Dong C, Tian P, Qin T, Yang X, Wang Z, Jing H, Shi Y, Wang L, Gao Y, Luo J (2021) Libfewshot: a comprehensive library for few-shot learning. ArXiv preprint arXiv:2109.04898
17. Liang M, Huang S, Pan S, Gong M, Liu W (2019) Learning multi-level weight-centric features for few-shot learning. ArXiv preprint arXiv:1911.12476
18. Odena A, Olah C, Shlens J (2017) Conditional image synthesis with auxiliary classifier gans. In: International conference on machine learning. PMLR, pp 2642–2651
19. Paszke A, Gross S, Massa F, Lerer A, Bradbury J, Chanan G, Killeen T, Lin Z, Gimelshein N, Antiga L, Desmaison A, Kopf A, Yang E, DeVito Z, Raison M, Tejani A, Chilamkurthy S, Steiner B, Fang L, Bai J, Chintala S (2019) Pytorch: an imperative style, high-performance deep learning library. In: Wallach H, Larochelle H, Beygelzimer A, d' Alché-Buc F, Fox E, Garnett R (eds) Advances in neural information processing systems, vol 32, pp 8024–8035
20. Raghu A, Raghu M, Bengio S, Vinyals O (2019) Rapid learning or feature reuse? towards understanding the effectiveness of maml. ArXiv preprint arXiv:1909.09157
21. Ren M, Triantafillou E, Ravi S, Snell J, Swersky K, Tenenbaum JB, Larochelle H, Zemel RS (2018) Meta-learning for semi-supervised few-shot classification. ArXiv preprint arXiv:1803.00676
22. Rusu AA, Rao D, Sygnowski J, Vinyals O, Pascanu R, Osindero S, Hadsell R (2018) Meta-learning with latent embedding optimization. ArXiv preprint arXiv:1807.05960
23. Santoro A, Bartunov S, Botvinick M, Wierstra D, Lillicrap T (2016) Meta-learning with memory-augmented neural networks. In: International conference on machine learning. PMLR, pp 1842–1850

24. Snell J, Swersky K, Zemel RS (2017) Prototypical networks for few-shot learning
25. Sun Q, Liu Y, Chua TS, Schiele B (2019) Meta-transfer learning for few-shot learning. In: Proceedings of the IEEE/CVF conference on computer vision and pattern recognition, pp 403–412
26. Sung F, Yang Y, Zhang L, Xiang T, Torr PH, Hospedales TM (2018) Learning to compare: relation network for few-shot learning. In: Proceedings of the IEEE conference on computer vision and pattern recognition, pp 1199–1208
27. Tian Y, Wang Y, Krishnan D, Tenenbaum JB, Isola P (2020) Rethinking few-shot image classification: a good embedding is all you need? In: European conference on computer vision. Springer, pp 266–282
28. Vinyals O, Blundell C, Lillicrap T, Kavukcuoglu K, Wierstra D (2017) Matching networks for one shot learning
29. Wang Y, Yao Q, Kwok JT, Ni LM (2020) Generalizing from a few examples: a survey on few-shot learning. ACM Comput Surveys (CSUR) 53(3):1–34
30. Wei-Yu Chen Jia-Bin Huang YCL. Baseline. https://github.com/wyharveychen/CloserLookFewShot
31. Wertheimer D, Tang L, Hariharan B (2020) Few-shot classification with feature map reconstruction networks. ArXiv preprint arXiv:2012.01506
32. Yonglong Tian Wang Y. Rfs. https://github.com/WangYueFt/rfs
33. Zhang Y, Tang H, Jia K (2018) Fine-grained visual categorization using meta-learning optimization with sample selection of auxiliary data. In: Proceedings of the European conference on computer vision (ECCV), pp 233–248
34. Zhao B Basics of few-shot learning with optimization-based meta-learning. https://towardsdatascience.com/basics-of-few-shot-learning-with-optimization-based-meta-learning-e6e9ffd4775a
35. Zhuang F, Qi Z, Duan K, Xi D, Zhu Y, Zhu H, Xiong H, He Q (2020) A comprehensive survey on transfer learning. Proc IEEE 109(1):43–76

A Fast and Efficient Methods for Eye Pre-processing and DR Level Detection

Shivendra Singh, Ashutosh D. Bagde, Shital Telrandhe, Roshan Umate, Aniket Pathade, and Mayur Wanjari

Abstract Diabetes mellitus causes diabetic retinopathy (DR), which is the primary cause of blindness worldwide. To solve this issue, early screening should be done to detect diabetic retinopathy in retina images to prevent vision impairments and vision loss. There are manual methods for diagnosis which is time taking and costly. Many deep learning models have been proposed for the detection of DR on fundus retina images but none are commercially used in India due to lack of robustness. These models are also facing problems to pre-process dark fundus retina images and have poor performance when such images are passed as input. In this paper, we proposed fast and efficient methods for image pre-processing to tackle dark images, balance lighting, and remove an uninformative area from the image and neural network architecture with 32 layers for DR detection and DR level classification. The proposed model is trained on 2929 images, and the training accuracy achieved was 98%; when validated on 733 test images, the accuracy of 96% was achieved.

Keywords Diabetic retinopathy (DR) · Image processing · Convolutional neural network · OpenCV

1 Introduction

Diabetic retinopathy is a serious health problem, affecting 6 million people in India and 93 million people worldwide [1], and by 2045, this number is estimated to climb to 700 million [2, 3]. DR leads to retina vessel expansion and spills liquid and blood. DR causes vision impairments in diabetic patients [4]. At the primary stages, there are no visible symptoms, and hence, screening or diagnosis at the early stage is required. Manual diagnosis is costly and time taking, and on the other side, an automated system for DR detection takes less time and makes the process easy. To address these issues, this paper proposed a fast and efficient deep learning architecture to detect DR. In this work, we focused on the image processing tasks to pre-process a

S. Singh (✉) · A. D. Bagde · S. Telrandhe · R. Umate · A. Pathade · M. Wanjari
Department of R & D, Jawaharlal Nehru Medical College, Datta Meghe Institute of Medical Sciences, Wardha, Maharashtra, India
e-mail: shivendra3497@gmail.com

M. Thakur et al. (eds.), *Soft Computing for Problem Solving*, Lecture Notes in Networks and Systems 547, https://doi.org/10.1007/978-981-19-6525-8_5

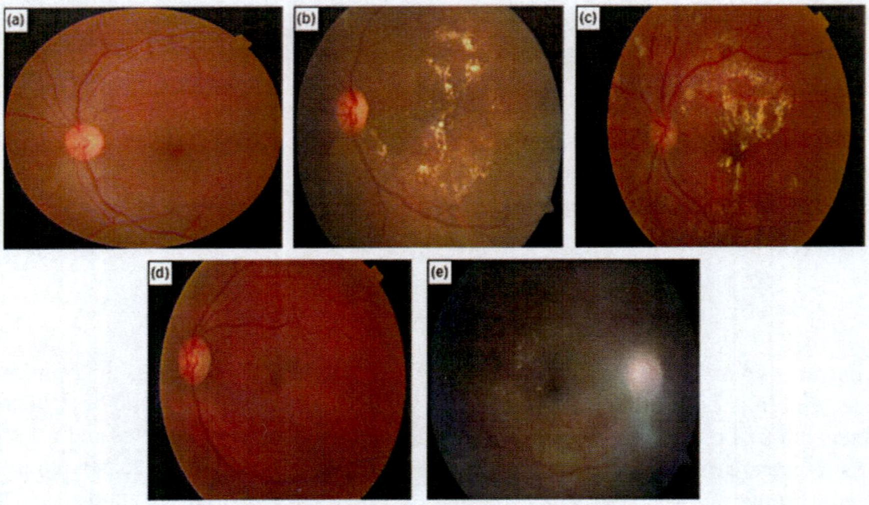

Fig. 1 Retina images with different diabetic retinopathy levels. **a** Normal, **b** Mild, **c** Moderate, **d** Severe, and **e** Proliferative [5]

variety of images taken under varying lighting conditions. The input images are preprocessed with image processing techniques such as Weighted Gaussian Filter and circle cropping, and all the implementation of methods is done using the OpenCV library and Python programming language with Visual Studio Code as editor. Our final model classifies the image dataset into five categories.

There are different levels of DR based on severity as shown in Fig. 1. The first level is considered as no DR, and the second level is considered a Mild DR. In this phase, small spherical swelling can be observed in the blood vessels and this occurs due to the presence of microaneurysms. The third level is considered Moderate DR. In this phase, the microaneurysm number and size increase and blood flow to the retina is obstructed. The fourth phase is considered Severe DR, and, in this phase, the obstructed blood vessel number increases and results in the blood supply loss to different regions of the retina. And the last phase which is the fifth is considered Proliferative DR which is the most extreme phase. In this phase, the number of novel blood vessels increases, and these vessels are weak and prone to injury. This led to constant leakage of blood and fluid inside the retina resulting in vision impairments.

2 Related Work

The medical practitioner in clinics and hospitals puts drops in the patient's eye to perform an extensive dilated eye examination; this step enables the practitioner to get a clear view of retina blood vessels to examine for abnormalities. Another diagnosis method is called fluorescein angiography where fluorescein also called yellow dye is injected into the patient's vein. A camera captures a picture as the dye flows through blood vessels to determine leaking fluid or clogged blood vessels [6]. Some research articles focus on clinical identification such as HbAIs and glucose for diabetic retinopathy diagnosis, and the authors also ranked these clinical identifications as the most significant risk factors [7]. Furthermore, some research articles presented deep learning models to classify and detect the level of diabetic retinopathy. Deperlioglu and Kose [8] presented image processing techniques to enhance the retinal fundus image and a deep learning approach to diagnose DR from these images. The image processing step includes methods such as HSV, histogram equalization, V transfer algorithm, and a low-pass Gaussian filter. The Kaggle's DR dataset was used to assess the proposed method, and the overall accuracy of the classification model was 96.33%. Shanthi et al. [9] worked on the Messidor dataset and trained modified AlexNet architecture on the data to detect DR and classify retinal fundus images with the validation set accuracy of 96.25%. Satwik et al. [5] utilized the transfer learning technique with SEResNeXt32 × 4d and EfficientNetb3 to classify retina images according to the severity. The accuracy with SEResNeXt32 × 4d architecture achieved was 85.15%, and the accuracy with EfficientNetb3 architecture achieved was 91.44%. To recognize DR, Wu et al. [10] utilized the transfer learning technique with VGG19, Resnet 50, Inception V3 with 150 epoch and 0.0001 as learning rate, Inception V3 with 300 epoch and 0.001 as learning rate, and Inception V3 with 300 epoch and 0.0001 as the learning rate, and accuracy achieved was 51%, 49%, 50%, 55%, and 61%, respectively. Xie et al. [11] worked on transformed deep learning network Resnet where aggregated residual transformation and follows Fb.resnet.torch [12] code for implementation. Pragathi et al. [13] proposed an integrated approach with support vector machine (SVM), Principal Component Analysis (PCA), and moth flame and compared various machine learning algorithms and achieved the highest accuracy of 67.7% with the SVM algorithm.

3 Dataset Description

The dataset is taken from Kaggle [14] and consists of large retina images which are captured using fundus photography. The images are labeled into five categories as shown in Table 1.

Table 1 Showing levels of diabetic retinopathy

0	No diabetic retinopathy
1	Mild DR
2	Moderate DR
3	Severe DR
4	Proliferative DR

The image dataset has noise in both images and labels, and some images contain artifacts. The images were collected from various towns by using different varieties of cameras, which introduce varying lighting conditions in the image.

4 Retina Image Pre-processing

4.1 Why Pre-processing is Required

To maximize the accuracy and performance of neural network architecture, the quality of images should be improved before feeding these images (retina fundus images) into the neural network. This step is necessary because of two reasons. First, the images taken using fundus photography have varying lighting conditions, and these are introduced in the images since the images were taken from different towns with different devices. Second, some retina images contain an uninformative area and we need to crop it. The algorithm used to pre-process an input image is described in Sect. 4.3.

4.2 The Methodology Used to Pre-process and Implementation

4.2.1 Ben's Pre-processing

To improve lighting conditions, Ben's pre-processing [15] method is used. Here, the sigma used for Gaussian blur is equal to 10; in this paper, the image quality is enhanced by using other values of sigma between 30 and 40.

4.2.2 Circle Cropping

The circle cropping is required to feed all images to the deep neural network in the same way. After visualizing the dataset, we found that there are different types of images present in the dataset such as rectangular images with vertical and horizontal cropping and square images with vertical and horizontal cropping.

4.3 The Algorithm Used to Pre-process an Input Image

Step 1: Start
Step 2: Declare variables N, sigma, resize the width, resize the height, and image.
Step 3: Set N = Total number of images.
Step 4: Repeat the steps until N = 0
    Step 4.1: Read an image from project directory
        Image = cv2.imread("input image")
    Step 4.2: If image Format = "RGB"
            GB = Gaussian Blur (Image)
            Cropped = circle Crop (Image)
                Return Cropped
        Else
            Image = cv2.cvtColor(image, cv2.COLOR_BGR2RGB)
            GB = Gaussian Blur (Image)
            Cropped = circle Crop (GB)
                Return Cropped
Step 5: Stop

5 Proposed Neural Network Architecture

The proposed model consists of 32 convolution layers. One convolution layer contains a batch normalization function, a ReLU activation function, and average pooling.

5.1 Batch Normalization

There is an internal covariate shift that means the dataset has different distributions in each mini-batch whenever the model's weights are updated. With the use of batch normalization, the input data is standardized.

5.2 Activation Function

The convolutional neural network contains neurons inside the network;, these neurons calculate the sum (weighted sum), and at last, bias is added. There are two flags (zero and one) for a neuron that decides, it should be fired or not. In our network model, an activation function called ReLU was used.

5.3 Average Pooling

Average pooling is used to downsample the input feature map. This solves the problem of feature maps in that the feature map's output is sensitive to the input image feature locations. The average pooling is performed by taking the average of each patch of the feature map.

Apart from these three, the last layer consists of a flattened layer and a dense layer with a SoftMax activation function. A SoftMax function is used to classify the input image into four classes.

5.4 Program Flowchart

Figure 2 of Sect. 5.4 describes the workflow of the project; as in the first step, the retina image data was collected, and in the second step, these images are sent for pre-processing to remove unwanted area of the image that does not contribute to the prediction model. At last, the image is resized and weighed Gaussian blur was used on the resized image to improve lightening. The sigma values used for this filter were between 30 and 40. In the third step, the pre-processed images are sent to the neural network model. And in the last step that is step 4, the input image is classified into one of the 5 classes No DR, Mild, Moderate, Severe, and Proliferative.

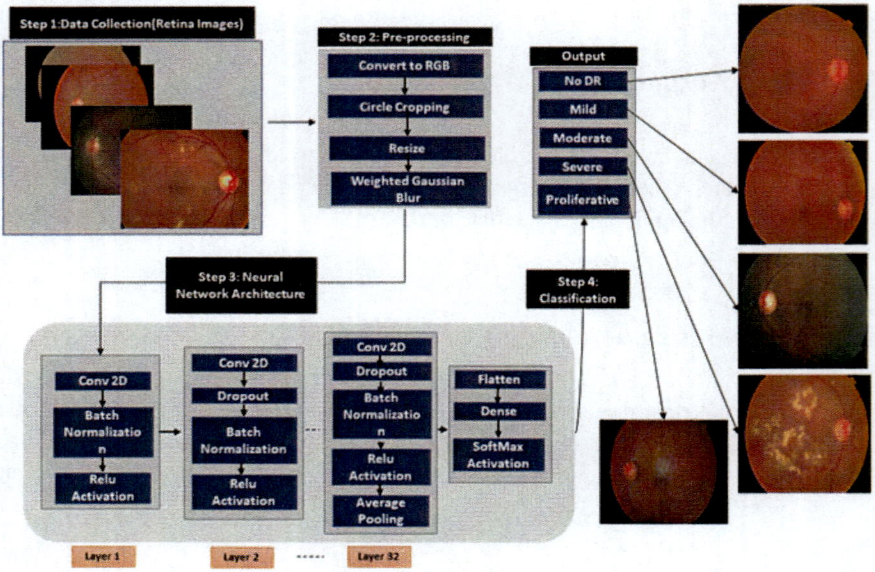

Fig. 2 Showing program flowchart, starting from image processing to DR level classification

Table 2 Showing parameter values such as learning rate, batch size, dropout rate, compression, number of filters, and function such as optimizer function, loss function, and metrics used for training of the neural network

Learning rate	1.4
Batch size	8
Dropout rate	000.1
Compression	0.5
Number of filters	32
Optimizer	Adam
Loss function	Cross-entropy
Metrics	Accuracy

6 Model Training

The dataset has a total of 3662 retina images, and these retina images are split into a training set and a test set. The neural network architecture is trained on the training dataset that has 2929 retina images and tested on 733 images. After the training phase, the training accuracy of 98% was achieved and performance accuracy of 96% was achieved when the model was validated on the test dataset. Parameter value used while training is shown in Table 2 where learning rate is equal to 1.4, dropout rate is equal to 000.1, compression is equal to 0.5, number of filters is equal to 35, and Adam function is used as optimizer, categorical cross-entropy as loss function, and accuracy as metrics. The learning rate is a parameter that determines how quickly the model adapts to the situation. Dropout is the method that helps to reduce overfitting by dropping the output nodes at the mentioned probability. The extension of the stochastic gradient is called the Adam optimizer, and it is used in the algorithm since it is easy to implement and computationally efficient and works well with large datasets. The confusion matrix (CM) and classification report (CR) after the model's training and testing are shown in Tables 3 and 4. The proposed model failed to work or has poor performance when the images that are blurred, cropped, and types of images in which vessels are not visible are passed as input to the model. The bias of the model is calculated, and its value is equal to 0.2 for class 0 or No DR class, 0.9 for class 1 or Mild DR class, 0.3 for class 2 or Moderate DR class, 0.8 for class 3 or Severe DR class, and 0.8 for class 4 or Proliferative DR class.

7 Conclusion

In this paper, we proposed an efficient method to pre-process the retina images and a 32-layer convolutional neural network to categorize these retina images into 5 classes such as No DR, Mild DR, Moderate DR, Severe DR, and Proliferative. The overall structure of the project is organized into four steps such as dataset acquisition, image pre-processing step, training of deep learning model, and classification. In the pre-processing step, circle cropping is used to remove the uninformative area and

Table 3 Confusion matrix

Actual class	Actual No DR class	**354**	5	2	0	0
	Actual Mild class	1	**69**	3	0	1
	Actual Moderate class	1	1	**195**	1	3
	Actual Severe class	2	1	1	**35**	1
	Actual Proliferative class	0	1	1	1	**54**
		Predicted No DR class	Predicted Mild class	Predicted Moderate class	Predicted Severe class	Predicted Proliferative class
	Predicted class					

Table 4 Final classification report

	Model precision (%)	Model recall (%)	Model f1-score (%)	Support (total images in level 0)
Level 0	99	98	98	361
Level 1	90	93	91	74
Level 2	97	97	97	200
Level 3	95	90	92	39
Level 4	92	92	92	59
Accuracy			96	733
Macro average	94	94	94	733
Weighted average	96	96	96	733

weighted Gaussian blur to improve lighting conditions. Different values of sigma were used in weighted Gaussian blur operation but the sigma values between 30 and 40 were best suited for the overall performance of the model. The trained model's performance is satisfactory and produced an excellent f1-score when tested on the validation dataset. The f1-score is 0.98, 0.91, 0.97, 0.92, and 0.92 for the classes 0, 1, 2, 3, and 4, respectively, and accuracy on validation set achieved is 96%. The system architecture used for training of the model includes 16 GB GEFORCE RTX GPU, 32 GB RAM, and 12th Generation Intel® Core™ i9-12900 K, 3187 MHz, 16 Core(s), and 24 Logical Processors.

References

1. Shukla UV, Tripathy K (2022) Diabetic retinopathy. (Updated 29 Aug 2021). In: StatPearls (Internet). Treasure Island (FL): StatPearls Publishing, Jan 2022. https://www.ncbi.nlm.nih.gov/books/NBK560805/. Last accessed 28 Apr 2022
2. International Diabetes Federation. International diabetes federation diabetes atlas. https://www.diabetesatlas.org/en/. Last accessed 24 Apr 2022
3. Tsiknakis N, Theodoropoulos D, Manikis G, Ktistakis E, Boutsora O, Berto A et al (2021) Deep learning for diabetic retinopathy detection and classification based on fundus images: a review. Comput Biol Med 135:104599
4. Bourne RRA, Stevens GA, White RA, Smith JL, Flaxman SR, Price H et al (2013) Causes of vision loss worldwide 1990–2010: a systematic analysis. Lancet Glob Health 1(6):e339–e349
5. Ramchandre S, Patil B, Pharande S, Javali K, Pande H (2020) A deep learning approach for diabetic retinopathy detection using transfer learning. In: 2020 IEEE international conference for innovation in technology (NEOCON) (Internet). IEEE, Bengaluru, India, Cited 6 Apr 2022, pp 1–5. Available from: https://ieeexplore.ieee.org/document/9298201/
6. Bora A, Balasubramanian S, Babenko B, Virmani S, Venugopalan S, Mitani A et al. (2021) Predicting risk of developing diabetic retinopathy using deep learning. p 40
7. Nneji GU, Cai J, Deng J, Monday HN, Hossin MA, Nahar S (2022) Identification of diabetic retinopathy using weighted fusion deep learning based on dual-channel fundus scans. Diagnostics 12(2):540
8. Deperlioglu O, Kose U (2018) Diagnosis of diabetic retinopathy using image processing and convolutional neural network. In: 2018 Medical technologies national congress (TIPTEKNO) (Internet). IEEE, Magusa, cited 6 Apr 2022, pp 1–4. Available from: https://ieeexplore.ieee.org/document/8596894/
9. Shanthi T, Sabeenian RS (2019) Modified Alexnet architecture for classification of diabetic retinopathy images. Comput Electr Eng 76:56–64
10. Wu Y, Hu Z (2019) Recognition of diabetic retinopathy based on transfer learning. In: 2019 IEEE 4th international conference on cloud computing and big data analysis (ICCCBDA) (Internet). IEEE, Chengdu, China, cited 6 Apr 2022. pp 398–401. Available from: https://ieeexplore.ieee.org/document/8725801/
11. Xie S, Girshick R, Dollár P, Tu Z, He K (2017) Aggregated Residual transformations for deep neural networks. ArXiv161105431 Cs (Internet). 10 Apr 2017, cited 6 Apr2022. Available from: http://arxiv.org/abs/1611.05431
12. Gross S, Wilber M (2016) Training and investigating residual nets. https://github.com/facebook/fb.resnet.torch. Last accessed 28 Apr 2022
13. Pragathi P, Nagaraja RA (2022) An effective integrated machine learning approach for detecting diabetic retinopathy. Open Comput Sci 12(1):83–91
14. Kaggle.com. 2015. Diabetic retinopathy detection | Kaggle. (Online). Available at: https://www.kaggle.com/competitions/diabetic-retinopathy-detection/data. Accessed 9 Mar 2022
15. Graham B (2015) Kaggle diabetic retinopathy detection competition report. The University of Warwick, 6 Aug 2015, pp 24–6

A Deep Neural Model CNN-LSTM Network for Automated Sleep Staging Based on a Single-Channel EEG Signal

Santosh Kumar Satapathy, Khelan Shah, Shrey Shah, Bhavya Shah, and Ashay Panchal

Abstract Sleep plays a vital role in human physiological behaviors. Sleep staging is a critical criterion for assessing sleep patterns. Therefore, it is essential to develop an automatic sleep staging algorithm. The present study proposes a deep neural network based on a convolutional neural network (CNN) and Long Short-Term Memory (LSTM) for automated sleep stage classification. We presented a deep neural CNN-LSTM network to model character-level information. In the proposed model, the CNN can extract high-level sleep signal features, and LSTM can realize sleep staging with high accuracy by combining the correlations among the sleep data in different sleep periods. Finally, we used the Sleep-EDF dataset for model assessment. On a single EEG channel (Fpz-Oz) from the Sleep-EDF dataset, the overall accuracy achieved 91.12%, according to the results. In most research, the data imbalance of training data exists, which has been solved in the proposed method. In addition, the overall accuracy of the proposed method was superior to those of the latest techniques based on Sleep-EDF. Hence eradicating the tedious work of sleep staging classification required by professionals. The proposed model helped achieve this accuracy level without using any hand-engineered features. The ability of the model to give such conspicuous results without using any handcrafted features makes it quite versatile and robust.

Keywords Electroencephalogram · Sleep stage · CNN-LSTM · Deep learning

1 Introduction

Sleep is considered a state of the brain and body that is differentiated based on elevated levels of consciousness and visual separation from the outer world. According to a survey, human newborns sleep around 75% of their lives. Sleep time and its quality play a significant role in a newborn's brain development, specifically for the initial one-two years of their life [1]. So, a sleep-deprived person can directly impact his

S. K. Satapathy (✉) · K. Shah · S. Shah · B. Shah · A. Panchal
Department of Information and Communication Technology, Pandit Deendayal Energy University (PDEU), Gandhinagar, India
e-mail: Santosh.Satapathy@sot.pdpu.ac.in

M. Thakur et al. (eds.), *Soft Computing for Problem Solving*, Lecture Notes in Networks and Systems 547, https://doi.org/10.1007/978-981-19-6525-8_6

mood and cognitive abilities. Maintaining good health for humans is an essential activity. Sleep deprivation and poor sleep quality interfere with regular exercise and cause physical and mental problems.

The ability to continuously monitor sleep quality helps identify sleep disorders quickly. A frequently used technique for sleep research, especially in diagnosing sleep disorders, is polysomnography (PSG), also known as a sleep test. It is a formal medical procedure that captures important biological signals for detecting sleep disorders. PSG consists of electroencephalogram (EEG), electrocardiogram (EOG), electromyogram (EMG), an electrocardiogram (ECG) [2]. The readings of EEG signals are taken by placing electrodes on the scalp.

They record the potential generated by nerve cells in the brain and open windows for studying nerve activity and brain function [3]. Therefore, efficient automated sleep scoring can save a lot of time and provide an objective assessment of sleep, regardless of expert subjective assumptions. The R&K criterion and the AASM's recommendation are the widely known high standards for evaluating sleep recordings. Each 30 s epoch of nocturnal sleep, as per R&K standards, corresponds to seven different stages: Wake, drowsiness, also known as S1, light sleep, also known as S2, S3, also known as deep sleep; S4, also known as deep or slow-wave sleep, rapid eye movement, and movement time MT. The rules by AASM were updated in 2007 to address and rectify certain newly invented flaws in the R&K. Stages such as S3 and S4 are mixed into a single-stage by AASM. After falling asleep, NREM stages 1–4 are attained within 45–60 min by an average individual. After the first episode of slow-wave sleep, the progression of non-REM sleep is reversed [4]. The first REM sleep usually occurs for 80 min or more. After falling asleep, REM sleep shortens with age. Attaining REM sleep in teenagers (especially if it lasts less than 30 min) may indicate a condition such as intrinsic depression, narcolepsy, circadian rhythm disorder, or withdrawal from the drug. Non-REM sleep and REM sleep change at night with an average 90–110 min cycle. The longer the sleep period, the lower the percentage of each process of slow-wave sleep and the higher the percentage of REM sleep. Overall, REM sleep accounts for approximately 20–30% of total rest, while non-REM sleep accounts for 45–60% (increased in the elderly) [5]. This research aims to devise an effective mechanism for identifying sleep pattern irregularities to improve sleep stage classification efficiency with different health conditions of subjects. The goal is to construct and create a more advanced deep learning approach for distinguishing sleep stages and accurately identifying sleep changes in behavior. For two-five sleep stages classification tasks, we will use Convolutional Neural Networks to retrieve time-invariant characteristics and bidirectional Long Short-Term Memory to automatically discover transition rules between sleep stages from EEG epochs in the proposed system. Hence the substantial motivation of research work is to improve the sleep staging classification accuracy and classify sleep stages by monitoring their behavior using improved deep learning models.

The flow of the paper consists of the following:

a) We surveyed various papers relevant to 'Sleep Staging Analysis' and have noted some of the remarkable ones in our literature Survey.

b) The Methodology section consists of an in-depth explanation of our proposed methodology, including its ins and outs, emphasizing our model structure and the data set we used to test our algorithm.
c) We have a detailed results section rife with charts from the algorithm's outcome.
d) The discussion section includes an extensive comparison of our model with previous models; this shows the strengths of our model.

2 Literature Survey

Yana et al. [6] proposed a sleep scoring method by using eight combinations of four modalities of polysomnography (PSG) signals. They used a Cyclic Alternating Pattern (CAP) PhysioNet database consisting of 232 characteristics, covering fractal characters, frequency characters, entropy characters, non-linear characters, time–frequency characters, and statistical characters. Machine learning approaches were performed for training the data in the presented study, in which the random forest classifier obtained the optimal accuracy of 86.24%.

Hence, the conclusion drawn by this study was that the proposed automated sleep scoring method was cost-effectual and also reduced the stress on medical practitioners. Apart from that, the limitation of this method was S1 stage is misidentified as W (wake) and Rapid Eye Movement by automated sleep scoring.

Shen et al. [7] worked on improved combination locality energy (LE), model-based essence features (IMBEFs), and DSSMs for sleep staging electroencephalogram signals in this research. The proposed work was performed on three public datasets: Sleep-EDF, Dreams Subjects, and ISRUC. Each EEG signal epoch is initially broken down into high-level and low-level sub-bands. Then the DSSM is estimated, and the LSB causes the LE calculation to be performed in the HSB. Finally, LE and DSSM are supplied to the suitable classifier for sleep staging classification. The accuracy of the Dreams Subjects database is approximately 78.92%, according to the R&K standard. According to the standards of the AASM, the accuracy of classification of the five classes in the ISRUC dataset was 81.65%, and the DS dataset was 79.90%. But the drawback of the proposed work was that the misclassified ratio was more for the S2 stage. Huanga et al. [8] showcased a novel feature screening and signaled preprocessing. They worked on the superposition of multi-channel signals to improvise the information in the actual movement. Sixty-two features were first selected for feature screening, including frequency-domain, time-domain, and non-linear characteristics. A Relief model was used to pick 14 factors substantially linked to the stages of sleep from the 62 parts. 2 redundant were removed using the Pearson correlation coefficient. The support vector machine (SVM) and 12 selected feature classifiers were employed for sleep staging based on 30 recordings using the above signal preprocessing procedure. But the proposed methodology gave an ineffective performance with the heterogeneous combinations of the signals. Ghimatgara et al. [9] proposed a feature selection algorithm in which various classic features were chosen from much data acquired during sleep electroencephalogram epochs. A random forest classifier was then used to classify EEG segments. Utilizing a leave-one-out validation plan, they obtained 79.4–87.4% accuracy. The only limitation of the presented work is that the classifier's performance is highly oblique on wake. Cooray et al. [10] provides a wholly automated RBD detection approach with automatic sleep staging and RBD identification. A restricted polysomnography montage was used to analyze 53 age-matched healthy controls and 53 patients with RBD. An RF classifier and 156 characteristics from EOG, EMG, and EEG signals classified sleep stages. An RF classifier was designed to detect RBD by integrating established techniques for measuring muscle atonia with extra variables such as the EMG fractal exponent and sleep architecture. This study surpasses existing measurements and shows that combining sleep architecture and transitions might help detect RBD. But the study gave more misclassification in the N2 stage. Diykh et al. [11] aimed to create a novel automatic approach for classifying EEG sleep stages using a weighted brain network and statistical model. As an outcome, a vector of characteristics represents every section. After that, the features vector is converted to a weighted, undirected network. Simultaneously, the network's characteristics are thoroughly examined. The network's properties are discovered to fluctuate depending on their sleep stages. The critical elements of each sleep stage's networks are best portrayed. The drawback of this method is imbalanced EEG data.

Cooray et al. [12] provides a wholly automated RBD detection approach that includes automatic sleep staging and RBD identification. A random forest Classifier was used, and a total of 156 characteristics were achieved from EOG, EMG, and EEG channels. Model attained Co hen's Kappa score of 0.62 while accuracy improved from 10 to 96% during manual sleep staging, but 92% accuracy was achieved during automated sleep staging. The advantage is despite the decline in incompetence for REM stage classification in RBD people, the specificity of REM stage detection remained high. Li et al. [13] proposed an approach named HyCLASS related to sleep stage classification using EEG signals of single-channel. 30 EEG signal characteristics, including frequency, temporal and non-linear variables, were used to train the Random Forest Classifier model from the Cleveland Family Study was used (CFS). The total accuracy and kappa coefficient applied to 198 patients were 85.95% and 0.8046, respectively. The benefit of this strategy is that it uses a Markov model to automatically address this problem by using existing sleep stage transition data from sleep stage sequences. Lai, et al. [14] is to discover sleep bruxism by evaluating the change in the electroencephalogram (EEG) domain during distinct sleep stages. The decision tree algorithm was used for sleep bruxism detection with the help of a mixture of A1 and P4 of C4 channels of scalp EEG. Welch's technique focused on the S1 sleep stage and rapid eye movement to detect bruxism. The database used for the method was Cyclic Alternating Pattern (CAP) PhysioNet Database, and the accuracy for C4-A1 and C4-P4 channels was 74.11% and 81.70%, respectively. In comparison, the average accuracy of the two channels was 81.25%. An individual result's specificity was lower when compared to the combined one. The limitation of the method is that the C4-A1 and C4-P4 channels could not trace all neuron channels. Zhang et al. [15] is to provide the findings in the form of a signal spectrum analysis of changes in the field of sleep stage. For a focus on two sleep stages, REM and

W channels ECG1-ECG2 and EMG1-EMG2 of the signal were mixed to diagnose bruxism with the help of power spectral density. The recordings of normal subjects and bruxism patients analyzed in the method were 95 and 149, respectively. During the REM and W sleep stages, the average normalized readings of the density of ECG1-ECG2 and EMG1-EMG2 are several times higher in bruxism than in normal sleep. The decision tree classifier for the power spectral density-based method shows an increase in the accuracy of sleep bruxism compared to previous ones. Diykh et al. [16] used the EEG signal divided into parts, and features were extracted and used to classify sleep stages. Two datasets named Sleep-EDF Database and Sleep Spindle Database were used for the whole process. The support vector machine (SVM) classifier was used. Experimental results reveal that the suggested method surpasses the other four methods and the SVM classifier regarding classification accuracy. The presented technique has an average classification accuracy of 95.93%. To identify sleep stages, graph theory is applied, and K-means with structural graph similarity mixed is used to reach a spike in the accuracy of classification that outperforms manual results. Still, this model fails to work in frequency-domain features.

In recent research developments, sleep studies are generally dependent on both machine learning (ML) [11, 17–34] and deep learning (DL) [16, 35–46] approaches by using different physiological signals. Most of the writers considered the EEG signal to monitor changes in sleep behavior. But it has been observed from the literature that most of the studies did not perform well in the case of multi-class sleep stages classification problems. Mainly sleep staging with ML approaches has a few limitations. First, it requires experts with prior knowledge to discriminate the features manually. Second, the extracted features cannot be fully observable concerning behavior changes in the individual sleep stages. Nowadays, this set of challenges is handled by using deep learning approaches. To examine the variations in hidden behavior of sleep over the individual stages of sleep concerning the changes in time and frequency levels. To enhance the accuracy of the classification of multi-class sleep staging classification problems, we propose a deep learning neural network to classify the various stages of sleep. First, we utilized the Single-channel EEG signals data as input and then moved to the preprocessing step to eliminate the muscle movement information and irrelevant noise compositions. Then, obtaining an LSTM and CNN for the five classification stages of sleep identifies the efficient extraction of hidden sleep behavior features and helps to get higher classification accuracy.

3 Methodology

Our proposed automatic sleep staging system mainly includes two steps: data processing and a CNN+LSTM, identifying the hidden relationship between the different sleep stages and classification. Figure 1 shows the detailed structure of our proposed method.

As shown in Fig. 1, after data were stored, we divided the EEG signals into 30 s epochs and performed necessary preprocessing and conversion operations. Moreover,

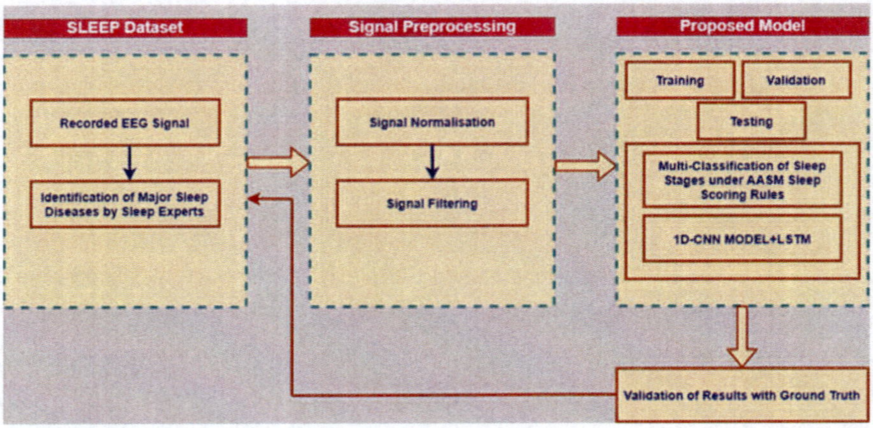

Fig. 1 The overall framework of the system

we present a hybrid neural network. A CNN was used to extract high-level data features, and LSTM was employed to combine the correlations among the sleep data in different periods. Finally, softmax classifiers were utilized to categorize sleep phases using the collected characteristics.

3.1 Dataset Used

The Sleep-EDF dataset was used in this paper. The Sleep-EDF dataset was used for parameter adjustment, training the model, and testing the model. Further, the self-recorded dataset was used to validate the generalization and reliability of the model.

3.2 Data Preprocessing

The number of samples of sleep recordings of each stage varied greatly. This imbalance in sleep stages data significantly affects classification performance. To overcome this imbalance, in terms of sleep data, we applied the method of equalized sampling for the training set in this research work. This process can be divided into four steps:

- Count the numbers of sleep samples for six types of sleep staging data, denoting them as e1, e2, e3, e4, e5, and e6, which correspond to the W, S1, S2, S3, S4, and REM epochs, respectively.
- Remove the most and least common values and calculate the average n of the remaining values.

- Under sample categories whose sample sizes are more significant than n, remove redundant data through random selection.
- Oversample the categories with sample sizes less than n and expand the dataset by random repeated sampling. The main aim of this paper was to achieve five-state sleep stages classification problems. Therefore, we merged S3 and S4 into N3. According to the corresponding sleep stage labels "W, N1, N2, N3, R" given by experts, the data were relabeled and transformed into "0, 1, 2, 3, 5", which corresponded to five classification sleep stages.

3.3 *Proposed Deep Neural Network Based on CNN-LSTM*

In this research work, the proposed deep neural network was mainly composed of a CNN architecture for automatic feature learning, an LSTM deep neural network architecture for decoding time information, and a multi-class classification component comprised a softmax layer.

The model first of extracting the feature automatically. This paper proposes a serial CNN architecture with four layers of convolutional units. An information input layer, a convolution operation layer, a normalization layer, a rectified linear unit (ReLU) activation layer, a pooling layer, and a dropout layer make up each convolution unit of the architecture. The sample tag field was mapped to the learned dispersed feature extraction. The obtained one-dimensional vector was normalized and activated to get the concise spatiotemporal features of the EEG signal. Figure 2 depicts the overall structure of the proposed CNN+LSTM model.

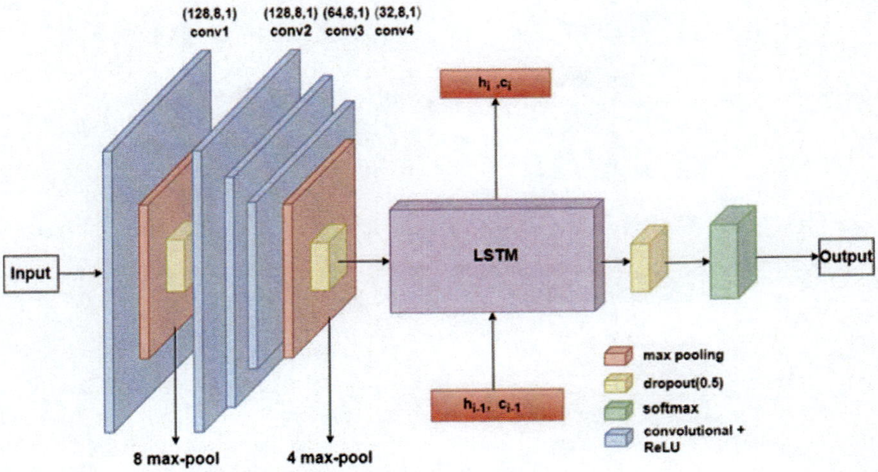

Fig. 2 The overall presentation of the proposed CNN+LSTM model

3.4 Model Specification

We have proposed a deep learning model for automatically evaluating sleep stages based on raw data from a single-channel EEG without using handmade features.

1. The algorithm segregates the data into three different parts, namely, train, validation, and test
2. We then assign specified weights to the weighted cross-entropy
3. The data is then sent to train our CNN model

We have used four-layered Convolution Neural networks; each encoded with 128, 128,128, and 128 filters, respectively. Each convolutional layer has a filter size of 8 and a stride of 1. Figure 3 presents the internal structure of the proposed CNN+LSTM model. After passing through each convolutional layer, the training batch is normalized and passed through the ReLU activation function. The advantages of ReLU are Sparse activation, Better Gradient Propagation: Reduces the problem of zero gradients compared to the sigmoid activation function that saturates in both directions.

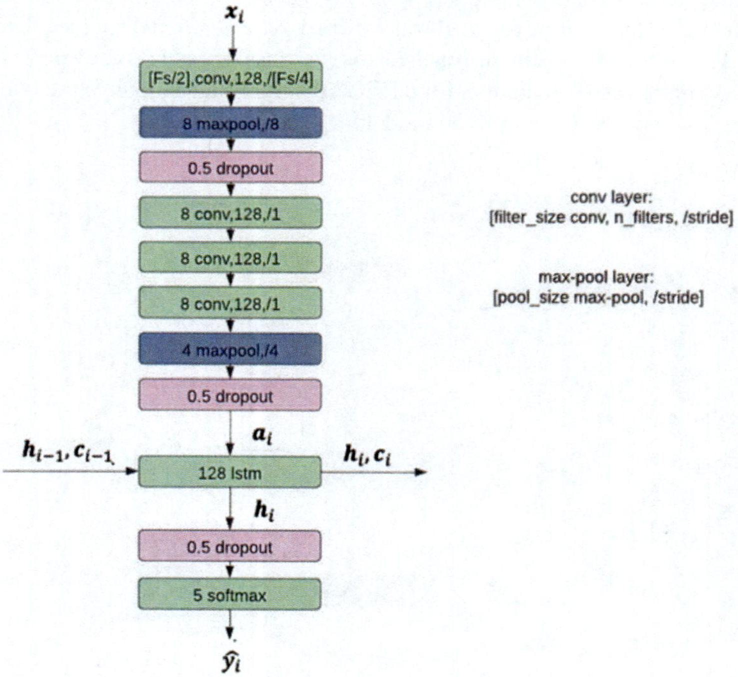

Fig. 3 Internal structure diagram of the CNN+LSTM model

3.5 *Evaluation Methodology*

K-fold cross-validation, a commonly used estimator of generalization performance, is widely used in automatic sleep stage classification. To assess the generalization ability of our method, this paper used five-fold cross-validation to obtain results. In each grouping result, (k-1) subclusters were utilized for training, but just one subcluster was used for testing. In this way, the overall accuracy was obtained by calculating the average accuracy of k predictions. The performance of our method was evaluated by using per-class precision (PR), recall (RE), F1-score (F1), and overall accuracy (ACC).

4 Experimental Results and Discussion

The research objective for this paper was a healthy population. Therefore, we selected a total of 100 groups of sleep data from a healthy population, of which 70% of data were used for the training purpose, 20% were used for the validation purpose, and 10% were used for the testing purpose to know about the performance of the model. After inputting the dataset into the automatic sleep stage classification system, it performed processing steps such as learning and feature extraction via the CNN-LSTM neural network and finally outputs the accuracy of each sleep stage through a softmax classifier. After using five-fold cross-validation, the average training accuracy was 90.31%, test accuracy was 91.12%, and validation accuracy was 89.74%. As we can see in Table 1, the Test F1 was 86.74%, and Val F1 was 82.74% for the W, N1, N2, N3, and REM stages. The training and validation loss of the five sleep stages was 0.28 and 0.33, respectively. Furthermore, to clearly show the automatic sleep classification results of the model, Fig. 4a, b c d and e demonstrates examples of the train, validation, and test accuracy, and Fig. 5a, b c d and e represents the F1 test and F1 validation. Figure 6a, b c d and e presented train and validation loss for five-fold cross-validation. Figure 7 illustrates the graph representation of the training accuracy and F1-score, and Fig. 8 represents the precision and recall for five-fold cross-validation.

5 Discussion

Sleep staging can reflect sleep structure and diagnose diseases, which is the basis for treating some conditions. Artificial sleep staging is a laborious and tedious task for sleep experts. Therefore, the research aims to obtain a high-accuracy automated sleep staging method. Here are some of the most advanced automatic scoring methods for sleep stages. Table 1 compares our suggested strategy to some advanced mechanical sleep stage research, which included single-channel EEG signals as inputs, and

Table 1 Comparison of the results obtained by our proposed work and other sleep staging methods of the Sleep-EDF dataset

Work	Input signals	Methodology used	Accuracy (%)
Ref [21]	Pz-Oz	Difference Visibility Graph	89.30
Ref [43]	Fpz-Cz+EOG	Multitask 1-max CNN	82.30
Ref [44]	Fpz-Cz+Pz-Oz+EOG+EMG+Resp+Tem	ReliefF+ICA	90.10
Ref [38]	Fpz-Cz+Pz-Oz	Complex Morlet Wavelets+L-BFGS	78.90
Ref [45]	Fpz-Cz+Pz-Oz	CNN+BiRNN	84.26
Ref [46]	Fpz-Cz	Wavelet+FFT+SVM	86.82
Our work	Fpz-Cz	CNN+LSTM	91.12

traditional machine learning, as the primary method and was classified according to AASM standards. In previous studies, many researchers have proposed different sleep staging methods. Table 1 summarizes the comparison between some cutting-edge automated sleep stage research and this study. In the existing research, most automatic sleep staging methods ignore the imbalance of training data, which may lead to poor classification effects. This problem has been well solved in the proposed way. Sleep-EDF is one of the most widely used datasets for sleep staging.

Table 1 compares the results and details of previously published works. The Sleep-EDF dataset was utilized in all of the studies described in Table 1, and the best accuracy is shown. The proposed method's average accuracy was 91.12%, as indicated. The results of our work reached the best accuracy among all the results of the compared studies, which shows that the implemented method has better performance than the existing method.

The main reasons for the high accuracy of our proposed method are. Initially, proposed and implemented an improved hybrid LSTM+CNN algorithm. The time structure information can be fully extracted to realize a more accurate feature extraction. The CNN model can receive the input of the graphic spectrum and adopts a relatively simple network structure to increase the training time of the model.

6 Conclusion

We proposed an automated sleep staging system mainly consisting of three steps: data processing, feature engineering, and a deep neural network based on CNN+LSTM, which achieved high classification performance in the public Sleep-EDF dataset of healthy subjects. We will develop the presented algorithm to make it suitable for more addition of physiological signals and several channels in the future. Furthermore, we

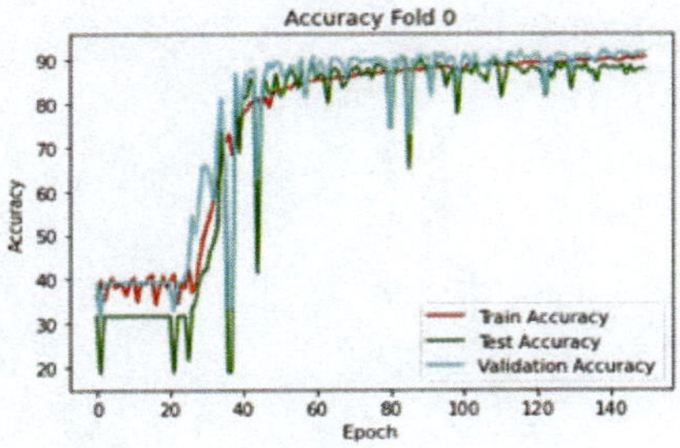

(a)

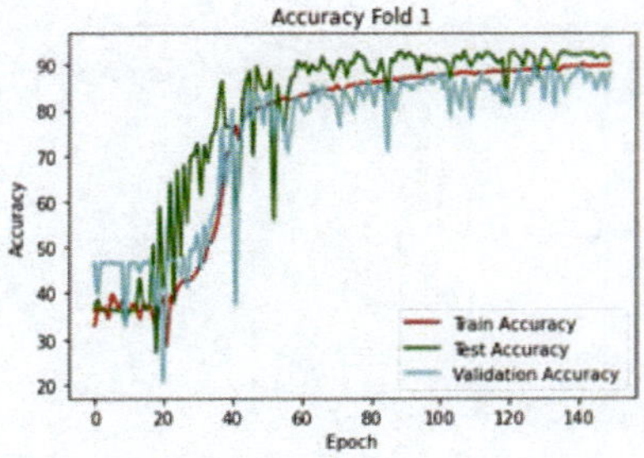

(b)

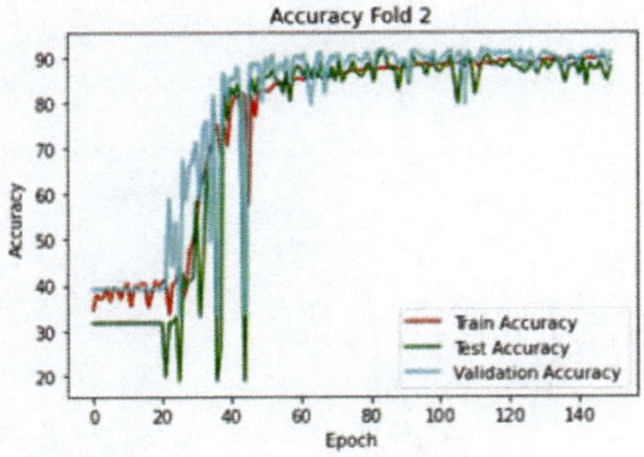

(c)

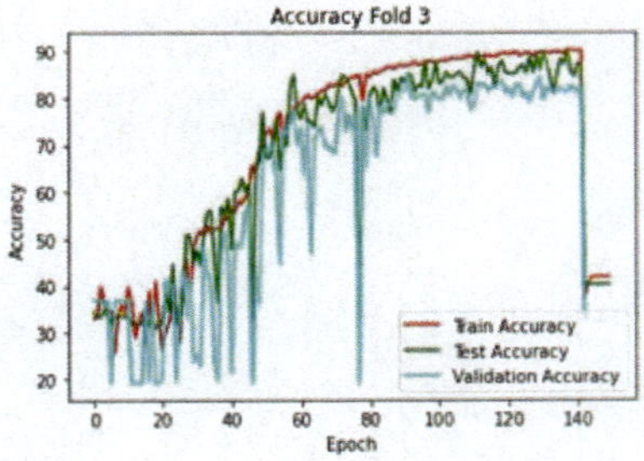

(d)

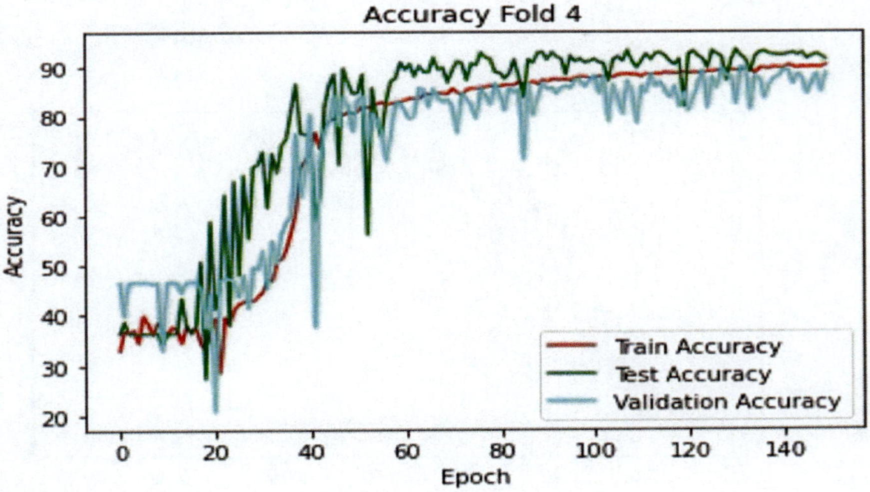

(e)

Fig. 4 Training, testing, and validation accuracy of the proposed model (CNN+LSTM)

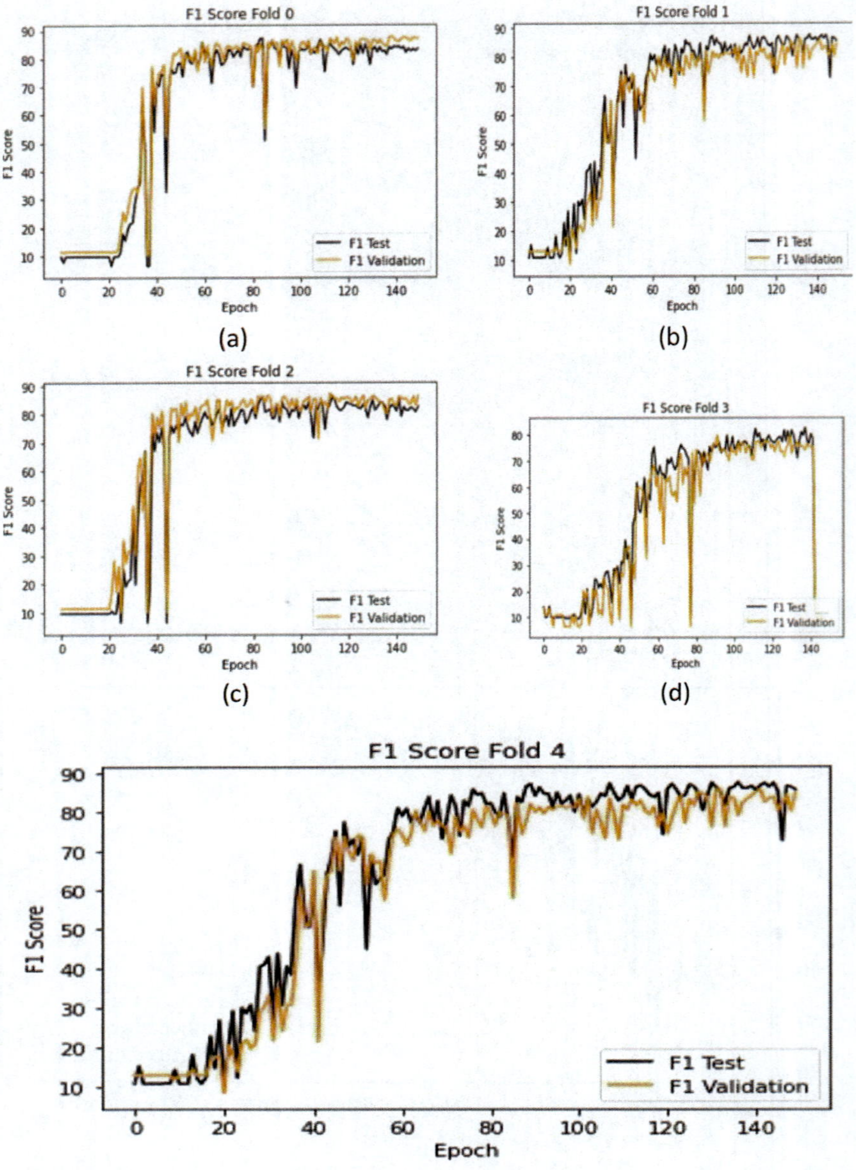

Fig. 5 F1 Test and F1 validation of the proposed model (CNN+LSTM)

will add the data augmentation and use transfer learning to improve the algorithm's sleep staging classification accuracy and generalization.

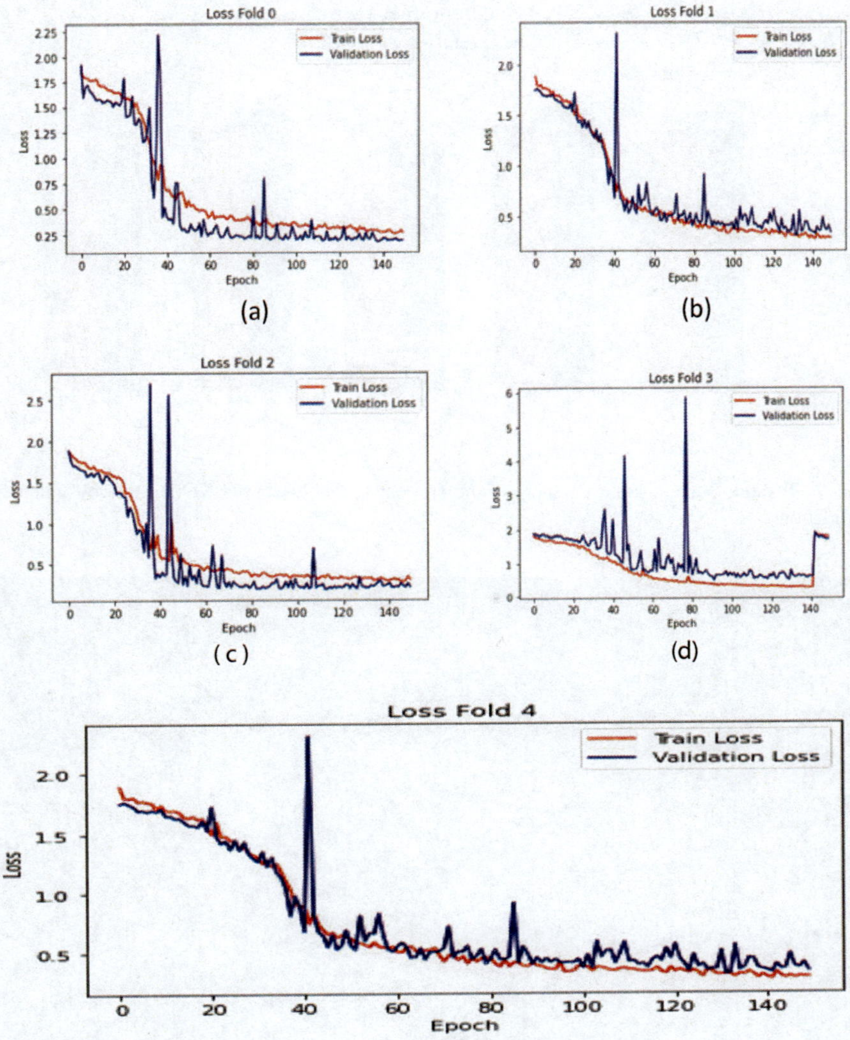

Fig. 6 Train loss and validation loss of the proposed model (CNN+LSTM)

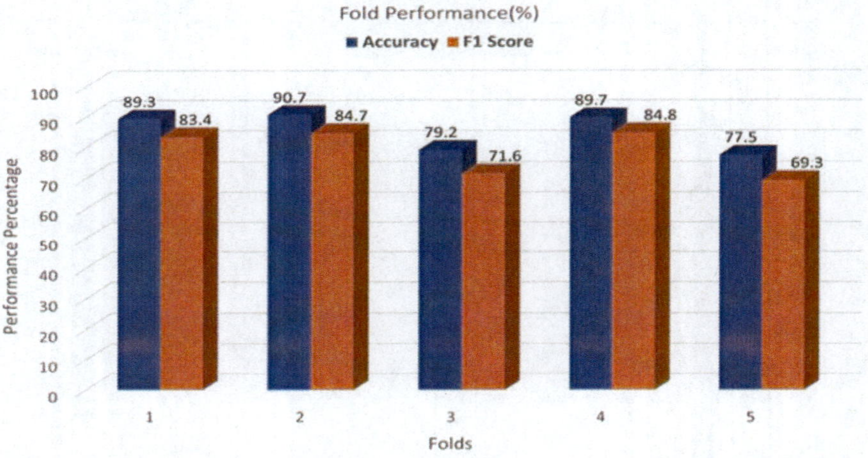

Fig. 7 Training accuracy and F1-score of the proposed model (CNN+LSTM) for five-fold cross-validation

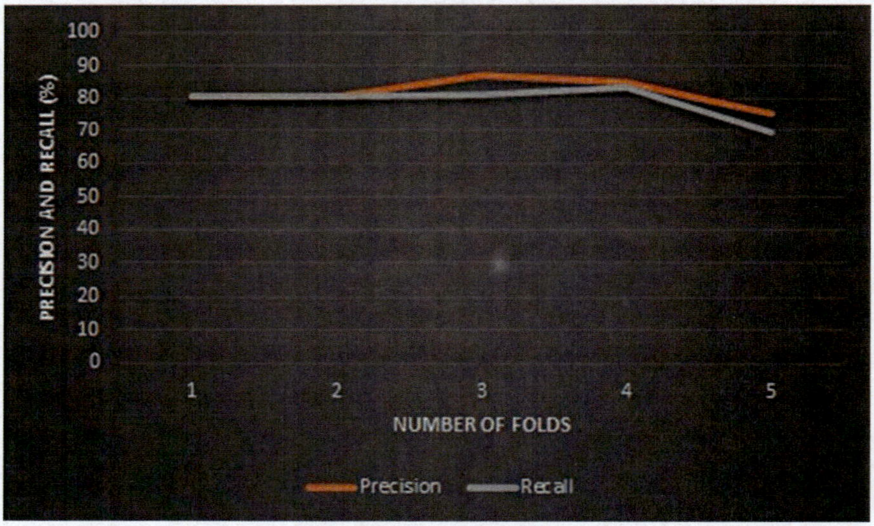

Fig. 8 Precision and recall of the proposed model (CNN+LSTM) for five-fold cross-validation

References

1. Panossian LA, Avidan AY (2009) Review of sleep disorders. Med Clin N Am 93:407–425
2. Smaldone A, Honig JC, Byrne MW (2007) Sleepless in America: inadequate sleep and relationships to health and well-being of our nation's children. Pediatrics 119:29–37
3. Hassan AR, Bhuiyan MI (2016) Automatic sleep scoring using statistical features in the EMD domain and ensemble methods. Biocybern Biomed Eng 36(1):248–255

4. Satapathy SK, Loganathan D, Narayanan P, Sharathkumar S (2020) Convolutional neural network for classification of multiple sleep stages from dual-channel EEG signals. In: 2020 IEEE 4th conference on information & communication technology (CICT), pp 1–16
5. Satapathy SK, Ravisankar M, Logannathan D (2020) Automated sleep stage analysis and classification based on different age specified subjects from a dual-channel of EEG signal. In: 2020 IEEE international conference on electronics, computing and communication technologies (CONECCT), pp 1–6
6. Satapathy SK, Loganathan D, Kondaveeti HK, Rath RK (2022) An improved decision support system for automated sleep stages classification based on dual channels of EEG signals. In: Mandal JK, Roy JK (eds) Proceedings of international conference on computational intelligence and computing. Algorithms for intelligent systems. Springer, Singapore
7. Shen H, Ran F, Xu M, Guez A, Li A, Guo A (2020) An automatic sleep stage classification algorithm using improved model based essence features. Sensors 20(17):4677
8. Satapathy SK, Loganathan D (2022) Automated accurate sleep stage classification system using machine learning techniques with EEG signals. In: Kannan SR, Last M, Hong TP, Chen CH (eds) Fuzzy mathematical analysis and advances in computational mathematics. Studies in fuzziness and soft computing, vol 419. Springer, Singapore
9. Ghimatgar H, Kazemi K, Helfroush MS, Pillay K, Dereymaeker A, Jansen K, De Vos M, Aarabi A (2020) Neonatal EEG sleep stage classification based on deep learning and HMM. J Neural Eng
10. Cooray N, Andreotti F, Lo C, Symmonds M, Hu MTM, De Vos M (2019) Detection of REM sleep behaviour disorder by automated polysomnography analysis. Clin Neurophysiol 130(4):505–514
11. Diykh M, Li Y, Abdulla S (2020) EEG sleep stages identification based on weighted undirected complex networks. Comput Methods Programs Biomed 184:105116
12. Satapathy SK, Kondaveeti HK (2021) Automated sleep stage analysis and classification based on different age specified subjects from a single-channel of EEG signal. In: 2021 IEEE Madras section conference (MASCON), pp 1–7
13. Li X, Cui L, Tao S, Chen J, Zhang X, Zhang G-Q (2017) HyCLASSS: a hybrid classifier for automatic sleep stage scoring. IEEE J Biomed Health Inform p 1
14. Heyat M, Lai D (2019) Sleep bruxism detection using decision tree method by the combination of C4-P4 and C4-A1 channels of scalp EEG. IEEE Access p 1
15. Lai D, Heyat M, Khan F, Zhang Y (2019) Prognosis of sleep bruxism using power spectral density approach applied on EEG signal of both EMG1-EMG2 and ECG1-ECG2 channels. IEEE Access p 1
16. Diykh M, Li Y, Wen P (2016) EEG sleep stages classification based on time domain features and structural graph similarity. IEEE Trans Neural Syst Rehabil Eng 24(11):1159–1168
17. Satapathy SK, Bhoi AK, Loganathan D, Khandelwal B, Barsocchi P (2021) Machine learning with ensemble stacking model for automated sleep staging using dual-channel EEG signal. Biomed Signal Process Control 69:102898
18. Memar P, Faradji F (2018) A novel multi-class EEG-based sleep stage classification system. IEEE Trans Neural Syst Rehabil Eng 26(1):84–95
19. Satapathy SK, Loganathan D (2021) Prognosis of automated sleep staging based on two-layer ensemble learning stacking model using single-channel EEG signal. Soft Comput 25:15445–15462
20. Satapathy S, Loganathan D, Kondaveeti HK, Rath R (2021) Performance analysis of machine learning algorithms on automated sleep staging feature sets. CAAI Trans Intell Technol 6(2):155–174
21. Zhu G, Li Y, Wen PP (2014) Analysis and classification of sleep stages based on difference visibility graphs from a single-channel EEG signal. IEEE J Biomed Health Inform 18(6):1813–1821
22. Satapathy SK, Loganathan D (2022) Multimodal multiclass machine learning model for automated sleep staging based on time series data. SN Comput Sci 3:276

23. Satapathy SK, Loganathan D (2022) Automated classification of multi-class sleep stages classification using polysomnography signals: a nine-layer 1D-convolution neural network approach. Multimed Tools Appl
24. Khalighi S, Sousa T, Santos JM, Nunes U (2016) ISRUC-sleep: a comprehensive public dataset for sleep researchers. Comput Methods Programs Biomed 124:180–192
25. Eskandari S, Javidi MM (2016) Online streaming feature selection using rough sets. Int J Approximate Reasoning 69:35–57
26. İlhan HO, Bilgin G (2017) Sleep stage classification via ensemble and conventional machine learning methods using single channel EEG signals. Int J Intell Syst Appl Eng 5(4):174–184
27. Sanders TH, McCurry M, Clements MA (2014) Sleep stage classification with cross frequency coupling. In: 36th Annual International Conference of the IEEE Engineering in Medicine and Biology Society (EMBC), pp 4579–4582
28. Bajaj V, Pachori R (2013) Automatic classification of sleep stages based on the time-frequency image of EEG signals. Comput Methods Programs Biomed 112(3):320–328
29. Hsu Y-L, Yang Y-T, Wang J-S, Hsu C-Y (2013) Automatic sleep stage recurrent neural classifier using energy features of EEG signals. Neurocomputing 104:105–114
30. Zibrandtsen I, Kidmose P, Otto M, Ibsen J, Kjaer TW (2016) Case comparison of sleep features from ear-EEG and scalp-EEG. Sleep Sci 9(2):69–72
31. Berry RB, Brooks R, Gamaldo CE, Hardsim SM, Lloyd RM, Marcus CL, Vaughn BV (2014) The AASM manual for the scoring of sleep and associated events: rules, terminology and technical specifications. American Academy of Sleep Medicine
32. Sim J, Wright CC (2005) The kappa statistic in reliability studies: use, interpretation, and sample size requirements. Phys Ther 85(3):257–268
33. Liang S-F, Kuo C-E, Kuo YH, Cheng Y-S (2012) A rule-based automatic sleep staging method. J Neurosci Methods 205(1):169–176
34. Satapathy SK, Kondaveeti HK (2021) Prognosis of sleep stage classification using machine learning techniques applied on single-channel of EEG signal of both healthy subjects and mild sleep effected subjects. In: 2021 International conference on artificial intelligence and machine vision (AIMV), pp 1–7
35. Satapathy SK, Pattnaik S, Rath R (2022) Automated sleep staging classification system based on convolutional neural network using polysomnography signals. In: 2022 IEEE Delhi section conference (DELCON), pp 1–10
36. Peker M (2016) A new approach for automatic sleep scoring: Combining Taguchi based complex-valued neural network and complex wavelet transform. Comput Methods Programs Biomed 129:203–216
37. Subasi A, Kiymik MK, Akin M, Erogul O (2005) Automatic recognition of vigilance state by using a wavelet-based artificial neural network. Neural Comput Appl 14(1):45–55
38. Tsinalis O, Matthews PM, Guo Y (2016) Automatic sleep stage scoring using time-frequency analysis and stacked sparse autoencoders. Ann Biomed Eng 44(5):1587–1597
39. Hassan AR, Bhuiyan MIH (2017) An automated method for sleep staging from EEG signals using normal inverse Gaussian parameters and adaptive boosting. Neurocomputing 219:76–87
40. Hassan AR, Bhuiyan MIH (2017) Automated identification of sleep states from EEG signals by means of ensemble empirical mode decomposition and random under sampling boosting. Comput Methods Programs Biomed 140:201–210
41. Diykh M, Li Y (2016) Complex networks approach for EEG signal sleep stages classification. Expert Syst Appl 63:241–248
42. Mahvash Mohammadi S, Kouchaki S, Ghavami M, Sanei S (2016) Improving time–frequency domain sleep EEG classification via singular spectrum analysis. J Neurosci Methods 273:96–106
43. Phan H, Andreotti F, Cooray N (2018) Joint classification and prediction CNN framework for automatic sleep stage classification. IEEE Trans Biomed Eng 66(5):1285–1296
44. Pan J, Zhang J, Wang F (2021) Automatic sleep staging based on EEG-EOG signals for depression Detection. Intell Autom Soft Comput 28(1):53–71

45. Mousavi S, Afghah F, Acharya UR (2019) SleepEEGNet: automated sleep stage scoring with sequence-to-sequence deep learning approach. PLoS ONE 14(5):e0216456
46. Chen T, Huang H, Pan J (2018) An EEG-based brain-computer interface for automatic sleep stage classification. In: 2018 13th IEEE conference on industrial electronics and applications (ICIEA). IEEE, pp 1988–1991

An Ensemble Model for Gait Classification in Children and Adolescent with Cerebral Palsy: A Low-Cost Approach

Saikat Chakraborty, Sruti Sambhavi, Prashansa Panda, and Anup Nandy

Abstract A fast and precise automatic gait diagnostic system is an urgent need for real-time clinical gait assessment. Existing machine intelligence-based systems to detect cerebral palsy gait have often ignored the crucial issue of performance and computation speed trade-off. This study, in a low-cost experimental setup, proposes an ensemble model by combining fast and deep neural networks. The proposed system demonstrates a competing result with an overall ≈82% of detection accuracy (sensitivity: ≈78%, specificity: ≈84%, and F1-score: ≈83%). Although the improvement in detection performance is marginal, the computation speed increased remarkably from state of the art. From the perspective of computation time and performance trade-off, the proposed model demonstrated to be competing.

Keywords Kinect(v2) · Gait · Cerebral palsy · ELM · LSTM · Ensemble model

1 Introduction

Cerebral palsy (CP) is a non-progressive developmental disorder of brain, causing postural instability and gait anomalies [21]. As per a recent report [22], globally, more than 4 children out of 1000 suffer from CP. Situation is more alarming in developing countries like India, where CP patients account for 15–20% out of all physically disabled children [4]. Therapeutic intervention is crucial to upgrade the gait quality of this population. The efficacy of an intervention is justified through precise and quantitative gait diagnosis [15].

The performance of shallow machine learning (ML)-based automatic gait diagnostic systems is basically dependent on expert knowledge for selecting salient fea-

S. Chakraborty (✉)
School of Computer Engineering, Kalinga Institute of Industrial Technology (KIIT), Bhubaneswar, Odisha, India
e-mail: saikat.sc@gmail.com

S. Chakraborty · S. Sambhavi · P. Panda · A. Nandy
Machine Intelligence and Bio-motion Research Lab, Department of Computer Science and Engineering, National Institute of Technology Rourkela, Rourkela, Odisha, India

M. Thakur et al. (eds.), *Soft Computing for Problem Solving*, Lecture Notes in Networks and Systems 547, https://doi.org/10.1007/978-981-19-6525-8_7

tures. In absence of a proper feature selection method, a noisy feature can drastically decrease the model performance [12]. On the other hand, deep learning (DL) models learn the objective function through automatic feature mapping [12]. It has shown great potential in dealing with high dimensional and nonlinear data [12, 20]. Furthermore, in some applications, DL approaches outperformed the classical ML methods [3]. In the case of time-series analysis, such as gait abnormality detection, long short-term memory (LSTM) has exhibited a promising performance [16]. Specifically for CP gait diagnosis, a few studies have used LSTM network and obtained a competing result [9]. But, LSTM suffers from slow computation speed [25], which could limit its applicability for real-time gait assessment tasks in clinic. A trade-off between computation speed and model performance is a vital issue that needs to be addressed while building a gait diagnostic system. But, studies have often ignored this crucial issue during decision-making. Recently, extreme learning machine (ELM) has earned popularity for its fast computation speed [14]. It has demonstrated remarkable performance in different application domains as well [1, 24]. Incorporation of this architecture during building a LSTM-based gait abnormality detection system could be beneficial to decreasing computation time. A LSTM+ELM ensemble model is proposed in this study which seems to be effective and fast compared to LSTM-based models in diagnosing CP gait. Another aspect is that, the existing diagnostic systems for CP gait is expensive due to the incorporation of high-end motion capture sensors [4]. These systems are unaffordable for most of the rehabilitation centers and clinics, especially in developing countries. A low-cost gait assessment system is a serious requirement for CP patients.

This study aimed to establish a novel computational model (in a low-cost environment) for fast as well as precise diagnosis of gait patterns for children and adolescents with CP (CAwCP). First, a low-cost motion sensor-based (Kinect v2) setup was constructed from which three-dimensional velocity of lower limb joints (i.e., ankles, knees, and hips of both limbs, and pelvis) was extracted. Second, an ensemble model (LSTM+ELM) was designed where the hidden layer of ELM was formed using LSTM units. The random feature mapping technique of ELM facilitates fast computation for the entire network. Finally, the proposed system was compared with state-of-the-art CP gait diagnostic techniques. The contributions of this work are summarized as follows:

- Developing a LSTM+ELM-based ensemble model where the hidden layer of ELM was constructed using a set of LSTM cells.
- Establishing a low-cost gait detection system based on the ensemble model.

Remaining of this paper is organized as follows: Sect. 2 demonstrates state-of-the-art studies. The experimental setup, data collection procedure, data analysis, and the proposed model are described in Sect. 3. Results and discussion are presented in Sect. 4. Conclusion and a future research direction have been given in Sect. 5.

2 Related Work

Different studies have tried to construct automated gait detection system for CP patients.

Kamruzzaman and Begg [15] examined different kernels of support vector machine (SVM) with normalized features to diagnose children with CP (CwCP). They reported SVM as a comparatively better classifier (accuracy: 96.80%) when normalized cadence and stride length are used as input features. Gestel et al. [23] checked the utility of Bayesian networks to detect gait abnormality in CwCP and obtained a competing result (88.4% accuracy) using gait features in the sagittal plane. Laet et al. [6] used logistic regression and Naive Bayes classifiers in addition to expert knowledge extracted from the Delphi-consensus study [18]. They observed a significant increase in performance after incorporating expert knowledge in feature selection. Though the above-mentioned studies have obtained promising results, they are biased to human experience for selecting salient gait features. Dobson et al. [7] and subsequently, Papageorgiou et al. [19] have reported that cluster-based gait assessment systems for CP patients were studied mostly in the literature. But, cluster-based gait diagnosis systems may form clinically irrelevant artificial groups [27]. On the other hand, despite the growing popularity of deep learning, very few studies have used it for CP gait diagnosis [10]. Alberto et al. [9] used LSTM and multilayer perceptron (MLP) to divide diplegic patients into 4 clinically relevant groups. In their experiment, LSTM marginally outperformed the MLP network. Notably, all the studies mentioned above represented a high-cost system with expensive sensors which is not affordable for most of the clinics. Although some studies have tried to provide a low-cost solution using single Kinect-based architecture [2, 8], but suffer to establish a clinically relevant walking track (i.e., 10 m or 8 m in length). As per the best of our knowledge, this work is the first one to adopt a deep learning technique, specifically an ensemble model, to diagnose CAwCP gait in a clinically relevant low-cost architecture.

3 Methods

3.1 Participants

Fifteen CAwCP patients (age (year): 12.55 ± 2.13, height (cm): 130.15 ± 14.87, male/female: 8/7, GMFCS level: I/II) were recruited from the Indian Institute of Cerebral Palsy (IICP), Kolkata. Fifteen age matched typically developed children and adolescent (TDCA) (age (year): 12.45 ± 3.51, height (cm): 132.06 ± 14.09, male/female: 9/6) were also recruited from REC School National Institute of Technology Rourkela Campus. Ethical approval for this study was given by competent authority.

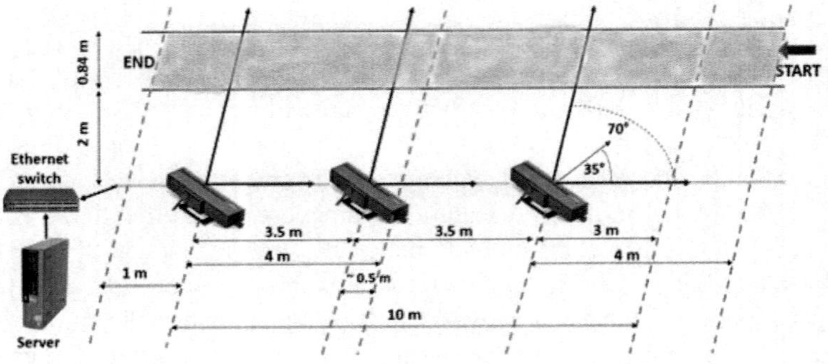

Fig. 1 Experimental setup [4]

3.2 Experimental Setup and Data Acquisition

A set of Microsoft Kinect v2 sensors were used to construct the data acquisition platform (see Fig. 1). Kinects were placed sequentially at a 35° angle to the walking direction. A client-server protocol was set up to control the 3 Kinects. This setup was observed to provide a walking track of the effective length 10m. For a detail description of the architecture, please see [4].

A mock-up practice was conducted for the subjects before starting the experiment. They were allowed to walk at a self-decided speed from a distance of 4m from the 1st Kinect. The system started to capture data after a 1m distance walk. Out of 12m distance of the path, data were collected within the 10m track (see Fig. 1). Extra distances at both ends of the path have been given to minimize the acceleration and deceleration effect on gait parameters. For each subject, five trials were taken. They were allowed to take 2 min gap between two consecutive trials. We have followed the protocol described by Geerse et al. [11] and Muller et al. [17] to record and combine signals, and removing noise from time-series data collected from Kinects.

3.3 Data Analysis

Feature Vector Representation: Body point time-series data combined from all the Kinects were converted to velocity using three-point estimation method. Then, gait cycles were extracted following the method described by Zeni et al. [26]. Each gait cycle was time normalized to 101 points (i.e., 0–100% cycle representation) like [5]. Across all the trials for all subjects, the mean value of the number of gait cycles (i.e., 6) was computed and was taken for further processing. Hence, for each subject, 30 gait cycles were considered for data analysis. So, for each gait cycle, the feature vector size was 21 (7 joints × 3 spatial directions).
Binary Classification: To distinguish between normal or CAwCP gait, an ensemble model consisting of ELM and LSTM units was established.

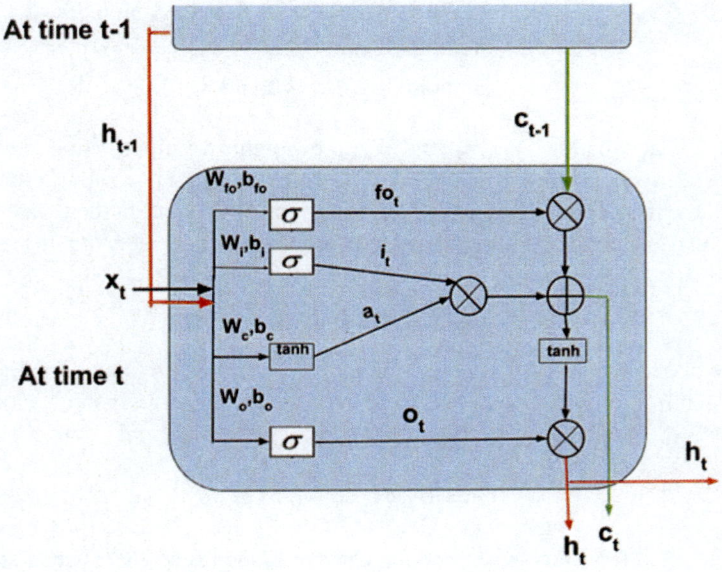

Fig. 2 A LSTM unit

(a) *Extreme Learning Machine*: ELM is a feedforward neural network which consists of only single hidden layer [14]. The output function on feature vector **f** can be written as:

$$\phi(\mathbf{f}) = \sum_{i=1}^{n} w_i \theta_i(\mathbf{f}) \tag{1}$$

In Eq. 1, w_i represents the weight associated with the ith hidden node to the output layer, whereas $\theta_i(\mathbf{f})$ represents nonlinear mapping of features from the ith hidden node to the output. $\theta(.)$ may vary across the hidden nodes. It can be written as:

$$\theta_i(\mathbf{f}) = \sigma_i(a_i, b_i, \mathbf{f}) \tag{2}$$

In Eq. 2, $\sigma_i(a_i, b_i, \mathbf{f})$ is a nonlinear piecewise continuous function, and (a_i, b_i) are the hidden node parameters (generated randomly) which speed up the learning procedure. In ELM, learning is basically a two-step procedure: (1) random mapping of features, and (2) solving the parameters linearly where the hidden to output layers weights are tuned by minimizing the output error [14].

(b) *Long Short-Term Memory*: LSTM is an updation of recurrent neural network (RNN) where the vanishing and exploding gradient problems were resolved at some extent [13]. A LSTM unit (at time instant t) (see Fig. 2) consists of an input (i_t), output (o_t) and forget gates (fo_t) which can be represented as:

$$
\begin{aligned}
i_t &= \text{sigmoid}(W_i.(x_t, h_{t-1}) + b_i) \\
o_t &= \text{sigmoid}(W_o.(x_t, h_{t-1}) + b_o) \\
fo_t &= \text{sigmoid}(W_{fo}.(x_t, h_{t-1}) + b_{fo})
\end{aligned}
\tag{3}
$$

In the above equation, W_i, W_o, and W_{fo} are weights for input, output, and forget gates, respectively. Similarly, b_i, b_o, and b_{fo} are biases for the input, output, and forget gates, respectively. x_t and h_{t-1} are input at time t and hidden state at time $(t-1)$, respectively. In addition, it also consists a candidate cell state (a_t) at time t;

$$a_t = \tanh(W_c.(x_t, h_{t-1}) + b_c) \tag{4}$$

In the above equation, W_c is the weight, and b_c is the corresponding bias. c_{t-1}, i_t, fo_t, and a_t are used to generate the current cell state (c_t) which act as internal memory:

$$c_t = \text{fo}_t \otimes c_{t-1} + i_t \otimes a_t \tag{5}$$

In Eq. 5, $\otimes$ refers entry-wise product. Current hidden state (h_t) is generated as:

$$h_t = o_t \otimes \tanh(c_t) \tag{6}$$

(c) *Proposed Model*: The ensemble model is demonstrated in Fig. 3. The hidden layer of ELM was constructed using a set of LSTM units. The input layer contains 21 neurons that correspond to 21 features. 35 LSTM units were selected empirically for the hidden layer. A single neuron in the output layer was used for binary classification.

Sigmoidal activation function was used between the layers of ELM. At time $(t+1)$, data input to a LSTM unit was defined as:

$$x_{t+1} = \mathbf{f}_t \cdot W_{ih}^T \tag{7}$$

where W_{ih} refers randomly initialized weight matrix from input layer to hidden layer.

Output function $(\phi(\mathbf{f}))$ can be written as:

$$\phi(\mathbf{f}) = \sigma(W_{ho} \cdot H) \tag{8}$$

where W_{ho} is weight matrix from hidden layer to output layer, and H refers the hidden state matrix computed from all LSTM units (H depends on the feature vector). In this experiment, sigmoid function was used for feature mapping.

The entire dataset was divided into 7:3 ratio (i.e., training : testing) (subjects in test and training were different). Eleven CAwCP and 10 normal (i.e., TDCA) subjects were included in the training set. Rest was used for the test data. Leave-one-out cross-validation was performed to reduce the error on test data. For comparison with state of the art, vanilla LSTM (or LSTM) [9], bi-directional LSTM (Bi-LSTM) [16] models along with normal ELM, and multilayer ELM (MELM) were tested on the

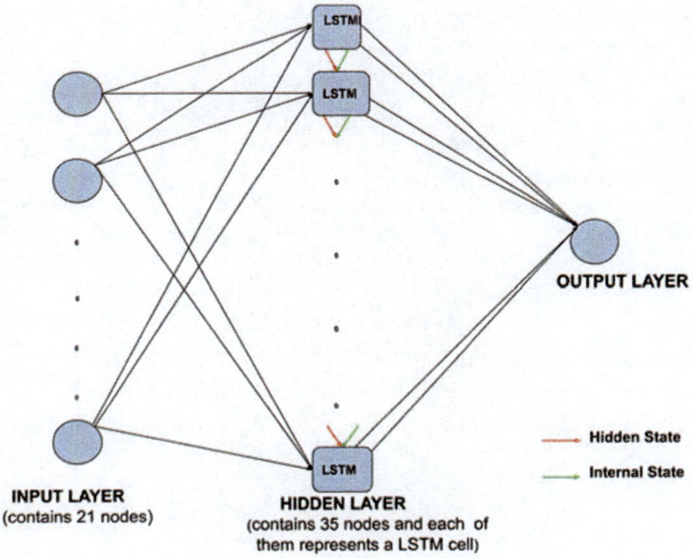

Fig. 3 Proposed model

Table 1 Hyperparameters of different models

Model	Hyperparameter
Vanilla LSTM	Epoch and batch size
Bi-LSTM	Epoch and batch size
ELM	Activation function, and hidden layer nodes
MELM	Activation function, hidden layers, and hidden layer nodes
Proposed	Epoch, batch size, hidden layer nodes, and activation function

same dataset. Grid search was used for hyperparameters (see Table 1) tuning. For MELM, 2–5 hidden layers were tested. Batch and epoch sizes (for Vanilla LSTM, Bi-LSTM, and the proposed model) were ranged between 300 and 1000 and 500 to 1000, respectively. For the proposed model, hidden layer nodes ranged between 20 and 50, while as activation function, sigmoidal, ReLU, and tanh were used. Mean square error was used for the loss function (for LSTM and Bi-LSTM). Models were learned until the validation accuracy stopped improving. The best performing models in the validation testing were selected for the testing phase. For example, MELM consisting of 2 hidden layers was observed to perform the best. The entire experiment was performed on a CPU having RAM: 8 GB, Intel(R) Core(TM) i7-4770 @ 3.40 GHz.

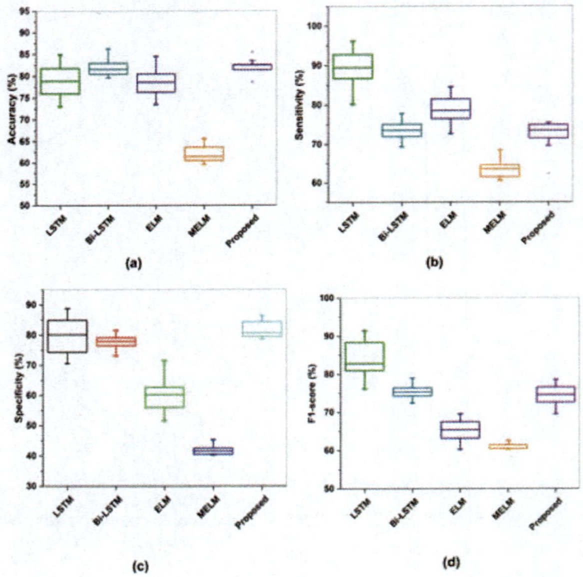

Fig. 4 Distribution of performance parameters for different models during leave-one-out cross-validation. **a** classification accuracies, **b** sensitivities, **c** specificities, and **d** F1-scores

4 Results and Discussion

The primary aim of this study was to setup a low-cost gait assessment system (for CAwCP patients) which should be able to provide precise result with faster computation. Accuracy, sensitivity, specificity, and F1-score were chosen to assess the performance of the models. Figure 4 demonstrates the distribution of performance metrics of different models during leave-one-out cross-validation. It can be seen that the overall performance of the proposed model is competing. In terms of accuracy, it exhibited the best performance with the lowest variance. Also for specificity, the distribution of the proposed model coincides with LSTM with comparatively lower variance. The range of distribution for sensitivity was comparatively lower than some of the other models. Figure 5 exhibits the comparative performance of different models on the testing dataset. Here also the proposed model outperformed the others in terms of specificity (84.77%), while accuracy and F1-score were marginally higher ($\approx$2% and $\approx$1% respectively) than LSTM. Notably, the model was trained much faster than LSTM and Bi-LSTM (see Table 2). Also, it exhibited faster computation than LSTM and Bi-LSTM on the testing data.

The comparatively lower sensitivity of the proposed model may be a result of the probability distribution for randomly generating hidden node parameters in ELM [14], which might be impacted the LSTM units to recognize the minor difference between the two populations. In terms of learning time, the proposed model outperformed LSTM and Bi-LSTM. The proposed model performance closely resembles LSTM, but the faster computation speed makes it more attractive for clinics. Random initialization of the weight matrix could be a determinant factor to increase

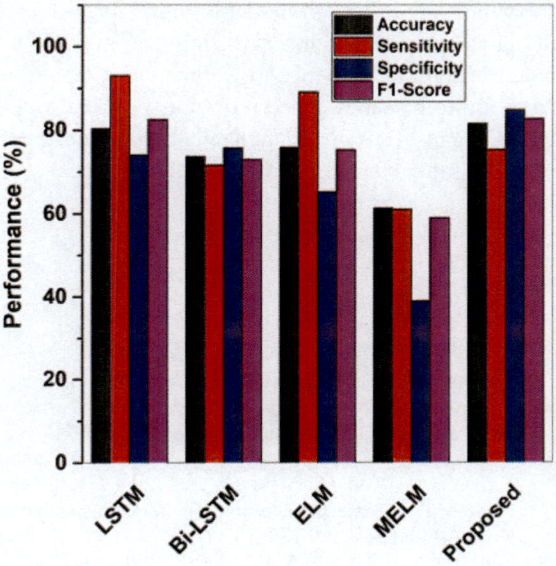

Fig. 5 Performance of different models on testing dataset

Table 2 Computation time for different models

	Vanilla LSTM	Bi-LSTM	Proposed
Validation time (s)	18,432.50	20,163.71	1186.87
Testing time (s)	532.78	549.61	**112.88**

the computation. The blending of LSTM and ELM has provided the precise and faster outcome. The trade-off between performance and computation speed makes this model more clinically significant.

In this study, data were captured by placing Kinects on a particular side of the walking direction. This may over or underestimate the gait features. Placing Kinects on both sides of walking direction may be a solution to this problem, but, it will increase the overall system cost as more Kinects will be required. Trade-off between the computation time and detection performance was maintained in the proposed model.

5 Conclusion

This study proposed an affordable automatic gait assessment system for CAwCP patients using an ensemble model. Data were collected from a low-cost multi-Kinect setup. The ensemble model was constructed using a fast neural network (i.e., ELM) and a deep neural network (i.e., LSTM). Along with the performance, the computation time (both validation and testing) was assessed for the proposed model. Results shows that the proposed model is competing with state of the art on the perspective of diagnostic performance and computation speed trade-off. Clinicians can take this model where performance as well as the computation speed are a point of concern. As a future research aspect, the utility of the proposed model in the diagnosis of CAwCP gait in a more challenged environment (i.e., uneven ground walking, backward walking, etc.) seems warranted.

References

1. Alharbi A (2020) A genetic-elm neural network computational method for diagnosis of the Parkinson disease gait dataset. Int J Comput Math 97(5):1087–1099
2. Bei S, Zhen Z, Xing Z, Taocheng L, Qin L (2018) Movement disorder detection via adaptively fused gait analysis based on Kinect sensors. IEEE Sens J 18(17):7305–7314
3. Camps J, Sama A, Martin M, Rodriguez-Martin D, Perez-Lopez C, Arostegui JMM, Cabestany J, Catala A, Alcaine S, Mestre B et al (2018) Deep learning for freezing of gait detection in Parkinson's disease patients in their homes using a waist-worn inertial measurement unit. Knowl-Based Syst 139:119–131
4. Chakraborty S, Thomas N, Nandy A (2020) Gait abnormality detection in people with cerebral palsy using an uncertainty-based state-space model. In: International conference on computational science. Springer, pp 536–549
5. Cui C, Bian G-B, Hou Z-G, Zhao J, Su G, Zhou H, Peng L, Wang W (2018) Simultaneous recognition and assessment of post-stroke hemiparetic gait by fusing kinematic, kinetic, and electrophysiological data. IEEE Trans Neural Syst Rehabil Eng 26(4):856–864
6. De Laet T, Papageorgiou E, Nieuwenhuys A, Desloovere K (2017) Does expert knowledge improve automatic probabilistic classification of gait joint motion patterns in children with cerebral palsy? PloS one 12(6):e0178378
7. Dobson F, Morris ME, Baker R, Graham HK (2007) Gait classification in children with cerebral palsy: a systematic review. Gait Posture 25(1):140–152
8. Dolatabadi E, Taati B, Mihailidis A (2017) An automated classification of pathological gait using unobtrusive sensing technology. IEEE Trans Neural Syst Rehabil Eng 25(12):2336–2346
9. Ferrari A, Bergamini L, Guerzoni G, Calderara S, Bicocchi N, Vitetta G, Borghi C, Neviani R, Ferrari A (2019) Gait-based diplegia classification using lsmt networks. J Healthcare Eng
10. Gautam R, Sharma M (2020) Prevalence and diagnosis of neurological disorders using different deep learning techniques: a meta-analysis. J Med Syst 44(2):49
11. Geerse DJ, Coolen BH, Roerdink M (2015) Kinematic validation of a multi-Kinect v2 instrumented 10-meter walkway for quantitative gait assessments. PloS one 10(10):e0139913
12. Halilaj E, Rajagopal A, Fiterau M, Hicks JL, Hastie TJ, Delp SL (2018) Machine learning in human movement biomechanics: best practices, common pitfalls, and new opportunities. J Biomech 81:1–11
13. Hochreiter S, Schmidhuber J (1997) Long short-term memory. Neural Comput 9(8):1735–1780
14. Huang G, Huang G-B, Song S, You K (2015) Trends in extreme learning machines: a review. Neural Netw 61:32–48
15. Kamruzzaman J, Begg RK (2006) Support vector machines and other pattern recognition approaches to the diagnosis of cerebral palsy gait. IEEE Trans Biomed Eng 53(12):2479–2490
16. Khokhlova M, Migniot C, Morozov A, Sushkova O, Dipanda A (2019) Normal and pathological gait classification lstm model. Artif Intell Med 94:54–66
17. Müller B, Ilg W, Giese MA, Ludolph N (2017) Validation of enhanced Kinect sensor based motion capturing for gait assessment. PloS one 12(4):e0175813
18. Nieuwenhuys A, Õunpuu S, Van Campenhout A, Theologis T, De Cat J, Stout J, Molenaers G, De Laet T, Desloovere K (2016) Identification of joint patterns during gait in children with cerebral palsy: a Delphi consensus study. Dev Med Child Neurol 58(3):306–313

19. Papageorgiou E, Nieuwenhuys A, Vandekerckhove I, Van Campenhout A, Ortibus E, Desloovere K (2019) Systematic review on gait classifications in children with cerebral palsy: an update. Gait Posture 69:209–223
20. Quisel T, Foschini L, Signorini A, Kale DC (2017) Collecting and analyzing millions of mhealth data streams. In: Proceedings of the 23rd ACM SIGKDD international conference on knowledge discovery and data mining, pp 1971–1980
21. Richards CL, Malouin F (2013) Cerebral palsy: definition, assessment and rehabilitation. In: Handbook of clinical neurology, vol 111. Elsevier, pp 183–195
22. Stavsky M, Mor O, Mastrolia SA, Greenbaum S, Than NG, Erez O (2017) Cerebral palsy-trends in epidemiology and recent development in prenatal mechanisms of disease, treatment, and prevention. Frontiers in pediatrics 5:21
23. Van Gestel L, De Laet T, Di Lello E, Bruyninckx H, Molenaers G, Van Campenhout A, Aertbeliën E, Schwartz M, Wambacq H, De Cock P et al (2011) Probabilistic gait classification in children with cerebral palsy: a Bayesian approach. Res Dev Disabil 32(6):2542–2552
24. You Z-H, Lei Y-K, Zhu L, Xia J, Wang B (2013) Prediction of protein-protein interactions from amino acid sequences with ensemble extreme learning machines and principal component analysis. In: BMC Bioinformatics, vol 14. Springer, p S10
25. Yu W, Li X, Gonzalez J (2019) Fast training of deep lstm networks with guaranteed stability for nonlinear system modeling. In: International symposium on neural networks, vol 422. Springer, pp 3–10
26. Zeni J Jr, Richards J, Higginson J (2008) Two simple methods for determining gait events during treadmill and overground walking using kinematic data. Gait and posture 27(4):710–714
27. Zhang Y, Ma Y (2019) Application of supervised machine learning algorithms in the classification of sagittal gait patterns of cerebral palsy children with spastic diplegia. Comput Biol Med 106:33–39

Imbalanced Learning of Regular Grammar for DFA Extraction from LSTM Architecture

Anish Sharma and Rajeev Kumar

Abstract In this work, we attempt to extract Deterministic Finite Automata (DFA) for a set of regular grammars from sequential Recurrent Neural Networks (RNNs). We have considered Long Short-Term Memory (LSTM) architecture, which is a variant of RNN. We have classified a set of regular grammars by considering their imbalances in terms of strings they accept and the strings they reject by using an LSTM architecture. We have formulated a set of the extended Tomita Grammar by adding a few more regular grammars. The different imbalance classes we introduce are Nearly Balanced (NB), Mildly Imbalanced (MI), Highly Imbalanced (HI), Extremely Imbalanced (EI). We have used L* algorithm for DFA extraction from LSTM networks. As a result, we have shown the performance of training an LSTM architecture for extraction of DFA in the context of the imbalances for a set of so formed regular grammars. We were able to extract correct minimal DFA for various imbalanced classes of regular grammar, though in some cases, we could not extract minimal DFA from the Network.

Keywords Deterministic Finite Automata (DFA) · Imbalance · Long Short-Term Memory (LSTM) Network · Sequential Neural Network · Extended Tomita Grammar Set (ETGS)

1 Introduction

Machine Learning is a challenging field in computer science that helps us to learn patterns from data and draw conclusions from such patterns. At present machine learning is still in its infancy for many researchers to understand many hidden treasures. Researchers have been trying to solve their problems using machine learning which provides a variety of algorithms that gives different ways to look at data and patterns from different perspectives leading to interpretation. Artificial neural net-

A. Sharma (✉) · R. Kumar
Data to Knowledge (D2K) Lab School of Computer and Systems Sciences, Jawaharlal Nehru University, New Delhi 110067, India
e-mail: anishsharma.ishu@gmail.com

M. Thakur et al. (eds.), *Soft Computing for Problem Solving*, Lecture Notes in Networks and Systems 547, https://doi.org/10.1007/978-981-19-6525-8_8

works are a class of machine learning algorithms which are inspired from human brain. They can only process static data. To address this drawback the extension of neural networks, Recurrent Neural Networks (RNNs) were introduced. They can process sequential data conveniently. RNN's also being black box model (i.e., we do not understand much of its internal working, which leads to its generalization and interpretation.) which makes interpreting the knowledge they have learned a challenging task. Some researchers like Jacobsson [10] have proposed extraction of deterministic finite automata in order to find underlying model from RNNs by processing input and output sequentially.

In early days of this field, Elman [6] introduced Elman RNN for time analysis. Giles et al. [7] introduced second order RNN which can easily learn to simulate DFA. Goudreau et al. [8] showed that second order machines perform better than the first order RNN. Omlin et al. [12] proposed extraction of Finite State Automata (FSA) using clustering with RNN. This method used second order RNN and each cluster here represented a state of FSA.

In this work, we are focused on a variant of RNN's architecture introduced by Hochreiter and Schmidhuber [9], known as Long Short-term memory unit (LSTM). This RNN is superior to second order RNN's and is widely used these days for various sequential learning. Early on Bengio et al. [2] suggested how gradient descent as an error optimization method is ineffective on simple RNN's whenever RNN's were required to handle long-term dependencies. To avoid these problems of simple RNN's we are using LSTM architecture for our research. We are training RNNs on an Extended Tomita Grammar Set (ETGS), which is extension of Tomita grammar and contains a set of regular grammars on alphabet Σ to extract Deterministic Finite Automata (DFA) from an RNN. We are extending the work done by Weiss et al. [14] for DFA extraction from RNN. They used the exact learning L* algorithm by Angluin [1] which answers the membership and equivalence queries for a set of regular languages. We are looking into the imbalance of regular grammar in terms of positive and negative binary strings that are associated with these regular grammars. The rest of the paper is organized as follows. The literature survey is included in Sect. 2. Problem Definition is formulated in Sect. 3. The proposed methodology is described in Sect. 4. Dataset and Preprocessing is included in Sect. 5. Results and Discussions are included in Sect. 6. Finally, we conclude the work in Sect. 7.

2 Related Work

Recurrent neural networks (RNN's) are an extension of neural networks which are able to process variable-length sequences. Since it is an extension of neural networks, it is also a black-box model. Hence it is unclear what they actually learn and how do they respond to unseen patterns. In order to understand the internal working of RNN's, researchers explored finite automata extraction from RNN with different approaches.

The simple RNN architectures were not able to train for the tasks involving long-term dependencies as observed by Bengio et al. [2] due to this the System would not be robust to input noise. Then different RNN architectures were proposed by researchers over the years. The Long short-term memory unit proposed by Hochreiter and Schmidhuber [9]. Later, Cho et al. [4] introduced the Gated Recurrent Unit (GRU) mechanism. The Chung et al. [5] showed that these two different architectures have significantly improved performance over RNN networks.

Early on, Omlin et al. [11] proposed dynamic state exploration and cluster analysis of multi-dimensional (N-dimensions) output space for an RNN. A simple partitioning procedure achieved this with parameter q for quantization levels. Each partition is treated as a cluster to apply K-means clustering, with different partitioning, there can be several minimal deterministic finite automata. This approach suffered from state space explosion problems and also required parameter tuning.

Cechin et al. [3] introduced another approach for the extraction of knowledge from RNN's. They tried K-means clustering and also fuzzy clustering because the network must be able to learn dynamic systems behavior. The clustering was applied to the neural activation space for construction of membership function in hidden layers. This approach required the setting of partitioning before extraction began. This method becomes challenging when we require to choose a suitable parameter value for extraction. Hence, it also requires parameter tuning.

In this approach, Weiss et al. [14] introduced a novel algorithm that uses exact learning for deterministic finite automata extraction. This method uses the L* algorithm, which focuses on answering membership and equivalence queries. It was successfully able to extract deterministic finite automata for any network. It also returns counter-examples so that it could point to incorrect patterns during extraction. This method requires no tuning of parameters and is not affected by hidden state size like other methods discussed above.

3 Problem Definition

In this work, we extend the set of regular grammars called Tomita Grammar, as proposed by Weiss et al. [14]. For this, we consider a set of regular grammars, which is a super-set of Tomita Grammar. Then, we examine this extended set of Tomita Grammar by generating automata using LSTM architecture. We analyze the set of strings which are accepted and rejected by this methodology. The strings which are accepted or rejected by the LSTM architecture are found to be imbalanced, making training challenging.

3.1 Tomita Grammar

Tomita grammar is one of most widely used set of regular grammar proposed by Tomita [13]. It consists of seven regular grammars, it consists of some of basic well known grammars. These grammars are chosen because for these grammars we also have ground truth DFA's. That means we can cross check the DFA extracted from RNN with ground truth DFA.

3.2 Extended Tomita Grammar

In this work, we introduce a few more well known regular grammars as an extension to Tomita grammar. These extended set of regular grammars are also among the well known regular grammars. For these grammars, we also have ground truth DFA. In this extension of Tomita grammar, we have added 5 regular grammars. We are calling them Extended Tomita Grammar.

3.3 Imbalancing

The regular grammars used in Tomita and Extended Tomita grammar have common alphabet $\Sigma = (0,1)$. These grammars generate binary strings. Among these binary strings, strings which are accepted by final state of DFA are treated as *Positive* strings and remaining are *Negative* Strings.

Imbalance (I) in regular grammars is established on the basis of modulus of difference between the number of Positive (S_+) and Negative (S_-) strings associated with the grammars. According to this we split our Grammars in different classes (Table 1).

$$I = |(S_+ - S_-)| \tag{1}$$

Table 1 Defined imbalance classes for extended Tomita grammar

#	Nature of grammar	Threshold
1	Nearly Balanced (NB)	$0 < I < 10$
2	Mildly Imbalanced (MI)	$10 < I < 50$
3	Highly Imbalanced (HI)	$50 < I < 90$
4	Extremely Imbalanced (EI)	$90 < I < 100$

- Nearly Balanced (NB Class): If the difference between (S_+) strings and (S_-) strings is very small.
- Mildly Imbalanced (MI Class): If the difference between (S_+) strings and (S_-) strings is in between the range of Highly Imbalanced and Balanced class.
- Highly Imbalanced (Class): If the difference between (S_+) strings and (S_-) strings is significantly large.
- Extremely Imbalanced (EI Class): If the difference between (S_+) strings and (S_-) strings is drastically large.

4 The Proposed Methodology

We are working toward the extraction of Deterministic Finite Automata from RNN. We know RNN's are black box models, so we do not know the internal workings of RNN. Hence we are extracting DFA from RNN to understand their working. Particularly LSTM architecture with 2 hidden layers each of size 10 dimensions and for loss function we are using Log loss which predicts between 0 and 1. We are using stop-threshold of 0.001 on train set. We have divided each grammar in batches of 20 and number of epochs we are using are 100. The train set and test set are split evenly between positive and negative strings for each string length. We have taken Tomita Grammar which is set of 7 regular grammars and introduced 5 more well known regular grammars to create Extended Tomita Grammar. For DFA extraction, we are using L* algorithm proposed by Angluin [1]. It helps us learning DFA from minimally adequate teacher. The L* algorithm basically works on membership and equivalence queries. It checks if the string belongs to the grammar in membership queries if not then reject them. In equivalence checking it compares the extracted DFA with ground truth DFA, whenever they disagree it generate a counter-example. This process eventually leads to a minimal DFA. Here, LSTM network is acting as teacher. The LSTM network is trained for classification of input sequences that are fed into the network. We check for membership and equivalence queries in order to see if the extracted DFA is correct. It generates counter-examples to check if the language accepted is correct (Table 2).

5 Datasets and Preprocessing

Tomita Grammar is one of widely used grammar for DFA extraction from RNN introduced by Tomita [13]. It is a set of 7 regular grammars. Grammars from 1 to 7 are called Tomita Grammar. We also introduced 5 new regular grammars to extend Tomita Grammar data-set to Extended Tomita Grammar. Grammars from 8 to 12 are added to Extend Tomita Grammar. All these grammars have common alphabet $\Sigma =$ (0,1) and generate infinite language over (0,1)*. Each grammar of Extended Tomita grammar has a set of binary strings where some strings are Positive strings and some are Negative strings.

Table 2 Extended Tomita grammar extension to Tomita grammar [13]

#	Description of grammar	Nature of grammar
Tomita grammar		
1	(1*)	Extremely imbalanced
2	(10)*	Extremely imbalanced
3	Odd no. of consecutive '1's followed by an even no. on consecutive 0's	Nearly balanced
4	Any string not containing "000" as a substring	Nearly balanced
5	Even number of 0's and even number of 1's	Nearly balanced
6	The difference between number of 0's and the number of 1's is a multiple of 3	Nearly balanced
7	(0*1*0*1*)	Nearly balanced
Extended Tomita grammar		
8	Last 2 bits are different	Nearly balanced
9	Start's and ends with same bit	Nearly balanced
10	'1' at every even position	Mildly imbalanced
11	Containing '01' as Sub-string	Highly imbalanced
12	No. of 1's in word are Mod 5 ==3	Nearly balanced

- Positive strings: Strings that are accepted by the DFA.
- Negative strings: Strings that are not accepted by the DFA.

We are classifying each grammar according to the imbalance of positive and negative strings among the binary strings generated by the grammars. This classification helps us to look at the grammars with different angle. The grammars 1 and 2 are classified in the category of Extremely Imbalanced (EI) Class because positive and negative strings associated to the grammar are in extremely disproportionate. The grammar 11 is in the category of Highly Imbalanced (HI) class. The number of positive and negative strings associated with this grammar are in between Mildly Imbalanced (MI) and Extremely Imbalanced (EI) classes.

Grammar 3, 4, 5, 6, 7, 8, 9, and 12 are in Nearly Balanced (NB) category. The positive and negative strings are mostly equal. Grammar 10 is in Mildly Imbalanced (MI) class. The positive and negative strings are somewhere in between the Highly Imbalanced (HI) and Nearly Balanced (NB) range.

6 Results and Discussion

In this work, we are assessing the results on our Extended Tomita grammar which is an addition of 5 new regular grammars to Tomita Grammar. We have divided our dataset into different classes to study their behavior. We will be focusing on the training score on LSTM Unit and DFA Extraction score and the quality of extracted DFA against ground truth DFA's. Here, LSTM score represents the training performance of LSTM network on particular grammar. We have observed that higher the training score the better our network will perform in DFA extraction. DFA Extraction score represents how accurate DFA we will get as compared to ground truth DFA's.

6.1 Experimental Setup

Our work on DFA extraction from LSTM architecture builds upon the work by Weiss et al. [14]. We extend the implementation made freely available by the authors.[1] This code is for extraction of DFA from Tomita grammar but we have extended it for Extended Tomita Grammar as explained in Sect. 5. The system we are using contains intel i5-2, 60GHz processor, 8GB RAM, Windows 10 operating system. We are using google colab environment for our python code with Python 3.7.13 as the current version (Table 3).

6.2 Results

We demonstrate how LSTM architecture of RNN's behaves since it is a black-box model. We introduced a new set of regular grammars to the already existing Tomita grammar. These grammars are classified into different categories according to their imbalances among positive and negative strings. These categories are NB, MI, HI, and EI. We look at overall performance of LSTM network regarding its training on these grammars and DFA extraction from network by presenting results in-view of our classification of set of regular grammars we call Extended Tomita Grammar. Our focus is primarily on how grammars of each class perform in extraction of DFA.

We have done bias analysis in Table 4 on our grammars we are running our training process in repetition for 30 times. Grammar 5 and 6 show high deviation in its LSTM training and DFA extraction score as expected since they extract wrong DFA from network. For grammars with correct minimal DFA there is no major deviation in their LSTM training and DFA extraction scores.

6.3 Discussion

In this work, we have extracted DFAs from LSTM network by training it on Tomita grammar and few more set of regular grammars that we introduced as Extended Tomita Grammar. We have divided these grammars into different classes. In Table 3, we have presented the information about the performance of training LSTM net-

[1] https://github.com/tech-srl/lstarextraction.

Table 3 LSTM training and DFA extraction score for extended Tomita grammar set with there Imbalance classes

Grammar type	Nature of grammar	Score	
		RNN (LSTM) (%)	Extracted DFA (%)
Tomita_1	Extremely imbalanced	100	100
Tomita_2	Extremely imbalanced	100	100
Tomita_3	Nearly balanced	99.96	99.96
Tomita_4	Nearly balanced	100	100
Tomita_5	Nearly balanced	72.74	55.6
Tomita_6	Nearly balanced	72.94	53.02
Tomita_7	Nearly balanced	100	100
Extended_Tomita_8	Nearly balanced	99.97	87.69
Extended_Tomita_9	Nearly balanced	100	100
Extended_Tomita_10	Mildly imbalanced	100	100
Extended_Tomita_11	Highly imbalanced	100	100
Extended_Tomita_12	Mildly imbalanced	99.96	99.9

Table 4 Bias and deviation score for every grammar

Grammar type	Bias and deviation score			
	Mean RNN(LSTM)	SD RNN(LSTM)	Mean DFA	SD DFA
Tomita_1	100	0	100	0
Tomita_2	99.96	0.10	99.79	0.62
Tomita_3	99.97	0.03	99.98	0.01
Tomita_4	99.99	0.01	89.14	21.54
Tomita_5	59.14	13.84	58.49	14.79
Tomita_6	67.26	22.18	65.38	21.32
Tomita_7	99.98	0.02	94.01	17.92
Extended_Tomita_8	99.98	0.01	99.98	0.01
Extended_Tomita_9	99.99	0.01	99.99	0.01
Extended_Tomita_10	100	0	100	0
Extended_Tomita_11	100	0	100	0
Extended_Tomita_12	99.92	0.05	95.12	14.62

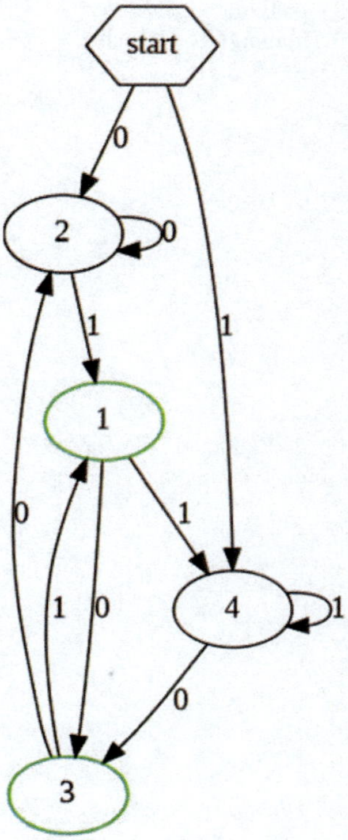

Fig. 1 Extracted DFA for Ex_Tomita_8 Grammar

work and DFA extraction from trained network in context of grammars associated to their respective classes. Here we have presented extracted DFA for Extended Tomita grammar 8, 10, and 2 in Figs. 1, 2, and 3. Grammar 8 is from Nearly Balanced class, Grammar 10 is from Mildly Imbalanced class and Grammar 10 is from Extremely Imbalanced class. Grammars which are associated with EI, HI, and MI classes have no trouble in training and also their extracted DFA's are minimal DFA's. Except Grammar 5 and 6, for all other grammars we are able to get correct minimal DFA from RNN. For grammar 5 and 6 our training loss is very high so it is affecting LSTM score and Extracted DFA score severely. These are product machine grammars. These are also bit more complex then other grammars we have used. Our LSTM network is able to train seemingly on EI, HI, and MI class of grammars with almost 100% score. For the grammars with correct minimal DFA we are getting results quickly compared to grammars with incorrect minimal DFA as expected.

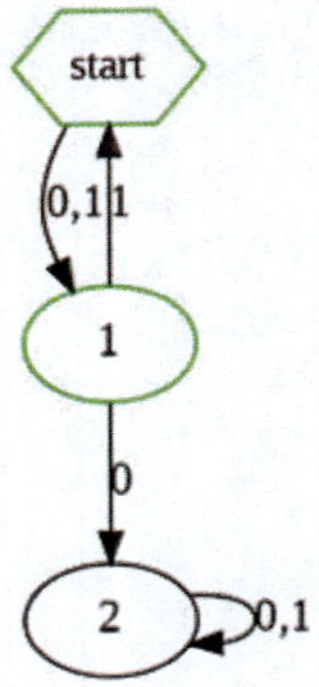

Fig. 2 Extracted DFA for Ex_Tomita_10 Grammar

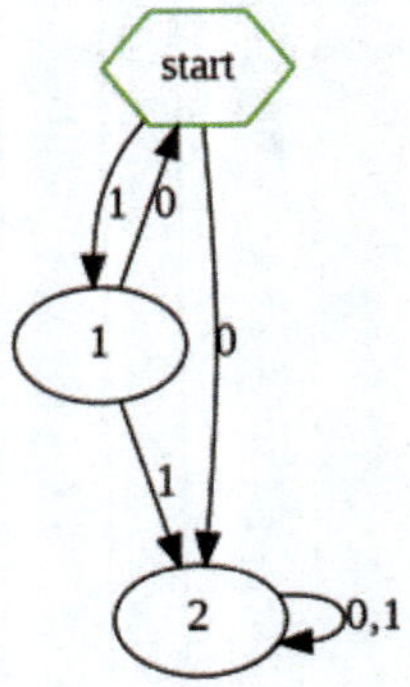

Fig. 3 Extracted DFA for Tomita_2 Grammar

7 Conclusion

In this work, we looked at the extraction of DFA for our classification of grammars we introduced regarding the imbalance of positive and negative strings. The LSTM network, we trained on different nature of grammars for DFA extraction. For grammars of categories EI, HI, and MI, we can extract minimal DFA from our network successfully but for some grammars in the NB category, we were not able to extract minimal DFA. The grammars from categories like EI, HI, MI, we can train our network for these grammars very effectively, and they accepted the language correctly. The extracted minimal DFA is also the same as ground truth DFA. In the case of grammars of the NB category, the DFA generated by the LSTM network contained more states than ground truth DFA. We may try to look into some more complex grammar for all different imbalanced classes in the future. We can also increase our data-set concerning the Imbalance classes we introduced in our work. We can also test our data-set on other RNN architectures such as Simple RNN and Gated Recurrent Unit.

Acknowledgements We thank the anonymous reviewers for their valuable feedback by which the readability of the paper is improved.

References

1. Angluin D (1987) Learning regular sets from queries and counterexamples. Inf Comput 75(2):87–106
2. Bengio Y, Simard P, Frasconi P (1994) Learning long-term dependencies with gradient descent is difficult. IEEE Trans Neural Netw 5(2):157–166
3. Cechin AL, Regina D, Simon P, Stertz K (2003) State automata extraction from recurrent neural nets using k-means and fuzzy clustering. In: Proceedings of 23rd international conference Chilean Computer Science Society (SCCC), IEEE, pp 73–78
4. Cho K, Van Merriënboer B, Gulcehre C, Bahdanau D, Bougares F, Schwenk H, Bengio Y (2014) Learning phrase representations using rnn encoder-decoder for statistical machine translation. arXiv preprint arXiv:1406.1078
5. Chung J, Gulcehre C, Cho K, Bengio Y (2014) Empirical evaluation of gated recurrent neural networks on sequence modeling. arXiv preprint arXiv:1412.3555
6. Elman JL (1990) Finding structure in time. Cognitive Sci 14(2):179–211
7. Giles C, Sun GZ, Chen HH, Lee YC, Chen D (1989) Higher order recurrent networks and grammatical inference
8. Goudreau MW, Giles CL, Chakradhar ST, Chen D (1994) First-order versus second-order single-layer recurrent neural networks. IEEE Trans Neural Netw 5(3):511–513
9. Hochreiter S, Schmidhuber J (1997) Long short-term memory. Neural Comput 9(8):1735–1780
10. Jacobsson H (2005) Rule extraction from recurrent neural networks: a taxonomy and review. Neural Comput 17(6):1223–1263
11. Omlin CW, Giles CL (1996) Extraction of rules from discrete-time recurrent neural networks. Neural Netw 9(1):41–52
12. Omlin C, Giles C, Miller C (1992) Heuristics for the extraction of rules from discrete-time recurrent neural networks. In: Proceedings of International Joint Conf Neural Networks (IJCNN), vol 1. IEEE, pp 33–38
13. Tomita M (1982) Dynamic construction of finite-state automata from examples using hill-climbing. In: Proceedings of 4th annual conference cognitive science society, pp 105–108
14. Weiss G, Goldberg Y, Yahav E (2018) Extracting automata from recurrent neural networks using queries and counterexamples. In: Proceedings of International Conference Machine Learning (ICML), PMLR, pp 5247–5256

Medical Prescription Label Reading Using Computer Vision and Deep Learning

Alan Henry and R. Sujee

Abstract One of the most crucial skills in a person's daily life is handwriting. When it comes to making scripts for various professions, doctors, on the other hand, have been familiar with their low-quality handwriting for decades. To solve this, we demonstrated a deep learning-based method for detecting drug names from doctor's prescriptions, which will benefit the public. The drug is first cropped from the image with reduced dimensions and then fed into two alternative architectures, CRNN alone and EAST + CRNN architecture. The cursive handwritten image is then converted to conventional text using these models. After obtaining the texts, the text is calculated using the CTC loss and the outcome is predicted.

Keywords Deep convolutional neural network · Connectionist temporal classification · Optical character recognition · Handwriting recognition

1 Introduction

Handwriting and reading are some of the key skills which will be acquired by a human in his childhood days. Even with the help of handwriting skills, a person can measure how other person's mindset will be. In brief, we can tell a lot about a person's personality just by looking at his handwriting. Handwriting can also facilitate help in memorizing complex concepts or ideas, which will be helpful for students to learn new concepts and studies rather than mugging up the concepts. So, handwriting plays an important role in day-to-day life in every human. Today, we can categorize handwriting into two categories. One is normal handwriting, and the other is typed text or printed text. Even though handwriting has a lot of benefits, it still has its demerits. One of the biggest concerns is that handwriting takes a lot of time, which

A. Henry (✉) · R. Sujee
Department of Computer Science and Engineering, Amrita School of Engineering, Amrita Vishwa Vidyapeetham, Amrita University, Coimbatore, India
e-mail: alanhenryhelda@gmail.com

R. Sujee
e-mail: r_sujee@cb.amrita.edu

M. Thakur et al. (eds.), *Soft Computing for Problem Solving*, Lecture Notes in Networks and Systems 547, https://doi.org/10.1007/978-981-19-6525-8_9

affects our efficiency and leaves us with less time to do other things. When we write for a longer duration, it produces pain in the finger. Handwriting can also cause other health problems such as headaches, neck pain, and back pain.

Even though there exist many tools on the market, people still prefer writing and reading on their own. Among these technologies, there is one such technology which is called optical character recognition (OCR). The main purpose of OCR is to recognize text which is present in an image or in a captured or scanned image and convert them into proper normal digital text. Here, what they mean as digital text is that the text is almost like the text produced by any printers or any electronic device, and the trend that we can see in the current world is that most of the functional machines are moving toward digitalized machines. Since OCR plays a vital role in digitizing the world, it can also lead to an increase in business production by reducing the labor costs and time of a person. Even though OCR problems have been solved by many researchers, it still becomes a tough part when there are more variations in handwritten texts. These variations are caused by different persons' different handwriting skills. It differs based on the person. Some may have good handwriting while others do not have good handwriting. The main applications of OCR can be applied in various industries like the banking, health care, and insurance sectors. Nowadays in some organizations, they used to enter all the details manually. Instead of typing manually, they use the OCR technique which will lead them to reduce their time and work. In insurance industry, daily they receive a huge number of claim documents. Manually entering each document is a tiresome process; instead, they can deploy OCR for ease of work. In banking sectors, OCR could be deployed to read the information which is present in the bank cheques. Also, it can be implemented in restaurants to extract invoice details from an invoice. Whereas in the healthcare industry, the OCR technique can be used to extract the medicines which are contained inside prescription or tablet strips.

So, in this project, the researchers are trying to implement an OCR-based deep learning model in the healthcare field. The main motivation to do this study is to analyze the doctor's handwriting. One of the main issues that arises in the case of medical prescriptions is when the handwriting of a doctor is not understandable. In most cases, only pharmacists can understand the handwriting written by doctors. Sometimes, even they cannot understand what is written inside it. So, what this research is trying to do is even a layman should understand the medicine name which is present in the prescription even without others' help. So, in that case we can use advanced technologies like OCR techniques, which will be useful for detecting the characters present inside a prescription. The OCR has a wide variety of applications. Even though there exist many handwriting recognition algorithms, recognizing handwriting is still a challenging part. Still, people are doing more research in this field to find out more proper and accurate algorithms. Earlier, they used models like CRNN to detect text in an image, which consisted of a set of convolutional and recurrent neural network layers. Nowadays, handwriting recognition is categorized into a general term called Intelligent Character Recognition (ICR). Currently, the trends are on Transformers, which can detect text more precisely than the previous algorithms. Also, there is Attention-OCR, but still, Transformers outperform them all.

Also, nowadays there are many inbuilt libraries available on the Internet like Pytesseract, to convert the text present in an image to digital form. Even Google has introduced an application programming interface called Cloud Vision API to detect texts, and Baidu engineers have released a deep learning model called PaddleOCR to do the same task. So, what we can understand is that these OCR techniques are in a more advanced state than we would have expected. In this paper, Sect. 2 of this paper explains why this initiative was started. Section 3 outlines the related study on extracting text from images. Section 4 describes how the experiment is carried out and which architectures are employed. The experimental results and implications from the experiments are described in Sect. 5. Section 6 describes about the inferences got from the experiments and how the research can be further enhanced.

2 Motivation

Predicting a medicine from a medical prescription written by a doctor is not a straightforward task. In the USA during 1999, the data has shown that around 44,000–98,000 people died because of this kind of medical errors. Among them, 7000 people were died because of the poor handwriting of the doctors [1]. In India also many events happened because of the illegible handwriting of doctors, which has cost many lives. The precise numbers are not well known since we do not have an ideal knowledge bank or correct written record.

Even though fewer systems with certain features exist, they are still not accurate and produce only results that are limited to text only. Also, the opposite aspect of this lousy handwriting is that even a layman cannot perceive what is written in a medical prescription which makes them difficult to understand. Thus, to avoid these problems, we can implement specific OCR techniques. Through OCR techniques, we can extract the texts on the region of interest. So, what this project tries to aim is in addition to conversion of the text, from the prescriptions. So that everyone will be able to recognize the medicine even without the help of the third person.

3 Related Work

In this analysis [2], what they have attempted to implement is a camera-based method to scan medical prescriptions for visually impaired people and extract the meaningful texts with the assistance of a region of interest and convert them into audio. The texts contained within medical prescriptions are scanned using OCR techniques. The various steps involved in the techniques include scanning the prescription and preprocessing, edge detection and segmentation, and also OCR through NLP. The result that they acquired when they scanned the prescription was smart and was able to convert the text into speech.

The technique involved [3] is a mobile-based application that detects a medicine dosage from a given prescription and ends up resulting in text which can be useful for both the patients and the pharmacists. The different steps involved in the process are preprocessing; after preprocessing, the images are then transferred into Convolutional Neural Network (CNN) that successively provides a major amount of the extracted details, and at the end, OCR is applied on the images with low quality. The suggested technique was able to accomplish 70% accuracy with the CNN model.

In analysis [4], they have performed a study on a medical prescription dataset which is collected from several clinics and hospitals and were able to recognize the proper texts from the cursive handwritten texts of doctors by applying Deep Convolutional Recurrent Neural Network. They have employed two CRNN models, out of which CRNN with batch normalization achieved a 76% accuracy and 72% accuracy in the validation set. This successful model is then hosted onto a mobile application.

In 2013, Roslay and Ruel [5] built an android application to determine the misinterpreted names within the medical slips with the help of an optical character recognition library such as tesseract and return them into a text. Usually in the above research papers, the region of interest is chosen by the system itself. But here, the user selects a particular region of interest and it is then converted into text. Instead of a dataset, they have used a separate database from where the drugs are checked whether they are matching are not and achieved significant results.

This study [6] examines the three different styles of OCR ways that will be helpful in scanning medical prescriptions through the OCR. The three different methods are classic computer vision techniques, standard deep learning methods, and specific-based deep learning. These methods are examined with 100 prescription medical image labels and are evaluated based on accuracy, speed, and resource. And they produced 76% accurate results.

The research [7] tries to identify the name of the capsule which is present in the blister strip. The result that they get is an audio output of the name of the capsule. The main aim of this project is to assist visually impaired people so that they will be able to recognize the name of the tablet even without a third person. First, they obtain the name of the tablet from the image, which is then converted to audio with the assistance of Google Text-to Speech (GTTS). To find the name found in the image, they have used SIFT algorithm along with the SQLite database which consists of the name of the capsules.

This project [8] tries to convert the handwritten into digital form. They have approached the problem with the IAM dataset along with CNN with various architectures and LSTM to build a setup of a bounding box around the characters. After this pass, the segmented characters are added to CNN for accurate classification.

The next paper [9] tries to implement a basic CRNN deep learning model to recognize text which is present inside a football match scene. Along with the CRNN architecture, they have added extra MFM layers to increase the contrast of the image. The model was employed on both public and manual datasets and achieved a significant result than the original model.

To convert the handwritten text to normal text, what this paper [10] has implemented is and what they have done is they classified their dataset into four different categories: printed, semi-printed, handwritten, and cursive handwritten. To train the printed texts, they have employed the Pytesseract model, but for handwritten images, they have applied the CRNN model. The overall accuracy of what they have got for printed text was 94.79%, 75.2% for handwritten images, and 65.7% for cursive handwriting.

On today's cell phones, there are a plethora of applications for personal event planning. The most popular modes of input for those programmers are text, voice input from the user, and E-Mail updates from some service providers. This paper [11] describes an event planner based on image processing that uses Google Calendar to arrange activities. The application recognizes text in photographs using optical character recognition (OCR), which is then utilized as an input to the scheduler, and Google Calendar is used to keep the event planner up to date.

Car owners' biggest problem when driving in the city is finding a parking spot for their vehicles. Most of the time, those waiting outside parking lots are informed at the last minute that parking is unavailable. The goal of this project [12] is to leverage GSM messaging services to offer real-time information to users about the number of available slots, as well as to automate parking lot administration and tariff calculation using optical character recognition and time stamping. Customers will wait substantially less time, and the number of staff required in the parking lot will be reduced as well.

The revolutionary Internet and digital technologies have necessitated the creation of a system to organize and categorize an abundance of digital photos for easy retrieval and categorization. The goal [13] is to create a semantic image search engine for the web. The overlay text embedded in photographs is used to do the search. Because the editor's aim can be adequately expressed by employing these embedded sentences, overlay text comprises crucial semantic hints in films as well as visual content analysis such as retrieval and summarization. Image feature extraction, representation, mapping of features to semantics, storage, and retrieval are all elements of the proposed Content-Based Image Retrieval (CBIR) system. We propose an architecture for picture retrieval based on the extracted text. The CBIR systems have a significant impact.

Allergic reactions to food can be influenced by a variety of circumstances, resulting in a wide range of proportional reactions. With such a wide spectrum of unpredictability, scientists have been working on identifying allergens and the rate at which they affect people for years. In this research [14], a 2-tab deep learning-based application to give nutrient and allergen content in fruits and vegetables, as well as to display allergen information in packaged food using OCR, to raise awareness of the food we consume, and the possible hazards it may pose. The image of the fruit or vegetable captured via an application is categorized and identified, as well as the nutritional facts and allergen information, using a unique Deep Learning Framework. On dataset, the fine-tuned deep learning model, which is deployed in the cloud, achieved a good accuracy of 97.37%. For packaged food, the application captures a picture of the Ingredient Index, and the allergen information is displayed once the text is detected using optical character recognition, which is done on a remote server.

The core of industrial digitization, often known as industry 4.0, is computer vision and its applications. Texts contained in photos are an excellent source of information about an object for automating a procedure. Because of the complicated background, size and spacing fluctuations, and uneven text arrangements, reading text from natural photos remains a difficult task. The main steps of reading in the outdoors are detection and recognition. Many researchers have developed approaches for identifying writings in photos in the last few years. These strategies work well with horizontal texts but not with irregular text arrangements. This research [15] focuses on a deep learning model for visual text recognition (DL-TRI). The model considers a variety of curved and perspective typefaces.

4 Design of the Proposed Work

4.1 Data Collection

As per architecture (see Fig. 1), data collection is the first step to acquiring the proper dataset for the project. For the first approach, the dataset (see Fig. 2) consisted of 50 images which were collected from several doctors and through the web. For the second approach, there is no particular dataset, but instead, we can pass the whole prescription image to the model. Most of the images are color images and are captured through a 12 mp camera.

4.2 Preprocessing

Since all images are captured through the phone, every image is in RGB format. To yield better accuracy, we have got to convert our RGB image into gray scaled images. After this step, in the first approach, since our model does not understand the character in strings, we have to convert the labels. To do this, we have to encode our labels into numerical; e.g., instead of character 'a,' we encode them as number 1. After this, we then normalize our images to yield good accuracy. Whereas in the second approach, the EAST and CRNN combined model, the captured image is only just converted to a gray scaled image and passed onto the EAST model to identify where the text is present inside a prescription. One of the main reasons to do preprocessing is to make images clearer and to get rid of unwanted noises which will make the images more suitable for model building and further processing.

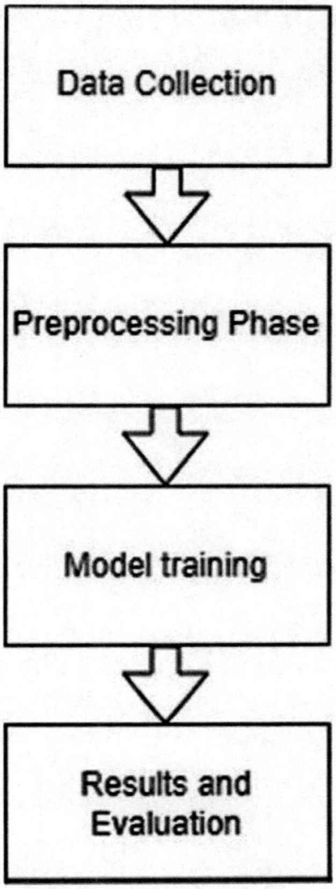

Fig. 1 Proposed architecture

4.3 *Training of Data Using Deep Learning*

In the first approach (see Fig. 3), the preprocessed data is then fed into a CRNN (Convolutional Recurrent Neural Network) deep learning model which accepts an input shape of width 128 and height 32. CNN has been used to extract vital information from preprocessed images. From the vital regions with help of bidirectional LSTM, the handwritten is then converted into a normal text along with the Connection Temporal Classification (CTC) loss. The model consists of full gated convolution and 7 convolution layers; after that, it is employed using ReLU, padding as same, and max pooling. After the CNN layer, the image is then fed into a bidirectional LSTM model for the prediction of the normal text along with the CTC loss. The main purpose of CTC loss is to avoid the spanning of the characters. For training purposes, TensorFlow, OpenCV, and Python were used.

In the second approach (see Fig. 4), the preprocessed image is fed into the EAST (Efficient and Scene Text) algorithm, which is a trained model on the ICDAR 2015 dataset and achieved an F-score of 0.7820. In the EAST text detection pipeline,

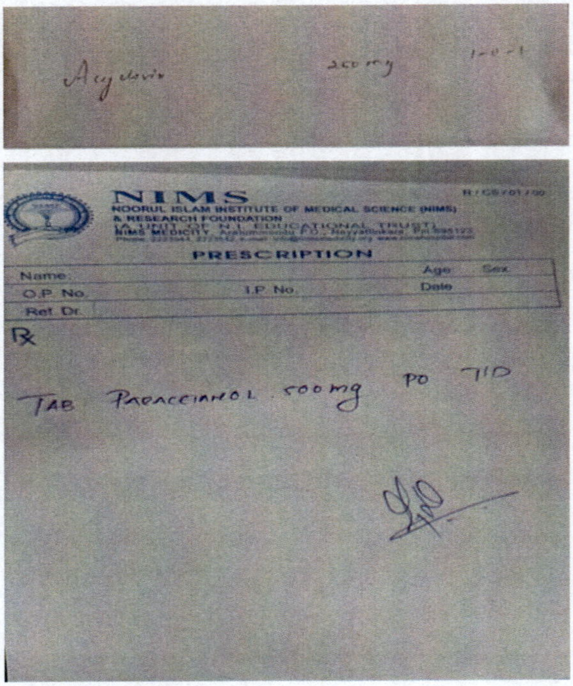

Fig. 2 Approach 1 and approach 2 dataset sample images

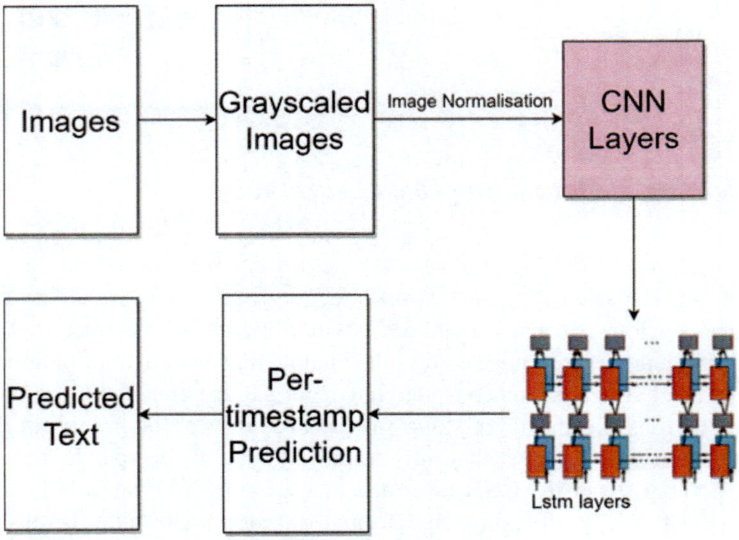

Fig. 3 Approach 1 (only CRNN) architecture

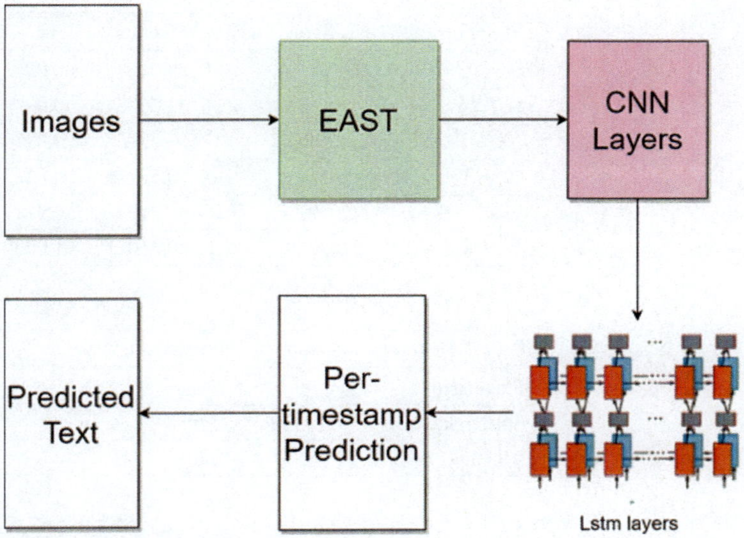

Fig. 4 Approach 2(EAST + CRNN) architecture

they are mainly divided into two stages. At the first stage, the preprocessed input image is fed into a multi-channel fully convolutional neural, which is a U-shaped network architecture, which detects the text with varied shapes and lengths, and as a result, they produce scores and geometry maps. After this, these maps are then converted into text regions, which are then post-processed using thresholding and non-max suppression to produce final text boxes. Hence, text is detected from the image. Afterward, to recognize the text inside an image, since our input is in image format and the output to be predicted is in text form, the CRNN model is employed, which is a combination of CNN and RNN. Here at CNN, the convolutional layers are converted into feature maps, which in turn are converted into a sequence of feature vectors. These extracted feature vectors are then fed into the bidirectional LSTM which solves the problem of vanishing gradient points. The output generated from bidirectional LSTM may consist of repeating string patterns. So, to avoid that we must use CTC loss. CTC tries all the possible combination and chooses the text based on the CTC mapping rule, high probability value, and best path decoding. The CRNN model that we have used is a trained model on the MySynth dataset.

4.4 Evaluation

To check the accuracy of the model, we have used metrics such as validation loss, training loss, and CTC loss.

Table 1 Results obtained based on approach 1(only CRNN)

No. of epochs	Training loss	Validation	Accuracy
10	27.3369	28.0733	0.00
100	26.1740	32.06784	0.00
200	1.5467	107.7516	0.9778
300	0.8582	30.82215	0.0222
350	2.1911	91.2187	0.9111
400	0.2048	30.61071	1.0000

Table 2 Medicine name predictions based on approach 1(only CRNN)

Original text	Predicted text
Fluconazole	Nyine
Spironolactone	Cyin
Albendazole	Cnine

5 Experimental Results

First, the images are scanned, and then, it is fed into the model to get the prediction. Based on comparing both the results produced by the different models, approach 2 (EAST + CRNN) generated much better results. Even though the results produced by the Model 2 (EAST + CRNN) are not accurate, it almost tried to predict a proper word. Approach 1 is only getting accuracy, but the prediction is not as expected, which might be because of fewer data. The hyperparameters for CRNN are learning rate = 0.01, batch size = 8, ReLU as the activation function, and Adam as the optimizer function.

Based on the Table 1 and Table 2 results that are obtained by approach 1 (CRNN), it is evident that the results that are predicted are not the same as the validated text or label, and also it is underperforming. Even though it showed a 100% accuracy at epoch 400, the results are not acceptable. So, it cannot be deployed for real use cases.

From the above results, it is noticeable that the medicines that are prescribed are in improper form. Also, compared to approach 1, approach 2 (CRNN) is much better (see Fig. 5). It also proved that this model works well in the case of printed text compared to handwritten text.

6 Conclusion and Enhancements

The models that are tried on this research are successfully implemented with the help of different datasets. But the results what the models have produced are not accurate or not satisfactory. It could be due to the author's unreadable handwriting, which is composed of various sizes, alignments, and shapes. It is difficult for the

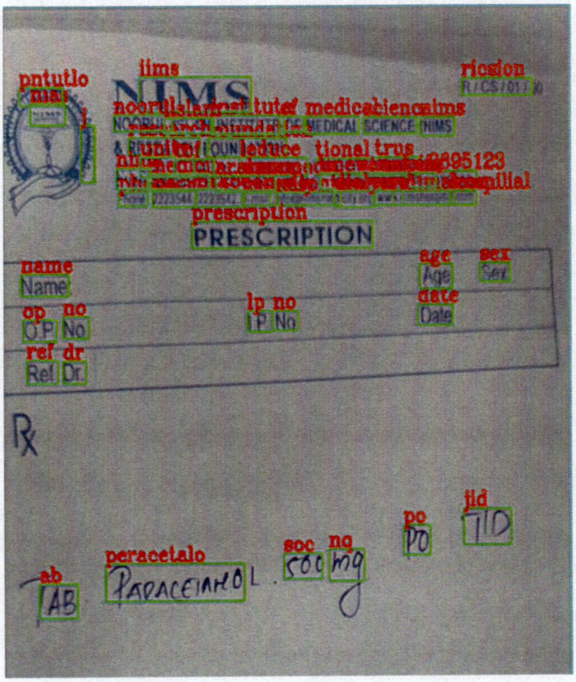

Fig. 5 Approach 2 (EAST + CRNN) result

model to learn to recognize proper texts from photos. In addition, as compared to handwritten writings, the results demonstrated good accuracy on printed texts. This work fails if the handwriting is too much curvy or in irregular shape. These models can be improved by including the Levenshtein distance as an evaluation metric at the last phase of the experiment, which helps the model to predict the nearest word.

In future, this study could be improved by adding highly advanced deep learning architectures like Attention-OCR, Transformers, and Visual Attention. Also further, we can add a text-to-speech converter, which converts the raw text into audio form. To achieve this, we can just use deep learning models like tacotron or web API like Google Text-to-Speech (GTTS). And it can be evaluated based on the Mean Opinion Score (MOS). Adding web-based applications or mobile-based applications could help the user to access the application from any part of the world where there is the Internet.

Acknowledgements The authors are thankful to Amrita Vishwa Vidyapeetham's Department of Computer Science and Engineering for providing us with the opportunity to work on medical prescriptions handwriting recognition.

References

1. Deccan Chronicle. https://www.deccanchronicle.com/nation/in-other-news/201018/its-time-to-totally-ban-handwritten-prescription.html
2. Sahu N, Raut A, Sonawane S, Shaikh R (2020) Prescription reading system for visually impaired people using NLP. Int J Eng Appl Sci Technol 4
3. Hassan E, Tarek H, Hazem M, Bahnacy S, Shaheen L, Elashmwai WH (2021) Medical prescription recognition using machine learning. In: 2021 IEEE 11th annual computing and communication workshop and conference (CCWC). IEEE, Jan 2021, pp 0973–0979
4. Fajardo LJ, Sorillo NJ, Garlit J, Tomines CD, Abisado MB, Imperial JMR, Fabito BS (2019) Doctor's cursive handwriting recognition system using deep learning. In: 2019 IEEE 11th international conference on humanoid, nanotechnology, information technology, communication and control, environment, and management (HNICEM). IEEE, pp 1–6
5. Alday RB, Pagayon RM (2013) MediPic: a mobile application for medical prescriptions. In: IISA 2013. IEEE, July 2013, pp 1–4
6. Bisiach J, Zabkar M (2020) Evaluating methods for optical character recognition on a mobile platform: comparing standard computer vision techniques with deep learning in the context of scanning prescription medicine labels
7. Shashidhar R, Sahana V, Chakraborty S, Puneeth SB, Roopa M (2021) Recognition of tablet using blister strip for visually impaired using SIFT algorithm. Indian J Sci Technol 14(23):1953–1960
8. Balci B, Saadati D, Shiferaw D (2017) Handwritten text recognition using deep learning. In: CS231n: Convolutional neural networks for visual recognition, Stanford University, Course Project Report. Spring, pp 752–759
9. Chen L, Li S (2018) Improvement research and application of text recognition algorithm based on CRNN. In: Proceedings of the 2018 international conference on signal processing and machine learning, Nov 2018, pp 166–170
10. Bagwe S, Shah V, Chauhan J, Harniya P, Tiwari A, Gupta V, Mehendale N (2020) Optical character recognition using deep learning techniques for printed and handwritten documents. Available at SSRN 3664620
11. Bhaskar L, Ranjith R (2020) Robust text extraction in images for personal event planner. In: 2020 11th International conference on computing, communication and networking technologies (ICCCNT). IEEE, July 2020, pp 1–4
12. Annirudh D, Kumar DA, Kumar ATSR, Chandrakala KV (2021) IoT based intelligent parking management system. In: 2021 IEEE Second international conference on control, measurement and instrumentation (CMI). IEEE, Jan 2021, pp 67–71
13. Hrudya P, Gopika NG (2012) Embedded text based image retrieval system using semantic web. Int J Comput Technol Appl 3(3):1183–1188
14. Rohini B, Pavuluri DM, Kumar LN, Soorya V, Aravinth J (2021) A framework to identify allergen and nutrient content in fruits and packaged food using deep learning and OCR. In: 2021 7th International conference on advanced computing and communication systems (ICACCS), vol 1. IEEE, Mar 2021, pp 72–77
15. Shrivastava A, Amudha J, Gupta D, Sharma K (2019) Deep learning model for text recognition in images. In: 2019 10th International conference on computing, communication and networking technologies (ICCCNT). IEEE, July 2019, pp 1–6
16. de Sousa Neto AF, Bezerra BLD, Toselli AH, Lima EB 2020 HTR-Flor: a deep learning system for offline handwritten text recognition. In: 2020 33rd SIBGRAPI conference on graphics, patterns and images (SIBGRAPI). IEEE, Nov 2020, pp 54–61
17. Nanonets. https://nanonets.com/BLOG/ATTENTION-OCR-FOR-TEXT-RECOGNTION/
18. AI learner. https://theailearner.com/2019/05/29/creating-a-crnn-model-to-recognize-text-in-an-image-part-1/
19. https://github.com/rajesh-bhat/spark-ai-summit-2020-text-extraction
20. https://github.com/GireeshS22/Handwriting-CNN-LSTM

Autoencoder-Based Deep Neural Architecture for Epileptic Seizures Classification

Monalisha Mahapatra, Tariq Arshad Barbhuiya, and Anup Nandy

Abstract This paper discusses a deep neural network architecture of Long Short-Term Memory (LSTM) with an autoencoder-based encoder-decoder scheme. Primarily, the proposed structure determines the time-domain features of electroencephalography (EEG) signals, which is subsequently trained to acquire reduced dimensions of EEG features. Later, these features are provided to a one-dimensional (1D) Convolutional Neural Network (CNN) for classification. The effectiveness of the proposed model is corroborated on the public benchmark Kaggle Epileptic Seizure Recognition dataset. The dataset consists of five classes corresponding to five different health states;comprising a seizure state (subjects with epileptic seizure) and four normal states (subjects without seizure). In this work, the binary classification task of epileptic seizure is performed. The outcomes exhibit that the proposed architecture attains a recognition accuracy of 92.47% on this task. Additionally, a relative study has been carried out with other standard neural models, specifically, deep neural networks (DNN), and CNN. Few machine learning models, namely, logistic regression (LR),random forest (RF), and K-Nearest Neighbors (KNN) are also studied for comparison. It further substantiates the dominance of the proposed model. The outcomes have proven the purpose of this study to demonstrate the efficacy of the proposed model in EEG epileptic seizure classification.

Keywords Long short-term memory · Convolutional neural network · Epileptic seizures

1 Introduction

Epilepsy disease (ED) is regarded as one such progressive diseases in cognitive functioning of brain for a certain months or years [1]. Seizure condition is the dominant general cause of ED. A sudden outbreak of extra electricity in the brain causes unusual actions bringing on unforeseen seizure attacks. It occurs instantaneously resulting

M. Mahapatra (✉) · T. A. Barbhuiya · A. Nandy
Machine Intelligence and Bio-motion Lab, Department of Computer Science and Engineering, National Institute of Technology, Rourkela, India
e-mail: mmahapatra435@gmail.com

M. Thakur et al. (eds.), *Soft Computing for Problem Solving*, Lecture Notes in Networks and Systems 547, https://doi.org/10.1007/978-981-19-6525-8_10

from brain abnormalities in absence of any signs. Frequent epileptic seizures tamper with the human brain, exerting influence on impaired memory, cerebral deterioration, and thereby increase in human mortality. Accordingly, it becomes critical to recognize epileptic seizure state and provide primary assistance early [2]. Epileptic seizure movements is recognized during the visual study of EEG recordings that demands considerable time and effort. EEG is a device for studying brain waves relevant to determining epileptic seizures. The brain neural signals are acquired by the electrodes positioned on the scalp at different situations. At the initial stages of EEG study, usual activity is detected, but, very significant amplitude and periodic activities are observed for a while. Subsequently, the signal again returns to original form. These periodic activities are termed as spikes throughout a seizure and are quite temporary. Throughout seizures, intricate spike and wave forms produced by the brain can be registered on the EEG recordings. Customary diagnosis of epileptic seizure and deciding the cause through visual inspection of EEG data is cumbersome and requires effort [3]. Thus, visual seizure detection has failed to be very systematic. Hence, it is essential to develop competent machine-driven seizure recognition schemes to enable the identification of epilepsy. Automated seizure classification in EEG is an arduous task and still demanding. Several authors have discussed their contributions on automatic epileptic seizure classification tasks, mostly using binary classes (epileptic and non-epileptic seizure EEG signal). The EEG dataset used in the current study may include some needless information (features). Different researchers have addressed several feature extraction techniques through standard frequency domain or time-domain analysis. Rabby et al. [4] proposed a unique method for feature extraction to separate the seizure and non-seizure classes of EEG signal. They introduced wavelet transform following petrosian and higuchi fractal dimension, and singular value decomposition entropy. Sayem et al. [5] discussed their work on epileptic seizure classification emphasizing feature extraction approach by applying discrete fourier and wavelet transformations. Further, to identify nonlinear features, they introduced multi-scale entropy with sample entropy. Hussain et al. [6] demonstrated the methods to extract only a marginal amount of extremely preferential features utilizing wavelet decomposition to classify epileptic seizures. In the past years, deep learning (DL) has advanced exceedingly for feature extraction and classification purpose. Shekokar et al. [7] implemented LSTM model on Bonn's dataset to classify epileptic seizures. The measure to verify the effectiveness of their proposed approach are accuracy, specificity and sensitivity. Xu et al. [8] proposed a complete framework for feature extraction by altering the traditional structure of CNN model. Next, the features obtained from 1D CNN are further processed to extract the temporal features using LSTM so as to improving the classification results of their introduced structure. And hence, the name of their proposed architecture is 1D CNN-LSTM. Regardless, determining an approach to instinctively extract features and maintain the classification performance is demanding. In the current study, we put forward an automated feature extraction approach by incorporating LSTM with autoencoders. In the current study, we put forward an automated feature extraction approach by incorporating LSTM with Autoencoders. Similarly, Ahmed et al. [9] presented a 1D convolutional autoencoder framework elucidating the feature extraction and dimensionality reduction concept. The standard structure of the encoding and decoding part is altered using stacked convolutional layers in lieu of fully connected layers. It incurs the overhead of feature extraction and high training time; thereby it turns out to be more analytical and burdensome to extract the significant features in the interest of upgrading their classification performance. Typically, the encoding and decoding layers of deep autoencoders are customized. In this work, we attempted to introduce the regular autoencoder structure comprising fully connected layers with LSTM layers as hidden units for the feature extraction purpose, rather than structuring more layers for encoding and decoding part. Predominantly, our proposed structure can accommodate the computational insufficiency. Therefore, we deliberately considered LSTM autoencoder for feature extraction approach. This introduced model can conform to different dataset types regardless the training time, relative to other models as discussed in the literature [8]. And to our knowledge, none of the earlier analysis accurately embodies autoencoder with any deep neural models in its primary structure. In the current work, we have developed LSTM autoenoder in its fundamental structure in order to weigh the importance of chosen features.

Primarily, DL models does not involve definite feature engineering and can execute directly with raw EEG data. However since, EEG waveform are series of times, an autoencoder-based deep model can consider the temporal factor of EEG data, comprehend the data well, and perform better encoding of the data. Therefore, in this study, a deep neural structure of LSTM incorporated with autoencoder-based encoder-decoder scheme is proposed. Primarily, the LSTM autoencoder extract features from time-domain interpretation of EEG signals. Later, it is further processed to acquire reduced dimensions of EEG features, that are provided to a 1D CNN for ultimate classification.

The current study is concerned with actively determining binary classified events found in EEG time series data. The intent of this study is to introduce an automated seizure recognition model for identifying subjects with epileptic and non-epileptic seizure states. The essential contributions of this work is as mentioned:

1. Primarily, a distinctive approach to determine the significant time-domain EEG features is introduced using an autoencoder-based LSTM model. Subsequently, a 1D CNN model architecture is designed, preferring these features for the ultimate classification of epileptic seizures.
2. A relative study has been carried out with other standard deep neural network models, in particular, DNN, and CNN.
3. Further,few machine learning models are put in use to substantiate the efficacy of the proposed approach.

The rest of the paper is organized as mentioned: Sect. 2 provides the details of the dataset used. The architecture of the proposed model is elucidated in Sect. 3. The model evaluation and result analysis is discussed in Sect. 4. Lastly, Sect. 5 provides conclusion and future work.

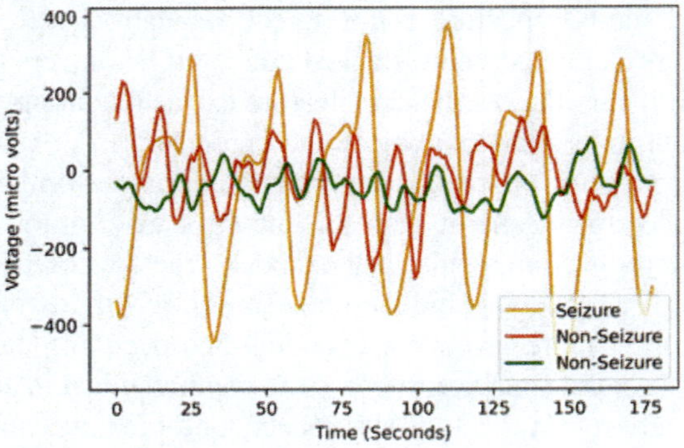

Fig. 1 EEG signals' visualization

2 Dataset Description

The dataset used in this study is pre-processed and a modified interpretation of epileptic seizure recognition dataset of UCI repository, available online in Kaggle [10]. The UCI dataset contains recordings of brain activity of 500 subjects sampled for 23.5 s including 4097 data points. Every data point corresponds to the value of EEG signals at various time points.

This modified dataset can be understood as every sample containing 4097 data values, distributed into 23 data segments with 178 data points of 1 s in every segment. Subsequently, the 23 data segments are reorganized. Ultimately, for every individuals, 11,500 data values are achieved.There are five health conditions for this dataset, comprising an epileptic state and others as single epileptic seizure status and other four as standard conditions. The information contains are in this way:

1. The subjects' recordings having epileptic seizures;
2. The subjects' recordings with opened eyes while EEG signals recording;
3. The recordings of individuals with eyes closed while recording of EEG signals;
4. The EEG signal recordings of subjects taken from sound brain region;
5. The EEG signal recordings of subjects' taken from tumor area in the brain.

The EEG signal visualization with one epileptic state and other non-epileptic seizure states is shown in Fig. 1. It depicts that seizure signals represent significant amplitude, while the non-seizure signals are low in amplitude. In this work, binary epileptic seizure recognition tasks is taken into consideration to measure the performance of the proposed model.

3 Proposed Approach

This section provides an absolute description of the proposed epileptic seizure recognition model. The EEG signals recorded in this dataset are previously pre-processed and represented in time-domain representation, and hence, do not necessitate any further signal preprocessing techniques.

3.1 Structure of Autoencoder-Based LSTM

An autoencoder consists of an encoder α and a decoder λ where

$$\alpha : Y \rightarrow Z,$$
$$\lambda : Z \rightarrow Y$$

Y and Z represents input and encoded space. The multi-dimensional features space is potentially transformed into smaller dimensional feature representations using LSTM autoencoder. Thereby, it promotes learning against high-dimensional data with low-sampled records.

The proposed approach performs sequence-to-sequence tasks, typically used for time series data. The sequenced data is passed to the network by LSTM autoencoder that regenerates it into latent space representation connected with LSTM as hidden units. The output is next given to the decoder component with the exact input from encoder hidden states to construct a different representation. The entire sequence is characterized as a fixed size embedding vector, thereupon the encoder is provided 'n' inputs in sequence meantime the decoder is provided with a transferred version with one time step.

3.2 1D CNN Structure

The basic structure of 1D CNN utilizes several filters to effectively perform the convolution operations. In this study, the filters and feature maps are one dimension so that it can correspond the 1D property of raw EEG signal data.

3.3 Proposed Architecture

The proposed LSTM autoencoder 1D CNN model is comprised of two basic blocks: (a) feature extraction, and (b) final Classification. The initial block indicates the feature extraction phase comprised of LSTM autoencoder neural network to recognize highly pertinent features which are next processed to the next block for final classification of epileptic seizures. The classification is achieved using the 1D CNN structure. The output of the encoder phase (the latent space) is taken as input by two 1D CNN layers succeeded by a fully connected layer for the classification.

In the current work, the sequential data with shape 178×1 is passed as input to the encoder part containing LSTM layer with 32 units. It transforms the original input signal of shape 178×1 into an abstract space representation Y of shape 16×1. The decoder part takes this 16 dimensional abstract space as input containing LSTM layer with 178 units and reconstructs back the original signal of shape 178×1.

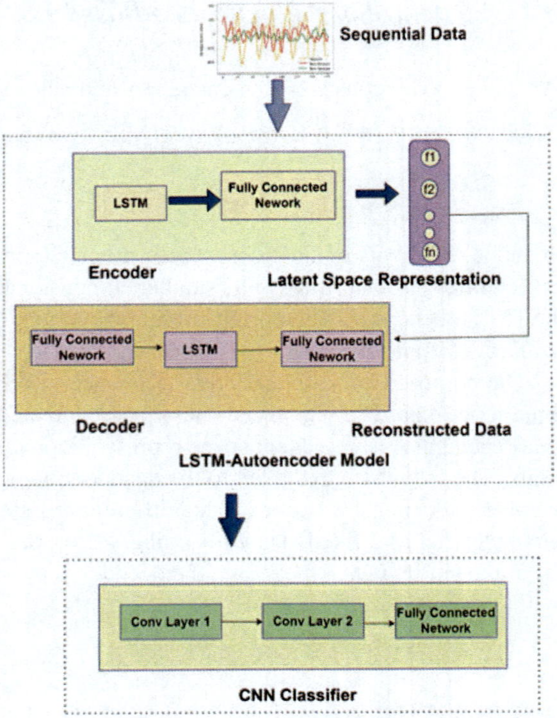

Fig. 2 Proposed model overview

Later, this autoencoder based LSTM model is followed by 1D CNN for the ultimate classification. The input layer contains the signal of shape 16×1 obtained from the encoder part of the autoencoder LSTM. Next, it is passed to the first convolutional layer with filter size 3 and kernel size 3 for further understanding of the data. Further, it is connected to the second convolutional layer with filter size 16 and kernel size 3. This layer passes its output to the fully connected layer, that classifies the epileptic and non-epileptic signals. Figure 2 provides an overview of the proposed approach and Fig. 3 displays the elaborate structure of the proposed model.

The Latent space representation is given as:

Abstract Space Representation, $Y = \text{en}(x)$
Reorganized EEG Data, $\hat{x} = \text{de}(y)$
where x = Original EEG Data
en = Encoder LSTM, and
de = Decoder LSTM.

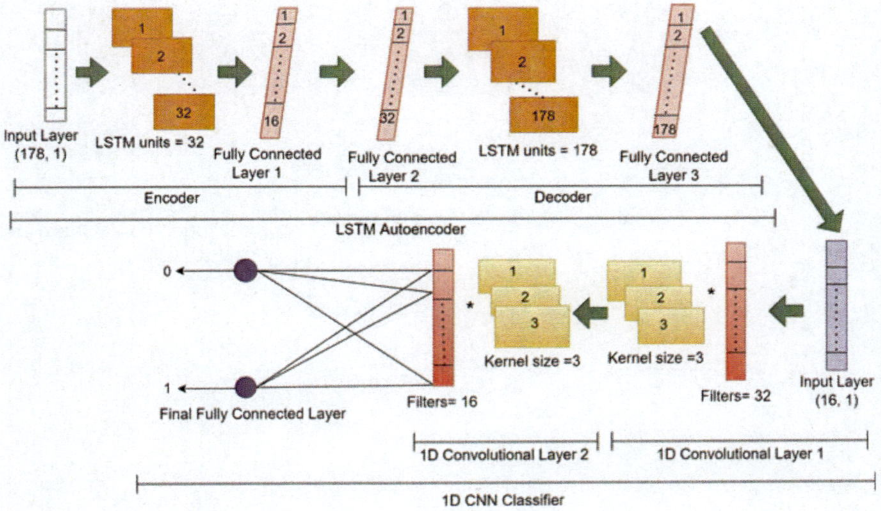

Fig. 3 Proposed model architecture

4 Model Evaluation and Results

This section discusses about the achievement of the proposed model on the experimental tests done on the epileptic seizure recognition dataset. The training and testing results of the proposed model are provided for the experiments carried out. Moreover, a comparative study with standard deep learning and machine learning models is realized to demonstrate its efficacy.

4.1 *Binary Classification Task*

The dataset includes five separate classes, however, in the current study, the binary classification is considered in favor of model evaluation, such as, Class 1 (epileptic condition) compared with the rest classes.

The experimental studies are carried out on google colab environment on python based Tensor Flow, a deep learning library. The model is evaluated by applying a single fold cross validation. The training was done on 70% data and validated on 10%, while rest 20% are used for testing purpose.

4.2 *Experimental Results and Discussion*

Primarily, the model loss curve of the proposed LSTM autoencoder 1D CNN model, is shown in Fig. 4a. Moreover, the two standard DL models, namely, DNN and CNN

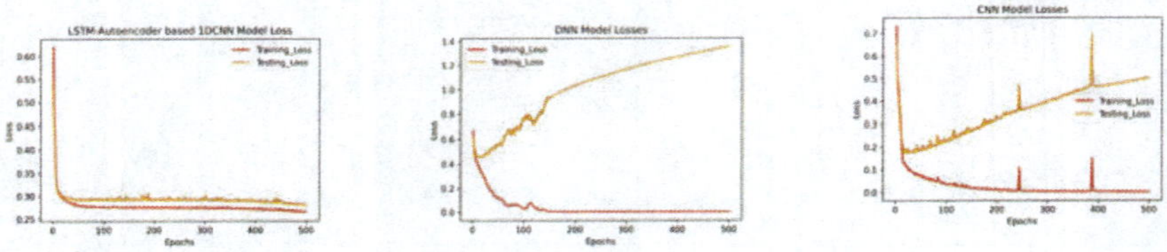

(a) Model Loss Curve of Proposed Model

(b) Model Loss Curve of DNN

(c) Model Loss Curve of CNN

Fig. 4 Model loss curves of proposed, DNN, and CNN models

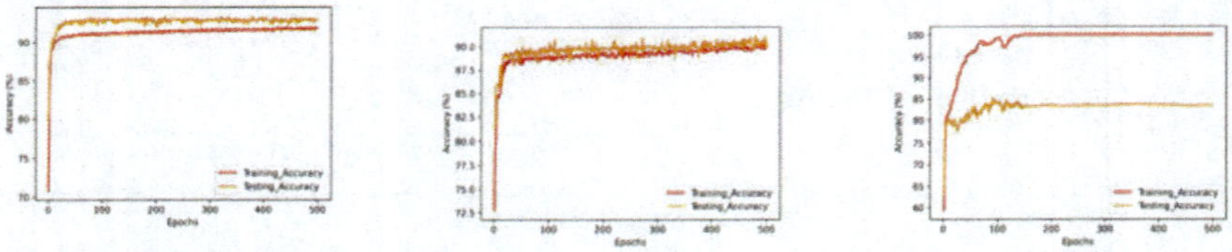

(a) Training and Testing Accuracy of Proposed Model

(b) Training and Testing Accuracy of CNN

(c) Training and Testing Accuracy of DNN

Fig. 5 Training and testing accuracies of proposed, DNN, and CNN models

are also considered for comparison purpose. Their training and testing loss graphs are also displayed in Fig. 4b and c. It can be observed from Fig. 4a that the error rate of the proposed model declines at values lower than DNN and CNN models. Additionally, even after attaining optimum parameter tuning, the error rate of DNN and CNN do not diminish. Therefore, it can be inferred that the proposed model attains better training performances over these two models. Further, the accuracies of these models while training and testing process are shown in Fig. 5.

The Adams optimizer and sigmoid activation function are set for the standard DNN and CNN models in order to tune the parameters, while the training and testing ratios are maintained in the same fashion as related to the proposed approach. Moreover, tanh activation function is utilized for the proposed LSTM autoencoder model.

Later, to measure the accomplishment of the proposed model in further detail, the testing accuracy graphs is obtained as depicted in Fig. 6. It can be observed that the proposed model attains the maximum testing accuracy against DNN and CNN. Moreover, accuracy, precision, recall, f1-score are estimated and measured to further assess the classification efficiency of such models as provided in Table 1. Further, the receiver operating characteristic (ROC) and area under curve (AUC) curve of proposed and other models are displayed in Figs. 7 and 8.

These performance measures are provided in brief as follows:

$$\text{accuracy} = \frac{\text{tpr} + \text{tnr}}{\text{tpr} + \text{tnr} + \text{fpr} + \text{fnr}} \quad (1)$$

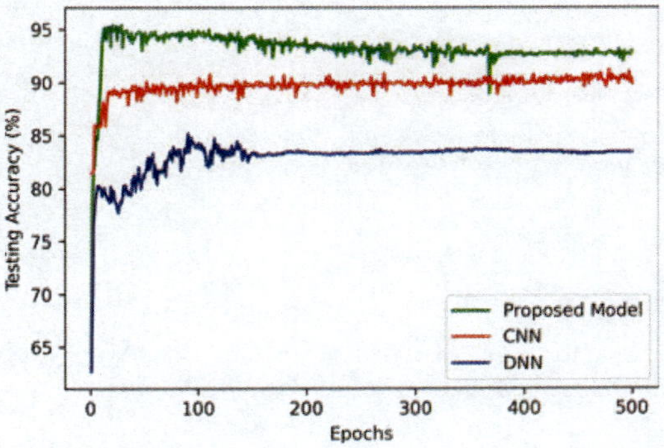

Fig. 6 Testing accuracies of proposed, DNN, and CNN models

Table 1 Binary classification performance of proposed, DNN, CNN, KNN, LR, and RF models

Approaches	Accuracy (%)	Precision	Recall	F1-score
DNN	81.56	0.73	0.72	0.72
CNN	90.47	0.88	0.82	0.85
KNN	88.43	0.90	0.81	0.77
LR	81.52	0.88	0.55	0.54
RF	89.65	0.92	0.75	0.80
Proposed model	**92.47**	**0.93**	**0.87**	**0.88**

$$\text{precision} = \frac{\text{tpr}}{\text{tpr} + \text{fpr}} \tag{2}$$

$$\text{recall} = \frac{\text{tpr}}{\text{tpr} + \text{fnr}} \tag{3}$$

$$\text{f1-score} = 2 \times \frac{\text{precision} \times \text{recall}}{\text{precision} + \text{recall}} \tag{4}$$

Here, tpr and fnr indicate the count of correct and incorrect classification respectively, provided a seizure identification task; tnr represents the count of not being classified to a class, provided a seizure identification tasks not included in this class; fpr denotes the count of a provided seizure identification task being wrongly categorized as this kind.

The distinction of the current work is validated and discussed using standard deep neural models, such as, DNN, CNN followed by few machine learning models as well. The precedence of the proposed model is demonstrated and evident in Table 1. It is apparent that the proposed model outperforms other models in terms of accuracy, precision, recall, and f1-score. In particular, compared to standard DNN and CNN models, the proposed model acquires an enhancement of 10.91 and 2% relating to accuracy, 20 and 5% concerning precision, 15 and 5% as for recall, and improve-

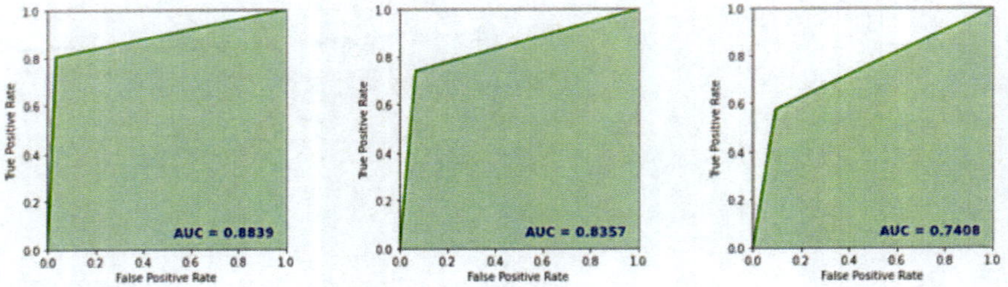

(a) Roc-AUC Curve of Proposed Model

(b) Roc-AUC Curve of CNN

(c) Roc-AUC Curve of DNN

Fig. 7 ROC-AUC curves of proposed, DNN, and CNN models

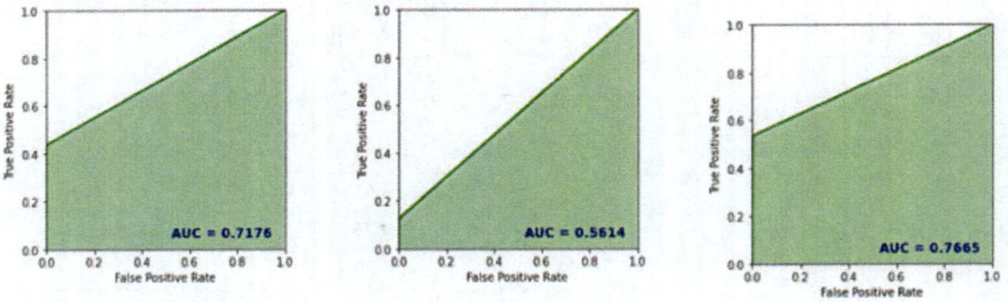

(a) Roc-AUC Curve of KNN

(b) Roc-AUC Curve of LGR

(c) Roc-AUC Curve of RF

Fig. 8 ROC-AUC curves of KNN, LGR, and RF models

ments in f1-score of 0.16 and 0.03. Moreover, as compared to KNN, LR, and RF, the proposed model excels in all the above metrics. Further, the ROC-AUC curve interprets as an actual standard of accuracy. It can be noted that the proposed model attains a significant AUC value of 0.8839 which is the finest among other models.

Additionally, Fig. 6 also corroborates the supremacy of the proposed model relative to standard DNN and CNN models.

5 Conclusions and Future Work

In this current study, an autoencoder-based deep learning framework incorporating LSTM neural structure is introduced for feature extraction from EEG signal data followed by a 1D CNN classification model for final classification. The results signify that the proposed architecture performs well in binary classification tasks on epileptic seizure recognition dataset with a prominent average accuracy of 92.47%. Additionally, the proposed model is validated against other standard deep neural architectures, specifically, CNN and DNN. Further, three machine learning models are studied to analyze the performance of the proposed model on this task. Regardless the substantial progress of the proposed model on the binary classification task of the epileptic seizure dataset, achieving a decent performance on multi-class classification task still remains worrisome. In view of this, the subsequent work emphasizes on altering the proposed model further by incorporating new neural structure to extract features and improving the 1D CNN structure in the interest of refining its achievement on the more challenging epileptic seizure classification tasks, which will enhance its recognition power on distinct datasets.

Acknowledgements We would like to be extremely thankful to the Department of Science and Technology (DST), Govt. of India to support this research work (File no. INT/Korea/P-53).

References

1. San-Segundo R, Gil-Martín M, D'Haro-Enríquez LF, Pardo JM (2019) Classification of epileptic EEG recordings using signal transforms and convolutional neural networks. Comput Biol Med 109:148–158
2. Tsubouchi Y, Tanabe A, Saito Y, Noma H, Maegaki Y (2019) Long-term prognosis of epilepsy in patients with cerebral palsy. Dev Med Child Neurol 61:1067–1073
3. Amin HU, Yusoff MZ, Ahmad RF (2019) A novel approach based on wavelet analysis and arithmetic coding for automated detection and diagnosis of epileptic seizure in EEG signals using machine learning techniques. Biomed Signal Proc Cont 56:1–10
4. Rabby MdK, Islam AKMK, Belkasim S, Bikdash MU (2021) Wavelet transform-based feature extraction approach for epileptic seizure classification. In: ACM SE '21: proceedings of the 2021 ACM southeast conference, pp 64–169
5. Sayem MA, Sarker MdSR, Ahad MAR, Ahmed MU (2021) Automatic epileptic seizures detection and EEG signals classification based on multi-domain feature extraction and multiscale entropy analysis. In: Ahad MAR, Ahmed MU (eds) Signal processing techniques for computational health informatics. Intelligent systems reference library, Springer, Cham, pp 315–333
6. Hussain SF, Qaisar SM (2021) Epileptic seizure classification using level-crossing EEG sampling and ensemble of sub-problems classifier. Elsevier 191:1–16
7. Shekokar K, Dour S, Ahmad G (2021) Epileptic seizure classification using LSTM. 2021 8th international conference on signal processing and integrated networks (SPIN). IEEE, Noida, India, pp 591–594
8. Gaowei X, Tianhe R, Yu C, Wenliang Ch (2020) A one-dimensional CNN-LSTM model for epileptic seizure recognition using EEG signal analysis. Front Neurosc 14:1–9
9. Abdelhameed AM, Daoud HG, Bayoumi M (2018) Epileptic seizure detection using deep convolutional autoencoder. In: 2018 IEEE international workshop on signal processing systems (SiPS), pp 223–228
10. Kaggle: your machine learning and data science community. https://www.kaggle.com/datasets

Stock Market Prediction Using Deep Learning Techniques for Short and Long Horizon

Aryan Bhambu

Abstract Long horizon forecasting in time series is a challenging task due to market volatility and stochastic nature. Traditional machine learning prediction models stated in the literature have several shortcomings in predicting long-term horizon time series forecasting. The deep learning algorithms are preferable to other existing algorithms as they can learn a time series's non-linear and non-stationary nature, reducing forecasting error. This research proposes a novel framework for recurrent neural network (RNN), long short-term memory (LSTM), gated recurrent unit (GRU), and bi-directional long short-term memory (Bi-LSTM) models. We presented a comparative study over the proposed models to predict short and long horizon time series forecasting. The computational results calculated over five time-series datasets demonstrate that the Bi-LSTM method with proper hyper-parameter tuning performs better than other deep neural networks.

Keywords RNN · LSTM · GRU · Bi-LSTM · Deep learning · Stock market prediction

1 Introduction

Stock market forecasting is a difficult task due to the market's variability and uncertainty orientation [1]. The stock market gets driven by many potential factors such as economic sustainability, investor sentiments, political upheaval, and natural calamities. These factors make the stock market very volatile and complex to anticipate accurately. In recent years, financial time series forecasting has gained significant attraction. Stock price prediction is vital for the growth of investors in a company's stock since it enhances the speculator's interest in investing in the company's stock. [1, 2]. A successful forecast of a stock's future price might result in a substantial profit. Recent research in deep learning (DL) algorithms evidenced that DL algorithms are able to learn the latent and non-linear patterns of the time series [3, 4]. Time series

A. Bhambu (✉)
Department of Mathematics, Indian Institute of Technology Guwahati, Assam, India
e-mail: a.bhambu@iitg.ac.in

M. Thakur et al. (eds.), *Soft Computing for Problem Solving*, Lecture Notes in Networks and Systems 547, https://doi.org/10.1007/978-981-19-6525-8_11

forecasting has become a demanding research topic as it recently attained enormous success in the domains such as stock price forecasting, cryptocurrency forecasting, and virus spread. Several statistical methods have already been used and published in the literature for time series prediction. Functional autoregressive (AR), autoregressive integrated moving average (ARIMA), and multiple linear regression models are examples of traditional statistical methodologies [2, 5, 6]. Roh [7] presented hybrid models for predicting the direction of movement of KOSPI 200 that used ANN and time series models. Exponentially weighted MA (EWMA), GARCH, and EGARCH were combined with NN for the stock price forecasting. The computational results reported that NN-EGARCH performs better when compared with NN and other hybrid models. Standard approaches are not preferable for such problems because there are not able to predict the non-periodic and non-stationary character of time series data [8]. Shen et al. [9] presented a methodology that utilized correlations across global markets and other items to anticipate the one-day-ahead trend prediction of stock prices called support vector machine (SVM). Later on, fuzzy Logic [10], k-nearest neighbors [11], and neural networks [12] also give better results for time series prediction as compared to statistical tools.

Neural networks are complex models that learn the hidden patterns in data and extract characteristics by hierarchically linking many artificial neurons (each executing a small computational task). Neural networks with one or more hidden layers in their architecture are known as deep neural networks. DNNs are better for independent input and of the same size. They are not an appropriate choice for the time series dataset because while dealing with sequential data, it is impossible to process the data at each time step and store the intermediate state [13]. Recurrent neural networks (RNNs) are introduced to deal with these tasks as they allow data to endure, which solves the problem of prior inputs being "forgotten" [13, 14]. Long short-term memory (LSTM) is a class of RNN that overcomes the vanishing and exploding gradient problem of RNNs. LSTMs use the gating mechanism comprised of three gates: input gate, forget gate, and output gate. LSTMs eliminate any error entering the self-recurring unit during backpropagation, and the vanishing gradient problem is no longer an issue [15]. A gated recurrent unit is a particular type of LSTM introduced to simplify the cell structure. GRU has two gates: the reset gate and the update gate [16, 17].

However, the new problem is to provide long horizon prediction, which entails predicting time series signals for several steps ahead. There are few applications for long-term forecasting of time series data due to increased uncertainty caused by many factors such as insufficient information and accumulation of errors [18, 19]. Recent research on non-linear time series has also found that multi-step ahead forecasting is more beneficial than predicting an intra-day or one-step ahead prediction. This research proposes a novel architecture for multi-day ahead forecasting associated with the current DL algorithms such as RNN, GRU, LSTM, and bi-LSTM using stock index datasets. Finally, the empirical experiments evidenced that bi-LSTM outperformed other DL models examined with the help of the performance metrics.

The contents of the article are summarized as follows. In Sect. 2, the literature survey related to the proposed methodologies has been done. In Sect. 3, we have briefly described the methodologies. The data description, pre-processing, and assessment metrics are given in Sect. 4. In Sect. 5, the simulation results of all methods over the short horizon and long horizon forecasting are compared and discussed. Section 6 is the conclusion.

2 Related Work

Financial time series forecasting attracts every investor because successful forecasting can substantially benefit stockholders. Artificial neural networks are used as they can learn the non-linear and hidden presentations in the datasets. Sezer et al. [20] used the multi-layer perceptron (MLP) over Dow-Jones Index and reported that it performs better and can be improved by fine-tuning of technical indicators. Roy et al. [21] studied the importance of LSTM over ARIMA and concluded that LSTM is a better choice than ARIMA. McCrae et al. [22] performed a comparative study between SVM and LSTM over DJI and commodity price dataset. The simulation outcomes indicated that LSTM outperformed SVM. Liu et al. [23] utilize the deep LSTM model for predicting the volatility of AAPL and S&P 500 datasets. LSTM better predicts the volatility of big data than the v-SVR model.

Li et al. [24] performed a comparative study between SVM, Naive Bayes, decision tree, MLP, RNN, and LSTM. This study used nine feature combinations and 23 technical indicators as input features to forecast one-day-ahead forecasting. The evaluated metrics evidenced that DL algorithms, MLP, RNN, and LSTM performed better than the other three models. Gao [25] studied the importance of LSTM on stock market forecasting. The proposed model is evaluated on hourly stock data for six different datasets. The empirical results evidenced that LSTM performs far better. Karmiani et al. [26] have done a comparative study between SVM, backpropagation neural network (BPNN), and LSTM over nine different stock markets. They have used six various technical indicators for feature extraction. The t-result analysis reported that LSTM performed far better than BPNN and SVM. Wang et al. [27] proposed a model based on recurrent LSTM one-day-ahead of photovoltaic power prediction. Zhou et al. [28] proposed a deep LSTM framework that predicts the multi-step air quality index.

Shah et al. [13] studied the suitability and proficiency of LSTM against DNN and found better for forecasting. Kumar et al. [29] presented a methodology to investigate the effectiveness of LSTM, GRU, and their hybrid variants. The models are trained on spark clusters for faster tuning of parameters for a better choice of the optimum model. The hybrid model performs better, having the least RMSE. Saiful et al. [30] provide a novel model for forecasting future FOREX currency closing values that combine GRU and LSTM. Namini et al. [31] studied the predictive capability of ARIMA and LSTM models over time series finance datasets. The empirical results evidenced that the LSTM model outperformed the ARIMA model and noticed that there is no effect if the number of epochs has been changed during the training of the models. Akita et al. [32] gathered data from scholarly publications to demonstrate the influence of the previous incidents on stock market opening prices. They devised a formula that handled numerical and textual inputs for the LSTM system to provide accurate forecasting. Yamak et al. [33] implicated the ARIMA, LSTM, and GRU models for predicting and found GRU performed better. Sunny et al. [2] reported that the proposed model using bi-LSTM generates lower RMSE than the LSTM model. Patel et al. [34] proposed a new hybrid model of GRU and LSTM for cryptocurrency price forecasting. The simulation results showed that the proposed model performs better than the LSTM network. Khaled et al. [35] used the importance of bi-LSTM, and empirical results revealed that Bi-LSTM networks have better outcomes for both short- and long-term predictions.

According to our literature survey, none of the research thus far has compared the RNN, LSTM, GRU, and bi-LSTM in stock price prediction over a short and long horizon. We have proposed a methodology for multi-day ahead forecasting by utilizing the performances of RNN, LSTM, GRU, and bi-LSTM models over finance datasets.

3 Methodology

RNN RNNs are a class of neural networks in which the links between the computational units form a directed graph [14]. Unlike feedforward networks, RNNs may handle any sequence of inputs using their internal memory. RNNs are derived from feedforward neural networks and can take different input sequences using internal memory. Each computing unit in an RNN has variable weights and a real-valued activation that change over time. The recurrent layers, also known as hidden layers in RNNs, are composed of recurrent cells whose states account for the information provided by the previous cell states and current input via feedback connections.

RNNs are typically networks consisting of regular recurrent cells like sigmoid (sig) and tan hyperbolic ($tanh$) cells. The standard recurrent sigmoid cell's mathematical formulae are as follows:

$$s_t = sig(W_s s_{t-1} + W_i X_t + B)$$
$$o_t = s_t$$

where x_t, s_t, and o_t signify the cell's input, recurrent information, and output at time step t, respectively; W_i and W_s are the weights, and B is the bias. However, recurrent neural networks cannot learn long-term relationships because assigning the priority to related input becomes more difficult as the gap increases.

LSTM Hochreiter and Schmidhuber presented LSTM in 1997, which soon gained popularity, considerably for solving time series prediction problems [15]. LSTM is a modified RNN algorithm that works well on many issues and is now frequently utilized. LSTMs are equipped with a gating mechanism that controls access to memory cells. They are well-suited to working with sequential data, such as time series data, since they can learn long-term dependencies.

The vanishing gradient problem addresses with LSTMs [33]. When the time step is considerable, the gradient becomes too tiny or enormous, resulting in a vanishing gradient problem. This problem happens when the optimizer propagates, causing the algorithm to execute even when the weights hardly change. Each LSTM cell consists of three gates: input, forget, and output. The gates allow the network to selectively write the information received from the output of the last cell, selectively read the information received from the intermediate step and forget the information that is not relevant. The cell state and hidden state are used for data collection and passing it to the next state. As a consequence, the problem of vanishing gradients gets resolved. Figure 1 shows the architecture of the LSTM block. The mathematical expression of the gates is as follows:

- Input gate: $i_t = sig\left(W_i s_{t-1} + U_i x_t\right)$
- Forget gate: $f_t = sig\left(W_f s_{t-1} + U_f x_t\right)$
- Output gate: $g_t = sig\left(W_g s_{t-1} + U_g x_t\right)$
- Intermediate cell state: $\tilde{H} = \tanh\left(W_h s_{t-1} + U_h x_t\right)$
- Cell state : $h_t = \left(i_t * \tilde{H}\right) + \left(f_t * h_{t-1}\right)$ (input to next memory)
- New state: $s_t = g_t * \tanh\left(h_t\right)$

Where h_t signifies the LSTM cell state, the weights are W_i, U_i, W_h, U_h, W_g, and U_g, the operator "." means dot product of two vectors, and the X_i, s_t input vector and output vector, respectively. The input gate decides the amount of the new information allowed in the cell state, and the output gate determines the output based on the past information received by the cell state while updating the cell state. The amount of data that will be erased from the cell state is determined by the forget gate. The information is retained when the forget gate, f_t, has a value of 1, and it is deleted when it has a value of 0.

GRU The LSTM cell has a higher learning capacity than the typical recurrent cell. The gated recurrent unit is proposed to eliminate the additional parameters that are increasing the computational cost [17]. The gated recurrent unit has similar structure to the LSTM unit. The mathematical expression for the cell is as follows:

$$\begin{aligned}
&\text{Update gate :} \quad n_t = sig\left(W_n s_{t-1} + U_n x_t\right)\\
&\text{Reset gate :} \quad r_t = sig\left(W_r s_{t-1} + U_r x_t\right)\\
&\text{Cell state :} \quad h_t = \tanh\left(W_h\left(s_{t-1} * r_t\right) + U_h x_t\right)\\
&\text{New state:} \quad s_t = \left(n_t * h_t\right) + \left(\left(1 - n_t\right) * s_{t-1}\right)
\end{aligned}$$

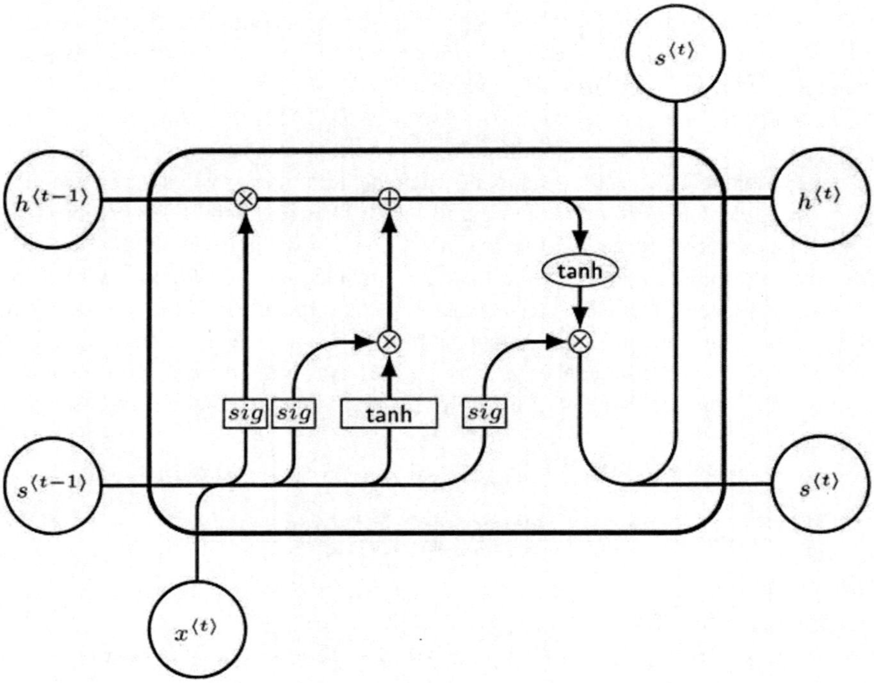

Fig. 1 Architecture of standard LSTM cell

GRU unit has two gates: update and reset gate. The update gate is the convex combination of input and forget gate of the LSTM cell. The reset gate processes and identifies how new input will integrate with the previously stored information. The GRU is just a forget-gated version of a vanilla LSTM. The single GRU cell is less potent than the LSTM since one gate is missing.

Bi-LSTM The traditional RNNs can only use information from one direction, i.e., from the initial to the final time step. Schuster and Paliwal [36] introduced the bi-directional RNN (BRNN) to overcome the limitation. The model simultaneously accounts for both time directions, i.e., forward and backward. Later, Graves and Schmidhuber [37] combine the BRNN and LSTM architecture to introduce bi-directional LSTMs. Bi-LSTM provides an advantage as it extracts the features from both directions, i.e., from the initial time step to the final time step with the help of the forward pass and from the last time step to the initial time step with the use of the backward pass. Figure 2 shows the connections and architecture of the bi-directional LSTM recurrent cell.

The forward layer connections in Fig. 2 are identical to those in the LSTM network, which computes sequences from time step $t-1$ to $t+1$. The hidden sequence and outputs for the backward layers are iterated from time step $t+1$ to $t-1$. The architecture's final output may be stated as

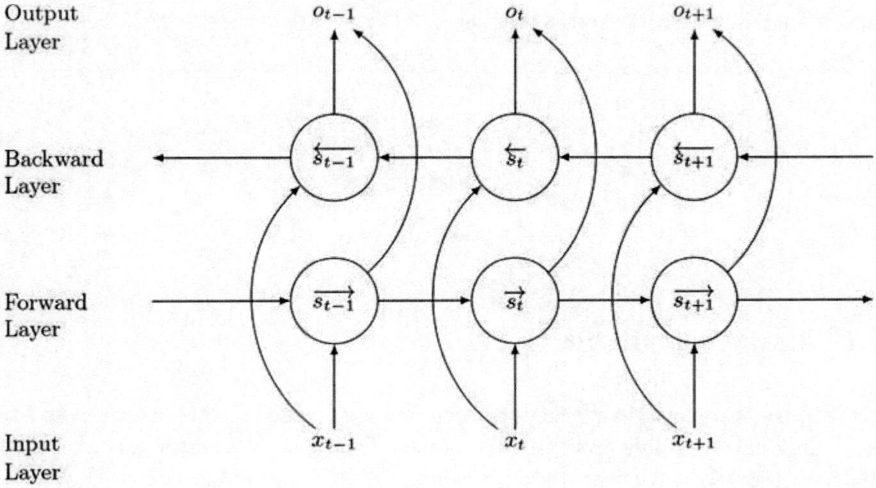

Fig. 2 Bi-directional LSTM layer

$$y_t = W_{\overrightarrow{s}} \overrightarrow{s_t} + W_{\overleftarrow{s}} \overleftarrow{s_t}$$

where the outputs of the forward and backward layers are $\overrightarrow{s_t}$ and $\overleftarrow{s_t}$, respectively.

4 Experiment

4.1 Data Description

The raw data is taken from Yahoo finance. This experiment is carried out on the various stock index datasets mentioned in Table 1 based from March 1, 2012, to March 1, 2022. The input data is in numeric format. The data is taken daily, including the stock's opening, high, low, and closing values for multi-day ahead prediction. The pre-processing of data has been done with the help of a min-max scalar. After that, the processed dataset is separated into two parts: training and testing. An initial 80% of the data is used for training and the rest 20% for testing of the models. A similar experimental setup is employed for all proposed hybrid models to generate a homogeneous setting for comparison.

The training data is passed through the networks with different tuning parameters to produce multi-dimensional output for closing price prediction. The outputs are obtained concerning each dataset and each proposed model. Then, the predictions are compared to testing datasets, and the evaluation metrics are calculated.

Table 1 Datasets used for financial forecasting

Index	Dataset
1	NIFTY 50
2	DJI
3	S&P 500
4	KOSPI
5	HSI

4.2 Assessment Metrics

There are many metrics to measure the accuracy. R-squared value, root mean squared error, and mean absolute error are among them. These are the performance metrics that are widely used for regression problems.

R-squared value: R-squared R^2 value is a measure that represents the proportion of variance for dependent variable calculated with the help of independent variables. It is defined as

$$\mathrm{R}^2 = 1 - \frac{\mathrm{RS}}{\mathrm{TS}}$$

where, RS is the sum of squares of residuals, and TS is the total sum of squares and residuals measures how far the regression line is from the data point.

Root Mean Squared Error: It is defined as the squared root of the standard deviation of the residuals. It is commonly denoted by "RMSE". Also,

$$\mathrm{RMSE} = \sqrt{\frac{RS}{N}}$$

where N is the number of total observations.

Mean Absolute Error: Mean absolute error (MAE) determines the regression problem's prediction accuracy or in other words, the average magnitude of error given a series of predictions. It is defined as

$$\mathrm{MAE} = \frac{1}{N}\sum_{i=1}^{N}(\hat{o}_i - o_i)$$

where o_i is the actual observation and $\hat{o}_i$ is the predicted observation of time series.

4.3 Experimental Setup

The input dataset containing the past information is passed through each DL algorithm, i.e., RNN, LSTM, GRU, and bi-LSTM. The model is trained with the backpropagation algorithm's help with the time step. While constructing a model, the choice of the loss function and optimization algorithm plays a major role. The loss function and optimization algorithm are "MSE" and "ADAM", respectively. For better generalization of the model, dropout is used as a regularization technique. The output is multi-dimensional, 5 to obtain 1, 2, 3, 4, and 5 days ahead prediction. The experiment is carried out using the Keras library and TensorFlow in Python.

The experimental study is conducted over five different time series datasets (mentioned in Table 1). The results (metrics) are evaluated with the help of different hyperparameters based on the nature of the dataset and proposed methods. All the metrics are calculated on the test data and are presented in the result section.

5 Results and Discussion

The assessment metrics are evaluated by employing different layers, different units in the hidden layers and dense layers. The results for one-day-ahead forecasting has discussed in the short horizon forecasting subsection. The results for prediction for 2, 3, 4, and 5 steps ahead are discussed in the long horizon forecasting subsection of section 5. The optimal value of prediction depends on many important parameters, and we will discuss the tuning process of the parameter in this section. The forecast could be improved after learning the time series pattern done with the help of the appropriate value of window size.

The dataset is divided into smaller parts, i.e., batch size, and then, the model is trained. DL algorithms employ gradient descent to enhance their models, passing the whole dataset through the model numerous (epochs) times with the purpose of updating the parameters and producing a stronger and more accurate prediction model makes sense. Each dataset has a different type of behavior; therefore, different epochs could be required to train the model precisely. It is observed that the bi-LSTM model converges very slowly due to its high model complexity in comparison with other proposed models, and the model is trained with a large number of epochs.

The hidden layer representation is crucial as it extracts the valuable feature representation from the dataset. As a result, the number of hidden layers, hidden dense layers, and the number of neurons in the layers are critical factors to consider while training the model. For developing a good framework, precise tweaking of the parameters in the hidden levels plays a vital role. It is found that the efficiency of the test data reduces with the increase in the number of hidden layers for the datasets (mentioned in Table 1), and the number of hidden layers are tuned by choosing 2-5. We have tried many combinations of neurons in the hidden layers in this framework, such as 128, 64, and 32.

Table 2 Experimental results of RNN, LSTM, GRU, and bi-LSTM models for one-day-ahead forecasting

Stock	Model	R-squared R^2	RMSE	MAE
NIFTY50	RNN	0.892	0.052	0.055
	LSTM	0.961	0.028	0.033
	GRU	0.941	0.037	0.041
	Bi-LSTM	0.978	0.019	0.025
DJI	RNN	0.845	0.053	0.057
	LSTM	0.951	0.034	0.037
	GRU	0.904	0.041	0.045
	Bi-LSTM	0.958	0.024	0.029
S&P 500	RNN	0.824	0.048	0.052
	LSTM	0.927	0.033	0.038
	GRU	0.862	0.052	0.0556
	Bi-LSTM	0.972	0.020	0.024
KOSPI	RNN	0.882	0.061	0.068
	LSTM	0.940	0.043	0.050
	GRU	0.953	0.012	0.040
	Bi-LSTM	0.944	0.042	0.048
HSI	RNN	0.930	0.010	0.027
	LSTM	0.940	0.007	0.026
	GRU	0.939	0.010	0.026
	Bi-LSTM	0.941	0.009	0.026

We found the best result corresponding to each model and calculated datasets and metrics. It is observed that the bi-LSTM model with 2 hidden layers with combinations of 128 x 64 and 2 dense layers gives the best result over the datasets.
Short Horizon Forecasting: Out of five time series datasets, bi-directional LSTM is giving better R^2 value, RMSE, and MAE on four of the time series. Table 2 shows that the bi-LSTM model performed better over the NIFTY 50 dataset having the least RMSE of 0.019, MAE of 0.025, and highest R^2 value of 0.978 among other models. For the KOSPI dataset, GRU performed better with R^2 value 0.953, RMSE of 0.012, and MAE of 0.040. Similarly, for DJI, S&P 500 and HSI datasets bi-LSTM performed better than all other networks with the metrics mentioned in Table 2.
Long Horizon Forecasting: The parameter obtained from short horizon forecasting has been used for long horizon forecasting. Long horizon predictions for NIFTY 50, DJI, S&P 500, KOSPI, and HSI are presented in Tables 3, 4, 5 and 6. The referenced tables have evaluated metrics associated with two, three, four, and five days ahead. The evaluated metrics clearly evidenced that vanilla RNN is not preferred due to fewer term dependencies or gradient problem. The LSTM and Bi-LSTM performed well over the long horizon. The empirical results evidenced that the Bi-LSTM model overall fit best to these datasets over the long horizon.

Table 3 Experimental results of RNN, LSTM, GRU, and bi-LSTM models for two day ahead forecasting

Stock	Model	R-squared R^2	RMSE	MAE
NIFTY50	RNN	0.883	0.053	0.057
	LSTM	0.959	0.029	0.033
	GRU	0.927	0.039	0.045
	Bi-LSTM	0.979	0.020	0.023
DJI	RNN	0.756	0.054	0.073
	LSTM	0.955	0.035	0.035
	GRU	0.909	0.041	0.043
	Bi-LSTM	0.956	0.025	0.029
S&P 500	RNN	0.413	0.053	0.112
	LSTM	0.915	0.034	0.042
	GRU	0.826	0.054	0.061
	Bi-LSTM	0.961	0.021	0.028
KOSPI	RNN	0.923	0.013	0.054
	LSTM	0.934	0.062	0.052
	GRU	0.929	0.044	0.055
	Bi-LSTM	0.927	0.043	0.055
HSI	RNN	0.898	0.010	0.034
	LSTM	0.913	0.007	0.031
	GRU	0.912	0.010	0.032
	Bi-LSTM	0.912	0.009	0.032

Table 4 Experimental results of RNN, LSTM, GRU, and bi-LSTM models for three day ahead forecasting

Stock	Model	R-squaredR^2	RMSE	MAE
NIFTY 50	RNN	0.887	0.055	0.057
	LSTM	0.969	0.030	0.029
	GRU	0.901	0.040	0.054
	Bi-LSTM	0.973	0.021	0.027
DJI	RNN	0.659	0.055	0.086
	LSTM	0.950	0.036	0.037
	GRU	0.821	0.042	0.062
	Bi-LSTM	0.941	0.026	0.034
S&P 500	RNN	0.785	0.054	0.069
	LSTM	0.910	0.035	0.043
	GRU	0.794	0.055	0.067
	Bi-LSTM	0.932	0.022	0.038
KOSPI	RNN	0.851	0.014	0.079
	LSTM	0.929	0.063	0.054
	GRU	0.885	0.044	0.070
	Bi-LSTM	0.933	0.043	0.053
HSI	RNN	0.875	0.011	0.037
	LSTM	0.892	0.007	0.035
	GRU	0.887	0.009	0.035
	Bi-LSTM	0.890	0.008	0.035

Table 5 Experimental results of RNN, LSTM, GRU, and bi-LSTM models for four day ahead forecasting

Stock	Model	R-squared R^2	RMSE	MAE
NIFTY 50	RNN	0.891	0.056	0.056
	LSTM	0.965	0.032	0.030
	GRU	0.915	0.041	0.049
	Bi-LSTM	0.966	0.023	0.030
DJI	RNN	0.692	0.056	0.082
	LSTM	0.932	0.038	0.044
	GRU	0.777	0.043	0.071
	Bi-LSTM	0.932	0.027	0.037
S&P 500	RNN	0.549	0.055	0.097
	LSTM	0.883	0.036	0.050
	GRU	0.674	0.056	0.085
	Bi-LSTM	0.937	0.023	0.037
KOSPI	RNN	0.680	0.015	0.116
	LSTM	0.936	0.064	0.051
	GRU	0.825	0.046	0.086
	Bi-LSTM	0.939	0.045	0.050
HSI	RNN	0.860	0.011	0.040
	LSTM	0.867	0.007	0.039
	GRU	0.865	0.009	0.039
	Bi-LSTM	0.869	0.008	0.038

Table 6 Experimental results of RNN, LSTM, GRU, and bi-LSTM models for five day ahead forecasting

Stock	Model	R-squared R^2	RMSE	MAE
NIFTY 50	RNN	0.843	0.057	0.068
	LSTM	0.925	0.033	0.046
	GRU	0.894	0.043	0.055
	Bi-LSTM	0.939	0.024	0.041
DJI	RNN	0.531	0.057	0.100
	LSTM	0.938	0.039	0.042
	GRU	0.762	0.044	0.072
	Bi-LSTM	0.898	0.028	0.046
S&P 500	RNN	0.063	0.056	0.141
	LSTM	0.823	0.037	0.062
	GRU	0.654	0.057	0.087
	Bi-LSTM	0.902	0.024	0.046
KOSPI	RNN	0.056	0.016	0.201
	LSTM	0.817	0.065	0.087
	GRU	0.810	0.047	0.088
	Bi-LSTM	0.928	0.046	0.054
HSI	RNN	0.829	0.011	0.044
	LSTM	0.843	0.007	0.042
	GRU	0.841	0.009	0.042
	Bi-LSTM	0.846	0.008	0.041

6 Conclusion

In this work, it is evident that deep learning algorithms have a considerable effect on current technology, notably in the construction of different time series-based prediction models. This study uses RNN, LSTM, GRU, and bi-directional LSTM models for short and long horizon forecasting of various time series datasets, NIFTY 50, DJI, S & P 500, KOSPI, and HSI datasets are used. It is attempted to determine the influence of the gap between historical data on predicting short- and long-term horizons. The usage of deep neural networks with adequate parameter modification is essential since prediction accuracy is highly dependent on these parameters. The proposed bi-LSTM method outperforms the other models for short and long horizons and can learn the information in both directions. In future, we intend to evaluate the performance of the different emerging DL algorithms over the long horizon by analyzing data from a broader range of stock markets.

References

1. Abu-Mostafa YS, Atiya AF (1996) Introduction to financial forecasting. Appl Intell 6(3):205–213
2. Sunny MAI, Maswood MMS, Alharbi AG (2020) Deep learning-based stock price prediction using LSTM and bi-directional LSTM model. In: 2020 2nd Novel Intelligent and Leading Emerging Sciences Conference (NILES). IEEE, pp 87–92
3. Vargas R, Mosavi A, Ruiz R (2017) Deep learning: a review
4. Sezer OB, Gudelek MU, Ozbayoglu AM (2020) Financial time series forecasting with deep learning: a systematic literature review: 2005–2019. Appl Soft Comput 90:106181
5. Li P, Jing C, Liang T, Liu M, Chen Z, Guo L (2015) Autoregressive moving average modeling in the financial sector. In: 2015 2nd International Conference on Information Technology, Computer, and Electrical Engineering (ICITACEE). IEEE, pp 68–71
6. Zhang G, Zhang X, Feng H (2016) Forecasting financial time series using a methodology based on autoregressive integrated moving average and Taylor expansion. Expert Syst 33(5):501–516
7. Roh TH (2007) Forecasting the volatility of stock price index. Expert Syst Appl 33(4):916–922
8. Hussain AJ, Ghazali R, Al-Jumeily D, Merabti M (2006) Dynamic ridge polynomial neural network for financial time series prediction. In: 2006 innovations in information technology. IEEE, pp 1–5
9. Shen S, Jiang H, Zhang T (2012) Stock market forecasting using machine learning algorithms. Stanford University, Stanford, CA, Department of Electrical Engineering, pp 1–5
10. Khemchandani R, Chandra S (2009) Regularized least squares fuzzy support vector regression for financial time series forecasting. Expert Syst Appl 36(1):132–138
11. Alkhatib K, Najadat H, Hmeidi I, Shatnawi MKA (2013) Stock price prediction using k-nearest neighbor (kNN) algorithm. Int J Bus Humanities Technol 3(3):32–44
12. Adhikari R (2015) A neural network based linear ensemble framework for time series forecasting. Neurocomputing 157:231–242
13. Shah D, Campbell W, Zulkernine FH (2018) A comparative study of LSTM and DNN for stock market forecasting. In: 2018 IEEE international conference on big data (big data). IEEE, pp 4148–4155
14. Mikolov T, Kombrink S, Burget L, Černocký J, Khudanpur S (2011) Extensions of recurrent neural network language model. In: 2011 IEEE international conference on acoustics, speech and signal processing (ICASSP). IEEE, pp 5528–5531

15. Hochreiter S, Schmidhuber J (1997) Long short-term memory. Neural Comput 9(8):1735–1780
16. Cho K, Van Merriënboer B, Gulcehre C, Bahdanau D, Bougares F, Schwenk H, Bengio Y (2014) Learning phrase representations using RNN encoder-decoder for statistical machine translation. arXiv preprint arXiv:1406.1078
17. Cho K, Van Merriënboer B, Gulcehre C, Bahdanau D, Bougares F, Schwenk H, Bengio Y (2014) Learning phrase representations using RNN encoder-decoder for statistical machine translation. arXiv preprint arXiv:1406.1078
18. Bala R, Singh RP (2019) Financial and non-stationary time series forecasting using lstm recurrent neural network for short and long horizon. In: 2019 10th International Conference on Computing, Communication and Networking Technologies (ICCCNT). IEEE, pp 1–7
19. Weigend AS (2018) Time series prediction: forecasting the future and understanding the past. Routledge, London
20. Sezer OB, Ozbayoglu AM, Dogdu E (2017) An artificial neural network-based stock trading system using technical analysis and big data framework. In: Proceedings of the southeast conference, pp 223–226
21. Mpawenimana I, Pegatoquet A, Roy V, Rodriguez L, Belleudy C (2020) A comparative study of LSTM and ARIMA for energy load prediction with enhanced data preprocessing. In: 2020 IEEE Sensors Applications Symposium (SAS). IEEE, pp 1–6
22. Lakshminarayanan SK, McCrae JP (2019) A comparative study of SVM and LSTM deep learning algorithms for stock market prediction. In: AICS, pp 446–457
23. Liu Y (2019) Novel volatility forecasting using deep learning-long short term memory recurrent neural networks. Expert Syst Appl 132:99–109
24. Li W, Liao J (2017) A comparative study on trend forecasting approach for stock price time series. In: 2017 11th IEEE international conference on Anti-counterfeiting, Security, and Identification (ASID). IEEE, pp 74–78
25. Gao Q (2016) Stock market forecasting using recurrent neural network (Doctoral dissertation, University of Missouri–Columbia)
26. Karmiani D, Kazi R, Nambisan A, Shah A, Kamble V (2019) Comparison of predictive algorithms: backpropagation, SVM, LSTM and Kalman Filter for stock market. In: 2019 Amity International Conference on Artificial Intelligence (AICAI). IEEE, pp 228–234
27. Wang F, Xuan Z, Zhen Z, Li K, Wang T, Shi M (2020) A day-ahead PV power forecasting method based on LSTM-RNN model and time correlation modification under partial daily pattern prediction framework. Energy Convers Manage 212:112766
28. Zhou Y, Chang FJ, Chang LC, Kao IF, Wang YS (2019) Explore a deep learning multi-output neural network for regional multi-step-ahead air quality forecasts. J Cleaner Prod 209:134–145
29. Kumar S, Hussain L, Banarjee S, Reza M (2018) Energy load forecasting using deep learning approach-LSTM and GRU in spark cluster. In: 2018 fifth international conference on Emerging Applications of Information Technology (EAIT). IEEE, pp 1–4
30. Islam MS, Hossain E (2020) Foreign exchange currency rate prediction using a GRU-LSTM hybrid network. Soft Comput Lett 100009
31. Siami-Namini S, Namin AS (2018) Forecasting economics and financial time series: ARIMA vs. LSTM. arXiv preprint arXiv:1803.06386
32. Akita R, Yoshihara A, Matsubara T, Uehara K (2016) Deep learning for stock prediction using numerical and textual information. In: 2016 IEEE/ACIS 15th International Conference on Computer and Information Science (ICIS). IEEE, pp 1–6
33. Yamak PT, Yujian L, Gadosey PK (2019) A comparison between arima, lstm, and gru for time series forecasting. In: Proceedings of the 2019 2nd international conference on algorithms, computing and artificial intelligence, pp 49–55
34. Patel MM, Tanwar S, Gupta R, Kumar N (2020) A deep learning-based cryptocurrency price prediction scheme for financial institutions. J Inf Security Appl 55:102583
35. Althelaya KA, El-Alfy ESM, Mohammed S (2018) Evaluation of bidirectional LSTM for short- and long-term stock market prediction. In: 2018 9th International Conference on Information and Communication Systems (ICICS). IEEE, pp 151–156

36. Schuster M, Paliwal KK (1997) Bidirectional recurrent neural networks. IEEE Trans Signal Process 45(11):2673–2681
37. Graves A, Schmidhuber J (2005) Framewise phoneme classification with bidirectional LSTM and other neural network architectures. Neural Netw 18(5–6):602–610

Improved CNN Model for Breast Cancer Classification

P. Satya Shekar Varma and Sushil Kumar

Abstract In this paper, we propose an improved convolutional neural network for the automatic classification of breast cancer pathological images, with the goal of achieving more accurate results. It is also proposed to use two different convolutional structures to improve the accuracy of pathological image recognition by the network, which is covered in greater detail in the paper. After constructing the foundation network from a deep residual network, octave convolution is used to replace the traditional convolutional layer during the feature extraction stage. This reduces the number of redundant features in the feature map and improves the effect of the detailed feature extraction, both of which are beneficial. As a result of the introduction of heterogeneous convolution into the network, a portion of the traditional convolutional layers can be replaced, resulting in a reduction in the number of parameters necessary for model training. It is possible to overcome the problem of overfitting caused by a small number of data samples available by employing an effective data enhancement method based on the concept of the image block. In the experiments, the accuracy of the network at the image level on the four-classification tasks of the network is shown to be 91.25%. This means that the network model that was proposed has a better recognition rate and is faster in real time.

Keywords Image processing · Pathological images · CNN · Residual networks · Octave convolution · Heterogeneous convolution

P. Satya Shekar Varma (✉) · S. Kumar
Department of Computer Science and Engineering, National Institute of Technology Warangal, Telangana, India
e-mail: ss721087@student.nitw.ac.in

S. Kumar
e-mail: kumar.sushil@nitw.ac.in

M. Thakur et al. (eds.), *Soft Computing for Problem Solving*, Lecture Notes in Networks and Systems 547, https://doi.org/10.1007/978-981-19-6525-8_12

1 Introduction

Breast cancer (BC) is one of the most widespread types of cancer among women. Transience of BC is very high when compared to other types of cancer [29]. It begins in the breast cells. Breast cancer is classified into several subtypes based on how the cell appears under a microscope [30]. The two primary types of breast cancer are invasive ductal carcinoma (IDC) and ductal carcinoma in situ (DCIS). DCIS grows slowly and has little effect on patients' daily lives. DCIS accounts for between 20% and 53% of all breast cancer cases. On the other hand, the IDC type is more dangerous because it spreads throughout the breast tissue. Around 80% of breast cancer patients fall into this category [1, 6].

Mammography, MRI, and X-ray imaging can accurately stage and grade breast cancer pathology images [26]. Computer-aided diagnosis increases diagnostic efficiency and objective judgment over manual approaches [18]. Overlapping nuclei modifies the characteristics of individual nuclei, resulting in nuclear traits that are not visible to the naked eye. Inconsistent staining complicates classification [25, 34]. Pathological pictures may have different nuclei, which affects their classification. Due to lack of contrast, noise, or visual attractiveness, instruments have been created to improve image processing [19]. AI, ML, and CNN are the healthcare industry's fastest-growing segments. AI and machine learning focus on constructing technology systems that can solve difficult issues without human intelligence [8, 33].

Deep learning uses neural networks (ANNs). ANNs were crucial to this field's progress. DNN, RNN, and deep belief networks (DBNs) are employed in computer vision, speech recognition, NLP, drug design, bioinformatics, and more[12, 13]. Medical image analysis, materials inspection, histology, and board games employ them. DL algorithms can increase cancer detection accuracy [4, 28].

To make high-quality images with digital pathology (DP), the slides are digitally digitized [15]. These digital images are used to find, separate, and classify things using image analysis techniques. Image categorization in deep learning (DL) with CNNs requires more steps, like digital staining [13].

The application of CNN in medical imaging research is not limited to the use of deep CNN to extract imaging characteristics but also includes other applications. Indeed, CNNs have the potential to be used to generate synthetic images that can be used in medical research applications [17, 22].

2 Related Works

Currently, public datasets are all small datasets, which pose great challenges to deep learning-based classification methods.

Spanhol et al. [31] classified pathological pictures of breast cancer using the BreaKHis dataset with an 80% to 85% accuracy rate. Spanhol and colleagues [30] used a modified AlexNet to categorize diseased photos on the same dataset; they attained an accuracy rate of 89.6%. Araujo et al. [1] used a CNN paired with SVM to classify the 2015 breast cancer dataset with 77.8% accuracy. Golatkar et al. [6] used the Inception-v3 network's transfer learning to preprocess and extract patches from 400 evenly divided images of four categories. Rakhlin et al. [26] combined deep neural network structures with gradient boosting tree (LightGBM) classifiers to achieve 87.2% accuracy on a four-classification assignment [6].

Kone et al. [18] used the idea of a binary tree to first classify the pathological images and then did a more detailed binary classification for each class to get to four classifications. They used the proposed hierarchical deep residual network to classify the three other classes. It takes a lot of work to build different networks. The accuracy of the four-classification task is 99%. Nazeri et al. [25] came up with a patch-based method that used two consecutive CNNs to classify four types of pathological images with a 94% accuracy rate. When Wang et al. used the SVM algorithm to classify 68 breast cancer pathological images, they got an accuracy rate of 96%. Krizhevsky et al. [19] used Google network (GoogLeNet) for transfer learning, and the recognition rate on the BreakHis dataset was 91%. The accuracy of some of these methods is not the same because they use different datasets and evaluation criteria, so they are not the same.

Using CNNs naively might not work because "medical images are more unique than normal images," Gravina et al. [7] said in a paper they wrote. Lesion segmentation has been shown to be a good source of information, because it can help both find shape-related structures and pinpoint the exact location of a lesion.

Desai and Shah [5] stated that considerable effort is expended comparing how each network operates and is constructed. Then, they assessed each network's ability to diagnose and categorize breast cancer to determine which was superior. CNN is slightly more accurate than MLP at diagnosing and identifying breast cancer.

Another researcher [27] searched for images of mitosis in breast histology, such as this one, using deep max pooling CNNs. The networks were able to sort the images according to their pixel count. Murtaza et al. [24] used an automated method to identify and study IDC tissue zones. Hossain [14] demonstrated how to classify breast WSIs into simple, ductal carcinoma in situ (DCIS), and invasive ductal carcinoma (IDC) using context aware stacked CNNs.

When the traditional method is used to extract features, it is difficult because cell features are very similar between cell types in the same class. This means that the feature extraction is difficult, and the accuracy of the algorithm is low. Even though the traditional CNN has a high accuracy, it will also have a high cost [12]. The number of convolution kernels in the transfer learning method will affect how well the features in the image are learned. Most of the training networks are single, so the learned feature weights cannot be used well in the organization. There is a low rate of recognition when pathological image features are taken out. To solve the problems above, this paper comes up with a deeper and more effective CNN, as well to make more data, which avoids overfitting when there is not enough data. This improves the model's ability to classify things.

3 Proposed Method

As an efficient recognition method in the field of image processing [8, 19, 32, 33], CNN has been widely used in many mainstream NN frameworks. Compared with the traditional residual network, the residual network [35] model used in the experiment has less computation and higher training accuracy. At the same time, the octave convolution (octave convolution) [4] is used to replace the traditional convolution layer, which can effectively extract pathological images. In addition, the heterogeneous convolution (heterogeneous convolution) [28] is introduced to reduce the training parameters and improve the classification accuracy of the model.

3.1 Network Architecture

Replacing the traditional convolutional layer with octave convolution, since the convolutional layer with the convolution kernel of 1×1 has the effect of increasing or reducing the dimension, replacing with octave convolution will only increase the training parameters and training time. Therefore, only the convolutional kernel is replaced as 3×3 convolutional layers. The convolution kernel of the basic convolution layer in the initial layer octave convolution adopts heterogeneous convolution to reduce the training parameters and improve the classification accuracy. The overall process structure of this network is shown in Fig. 1, where Conv2D represents two-dimensional convolution.

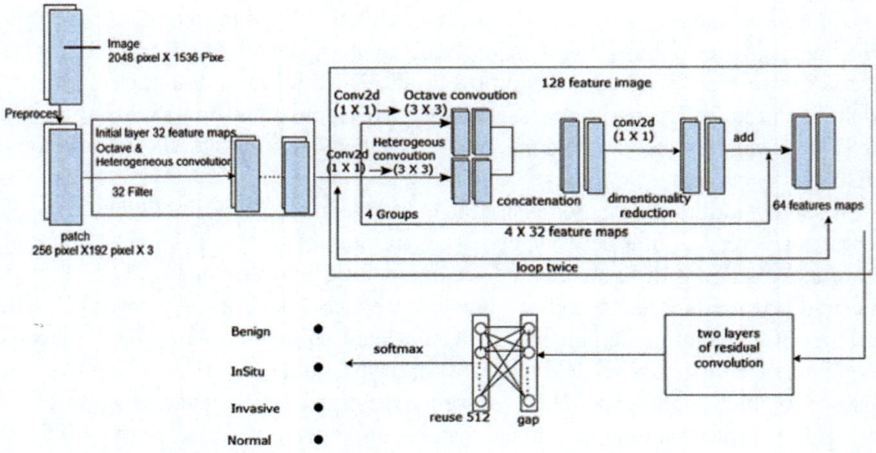

Fig. 1 Architecture of proposed algorithm

3.1.1 Initial Layer

The initial layer is a convolutional layer with 32 convolution kernels with a size of 3×3 and a stride of 1. The input size is a tensor with a size of 256pixel$\times$192pixel$\times$3, where 3 is the number of channels. Batch normalization (BN) [16] is then performed, and activation is performed with a linear rectification function (ReLU) [36] for preliminary feature extraction.

Octave Convolution Module: The high-frequency portion of the image provides a great deal of information, whereas the low-frequency portion of the image represents the overall information of the image and contains less definitive information.

Chen et al. [4] substituted octave convolution for the convolutional layer in typical CNNs, separated the feature map into high- and low-frequency channels, and then divided the low-frequency channel. The channel's feature map is halved; that is, the feature map is separated into high- and low-frequency segments, as illustrated in Fig. 2.

The input feature map X is partitioned into high-frequency X^{H} and low-frequency X^{L} in Fig. 2. To begin, the high-frequency segment is subjected to a convolution operation from pooling (AvgPooling), which reduces the size of the feature map by half. Next, the high-frequency to high-frequency segment is generated, yielding the feature map $Y^{\mathrm{L}\to\mathrm{L}}$.Then, average a feature map with the same number of low-frequency channels through convolution, low frequency is performed on the low-frequency part, and the feature map $Y^{\mathrm{H}\to\mathrm{L}}$ is and obtain the final feature map $Y^{\mathrm{H}\to\mathrm{L}}$. Convolution operation from low frequency to obtained. Then perform up-sampling operation on XL to obtain a feature map with thethrough convolution. Add $Y^{\mathrm{H}\to\mathrm{H}}$ and $Y^{\mathrm{L}\to\mathrm{H}}$, $Y^{\mathrm{H}\to\mathrm{L}}$ and $Y^{\mathrm{L}\to\mathrm{L}}$ to obtain high and lowsame size as the high-frequency channel, and then obtain the final feature map $Y^{\mathrm{L}\to\mathrm{H}}$ frequency feature maps Y^{H} , Y^{L} which can be expressed as

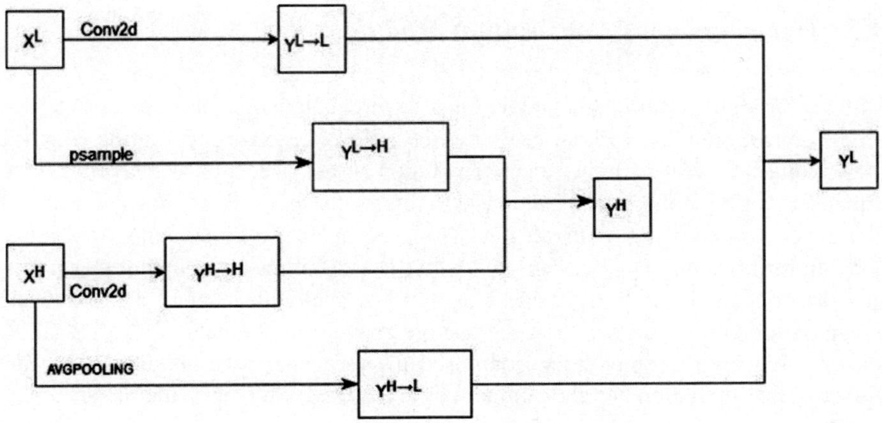

Fig. 2 Separation of feature map by transition layer

$$Y^{H} = Y^{H \to H} + Y^{L \to H} \quad (1)$$

$$Y^{L} = Y^{L \to L} + Y^{H \to L} \quad (2)$$

Similar types of pathological images have many same features, and it is difficult to extract their detailed features. Therefore, the octave convolution network module is introduced to improve the extraction efficiency of high-frequency information, reduce redundant low-frequency information, and improve the accuracy of the image. The designed octave convolution structure includes an initial layer, a transition layer, and an output layer. Among them, the initial layer is single input and double output, which is responsible for receiving the input feature map. The original image is passed through a convolutional layer with a convolution kernel size of 3 × 3 to output a high-frequency feature map (X^{H}), and the original image is averagely pooled and then passed through the same convolutional layer to output a low-frequency feature map (X^{L}).

The number of low-frequency channels is $F_{\text{filters}} \times \alpha$, and the number of high-frequency channels is $1 - F_{\text{filters}} \times \alpha$, where $F_{filters}$ is the number of input channels, and the experiment takes 64. In order to reduce the redundancy of low-frequency features, the value range of parameter α is 0 0.5, and α is an integer multiple of 0.125. After comparative experiments, it is finally set to 0.25. The transition layer is a double input and double output, with X^{H} and X^{L} as inputs, high-frequency features X^{H} and low-frequency features X^{L} go through a convolutional layer and perform down-sampling and up-sampling operations, respectively. Output Y^{H}, Y^{L}; the output layer is double input and single output, the input is Y^{H}, Y^{L}, the output Y^{L} of the transition layer is upsampled after a layer of convolution layer, and then the feature map obtained after a layer of convolution layer with Y^{H}. The addition operation is performed to obtain the output feature map of the module.

3.2 Heterogeneous Convolution Module

Different from the traditional convolution structure, heterogeneous convolution acts on the convolution layer with a new filter idea. Homogeneous convolution consists of convolution layers with the same convolution kernel size, such as a layer of two-dimensional (2D) convolution contains 256 filters of size 3 × 3, as shown in Fig. 3a. Heterogeneous convolution divides the $F_{filters}$ convolution kernels in a traditional convolution layer into $F_{filters}/P$ groups, where P is the number of convolution kernels in each group, and there is only one convolution kernel with a size of 3 × 3 in each group of convolution kernels, and the rest are convolution kernels of size 1 × 1, as shown in Fig. 3b. This convolution composed of heterogeneous kernels can reduce the amount of computation and the number of parameters, while ensuring the accuracy of training.

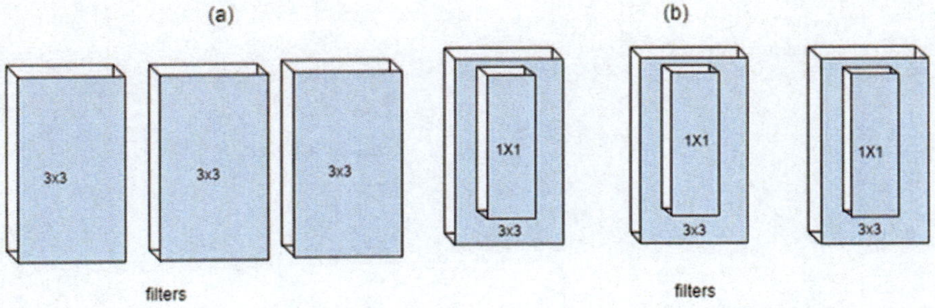

Fig. 3 Convolutions with different structures. **a** Traditional convolution with 3 × 3; **b** heterogeneous convolution with 3 × 3 and 1 × 1

The size of the convolution output feature map is $D_O \times D_O$, the number of output channels is N, the size of the convolution kernel is K × K, the number of convolution kernels with the size of K × K is $F_{filters}/P$, and the number of the remaining convolution kernels is $F_{filters}/P$. The size is 1×1, then the total calculation amount of each layer of convolution (F_{all}) is the calculation amount of each convolution kernel of size K×K (F_K) and 1×1 convolution kernel calculation amount (F_1) and can be expressed as

$$F_K = D_0 \times D_0 \times F_{\text{filers}} \times N \times K \times K \tag{3}$$

$$F_1 = (D_0 \times D_0 \times N) \times \left(1 - \frac{1}{P}\right) \times F_{\text{filters}} \tag{4}$$

$$F_{\text{all}} = F_K + F_1 \tag{5}$$

The convolution kernel of the initial layer is designed in the form of heterogeneous kernel, and the algorithm designed according to the heterogeneous convolution principle is used to replace the traditional convolution kernel to achieve the purpose of heterogeneous convolution kernel. The flow of the algorithm is shown in Fig. 4, where $P \geqslant 2$, the size of the first convolution kernel in each group is 3×3, and the size of the remaining P-1 convolution kernels is 1×1. In the experiment, P takes 2 and 4 to the power of 2. After many experiments, it is found that when $P = 2$, the effect of the network is the best.

Residual layer: Residual network [18] is a highly modular network structure. The experimentally designed residual network module stacks convolutional layers with convolution kernels of 1 × 1 and 3 × 3. The convolutional layer with the product kernel of 1×1 is used as the last layer of the residual network module, and then the output of the last layer and the input of the residual network module are used for the add operation, and finally a complete residual network module is formed, as

Fig. 4 Flowchart of the heterogeneous convolution algorithm

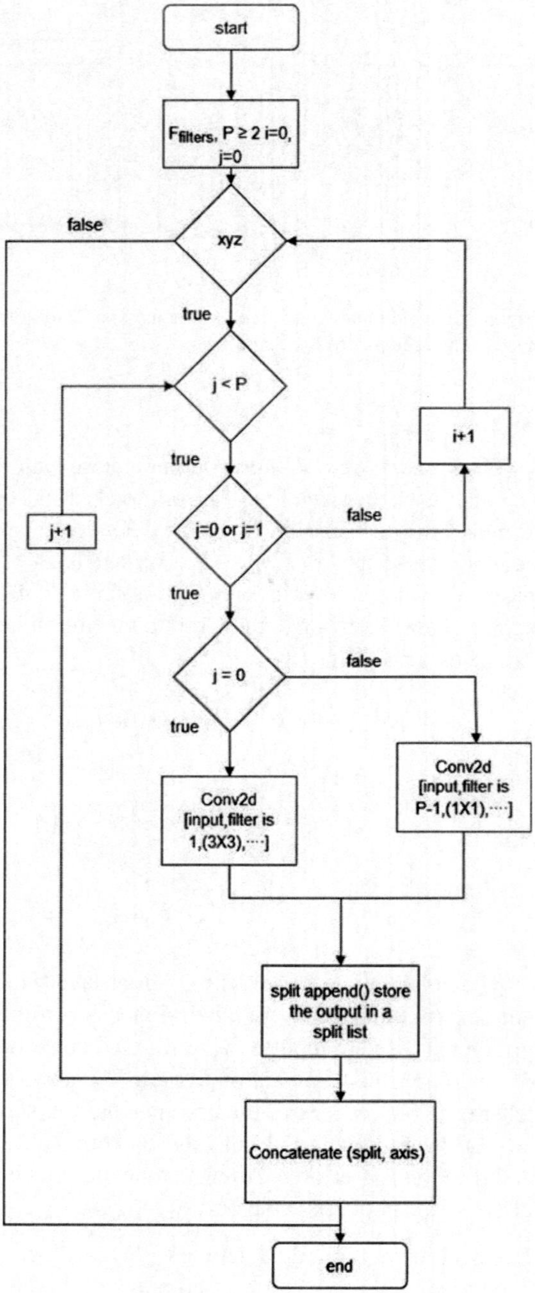

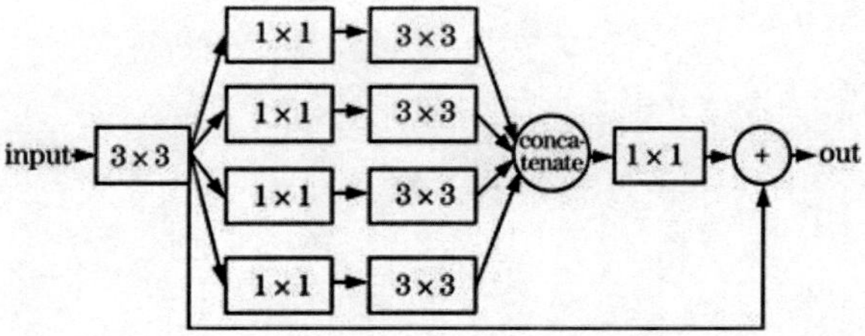

Fig. 5 Structure of the residual network module

shown in Fig. 5. The main part of this network model is composed of three residual network modules, and the number of output filters is 64, 128, and 256. Subsequent experiments are based on this model to improve performance.

Global Average Pooling Layer: The traditional fully connected layer is replaced by a global average pooling (GAP) layer [20], which is followed by a dense layer with 512 nodes. The dense layer can convert the output features of the previous layer into a 1-dimensional vector of N × 1 and synthesize the obtained image features with high-level meanings. In order to avoid overfitting, a dropout layer [31] is added after the dense layer, which is output by the softmax classifier. The rectified Adam (RAdam) optimizer is selected for optimization [22]. On the one hand, the experimental dataset is relatively small. Compared with the Adam optimizer, the RAdam optimizer can save the warm-up step [17]. On the other hand, the RAdam optimizer is more robust to the learning rate and has the same convergence speed as the Adam optimizer, which can avoid falling into a local optimal solution.

3.3 Data Preprocessing

Dataset: Experiments were performed on the BACH public dataset [2] of the Grand Challenge, which included four types of normal tissue (normal), benign lesions (benign), carcinoma in situ (in situ), and invasive carcinoma (invasive) after H&E (Hematoxylin) -Eosin) stained breast pathological microscope image, the size of the image is 2048pixel × 1536pixel, all are red, green,blue (RGB) images, each pixel covers 0.42 μm × 0.42 μm of tissue. The dataset contains annotations jointly given by two pathologists, and the disagreeing images have been discarded. In order to make the data uniform, 100 images were selected from each type of images for the experiment.

Preprocessing: Data preprocessing [9, 16, 21] is an essential step in image processing. Due to the limited number of selected dataset samples, in order to prevent model training from overfitting, small patches (patches) are extracted from each image to increase the number of samples in the dataset, while only retaining patches with nuclei, discarding patches without nuclei or with fewer nuclei.

Since the size and shape of the nucleus and its surrounding tissue structure are the main features of the classification [3, 10, 11, 23, 37]. Among them, normal tis-

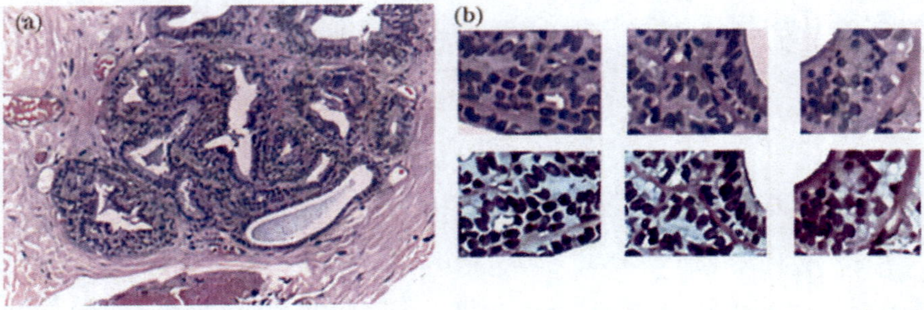

Fig. 6 Images of the benign class in the validation set **a** complete image **b** small patch

sues had larger cytoplasmic regions and dense nucleation clusters after H&E staining. Benign lesions consist of multiple adjacent nuclear clusters; carcinoma in situ presents with enlarged nuclei and prominent nucleoli, but all in a circular cluster; invasive carcinoma breaks the cluster form of carcinoma in situ, and the nuclei spread to nearby areas and the nuclear density was high, and the nuclei were arranged in a disorderly manner. Therefore, the experimental extraction area is 256pixel×192pixel, which can well contain the outline of the cluster, the nucleus, and its surrounding structures. In order to obtain more comprehensive image feature information, each patch needs to cover 50% of the area of the extracted image, that is, the step size S_{weight} in width is 128 pixels, the step size in S_{height} is 96 pixels, and the number of patches that can be extracted in width is W_T , the number of $patches$ that can be extracted in height is H_T, which can be expressed as

$$W_{\mathrm{T}} = \frac{2048 - 256}{S_{\mathrm{weight}}} + 1 \tag{6}$$

$$H_{\mathrm{T}} = \frac{1536 - 192}{S_{\mathrm{height}}} + 1 \tag{7}$$

$$T_{\mathrm{all}} = W_{\mathrm{T}} \times H_{\mathrm{T}} \tag{8}$$

where T_{all} is the patch extracted from each complete image. The experiment does not use all the patches, only retains the patches with high kernel density, discards the patches with sparse kernel density, retains and discards the standard reference [6], and extracts the labels of all patches consistent with the labels of the original image. The edge features of the nuclei are not prominent in the pathological images stained by H&E, therefore, contrast stretch processing was performed on all patches to make the nuclei and their surrounding features more obvious, as shown in Fig. 6. The study found that training on contrast-stretched data resulted in higher network accuracy than unprocessed data.

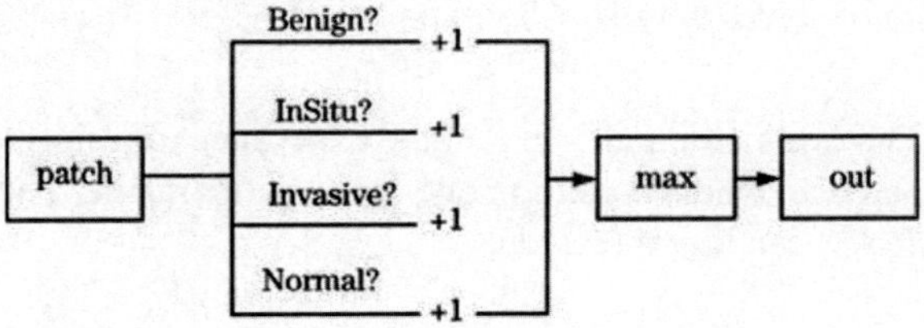

Fig. 7 Principle of the majority voting algorithm

Majority Voting Principle: Any patch of each image will get the output of one category after passing through the softmax classifier, and the number of corresponding categories will be increased by 1. When the number of patch samples belonging to a certain category in all patches extracted from the image is relatively large, the image is judged as this category, and the classification principle is shown in Fig. 7.

4 Results and Analysis

4.1 Experimental Environment

The experiment uses Python language to program based on Keras framework, and the experimental platform is DGX Ubuntu system NVIDIA v100. Data preprocessing is performed in the CPU environment, and the CNN model is trained on the GPU to speed up the parallel computing of the data and improve the experimental efficiency.

4.2 Training Strategy

Each class of the dataset is evenly divided into training set (60%), validation set (20%), and test set (20%), and the extracted patches are used as the final training set, validation set, and test set. Among them, the training set is used for model training and parameter learning; the validation set is used to verify the model, the generalization ability of the model is continuously verified through training, and the parameters are automatically fine-tuned to save the best model at any time; the test set is used to test the recognition rate and generalization of the model. All training data is scrambled and then processed during training.

Training is performed under a four-class task, namely normal, benign, carcinoma in situ, and invasive carcinoma. Training strategy: first train the model in the original residual network; then replace the 3×3 convolutional layers in residual network with octave convolution to retrain. Finally, the traditional convolution kernel of the initial layer is replaced with heterogeneous convolution and retrained. The data obtained from the training of the former model can provide an effective judgment basis for the optimization of the latter model and serve as a comparative experiment.

4.3 Evaluation Criteria

The evaluation indicators of the experimental results are the recognition rate of the patch and the recognition rate of the entire image. The recognition rate of the patch P_{patch} can be expressed as

$$P_{patch} = \frac{N_{\text{right}}}{N_{\text{sum}}} \tag{9}$$

In the formula, N_{right} is the number of correct patches identified in the test set, and N_{sum} is the total number of patches in the test set. The recognition rate I of the whole image can be expressed as

$$I = \frac{N_{\text{rp}}}{N_{\text{all}}} \tag{10}$$

where N_{rp} is the number of successfully classified images in the test set, and N_{all} is the total number of images in the test set.

4.4 Experimental Results and Analysis

The performance of this network model is evaluated by the accuracy of the training patch and the classification accuracy of the entire image (the number of images of each type in the test set is 20), and the accuracy of the former greatly affects the accuracy of the latter. Experiments are carried out in turn in the three improved networks, and the experimental results are gradually increased. The experimental results are analyzed by the accuracy of the training set(acc), the accuracy of the validation set (val_acc), and the confusion matrix of the entire image of the test set.

Residual Network: Residual network consists of a convolutional layer (input layer) containing 32 filters, three residual network modules, a GAP layer, a dense layer, a dropout layer, and a softmax classifier. The overall structure is shown in Fig. 1. When training to the 49th round, the generalization ability of the model performance is the best, the accuracy rate of the training set reaches 90.97%, and the accuracy rate of the validation set reaches 71.92% (based on patch level), as shown in Fig. 8. The difference in accuracy between the two is about 19% points, which indicates that the model has serious overfitting and needs to be improved.

Since the experiment adopts the principle of majority decision-making, the accuracy of the validation set cannot represent the discrimination accuracy of the entire image. The accuracy of the validation set is based on the patch level, while the image-level accuracy is determined by the maximum number of categories of all patches in the image. The final result is shown in Table 1, where the row represents the predicted value and the column represents the true value, and the final image-level classification accuracy is 82.5%. It can be seen that there are relatively more wrong images

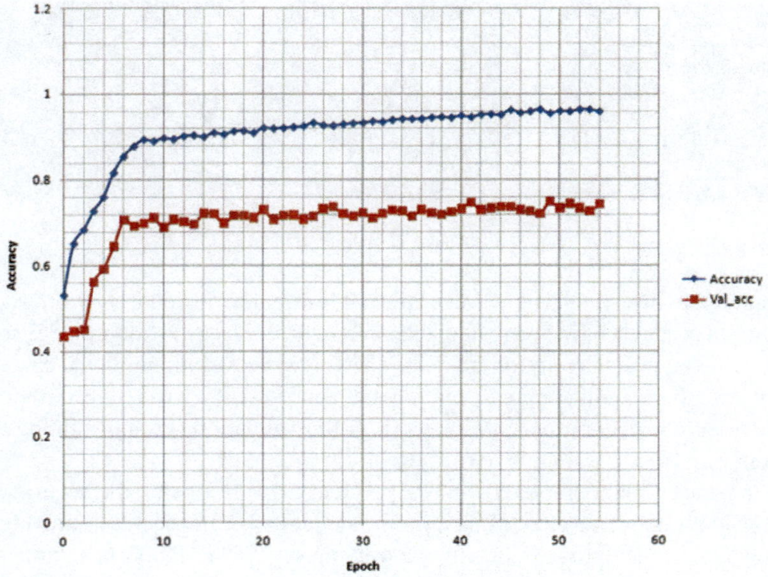

Fig. 8 Training accuracy and validation accuracy of the residual network model

Table 1 Confusion matrix of the residual network model

	Benign	In situ	Invasive	Normal
Benign	18	1	1	3
In situ	1	18	4	0
Invasive	0	1	14	1
Normal	1	0	1	16

in the two categories of invasive and normal, because normal images and benign images, carcinoma in situ and invasive carcinoma images have similar features, and the model does not learn deeper features.

Figure 9 is an image of part of the wrong prediction. Figure 9a belongs to the invasive class, but it is misjudged as the in situ class. Figure 9b belongs to the in situ class. It can be found that the general features of the two images are very similar, and there are basically no clusters, but the nucleus density in Fig. 9a is greater than that in Fig. 9b. Figure 9c is also an in situ class, which is misjudged as an invasive class, because the nucleus density in some regions of Fig. 9c is relatively high. This shows that the network is not sensitive to the feature recognition of kernel density and cannot extract the detailed features of the image. In order to extract more detailed features and reduce feature redundancy, the octave convolution module is introduced into the model.

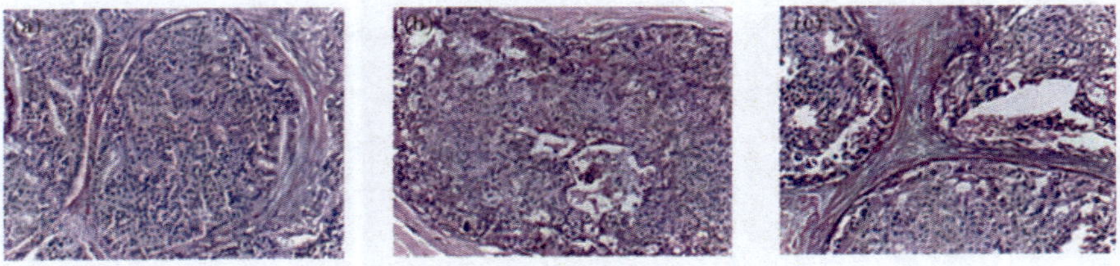

Fig. 9 Partially misjudged images. **a** Invasive, **b** in situ1, **c** in situ2

Residual Network + Octave Convolution Model: The residual network+octave convolution model is based on residual network and replaces the traditional convolution with the octave convolution module. The octave convolution module can effectively extract high-frequency information and appropriately weaken low-frequency information. The accuracy of the network in the validation set and training set is shown in Fig. 10.

It can be seen that compared with the residual network model, the accuracy rate of the residual network+octave convolution model has been greatly improved. The accuracy rate of the optimal model in the training set reaches 97.38%, and the accuracy rate in the validation set reaches 81.73%. The difference between the two is about 16%. Compared with the residual network model, the generalization ability is also improved, which indicates that the patch accuracy affects the image-level recognition results.

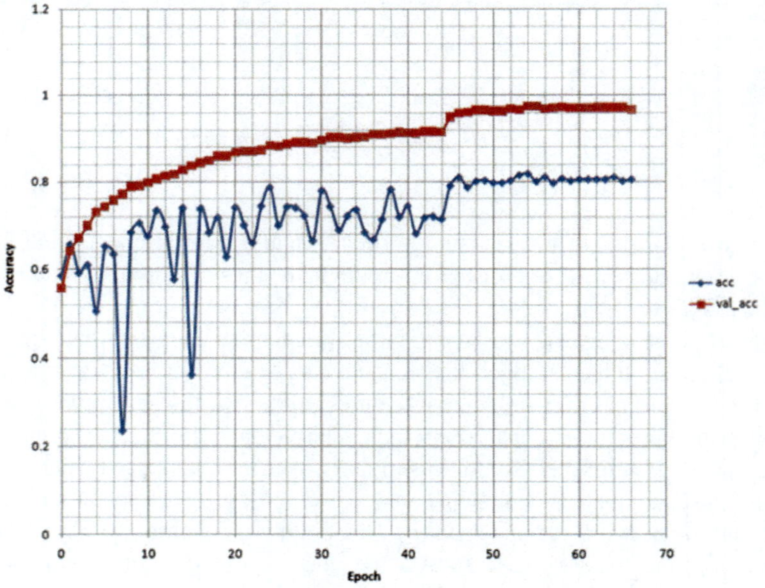

Fig. 10 Training accuracy and validation accuracy of the residual network + octave convolution model

Table 2 Confusion matrix of residual network + octave convolution model

	Benign	In situ	Invasive	Normal
Benign	18	1	0	1
In situ	1	18	3	0
Invasive	0	0	17	0
Normal	1	1	0	19

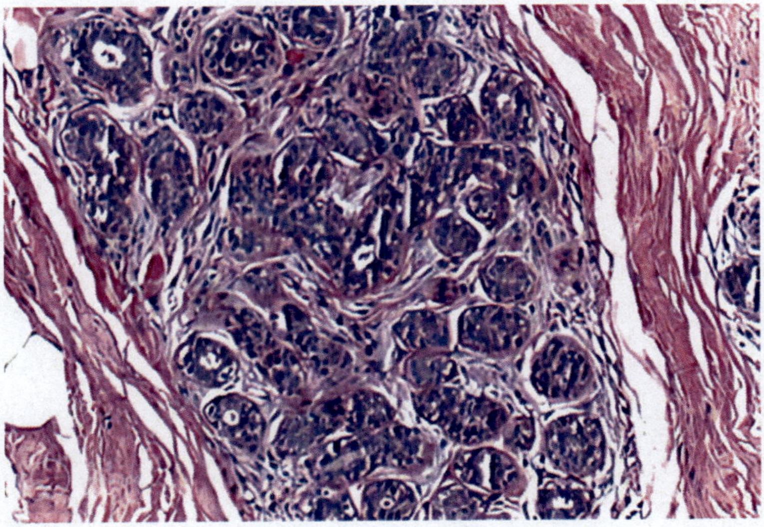

Fig. 11 Image of the normal class

The image-level confusion matrix of the residual network+octave convolution model is shown in Table 2. It can be found that the accuracy rate of the model at the image level can reach 90%, and the number of correct images for the invasive and normal categories has increased. The recognition rate of the normal class with similar classes is greatly improved, and the recognition rate of the invasive class is also improved. Figure 9c can be successfully recognized, which shows that the octave convolution module can extract more detailed features in the image.

Figure 11 shows the wrong image of normal class recognition. The main reason is that the image is dyed unevenly, which makes the edge features of the nucleus blurred, and the features extracted during the test are not obvious enough. The experimental results show that octave convolution has strong robustness to the recognition of similar categories.

Residual Network + Octave Convolution + Heterogeneous Convolution model: Due to the special structure of the octave convolution module, replacing the original traditional convolution with it will greatly increase the number of training parameters, so that the training time for one round is about twice that of the residual network

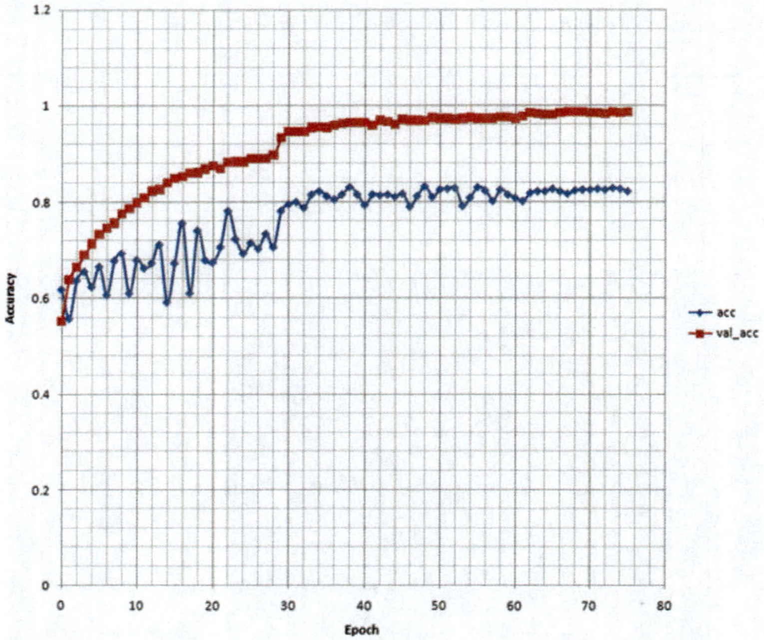

Fig. 12 Training accuracy and validation accuracy of the residual network + octave convolution + heterogeneous convolution model

model. In order to reduce the training time, the heterogeneous convolution module is introduced, and the model performance can be improved at the same time. The network introduces the heterogeneous convolution ($P = 2$) structure based on the residual network+octave convolution network, replacing the traditional convolutional layer in the octave convolution module of the initial layer. The accuracy of the network in the validation set and training set is shown in Fig. 12. It can be seen that the experiment only introduces the heterogeneous convolution module to the initial layer, reducing 37,632 training parameters, but the training accuracy and validation accuracy are good. It is better than the residual network+octave convolution model, and the curve fluctuation in the first 30 rounds is smaller, the training accuracy rate of the optimal model reaches 97.07%, and the accuracy rate of the validation set reaches 83.04%. The image-level confusion matrix of the residual network+octave convolution+heterogeneous convolution model is shown in Table 3. The recognition of Fig. 11 by the network is still wrong, which may be caused by the blurring of the nucleus edges caused by image staining.

From Table 3, it can be found that the final accuracy of the residual network+octave convolution+heterogeneous convolution model at the image level is 91.25%. Comparing Tables 2 and 3, it can be seen that the recognition accuracy of the model on the invasive class has been improved. In the end, only one image in the normal class was identified as the benign class. The image-level confusion matrix of the literature

Table 3 Confusion matrix of residual network + octave convolution + heterogeneous convolution model

	Benign	In situ	Invasive	Normal
Benign	18	1	0	1
In situ	0	18	2	0
Invasive	0	0	18	0
Normal	2	1	0	19

Table 4 Confusion matrix of Golatkar et al. [6]

	Benign	In situ	Invasive	Normal
Benign	23	1	1	4
In situ	1	20	2	1
Invasive	0	1	22	0
Normal	1	3	0	20

[4] is shown in Table 4. Compared with the normal class in Table 4, the number of images in this class that was misjudged as the benign class obviously decreases. Use the confusion matrix of the two methods to calculate the recall rate (recall), precision rate (precision), and final accuracy rate (accuracy) of each category. The results are shown in Table 5. Among the positive samples of a certain type to be predicted, the number of samples that are predicted to be correct is represented by T_P, and the number of samples that are predicted to be wrong is represented by F_N. Conversely, the number of samples predicted to be correct in a certain class of negative samples is denoted by T_N, and the number of samples predicted to be wrong is denoted by F_P. In the experiment, each class is its own positive sample, and the other classes are negative samples.

The recall rate refers to the probability of being predicted as a positive sample in the actual positive sample, which can be expressed as

$$X_{\text{Recall}} = \frac{X_{\text{TP}}}{X_{\text{TP}} + X_{\text{FN}}} \tag{11}$$

Precision refers to the probability that all predicted positive samples are actually positive samples, which can be expressed as

$$X_{\text{Precision}} = \frac{X_{\text{TP}}}{X_{\text{TP}} + X_{\text{FP}}} \tag{12}$$

The accuracy rate refers to the probability that the correct prediction results account for the total number of samples, which can be expressed as

Table 5 Comparison of recall, precision, and accuracy

Method		Recall	Precision	Accuracy
Proposed method	Benign	90.00	90.00	91.25%
	In situ	90.00	90.00	
	Invasive	90.00	100.00	
	Normal	95.00	86.36	
Golatkar et al. [6]	Benign	90.00	79.31	85.0.0%
	In situ	80.00	83.33	
	Invasive	88.00	95.65	
	Normal	80.00	83.33	

$$X_{\text{Accuracy}} = \frac{X_{\text{TP}} + X_{\text{TN}}}{X_{\text{TP}} + X_{\text{TN}} + X_{\text{FP}} + X_{\text{FN}}} \quad (13)$$

The recall rate can reflect the proportion of correctly judged positive examples in the total positive examples. It can be seen from Table 5 that, except for the Benign class, the accuracy of this method for other classes is significantly higher than that of the transfer learning method. This shows that the method can extract features that can distinguish similar classes and improve the performance of the model while reducing the training parameters of the model. In addition, this method reduces the actual test time of the test set through the offline trained model. For 80 RGB images with a size of 2048pixel×1536pixel in the test set, the test time is 562 s, and the test time of each image is 7.025s. It has good performance and can meet the needs of practical applications.

The parameter P in the heterogeneous convolution structure is obtained through experiments, when $P = 4$, it is a structure composed of 3×3 and three 1×1 convolution kernels, and the generalization ability of the model is weak. Therefore, choose the heterogeneous convolution structure with $P = 2$.

4.4.1 Comparison of Experimental Results:

In order to verify the effectiveness of this method, the recognition rates of four-classification tasks of different models are compared with the same dataset, and the results are shown in Table 6. It can be found that the patch-level accuracy (patch-accuracy) and the image-level accuracy (image accuracy) of the final model obtained in the experiment are higher than the transfer learning method in the literature [4], and this method is more effective for similar categories of the recognition rate has been greatly improved.

Table 7 is the four-category comparison results between this method and other methods. Among them, the traditional machine learning method in the literature [1] uses three different machine learning algorithms, and the extraction of artificial features has limitations. The final accuracy rate obtained is low; the method in the

Table 6 Classification rates of different models

Method	Residual convolution	Residual + octave convolution	Residual + octave + heterogeneous convolution nv $p = 2$ ($p = 4$)	Golatkar et al. [6]
Patch accuracy	71.92	81.73	83.04 (78.12)	79.00
Image accuracy	82.50	90.00	91.25 (88.75)	85.00

Table 7 Experimental results obtained by different methods

Method	Accuracy (%)
Traditional ML	80.00–85.00
Alex Net	89.60
CNN + SVM	77.80
Inception transfer learning	85.00
LightGBM	87.20
Proposed method	91.25

literature [2] is improved on the basis of AlexNet, using the advanced texture descriptor, the accuracy rate under the same dataset is improved; the method in the literature [3]. Combining CNN and SVM, the accuracy rate is low in multi-classification tasks; the method in the literature [4] uses the inception network for migration learning, and the accuracy rate is also low; the method in the literature [5] combines three different CNNs, and the LightGBM classifier, with a single network structure, cannot extract the deep features of the image and has a low recognition rate in multi-classification tasks. The method in the literature [6] transforms a four-classification task into a simple two-classification task, that is, using the binary tree idea to gradually carry out two-classification to achieve the purpose of four-classification, the recognition rate is high, but human intervention is required in the experiment for training and classification. The three models tested have poor real-time performance. To sum up, artificial feature extraction will bring subjectivity and limitations, and traditional CNN will lead to single feature and feature redundancy, which will affect the recognition rate to a certain extent. However, this method uses the improved residual network, which has a deeper network structure, can effectively reduce the redundancy of feature space, and has a higher recognition rate for similar classes.

5 Conclusions

The CNN is used to automatically classify breast cancer histopathological images, and the network has a deeper network structure because of the improved deep CNN model, which reduces the number of training parameters while increasing the accuracy rate of the classification. In this paper, the contrast stretching method is used for the data preprocessing in order to increase the recognizability of the nuclei in the image while also overcoming the problem of overfitting caused by the small amount of data available. As demonstrated by the experimental results, the proposed improved CNN model improves the classification rate, reduces the redundancy of extracted features, and reduces the impact of redundancy on recognition and computation consumption. Further, the proposed method does not give biased results. Further, the major limitation of this work is that the proposed model is significantly slower due to an operation such as AvgPool. This method is also more sensitive to identifying the detailed features of similar categories and has better robustness and real-time performance, which can meet the needs of clinical applications to a certain extent when used in conjunction with them.

References

1. Araújo T, Aresta G, Castro E, Rouco J, Aguiar P, Eloy C, Polónia A, Campilho A (2017) Classification of breast cancer histology images using convolutional neural networks. PloS one 12(6):e0177544
2. Aresta G, Araújo T, Kwok S, Chennamsetty SS, Safwan M, Alex V, Marami B, Prastawa M, Chan M, Donovan M et al (2019) Bach: grand challenge on breast cancer histology images. Med Image Anal 56:122–139
3. Bardou D, Zhang K, Ahmad SM (2018) Classification of breast cancer based on histology images using convolutional neural networks. IEEE Access 6:24680–24693
4. Chen Y, Fan H, Xu B, Yan Z, Kalantidis Y, Rohrbach M, Yan S, Feng J (2019) Drop an octave: reducing spatial redundancy in convolutional neural networks with octave convolution. In: Proceedings of the IEEE/CVF international conference on computer vision, pp 3435–3444
5. Desai M, Shah M (2021) An anatomization on breast cancer detection and diagnosis employing multi-layer perceptron neural network (mlp) and convolutional neural network (cnn). Clinical eHealth 4:1–11
6. Golatkar A, Anand D, Sethi A (2018) Classification of breast cancer histology using deep learning. In: International conference image analysis and recognition. Springer, Heidelberg, pp 837–844
7. Gravina M, Marrone S, Sansone M, Sansone C (2021) Dae-cnn: exploiting and disentangling contrast agent effects for breast lesions classification in dce-mri. Pattern Recogn Lett 145:67–73
8. Gu Y, Lu X, Yang L, Zhang B, Yu D, Zhao Y, Gao L, Wu L, Zhou T (2018) Automatic lung nodule detection using a 3d deep convolutional neural network combined with a multi-scale prediction strategy in chest cts. Comput Biol Med 103:220–231
9. Gu Y, Lu X, Zhang B, Zhao Y, Yu D, Gao L, Cui G, Wu L, Zhou T (2019) Automatic lung nodule detection using multi-scale dot nodule-enhancement filter and weighted support vector machines in chest computed tomography. PLoS One 14(1):e0210551
10. Gu Y, Lu X, Zhao Y, Yu D (2015) Research on computer-aided diagnosis of breast tumors based on pso-svm. Comput Simulation 05:344–349
11. Gupta V, Bhavsar A (2017) Breast cancer histopathological image classification: is magnification important? In: Proceedings of the IEEE conference on computer vision and pattern recognition workshops, pp 17–24
12. Hayat MJ, Howlader N, Reichman ME, Edwards BK (2007) Cancer statistics, trends, and multiple primary cancer analyses from the surveillance, epidemiology, and end results (seer) program. The Oncologist 12(1):20–37
13. He K, Zhang X, Ren S, Sun J (2016) Deep residual learning for image recognition. In: Proceedings of the IEEE conference on computer vision and pattern recognition, pp 770–778

14. Hossain MS (2017) Cloud-supported cyber-physical localization framework for patients monitoring. IEEE Syst J 11(1):118–127
15. Ioffe S, Szegedy C (2015) Batch normalization: accelerating deep network training by reducing internal covariate shift. In: International conference on machine learning, PMLR, pp 448–456
16. Khan S, Islam N, Jan Z, Din IU, Rodrigues JJC (2019) A novel deep learning based framework for the detection and classification of breast cancer using transfer learning. Pattern Recogn Lett 125:1–6
17. Kingma DP, Ba J (2014) Adam: a method for stochastic optimization. arXiv preprint arXiv:1412.6980
18. Koné I, Boulmane L (2018) Hierarchical resnext models for breast cancer histology image classification. In: International conference image analysis and recognition. Springer, Heidelberg, pp 796–803
19. Krizhevsky A, Sutskever I, Hinton GE (2012) Imagenet classification with deep convolutional neural networks. In: Advances in neural information processing systems, vol 25
20. Lin M, Chen Q, Yan S (2013) Network in network. arXiv preprint arXiv:1312.4400
21. Linlin Guo YL (2018) Histopathological image classification algorithm based on product of experts. Laser Optoelectronics Progr 55:021008
22. Liu L, Jiang H, He P, Chen W, Liu X, Gao J, Han J (2019) On the variance of the adaptive learning rate and beyond. arXiv preprint arXiv:1908.03265
23. Melekoodappattu JG, Subbian PS (2019) A hybridized elm for automatic micro calcification detection in mammogram images based on multi-scale features. J Med Syst 43(7):1–12
24. Murtaza G, Shuib L, Abdul Wahab AW, Mujtaba G, Nweke HF, Al-garadi MA, Zulfiqar F, Raza G, Azmi NA (2020) Deep learning-based breast cancer classification through medical imaging modalities: state of the art and research challenges. Artif Intell Rev 53(3):1655–1720
25. Nazeri K, Aminpour A, Ebrahimi M (2018) Two-stage convolutional neural network for breast cancer histology image classification. In: International conference image analysis and recognition. Springer, Heidelberg, pp 717–726
26. Rakhlin A, Shvets A, Iglovikov V, Kalinin AA (2018) Deep convolutional neural networks for breast cancer histology image analysis. In: International conference image analysis and recognition. Springer, Heidelberg, pp 737–744
27. Rezaeilouyeh H, Mollahosseini A, Mahoor MH (2016) Microscopic medical image classification framework via deep learning and shearlet transform. J Med Imaging 3(4):044501
28. Singh P, Verma VK, Rai P, Namboodiri VP (2019) Hetconv: heterogeneous kernel-based convolutions for deep cnns. In: Proceedings of the IEEE/CVF conference on computer vision and pattern recognition, pp 4835–4844
29. Spanhol FA, Oliveira LS, Petitjean C, Heutte L (2015) A dataset for breast cancer histopathological image classification. IEEE Trans Biomed Eng 63(7):1455–1462
30. Spanhol FA, Oliveira LS, Petitjean C, Heutte L (2016) Breast cancer histopathological image classification using convolutional neural networks. In: 2016 International Joint Conference on Neural Networks (IJCNN). IEEE, pp 2560–2567
31. Srivastava N (2013) Improving neural networks with dropout. Univ Toronto 182(566):7
32. Sumei L, Guoqing LRF (2019) Depth map super-resolution based on two-channel convolutional neural network. Acta Optica Sinica 38:081001
33. Ting M, Yuhang LKZ (2019) Algorithm for pathological image diagnosis based on boosting convolutional neural network. Acta Optica Sinica 39:081001
34. Wang Z, You K, Xu J, Zhang H (2014) Consensus design for continuous-time multi-agent systems with communication delay. J Syst Sci Complexity 27(4):701–711
35. Xie S, Girshick R, Dollár P, Tu Z, He K (2017) Aggregated residual transformations for deep neural networks. In: Proceedings of the IEEE conference on computer vision and pattern recognition, pp 1492–1500
36. Xu B, Wang N, Chen T, Li M (2015) Empirical evaluation of rectified activations in convolutional network. arXiv preprint arXiv:1505.00853
37. Xueying H, Zhongyi H, Benzheng W (2018) Breast cancer histopathological image auto-classification using deep learning. In: Computer engineering and applications

Performance Assessment of Normalization in CNN with Retinal Image Segmentation

Junaciya Kundalakkaadan, Akhilesh Rawat, and Rajeev Kumar

Abstract Retinal vessel segmentation segmentizes the blood vessels from retinal fundus images; this helps detect retinal diseases. Normalization techniques such as group normalization, layer normalization, and instance normalization were introduced to replace batch normalization. This paper evaluates the performance of these normalization techniques in a convolutional neural network (CNN) on retinal vessel segmentation: how it helps in improving the generalization ability of the model. The digital retinal images for vessel extraction (DRIVE), a publicly available dataset, are used for this experiment. Accuracy, F1 score, and Jaccard index of models with these normalization techniques were calculated. By empirical experiments, it is observed that the batch normalization outperforms its peers in CNN in terms of its accuracy. However, group normalization gives better convergence than other normalization techniques in terms of the validation error and results in a better generalized architecture for this segmentation task.

Keywords CNN · Generalization · Normalization · Batch normalization · Group normalization · Layer normalization · Instance normalization · Retinal fundus

1 Introduction

Computer-aided diagnosis (CAD) helps the medical field diagnose the problem quickly with lesser human effort. Retinal vessel segmentation extracts blood vessels from retinal fundus images; these images are helpful in detecting diabetic retinopathy diseases [5, 11]. Any changes in blood vessels signal the retinal disease in the

J. Kundalakkaadan (✉) · A. Rawat · R. Kumar
Data to Knowledge (D2K) Lab, School of Computer and Systems Sciences,
Jawaharlal Nehru University, New Delhi 110067, India
e-mail: junaci66_scs@jnu.ac.in

A. Rawat
e-mail: akhile62_scs@jnu.ac.in

M. Thakur et al. (eds.), *Soft Computing for Problem Solving*, Lecture Notes in Networks and Systems 547, https://doi.org/10.1007/978-981-19-6525-8_13

patient. As demand for computing power to detect diseases increases in the medical field due to the complex structure of data, neural network-based techniques are gaining interest in this area due to robustness.

Convolutional neural network (CNN), the deep neural network, has become state of the art in many applications such as image classification[7]. CNN has been successfully used in several medical image classification applications [15]. The deep neural network has great importance in medical applications because of its adaptability to complex structures. Supervised machine learning algorithms are based on the statistical assumption that both training data and test data follow the same distribution [23]. The model is trained to learn the patterns hidden in this training data, and based on these patterns learned from training data, it makes predictions on a test data: the data which the model has never seen during training and sampled from the same population of the training data. The key to the success of deep neural networks is their ability to predict well on this unseen data [24]. If a CNN model can be developed by reducing the gap between training error and testing error, it will lead to better generalization, which will help in superior detection of disease without much human intervention.

Since overfitting leads to poor generalization [18], regularizers like weight decay, data augmentations, and dropout are used for better performance of the neural network [18]. Normalization technique like batch normalization has been widely used as an implicit regularizer in CNN models for reducing generalization error [24]. Recently, many researchers have carried out their work on retinal vessel segmentation with batch normalization [1, 4, 8, 22]. Group normalization, layer normalization, and instance normalization have been introduced as a replacement for batch normalization. These normalization techniques are performed differently in different architectures. The majority of studies were carried out using batch normalization in CNN. Layer normalization performs well in recurrent neural network (RNN) [2] and instance normalization does better in image generation [19].

Batch normalization has a significant influence on many existing systems. However, this paper analyzes the performance of different normalization techniques. These techniques are applied in a CNN on segmenting retinal blood vessels to evaluate how these techniques are helping a CNN to predict well on unseen data and to improve the performance. Jaccard index is used to measure the similarity between the predicted image and the ground truth image. F1 scores are calculated, and the accuracy graph of each normalization is plotted to evaluate the performance of the normalization technique on retinal vessel segmentation.

The rest of the paper is organized as follows: the literature review, problem definition, and proposed model are included in Sects. 2, 3, and 4, respectively. Datasets used, results, and discussions are included in Sect. 5. Finally, the work is concluded in Sect. 6.

2 Literature Review

Despite the great potential of neural networks, training neural networks by keeping their generalization error minimal is a challenging task. Neural networks are prone to overfitting. Regularization is a fundamental technique to prevent overfitting, and it helps in improving generalization [3]. Several authors studied the impact of regularizers like data augmentation and weight decay; they found that data augmentation and weight decay dropped the performance of some classes [3].

Zhang et al. [23] argued that explicit regularizers such as dropout and weight decay are not essential and sufficient for generalization. The authors argued the necessity of regularization in training an over-parameterized network. The authors observed that test time performance remained strong while turning off the regularizers like data augmentation, weight decay, and dropout.

Ioffe and Szegedy [9] introduced batch normalization; they explored the possibility of batch normalization. The model's generalization performance is illustrated in the training of neural networks utilizing batch normalization, dropout, and data augmentation in [18]. Zhang et al. [23] categorized the batch normalization as an implicit regularizer. Wu and He [21] observed that when the batch size becomes smaller, batch normalization rapidly increases errors which leads to the introduction of group normalization (GN).

A layer normalization is applied to recurrent neural networks (RNNs) by Ba et al. [2]. It was not obvious how to apply batch normalization to RNN. Authors showed that RNN is benefited from layer normalization, especially for long sequences and mini-batches. It improves both generalization and training time in an RNN model. Batch normalization is replaced with instance normalization (IN) by Ulyanov et al.[19]. Authors showed that the instance normalization improved the performance in generating images using a certain deep neural network.

Murugan and Roy [11] suggested a CNN architecture for detecting microaneurysms (MA). For their experiment, they employed retinal fundus images from the ROC dataset. They used regularization techniques to deal with overfitting in their model, but they did not use any normalizing techniques, which may provide a more generalized model for their experiment. To obtain improved generalization, many researchers have employed normalization techniques such as batch normalization [14] and group normalization [12].

An empirical study is conducted to analyze the performance of dropout and batch normalization on training the multilayered dense neural network and convolutional neural network (CNN) by Garbin et al. [6]. Authors found that adding dropout to CNN has reduced the accuracy, whereas adding batch normalization to CNN has increased the accuracy of the model. Jifara et al. [10] used batch normalization as a regularization model. They applied batch normalization between the convolution layer and rectified linear unit (ReLU) activation. The experiments result showed that the accuracy of the model and training speed were increased. Chen et al. [4] did a review of retinal vessel segmentation based on deep learning. Batch normalization is used before the activation function to accelerate the training process. Various models such as FCN, U-Net, multi-modal network, and generative adversarial network (GAN) for retinal vessel segmentation are analyzed in this paper.

A multichannel U-Net (M-U-Net) to separate blood vessels from a retinal fundus image employed by authors in [5]. To enhance the model's accuracy, batch normalization is used in this study with a batch size of 3. A convolutional network known as Sine-Net is developed by Atli et al. [1]. Under this study, all inputs are batch normalized and scaled between zero and one to minimize the complexity. [8, 22] utilized batch normalization with a mini-batch size of 32 to detect retinal blood vessels using U-Net. Recent studies are used batch normalization for retinal vessel segmentation using CNN.

In summary, convolutional neural network (CNN) imposes a low computational cost [7] on image classification. Many researchers have started using batch normalization along with a regularizer to improve the performance of the model. Other normalization techniques like group normalization, layer normalization, and instance normalization have been introduced as a replacement for batch normalization. How these normalization techniques are performing in CNN which is needed to be evaluated.

3 Problem Definition: Research Questions

How CNN performs on test data without using any normalization techniques is to be studied. How normalization techniques such as batch normalization, group normalization, layer normalization, and instance normalization are performing in CNN which has to be observed. The possibility of improving the generalization ability of CNN when applying these newly introduced normalization techniques as a replacement to batch normalization is to be analyzed.

4 The Proposed Methodology

4.1 CNN Architecture

This study implements a U-Net architecture (Fig. 1) for biomedical image segmentation provided in [13] with various normalization strategies. A contracting path and an expansive path are included in this design. Two 2×2 convolutions are applied, with each convolution being followed by a normalization layer and the ReLU activation function. In the first convolutional block, a dropout is used, followed by the ReLU. For downsampling, two repeated convolutions are followed by a max-pooling layer with stride 2. The number of feature channels doubles with each downsampling step. The expansive path consists of upsampling followed by a convolution layer that

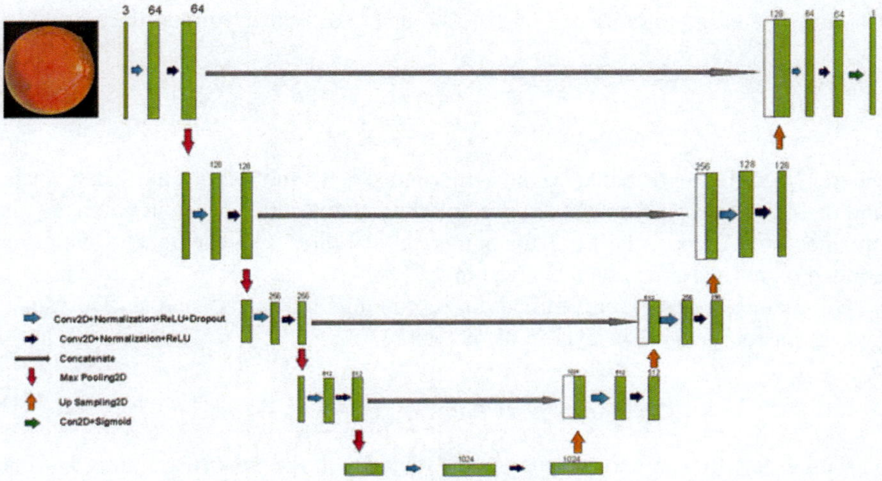

Fig. 1 U-Net Architecture for retinal image segmentation

halves the number of feature channels, a concatenative skip connection from the contracting path for an alternative way for feature reusability of the same dimension, and two 2 × 2 convolutions each followed by normalization and ReLU. Dropout is used in the first convolution block. A sigmoid activation is used in the last 1 × 1 convolutional layer, which gives the final output representing pixel-wise classification. We assessed the performance of normalization techniques in deep neural networks for medical applications using this architecture.

4.2 Normalization Techniques

Normalization techniques are used to standardize inputs, which help in improving the system's performance[9, 21]. In this study, the CNN has experimented with each normalization technique, such as batch normalization, group normalization, layer normalization, and instance normalization. Batch normalization deals with the internal covariance shift [9].

Consider a batch of size n, the batch mean (μ_B), and the batch variance ($\sqrt{\sigma^2{}_B}$) are calculated using the Eq. 1 and the Eq. 2, respectively, for values of x over the batch B [9].

$$\mu_B \leftarrow \frac{1}{n}\sum_{i=1}^{n} x_i \tag{1}$$

$$\sigma_B^2 \leftarrow \frac{1}{n}\sum_{i=1}^{n} (x_i - \mu_B)^2 \tag{2}$$

Using the mean and variance of mini-batches, it normalizes each dimension as follows

$$\hat{x}_B \leftarrow \frac{x_B - \mu_B}{\sqrt{\sigma^2{}_B + \epsilon}} \tag{3}$$

where $\hat{x}_B$ is the input batch which is normalized by subtracting the batch mean and then divided using batch variance and thus, the normalized input batch, $\hat{x}_B$, is obtained. ϵ is used in the Eq. 3 for numerical stability. A linear transformation is applied on normalized values as given in the Eq. 4.

Parameters, γ and β, are learned during training and scaled, and shifted values (y_i) are passed to the next layer in the network.

$$y_i = \gamma \hat{x}_i + \beta \tag{4}$$

The mini-batch mean and variance give efficiency at the time of training. During the inference, the population mean and variance are used in Eq. 3 to normalize the output.

Group normalization divides input channels into groups [21]. The mean and variance of these groups are calculated for normalization. Thus, group normalization is batch-independent. The parameters γ and β are used to scale and shift the normalized value as we did in Eq. 4.

Layer normalization [2] is computed over all hidden units in a layer. Mean and variance are calculated across the same layer. Normalization is performed for a specific channel within the sample in CNN. As given in Eq. 4, learning parameters are used for scaling and shifting.

In instance normalization [19], every channel is considered. Channel represents the number of features in the input. The mean and the variance of each feature are calculated for normalization as shown in Eq. 3. γ and β are used for scale and shift as in Eq. 4.

5 Results and Discussion

5.1 Datasets and Preprocessing

The digital retinal images for vessel extraction (DRIVE), a publicly available dataset, was introduced [17]. It includes 40 JPEG images which are equally divided into 20 images for training and 20 images for testing of 584 × 565 resolutions with three RGB color channels. These images were collected from diabetic retinopathy patients in the Netherlands. Data augmentation (DA) techniques like horizontal flip, vertical flip, elastic transform, grid distortion, and optical distortion are applied to generate 120 images, resized to 512 × 512, each for training and testing. Pixel values of both

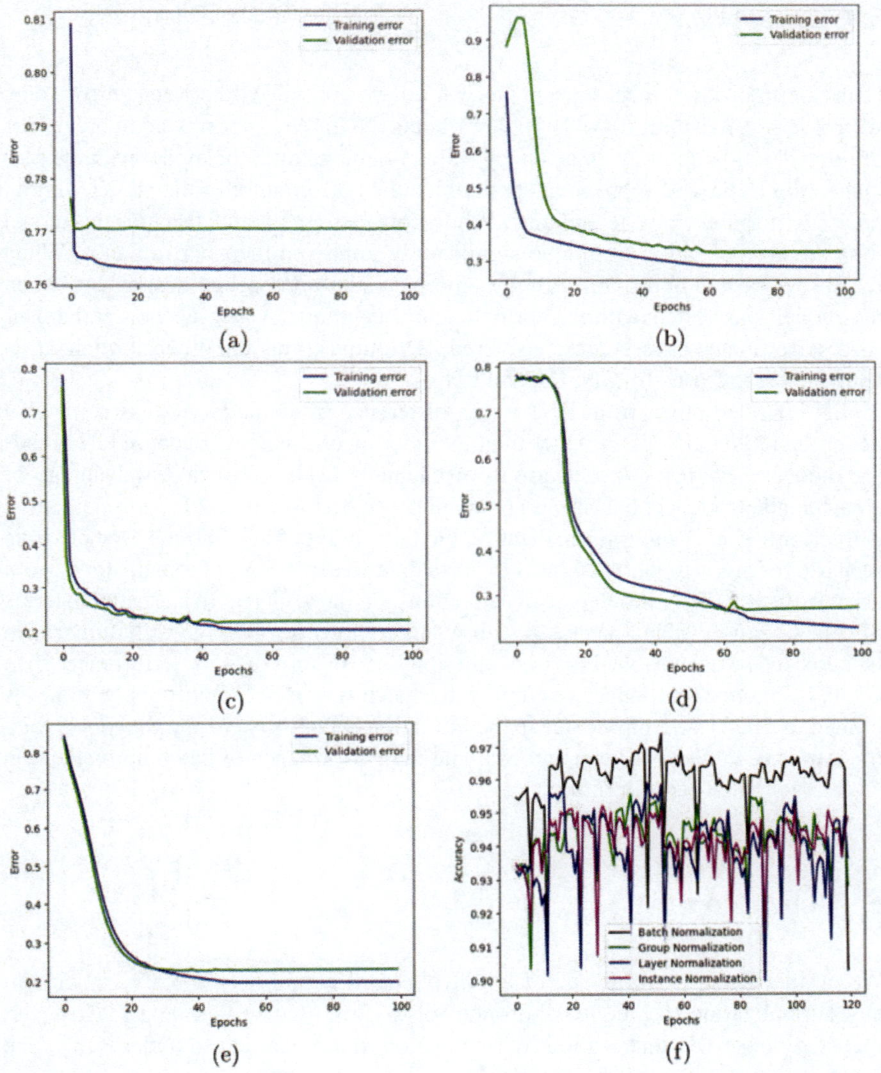

Fig. 4 **a** Without normalization, **b** batch normalization, **c** group normalization, **d** layer normalization, **e** instance normalization, and **f** accuracy of all normalization techniques

accuracy of the validation set is plotted, which shows batch normalization outperforms the other model in terms of accuracy. But group normalization gives better convergence than other normalization techniques in terms of validation error and works as better generalized architecture for this segmentation task. In Fig. 4, group normalization (Fig. 4) performs with 71% lesser validation error without normalization (Fig. 4) and with 31% lesser validation error with batch normalization (Fig. 4).

5.4 Discussion

Using normalization techniques, pixel values are normalized between zero to one, which helps the neural network to work better. In Table 1, norm is an abbreviation of normalization. In terms of accuracy, Table 1 demonstrates the model with the normalization technique approximately doubled the performance of the model without applying normalization techniques. While considering F1 and Jaccard, the model with the normalization technique significantly improved their performance. When recall approached approximately one, the precision approached nearly zero when the model was tested without normalization technique. When accuracy and Jaccard performance criteria are considered, BN outperforms the other normalization approaches, followed by GN, IN, and LN.

GN is scaled down from BN, with a difference of about 2%. GN is somewhat better than LN and slightly better than IN in terms of accuracy. Because GN allows the model to learn from each group of channels [21], and it takes advantage of channel independence, but LN and IN do not. The performance of LN normalization is inferior to that of other normalization methods in Fig. 4 and Table 1. One possible scenario for this is that LN attempts to normalize all activations of a single layer from a batch; it normalizes the average of activations. In Fig. 4, GN performs better in terms of generalization. When batch sizes grow larger, however, BN may still outperform GN. Because when the batch size becomes higher, BN improves its performance than GN [21]. In our experiment, we employed a batch size of one. Our investigation was limited to a batch size of one due to the limitations of our system and the time it took for it to execute. We will test and evaluate the performance of batch normalization with larger batch sizes in future.

6 Conclusion

We evaluated the performance of a convolutional neural network (CNN) without any normalization technique. A dropout of 0.25 is used in this model. However, this study observed that without normalization, the model led to underfitting. The

Table 1 Performance assessment of normalization techniques

	Without norm.	Batch norm.	Group norm.	Layer norm.	Instance norm.
Accuracy	0.47123	0.96194	0.94281	0.93979	0.94134
F1	0.09506	0.56982	0.51351	0.49041	0.50494
Jaccard	0.04999	0.40008	0.34654	0.36237	0.3392
Recall	0.84780	0.80628	0.92792	0.89292	0.93274
Precision	0.05044	0.45306	0.35683	0.34194	0.34872

model performed with an accuracy of 47%. We also evaluated the performance of the model with normalization techniques. Normalization techniques such as batch normalization, group normalization, layer normalization, and instance normalization are applied to assess the performance of a CNN. We observed that the batch normalization outperformed the other normalization techniques with an accuracy of 96%, in CNN. Group normalization, on the other hand, provides better convergence than other normalization techniques in terms of validation error and works as a better generalized architecture for this segmentation task.

Acknowledgment We thank the anonymous reviewers for their valuable feedback by which the readability of the paper is improved.

References

1. Atli I, Gedik OS (2021) Sine-Net: a fully convolutional deep learning architecture for retinal blood vessel segmentation. Eng Sci Technol Int J 24(2):271–283
2. Ba JL, Kiros JR, Hinton GE (2016) Layer normalization. arXiv preprint arXiv:1607.06450
3. Balestriero R, Bottou L, LeCun Y (2022) The effects of regularization and data augmentation are class dependent. arXiv preprint arXiv:2204.03632
4. Chen C, Chuah JH, Raza A, Wang Y (2021) Retinal vessel segmentation using deep learning: a review. IEEE Access
5. Dong H, Zhang T, Zhang T, Wei L (2022) Supervised learning-based retinal vascular segmentation by M-UNet full convolutional neural network. In: Signal, image & video processing, pp 1–7
6. Garbin C, Zhu X, Marques O (2020) Dropout vs. batch normalization: an empirical study of their impact to deep learning. Multimedia Tools Appl 79(19):12777–12815
7. Guo T, Dong J, Li H, Gao Y (2017) Simple convolutional neural network on image classification. In: Proceedings of the IEEE 2nd International Conference Big Data Analysis (ICBDA. IEEE, pp 721–724
8. Hakim L, Kavitha MS, Yudistira N, Kurita T (2021) Regularizer based on euler characteristic for retinal blood vessel segmentation. Pattern Recogn Lett 149:83–90
9. Ioffe S, Szegedy C (2015) Batch normalization: accelerating deep network training by reducing internal covariate shift. In: Proceedings of the International Conference Machine Learning (ICML), PMLR, pp 448–456
10. Jifara W, Jiang F, Rho S, Cheng M, Liu S (2019) Medical image denoising using convolutional neural network: a residual learning approach. J Supercomput 75(2):704–718
11. Murugan R, Roy P (2022) MicroNet: microaneurysm detection in retinal fundus images using convolutional neural network. In: Soft computing, pp 1–10
12. Myronenko A (2018) 3D MRI brain tumor segmentation using autoencoder regularization. In: Proceedings of the international MICCAI Brainlesion workshop. Springer, Heidelberg, pp 311–320
13. Ronneberger O, Fischer P, Brox T (2015) U-net: convolutional networks for biomedical image segmentation. In: Proceedings of the international conference on medical image computing and computer-assisted intervention. Springer, Heidelberg, pp 234–241
14. Saranya P, Prabakaran S, Kumar R, Das E (2021) Blood vessel segmentation in retinal fundus images for proliferative diabetic retinopathy screening using deep learning. In: The visual computer, pp 1–16
15. Sarvamangala D, Kulkarni RV (2021) Convolutional neural networks in medical image understanding: a survey. In: Evolutionary intelligence, pp 1–22

16. Soomro TA, Afifi AJ, Gao J, Hellwich O, Paul M, Zheng L (2018) Strided U-Net model: retinal vessels segmentation using dice loss. In: Digital Image Computing: Techniques and Applications (DICTA). IEEE, pp 1–8
17. Staal J, Abràmoff MD, Niemeijer M, Viergever MA, Van Ginneken B (2004) Ridge-based vessel segmentation in color images of the retina. IEEE Trans Med Imaging 23(4):501–509
18. Thanapol P, Lavangnananda K, Bouvry P, Pinel F, Leprévost F (2020) Reducing overfitting and improving generalization in training convolutional neural network under limited sample sizes in image recognition. In: Proceedings of the 5th International Conference on Information Technology (InCIT). IEEE, pp 300–305
19. Ulyanov D, Vedaldi A, Lempitsky V (2016) Instance normalization: the missing ingredient for fast stylization. arXiv preprint arXiv:1607.08022
20. Wang C, Zhao Z, Yu Y (2021) Fine retinal vessel segmentation by combining Nest U-net and patch-learning. Soft Comput 25(7):5519–5532
21. Wu Y, He K (2018) Group normalization. In: Proceedings of the European Conference on Computer Vision (ECCV), pp 3–19
22. Xiancheng W, Wei L, Bingyi M, He J, Jiang Z, Xu W, Ji Z, Hong G, Zhaomeng S (2018) Retina blood vessel segmentation using a u-net based convolutional neural network. In: Procedia Computer Science: Proceedings of the International Conference Data Science (ICDS 2018), pp 8–9
23. Zhang C, Bengio S, Hardt M, Recht B, Vinyals O (2021) Understanding deep learning (still) requires rethinking generalization. Commun ACM 64(3):107–115
24. Zheng Q, Yang M, Yang J, Zhang Q, Zhang X (2018) Improvement of generalization ability of deep CNN via implicit regularization in two-stage training process. IEEE Access 6:15844–15869

A Novel Multi-day Ahead Index Price Forecast Using Multi-output-Based Deep Learning System

Debashis Sahoo, Kartik Sahoo, and Pravat Kumar Jena

Abstract Movement in the stock market or equity market can have unfathomable consequences on the economy and individual investors. A collapse in the stock market especially in the indexes has the potential to cause extensive economic disruption. Today's smart AI has the capability to capture extreme fluctuations, irrational exuberance, and episodes of very high volatility. These sophisticated AI driven systems can detect such non-linearity with much-improved forecast results compared to conventional statistical methods. Prediction and analysis of index prices have greater importance in today's economy. In this piece of work, we have experimented with three types of deep learning architectures, a simple feed-forward neural network (ANN), a long short-term memory network (LSTM), and a blend of convolutional neural networks with LSTMs (CNN-LSTMs). Along with open, high, low, close (OHLC) data, a set of 55 technical indicators have been considered based on their importance in technical analysis to predict the daily price for 5 different global indices. A random forest-based recursive feature elimination has also been used to obtain the most important technical indicators, and these results have been compared with all deep learning models for a horizon of 5 days ahead index price forecast.

Keywords Deep learning · Multi-day ahead · Index price forecast · Multi-in multi-out model · Recursive feature elimination · Random forest · Feature importance algorithm · LSTM · CNN-LSTM · Nifty 50 · Dow Jones · SP 500 · Nasdaq

D. Sahoo (✉) · K. Sahoo
Indian Institute of Technology Mandi, Mandi, Himachal Pradesh, India
e-mail: S18020@students.iitmandi.ac.in

K. Sahoo
e-mail: D17045@students.iitmandi.ac.in

P. K. Jena
University of Petroleum and Energy Studies, Dehradun, India
e-mail: pravat.jena@ddn.upes.ac.in

M. Thakur et al. (eds.), *Soft Computing for Problem Solving*, Lecture Notes in Networks and Systems 547, https://doi.org/10.1007/978-981-19-6525-8_14

1 Introduction

For any developed economy, the stock market plays a major sentiment indicator, and investors consider the financial market a highway for superlative investment return and an outstanding opportunity for wealth creation. There are many ways investors do analysis in the stock market, based on the fair value of the share by looking at companies sales, income statement, balance sheet, etc., and this kind of investment generally is done for a longer period of time (long-term investment), also looking some technical indicators like moving average, Bollinger bands, relative strength index (RSI) values, and many more, investors take a decision either for shorter period or for longer period of time. While technologies are rapidly changing and growing, nowadays, artificial intelligence (AI) is also playing a significant role in financial industries. To begin with, wall street statisticians to many famous investment bankers, fund managers, high net-worth individuals have adapted many applications of statistics, machine learning, and deep learning model-based systems to their banking and trading systems.

Since stock and future prices are highly non-linear and stochastic in nature, so mathematical models have the capabilities to analyze and capture the linear, non-linear, and complex statistical and mathematical structures. Time series forecasting plays a major role here since price movements are observed in sequence and form a time series. Time series analysis can be uni-variate and multivariate analysis, depending upon the number of independent variables. To determine our target or dependent variable if we consider a single independent variable or feature or attribute, then this is uni-variate time series, and if the dependent variable is calculated with help of many factors or independent variables, then it is a multivariate time series analysis. Many statistical moving average-based linear models like auto-regressive (AR), ARMA, auto-regressive integrated moving average (ARIMA) [1], and its variations have been used to deal with uni-variate time series forecasting.

In recent decade, several machine learning and deep learning models have been developed to deal with stock market signal generation to buy sell any particular asset, trend forecasting for both short, medium, and longer period of time. Artificial neural networks (ANNs) have capability to approximate any non-linear functions to certain arbitrary precision [2, 3], and with huge availability of structured data and computer resources like high-end CPUs and graphical processing like GPUs, ANNs can be scaled with more number of layers and vast number of nodes per layer, and they are capable to handle and learn from huge amount of data and to produce accurate signals, good alpha by exploiting inefficiencies in the market, which is almost impossible for human beings.

While a simple neural network can produce wonders, there are more complex and advanced deep learning models like recurrent neural networks (RNNs) which are especially developed for sequence data and long short-term memory networks (LSTMs) have memory cells, and these models have performed better for time series forecasting [4–6]. Time series forecasting can be one-step ahead or single-step prediction, where target value is predicted for next time step or a time series forecasting can be of multistep ahead, where target value is predicted for several steps ahead of current time.

In this study, for 5-day ahead index price forecasting with help of various technical indicators, and random forest-based robust feature engineering have been experimented with state-of-the-art deep learning models. For the purpose of our comparisons, open, high, low, and close (OHLC) data from top five global indices have been taken, and details about data are described in Sect. 4. The rest of the paper is organized as follows: research related to multistep ahead for stock market and index prices are discussed in Sect. 2. In the next Sect. 3, some background about different deep learning models like ANNs, LSTMs, and CNN-LSTMs have been briefed. Section 4 talks about the dataset and pre-processing techniques. Details about proposed models and model calibrations are discussed in Sect. 5. Results are presented in Sect. 6 and conclusions and future works in Sect. 7, followed by references.

2 Related Works

Price forecasting using deep learning models Different machine learning and deep learning models have been used in past. A hybrid model using ARIMA and multilayer perceptron (MLP) for stock price forecasting [7], using random forest(RF), support vector machine (SVM), and comparison of price forecasting results with ANN [8, 9], in recent years LSTMs and bi-directional LSTMs has been compared with other models like ANN, SVM, gated recurrent unit (GRU) for price forecasting in global indices and predicting Bitcon price [10–13]. Genetic fuzzy systems integrated with ANNs to improve results in forecasting price in IT and Airlines sector [14, 15]. Phase space reconstruction method for time series reconstructions along with LSTMs [16–18]. In some work, CNNs, which are well known for feature extraction, have been compared with other models like MLP, RNN, LSTMs for stock price prediction.

Multistep ahead forecasting Due to future uncertainty, multistep ahead forecasting is more difficult and error prone as compared to single-step forecasting. Various methods have been proposed in literature for multistep ahead forecasting using ANNs, LSTMs, e.g., direct method, multi-input multi-output (MIMO) method, recursive methods [4, 19–21]. According to literature study, multi-input and multi-output method are best among all other methods due to computational in-expensiveness like direct method and errors produced in the previous steps are not propagated as it is in recursive methods. Many works have been done in past on index price forecasting using RNNs, LSTMs [22], encoder–decoder frameworks [23, 24], and some of the works also used attention mechanism along with encoder–decoder framework for index price forecasting [25–27].

Dimension reduction and feature selection Index price depends on many factors, after all many features are taken into considerations while building a model, and on the other hand, all features which are given as input may not have equally significant role, so various feature selection and dimension reduction techniques have been used along with deep networks. In many cases to predict stock price, gold and crude oil have been used as extra indicators [28], in some of the works, 43 technical indicators as input features along with LSTM to forecast price [29], and in other work, 715 technical indicators suggested by experts have been used as input to forecast stock price [30]. Principal component analysis (PCA), a dimension reduction technique, used to extract a new set of variables from actual large set of features, has been used along with LSTMs or ANNs for stock price predictions [31–33].

3 Methodology

Before introducing our proposed framework, in this section, we will review the artificial neural network, long short-term memory network, and convolutional neural networks.

3.1 Artificial Neural Networks (ANNs)

Artificial neural networks are biological inspired computational networks, evolved from idea of mimicking the human nervous system that allows learning by examples from representative data. The ANN was first introduced by McCulloch and Pitts [34, 35] based on threshold logic, and later backpropagation algorithm triggered the interest and enabled practical applications and equipped training of multi-layer networks. An ANN consists of an input layer, one or more hidden layers, and an output layer, there are connection between every two neurons of successive layers and a weight associated to each connections. Input passes all the way through the hidden layers and backpropagation distribute the errors back through the layers by modifying the weights in every node. A neural network works by learning the non-linear relationship between dependent and independent variables. According to universal function approximation theorem, a neural network with a single hidden layer and finite number of nodes can learn any continuous function and under assumption on activation functions to achieve higher order non-linearity. With the help of optimizers, the network tries to minimize the cost function. The weights are initialized randomly, errors are calculated after all the computations, and then gradient is calculated, i.e., derivative of error with respect to current weight. New weights are calculated as shown in Eqn. 1, where η is the learning rate, w^{n+1}, w^n are new and old weights, respectively, and $\frac{\partial \text{Error}}{\partial w^n_{jk}}$ is derivative of error with respect to weight.

$$w^{n+1}_{jk} = w^n_{jk} - \eta \frac{\partial \text{Error}}{\partial w^n_{jk}} \tag{1}$$

3.2 Long Short-term Memory Networks (LSTMs)

Long short-term memory network (LSTM) is an artificial recurrent neural network that allows information to persist, useful for sequential information, introduced by Hochreiter and Schmidhuber [36]. Being equipped with gated architecture (forget gate, input gate, output gate), LSTMs also have feedback connections which makes them different than traditional feed-forward neural network, this delegates LSTMs to process long sequence of data without treating every points in the sequence independently, and also it retains or memorize useful information from the previous time stamps through its memory cell. These very properties empower LSTMs to handle vanishing gradient faced by recurrent neural networks (RNNs). The cell state is updated in each time step, the input gate regulates the amount of information to consider at current time step t, the forget gate takes previous hidden state h_{t-1} and current input x_t to determines what are the information to be preserved from previous time step $t-1$. Finally, the "output gate" takes input from updated cell state, previous hidden state h_{t-1} and new input data x_t and produces the new hidden state h_t. Summarizing, Eqs. 2–6 briefly describes the operations performed by an LSTM unit.

$$\text{[Input gate] } i_t = \sigma(V_{gi}x_t + W_{zi}h_{t-1} + b_i), \tag{2}$$

$$\text{[Forget gate] } f_t = \sigma(V_{gf}x_t + W_{zf}h_{t-1} + b_f) \tag{3}$$

$$\text{[Current Memory cell] } c_t^* = \tanh(V_{gc}x_t + W_{zc}h_{t-1} + b_c) \tag{4}$$

$$\text{[Updated Memory cell] } c_t = f_t \odot c_{t-1} + i_t \odot c_t^* \tag{5}$$

$$\text{[Output gate] } o_t = \sigma(V_{go}x_t + W_{zo}h_{t-1} + b_o) \tag{6}$$

where x_t denotes the input, W_*, V_* are the weight matrices, b_* is the bias term. Finally, the hidden state h_t that comprises the output of LSTM memory cell is calculated by Eq. 7. When more than one LSTM layers are stacked together, the state of memory c_t and hidden state h_t of each LSTM layer are forwarded to next layer.

$$\text{[Hidden state] } h_t = o_t \odot \tanh(c_t) \tag{7}$$

Figure 1 shows an architecture of a LSTM cell with cell state, input, hidden states and input, output, and forget gates.

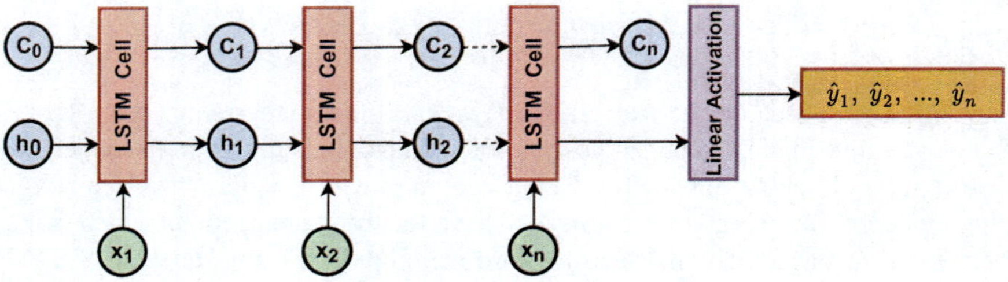

Fig. 1 Representation of LSTM with input, output, hidden, and memory units

3.3 Proposed Hybrid Model (CNN-LSTM)

A convolutional neural network (CNN) is a class of artificial neural network, proposed by LeCun et al. in 1989 [37], most commonly applied in 2D image data, to find patterns in images, recognize objects. CNNs have characteristics of paying attention to important features from the input. It uses a shared weight architecture called filters that slides along input features and produces a response as feature map. CNN learns spatial hierarchies of features using multiple building blocks such as convolution layer, pooling layer, and fully-connected (FC) layer.

Convolution layer—the kernels The convolution layer is the core building block of CNN, which contains multiple convolution kernels or filters. The convolution operation extracts the features and due to high dimension of the convolved features, a pooling layer is used to reduce the feature dimension and to extract the dominant features. A convolution is a linear mathematical operation, where the weights are multiplied with the input, and these weights are named to be filters or kernels. The filter performs the dot product or element-wise multiplication, filter applied to each overlapping part orderly left to right, top to bottom, and the results of kernel operations are stored in the feature map [38, 39]. Once the feature map is obtained, it is passed with a non-linear activation function like ReLU or Sigmoid. This convolution operation for a time series can be thought of a cross-correlation among different features.

Multiple filters and stacked layers Multiple filters in parallel are used to learn multiple features in parallel for a given input. For example, if 8 filters are used, then from a given input, 8 different sets of features are extracted, this diverseness empowers specialization and helps to extract salient features from multivariate inputs. Stacking of convolution layers can be leveraged for hierarchical decomposition of the multivariate input data [40]. Filters that operates on initial layers can learn the higher level feature, and the filters at deeper layers can learn the lower level features from the input.

Pooling layer Pooling layer is responsible for reducing the spatial size and information filtering by acquiring a concise static of the neighboring outputs, so the output feature map is passed to the pooling layer after each convolution operation [41].

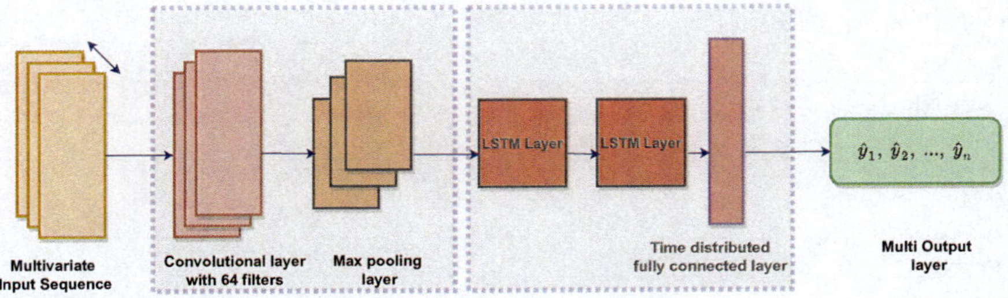

Fig. 2 Architecture of proposed hybrid model with different layers

In literature, there are many types of pooling layers: maximum pooling, average pooling, L2 norm of rectangular neighborhood, a weighted average based on the distance from the central pixel [38, 39]. Average pooling takes the average value, and maximum pooling takes the maximum value in the area.

In Eq. 8, x_t is the input, w_t is the weight of convolution filter, b_t is the bias, and conv_t represents the output value after convolution operation.

$$\text{conv}_t = \tanh(x_t * w_t + b_t) \tag{8}$$

Depending upon the complexity of data, a number of convolution and pooling layers can be used to capture both high level and lower level details even further.

4 Data Pre-processing and Feature Engineering

In this section, we will be discussing about different technical indicators used for our experiment and different feature selection procedures.

4.1 Dataset

To perform comparative study of models, experiments have been performed on five major indices from different parts of the world (listed in Table 1). Daily historical open, high, low, and close (OHLC) data have been considered from the period of January 2010 to December 2021. Since index price movements are ambivalent in nature and looking only the OHLC data, and predicting the closing price may or may not produce better results. So based on previous study [42, 43] and huge application in technical analysis [44, 45], 55 technical indicators and various oscillators have been computed and used as input features.

Table 1 Market indices

Symbol	Index name	Country
Nifty 50	NSE	India
DJIA	Dow Jones Industria Average	USA
IXIC	Nasdaq Composite	USA
INX	S&P 500	USA
NI225	Nikkei 225	Japan

4.2 Technical Indicators

Technical indicators have been selected based on popularity and the previous study [29, 30, 42] and suggestion from market experts. Few technical indicators based on historical prices that is used for this work are briefed here. Simple moving averages (SMA) with different time frame like SMA5, SMA10, SMA15, SMA20, exponential moving average (EMA) along with different time frames EMA5, EMA10, EMA15, EMA20, Kaufman's adaptive moving average (KAMA), KAMA10, KAMA20, KAMA30, stop or reversal (SAR), triangular moving average (TRIMA), TRIMA5, TRIMA10, TRIMA20, average directional index (ADX), absolute price oscillator (APO), commodity channel index (CCI), moving average convergence divergence (MACD), money flow index (MFI), momentum indicator (MOM), rate-of-change (ROC), percentage price oscillator (PPO), relative strength index (RSI) are considered.

4.3 Random Forest-Based Feature Importance

Along with OHLC data, we have added many technical indicators value to our input, as discussed in the previous section. But all the features may or may not have significant importance to our multi-day ahead price forecasting model. So the question always behind the scene—is "Which features should be used to create a good predictive machine learning model". More features include complexity to the model which leads to longer training time, harder to interpret, and may initiate noise in the model. Random forest-based feature importance is one of the ways of doing feature selection [46–48]. A random forest is a supervised model consisting of a set of decision trees and a bagging method. According to the pureness of the leaves, each individual tree calculates the importance of the feature, and the importance of the feature is decided by the higher value of increment in leave purity. This is calculated for every tree and further averaged and normalized, such that the importance scores are summed to be one.

Recursive feature elimination (RFE) Once we have got the importance of each individual feature, we perform a recursive feature elimination with the use of k-fold

Table 2 Number of RFE features

Dataset	Total features	RFE features
Nifty 50	64	35
DJIA	64	27
IXIC	64	31
INX	64	42
NI225	64	28

cross-validation to finalize our set of features. RFE starts with all the features in the training set and successively removes features until the desired number remains. Random forest algorithms have been used to rank the importance of the features, discarding the least important features and again fitting the model, this process is repeated until a desired number of features remain. Here, fivefold cross-validation is used with RFE to score different feature subsets and to select the best scoring collection of features.

For our experiments, original dataset has 4 features, open, high, low, and close. Along with that we have included 5 previous day close price and 55 technical indicators which in total are 64 features, and using random forest-based RFE, we have different number of features for different datasets shown in Table 2.

4.4 Scaling the Training Set

Since in our experiment along with OHLC data, we are also using various technical indicators, so all the independent variables are having features in different range. Through feature scaling, we have normalize the range of independent variables to same scale or fixed range. To avoid numerical difficulties during calculations of step size of the gradient descent, as the steps of the gradient descent are updated in the same rate for every features and to ensure the gradient descent moves smoothly toward minima, it is required to scale the input data before feeding in to the model. Other main advantage of feature scaling is to avoid dominating features having higher numeric ranges to the features with smaller numeric ranges. For our experiment, input features are normalized (also called min-max scaling), where feature values are rescaled to the range between 0 and 1 as shown in the Eq. 9, where x is original input, $x_{\min}$ and $x_{\max}$ are the minimum and maximum values of corresponding feature, and x' is the normalized value.

$$x' = \frac{x - x_{\min}}{x_{\max} - x_{\min}} \tag{9}$$

5 Proposed Price Forecasting Framework

The contribution of this research is the development of a multi-day ahead index price forecasting using RFE features and exploiting deep learning techniques. Having a deep learning model for multivariate time series data, it has to deal with two axis of difficulty, and it needs to learn the temporal relationship so as to grasp changes of values over time and spatial relationships to know how independent variables impact on one and all. Having said that, we experimented with ANNs and LSTMs which are good for sequential time series data, but a novel CNN with stacked LSTMs combined is capable of learning both spatial and temporal relationships from input features. CNN by nature acts like a trainable feature extractor for spatial signals, with different filters and later LSTM receives a sequence of high-level representation, and again, it learns the temporal relations to generate the desired outputs. Details about each individual models have been discussed in the methodology section.

Along with feature selection procedures having data from 5 global indices, we evaluate our deep learning models ANNs, LSTMs, and CNN-LSTMs for following experiments for 5-day ahead forecasting:

- Prediction using OHLC data.
- Prediction using a combination of OHLC data and technical indicators and oscillators.
- Prediction using most important features obtained from RFE methods.

5.1 Model Calibration

In our implementation, we utilized the versions of our proposed models. For CNN-LSTM, the convolution layers of 64 filters of size 3 followed by a pooling layer. A CNN layer has been used to extract the most prominent features, and then, this is passed to two stacked LSTM layers to learn the temporal relationship among input, followed by a time distributed dense layer to keep one-to-one relations on input and output, and the output is produced in the output layer as shown in Fig. 2.

For MLP and LSTM models, number of layers, number of neurons in each individual layer, batch size, and number of epochs were varied. Models have build with two, three, and four layers, and for number of nodes, we have experimented with 16, 32, 64, 128, and 256 number of nodes in each layers as shown in Table 3. Algorithms were calibrated using grid search techniques. For optimizer adaptive moment estimation (Adam) optimizer, which is an extension of stochastic gradient descent was used.

Table 3 Possible models for ANN and LSTM

Number of layers	Nodes in each layer	Possible model
2	16, 32, 64, 128, 256	[16, 16], [32, 32], [64, 64], [128, 128], [256, 256]
3	16, 32, 64, 128, 256	16, 16, 16], [32, 32, 32], [64, 64, 64], [128, 128, 128], [256, 256, 256]
4	16, 32, 64, 128, 256	16, 16, 16, 16], [32, 32, 32, 32], [64, 64, 64, 64], [128, 128, 128, 128], [256, 256, 256, 256]

5.2 Model Evaluation

Machine learning models cannot have results with cent percent accuracy; otherwise, the model is known to be a biased model. For 5-day ahead multistep forecasting, models were evaluated based on root mean squared error (RMSE) and R squared (R^2 score), which are defined in Eqs. 10 and 11. In Eq. 11, SSR is the sum squared regression which is the sum of the residuals squared, and SST is the total sum of squares, i.e., the distance of the data point away from the mean of all squared. A model with low RMSE and a high R^2 score is desired. R^2 score tells how well the model has performed, and its value ranges from 0 to 1, while mean squared error (MSE) states the average of the squared difference between actual and predicted values, and the root of that is RMSE.

$$\text{RMSE} = \sqrt{\frac{\sum_{t=1}^{T}(x_t - \hat{x}_t)^2}{T}} \tag{10}$$

$$R^2 = 1 - \frac{\text{SSR}}{\text{SST}} = 1 - \frac{\sum_{t=1}^{T}(x_t - \hat{x}_t)^2}{\sum_{t=1}^{T}(x_t - \bar{x}_t)^2} \tag{11}$$

6 Experimental Results

Comparison results of all experiments with RMSE and R^2 score for both train and test data are shown in Table 4 and Fig. 3.

By comparing ANNs and LSTMs, our proposed hybrid model has least RMSE and best R^2 score, followed by LSTMs due to inherent memory and gated architecture in LSTMs and CNNs along with stacked LSTMs incubates an extra leverage for being extracted the spatial information helps the model to learn better from input features. We also experimented with only OHLC data, OHLC data with all technical indicator and OHLC data with RFE features, and we observed that OHLC with all technical indicators has performed well in maximum cases, but when important features are extracted and along with OHLC data are provided to the models input, they actually

Table 4 Results for 5 d ahead index price forecasting in RMSE and R^2 Score for major global indices.

Indices	Nifty 50		NASDAQ		Dow Jones (DJI)		SP 500		Nikkei_225	
Methods for prediction	RMSE (train/test)	R^2 score (train/test)	RMSE (train/test)	R^2 score (train/test)	RMSE (train/test)	R^2 score (train/test)	RMSE (train/test)	R^2 score (train/test)	RMSE (train/test)	R^2 score (train/test)
ANN + OHLC	0.017/0.030	0.95/0.86	0.018/0.032	0.95/0.85	0.031/0.04	0.95/0.91	0.013/0.022	0.97/0.95	0.028/0.035	0.96/0.95
LSTM + OHLC	0.014/0.022	0.96/0.88	0.013/0.022	0.97/0.89	0.026/0.038	0.97/0.96	0.01/0.02	0.97/0.95	0.021/0.03	0.97/0.96
CNN + LSTM + OHLC	0.014/0.02	0.97/0.89	0.011/0.017	0.98/0.92	0.021/0.03	0.98/0.96	0.008/0.02	0.98/0.96	0.0017/0.028	0.098/0.96
ANN + Indicators	0.015/0.027	0.97/0.89	0.016/0.029	0.95/0.83	0.03/0.04	0.95/0.9	0.015/0.023	0.97/0.95	0.029/0.38	0.96/0.95
LSTM + Indicators	0.012/0.018	0.99/0.91	0.011/0.02	0.97/0.92	0.023/0.037	0.96/0.95	0.011/0.022	0.97/0.96	0.022/0.031	0.97/0.96
CNN + LSTM + Indicators	0.01/0.016	0.99/0.93	0.01/0.018	0.98/0.9	0.017/0.026	0.98/0.97	0.007/0.019	0.98/0.96	0.017/0.03	0.99/0.98
ANN + RFE	0.014/0.023	0.98/0.9	0.014/0.028	0.96/0.88	0.027/0.037	0.96/0.92	0.013/0.021	0.98/0.96	0.25/0.033	0.97/0.95
LSTM + RFE	0.011/0.013	0.99/0.93	0.01/0.017	0.98/0.93	0.02/0.033	0.98/0.94	0.007/0.018	0.98/0.96	0.019/0.028	0.98/0.97
CNN + LSTM + RFE	0.009/0.011	0.099/0.95	0.008/0.013	0.99/0.96	0.016/0.24	0.99/0.97	0.004/0.011	0.99/0.98	0.014/0.026	0.99/0.97

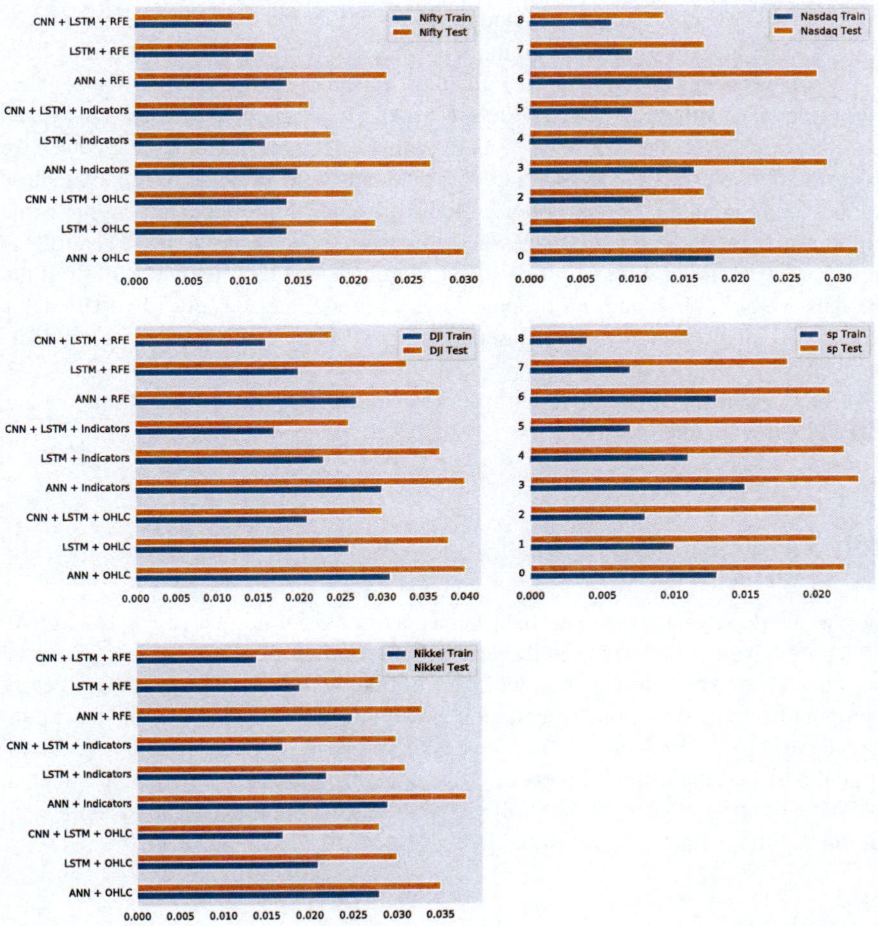

Fig. 3 RMSE for 5 major indices (Nifty 50, NASDAQ, Dow Jones, S&P 500, and Nikkei 225) for 5-day ahead forecasting

outperformed other results, hence this summarizes with all 55 technical indicators, there must be some indicators which are not much important for this 5 step ahead index price forecasting, and a good feature engineering is utmost important.

For multi-day ahead price forecasting, feature engineering plays a vital role. So along with OHLC data, we included 55 technical indicators, and also extensively performed experiments including all 55 indicators along with OHLC data and 5 previous day closing price, and also through recursive feature engineering, prominent features were selected. For every dataset, different number of features were picked by random forest-based RFE algorithm as shown in Table 2, and we observed through our experiment that, model with RFE features as input performed better than models having input features as either OHLC data or technical indicators.

For Nifty 50 index data, LSTM and CNN-LSTM, both the models have produced good RMSE and R^2 scores. When models were given only OHLC input, they have captured the trend well, but with RFE features, especially proposed hybrid model has performed best among all other models. For Dow Jones IA, S&P500 and Nikkei225 datasets, we have observed that models with OLHC and 55 indicator data have performed worse than models with OHLC data only. One possible reason may these models were not able to learn better with the presence of all indicators, rather some irrelevant features produce unnecessary adverse impact or act as noise, and those things were rectified with help of RFE algorithms. And with RFE features, all the models ANN, LSTM, and our proposed hybrid model have produced good results, and among all the models, our proposed hybrid model has outperformed other models in every other dataset too.

RMSE values have been plotted in Fig. 3, and RMSE along with R^2 scores has been detailed in Table 4.

6.1 Generalizability

For our work, every experiments have been performed 5 times, and mean RMSE and R^2 score have been reported in Table 4. While training our deep learning models, we also have taken care of the model over fittings with help of built-in Keras early stopping to stop training once the model performance stops improving on the validation dataset [49]. For 5-day ahead index price forecast, we have contemplate with 5 global indices from different parts of globe, and through our extensive experiments, we have observed that, with help of REF features, our proposed hybrid model can perform good result for other global indices too.

7 Conclusion and Future Work

While most of the experiments in the literature have shown promising results using only OHLC data for index price forecast for single-step price forecast, but multi-day ahead price forecast is much required for long-term investments and pair tradings. So in this work, we have performed experiments both with OHLC data and other indicators. Later, we found that arbitrarily adding indicators is not going to help to our prediction results, so we took help of random forest-based RFE algorithm, to eliminate trivial and insignificant features from our input, and later observed that RFE features worked better for every models in terms of RMSE and R^2 score.

The experimental outcomes proclaim that our proposed hybrid model has highest accuracy (least RMSE and best R^2 score) compared with ANNs and LSTMs. While LSTMs are good for time series, where it learns the temporal relationship, and CNNs are good for extracting the features, so with CNN-LSTMs, from a REF multivariate data, models have learnt both temporal pattern of the data points and spatial relationships among the input features. One more advantage with this model hybridization is convolutional neural nets can be parallelized, the weight sharing of CNN can lessen the number of parameters, and CNN is computationally cheaper than RNNs. So with our proposed model, we are getting better accuracy without compromising or increasing time complexity. The scope of this work can be extended to take decision in swing tradings, long-term investments and pair tradings. It is also worth noting that this framework can be diversified to other scientific areas of time series forecasting applications such as gold price forecasting, cryptocurrency, predicting rainfall, and even in network domains. In future, we may experiment more with feature engineering with help of auto-encoders and also try to include market sentiments or social media information like Twitter feeds to our multistep ahead index price forecasting.

Acknowledgements We would like to show our gratitude to Dr. Dileep A. D., Associate Professor School of Computing and Electrical Engineering, IIT Mandi and Dr. Manoj Thakur, Professor, School of Mathematical and Statistical Sciences, IIT Mandi, for sharing their pearls of wisdom with us during the course of this research.

References

1. Zhang GP (2003) Time series forecasting using a hybrid ARIMA and neural network model. Neurocomputing 50:159–175. https://doi.org/10.1016/S0925-2312(01)00702-0
2. Yegnanarayana B (2009) Artificial neural networks. PHI Learning Pvt. Ltd. https://www.google.co.in/books/edition/ARTIFICIAL_NEURAL_NETWORKS/RTtvUVU_xL4C
3. Jain AK, Mao J, Mohiuddin KM (1996) Artificial neural networks: a tutorial. Computer 29(3):31–44. https://doi.org/10.1109/2.485891
4. Sahoo D, Sood N, Rani U, Abraham G, Dutt V, Dileep AD (2020) Comparative analysis of multi-step time-series forecasting for network load dataset. In: 2020 11th International conference on computing, communication and networking technologies (ICCCNT). IEEE, pp 1–7. https://doi.org/10.1109/ICCCNT49239.2020.9225449
5. Yadav A, Jha CK, Sharan A (2020) Optimizing LSTM for time series prediction in Indian stock market. Procedia Comput Sci 167:2091–2100. https://doi.org/10.1016/j.procs.2020.03.257
6. Kim HY, Won CH (2018) Forecasting the volatility of stock price index: a hybrid model integrating LSTM with multiple GARCH-type models. Expert Syst Appl 103:25–37. https://doi.org/10.1016/j.eswa.2018.03.002
7. Khashei M, Hajirahimi Z (2019) A comparative study of series arima/mlp hybrid models for stock price forecasting. Commun Stat Simul Comput 48(9):2625–2640. https://doi.org/10.1080/03610918.2018.1458138
8. Tsang PM, Kwok P, Choy SO, Kwan R, Ng SC, Mak J, Wong TL (2007) Design and implementation of NN5 for Hong Kong stock price forecasting. Eng Appl Artif Intell 20(4):453–461. https://doi.org/10.1016/j.engappai.2006.10.002
9. Nikou M, Mansourfar G, Bagherzadeh J (2019) Stock price prediction using DEEP learning algorithm and its comparison with machine learning algorithms. Intell Syst Acc Finan Manag 26(4):164–174. https://doi.org/10.1002/isaf.1459
10. Wang Y, Liu Y, Wang M, Liu R (2018) LSTM model optimization on stock price forecasting. In: 2018 17th International symposium on distributed computing and applications for business engineering and science (DCABES). IEEE, pp 173–177. https://doi.org/10.1109/DCABES.2018.00052

11. Ding G, Qin L (2020) Study on the prediction of stock price based on the associated network model of LSTM. Int J Mach Learn Cybern 11(6):1307–1317. https://doi.org/10.1007/s13042-019-01041-1
12. Althelaya KA, El-Alfy ESM, Mohammed S (2018) Stock market forecast using multivariate analysis with bidirectional and stacked (LSTM, GRU). In: 2018 21st Saudi computer society national computer conference (NCC). IEEE, pp 1–7. https://doi.org/10.1109/NCG.2018.8593076
13. Dutta A, Kumar S, Basu M (2020) A gated recurrent unit approach to bitcoin price prediction. J Risk Finan Manag 13(2):23. https://doi.org/10.3390/jrfm13020023
14. Huang Y, Gao Y, Gan Y, Ye M (2021) A new financial data forecasting model using genetic algorithm and long short-term memory network. Neurocomputing 425:207–218. https://doi.org/10.1016/j.neucom.2020.04.086
15. Hadavandi E, Shavandi H, Ghanbari A (2010) Integration of genetic fuzzy systems and artificial neural networks for stock price forecasting. Knowl Based Syst 23(8):800–808. https://doi.org/10.1016/j.knosys.2010.05.004
16. Ramadhan NG, Atastina I (2021) Neural network on stock prediction using the stock prices feature and Indonesian financial news titles. Int J Inf Commun Technol (IJoICT) 7(1):54–63. https://doi.org/10.1007/s00521-019-04212-x
17. Mehtab S, Sen J (2020) Stock price prediction using convolutional neural networks on a multivariate timeseries. arXiv preprint arXiv:2001.09769. https://doi.org/10.48550/arXiv.2001.09769
18. Gao P, Zhang R, Yang X (2020) The application of stock index price prediction with neural network. Math Comput Appl 25(3):53. https://doi.org/10.3390/mca25030053
19. Taieb SB, Bontempi G, Atiya AF, Sorjamaa A (2012) A review and comparison of strategies for multi-step ahead time series forecasting based on the NN5 forecasting competition. Expert Syst Appl 39(8):7067–7083. https://doi.org/10.1016/j.eswa.2012.01.039
20. Cheng H, Tan PN, Gao J, Scripps J (2006) Multistep-ahead time series prediction. In: Pacific-Asia conference on knowledge discovery and data mining. Springer, Berlin, pp 765–774. https://doi.org/10.1007/11731139_89
21. Sorjamaa A, Hao J, Reyhani N, Ji Y, Lendasse A (2007) Methodology for long-term prediction of time series. Neurocomputing 70(16–18):2861–2869. https://doi.org/10.1016/j.neucom.2006.06.015
22. Hussein S, Chandra R, Sharma A (2016) Multi-step-ahead chaotic time series prediction using coevolutionary recurrent neural networks. In: 2016 IEEE Congress on evolutionary computation (CEC). IEEE, pp 3084–3091. https://doi.org/10.1109/CEC.2016.7744179
23. Alghamdi D, Alotaibi F, Rajgopal J (2021) A novel hybrid deep learning model for stock price forecasting. In: 2021 International joint conference on neural networks (IJCNN). IEEE, pp 1–8. https://doi.org/10.1109/IJCNN52387.2021.9533553
24. Sunny MAI, Maswood MMS, Alharbi AG (2020) Deep learning-based stock price prediction using LSTM and bi-directional LSTM model. In: 2020 2nd Novel intelligent and leading emerging sciences conference (NILES). IEEE, pp 87–92. https://doi.org/10.1109/NILES50944.2020.9257950
25. Liu G, Wang X (2018) A numerical-based attention method for stock market prediction with dual information. IEEE Access 7:7357–7367. https://doi.org/10.1109/ACCESS.2018.2886367
26. Fan C, Zhang Y, Pan Y, Li X, Zhang C, Yuan R, Wu D, Wang W, Pei J, Huang H (2019) Multi-horizon time series forecasting with temporal attention learning. In: Proceedings of the 25th ACM SIGKDD international conference on knowledge discovery & data mining, pp 2527–2535. https://doi.org/10.1145/3292500.3330662
27. Zhang H, Li S, Chen Y, Dai J, Yi Y (2022) A novel encoder-decoder model for multivariate time series forecasting. Comput Intell Neurosci. https://doi.org/10.1155/2022/5596676
28. Chen YC, Huang WC (2021) Constructing a stock-price forecast CNN model with gold and crude oil indicators. Appl Soft Comput 112:107760. https://doi.org/10.1016/j.asoc.2021.107760

29. Park HJ, Kim Y, Kim HY (2022) Stock market forecasting using a multi-task approach integrating long short-term memory and the random forest framework. Appl Soft Comput 114:108106. https://doi.org/10.1016/j.asoc.2021.108106
30. Song Y, Lee JW, Lee J (2019) A study on novel filtering and relationship between input-features and target-vectors in a deep learning model for stock price prediction. Appl Intell 49(3):897–911. https://doi.org/10.1007/s10489-018-1308-x
31. Yu L, Wang S, Lai KK (2009) A neural-network-based nonlinear metamodeling approach to financial time series forecasting. Appl Soft Comput 9(2):563–574. https://doi.org/10.1016/j.asoc.2008.08.001
32. Gao T, Chai Y (2018) Improving stock closing price prediction using recurrent neural network and technical indicators. Neural Comput 30(10):2833–2854. https://doi.org/10.1162/neco_a_01124
33. Wen Y, Lin P, Nie X (2020) Research of stock price prediction based on PCA-LSTM model. IOP Conf Ser Mater Sci Eng 790(1):012109. http://iopscience.iop.org/article/10.1088/1757-899X/790/1/012109/meta
34. Jain AK, Mao J, Mohiuddin KM (1996) Artificial neural networks: a tutorial. Computer 29(3):31–44. https://doi.org/10.1109/2.485891
35. Lettvin JY, Maturana HR, McCulloch WS, Pitts WH (1959) What the frog's eye tells the frog's brain. Proc IRE 47(11):1940–1951. https://doi.org/10.1109/JRPROC.1959.287207
36. Hochreiter S, Schmidhuber J (1997) Long short-term memory. Neural Comput 9(8):1735–1780. https://doi.org/10.1162/neco.1997.9.8.1735
37. LeCun Y, Boser B, Denker JS, Henderson D, Howard RE, Hubbard W, Jackel LD (1989) Backpropagation applied to handwritten zip code recognition. Neural Comput 1(4):541–551. https://doi.org/10.1162/neco.1989.1.4.541
38. Wu JMT, Li Z, Herencsar N, Vo B, Lin JCW (2021) A graph-based CNN-LSTM stock price prediction algorithm with leading indicators. Multimedia Syst 1–20. https://doi.org/10.1007/s00530-021-00758-w
39. Chandra R, Goyal S, Gupta R (2021) Evaluation of deep learning models for multi-step ahead time series prediction. IEEE Access 9:83105–83123. https://doi.org/10.1109/ACCESS.2021.3085085
40. Lu W, Li J, Li Y, Sun A, Wang J (2020) A CNN-LSTM-based model to forecast stock prices. Complexity. https://doi.org/10.1155/2020/6622927
41. Livieris IE, Pintelas E, Pintelas P (2020) A CNN-LSTM model for gold price time-series forecasting. Neural Comput Appl 32(23):17351–17360. https://doi.org/10.1007/s00521-020-04867-x
42. Kumar D, Meghwani SS, Thakur M (2016) Proximal support vector machine based hybrid prediction models for trend forecasting in financial markets. J Comput Sci 17:1–13. https://doi.org/10.1016/j.jocs.2016.07.006
43. Kim KJ (2003) Financial time series forecasting using support vector machines. Neurocomputing 55(1–2):307–319. https://doi.org/10.1016/S0925-2312(03)00372-2
44. Nti IK, Adekoya AF, Weyori BA (2020) A systematic review of fundamental and technical analysis of stock market predictions. Artif Intell Rev 53(4):3007–3057. https://doi.org/10.1007/s10462-019-09754-z
45. Huang CL, Tsai CY (2009) A hybrid SOFM-SVR with a filter-based feature selection for stock market forecasting. Expert Syst Appl 36(2):1529–1539. https://doi.org/10.1016/j.eswa.2007.11.062
46. Breiman L (2001) Random forests. Mach Learn 45(1):5–32. https://doi.org/10.1023/A:1010933404324
47. Genuer R, Poggi JM, Tuleau-Malot C (2010) Variable selection using random forests. Pattern Recogn Lett 31(14):2225–2236. https://doi.org/10.1016/j.patrec.2010.03.014
48. Strobl C, Boulesteix AL, Zeileis A, Hothorn T (2007) Bias in random forest variable importance measures: illustrations, sources and a solution. BMC Bioinf 8(1):1–21. https://doi.org/10.1186/1471-2105-8-25
49. Keras callback API—early stopping. https://keras.io/api/callbacks/. Last accessed 15 Apr 2022

Automatic Retinal Vessel Segmentation Using BTLBO

Chilukamari Rajesh and Sushil Kumar

Abstract The accuracy of retinal vessel segmentation (RVS) is crucial in assisting physicians in the ophthalmology diagnosis or other systemic diseases. However, manual segmentation needs a high level of knowledge, time-consuming, complex, and prone to errors. As a result, automatic vessel segmentation is required, which might be a significant technological breakthrough in the medical field. We proposed a novel strategy in this paper, that uses neural architecture search (NAS) to optimize a U-net architecture using a binary teaching learning-based optimization (BTLBO) evolutionary algorithm for RVS to increase vessel segmentation performance and reduce the workload of manually developing deep networks with limited computing resources. We used a publicly available DRIVE dataset to examine the proposed approach and showed that the discovered model generated by the proposed approach outperforms existing methods.

Keywords Retinal vessel segmentation · Binary teaching learning-based optimization · Neural architecture search · Convolutional neural network

1 Introduction

Retinal vessel segmentation (RVS) plays a crucial role in the field of medical image processing to detect the presence of pathological abnormalities in retinal blood vessels that might reflect diseases of ophthalmology and various systemic diseases like diabetes, arteriosclerosis, high blood pressure, etc. The fundus examination is now considered a regular clinical examination by ophthalmologists and other physicians [1]. The retinal vasculature tree has a complex structure with many tiny blood vessels and several interconnected blood vessels. The fundus images are prone to noise, uneven illumination, and the distinction between vascular zone and background is

Ch. Rajesh (✉) · S. Kumar
Department of Computer Science and Engineering, National Institute of Technology Warangal, Telangana 506004, India
e-mail: cr720084@student.nitw.ac.in

S. Kumar
e-mail: kumar.sushil@nitw.ac.in

M. Thakur et al. (eds.), *Soft Computing for Problem Solving*, Lecture Notes in Networks and Systems 547, https://doi.org/10.1007/978-981-19-6525-8_15

quite delicate. Therefore, segmenting the retinal vascular trees from fundus images has become a difficult task. Various manually developed neural network structures have been developed for RVS. In these, many of the existing neural network methods are unable to capture the vascular trees in complex fundus images. Consequently, in these situations, an automated optimized neural network model is required for more precise feature extraction from the complex vascular tree.

The objective of the proposed work is to apply NAS with a BTLBO evolutionary algorithm as an optimization method for RVS. Inspired by the encoder–decoder framework of U-net [2], we designed a search space with optimized neural network architecture to increase the RVS performance. Furthermore, we employed a non-redundant and fixed-length binary code encoding approach to represent the structure of the neural network for macro-architecture search. The teaching and learning phases of the BTLBO can produce more competitive neural network structures under the specified search space with limited computational resources during the evolution process as demonstrated in Fig. 1.

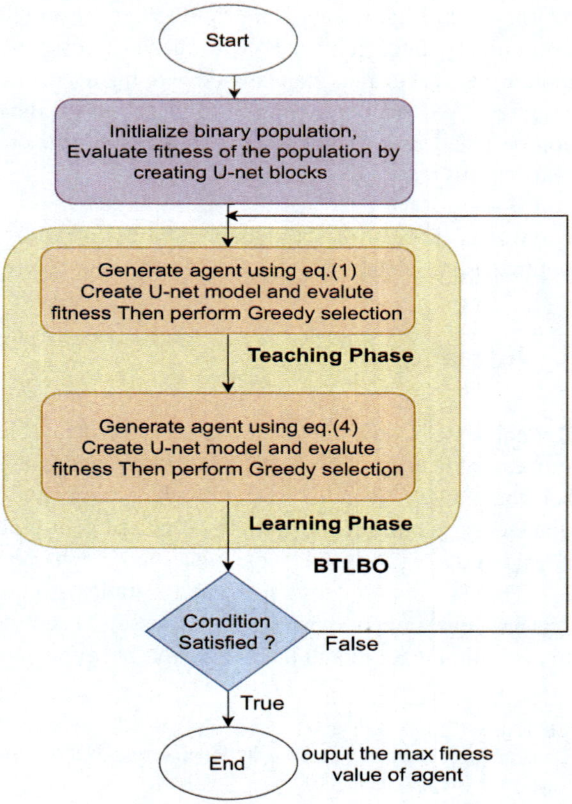

Fig. 1 Proposed approach flowchart

The contributions are:

- A novel automated U-net framework-based model has been proposed by using an evolutionary BTLBO algorithm for retinal vessel segmentation.
- A specific search space is designed to optimize the block structures in a U-net.
- The discovered U-net models significantly perform well on the public dataset DRIVE.

2 Related Work

2.1 Retinal Vessel Segmentation

Since the development of U-net [2] and FCN [3], image segmentation techniques based on fully convolutional neural networks have been popular due to their promising results. U-net varied neural network models have recently dominated new state-of-the-art models for RVS. Yan et al. [4] used a joint loss with pixel-wise and segment-level loss to provide supervision information for U-net. The joint loss can improve the model's capabilities by balancing the segmentation between thin and thick vessels. Jin et al. [5] designed DU-Net used instead of the traditional convolution to capture different vascular tree morphologies effectively. To make use of multi-scale features, a new inception-residual block with four supervision paths of varied convolution filter sizes for a U-shaped network was proposed by Wu et al. [6]. R2U-Net [7] was proposed to capture the features by using a recursive residual layer with cyclical convolutions. Dual U-net with two encoders, where one encoder extracts the spatial information and the other collects the context information, was proposed by Wang et al. [8]. Gu et al. [9] designed CE-Net to retain spatial information and capture more high-level information by using several convolution branches with distinct receptive fields for segmentation. To increase the model's hierarchical representation capture capabilities, a self-attention mechanism in the U-shape encoder–decoder was implemented by Mou et al. [10]. The networks mentioned above are notoriously complex and were developed manually. Therefore, some networks must be developed automatically to improve the feature extraction capacity. In this paper, we design the network structures automatically that can extract features effectively from fundus images.

2.2 Neural Architecture Search (NAS)

NAS is a powerful approach for assisting end-users to automatically create efficient deep neural networks. The NAS can be categorized into three types depending on the search method: reinforcement learning, evolutionary algorithms, and differential architectural search. NAS is defined as a Markov decision process by the reinforcement learning technique [11]. The performance of a model is used as reward feedback to the controller, which is used to sample the design neural network topologies to develop better structures through continuous trial and error. NAS is formulated as an optimization problem and encodes the structures by evolutionary algorithm [12]. Then perform some operations (such as mutation and crossover) on the neural network architecture to generate offspring. Then, continuously adjust the neural network structures through generations based on the survival of the fittest principle to get the optimized model. Weight coefficient is assigned to each cell operation in the differential architecture search [13]. Then, a gradient descent is used to update the neural network's parameter weight as well as the weight of each operation. After convergence, choose the operation with the highest weight to get the optimal model. NAS has applied and achieved good results in medical image segmentation [14–17]. In [15, 16], optimized the building blocks layer operations and hyperparameters, but the structure of the block was kept fixed. Some block type operations were optimized in [14, 17], then constructed architecture by stacking them repeatedly. Recently, Rajesh et al. [18] used differential evolution algorithm for finding optimal block structures for medical image denoising.

2.3 BTLBO

Metaheuristic algorithms have successfully tackled many optimization problems in the recent decades. TLBO was proposed by Rao et al. [19] which is a popular social-inspired algorithm and has been broadly applied in solving various optimization problems in a variety of domains and real-world applications [20]. Initially, TLBO was designed to solve problems of continuous optimization. By adjusting the operators to TLBO, we can solve binary optimization problems like feature selection problems. To design an array of plasmonic nanopyramids with the maximum absorption coefficient spectrum, a binary TLBO algorithm was developed by Akhlaghi et al. [21]. Later, the approach was extended to optical applications like plasmonic nanoantennas [22]. Furthermore, Sevinç and Dökerolu [23] presented TLBO-ELM, which is extreme learning machines (ELM) with a TLBO feature selection technique.

3 Methodology

In this section, we have given brief information on U-shaped architecture. Then, we have explained the proposed search space of node sequence operations, designing U-net block inter-node connections from the binary agent that BTLBO generates, and finally BTLBO algorithm.

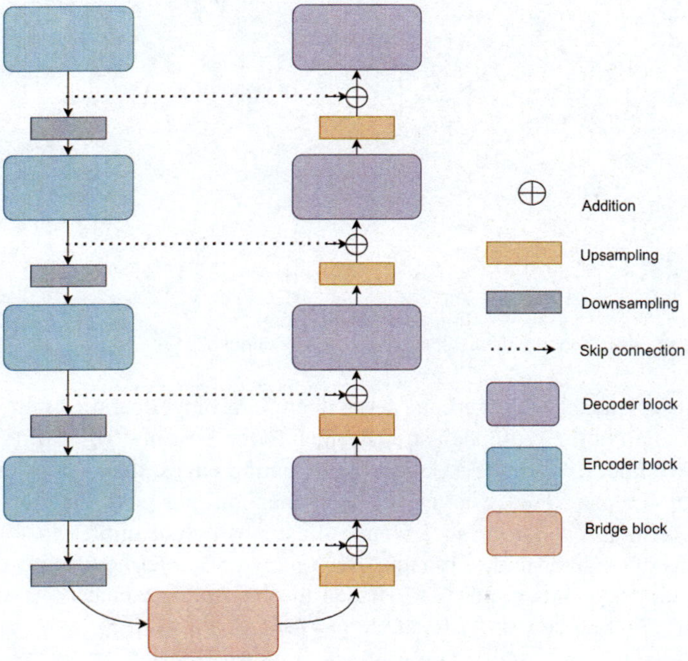

Fig. 2 U-shaped architecture

3.1 U-net

In medical image segmentation, U-shaped neural network frameworks are quite a popular choice due to their remarkable performance and transferability. The U-shaped neural network's design contains encoders (down-sampling) and decoders (up-sampling). Features of the image at various scales are collected by encoders, while the decoders restore the encoder's extracted features to the size of an original image and classify each pixel in an original image. Therefore, the proposed model also adopted a U-shaped architecture with one bridge block, four encoder blocks, and four decoder blocks, as shown in Fig. 2.

3.2 Search Space and Encoding

Manually designed U-net variants [7, 9, 24] typically enhance the performance by changing the internal structure of the block (such as InceptionNet block [25], DenseNet block [26], and ResNet block [27]). Whereas, in our work, we have considered each block as a directed acyclic network with nodes and edges, which is similar to genetic CNN [12]. Here, the connections between nodes are indicated by

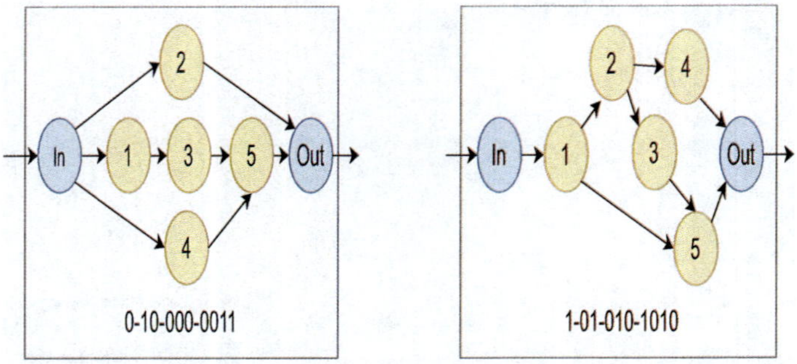

Fig. 3 Examples of encoding the block inter-node connections

edges, and each node can represent a set of operations or an operation unit. The output feature map of the pre-node is transformed to the post-node by a directed edge between two nodes. If a node has multiple incoming edges, then its corresponding feature maps can be added. The inter-node connections are represented by a binary encoding string as shown in Fig. 3, which illustrates two example block encoding connections of the inter-nodes. If a maximum number of intermediate nodes are M, then the number of bits required to encode the inter-node connections is $\frac{M(M+1)}{2}$ $(1 + 2 + 3 + \ldots + (M + 1) = \frac{M(M+1)}{2})$. The proposed model considers 5 nodes, so we require 10 bits to build the network structure as shown in Fig. 3. The first bit represents the link between (node1, node2), the next two represents the link between (node1, node3) and (node2, node3), and so on. The two nodes are connected if the corresponding bit is 1. The input node (represented as "In") takes input from the previous pooling layer and transfers it to the successor nodes. The output node (represented as "Out") takes input from the predecessor nodes and forwards it to the next pooling layer in a U-net.

As illustrated in Table 1, we have considered thirty-two operation sequences, and our main aim is to search for the most effective operation sequences of the optimal U-net block structures for retinal vessel segmentation. Every operation sequence can have distinct ID and contains several basic operational units (convolutional kernel size such as conv (1 * 1), conv (3 * 3), and conv (5 * 5), activation functions (pre-activation or post-activation), such as ReLU [28], and Mish [29], normalization such as instance normalization (IN) [30], and batch normalization (BN) [31]). As a result, the operation sequence is represented using five-bit binary encoding (to represent thirty-two alternative operation sequences). ReLU is a nonlinear activation function used in many tasks due to its performance. Mish is a recently proposed activation function that resembles ReLU but adds continuous differentiability, non-monotonicity, and other features and outperforms ReLU in various applications. As a result, both are accounted for in the search space. The U-net architecture is made up of nine blocks; each block needs 15 bits (the first 5 bits for getting the operation ID from the table 1 and the remaining 10 bits for building the structure of the network as shown in Fig. 3). The designed search space is quite adaptable, where each block can have distinct structures and operations.

Table 1 Node sequence operations

ID	Sequence of operation	ID	Sequence of operation
0	"conv 1 * 1 → ReLU"	16	"ReLU → conv 1 * 1"
1	"conv 1 * 1 → Mish"	17	"Mish → conv 1 * 1"
2	"conv 1 * 1 → IN"	18	"IN → conv 1 * 1"
3	"conv 1 * 1 → BN"	19	"BN → conv 1 * 1"
4	"conv 3 * 3 → ReLU"	20	"ReLU → conv 3 * 3"
5	"conv 3 * 3 → Mish"	21	"Mish → conv 3 * 3"
6	"conv 3 * 3 → IN → ReLU"	22	"IN → ReLU → conv 3 * 3"
7	"conv 3 * 3 → IN → Mish"	23	"IN → Mish → conv 3 * 3"
8	"conv 3 * 3 → BN → ReLU"	24	"BN → ReLU → conv 3 * 3"
9	"conv 3 * 3 → BN → Mish"	25	"BN → Mish → conv 3 * 3"
10	"conv 5 * 5 → ReLU"	26	"ReLU → conv 5 * 5"
11	"conv 5 * 5 → Mish"	27	"Mish → conv 5 * 5"
12	"conv 5 * 5 → IN → ReLU"	28	"IN → ReLU → conv 5 * 5"
13	"conv 5 * 5 → IN → Mish"	29	"IN → Mish → conv 5 * 5"
14	"conv 5 * 5 → BN → ReLU	30	"BN → ReLU → conv 5 * 5"
15	"conv 5 * 5 → BN → Mish"	31	"BN → Mish → conv 5 * 5"

3.3 BTLBO

A TLBO is a popular and effective metaheuristic technique representing a global solution as a population. It is modeled around a student's learning process in the classroom. The goal is to design an effective segmentation model by selecting the optimal node structures. This approach is used to choose the optimal U-net blocks for segmenting retinal vessel images with a maximum f1 score. The advantage of TLBO is that it needs only two tuning parameters where one is the population size, and the other is a number of iterations that can be used as stopping criteria. The teacher phase and the learner phase are the two key parts of this algorithm.

3.3.1 Teacher Phase

This phase entails gaining knowledge from the teacher. Generally, a teacher's role is to improve learners' skills based on their comprehending capability and the teaching method. All learners in a classroom are considered as a population which is represented as X. In our case, population size (N) initialized to 20, i.e., the number of learners in a classroom. The size of each learner (subject) is 145 (9 blocks, and each block takes 15 blocks). We have chosen the fitness function as F1 score Eq. 7. The proposed model used a combination of dice loss and binary cross-entropy as a loss function [32].

In the teacher phase, the most knowledgeable learner is designated as the teacher (T) in the classroom for each iteration. The teacher tries to improve the class's mean result by bringing other learners' knowledge up to his/her level, depending on their capabilities. The class mean position can be determined by calculating the mean score of all learners, which is denoted as X_{mean}. The teacher instructs other learners by the following Eq. 1

$$X_i^t = X_i^t + (r * (T - (T_f * X_{\text{mean}}))) \tag{1}$$

where $i = 1, 2, 3 \ldots N$, where N is a classroom learners size. t is an iteration number; in our case, t is chosen as 10. A random number r has been chosen from 0 to 1. The teaching factor (T_f) value can be either 1 or 2 chosen randomly for each iteration. X_{mean} is the mean value of learners' scores for a certain subject. Now, we evaluate the fitness value of the evaluated new teacher phase X_i^t by converting it to a binary agent using Eq. 2, 3. Then performs greedy selection between calculated teacher phase X_i^t and old X_i^t, whichever has the highest fitness value that will add to population X.

$$X_i^t = \begin{cases} 1, & \text{Sigmoid}(X_i^t) \geq \text{rand}(0, 1) \\ 0, & \text{otherwise} \end{cases} \tag{2}$$

$$\text{Sigmoid}(x) = \frac{1}{1 + e^{-x}} \tag{3}$$

3.3.2 Learner Phase

In this phase, a learner can learn from the other learners by interacting with them. A learner X_i tries to improve his knowledge by randomly choosing partner X_p from the population where $i \neq p$, and $i, p \in [1, .., N]$ based on their fitness value as given in Eq. 4

$$X_i^t = \begin{cases} X_i^t + r * (X_i^t - X_p^t), & f(i) \leq f(p) \\ X_i^t - r * (X_i^t - X_p^t), & \text{otherwise} \end{cases} \tag{4}$$

where a random number r has been chosen in the range of 0 and 1. Then, we evaluate the fitness value by converting the calculated learner phase X_i^t to a binary agent using Eq. 2, 3. In greedy selection, the calculated learner phase X_i^t and teacher phase X_i^t can be compared; the highest fitness valued learner will add to the population X.

4 Experiments and Results

4.1 Dataset

In our experiments, we used a publicly available DRIVE dataset. Total 40 colored fundus images of resolution 565 * 584 are present in the DRIVE dataset. The dataset is divided into a training set and a test set of size 20 images. To avoid overfitting, we have used data augmentation techniques such as vertical flipping, random horizontal, and as well as random rotation with the range between [−180°, 180°], as we have a small training dataset. Furthermore, we crop the original images to 512 * 512 resolution and give them as input to the models.

4.2 Metrics

Retina vessel segmentation is a binary classification problem that predicts whether a pixel is positive (vessel) or negative (non-vessel) in retina vessel images. The five metrics we have chosen as evaluation metrics which are given in Eqs. 5, 6, 7, 8, and 9. All of these measurements are calculated using true negative (TN), true positive (TP), false negative (FN), and false positive (FP). The global threshold value is set to 0.5 in our work, while computing FP, TP, TN, and FN values.

$$\text{Accuracy} = \frac{\text{TN} + \text{TP}}{\text{TN} + \text{FN} + \text{FP} + \text{TP}} \tag{5}$$

$$\text{Specificity} = \frac{\text{TN}}{(\text{TN} + \text{FP})} \tag{6}$$

$$\text{F1 Score} = \frac{\text{TP} * 2}{\text{FP} + \text{FN} + (\text{TP} * 2)} \tag{7}$$

$$\text{Sensitivity} = \frac{\text{TP}}{(\text{FN} + \text{TP})} \tag{8}$$

$$\text{Precision} = \frac{\text{TP}}{(\text{TP} + \text{FP})} \tag{9}$$

4.3 Results

We are comparing the results of the discovered architecture using proposed approach with the U-net [2], and FC-Densenet [33], which are also U-shaped networks. We trained the U-net, FC-Densenet, and our model in the same environment on DRIVE dataset. Table 2 gives that the discovered architecture outperforms U-net and FC-Densenet in terms of Accuracy, Sensitivity, and F1 Score. However, the discovered architecture outperforms U-net and closely comparable to FC-Densenet in terms of Specificity and Precision. Figure 4 shows examples of the resulting segmentation output images.

Table 2 Comparing results with the existing models

Model	Accuracy	Sensitivity	Specificity	F1 score	Precision
U-net	0.9550	0.7715	0.9789	0.7950	0.8242
FC-Densenet	0.9588	0.7485	**0.9860**	0.8141	**0.8734**
Proposed	**0.9617**	**0.8029**	0.9825	**0.8259**	0.8546

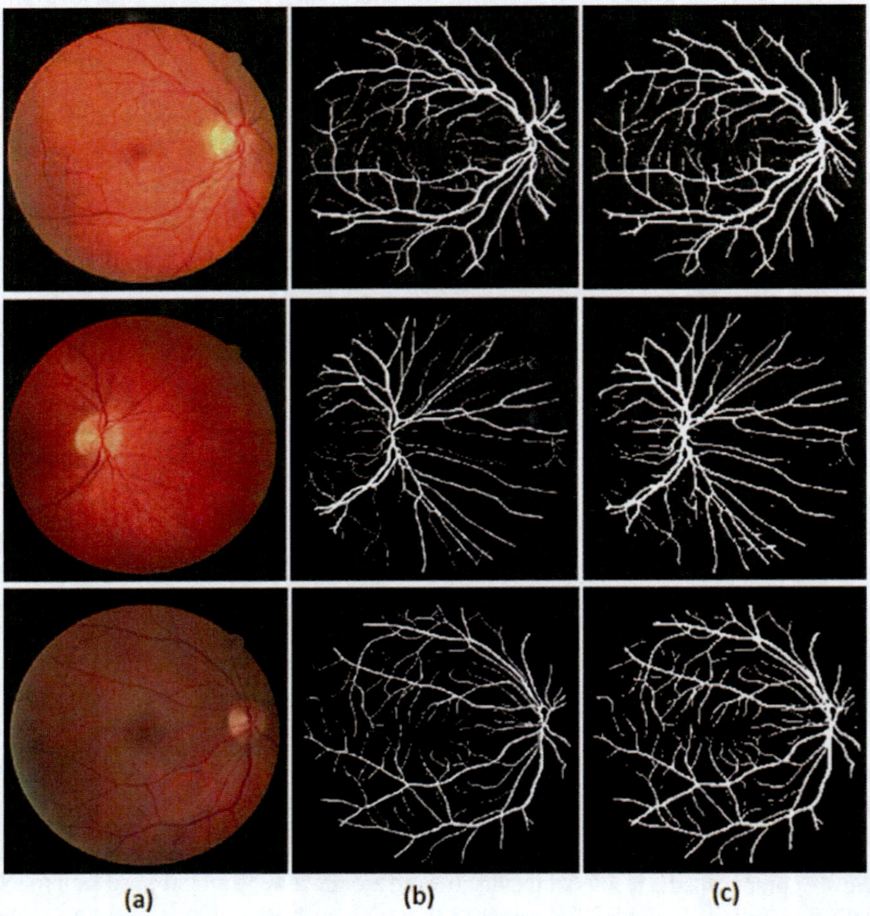

Fig. 4 Segmentation results of proposed model on the DRIVE dataset. **a** Original images, **b** Groundtruth mask, **c** Predicted mask

5 Conclusion

In this paper, we adopt a binary teaching learning-based optimization evolutionary strategy for finding the optimal blocks in a U-net framework based on the proposed specific search space for retinal vessel segmentation. The proposed automated model can capture the more complicated vascular trees features from the fundus images and produce optimal segmentation results than the other manually designed models. In the future, the proposed model can be transferred to other vessel segmentation datasets to show the model's potential in clinical applications.

References

1. Chatziralli IP, Kanonidou ED, Keryttopoulos P, Dimitriadis P, Papazisis LE (2012) The value of fundoscopy in general practice. Open Ophthalmol J 6:4
2. Ronneberger O, Fischer P, Brox T (2015) U-net: convolutional networks for biomedical image segmentation. International conference on medical image computing and computer-assisted intervention. Springer, Cham, pp 234–241
3. Long J, Shelhamer E, Darrell T (2015) Fully convolutional networks for semantic segmentation. In: Proceedings of the IEEE conference on computer vision and pattern recognition, pp 3431–3440
4. Yan Z, Yang X, Cheng KT (2018) Joint segment-level and pixel-wise losses for deep learning based retinal vessel segmentation. IEEE Trans Biomed Eng 65(9):1912–1923
5. Jin Q, Meng Z, Pham TD, Chen Q, Wei L, Su R (2019) DUNet: a deformable network for retinal vessel segmentation. Knowl-Based Syst 178:149–162
6. Wu Y, Xia Y, Song Y, Zhang D, Liu D, Zhang C, Cai W (2019) Vessel-Net: retinal vessel segmentation under multi-path supervision. International conference on medical image computing and computer-assisted intervention. Springer, Cham, pp 264–272
7. Alom MZ, Yakopcic C, Hasan M, Taha TM, Asari VK (2019) Recurrent residual U-Net for medical image segmentation. J Med Imag 6(1):014006
8. Wang B, Qiu S, He H (2019) Dual encoding u-net for retinal vessel segmentation. International conference on medical image computing and computer-assisted intervention. Springer, Cham, pp 84–92
9. Gu Z, Cheng J, Fu H, Zhou K, Hao H, Zhao Y, Zhang T, Gao S, Liu J (2019) Ce-net: context encoder network for 2d medical image segmentation. IEEE Trans Med Imag 38(10):2281–2292
10. Mou L, Zhao Y, Fu H, Liu Y, Cheng J, Zheng Y, Su P, Yang J, Chen L, Frangi AF, Akiba M (2021) CS2-Net: deep learning segmentation of curvilinear structures in medical imaging. Med Image Anal 67:101874
11. Zoph B, Le QV (2016) Neural architecture search with reinforcement learning. ArXiv preprint arXiv:1611.01578
12. Xie L, Yuille A (2017) Genetic CNN. In: Proceedings of the IEEE international conference on computer vision, pp 1379–1388
13. Liu H, Simonyan K, Yang Y (2018) Darts: differentiable architecture search. ArXiv preprint arXiv:1806.09055
14. Weng Y, Zhou T, Li Y, Qiu X (2019) Nas-unet: neural architecture search for medical image segmentation. IEEE Access 7:44247–44257
15. Mortazi A, Bagci U (2018) Automatically designing CNN architectures for medical image segmentation. International workshop on machine learning in medical imaging. Springer, Cham, pp 98–106

16. Zhu Z, Liu C, Yang D, Yuille A, Xu D (2019) V-nas: neural architecture search for volumetric medical image segmentation. In: 2019 International conference on 3d vision (3DV). IEEE, pp 240–248
17. Kim S, Kim I, Lim S, Baek W, Kim C, Cho H, Yoon B, Kim T (2019) Scalable neural architecture search for 3d medical image segmentation. International conference on medical image computing and computer-assisted intervention. Springer, Cham, pp 220–228
18. Rajesh Ch, Kumar S (2022) An evolutionary block based network for medical image denoising using Differential Evolution. Appl Soft Comput 121:108776
19. Rao RV, Savsani VJ, Vakharia DP (2011) Teaching-learning-based optimization: a novel method for constrained mechanical design optimization problems. Comput-Aided Des 43(3):303–315
20. Zou F, Chen D, Xu Q (2019) A survey of teaching-learning-based optimization. Neurocomputing 335:366–383
21. Akhlaghi M, Emami F, Nozhat N (2014) Binary TLBO algorithm assisted for designing plasmonic nano bi-pyramids-based absorption coefficient. J Mod Opt 61(13):1092–1096
22. Kaboli M, Akhlaghi M (2016) Binary teaching-learning-based optimization algorithm is used to investigate the super scattering plasmonic Nano disk. Opt Spectrosc 120(6):958–963
23. Sevinc E, DÖKEROGLU T, (2019) A novel hybrid teaching-learning-based optimization algorithm for the classification of data by using extreme learning machines. Turkish J Electr Eng Comput Sci 27(2):1523–1533
24. Guan S, Khan AA, Sikdar S, Chitnis PV (2019) Fully dense UNet for 2-D sparse photoacoustic tomography artifact removal. IEEE J Biomed Health Inf 24(2):568–576
25. Szegedy C, Liu W, Jia Y, Sermanet P, Reed S, Anguelov D, Erhan D, Vanhoucke V, Rabinovich A (2015) Going deeper with convolutions. In: Proceedings of the IEEE conference on computer vision and pattern recognition, pp 1–9
26. Huang G, Liu Z, Van Der Maaten L, Weinberger KQ (2017) Densely connected convolutional networks. In: Proceedings of the IEEE conference on computer vision and pattern recognition, pp 4700–4708
27. He K, Zhang X, Ren S, Sun J (2016) Deep residual learning for image recognition. In: Proceedings of the IEEE conference on computer vision and pattern recognition, pp 770–778
28. Nair V, Hinton GE (2010) Rectified linear units improve restricted Boltzmann machines, In Icml
29. Misra D (2019) Mish: a self regularized non-monotonic activation function. ArXiv preprint arXiv:1908.08681
30. Ulyanov D, Vedaldi A, Lempitsky V (2016) Instance normalization: the missing ingredient for fast stylization. ArXiv preprint arXiv:1607.08022 (2016)
31. Ioffe S, Szegedy C (2015) Batch normalization: accelerating deep network training by reducing internal covariate shift. In: International conference on machine learning. PMLR, pp 448–456
32. Zhang M, Li W, Chen D (2019) Blood vessel segmentation in fundus images based on improved loss function. In: 2019 Chinese automation congress (CAC). IEEE, pp 4017–4021
33. Jégou S, Drozdzal M, Vazquez D, Romero A, Bengio Y (2017) The one hundred layers tiramisu: fully convolutional densenets for semantic segmentation. In: Proceedings of the IEEE conference on computer vision and pattern recognition workshops, pp 11–19

Exploring the Relationship Between Learning Rate, Batch Size, and Epochs in Deep Learning: An Experimental Study

Sadaf Shafi and Assif Assad

Abstract Deep learning has promised us great outcomes when enough data are fed to it. Deep learning is a branch of artificial intelligence which employs artificial neural networks to learn. The quality of the performance of these ANNs majorly depends on the data fed to it, architecture of the ANN and hyperparameters. The hyperparameters are the parameters whose values control the process of learning, which in turn controls the performance of ANNs. These hyperparameters are assigned different values usually using hit and trial methods. Hyperparameters such as learning rate, batch size, and epochs are assigned some values independent of each other before training the ANN model. In this study, we introduce a novel method of allowing the learning rate to be a function of batch size and epoch, thereby reducing the number of hyperparameters to be tuned. We later on introduce some randomness to the learning rate to see the effects on accuracy. It was found that the proposed strategy helped increase accuracy by more than 2% in certain cases, when compared to the existing methods.

Keywords Learning rate · Batch size · Random learning rate

1 Introduction

Deep learning is a branch of artificial intelligence which employs artificial neural networks to learn the patterns from the data fed to it. Deep learning is widely used in the fields of NLP, data mining, computer vision, etc.

The most difficult part after data collection and cleaning is training these neural networks in a way that they generalize well. Therefore, the learning process is controlled in a way that DL model learns the best out of the data fed to it. The learning process is controlled by tuning some variables called hyperparameters. Some of the

S. Shafi (✉) · A. Assad
Islamic University of Science and Technology, Awantipora, J&K, India
e-mail: Sadaf.shafi@islamicuniversity.edu.in

A. Assad
e-mail: assif.assad@islamicuniversity.edu.in

M. Thakur et al. (eds.), *Soft Computing for Problem Solving*, Lecture Notes in Networks and Systems 547, https://doi.org/10.1007/978-981-19-6525-8_16

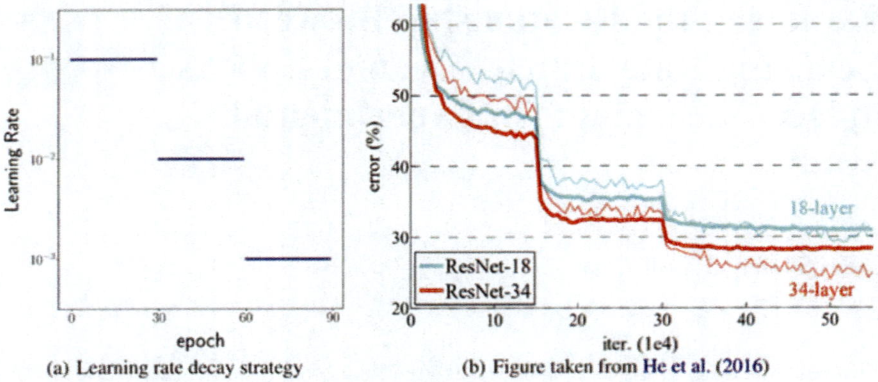

Fig. 1 **a** Shows step-wise learning rate decay and **b** shows error rate against it [4, 15, 16]

hyperparameters are batch size, learning rate, and epoch. Batch size is the number of training examples utilized in one iteration. Learning rate is the size of the step taken in the optimization algorithm in each iteration. Epochs on the other hand are the number of passes of the entire training dataset the machine learning algorithm has completed.

In all the studies conducted up until now, it was seen that these three hyperparameters are treated as independent variables, where learning rate is usually decayed [1, 9, 10] and batch size kept constant in most of the cases and so is the epoch. While training by this method, having bigger batch sizes is usually preferred in order to have only fewer parameter updates. Also, when we increase batch size, test accuracy seems to decrease [2, 11–13] (Fig. 1).

It is also shown that on increasing the batch size while keeping the learning rate constant, model accuracy comes out to be the way it would have been if batch size was constant, and learning rate was decaying [5, 14, 17, 18].

It has also been observed in the deep learning practitioners' community that the learning rate is almost always chosen without considering its effect on batch size and vice versa; however, the learning rate is chosen while keeping the number of epochs in view; this is accomplished usually through intuition of the practitioner. For example, if the learning rate is to be decayed, it'd decay in a way that in the last epoch it is very small, for example, up to 0.00001. Usually, this decay is designed using some mathematical relation of which the epoch is a part. This method has been proved to be one of the most effective methods and simple to implement.

However, this practice of setting one or more hyperparameters as the function of other ones has not been a common practice at all. Hence, this study explores the possibility of optimizing the model accuracy by choosing the hyperparameters a bit differently, i.e., making them a function of one another and then in another method subjecting the decaying learning rate to some randomness.

In this study, we explore the area where these hyperparameters can be treated as dependent variables and record the influence of the synergy on the generalization ability of the DL model. In this manuscript, we build a synergy between the mentioned hyperparameters, namely batch size, learning rate, and epoch, in which learning rate is a function of batch size and epochs. After conducting experiments on a wide array of datasets, it was observed the training and validation accuracy increased when compared to the commonly practiced methods of training an ANN.

The learning rate is also subjected to some randomness, and as a result, increase in accuracy is observed in a few cases.

Another advantage of the proposed strategy is that only two hyperparameters need to be tuned, i.e., batch size and epoch; as a result, learning rate would be set to the required values by the synergy itself.

The source code is also given along with this manuscript.

2 Proposed Methodology

This study proposes a novel method of fine-tuning the following hyperparameters, learning rate, batch size, and epochs, where learning rate is a function of batch size and epochs. Learning rate is directly proportional to batch size and inversely proportional to epochs. The synergy used is as follows.

$$\mathrm{Lr} = (B/(E+1)C_1 * C_2)/C_3 \tag{1}$$

where

Lr = learning rate
B = batch size, which goes as. 5, 10, 15, 20, and so on
E = epoch, which goes as 1, 2, 3, and so on
where c_1, c_2, c_3 are constants and are set to values such that Lr decays.

The values we used for the mentioned constants are 3/2, 80, 8, respectively.

The learning rate we got as a result of this synergy with the mentioned values of the constants is shown in Fig. 2.

The other synergy used to introduce randomness to the decreasing learning rate is as follows.

$$\mathrm{Lr} = R * E/C_1 \tag{2}$$

where

Lr = learning rate
R = random number
E = epoch, which goes as 1, 2, 3, and so on
where C_1 we used to be 10,000.

The learning rate obtained using (2) is shown in Fig. 3.

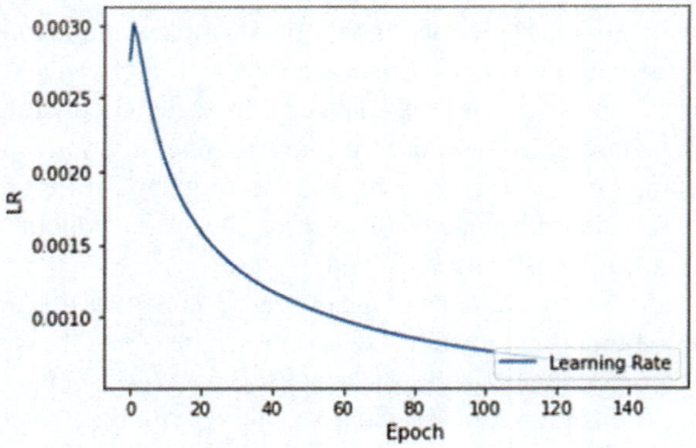

Fig. 2 Learning rate according to (1)

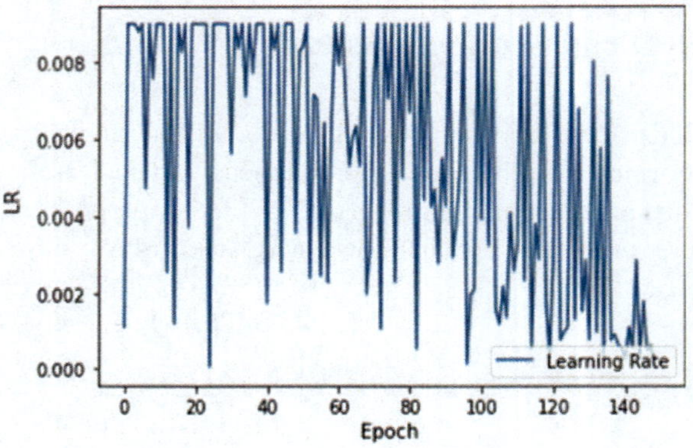

Fig. 3 Learning rate according to (2)

3 Datasets

The datasets utilized for the experimentation of the proposed method are as follows:

1. Audio

   a. **Speech Commands Dataset**

   This dataset hosts audio recordings of 10 commands in different voices which are as follows: "go," "up," "stop," "yes," "right," "no," "down," "left."

2. Structured Data

   a. **Pet Finder Dataset**

   This dataset includes tabular entries about the animals, like their color, breed, and age.

3. Image

   a. **MNIST**

MNIST ("Modified National Institute of Standards and Technology") is the de facto "hello world" dataset of computer vision.

This dataset is buildup of images of handwritten digits from 0 to 9.

b. **Fashion-MNIST**

Fashion-MNIST is a dataset of Zalando's article images—consisting of a training set of 60,000 examples and a test set of 10,000 examples.

c. **Cats and Dogs**

It contains 25,000 images of dogs and cats.

d. **Cifar10**

The CIFAR-10 dataset consists of 60,000 32 × 32 color images in 10 classes, with 6000 images per class. There are 50,000 training images and 10,000 test images.

4 Results

4.1 Using the Proposed Synergy Between Learning Rate, Batch Size, and Epochs

In order to test the performance of the model, it was trained by the proposed strategy, i.e., using (1) to choose the values of learning rate, batch size, and epoch. Then, the results were compared with widely practiced methods of setting these hyperparameters which are, setting learning rate and batch size to any constant value, setting batch size to any constant value, i.e., changing learning rate as epochs increase, setting learning rate to any constant value, i.e., changing batch size as epochs increase. In this subsection, we produce the results of the proposed series of experiments on Audio dataset and Pet Finder dataset, which is of structured numeric type.

The following plot (Fig. 4) is the comparison of the proposed strategy with three widely practiced strategies which are.

a. Keeping learning rate and batch size constant.
b. Keeping learning rate constant.
c. Keeping batch size constant.
d. Keeping learning rate proportional to batch Size and inversely proportional to epochs, the proposed strategy.

The plot in Fig. 4 shows the experiment on audio data. The model used in this experiment had 2 convolution layers, a max pooling layer, 2 dropouts, and 2 dense layers. After training in all the mentioned methods, the proposed method increased the validation accuracy by more than 2%, which is evident in the following plot.

The following plot in Fig. 5 corresponds to the experiments on structured data Pet Finder dataset; the accuracy in the proposed strategy is greater than the rest of the strategies.

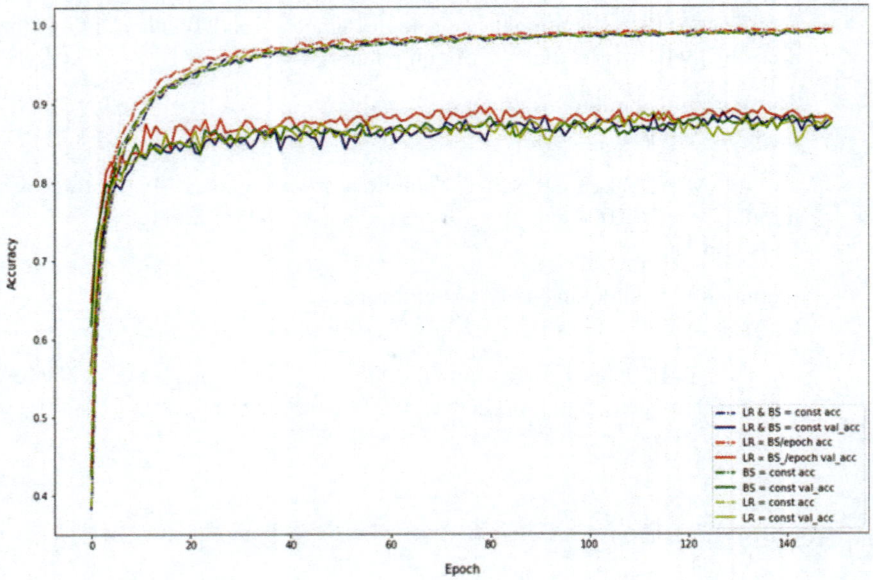

Fig. 4 Training and validation accuracy of the models trained by values of learning rate and batch size on audio dataset using (1)

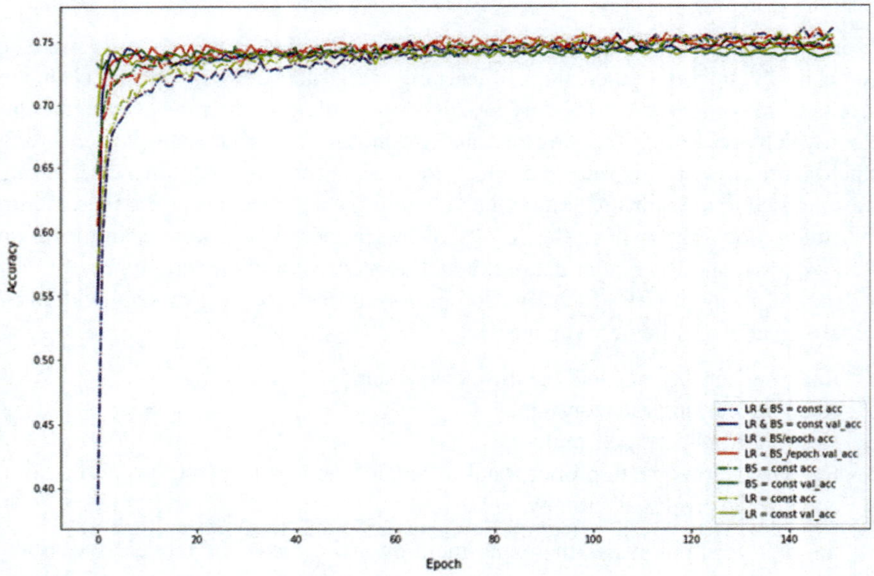

Fig. 5 Training and validation accuracy of the models trained by values of learning rate and batch size on Pet Finder dataset using (1)

4.2 *Introducing Some Randomness in Learning Rate*

In this subsection, we explore the utilization of (2) for choosing our learning rate, which is random decay, i.e., with the increasing epochs, the learning rate over all decays, but in each epoch, the learning rate is multiplied with a random number as is evident in Fig. 2. The outcome is then compared with three widely practiced methods of setting values to learning rate and batch size. These widely practiced methods to which the proposed strategy was compared to areas, setting learning rate and batch size to any constant value, setting batch size to any constant value, i.e., changing learning rate as epochs increases, setting learning rate to any constant value, i.e., changing batch size as epochs increase. All the mentioned datasets were used for this experiment with a simple architecture given in the source code; the results shared here correspond to the Pet Finder dataset which is a structured numeric dataset. The plot in Fig. 5 corresponds to the experiments on structured data Pet Finder dataset. After introducing randomness to the learning rate as mentioned in the experimental setup section, the validation accuracy increased a little more that 1.5% which can be seen from the Fig. 6 that is displayed.

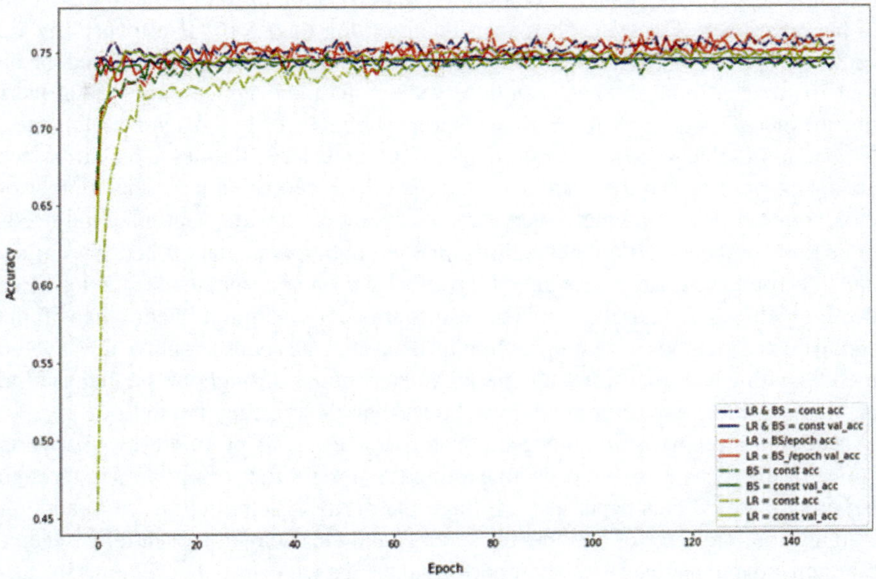

Fig. 6 Training and validation accuracy of the models trained by values of learning rate and batch size on Pet Finder dataset using (2)

4.3 Experiments on Other Datasets

The performance of the models for classifying image datasets was not as good as on audio and structured numeric datasets. The performance on them was quite opposite, i.e., the proposed method performed the worst when compared with other existing methods. Also, within the domain of image classification, it was observed that as the complexity of the image dataset increased the proposed technique performed more poorly which implies that as the complexity of the data decreases the proposed method performs better than the existing methods. The other behavior observed in case of image data classification was that using the proposed method fitting of the model was relatively better than the existing methods.

5 Conclusion

In the deep learning world, the focus usually goes either toward the model to train or the data to train on when it comes to optimizing the performance of the deep learning model. This paper intends to prompt the research community to explore the world of hyperparameters as well. The paper tune it or don't use it [6] also focuses on the same and explores the wide possibility of better performance if more work is done on hyperparameter tuning. It has been observed among the deep learning practitioners that the model accuracy is generally seen proportional to log of the number of images, which implies that if the number of images is not increased 10 times, a big difference cannot be observed in the accuracy. There is only a certain limit of the number of data points which can widely increase the accuracy; after that optimization needs to be done in the area of model architecture and hyperparameters if better accuracy has to be observed. Brigato et al. [7] have focused on hyperparameter tuning when dealing with small datasets, and their results are overwhelmingly better than all the sophisticated methods of dealing with small datasets. They chose a simple few layered models with CNN, focusing on hyperparameter tuning throughout the manuscript, and their results were surprisingly way better than the existing methods.

Building a relationship among the hyperparameters, as proposed by this work, is one of the many possible ways to explore the possibilities of optimizing through hyperparameters. This paper also explored the decay of learning rate in a randomized fashion. This idea of building the synergy and the decay of the learning rate in a randomized manner lead to a better performance in a wide array of problems in classification of audio datasets and structured numeric datasets. When using the method of building the synergy among the hyperparameters, the problem of assigning value of one or more hyperparameter can be reduced since one or more hyperparameter is made a function of other parameters, whose values determine the value of the other one, for example, in case of the proposed synergy learning rate is a function of batch size and epoch, so a deep learning practitioner would not have to assign the value to learning rate; only, batch size needs to be taken care of as the number of epochs increase.

References

1. You K, Long M, Wang J, Jordan MI (2019) How does learning rate decay help modern neural networks? arXiv preprint arXiv:1908.01878
2. Keskar NS, Mudigere D, Nocedal J, Smelyanskiy M, Tang PTP (2016) On large-batch training for deep learning: generalization gap and sharp minima. arXiv preprint arXiv:1609.04836
3. Smith SL, Kindermans PJ, Ying C, Le QV (2017) Don't decay the learning rate, increase the batch size. arXiv preprint arXiv:1711.00489
4. He K, Zhang X, Ren S, Sun J (2016) Deep residual learning for image recognition. In: CVPR 2016. arXiv preprint arXiv:1512.03385
5. Goyal P, Dollár P, Girshick R, Noordhuis P, Wesolowski L, Kyrola A, He K et al (2017) Accurate, large minibatch SGD: Training ImageNet in 1 hour. arXiv preprint arXiv:1706.02677
6. Brigato L, Barz B, Iocchi L, Denzler J (2021) Tune it or don;t use it: benchmarking data-efficient image classification. In: Proceedings of the IEEE/CVF international conference on computer vision, pp 1071–1080
7. Brigato L, Barz B, Iocchi L, Denzler J (2022) Image classification with small datasets: overview and benchmark. IEEE Access
8. Source code. https://github.com/SadafShafi/Experiment-LR. Last accessed 24 June 2022
9. Liu L, Jiang H, He P, Chen W, Liu X, Gao J, Han J (2019) On the variance of the adaptive learning rate and beyond. arXiv preprint arXiv:1908.03265
10. Jacobs RA (1988) Increased rates of convergence through learning rate adaptation. Neural Netw 1(4):295–307
11. Yu XH, Chen GA, Cheng SX (1995) Dynamic learning rate optimization of the backpropagation algorithm. IEEE Trans Neural Netw 6(3):669–677
12. Luo L, Xiong Y, Liu Y, Sun X (2019) Adaptive gradient methods with dynamic bound of learning rate. arXiv preprint arXiv:1902.09843
13. Radiuk PM (2017) Impact of training set batch size on the performance of convolutional neural networks for diverse datasets
14. Masters D, Luschi C (2018) Revisiting small batch training for deep neural networks. arXiv preprint arXiv:1804.07612
15. Yao Z, Gholami A, Arfeen D, Liaw R, Gonzalez J, Keutzer K, Mahoney M (2018) Large batch size training of neural networks with adversarial training and second-order information. arXiv preprint arXiv:1810.01021
16. Devarakonda A, Naumov M, Garland M (2017) Adabatch: adaptive batch sizes for training deep neural networks. arXiv preprint arXiv:1712.02029
17. Smith LN (2018) A disciplined approach to neural network hyper-parameters: Part 1—learning rate, batch size, momentum, and weight decay. arXiv preprint arXiv:1803.09820
18. Hoffer E, Hubara I, Soudry D (2017) Train longer, generalize better: closing the generalization gap in large batch training of neural networks. In: Advances in neural information processing systems, p 30

Encoder–Decoder (LSTM-LSTM) Network-Based Prediction Model for Trend Forecasting in Currency Market

Komal Kumar, Hement Kumar, and Pratishtha Wadhwa

Abstract Trend prediction of exchange rates has been a challenging topic of research. This problem is studied using non-stationary pattern recognition techniques. Several statistical, traditional machine learning, and deep learning techniques have been utilized for trend forecasting in the currency market. A novel encoder–decoder network-based model is proposed for trend prediction in this work. Furthermore, a comparative analysis is drawn with existing models using the Wilcoxon test for significant differences and model performance metrics to evaluate the proposed model.

Keywords Encoder–decoder · Trend prediction · Series analysis (TSA) · Currency market

1 Introduction

The currency market is the biggest financial market in the world. The currency market is functional 24 h a day [12], but the time for trading is classified into four time zones, namely the Australian zone, North American Zone, Asian Zone, and European Zone [16]. Currencies are traded in pairs, and most traded currencies are priced against the USD [17]. Currencies are coded using three letter symbols. The rates of a given currency pair are indexed by time forming a time series. Currency pair trend prediction becomes a problem of interest due to the size of the currency market and the variety of players in the market. This trend prediction can assist the investors with useful information for decision-making to lower the risk and maximize the return.

Modeling time series is one of the fundamental academic research fields in machine learning applications that solves two problems of time series, namely regression and classification analysis. Regression analysis predicts the Y_t^{th} term of time series given $Y_1, Y_2, \ldots Y_{t-1}$ terms, and classification problem predicts the class of

K. Kumar (✉) · H. Kumar · P. Wadhwa
Indian Institute of Technology Mandi, Himachal Pradesh 175005, India
e-mail: suryavansi8650@gmail.com

M. Thakur et al. (eds.), *Soft Computing for Problem Solving*, Lecture Notes in Networks and Systems 547, https://doi.org/10.1007/978-981-19-6525-8_17

Y_t^{th} given $Y_1, Y_2, \ldots Y_{t-1}$ terms. Application field of modeling time series data started with a few applications such as modeling climate [8], Convergence of Mortal and AI in Medicine (TSA) [21], and volatility forecasting (financial market analysis, risk management analysis, and high-frequency forecasting) [1].

Parameter estimation methods for modeling time series have been widely adopted by field experts. These methods include autoregressive integrated moving average (ARIMA) [3], exponential smoothing [6], and forecasting and structural TSA models [10]. Furthermore, there exists a rich amount of literature on time series modeling including time series automatic forecasting models: the forecast package for R [13], price prediction via hidden Markov model (HMM) using previous patterns [11], and using maximizing the posterior approach [9].

In traditional machine learning approaches, most of the used attributes require to be specified by domain expertise to reduce the complexity of the problem and construct ways more visual for ML models to perform well. Traditional ML models are used for predicting the trend. The previous works include support vector machine (SVM) in time series Landsat [24], trading support system with hybrid feature selection methods (hybrid) [19], forex mentoring using SVM [20], and PSVM in financial time series forecasting [14].

Deep learning can deal with high-dimensional data [2] with better accuracy as it has tremendous power and flexibility by memorizing to represent the problem as a nested scale of concepts whose subconcept defined concerns simpler concepts. The key advantage of the deep learning models is that they try to learn high-level features from the given set of information in an accumulative way that does not require domain expertise and hardcore feature extraction. In the accumulation of traditional ML, Gaussian processes [23] with the latest extensions include deep Gaussian [4] and conditional neural for stochastic processes [7]. Furthermore, the neural network has been widely used historically in time series modeling as shown in [22] and [18].

This paper proposes an encoder–decoder network-based model for trend prediction. Also, the results are compared with HMM, LSTM, GRU, CNN, and some variants of SVM models. This comparison is performed using the performance metrics and statistical test to evaluate the significance of the difference in the models.

The remaining paper is structured in the following way. Section 2 provides the theoretical background for the model that is proposed. The specifications of the data set used are mentioned in Sect. 3. The implementation workflow and proposed model are discussed in Sect. 4. Measures of performance are given in Sect. 5. Section 6 contains the factual findings of the comparative analysis. The conclusion is stated in Sect. 7.

2 Methodology

We start by summarizing the LSTM block in Sect. 2.1 and then discuss encoder–decoder-based model in Sect. 2.2 followed by encoder and decoder layer in Sects. 2.3 and 2.4, respectively. Finally, we develop encoder–decoder-based architectures for trend forecasting in Sect. 2.5.

2.1 LSTM Block

LSTM [5] consists of a chain arrangement that includes four networks and different recollection blocks which are known as cells. The info is possessed by cells, and the remembering manipulations are accomplished by some gates. LSTM consists of three gates:

- **Forget gate (selectively forget)**: $f_k = \sigma(x_k U_f + h_{k-1} W_f)$, here x_k is the current at current timestep k input and and h_k is the output of current hidden layer containing the output of all LSTM block, and U_f, W_f are the weighs (ignoring bias for now).
- **Input gate (selectively write)**: $i_k = \sigma(x_k U^i + h_{k-1} W^i)$ The candidate value $(\tilde{C}) = \tanh(W_k U^g + h_{k-1} W^g)$ is also responsible for updating the cell state. and the state of the block is $C_k = \sigma(f_k * C_{k-1} + i_k * \tilde{C}_k)$ at the kth step.
- **Output gate (selectively read)**: The output gate $(O_t) = \sigma(x_t U^o + h_{t-1} W^o)$ decides how to update the value of hidden nodes, and then the value of the hidden nodes is computed as $h_t = \tanh(C_t) * O_t$

The equation for the candidate is $\tilde{C}_t = \sigma(W(O_t \odot C_{t-1}) + x_t U + b)$, and the state value is $C_t = (1 - i_t) \odot C_{t-1} + i_t \odot \tilde{C}_t$, where $\odot$ is the Hadamard product.

2.2 Encoder–Decoder Network

Consider the problem of time series data modeling. Informally, given $t - 1$ trend we are interested in predicting the ith trend. More generally, for given $y_1, y_2, \ldots y_{t-1}$, we want to predict $y^\star = argmax(y_t | y_1, y_2, \ldots y_{t-1})$. The desired model is a combination of an encoder and a decoder model. An encoder takes the data (matrix of closing price in our case), and the decoder outputs the cleaned and rebuilt data. So, the simplified interpretation of the total system can be described through the following equations.

$$s_t = f_{\text{Encoder}}(x_t) \tag{1}$$

$$y_t = f_{\text{decoder}}((y_1, y_2, \ldots, y_{t-1}) + s_t) \tag{2}$$

where s_t is the current state of the input data, $y_1, y_2, \ldots y_{t-1}$ are the previous trends, and y_t is the trend predated trend by the model. So the motivation behind this model is that the encoder reads the corrupted input data and constructs a state vector where it is presumed to have the full details of the input data in the state vector. The stack of encoder and decoder layer is shown in Fig. 3.

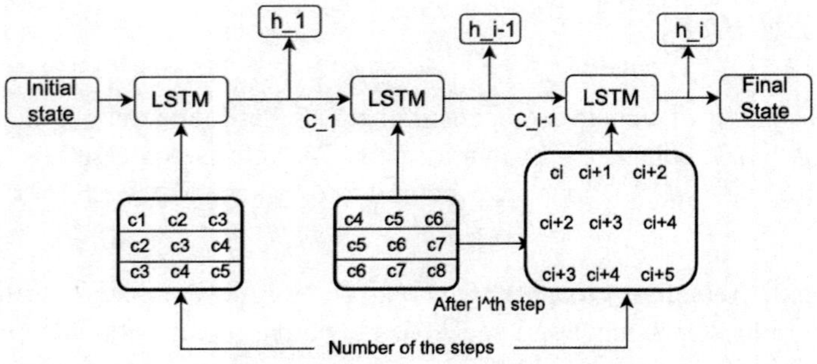

Fig. 1 Encoder as LSTM layer

2.3 *Encoder Layer*

Encoder consists of an LSTM neural network, reads the input data, and constructs a state vector where it is presumed to have the full details of the input data in the state vector. Here encoder has the full privilege to encode anything which will help the model to reduce the loss, so it encodes a state version of the input data in the vector as shown in Fig. 1.

2.4 *Decoder Layer*

Decoder consists of an LSTM whose output is a fully connected layer. Inp1 is the matrix of the closing price of the data by which decoder outputs the state of the current trend. Initial state of decoder is the state of the output from encoder, and the $inp2$ is the sequence $y_1,\ y_2, \dots y_{t-1}$ in Fig. 3. $inp2$ and the state from the encoder from Fig. 1 feed into the decoder shown in Fig. 2 that outputs the probabilities of uptrends and downtrends. Decoder takes the state vector from encoder and tries to produce a feature extracted data where it has the access to that related row of the input data. So overall, when the decoder reads a trend (uptrend or downtrend) it has the report about some current dependences of that trend, the state vector from the encoder, and what was the last output by the decoder, so when tth trend of a second input is demanded it can produce that by its memory.

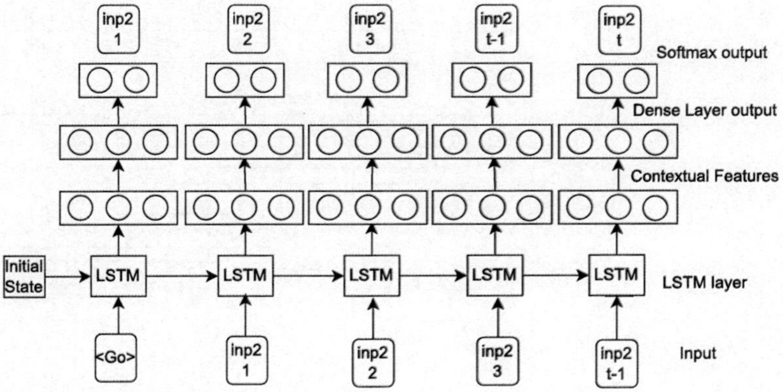

Fig. 2 Encoder as LSTM layer

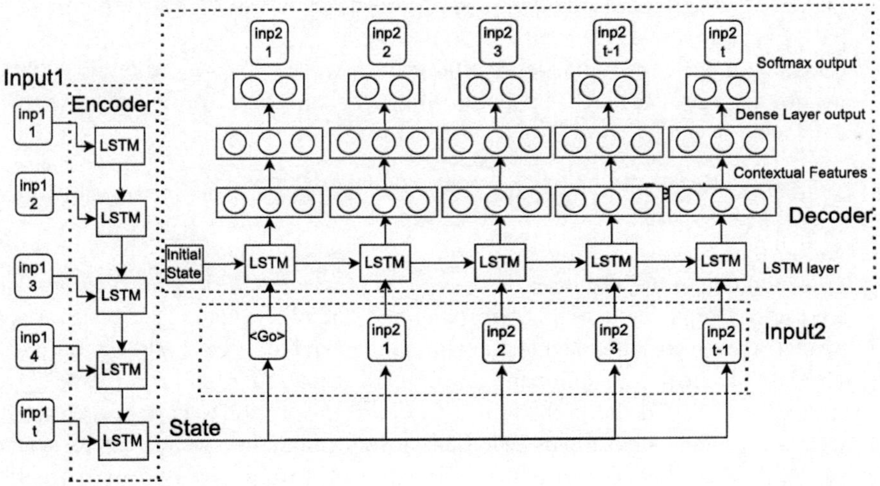

Fig. 3 Encoder–decoder complete architecture

2.5 Combination of Encoder and Decoder Architectures

See Fig. 3.

3 Model Formulation and Implementation

1. **Data Preprocessing**: We are interested in finding the trend on the closing price of currency pair data so suppose $c_1,\ c_2,\ c_3, \ldots, c_n$ is the closing price where n is the number of trading days. We are getting a high correlation coefficient for three

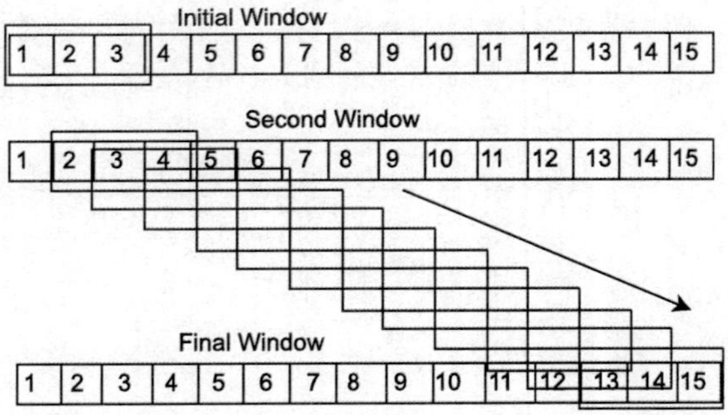

Fig. 4 Data prepossessed using sliding window technique

days, so our model will predict the $(i + 3)$th trend of closing price by looking back up to 3 days data, and then the input preprocessing is given in Fig. 4.
In the output preprocessing, if $c_{i+3} < c_{i+4}$ then uptrend so assign 1 else assign 0 for i from 0 to $n - 4$.
Encoding time series to images by converting $A_1, A_2, \ldots, A_t$ from the sliding window technique in Fig. 4 into matrix with step size = 3. So the final input format to encoder is given below:

$X_i = \begin{bmatrix} c_i & c_{i+1} & c_{i+2} \\ i+1 & c_{i+2} & c_{i+3} \\ i+3 & c_{i+4} & c_{i+5} \end{bmatrix}$ for i from 1 to $n - 5$, where n is the number of forex trading days.
$= X_i$ and inp2 $= (< Go >, y_1, y_2, \ldots, y_{t-1})$ where $< Go >$ indicates to start see Fig. 3. Output is the trend corresponding trend of X_i that is $y_1, y_2, \ldots, y_t$.

2. **Model Architecture**: Model architecture of encoder–decoder is shown in Fig. 3. We will model the conditional probability distribution $p(y_t|y_{t-1})$. More generally, we will find the probability of trend y_t given previous trends. What if we want to generate a trends given all closing prices? So we are interested in $p(y_t|y_{t-1}, C_t)$ where C_t has all the information of closing price. We would model $p(y_t = j|FC(y_{t-1}, C_t))$ where FC is the fully connected layer and $p(y_t = j)$ is the probability of being uptrend. There are many ways of making $p(y_t = j)$ conditioned on $FC(y_{t-1})$. Our proposed model is, Encoder = LSTM(x_i), Decoder = LSTM(C_{t-1}, y_{t-1}),
3. **Loss Function**: $L_t = -\log p(y_t = j|y_{t-1}, s_t)$, and training is done using Adam optimizer with default parameters.

4 Description of Experimental Data

All techniques (existing and proposed) are applied to 12 currency pairs shown in Table 1 from the currency market to evaluate the models. Historical open, close, high, and low price data for all the 12 currency pairs are extracted from Yahoo

Table 1 Currency pairs

Code	Currency pair
AUDUSD	Australia Dollar—United States Dollar
EURUSD	Euro member Euro—United States Dollar
GBPUSD	United Kingdom Pond—United States Dollar
JPYINR	Japan Yen—Indian Rupee
NZDUSD	New Zealand Dollar—United States Dollar
USDCAD	United States Dollar—Canada Dollar
USDJPY	United States Dollar—Japan Yen
USDKRW	United States Dollar—Korea (South) Won
USDSEK	United States Dollar—Sweden Krona
USDSGD	United States Dollar—Singapore Dollar
USDTHB	United States Dollar—Thailand Baht
USDZAR	United States Dollar—South Africa Rand

Finance for January 2007–March 2022. Each data set is split into two—the training data set (January 2007 to January 2017) and the testing data set(February 2017 to March 2022).

5 Performance Measure and Implementation of Prediction Model

Traders can buy as well as short the currency pair concerning upward and downward movement, respectively, to gain profit. Hence, for currency pair rates, trend prediction recall and precision for currency pair rates' increase and decrease are of equal importance. Therefore, to compare the performance of the models, several measures that are calculated from the confusion matrix are used.

Confusion matrix visualizes the classification results. It is a 2×2 matrix whose diagonal elements show the number of correct classification, and other two elements show the number of incorrect classification. The entry of confusion matrix will be confusion_matrix$[i][j]$ = predicted in ith class but actually belongs to jth class. From the confusion matrix for uptrend and downtrend class, we have

1. True Up (T_U): Number of uptrend class predicted correctly.
2. True Down (T_D): Number of downtrend class predicted correctly.
3. False Up (F_U): Number of uptrend class predicted incorrectly .
4. False Down (F_D): Number of downtrend class predicted incorrectly.

5.1 Recall

The recall is the proportion of the predicted uptrend (or downtrend) out of the total uptrend (or downtrend).

Recall of upward trend $(R^U) = \frac{T_U}{T_U + F_D}$

Recall of downward trend $(R^D) = \frac{T_D}{T_D + F_U}$

5.2 Precision

Precision is the fraction of true uptrend (or downtrend) out of all the uptrend (or downtrend prediction).

Precision of upward trend $(P^U) = \frac{T_U}{T_U + F_U}$

Precision of downward trend $(P^D) = \frac{T_D}{T_D + F_D}$

5.3 F_1-Score

F_1-score gives the harmonic mean of recall and precision. It is defined as follows:

F_1 score of Upward Trend $(F_1^U) = \frac{2 \times R^U \times P^U}{R^U + P^U}$

F_1 score of Downward Trend $(F_1^D) = \frac{2 \times R^D \times P^D}{R^D + P^D}$

Joint prediction error (JPE) [15] is used to assess the combined effect of F_1^U and F_1^D. Using JPE, the JPE coefficient is defined below:

$$\text{JPE coefficient} = \frac{(1 - F_1^U)^2 + (1 - F_1^D)^2}{2}$$

JPE coefficient will be 0 for the best prediction model and 1 for the worst prediction model.

6 Result and Discussion

To build a comparative analysis for the model, the performance measures (discussed above in Sect. 5) are computed for all the 12 currency pairs (given in Table 1) and for all the models stated in Sect. 2. The results are reported in Table 2. Table 3 shows

Table 2 Comparison of models

Currency pair	Metrics	HMM	SVM	LSSVM	PSVM	LSTM	GRU	CNN	ED
AUDUSD	$\mathbf{P}^D$	0.50	0.85	0.86	0.86	0.94	0.64	0.85	**0.90**
	$\mathbf{P}^U$	0.52	0.72	0.71	0.72	0.60	0.93	0.93	**1.00**
	$\mathbf{R}^D$	0.67	0.66	0.63	0.65	0.33	0.91	0.92	**1.00**
	$\mathbf{R}^U$	0.35	0.88	0.90	0.89	0.98	0.72	0.87	**0.91**
EURUSD	$\mathbf{P}^D$	0.50	0.78	0.65	0.65	0.79	0.57	0.81	**1.00**
	$\mathbf{P}^U$	0.50	0.83	0.91	0.93	0.80	0.95	0.78	**0.90**
	$\mathbf{R}^D$	0.63	0.84	0.95	0.96	0.82	0.92	0.79	**0.90**
	$\mathbf{R}^U$	0.37	0.77	0.49	0.48	0.77	0.66	0.81	**1.00**
GBPUSD	$\mathbf{P}^D$	0.51	0.78	0.78	0.78	0.97	0.34	**0.82**	1.00
	$\mathbf{P}^U$	0.52	0.83	0.83	0.82	0.59	0.98	**0.82**	0.62
	$\mathbf{R}^D$	0.63	0.84	0.84	0.83	0.33	0.94	**0.84**	0.72
	$\mathbf{R}^U$	0.39	0.77	0.77	0.78	0.99	0.59	**0.83**	1.00
JPYINR	$\mathbf{P}^D$	0.52	0.78	0.79	0.83	0.68	0.74	0.89	**1.00**
	$\mathbf{P}^U$	0.50	0.84	0.82	0.78	0.90	0.70	0.92	**0.88**
	$\mathbf{R}^D$	0.74	0.88	0.84	0.78	0.94	0.87	0.93	**0.90**
	$\mathbf{R}^U$	0.28	0.73	0.75	0.83	0.54	0.85	0.89	**1.00**
NZDUSD	$\mathbf{P}^D$	0.51	0.86	0.88	0.85	0.87	0.79	0.99	**0.97**
	$\mathbf{P}^U$	0.51	0.72	0.69	0.73	0.70	0.82	0.89	**1.00**
	$\mathbf{R}^D$	0.69	0.64	0.59	0.67	0.60	0.81	0.89	**1.00**
	$\mathbf{R}^U$	0.33	0.90	0.92	0.88	0.91	0.80	0.99	**0.97**
USDCAD	$\mathbf{P}^D$	0.49	0.76	0.77	0.79	0.68	0.85	0.83	**0.62**
	$\mathbf{P}^U$	0.50	0.79	0.77	0.70	0.89	0.74	0.71	**1.00**
	$\mathbf{R}^D$	0.65	0.79	0.76	0.64	0.93	0.77	0.74	**1.00**
	$\mathbf{R}^U$	0.34	0.75	0.77	0.84	0.55	0.83	0.81	**0.73**
USDJPY	$\mathbf{P}^D$	0.49	0.81	0.82	0.90	0.79	0.82	0.82	**0.79**
	$\mathbf{P}^U$	0.50	0.81	0.80	0.67	0.82	0.80	0.97	**1.00**
	$\mathbf{R}^D$	0.65	0.80	0.79	0.52	0.82	0.80	0.96	**1.00**
	$\mathbf{R}^U$	0.34	0.82	0.84	0.94	0.79	0.83	0.85	**0.83**
USDKRW	$\mathbf{P}^D$	0.50	0.79	0.81	0.85	0.59	0.96	0.95	**1.00**
	$\mathbf{P}^U$	0.50	0.87	0.86	0.71	0.95	0.53	0.91	**0.94**
	$\mathbf{R}^D$	0.49	0.88	0.87	0.60	0.98	0.66	0.91	**0.94**
	$\mathbf{R}^U$	0.52	0.77	0.80	0.88	0.35	0.94	0.96	**1.00**
USDSEK	$\mathbf{P}^D$	0.50	0.83	0.81	0.82	0.73	0.86	0.76	**0.88**
	$\mathbf{P}^U$	0.50	0.77	0.75	0.76	0.91	0.77	0.80	**1.00**
	$\mathbf{R}^D$	0.52	0.75	0.73	0.74	0.93	0.79	0.79	**1.00**
	$\mathbf{R}^U$	0.48	0.84	0.83	0.84	0.66	0.85	0.77	**0.89**
USDSGD	$\mathbf{P}^D$	0.51	0.74	0.74	0.76	0.51	0.82	0.88	**0.98**
	$\mathbf{P}^U$	0.46	0.84	0.84	0.82	0.96	0.78	0.86	**1.00**
	$\mathbf{R}^D$	0.69	0.88	0.87	0.85	0.99	0.79	0.87	**1.00**
	$\mathbf{R}^U$	0.28	0.67	0.68	0.72	0.30	0.81	0.87	**0.98**

(continued)

Table 2 (continued)

Currency pair	Metrics	HMM	SVM	LSSVM	PSVM	LSTM	GRU	CNN	ED
USDTHB	$\mathbf{P}^D$	0.53	0.71	0.82	0.84	0.83	0.95	0.89	**0.97**
	$\mathbf{P}^U$	0.47	0.89	0.75	0.55	0.81	0.56	0.94	**0.95**
	$\mathbf{R}^D$	0.50	0.93	0.73	0.25	0.83	0.71	0.94	**0.96**
	$\mathbf{R}^U$	0.49	0.60	0.83	0.95	0.80	0.91	0.88	**0.97**
USDZAR	$\mathbf{P}^D$	0.53	0.80	0.81	0.76	0.58	0.93	0.93	**0.99**
	$\mathbf{P}^U$	0.49	0.81	0.80	0.85	0.99	0.68	0.42	**0.99**
	$\mathbf{R}^D$	0.51	0.83	0.82	0.89	0.99	0.76	0.63	**0.99**
	$\mathbf{R}^U$	0.52	0.77	0.79	0.70	0.23	0.90	0.84	**0.99**

Table 3 Wilcoxon ranked sign test results

Algorithm	Measures	HMM	SVM	LSSVM	PSVM	LSTM	GRU	CNN
SVM	z-statistics	**−7.95**						
	p	**0.00**						
LSSVM	z-statistics	**−7.95**	−1.76					
	p	**0.00**	0.08					
PSVM	z-statistics	**−7.68**	**−2.28**	−0.84				
	p	**0.00**	**0.02**	0.40				
LSTM	z-statistics	**−7.00**	**−2.81**	**−1.96**	−0.89			
	p	**0.00**	**0.00**	**0.05**	0.37			
GRU	z-statistics	**−7.89**	−0.35	−0.46	−1.73	−1.60		
	p	**0**	0.72	0.64	0.08	0.11		
CNN	z-statistics	**−7.95**	**−4.99**	**−5.24**	**−5.22**	**−4.84**	**−4.34**	
	p	**0**	**6E−07**	**2E−07**	**2E−07**	**1.3E−06**	**1.4E−05**	
ED	z-statistics	**−7.96**	**−7.35**	**−7.25**	**−7.15**	**−7.46**	**−6.77**	**−4.34**
	p	**0**	**0**	**0**	**0**	**0**	**0**	**1.4E−05**

that the maximum test accuracy (highlighted in bold) in eleven from twelve currency pairs is achieved by the proposed encoder–decoder model. The performance of the models based only on accuracy is ambiguous.

We have calculated recall and precision for uptrend and downtrend. ED has given approximately equal values for uptrend and downtrend in both (recall and precision), whereas other models are biased toward downtrend compared to CNN which has also given approximately equal priority to uptrend and downtrend.

To evaluate the joint effect of F_1^U and F_1^D, JPE coefficient is calculated. (discussed in Sect. 5). Under traditional machine learning techniques, PSSVM has the highest JPE coefficient for seven currency pairs, LSSVM has the highest JPE coefficient for four currency pairs, and SVM performs best for most of the currency pairs as shown in Fig. 5.

Under the deep learning techniques, LSTM has the highest JPE coefficient for nine currency pairs and GRU has the highest JPE coefficient for three currency pairs

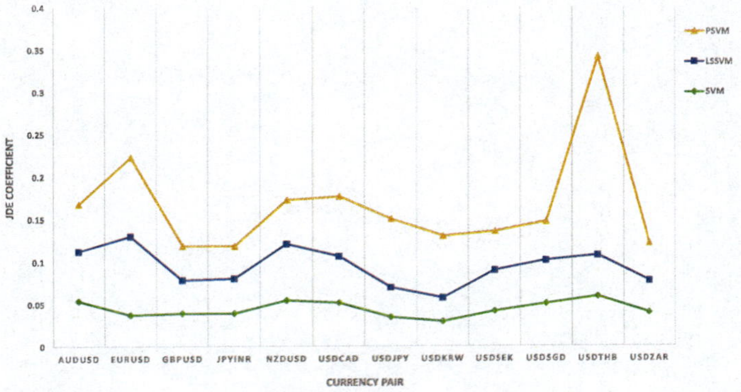

Fig. 5 JPE coefficient comparison of LSSVM, PSVM, and SVM models

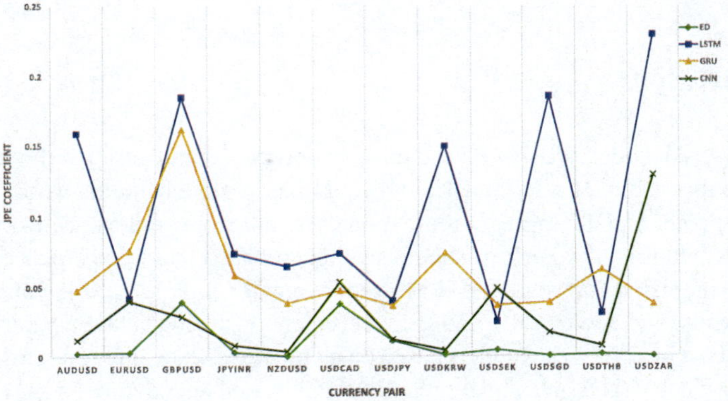

Fig. 6 JPE coefficient comparison of CNN, ED,LSTM, and GRU models

as depicted in Fig. 6. For most of the data sets, the proposed model performs better compared to other deep learning techniques.

HMM has the maximum JPE coefficient. Therefore, HMM has worst performance, and the proposed model outperforms all other models in terms of the JPE, as depicted in Fig. 7.

Using a wide range of performance metrics, a comparison is drawn for the models. But, it is not known if these models are significantly different statistically or not. To assess that which model performs better than other models significantly, the Wilcoxon rank sum test is carried out at a 5 percent level of significance. Table 3 contains the result of the test for all the currency pairs corresponding to the testing data for all the models. Table 3 depicts that most models are significantly different. Especially, the encoder–decoder model is significantly different from the rest of the models.

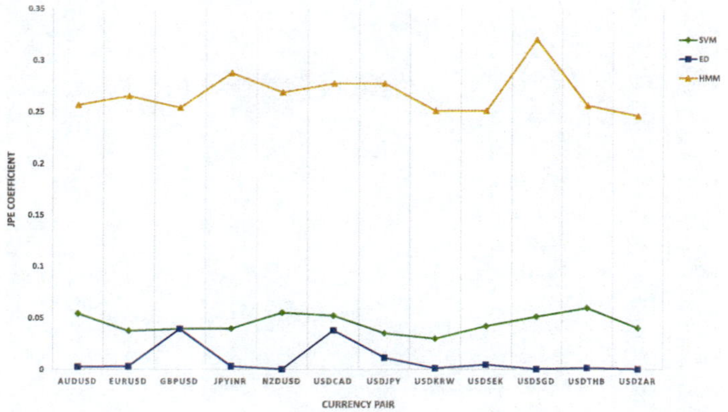

Fig. 7 JPE coefficient comparison of ED, HMM, and SVM models

7 Conclusion

The encoder–decoder network-based model is proposed and compared with HMM, SVM and its variants, LSTM, GRU, and CNN-based models in this work. To check the performance and flexibility of these models, twelve forex indexes across the globe are studied empirically. Performance is evaluated based on various metrics such as accuracy, precision (uptrend and downtrend), recall (uptrend and downtrend), F_1-score (uptrend and downtrend), and JPE coefficient on testing data. Encoder–decoder performs best in all of these models based on all measures except for one currency pair GBPUSD where CNN gives better results. In CNN and encoder–decoder, a five-day window is used which is one of the reasons for their outstanding performance. The strength of the ED model is that encoder gives the state of the input from the previous five days' data, and the decoder predicts the uptrend or downtrend based on given previous trends. The proposed encoder–decoder model not only outperforms the existing models but is also statistically different from them. The comparative analysis indicates that deep learning methods give predictions with greater accuracy. The HMM that produced good results in the past is now obsolete for the given data sets. SVM still provides a good prediction when compared with its variants and HMM. In the future, this problem can be converted to sequence-to-sequence modeling for multi-span identification tasks based on the dynamics of markets and user behavior on social media where we can use NLP models to encode user's comments.

References

1. Andersen TG, Bollerslev T, Christoffersen P, Diebold FX (2005) Volatility forecasting
2. Bengio Y, Courville A, Vincent P (2013) Representation learning: a review and new perspectives. IEEE Trans Pattern Anal Mach Intell 35(8):1798–1828

3. Box GE, Jenkins GM, Reinsel GC, Ljung GM (2015) Time series analysis: forecasting and control. Wiley
4. Damianou A, Lawrence ND (2013) Deep gaussian processes. In: Artificial intelligence and statistics. PMLR, pp 207–215
5. Fischer T, Krauss C (2018) Deep learning with long short-term memory networks for financial market predictions. Eur J Operat Res 270(2):654–669
6. Gardner ES Jr (1985) Exponential smoothing: the state of the art. J Forecast 4(1):1–28
7. Garnelo M, Rosenbaum D, Maddison C, Ramalho T, Saxton D, Shanahan M, Teh YW, Rezende D, Eslami SA (2018) Conditional neural processes. In: International conference on machine learning. PMLR, pp 1704–1713
8. Giorgi F (2019) Thirty years of regional climate modeling: where are we and where are we going next? J Geophys Res: Atmos 124(11):5696–5723
9. Gupta A, Dhingra B (2012) Stock market prediction using hidden Markov models. In: 2012 Students conference on engineering and systems. IEEE, pp 1–4
10. Harvey AC (1990) Forecasting, structural time series models and the Kalman filter
11. Hassan MR, Nath B (2005) Stock market forecasting using hidden Markov model: a new approach. In: 5th international conference on intelligent systems design and applications (ISDA'05). IEEE, pp 192–196
12. Huang RD, Masulis RW (1999) Fx spreads and dealer competition across the 24-hour trading day. Rev Financ Stud 12(1):61–93
13. Hyndman RJ, Khandakar Y (2008) Automatic time series forecasting: the forecast package for r. J Stat Softw 27:1–22
14. Kumar D, Meghwani SS, Thakur M (2016) Proximal support vector machine based hybrid prediction models for trend forecasting in financial markets. J Comput Sci 17:1–13
15. Kumar D, Meghwani SS, Thakur M (2016) Proximal support vector machine based hybrid prediction models for trend forecasting in financial markets. J Comput Sci 17:1–13
16. Masry S, Dupuis A, Olsen R, Tsang E (2013) Time zone normalization of fx seasonality. Quant Financ 13(7):1115–1123
17. Ozturk M, Toroslu IH, Fidan G (2016) Heuristic based trading system on forex data using technical indicator rules. Appl Soft Comput 43:170–186
18. Sen R, Yu H-F, Dhillon IS (2019) Think globally, act locally: a deep neural network approach to high-dimensional time series forecasting. Adv Neural Inf Proc Syst 32
19. Thakur M, Kumar D (2018) A hybrid financial trading support system using multi-category classifiers and random forest. Appl Soft Comput 67:337–349
20. Thu TNT, Xuan VD (2018) Supervised support vector machine in predicting foreign exchange trading. Int J Intell Syst Appl 11(9):48
21. Topol EJ (2019) High-performance medicine: the convergence of human and artificial intelligence. Nat Med 25(1):44–56
22. Wan EA et al (1993) Time series prediction by using a connectionist network with internal delay lines. In: Santa FE Institute Studies in the Sciences of Complexity-Proceedings, vol 15. Addison-Wesley Publishing co., pp 195–195
23. Williams C, Rasmussen C (1995) Gaussian processes for regression. Adv Neural Inf Proc Syst 8
24. Zheng B, Myint SW, Thenkabail PS, Aggarwal RM (2015) A support vector machine to identify irrigated crop types using time-series Landsat NDVI data. Int J Appl Earth Observat Geoinf 34:103–112

Histopathological Nuclei Segmentation Using Spatial Kernelized Fuzzy Clustering Approach

Rudrajit Choudhuri and Amiya Halder

Abstract Image segmentation is a crucial image processing step in many applications related to biomedical image analysis. One such key sector is high-resolution histopathological image segmentation for nuclei detection that aids in high-quality feature extraction for meticulous analysis in the domain of digital pathology for disease diagnosis. Manual nuclei detection and segmentation require domain expertise and are rigorous and time-consuming. The existing automated analytical tools are capable of nuclei segmentation, but there is a granulated segmentation technique selection and configuration management requirement for each analysis due to a wide variation in nuclei structures, along with overlapping and highly correlated image regions. In this paper, an unsupervised spatial kernelized fuzzy segmentation algorithm is presented for automated nuclei segmentation of light microscopy images of stained nuclei. The algorithm has stable performance across a wide gamut of image types without the need for experiment dependent adjustment of segmentation parameters. For performance analysis and rigorous use-case testing, a highly standardized dataset is obtained from the Data Science Bowl. The proposed algorithm manages to achieve segmentation accuracies in the range of 95–96% across varied image types which defends the robustness of the technique along with visual indications obtained from qualitative results.

Keywords Iterative optimization · Statistical fuzzy clustering · Nuclei segmentation · Kernel methods

1 Introduction

Segmentation of medical images with the intent of partitioning it into several consistent non-overlapping regions with similar texture, intensity, and structure is an important processing task in the domain of medical image analysis. It plays a vital role in biomedical applications of image processing and computer vision and helps in

R. Choudhuri · A. Halder (✉)
St. Thomas College of Engineering and Technology, 4-D. H. Road, Kolkata, India
e-mail: amiya.halder77@gmail.com

M. Thakur et al. (eds.), *Soft Computing for Problem Solving*, Lecture Notes in Networks and Systems 547, https://doi.org/10.1007/978-981-19-6525-8_18

quantification of various regions, partitioning, and automated diagnosis. Radiological techniques along with medical image analysis accelerate the diagnostic process of detecting abnormalities in various organs of the human body (brain, lungs, heart, and many more). With the advancement in digital whole slide imaging (WSI) acquisition techniques, histopathological image analysis has been an effective tool in cancer diagnosis [8] and is considered a standardized result for computer aided cancer prognosis in a wide variety of clinical protocols. Identifying the structure and shape of cell nuclei along with its distribution in pathology images is critically useful in prognosis such as cell and tissue determination along with cancer identification, cancer type detection, and grade classification.

As a first crucial step, precise histopathological image segmentation for nuclei identification is necessary for qualitative and quantitative analysis. The manual process of nuclei segmentation is rigorous, expensive, error-prone, time-consuming, requires expertise and domain knowledge, and has limited reproducibility. Therefore, efficient automated nuclei detection and segmentation methods are the need of the hour in the clinical domain for improving upon the resiliency, fault tolerance, and scalability of histopathological image analysis. Around 30 trillion cells are present in a human body each containing nuclei full of DNA. Accurate automated nuclei identification has the potential in aiding experts in the observation of cell reactions corresponding to various treatments, thus ensuring and accelerating patient treatment and drug discovery. Also, computer-aided pathology [17] along with microscopy imaging plays a crucial role in proving detailed information, reduces interobserver variations [7], experimental bias, produces meticulous results of pathological image features, and enables better analysis which benefits scientists, doctors, pathologists, and patients alike.

Resilient automated nuclei segmentation is fruitful, but is equally challenging for many reasons. Firstly, the histopathological and microscopy images consist of intensity inhomogeneities, noise corruptions, lousy depth resolutions, and other artifacts owing to faults in image acquisition. Secondly, the images have a disarranged background along with low contrast changes between the background and the foreground. Thirdly, there exists a huge spectrum of appearance variations including structural differences, shapes, sizes, and intensity variations within the cell depending on the histological grade and the cell and disease type. Also, there are strong correlations in the image and often there may exist overlapping nuclei regions. Finally, tissue preparation procedures can lead to inconsistent and distorted tissue appearance which jeopardizes image processing and analysis tasks due to the artifacts.

Over the years, researchers have adapted several techniques for tackling automated nuclei identification, detection, classification, and segmentation techniques from pathology images [2, 12–14]. These techniques can be broadly divided into supervised, weakly supervised, and unsupervised paradigms. Several supervised detection [6, 18, 20], and segmentation [1, 4, 10, 11, 16] techniques have been designed and implemented for applications in pathological image analysis. These techniques are hugely dependent on an enormous amount of standardized annotated data and require heavy computation power for model training. Also, meticulous annotated data acquisition for model training is a time-consuming task, and the trained model often lacks generality and scalability across different data and use cases. Weakly supervised techniques [5, 15] are able to relax the requirements up to an extent, but they still require heavy computation power along with a large annotated dataset requirement to yield satisfactory results. Unsupervised techniques [9, 19] are scalable across datasets but often fail on complex samples due to the lack of spatial feature consideration during segmentation. The complex samples usually correspond to degenerated use cases caused due to cancer. These use cases are critical for clinical diagnosis, and although the methods are scalable, they lack efficiency for these specific cases. Also, the lack of spatial information consideration makes these algorithms sensitive to noise and inhomogeneities.

To overcome the mentioned challenges and to tackle the current drawbacks in literature, a spatial circular kernelized fuzzy clustering algorithm is proposed for robust unsupervised nuclei segmentation. The proposed method is substructured upon an iterative optimization paradigm, and it assimilates fuzzy logic along with kernelized mapping technique in its objective function. It also incorporates local neighborhood feature consideration for effective data point clustering which also compensates for corruptions and inhomogeneities in the image. The amalgamation of fuzzy logic in the algorithm takes care of ambiguous, vague, and overlapping image regions and ensures an efficient segmentation for highly correlated and complex degenerated data samples. For quantitative and qualitative performance evaluation, the benchmark histopathological dataset: Data Science Bowl [3] has been used. The dataset consists of a wide variety of high resolution data samples with detailed annotated masks and encompasses necessary and critical use cases that may arise during histopathological image analysis for nuclei identification. Experimental results defend the reliability and the resiliency of the presented technique across multiple use cases and uphold the superior performance of the method when compared to the existing state-of-the-art methods. The technique manages to achieve robust results for both common and critical samples and proves its exemplary robustness in the domain of nuclei segmentation.

The paper is organized into six sections: The related works in literature are discussed in Sects. 2 and 3 which summarize the relevant background concepts required for comprehension of the proposed technique, and Sect. 4 presents the proposed algorithm for nuclei segmentation. Experimental results and performance comparison are provided in Sect. 5, and finally conclusion is derived in Sect. 6.

2 Related Work

Computer-aided automated nuclei detection, classification, and segmentation from pathological images [2–13] have gained quite some popularity in the last decade owing to the challenges and problems of the manual process. Several deep learning-based methods have shown notable performance in the domain of nuclei detection. One of the early state-of-the-art approaches was proposed by Xu et al. [20], where a stacked sparse autoencoder was introduced for applications in breast cancer histopathological image analysis. Two years later, a local feature sensitive deep learning-based architecture [18] came into existence which initiated the efforts in tackling nuclei detection and classification from colon cancer pathology images. In recent years, convolutional neural net-based architectures [11–16] have shown notable performance and have improved upon the benchmark of supervised automated nuclei segmentation approaches. These techniques incorporate optimization paradigms, for instance, some are based on network architecture optimization and some rely on stepwise image contour segmentation. An improved mask region-based convolutional neural net architecture [6] is one of the recent methods for stained cell detection and segmentation. Another recent method is structured on a double U-Net-based architecture [10] which tackles the problem of medical image segmentation. The supervised techniques have a decent performance, but when it comes to computation power requirement, scalability, and reproducibility, these methods do not uphold the established benchmark. Deep Adversarial-based multi-organ nuclei segmentation [15] and Mutual Complementing Framework for pathology image segmentation [5] are some of the state-of-the-art weakly supervised segmentation techniques present in literature. Although they achieve stable performance, the model architectures are not very straightforward to implement, at the same time being computationally expensive and highly dependent on enormous detailed annotated data, which makes them lose scalability and compatibility with hardware systems for real time image segmentation. A fuzzy unsupervised clustering algorithm was proposed in 2015 [19] for leukemia detection. In 2019, Le Hou et al. proposed a robust unsupervised sparse autoencoder-based histopathological nuclei detection technique [9]. These techniques do not have enormous high resolution data dependence, and they maintain a stable performance across use cases. However, the techniques do not perform well on complex degenerated samples as the spatial information consideration is not efficiently taken into account. These complex samples are necessary for an automated tool to detect as these correspond to critical clinical and pathological use cases. Therefore, although being robust on multiple use cases, the techniques are not that reliable for complex sample analysis.

3 Background

3.1 Fuzzy C-Means Clustering

Fuzzy C-means is an iterative optimization algorithm that clusters data points incorporating fuzzy set logic in order to calculate membership degrees that are needed to assign a category to a data point. Consider $\mathbf{P} = (P_1, P_2, \ldots, P_n)$ to be an image consisting of n pixels which are to be segmented into C clusters, where multispectral features are denoted by image pixels in P. The cost function (objective function) that is intended to be optimized is defined in Eq. (1).

$$\mathbf{J} = \sum_{k=1}^{C} \sum_{i=1}^{N} \mu_{ki}^{q} ||P_i - v_k||^2 \quad (1)$$

where μ_{ki} represents the membership value of pixel P_i corresponding to the kth cluster, v_k corresponds to the kth cluster centroid, $||.||$ signifies the Euclidean norm function, and q is a fuzziness control parameter for the resultant partition. The value of q is set to 2 for all experiments in this paper. The algorithm minimizes the cost function by assigning high membership values to pixels close to the cluster centroid and vice versa. The degree of membership is a probabilistic measure to represent the probability that a pixel belongs in a particular cluster. In the conventional fuzzy c-means, this probability measure is solely dependent on the Euclidean distance between the concerned pixel and each of the individual cluster centroids. This hinders the algorithm from considering spatial information while segmenting data points. The fuzzy membership is calculated using Eq. (2).

$$\mu_{ki} = \frac{1}{\sum_{j=1}^{C} \left(\frac{||P_i - v_k||}{||P_j - v_j||} \right)^{\frac{2}{q-1}}} \quad (2)$$

It must be noted that the membership values should satisfy the constraints $\sum_{k=1}^{c} \mu_{ki} = 1$, $\mu_{ki} \in [0, 1]$.

For calculation of the cluster centroids, the membership values are incorporated and the centroid is calculated using Eq. (3).

$$v_k = \frac{\sum_{j=1}^{n} \mu_{kj}^{q} P_j}{\sum_{j=1}^{n} \mu_{kj}^{q}} \quad (3)$$

FCM starts off with an initial assumption for each cluster centroids and iteratively converges to give solutions for v_k which represents a saddle or a minima point of the defined cost function. The data points (pixels in this case) are thus segmented into C clusters. The change detected in the membership and centroid values in successive iterations demonstrates the convergence of the approach.

3.2 Kernel Methods and Functions

Kernel methods are a particular category of algorithms used in pattern recognition. The main idea behind the approach is to easily structure or separate the data points by mapping them to a higher-dimensional space. The approach finds its best usage in support vector machines (SVMs). There is no need for rigorous computation of the mapping function, and the job can be done by incorporating vector algebra. This is a powerful trick, and it bridges the gap between linearity and nonlinearity for any algorithm which can be defined in terms of scalar products between two vectors. The basic intuition behind this is that after mapping the data into a higher-dimensional space, any linear function in the augmented space acts as a nonlinear one in the original space. The trick works by replacing the scalar product of the two vectors with the scalar product from a substitute suitable space, i.e., by replacing the scalar product with a Kernel function. A kernel function represents a scalar product in a feature space and is of the form defined in Eq. (4).

$$\mathbf{K(x,y)} = \langle \Psi(x), \Psi(y) \rangle \tag{4}$$

where $\langle \Psi(x), \Psi(y) \rangle$ represents the product between the vectors.

This technique can be used in the domain of image segmentation as well. Commonly, the cluster centroids are represented as a sum of a linear combination of all $\Psi(P_i)$, which basically implies that all centroids lie in the feature space.

4 Proposed Methodology: Spatial Circular Kernel Based Fuzzy C-Means Clustering Algorithm (SCKFCM)

In this section, a spatial circular kernel-based fuzzy C-means clustering algorithm (SCKFCM) is proposed. The objective function that the proposed algorithm tries to optimize is defined in Eq. (5).

$$J_{\text{SCKFCM}} = \sum_{k=1}^{C} \sum_{i=1}^{n} \mu_{ki}^{q} ||\Psi(S_i) - \Psi(v_k)||^2 + \sum_{k=1}^{C} \lambda \sum_{i=1}^{n} (1 - \mu_{ki}) \tag{5}$$

where Ψ represents a nonlinear mapping, S_i represents the immediate spatial neighborhood information around the pixel P_i, and λ is a cluster influence control parameter. The presented approach considers the spatial information around an image pixel along with its gray level intensity as a feature. For this amalgamation to take place, the algorithm picks a 3×3 window centered at a pixel (P_i) and computes the average pixel intensity S_i of the window, and finally uses the value as a data point. This spatial substructure makes sure that the data point not only contains the concerned pixel information but also its neighborhood information. This accounts for inhomogeneities and correlations in the image and ensures an efficient segmentation. $\Psi(v_k)$ is not represented as a linear combination sum of $\Psi(P_i)$ in this case but is still viewed as a mapping point. Simplifying the Euclidean norm term in Eq. (5) using the kernel substitution mapping we get.

$$
\begin{aligned}
||\Psi(S_i) - \Psi(v_k)||^2 &= [\Psi(S_i) - \Psi(v_k)]^T [\Psi(S_i) - \Psi(v_k)] \\
&= \Psi(S_i)^T \Psi(S_i) - \Psi(v_k)^T \Psi(S_i) - \Psi(S_i)^T \Psi(v_k) + \Psi(v_k)^T \Psi(v_k) \\
&= K(S_i, S_i) - 2K(S_i, v_k) + K(v_k, v_k)
\end{aligned} \tag{6}
$$

The scalar product of the vectors is replaced using kernel substitution in this case. The algorithm uses a circular kernel as defined in Eq. (7). In a higher-dimensional space, the circular kernel is able to encompass all data points in a mentioned radius without missing out on near vicinity points while at the same time being bound by constraints.

$$
\mathbf{K(x,y)} = \frac{2}{\pi} \cos^{-1}\left(\frac{-||x - y||}{\sigma}\right) - \frac{2}{\pi} \frac{||x - y||}{\sigma} \sqrt{1 - \left(\frac{||x - y||}{\sigma}\right)^2} \tag{7}
$$

where σ is the tuning parameter for adjusting the kernel. Using the circular kernel, we get K(m, m) = 1. Simplifying Eqs. (5) and (6), the objective function can be defined as:

$$
J_{\text{SCKFCM}} = 2\sum_{k=1}^{C}\sum_{i=1}^{n} \mu_{ki}^{q}(1 - K(S_i, v_k)) + \sum_{k=1}^{C} \lambda \sum_{i=1}^{n}(1 - \mu_{ki}) \tag{8}
$$

For optimization, equating the partial derivative of the objective function with respect to the membership function to zero we get:

$$
\begin{aligned}
&\tfrac{\partial J_{\text{SCKFCM}}}{\partial \mu_{ki}} = 0 \\
&=> 2 \times q \times \mu_{ki}^{q-1} \times (1 - K(S_i, v_k)) = 0 \\
&=> q\mu_{ki}^{q-1}(1 - K(S_i, v_k)) - \lambda_i = 0 \\
&=> \mu_{ki} = (\tfrac{\lambda_i}{2q})^{\frac{1}{q-1}} \times \left(\tfrac{1}{1-K(S_i,v_k)}\right)^{\frac{1}{q-1}}
\end{aligned} \tag{9}
$$

As $\sum_{j=1}^{C} \mu_{ji} = 1$ is a boundary constraint,

$$
\begin{aligned}
&\textstyle\sum_{j=1}^{C} \left(\tfrac{\lambda_i}{2q}\right)^{\frac{1}{q-1}} \times \left(\tfrac{1}{1-K(S_i,v_j)}\right)^{\frac{1}{q-1}} = 1 \\
&=> \left(\tfrac{\lambda_i}{2q}\right)^{\frac{1}{q-1}} = \frac{1}{\sum_{j=1}^{C}\left(\frac{1}{1-K(S_i,v_j)}\right)^{\frac{1}{q-1}}}
\end{aligned} \tag{10}
$$

From Eqs. (9) and (10), the membership function obtained is

$$\mu_{ki} = \frac{(1 - K(S_i, v_k))^{-\frac{1}{q-1}}}{\sum_{j=1}^{C}(1 - K(S_i, v_j))^{-\frac{1}{q-1}}} \tag{11}$$

Again, for optimizing the centroid values, the partial differentiation of the objective function with respect to the centroid computation function is equated to 0 and we get

$$\begin{aligned} &\frac{\partial J_{\text{SCKFCM}}}{\partial v_k} = 0 \\ &=> \sum_{i=1}^{n} \mu_{ki}^{q} \times -K(S_i, v_k) \times (S_i - v_k) \times (-1) + 0 = 0 \\ &=> \sum_{i=1}^{n} \mu_{ki}^{q} K(S_i, v_k) v_k = \sum_{i=1}^{n} \mu_{ki}^{q} S_i K(S_i, v_k) \end{aligned} \tag{12}$$

Therefore, the obtained centroid function is

$$v_k = \frac{\sum_{i=1}^{n} \mu_{ki}^{q} S_i K(S_i, v_k)}{\sum_{i=1}^{N} \mu_{ki}^{q} K(S_i, v_k)} \tag{13}$$

Given the computation functions for the pixel membership values and the centroids, the proposed algorithm iteratively converges to obtain optimum cluster centroids which represent a saddle point for the defined cost function, thereby segmenting the input image pixels into the required number of clusters. In subsequent iterations, the change noticed in the membership and centroid values highlight the convergence of the algorithm. The proposed algorithm is summarized in Algorithm 1.

Algorithm 1: PROPOSED ALGORITHM

Input: Light Microscopic Image of Stained Nuclei
Output: Nuclei Segmented Image

```
Assign number of clusters(C), maximum iterations (M), iterator (R), fuzziness control
parameter (q) [q > 1], and threshold T > 0
Set initial membership values μ^0 as 0 and set R to be 1.
while R < M do
    Compute membership values μ_ki^R for every cluster k with respect to each pixel P_i using
    Eqn. (11)
    for each pixel P_i do
        Find the maximum membership value μ_ki^R along with its corresponding cluster j
        Assign the pixel to cluster j
    Compute cluster centroids v_k^R based on Eqn. (13).
    if |v_j^R − v_j^(R−1)| < T, ∀j ∈ {1, C} then
        break
    R = R + 1
After iterative optimization, each pixel is assigned a cluster (c) and the pixel intensity P_i is
set to the value of the corresponding cluster centroid v_c.
```

5 Results

In this section, the performance of the proposed spatial circular kernel-based fuzzy c-means clustering algorithm is evaluated on the benchmark Data Science Bowl dataset, and after rigorous experiments, the qualitative and the quantitative results are presented. For a fair comparison of the performances, state-of-the-art segmentation techniques including improved mask-rcnn (IMRCNN) [6], convolutional neural net-based recurrent residual U-net (R2UNET) [1], and double U-net (DUNET) [10], Mutual Complementing Framework (MCF) [5], Conditional Deep Adversarial Net (CGAN) [15], sparse autoencoder-based segmentation (SABS) [9], and morphological contour-based fuzzy c-means (MFCM) [19] are also implemented.

The mentioned algorithms were implemented in Python using Google Colaboratory with NVIDIA Tesla K80 GPU acceleration. The proposed algorithm has entirely been implemented in C without any external library dependence. Dev C++ served as the development environment, and the implementation is done on a single CPU machine with 8GB RAM and an Intel i5 processor.

5.1 Dataset

For performance evaluation, the implemented algorithms are tested using the 2018 Data Science Bowl Grand Challenge dataset [3]. It consists of 735 light microscopy histopathological images of stained nuclei with varying multi-modalities with highly annotated manual nuclei masks. The dataset consists of images collected from more than 30 varied experiments with different research facilities across various samples under varying imaging conditions, staining protocols, microscopic instruments, and cell lines. The detailed annotations were manually curated by domain experts and biologists, where each annotation done by an expert was peer reviewed by multiple experts. The dataset encompasses typical and crucial cases encountered in pathology. The typical normal samples are common in pathological and clinical domains. There also exists a wide variety of complex data samples which correspond to critical cases.

Standard data augmentation is applied on the dataset before proceeding toward performance comparison. The data samples are resized to 128×128. Image darkening, Gaussian blurring, flipping, and rotation are also performed to further widen and mimic the varying imaging and acquisition conditions.

5.2 Quantitative Evaluation Metrics

For analyzing and comparing the segmentation performances, evaluation metrics including dice coefficient (DC) and root mean squared error (RMSE) are used. Dice coefficient measures the similarity between the generated segmentation output and the ground truth annotated masks and judges the reliability of the segmentation technique in precise labeling of pixel classes. RMSE is useful for measuring the deviation between the actual expected result and the generated results, thereby signifying the resiliency flaw of a technique. The metric calculations are performed based on the Eq. (14).

$$DC = 2\frac{|P_G \cap P_S|}{|P_G| + |P_S|}; \qquad RMSE = \sqrt{\frac{1}{n}\sum_{i=1}^{n}(P_{G_i} - P_{S_i})^2} \tag{14}$$

where P_G and P_S correspond to the ground truth mask and the generated result, respectively.

5.3 Performance Evaluation

Table 1 presents the quantitative evaluation metrics for performance comparison between the existing methods and the proposed segmentation technique. As noticed from the table, the dice coefficient for the proposed method is much higher than the other methods signifying its reliability and accuracy across use cases in nuclei segmentation. The algorithm has a 95.89% average segmentation accuracy value over all the different types of images in the dataset. The performance for accurate segmentation of nuclei from pathology images is stable for both common and critical samples. Furthermore, the average RMSE loss is the lowest for the presented

Table 1 Quantitative nuclei segmentation results corresponding to different techniques on Data Science Bowl dataset (average values obtained over all the images in the dataset)

Algorithm	RMSE	DC
MFCM	82.8186	0.8521
R2UNET	74.0184	0.9108
CGAN	76.4867	0.8921
SABS	78.8764	0.8778
IMRCNN	72.1291	0.9202
DUNET	71.9843	0.9211
MCF	71.5221	0.9234
Proposed method	66.1414	0.9589

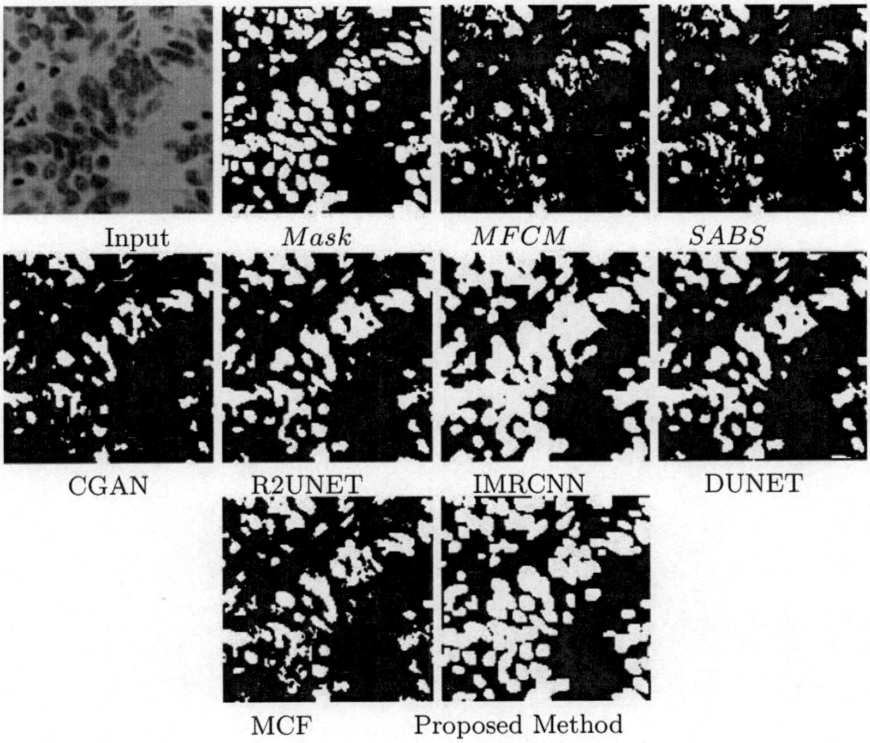

Fig. 1 Qualitative results for densely populated nuclei sample

algorithm (average loss being 66.14), again defending the resiliency of the approach. Qualitative indications suggested in Figs. 1, 2, and 3 also highlight the robustness of the algorithm. It can be seen from Figs. 2 and 3 that the algorithm performs better than its peers in nuclei segmentation on common samples, i.e., detecting medium and sparsely populated nuclei in pathology images. Also, from Fig. 1, it is noticed that the algorithm is stable in segmentation of small and dense cluttered nuclei (which is most rigorous and error-prone during manual segmentation). Overall, the algorithm has a straightforward implementation, and the performance is much better than the existing algorithms across various test cases which makes it a new benchmark for nuclei segmentation.

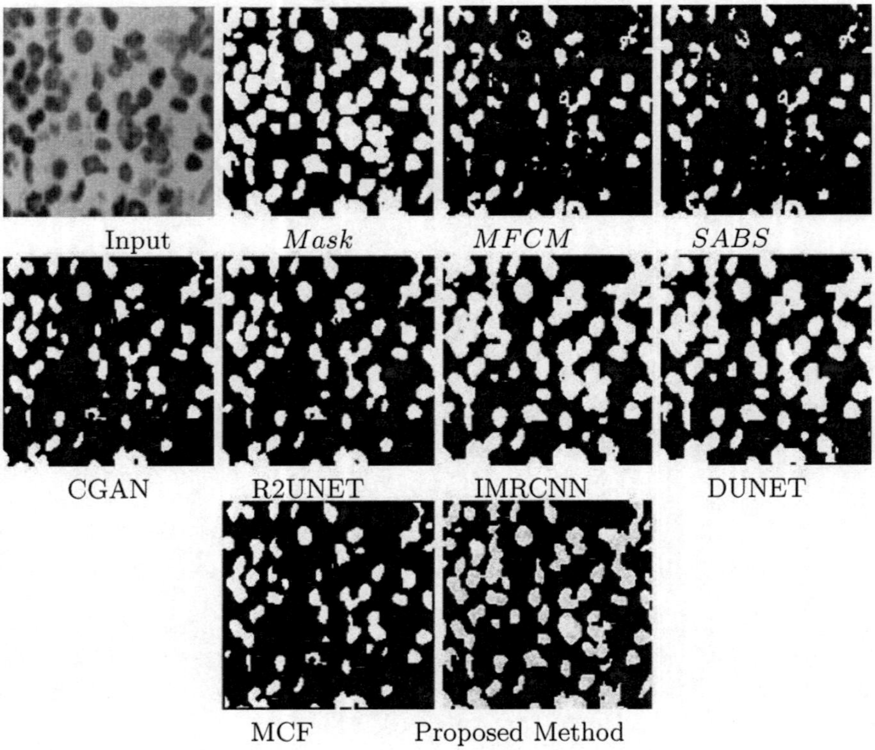

Fig. 2 Qualitative results for medium populated nuclei sample

6 Conclusion

In this paper, a spatial kernelized fuzzy c-means clustering algorithm has been proposed for unsupervised segmentation of nuclei from histopathological images. Experimental results prove the robustness and the efficiency of the proposed approach in the domain of light microscopy medical image segmentation. The algorithm is straightforward, has a low computation power requirement, and is scalable across domains of medical image segmentation. Owing to its simplicity and high reproducibility, it can be integrated with hardware to form an embedded system for real-time segmentation leading to computer-aided analysis and diagnosis. In future, a conditional local feature-based tuning parameter can be introduced into the objective function along with an adaptive window consideration to further enhance the performance of the segmentation algorithm.

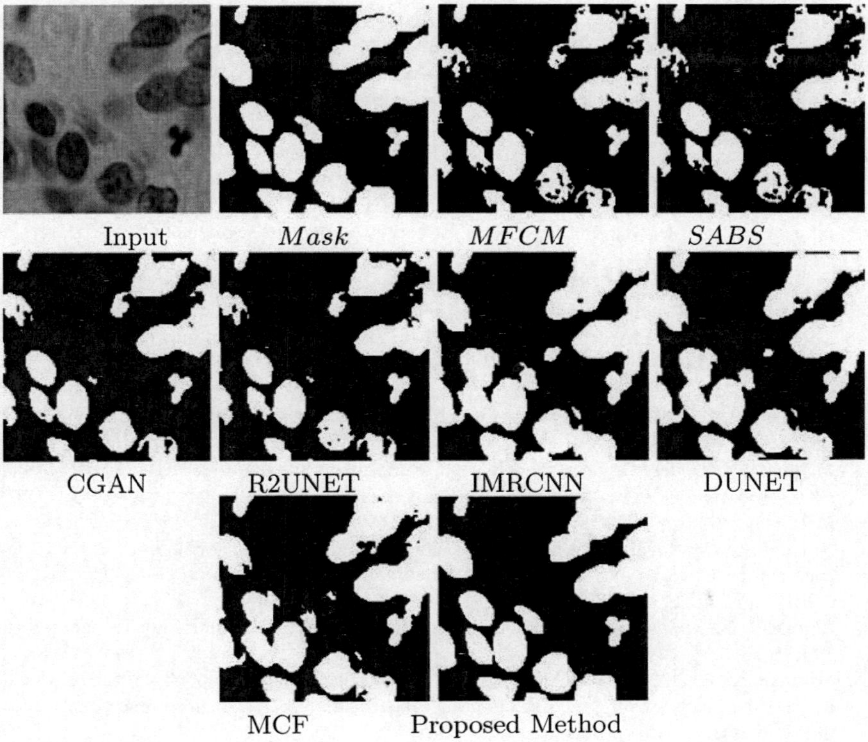

Fig. 3 Qualitative results for sparsely populated big nuclei sample

References

1. Alom MZ, Yakopcic C, Taha TM, Asari VK (2018) Nuclei segmentation with recurrent residual convolutional neural networks based unet (r2unet). In: IEEE national aerospace and electronics conference. IEEE, pp 228–233
2. Belsare A, Mushrif M (2012) Histopathological image analysis using image processing techniques: an overview. Signal Image Proc 3(4):23
3. Caicedo JC, Goodman A, Karhohs KW, Cimini BA, Ackerman J, Haghighi M, Heng C, Becker T, Doan M, McQuin C et al (2019) Nucleus segmentation across imaging experiments: the 2018 data science bowl. Nat Methods 16(12):1247–1253
4. Cui Y, Zhang G, Liu Z, Xiong Z, Hu J (2019) A deep learning algorithm for one-step contour aware nuclei segmentation of histopathology images. Med Biolog Eng Comput 57(9):2027–2043
5. Feng Z, Wang Z, Wang X, Mao Y, Li T, Lei J, Wang Y, Song M (2021) Mutual-complementing framework for nuclei detection and segmentation in pathology image. In: Proceedings of the IEEE/CVF international conference on computer vision, pp 4036–4045
6. Fujita S, Han XH (2020) Cell detection and segmentation in microscopy images with improved mask r-cnn. In: Proceedings of the Asian conference on computer vision
7. Garcia Rojo M, Punys V, Slodkowska J, Schrader T, Daniel C, Blobel B (2009) Digital pathology in europe: coordinating patient care and research efforts. In: Medical informatics in a united and healthy Europe. IOS Press, pp 997–1001

8. Gurcan MN, Boucheron LE, Can A, Madabhushi A, Rajpoot NM, Yener B (2009) Histopathological image analysis: a review. IEEE Rev Biomed Eng 2:147–171
9. Hou L, Nguyen V, Kanevsky AB, Samaras D, Kurc TM, Zhao T, Gupta RR, Gao Y, Chen W, Foran D et al (2019) Sparse autoencoder for unsupervised nucleus detection and representation in histopathology images. Pattern Recogn 86:188–200
10. Jha D, Riegler MA, Johansen D, Halvorsen P, Johansen HD (2020) Doubleu-net: a deep convolutional neural network for medical image segmentation. In: 2020 IEEE 33rd international symposium on computer-based medical systems (CBMS). IEEE, pp 558–564
11. Johnson JW (2018) Adapting mask-rcnn for automatic nucleus segmentation. arXiv preprint arXiv:1805.00500
12. Komura D, Ishikawa S (2018) Machine learning methods for histopathological image analysis. Comput Struct Biotech J 16:34–42
13. Kothari S, Phan JH, Stokes TH, Wang MD (2013) Pathology imaging informatics for quantitative analysis of whole-slide images. J Am Med Inf Assoc 20(6):1099–1108
14. Litjens G, Kooi T, Bejnordi BE, Setio AAA, Ciompi F, Ghafoorian M, Van Der Laak JA, Van Ginneken B, Sánchez CI (2017) A survey on deep learning in medical image analysis. Med Image Anal 42:60–88
15. Mahmood F, Borders D, Chen RJ, McKay GN, Salimian KJ, Baras A, Durr NJ (2019) Deep adversarial training for multi-organ nuclei segmentation in histopathology images. IEEE Trans Med Imag 39(11):3257–3267
16. Naylor P, Laé M, Reyal F, Walter T (2017) Nuclei segmentation in histopathology images using deep neural networks. In: 2017 IEEE 14th international symposium on biomedical imaging (ISBI 2017). IEEE, pp 933–936
17. Rojo MG (2012) State of the art and trends for digital pathology. Stud. Health Technol Inform 179:15–28
18. Sirinukunwattana K, Raza SEA, Tsang YW, Snead DR, Cree IA, Rajpoot NM (2016) Locality sensitive deep learning for detection and classification of nuclei in routine colon cancer histology images. IEEE Trans Med Imag 35(5):1196–1206
19. Viswanathan P (2015) Fuzzy c means detection of leukemia based on morphological contour segmentation. Proc Comput Sci 58:84–90
20. Xu J, Xiang L, Liu Q, Gilmore H, Wu J, Tang J, Madabhushi A (2015) Stacked sparse autoencoder (SSAE) for nuclei detection on breast cancer histopathology images. IEEE Trans Med Imag 35(1):119–130

Tree Detection from Urban Developed Areas in High-Resolution Satellite Images

Pankaj Pratap Singh, Rahul Dev Garg, and Shitala Prasad

Abstract Preserving trees is a challenging area which indeed an automated method to analyze the percentage of trees area in respect of total land area. In this regard, a good level of extraction approach is required for finding trees area. Initially, three image segmentation approaches have implemented for detection of tree areas in urban developed regions, basic color thresholding, automatic thresholding, and region growing segmentation methods. A semi-automatic approach is proposed for detecting tree areas from high-resolution satellite images (HRSI) of urban developed area in this paper. Initially, a pixel-level classifier will train to assign into two-class label {tree and non-tree} to each pixel in a HRSI and later as pixels group. The pixel-level classification is then refined by region growing method in an image to accurately segmentation of tree and non-tree regions. Therefore, this refined segmentation results will show the tree crowns with natural shape. The proposed approach will be trained on an aerial image of different urban developed area. Finally, the outcomes show tree detection results as well as good scalability of this approach.

Keywords Image segmentation · Gray level · Binary image · Image thresholding · Automatic thresholding · Region growing

1 Introduction

In the current scenario of environment, rapid changes are found due to human intervention, but the better classification is still a challenging task to provide the particular

---

P. P. Singh (✉)
Department of Computer Science and Engineering, Central Institute of Technology Kokrajhar, Kokrajhar, Assam, India
e-mail: pankajp.singh@cit.ac.in

R. D. Garg
Geomatics Engineering Group, Department of Civil Engineering, Indian Institute of Technology Roorkee, Roorkee, Uttarakhand, India

S. Prasad
Institute for Infocomm Research, A*Star, Singapore, Singapore

M. Thakur et al. (eds.), *Soft Computing for Problem Solving*, Lecture Notes in Networks and Systems 547, https://doi.org/10.1007/978-981-19-6525-8_19

region such as trees in automated manner. Since long, scientists and engineers have been using remote sensing technology to collect distant information about an object or class [1]. In respective of forest environment, an autonomous navigation is done for detection of tree regions based on the image color and texture features [2]. Image segmentation methods are quite useful to extract the tree region based on thresholding criteria [3]. Level-set method is used in medical image segmentation in successful manner [5–7]. Land cover classification is also quite helpful to identify tree regions, but it also includes the other vegetation regions [8]. Active contour method is also used to identify targeted regions effectively for medical diagnosis in image processing [9].

Identification of tree regions is still a challenging task, due to their irrespective shapes and also their boundary region. One of the major causes can be little spectral variation among the different kind of tree regions which can be matched with other vegetation class. In this regard, to achieve this challenge, the existing classification approaches emphasize to identify only those classes which keep low spectral similarity in high-resolution satellite image (HRSI). Expert system is also used to segment tree regions from the images based not only spectral but also shape feature. In addition, the edge information of tree regions is also quite reasonable to segregate from other image objects [10–13]. Individual tree detection was detected in the orchard using a Gaussian blob model in two-step modeling from VHR satellite images [14]. In the next Sect. 2, a detailed description of the proposed framework for tree region detection is discussed. In the Sect. 3, results are shown with a detailed analysis, and lastly, it is concluded with the proposed approach advantages and also the limitation for future scope.

2 A Designed Framework for Tree Region Detection Using Thresholding Approach

The proposed framework explains thresholding-based approach for detecting the tree regions in HRSI. The separation of two regions (tree and non-tree) is an essential task for achieving the perfect object recognition algorithm in case of trees. This kind of challenge can only be resolved by applying image segmentation on these images. Thus, image segmentation is an important step for recognizing an object. The main purpose of utilizing image segmentation is to split the image into different segments that correspond to substantial objects, or portions of an object, existing in the image. More precisely, image segmentation is the method of assigning tags to every pixel of an image, on the basis of certain matching criteria such that the pixels with the same tag combine to form a relevant region of the image with similar spectral behavior properties to decrease the computational complexity in this way. Hence, this proposed approach is adjustable and utilized for identifying the different objects in satellite images.

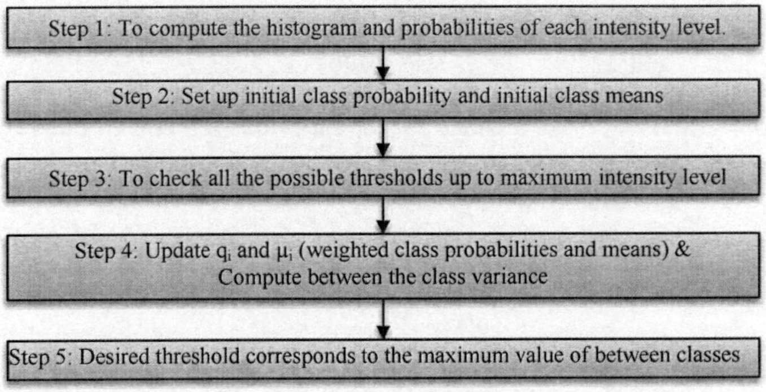

Fig. 1 Detailed step for detection of tree region using automatic thresholding

2.1 *Automatic Thresholding-Based Tree Region Detection in the Satellite Images*

In the regarding of thresholding-based approach, Otsu proposed a method which shows the maximum variance between classes [4]. Due to its simplicity, stability, and effectiveness, it is still in use with various kind of applications. It also performed well due to the automatic selection of threshold, and its time complexity is significantly less in compare to other thresholding methods. This method consists an important properties like the high inter-class variance between tree object and background as a principle to choose the best segmentation threshold. Otsu method chooses the optimal threshold by maximizing the variance between classes, which is equivalent to reducing the variance within class since the total variance (the addition of the intra- and inter-class variances) is constant for different regions. It functions directly on the gray-level histogram, but this method agonizes in presence of noises in an image; it provides acceptable outputs in such kind of scenario. Figure 1 shows detailed steps which are used in the automatic thresholding for detecting tree region.

The stopping criterion for this proposed approach is determined from the Otsu's adaptive thresholding method. In this connection, the threshold value is a decidable criteria which has to be more than the distance between labeled pixel and non-labeled pixel, then can say that both pixels belong to the same type of class region.

2.2 *Region Growing-Based Tree Region Detection in the Satellite Images*

In this proposed approach, region growing (RG) method exploits for extracting the tree regions only with excluding non-tree region. It is based on single-seeded region growing-based algorithm, and seed selection is the initial step of this RG technique. The proposed structure of region growing approach for tree detection in the satellite images is shown in Fig. 2. Initially, a seed selection as an image pixel is done for applying the next step that is based on proximity criteria. It selects the neighboring pixels having similar pixel value as a seed pixel in the HRSI. Thereafter, this region growing approach using these mentioned steps provides segmented tree regions. The key function of RG approach is to segment the image into non-overlapped regions.

It takes seeds as an input, and later, merge pixel based on the similarity criteria and also provide a region correspond to each seed. The result of RG method must follow the given constraints in Eq. (1).

$$\bigcup_{i=1}^{L} R_i = I \tag{1}$$

RC_i shows the connected region as $i = 1, 2, 3, \ldots, n$, where n signifies the number of regions. Now, Eq. (2) delivers mutual exclusion between the region RC_i and RC_j,

$$RC_i \cap RC_j = \text{Null}\, \forall\, i \neq j \tag{2}$$

In this method, proximity pixels are those one which are quite close together which signifies their similar spectral (pixel color) values. Region is grown from the arbitrarily chosen pixel p (single band) by adding in neighboring pixels p* that are similar in region R_{Trees}, increasing the size of the region. It is used to resolve the automatic thresholding (Otsu's method based) limitations. RG techniques start with a single pixel for the target class in a potential region and also growing it by accumulation of adjacent pixels. This process continues till the pixels being compared which are satisfying dissimilarity criteria, and later, it is also classified with the help

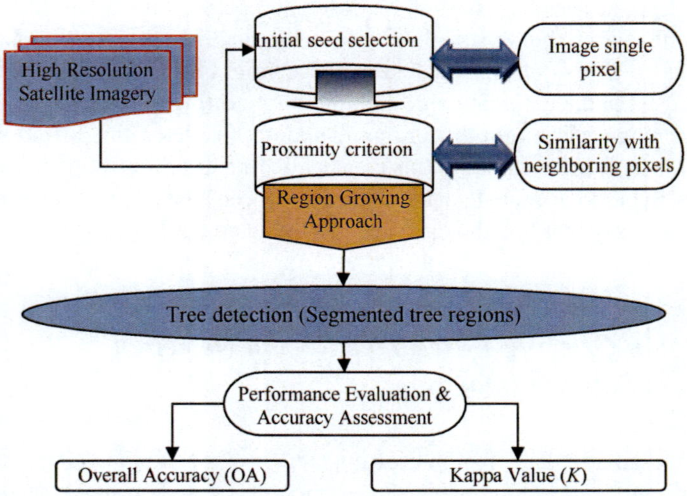

Fig. 2 Detailed framework for detection of tree region in HRSI

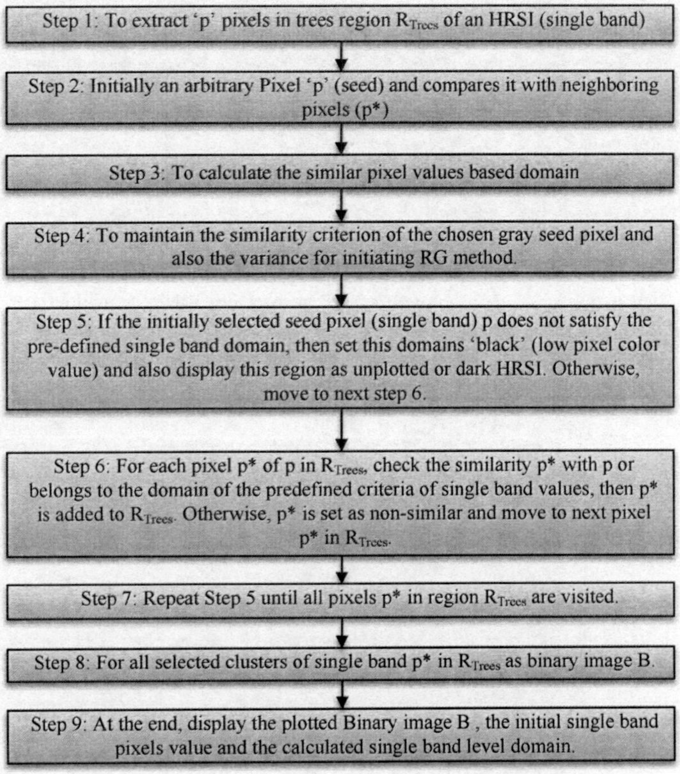

Fig. 3 Detailed steps description for tree region detection using RG approach

of this pixel-based segmentation method. The RG approach has been adapted in this proposed approach, and outputs have been achieved; furthermore, it provides good segmentation results in spectrally similar classes also. Each step of this RG approach for tree region detection is described in detail with the successive steps. Tree detection steps using RG method in HRSI are shown in Fig. 3.

3 Results and Discussion

In this paper, results are evaluated with the satellite images having 1-m spatial resolution which is acquired from Wikimapia. In this present experimental work, the date and place are not the significance factors since the extraction of the objects in the image is the key objective. Figure 4 shows the extracted tree pixels using color thresholding method in HRSI. Figure 5 shows the extracted tree areas (black color pixels) using Otsu's automatic thresholding method in Fig. 5c, d output images (Segmented trees) corresponding to Fig. 5a, b input HRS image.

Fig. 4 Extracted tree pixels using color thresholding method in HRSI: **a** input HRS image, **b** selected pixels from the HRS image, **c** output image (segmented tree region with white color)

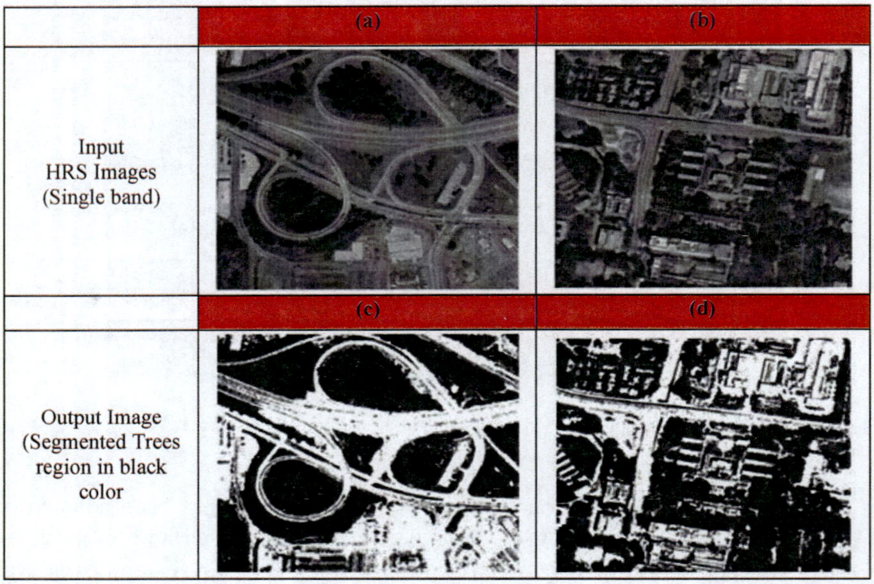

Fig. 5 Extracted tree areas (black color pixels) using Otsu's automatic thresholding method: **a**, **b** input HRS image; **c**, **d** output images (segmented trees)

Figure 6c, d shows the extracted tree areas (white color pixels) using region growing method for the input HRS image in Fig. 6a, b, respectively. Input images are also same in Figs. 5 and 6 which gives a comparison between the results of two methods. In Fig. 5, tree regions are shown with black color, but white color is used to show tree regions in Fig. 6. Finally, a segmented satellite image shows the tree regions, and performance evaluation is explained in next section.

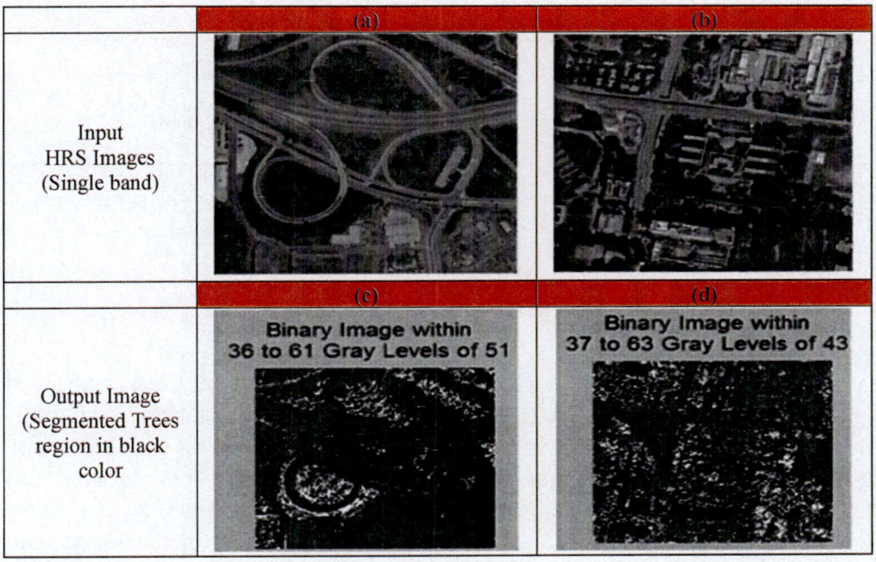

Fig. 6 Extracted tree areas (white color pixels) using region growing method: **a**, **b** input HRS image; **c**, **d** output images (segmented trees region)

3.1 Accuracy Assessment

The segmented tree regions performance is evaluated using the kappa coefficient (κ) and overall accuracy (OA) which is evaluated with the help of fuzzy error matrix corresponding to the results in Figs. 5 and 6. Tables 1 and 2 show the fuzzy error matrices corresponding to Fig. 5c, d as output images (Segmented trees). Table 3 illustrates the comparison results for segmented trees regions in HRSI in Figs. 5 and 6 with the help of kappa coefficient (κ) and overall accuracy (OA).

Table 1 Fuzzy error matrix of segmented trees region image in Fig. 5c

		Soft classified data	
		Trees	Non-trees
Reference data	Trees	0.2760	0.2560
	Non-trees	0.2602	0.2712

Table 2 Fuzzy error matrix of segmented trees region image in Fig. 5d

		Soft classified data	
		Trees	Non-trees
Reference data	Trees	0.2660	0.2420
	Non-trees	0.2532	0.2634

Table 3 Kappa coefficient and overall accuracy of classification of HRSI

Input HRS images	Overall accuracy (OA)		Kappa coefficient (κ)	
	Automatic thresholding method (%)	RG method (%)	Automatic thresholding method	RG method
Figures 5 and 6a	69.35	83.25	0.5426	0.7426
Figures 5b and 6b	61.13	82.73	0.5168	0.7103

4 Results and Discussion

In this paper, our focus was on segmentation and detection between two classes such as trees and non-trees areas in HRSI related to urban developed region. Initially, the performance of existing Otsu's-based method is evaluated for satellite images having thresholding phenomenon which is compared with the proposed RG method. Segmentation methods reflected the limitation for detection of tree areas in urban developed regions which are based on traditional thresholding methods, and later, region growing method is used to improve the segmentation results. The extracted tree region shows the satisfactory results. It implies that the proposed methodologies can also be used to accomplish good image segmentation from HRSI and can also be used in various types of object extraction from the satellite images. It is observed from the segmented results using RG method that the tree region (pervious surface) in different kind of areas yields the higher accuracy values than the thresholding approach. In the future, these results can be used as preprocessed results for object-based classification which can also be more effective in terms of accuracy and interpretation of results. It can also improve the automatic detection of tree crown.

References

1. Jensen JR (2000) Remote sensing of the environment: an earth resource perspective. Prentice-Hall Pub., p 544
2. Ali W (2006) Tree detection using colour, and texture cues for autonomous navigation in forest environment. Master's thesis report, Umeå University, Department of Computing Science, Sweden, June 2006
3. Kamdi S, Krishna RK (2012) Image segmentation and region growing algorithm. Int J Comput Technol Electron Eng 2(1)
4. Otsu N (1979) A threshold selection method from gray level histograms. IEEE Trans Syst Man Cyber 9(1):62–66
5. Shi Y, Karl WC (2005) A fast implementation of level set method without solving partial differential equations. Technical report number, Department of Electrical and computer engineering, Boston University, ECE-2005-02
6. Lankton S (2009) Sparse field methods. Technical Report, 6 July 2009
7. Airouche M, Bentabet L, Zelmat M (2009) Image segmentation using active contour model and level set method applied to detect oil spills. In: Proceedings of world congress of engineering, vol 1

8. Singh PP, Garg RD (2011) Land use and land cover classification using satellite imagery: a hybrid classifier and neural network approach. In: Proceedings of international conference on advances in modeling, optimization and computing. IIT Roorkee, India, pp 753–762
9. Coste A (2012) Active contours models. Image Processing Final Project, December 2012
10. Singh PP, Garg RD (2013) A hybrid approach for information extraction from high resolution satellite imagery. Int J Image Graph 13(2):1340007(1–16)
11. Singh PP, Garg RD (2013) Information extraction from high resolution satellite imagery using integration technique. In: Agrawal A, Tripathi RC, Do EYL, Tiwari MD (eds) Intelligent interactive technologies and multimedia, CCIS, vol 276. Springer, Berlin, Heidelberg, pp 262–271
12. Singh PP, Garg RD, Raju PLN (2013) Classification of high resolution satellite imagery: an expert system based approach. In: 34th Asian conference on remote sensing. Bali, Indonesia, pp SC02(725–732)
13. Singh PP, Garg RD (2016) Extraction of image objects in very high resolution satellite images using spectral behaviour in LUT and color space based approach. In: IEEE technically sponsored SAI computing conference. IEEE, London, UK, pp 414–419
14. Mahour M, Tolpekin V, Stein A (2020) Automatic detection of individual trees from VHR satellite images using scale-space methods. Sensors (Basel) 20(24):7194

Emotional Information-Based Hybrid Recommendation System

Manika Sharma, Raman Mittal, Ambuj Bharati, Deepika Saxena, and Ashutosh Kumar Singh

Abstract In this technology-driven times, recommender systems play a crucial role in providing better user experience and attracting as many users as possible on the website. We propose a hybrid approach for selecting movies that incorporates emotional data in this study. The model will include the benefits of both content-based filtering and collaborative filtering to make up a hybrid model, and then, an additional parameter which is emotions further enhances the accuracy and efficiency of the model. The model is evaluated and compared with some other approaches, and it appears to perform better in a real-time environment. The model also tries to eliminate some of the existing limitations up to some extent.

Keywords Collaborative filtering · Content-based filtering · Emotional information · Hybrid model · Recommendation system

1 Introduction

Almost every platform uses a recommendation system to recommend items to their users. These are systems that analyze a user's behavior, including information on previous preferences, to predict what they need. These systems make it easier for consumers to obtain items they might be interested in and would not have known about otherwise. The most common application areas of recommendation systems include e-commerce, electronic media, and many more. E-commerce organizations

M. Sharma · R. Mittal · A. Bharati (✉) · D. Saxena · A. K. Singh
Department of Computer Application, National Institute of Technology Kurukshetra, Kurukshetra, Haryana 136119, India
e-mail: ambuj_52010039@nitkkr.ac.in

M. Sharma
e-mail: manika_52010047@nitkkr.ac.in

R. Mittal
e-mail: raman_52010026@nitkkr.ac.in

A. K. Singh
e-mail: ashutosh@nitkkr.ac.in

M. Thakur et al. (eds.), *Soft Computing for Problem Solving*, Lecture Notes in Networks and Systems 547, https://doi.org/10.1007/978-981-19-6525-8_20

and streaming platforms such as Amazon and Netflix are best placed to make accurate recommendations because they have millions of customers and information on their online platform. Media companies, similar to e-commerce, provide recommendations on their online platform. It is uncommon to stumble across a news site without a recommendation system. Recommendation systems increase the user performance on the platform. Customer familiarity can be beneficial by getting users to spend longer time on the website, improving their chance of making future purchases and thus increasing the sales. So, both users and companies benefit by using recommendation systems.

The most common approaches used for generating recommendations are content-based filtering, collaborative filtering, knowledge-based filtering, and hybrid filtering. To deliver suggestions to users, these systems employ both implicit and explicit data, such as the user's browsing history and purchases, as well as user ratings. So, the recommendations rely totally on the data provided by the users and their similarity which might not work accurately in certain scenarios. The biggest challenge is the changing needs of users. One day a person might want to watch an action movie but the next day he might be interested in some documentary. Past behavior of users is not entirely an efficient approach because the trends are always changing and users' needs are dynamic especially in times like today when a lot of data is available online for the user to choose. With a large amount of data, the systems need to be highly scalable as well. As the new items and users enter into the system, there is no previous history with the help of which products can be recommended to the users. This poses a main challenge to the recommendation system, and the problem is known as the cold-start problem. Lastly, the user might not always provide ratings to the items. However, if a large number of customers buy the same product but do not leave any comments or ratings, the recommendation engine will find it difficult to suggest that product. This is known as data sparsity problem. In general, data from a system like MovieLens is represented as a user-item matrix populated with movie ratings, so matrix dimensions and sparsity rise as the number of users and movies grows. Data sparsity gives a negative effect on the quality of recommendations that is given by traditional collaborative filtering algorithms.

In this paper, we have taken movies as the domain of recommendation systems. We have proposed a hybrid model of a recommender system combined with an additional parameter which is the emotions of the user at any particular moment of time. Combining users' past behavior with the current emotions, the system can provide the possible list of movies that the user might like. This approach seeks to increase the efficiency and accuracy of recommendation systems to some level while also attempting to resolve any issues that may emerge. As an example, if a new user joins the system, the recommendation algorithm, despite the fact that the user is new and has no browsing history, will suggest movies based on current emotion. Similarly, if a new item enters or there is an item which is not rated by the user, it can still be recommended based on the user's current emotion.

The rest of this paper is structured as follows. In Sect. 2, we go over the related work that has been done with the various primary strategies for recommendation systems. Section 3 explains the proposed model, which is also a contribution of this research. Following that, in Sect. 4, we presented the experiments and results, in which we compared our model's outcomes to previously published results. Finally, Sect. 5 concludes our paper.

2 Related Work

Recommender systems may make personalized and specialized recommendations for its users. Various methodologies for developing recommendation systems have recently been developed. In [1], Kim et al. introduced a recommendation system model that captures six human emotions. This model was created by combining collaborative filtering with user-provided speech emotional information recognition in real time. It comprises primarily an emotion categorization module, a collaborative emotion filtering module, and a mobile application. The emotional model used in this model is Thayer's extended 2-dimensional emotion model. The SVM classifier was also used to identify patterns in the emotion data in the optimized featured vectors. Because of the presence of emotion data, this model provides more accurate recommendations to users.

Phorasim et al. [2] used collaborative filtering to create a movie recommender system that uses the K-means clustering technique to categorize users based on their interests and then finds similarities between the users to generate a recommendation for the active user. The proposed methodology aims to reduce the time it takes to recommend a product to a consumer. Juan et al. [3] developed a hybrid collaborative strategy for overcoming data sparsity and cold-start concerns in customized recommendation systems by using the scores predicted by the model-based personalized recommendation algorithm as features. The idea is to produce a new mode by learning from past data.

Geetha et al. in [4] introduced a movie recommendation system that addresses cold-start problems. Collaborative, content-based, demographics-based, and hybrid techniques are the most common techniques to build recommendation models, and this study tries to address the limitations in each technique separately. To improve the performance of recommender systems, [5] presented a hybrid model that combines content and collaborative filtering with association mining techniques. Other hybridization procedures are studied in the paper, such as the weighted method, which is used to partially address the limitations of prior methods. Additionally, it addresses issues such as cold-start issues also.

Wang et al. [6] is a Sentiment-Enhanced Hybrid Recommender System that focuses on extending the hybrid model by performing sentiment analysis on the output. The algorithm can make informed selections about which product to recommend by understanding the sentiments underlying the user reviews. When compared to existing hybrid models, this strategy yields a model with a high efficiency. Unlike prior methodologies, research in [7] does not solely rely on content or collaborative methods, but rather considers their benefits in order to create a hybrid model. This research combines K-nearest neighbors (KNN) and frequent pattern tree (FPT) to provide good recommendations to researchers, overcoming the drawbacks of existing methodologies. The system solves the problem of a cold start.

MovieMender is a movie suggestion system created by Author in [8] with the objective of helping users in finding movies that match their interests without effort. A user-rating matrix is created once the dataset has been preprocessed. In order to generate a matrix, each user-rating pair is subjected to content-based filtering. In order to produce suggestions for a user that is currently active, collaborative filtering uses matrix factorization to determine the relationship between items and user entities. Pawl Tarnowski et al. presented a paradigm for understanding seven major emotional states such as neutral, joy, surprise, anger, sadness, fear, and disgust based on facial expressions in [9]. The characteristics which are the elements of facial expressions were subsequently classified using the K-nearest neighbor classifier and the MLP neural network. This model delivers good classification results with a 96% (KNN) and 90% (KNN) accuracy for random data division (MLP). In [10], a model for a recommendation system is proposed that will suggest material to users based on their present mood or emotion. In this model's facial expression recognition approach, a Convolutional Neural Network (CNN) captures features from the face picture, which is subsequently followed by Artificial Neural Networks. This model corrects a fault in the old approach and enhances its accuracy by adding one more real-time variable to the system.

3 Proposed Model

Several approaches and methods have already been in use as discussed in the literature part of the paper, and they have their own challenges but the most important thing to observe is that almost all the methods fully rely on the user's previous actions ignoring the present mood of the user. So, in the proposed model we have introduced the emotional information as the new feature in addition to the other features in order to get better and user-friendly results.

As we can see in the process flow diagram illustrated in Fig. 1, the model is feeded with the movie data and requires user id and movies the user has already seen as the input, and in addition to that, it also detects the user's current emotion.

Then, the model uses cosine similarity and Singular Value Distribution methods in order to perform content-based and collaborative filtering. Finally, the emotional information is mapped to the possible genres and combined with the output given by the above methods in order to generate the result which not only relies on the previous action of the user but also takes care of the current mood of the user.

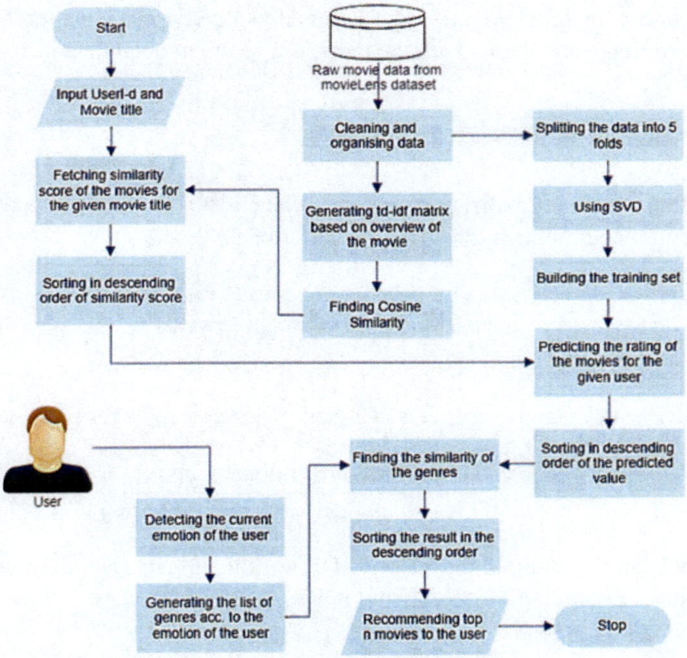

Fig. 1 Process flow diagram of the proposed model

3.1 Content-Based Method

For recommendation, content-based techniques must analyze the products and the user profile. It suggests information based on the user's surfing history, number of clicks, and products seen. This method can suggest unrated things and is entirely dependent on the user's rating. The content of an object might be a pretty abstract concept, so we have a lot of variables to choose from. For instance, when considering a film, we can think about the genre, the actors, movie reviews, and so on. In our algorithm, we can employ only one or a mix of them. Once we have decided which qualities to employ, we will need to convert all of this information into a Vector Space Model, which is an algebraic representation of text documents. Term Frequency and Inverse Document Frequency are concepts that have long been utilized in information retrieval systems, and they are now being employed in content-based filtering recommenders as well. Term Frequency and Inverse Document Frequency are abbreviated as TF-IDF. The TF-IDF can help you figure out how important a word is in a manuscript. They may be used to determine the relative value of a document or a movie, for example. Another crucial notion is the similarity metric, which may determine how similar objects are to one another. One of the most well-known examples in this regard is cosine similarity. The cosine similarity was characterized as follows by Ashwini et al. in [11].

Let us start with Term Frequency, which is illustrated in the equation below and indicates the frequency with which the term 't' appears in document d.

$$f_{t,d} = \sum_{t \in d} f(t, d) \tag{1}$$

Term Frequency normalization is used when a word or term appears more frequently in longer publications than in shorter ones.

$$\text{TF}n = \frac{\text{No. of times the word } t \text{ is used in a document}}{\text{The document's total no. of terms}} \tag{2}$$

where n is normalized.

Now, IDF which stands for Inverse Document Frequency is formally defined as:

$$\text{IDF} = \log \frac{(\text{Total no. of documents})}{(\text{The document's total no. of terms})} \tag{3}$$

The final step is to obtain the TF and IDF weight. TF and IDF are combined to form a matrix. The weight of the TF and IDF is thus expressed as:

$$\text{TF-IDF Weight} = \text{TF}(t, d) * \text{IDF}(t, D) \tag{4}$$

Following the calculation of the TF-IDF weight, the next step is to determine the similarity measure using that weight. The similarity metrics can be shown on a plot, with the coordinates indicating each user (or object). The similarity between two coordinates is determined by the distance between them. The higher the resemblance, the shorter the distance. The first stage is to locate users (or objects) who are similar, which is done using the ratings supplied by the users. We have employed the following methods, which are among the most common: cosine similarity.

The angle between two n-dimensional vectors in the vector space is used to calculate cosine similarity, which estimates the distance between them. When applying this to the recommender system, we consider the item user to be an n-dimensional vector and the resemblance between them to be an angle. The smaller the angle, the closer the pieces are (or users). The dot product is important for defining similarity because it is directly related to it. The ratio of their dot product and the product of their magnitudes is used to determine how similar two vectors u and v are.

$$\text{similaritiy} = \cos(\theta) = \frac{u.v}{\|u\| \|v\|} \tag{5}$$

If the two vectors are the same, this will be one, and if they are orthogonal, it will be zero, according to the concept of similarity. In other words, the similarity is a number between 0 and 1 that represents how similar the two vectors are. The pseudo-code for our implementation of content-based method is shown below:

```
INPUT: movie_title, movie_dataset
v1: = movie_title.genres
for x in movie_dataset:
  v2: = x.genres
  cos_sim: = (v1*v2)/|v1||v2|
  movie_dataset['similarity"]: = cos_sim
sort (movie_dataset, key: = similarity, reverse: =true)
OUTPUT: movie_dataset
```

3.2 Collaborative Filtering Method

Collaborative methods work by identifying commonalities between users and proposing things they use. Memory-based approach and model-based approach are two basic classes of collaborative approaches. The memory-based approach follows a three-step process: determining the degree of similarity between the training users and the target user, finding the target user's nearest neighbors (i.e., users who are very similar to the target user), and generating a final list of recommendations instead of using the data directly, and the user-rating behavior is used to extract the parameters for the model, resulting in improved accuracy and performance.

NormalPredictor and BaselineOnly are two extremely simple algorithms that may be utilized for collaborative filtering. The NormalPredictor method predicts a random rating based on the training set's distribution, which is deemed normal. For a given user and item, the BaselineOnly method forecasts the baseline estimate. There are other KNN-based algorithms as well. The KNN approach is slow, a technique of non-parametric learning. It makes predictions for fresh samples using a database in which the data points are divided into many clusters. Rather than making any assumptions about the underlying data distribution, KNN relies on item feature similarity. When KNN infers about a song, for example, it calculates the distance between the target song and every other song in its database, ranks the distances, and returns the top k most related song choices.

All non-negative matrix-factorization approaches are based on matrix factorization. SVD stands for Singular Value Decomposition with an implicit rating. Probabilistic matrix factorization is the SVD algorithm's counterpart. In this study, we have used the SVD algorithm. The SVD algorithm is described by Aggarawal in [12] as follows: SVD is a matrix-factorization approach that reduces the number of features in a data collection by lowering space dimensions from *N* to *K*, where *K* is less than *N*. However, for recommendation systems, the component of the matrix factorization that maintains the dimensionality constant is all that matters. The user-item rating matrix is used to factorize the matrix. Matrix factorization may be thought of as the process of finding two matrices whose product is the original matrix. A vector 'q_i'

can be used to represent each object. Similarly, each user may be represented by a vector 'p_u', with the predicted rating being the dot product of those two vectors.

$$\hat{r}_{ui} = p_u^T q_i + b_{ui} \tag{6}$$

Here, $\hat{r}_{\text{ui}}$ is the predicted rating, p_u is the user matrix, q_i is the item matrix, and b_{ui} is baseline prediction. We use stochastic gradient descent to build our output matrices because our input matrix is sparse. We iterate through the supplied ratings, minimizing the RMSE with each iteration.

$$\min\left(\sum_{(u,i)\in\kappa}\left(r_{ui} - \mu - b_u - b_i - p_u^T q_i\right)^2 + K\left(b_u^2 + b_i^2 + ||p_u||^2 + ||p_i||^2\right)\right) \tag{7}$$

Here, κ is the set of all present ratings and K is the regularization constant. The pseudo-code for our implementation of collaborative filtering method is shown below:

```
INPUT: user_ID, movie_dataset,ratings_dataset
title = movie_dataset.title
list_movie_ID = ratings.movieID
for x in list_movie_ID:
  pred_ratings = svd.predict(user_ID,x)
  movie_dataset['est_pred'] = pred_ratings
sort (movie_dataset, key: = est_pred, reverse: =true)
OUTPUT: movie_dataset
```

3.3 *Methods for Evaluating the Models*

Now, we will look at some methods for determining if a model overfits or underfits. Dietmar et al. in [13] says any model's ultimate objective is to perform well for any new data. So, what are our options? The dataset utilized is split into two parts: The first part is training, and the other one is testing. The model is trained using train data, and it is evaluated using test data. In an ideal circumstance, we would divide the dataset in an 8:2 ratio, with 80% of the data utilized to train the model and 20% to test the model. Allowing nonlinear data to be scattered and fitting it with a linear model can result in underfitting, making this method useless for training data. Overfitting, on the other hand, performs well with training data but not so well with test data. In this situation, the model is well suited to the data distribution area. The following are some of the evaluation approaches we have used.

Root Mean Square Error (RMSE)
The root mean square error is a typical means of calculating a model's error in predicting quantitative data. Its formal definition is as follows:

$$\text{RMSE} = \sqrt{\sum_{i=1}^{n} \frac{(\hat{y}_i - y_i)^2}{n}} \tag{8}$$

Here, $\hat{y}_i$ is predicted value, y_i is observed value, and n is no. of observations.

Mean Absolute Error (MAE)
MAE is one of the several measures for describing and evaluating a machine learning model's quality. The MAE is the mean of all recorded absolute errors, and error refers to the difference between the predicted value and actual value:

$$\text{MAE} = \frac{\sum_{i=1}^{n} |y_i - x_i|}{n} \tag{9}$$

Here, y_i is the predicted value, x_i is the actual value, and n is the no. of observations.

Qualitative and Quantitative Analysis
This study compares different systems with two distinct qualities. Metrics like RMSE and MAE, which were discussed in the preceding subsections, are used in the quantitative component. The qualitative element, on the other hand, is determined by the quality of the recommendation, which we assess by looking at the created recommendation.

4 Experimentation and Results

4.1 Setup

The experimental setup includes Anaconda which is a well-known machine learning and data science tool. It is a Python and R language distribution that is free and open source. We have used Jupyter Notebook and Python language for coding purposes.

4.2 Dataset Used

We used the 'MovieLens 1M Dataset' for this research. Since the site's launch in 2000, 1,000,209 anonymous reviews of about 3900 films have been submitted by 6040 MovieLens members. We specifically used two files: ratings and movies. UserID, MovieID, Rating, and Timestamp were the four fields present in the rating file. After analyzing the rating file, we found that UserIDs vary from 1 to 6040. Ratings are given on a 5-star scale (only whole-star ratings are accepted), the date is in seconds since the epoch, and each user has at least 20 ratings, while MovieIDs range from 1 to 3952.

In the movie file, there were three fields. They are MovieID, Title, and Genres, and we discovered that Titles are comparable to IMDB titles (containing year of premiere), and Genres are pipe-separated and picked from the following genres: Adventure, Animation, Children's, Comedy, Crime, Documentary, Drama, Fantasy, Film-Noir, Horror, Musical, Mystery, Romance, Science Fiction, Thriller, War, and Western are only some of the genres represented.

On the datasets, we did some preliminary exploratory analysis. Figure 2 shows a histogram of average ratings provided by users. As can be observed, this plot resembles a normal distribution with a left heavy tail. Between 3.5 and 4 stars is the average user rating.

Figure 3 shows the histogram of average ratings for each item. This chart resembles a normal distribution with a lengthy left tail. In this scenario, though, the values are more evenly distributed. The majority of the goods have been given a rating of 3–4.

The histogram of ratings is shown in Fig. 4. The most common ratings are 4 and 3 correspondingly, which is similar with the preceding two graphs.

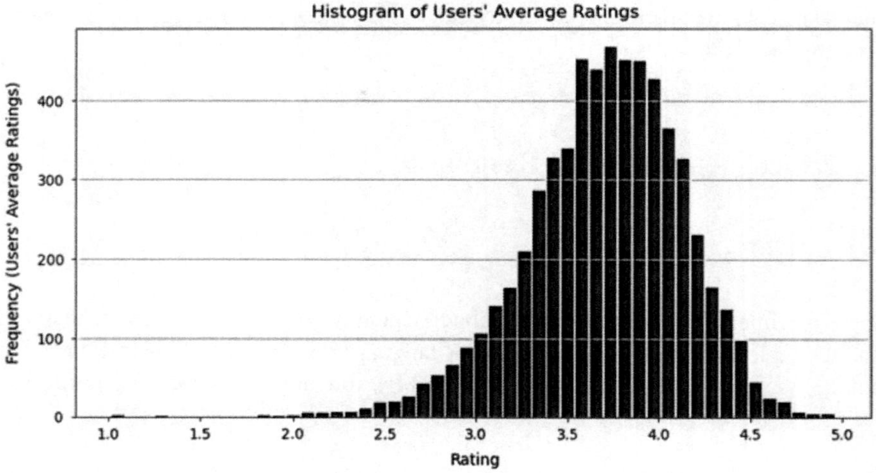

Fig. 2 Histogram of users' average ratings

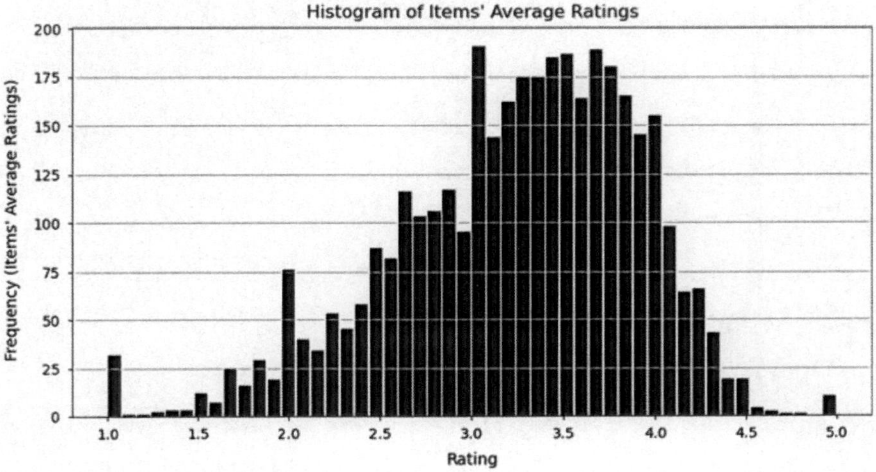

Fig. 3 Average rating of items (histogram)

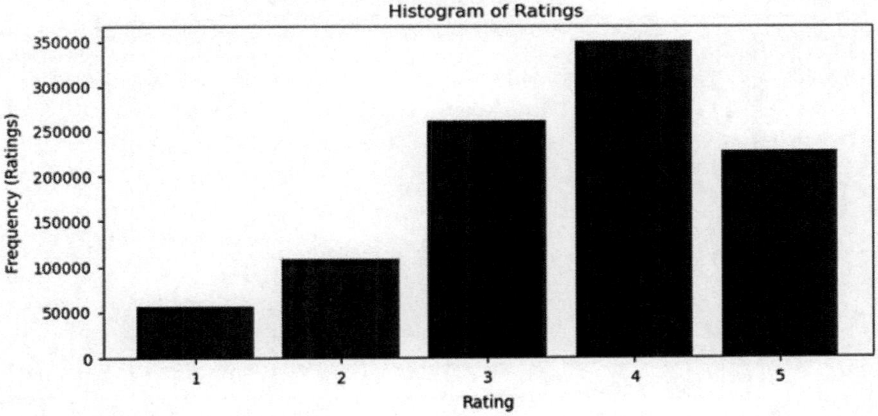

Fig. 4 Histogram of ratings

Figures 5 and 6 illustrate the histograms of objects rated by users and users who rated items. As can be seen from these two graphs, most users only rank a few items in result analysis.

4.3 Quantitative Analysis

We begin by comparing the RMSE and MAE errors between a collaborative filtering and a hybrid system. We will address the content-based filtering approach and

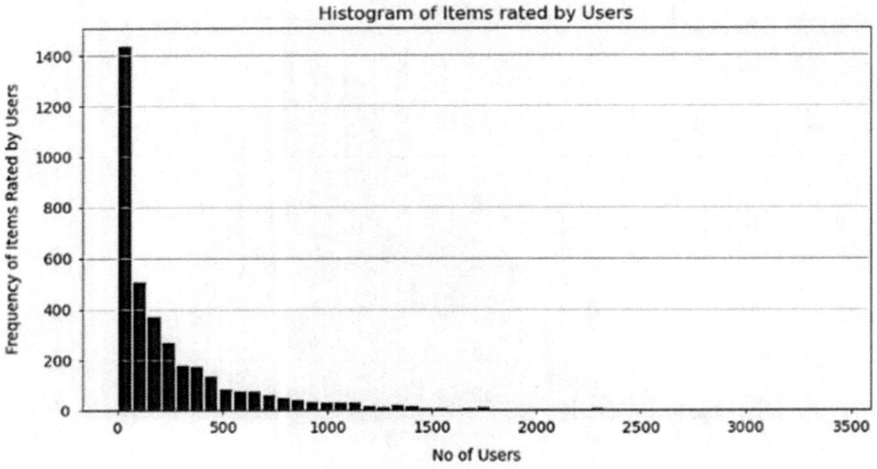

Fig. 5 Product rated by users (histogram)

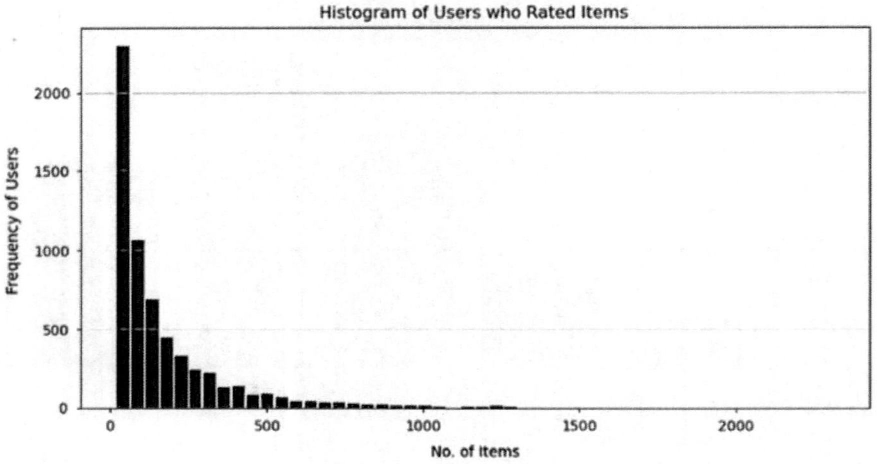

Fig. 6 Histogram of users who rated items

emotional information-based hybrid system in the next section because they only have a qualitative property. We select top-recommended movies for ten users from both systems and calculate RMSE errors for each system to compare. The RMSE plot for ten users is shown in Fig. 7, demonstrating that the hybrid system has a reduced overall RMSE. The hybrid system's superiority is also demonstrated by the average RMSE plot in Fig. 8.

We next do the same analysis for MAE, and as shown in Figs. 9 and 10, the hybrid recommendation system has a lower MAE, implying greater accuracy.

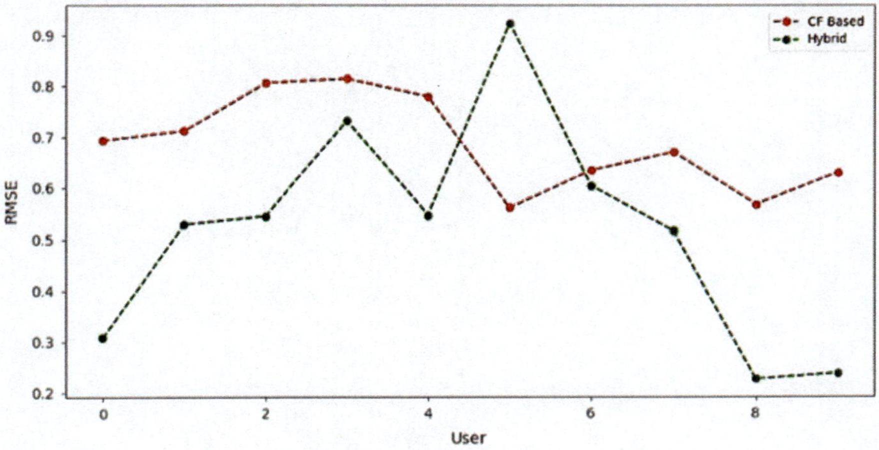

Fig. 7 RMSE of hybrid recommendation system and collaborative filtering based

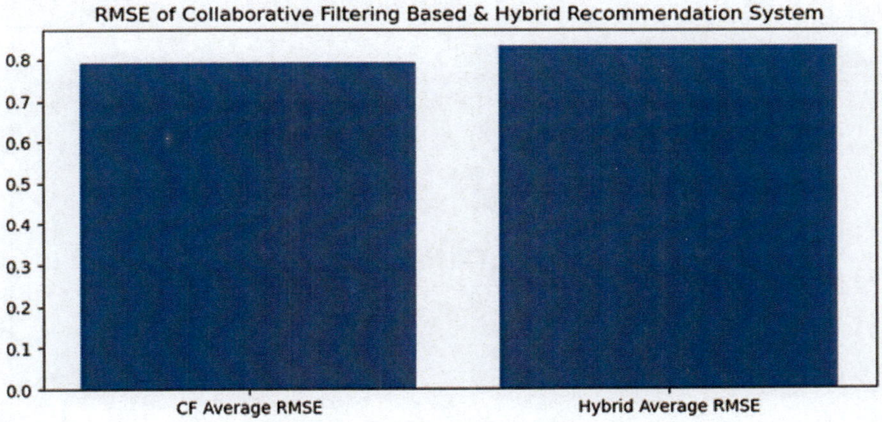

Fig. 8 Average RMSE of hybrid recommendation and collaborative filtering based

4.4 Qualitative Analysis

Collaborative filtering can forecast which movies a user is more likely to enjoy, as seen in Table 1. It does not, however, have a way of proposing similar films to a specific one depending on the user's interests. The genre column reveals that the genres are all over the place. In this scenario, we will look at User 2 and recommend the top ten movies he will likely enjoy.

A content-based system, on the other hand, can find the most similar movies to a given one (see Table 2), but it has no way of knowing if a user will enjoy it. Here, we look at the movie Chicken Run (2000), which has a Movie ID of 3751, and suggest the top ten movies that are comparable to it. Chicken Run is an animated film that

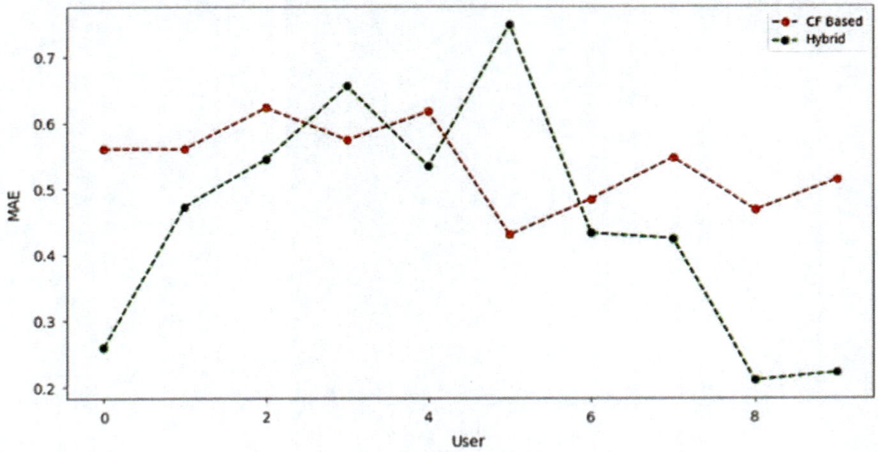

Fig. 9 MAE of hybrid recommendation system and collaborative filtering based

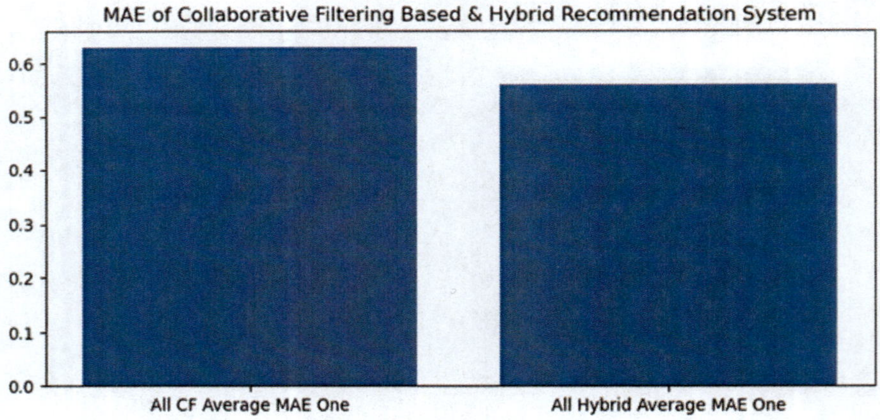

Fig. 10 Average MAE of hybrid recommendation and collaborative filtering based

falls within the comedy genre; thus, the genres of the recommended films are quite similar to the genre of the Chicken Run, as shown in Table 2.

With a hybrid system, we get the perfect blend. Table 3 shows how it can propose similar films to a certain one that the customer is likely to appreciate. Using User ID 1 and Movie ID 3751, We have compiled a list of the top ten movies that are similar to Chicken Run and are likely to receive good ratings from User 1.

In addition, we are attempting to improve the results of these models by including a new component, emotional information. The recommendation system must recognize and reflect the user's unique qualities and situations, such as personal preferences and moods, in order to increase user pleasure. As shown in Table 5, the hybrid system's results for the movie Chicken Run and User 1 are reordered based

Table 1 Collaborative filtering's top ten recommended movies for a certain user

Movie Id	Estimated rating	Title	Actual rating	Genres
914	4.725468	[My Fair Lady (1964)]	[]	[["Musical", "Romance"]]
1148	4.655111	[Wrong Trousers, The (1993)]	[]	[["Animation", "Comedy"]]
2905	4.593890	[Sanjuro (1962)]	[]	[["Action", "Adventure"]]
1223	4.572039	[Grand Day Out, A (1992)]	[]	[["Animation", "Comedy"]]
2565	4.559226	[King and I, The (1956)]	[]	[["Musical"]]
50	4.549067	[Usual Suspects, The (1995)]	[]	[["Crime", "Thriller"]]
3030	4.520446	[Yojimbo (1961)]	[4]	[["Comedy", "Drama", "Western"]]
745	4.519310	[Close Shave, A (1995)]	[]	[["Animation", "Comedy", "Thriller"]]
1784	4.515350	[As Good As It Gets (1997)]	[5]	[["Comedy", "Drama"]]
3578	4.491399	[Gladiator (2000)]	[5]	[["Action", "Drama"]]

on the similarity score between the genres of the suggested movies and the genres determined by the users' emotions. To use emotions as a parameter in our model, we must map emotions with the genres of the movies. We tried to collect the correspondence of human emotions with the genres through a research survey, and the result of the survey is shown in Table 4. With the help of that survey, we got to know the user's preferred genre when confronted with any emotion. The survey included people majorly in the age group of 18–25 years. The result shows that around 29% of people would like to watch comedy movies when faced with an emotion of anger, 32% of people would like to watch comedy movies and 25% would like to watch horror movies when faced with an emotion of fear, 32% of people would like to watch comedy movies when sad, 22% of people would like to watch comedy movies when feeling disgusted, 20% of people would like to watch comedy movies when feeling happy, and 26% of people would like to watch thriller movies when surprised.

4.5 Result Comparison

We used the findings from a model called "hybrid recommendation system" built by researchers at IIT Kanpur's Department of Computer Science and Mathematics [14], who conducted a comparative examination of algorithm performance using the same dataset. The findings achieved by them and the findings achieved by our model

Table 2 The top ten recommended movies for a certain film according to a content-based system

Movie index	Similarity score	Title	Movie Id	Genres
1050	1.000000	Aladdin and the King of Thieves (1996)	1064	[["Animation", "Children's", "Comedy"]]
2072	1.000000	American Tail, An (1986)	2141	[["Animation", "Children's", "Comedy"]]
2073	1.000000	American Tail: Fievel Goes West, An (1991)	2142	[["Animation", "Children's", "Comedy"]]
2285	1.000000	Rugrats Movie, The (1998)	2354	[["Animation", "Children's", "Comedy"]]
2286	1.000000	Bug's Life, A (1998)	2355	[["Animation", "Children's", "Comedy"]]
3045	1.000000	Toy Story 2 (1999)	3114	[["Animation", "Children's", "Comedy"]]
3542	1.000000	Saludos Amigos (1943)	3611	[["Animation", "Children's", "Comedy"]]
3682	1.000000	Chicken Run (2000)	3751	[["Animation", "Children's", "Comedy"]]
3685	1.000000	Adventures of Rocky and Bullwinkle, The (2000)	3754	[["Animation", "Children's", "Comedy"]]
236	0.869805	Goofy Movie, A (1995)	239	[["Animation", "Children's", "Comedy", "Romance"]]

are shown in Table 6, and we can clearly see that the RMSE value of our model is lower and lower RMSE value implies better accuracy of the model.

As a result, we may conclude that a hybrid recommendation system outperforms a single collaborative filtering or content-based filtering system in both qualitative and quantitative aspects. Furthermore, the emotional information-based hybrid model outperforms the hybrid models statistically for a specific user.

4.6 Future Insights

As the amount and quality of data grow, current algorithms will need to scale effectively. Further study in the area of recommendation systems may reveal some

Table 3 Top ten movies via hybrid recommendation system for a specific user

Movie index	Similarity score	Title	Movie Id	Estimated rating	Actual rating	Genres
2285	1.000000	Rugrats Movie, The (1998)	2354	4.265341	[]	[["Animation", "Children's", "Comedy"]]
1050	1.000000	Aladdin and the King of Thieves (1996)	1064	3.923892	[]	[["Animation", "Children's", "Comedy"]]
2073	1.000000	American Tail: Fievel Goes West, An (1991)	2142	3.917660	[]	[["Animation", "Children's", "Comedy"]]
3685	1.000000	Adventures of Rocky and Bullwinkle, The (2000)	3754	3.891245	[]	[["Animation", "Children's", "Comedy"]]
3682	1.000000	Chicken Run (2000)	3751	3.778759	[]	[["Animation", "Children's", "Comedy"]]
3542	1.000000	Saludos Amigos (1943)	3611	3.504144	[]	[["Animation", "Children's", "Comedy"]]
3045	1.000000	Toy Story 2 (1999)	3114	3.224993	[]	[["Animation", "Children's", "Comedy"]]
2072	1.000000	American Tail, An (1986)	2141	3.137429	[]	[["Animation", "Children's", "Comedy"]]
2286	1.000000	Bug's Life, A (1998)	2355	2.968066	[]	[["Animation", "Children's", "Comedy"]]
236	0.869805	Goofy Movie, A (1995)	239	3.756252	[]	[["Animation", "Children's", "Comedy", "Romance"]]

approaches to deal with this expanding volume of data that will solve the existing recommender systems' scalability problem. The strategy employed in this study can be applied to a variety of fields, including e-commerce, music, books, and many more.

The present emotion of the user is detected using facial recognition in this research. It can be further enhanced by incorporating speech recognition or semantic analysis to obtain data on the user's current area of interest. Many such approaches can be identified in the near future that could result in the increased reliability and efficiency of the recommender systems.

Table 4 Survey result on the correspondence of human emotion with the genres

Genres\emotions	Angry	Fear	Sad	Disgust	Happy	Surprise
Action	21.7	0.0	1.4	7.1	7.1	0.0
Adventure	4.3	4.3	7.1	5.7	5.7	5.7
Animation	1.4	7.2	5.7	7.1	2.9	5.7
Comedy	27.5	37.7	15.7	22.9	14.3	4.3
Crime	11.6	2.9	7.1	2.9	2.9	2.9
Documentary	4.3	2.9	10.0	1.4	7.1	8.6
Drama	5.8	1.4	10.0	12.9	2.9	5.7
Fantasy	4.3	0.0	4.3	5.7	11.4	7.1
Horror	1.4	27.1	2.9	8.6	4.3	2.9
Mystery	0.0	5.7	2.9	5.7	7.1	11.4
Romance	7.2	5.7	24.3	12.9	20.0	11.4
Sci-Fi	1.4	2.9	2.9	2.9	8.6	10.0
Thriller	8.7	2.9	5.7	4.3	5.7	24.3

Table 5 Top ten movies via emotion-based hybrid recommendation system for a specific user

Title	Genres	Similarity
Goofy Movie, A (1995)	[["Animation", "Children's", "Comedy", "Romance"]]	40.1
American Tail: Fievel Goes West, An (1991)	[["Animation", "Children's", "Comedy"]]	20.1
Rugrats Movie, The (1998)	[["Animation", "Children's", "Comedy"]]	20.1
Aladdin and the King of Thieves (1996)	[["Animation", "Children's", "Comedy"]]	20.1
Adventures of Rocky and Bullwinkle, The (2000)	[["Animation", "Children's", "Comedy"]]	20.1
Chicken Run (2000)	[["Animation", "Children's", "Comedy"]]	20.1
Saludos Amigos (1943)	[["Animation", "Children's", "Comedy"]]	20.1
Toy Story 2 (1999)	[["Animation", "Children's", "Comedy"]]	20.1
American Tail, An (1986)	[["Animation", "Children's", "Comedy"]]	20.1
Bug's Life, A (1998)	[["Animation", "Children's", "Comedy"]]	20.1

Table 6 RMSE values for different models by IITK

IITK model's result			Our model's result	
S. No	Method	RMSE value	RMSE value	MAE value
1	Collaborative filtering (SVD)	0.942863	0.685817	0.536795
2	Hybrid system	0.915856	0.516602	0.450063

5 Conclusion

Due to the constantly changing needs of users, it has become important to recommend items to the user as per their needs and preferences. The various approaches used in recommender systems suggest items on the basis of the user's past records and behavior. So, combining the existing approaches we introduced a hybrid model in this paper which proved to have less error in comparison with the individual methods. Emotional information of the user further enhances the hybrid recommender system with even a lower error score than the hybrid method. A survey has also been conducted to know how user's behavior and the movie's genre are correlated with each other. It is expected that the model will try to eliminate some current limitations of the recommender system and increase its efficiency.

References

1. Kim T-Y, Ko H, Kim S-H, Kim H-D (2021) Modeling of recommendation system based on emotional information and collaborative filtering. Sensors 21:1997. https://doi.org/10.3390/s21061997
2. Phorasim P, Yu L (2017) Movies recommendation system using collaborative filtering and k-means. Int J Adv Comput Res 7:52–59. https://doi.org/10.19101/IJACR.2017.729004
3. Juan W, Yue-xin L, Chun-ying W (2019) Survey of recommendation based on collaborative filtering. J Phys Conf Ser 1314:012078. https://doi.org/10.1088/1742-6596/1314/1/012078
4. Geetha G, Safa M, Fancy C, Saranya D (2018) A hybrid approach using collaborative filtering and content based filtering for recommender system. J Phys Conf Ser 1000:012101. https://doi.org/10.1088/1742-6596/1000/1/012101
5. Shah JM, Sahu L (2015) A hybrid based recommendation system based on clustering and association. Binary J Data Mining Netw 5:36–40
6. Wang Y, Wang M, Xu W (2018) A sentiment-enhanced hybrid recommender system for movie recommendation: a big data analytics framework. Wirel Commun Mob Comput 2018:1–9. https://doi.org/10.1155/2018/8263704
7. Bhatt B, Patel PJ, Gaudani H (2014) A review paper on machine learning based recommendation system. IJEDR 2
8. Hande R, Gutti A, Shah K et al (2016) MOVIEMENDER—a movie recommendation system. IJESRT 5:469–473. https://doi.org/10.5281/zenodo.167478
9. Tarnowski P, Kołodziej M, Majkowski A, Rak RJ (2017) Emotion recognition using facial expressions. Procedia Comput Sci 108:1175–1184. https://doi.org/10.1016/j.procs.2017.05.025
10. Iniyan S, Gupta V, Gupta S (2020) Facial expression recognition based recommendation system. Int J Adv Sci Technol 29:5669–5678
11. Lokesh A (2019) A comparative study of recommendation systems. Thesis, Western Kentucky University
12. Aggarwal CC (2016) Recommender systems. Springer International Publishing, pp 113–117
13. Jannach D, Zanker M, Felfernig A, Friedrich G (2011) Recommender systems: an introduction. Cambridge University Press, pp 166–188
14. Patel K, Sachdeva A, Mukerjee A (2014) Hybrid recommendation system

A Novel Approach for Malicious Intrusion Detection Using Ensemble Feature Selection Method

Madhavi Dhingra, S. C. Jain, and Rakesh Singh Jadon

Abstract Nowadays, machine learning-based intrusion detection is a hot topic of research. Analysing the nodes in the network or the underlying traffic can both be used to identify network behaviour. In recent years, malicious traffic has become a big issue. The process of detecting malicious traffic using machine learning algorithms is described in this study. The characteristics of network traffic can be studied to monitor its behaviour. Because working with a large number of features is time consuming, the proposed feature selection approach is used on a standard dataset. For categorising hostile traffic, the suggested work uses an ensemble feature selection approach and multiple machine learning classification algorithms. When the accuracy of the model developed using existing common feature selection techniques and the proposed ensemble-based technique is compared, it is discovered that the ensemble method produces more promising outcomes.

Keywords Intrusion detection · Malicious · Network attack · Feature engineering

1 Introduction

Behaviour of node in the wireless network is attracting many researchers in the field of wireless security. Identification of node as malicious or non-malicious is an important task which can prevent the intrusions in the network. As a result, determining the node's normal and malicious behaviour is critical. Any malicious node violates the triangle of security principles: availability, confidentiality, integrity and non-repudiation. An attacker can take advantage of these security flaws to compromise the network's security information. By periodically replaying, reordering, or

M. Dhingra (✉) · S. C. Jain
Amity University Madhya Pradesh, Maharajpura Dang, Gwalior, MP 474005, India
e-mail: madhavi.dhingra@gmail.com

S. C. Jain
e-mail: scjain@gwa.amity.edu

R. S. Jadon
MITS, Gwalior, MP, India

M. Thakur et al. (eds.), *Soft Computing for Problem Solving*, Lecture Notes in Networks and Systems 547, https://doi.org/10.1007/978-981-19-6525-8_21

discarding packets, as well as providing bogus routing messages, the attacker can launch a variety of denial-of-service (DoS) assaults.

The wireless networks have two types of nodes based on their behaviour [1].

Normal node—a normal node performs the task according to the defined protocol while maintaining the network's security.

Malicious node—a malicious node is the one that violates any of the security standards. Such a node might have a negative impact on the network, lowering its performance. As a result, different methods were utilised to identify and remove the rogue node. Malicious or vulnerable behaviour can take many forms, including packet loss, battery depletion, bandwidth consumption, linking issues, resource denial, data manipulation and the insertion of duplicate packets, among others.

This paper has proposed a new ensemble feature selection method based on which classification of malicious nodes is performed. The outline of the paper is such that Sect. 2 highlights the literature review in the corresponding domain. Section 3 describes the proposed work. Sections 4 and 5 show the experimental results and discussion followed by the last section of conclusion.

2 Related Work

Intrusion detection examines the behaviour of both the typical user and the invader, assuming that they operate in distinct ways [2]. According to their context, IDS is divided into anomaly and misuse detection systems [3, 4]. Classification-based detection is a kind of intrusion detection in which statistical significances are used to distinguish between normal with misbehaving nodes. It looks for any deviation between the actual value and the predicted value of the selected features and identifies the malicious action on the basis of it. The performance of the classification method depends on the following:

- selection of proper set of features,
- selection of proper classifier that categorises the calculated values of a feature into defined classes and
- training the classifier over a wide range of scenarios.

Intelligent systems have always aided decision-making and the verification of various constraints. Since the last few years, intelligent intrusion detection systems have grown, analysing both the network and the host to produce a variety of outputs [5, 6]. Such systems operate as a rule-based system that generates results based on the rules [7]. Intelligent IDS is created with the use of intelligent preprocessing and categorisation techniques.

In preprocessing techniques, feature selection is one of the major steps performed while designing intelligent IDS. It has several benefits like improving the accuracy of machine learning algorithm, knowing the data in a more efficient manner and helping in analysing it. It also minimises the storage and reduces the computation costs [8, 9].

Classification techniques work in two phases: first is training part, where the learning model is developed on the training dataset, and then second is testing part, where the selected learning model is tested on the test data and the instance is defined as normal or abnormal [10, 11]. Classification methods work both for one-class classification and multi-class classification. Several intelligent classification methods like neural networks, decision tree, Naive Bayes, etc., exist and are used extensively in the previous researches [12–14].

3 Proposed Work and Implementation

Data from the real-time network must be obtained for the building of a machine learning model for intrusion detection. The process of collecting real-time data in order to access the attacks is quite costly. The study activity necessitates the use of real databases on which data analysis can be carried out in a systematic manner, and some significant findings may be obtained. The results of a small database cannot be generalised. As a result, typical genuine datasets of good quality are employed to construct a learning model.

UNSW-NB 15 dataset [15] has total of 49 features containing a class feature which determines that the traffic category is normal or malicious [16, 17]. Forty-nine features are categorised into five groups: flow, basic, content, time, and additionally generated.

The attributes of the dataset are divided into flow-based features, basic features, content-based features, time-based features, connection-based features and class-based features. The research work is done by loading the dataset into the WEKA platform [18].

3.1 Proposed Ensemble-Based Feature Selection

An ensemble-based feature selection (EFS) approach is used in the proposed work to execute feature selection using the three feature selection methods that were chosen. The features are reduced to 17 in the proposed work, including four derived features and two class features.

The first fifteen ranking features of the selected filter methods are used in this phase. The filter methods used are correlation, gain ratio and information gain. To establish the ranks of the features, attribute filter methods are applied to the training dataset. The result of the three filters is transferred to the final feature generating process, which uses the threshold value to construct the output feature set. The value of the minimal threshold is set to two. The common characteristics of the three filter techniques depending on the minimal threshold value are combined in the final output feature set, which is utilised for training and testing set classification.

Ensemble Feature Selection (EFS) Algorithm

Step-1: Begin.
Step-2: Let $Xi = X1, X2, \ldots . X39$ and $Ci = C1, C2,$
where $Xi = featuresofdataset, Ci = class$
Step-3: For each $Fi = F1, F2, \ldots$ do
$Fi' = X1', X2', \ldots, X15'$, based on $Ri = Rank(i = 1to15)$
Step-4: Combine $Fi''\ asFi'' = F1', F2', \ldots$
Step-5: Calculate Threshold by FC, feature count
Step-6: $ForeachXi'\ inFi''$ '
$If(n(Xi') >= FC)Then$
$FSi = add(Xi'),$
Else $drop(Xi')$, where$FSi(FinalFeatureSet)$,
Step-7: FSi is the outcome set of features.
Step-8: End.

Following the feature selection result, the following preprocessing procedures are conducted, which include the insertion of new features and the removal of redundant features.

The first feature is tput(throughput), which is calculated using the algorithm below. tput = (a6+a7)/a2, where a6 represents spkts, a7 represents dpkts, and a2 represents dur. In order to calculate the tput, a2 must never be 0. As a result, data preprocessing is performed by utilising Weka's RemovewithValue filter to remove any rows with dur = 0.

The second feature, ploss (Packet Loss), is calculated as ploss = a15+a16, where a15 represents sloss and a16 represents dloss.

tjitter = a19+a20, where a19 is sjit and a20 is djit, is the third feature.

After computing some new features, the previous features become irreversible, such as a6 and a7 being eliminated after calculating tput. a13, a14, a15, a16, a19 and a20 are also eliminated.

tcprtt is the result of adding snack and ackdat, i.e. a17 = a18+a19 (given in dataset). As a result, a18 and a19 are eliminated from the dataset as well.

Following these five processes, the 11 most significant traits are selected in order to predict the correct assault class. The four derived features and two class features are added to these 11 features, resulting in a new training dataset of 17 features, as illustrated in Table 1.

3.2 Training Process

The UNSW-NB 15 dataset training dataset is subjected to the EFS, which reduces the number of features from 49 to 17, including a class feature. There are 14,325 instances in total. Figure 1 depicts the display of a reduced training dataset.

Table 1 Reduced features of the training dataset

S. No.	Features	S. No.	Features	S. No.	Features
1	id	8	tcprtt	15	ploss
2	service	9	smean	16	tjitter
3	sbytes	10	dmean	17	tload
4	dbytes	11	ct_state_ttl		
5	sttl	12	attack_cat		
6	dttl	13	label		
7	swin	14	tput		

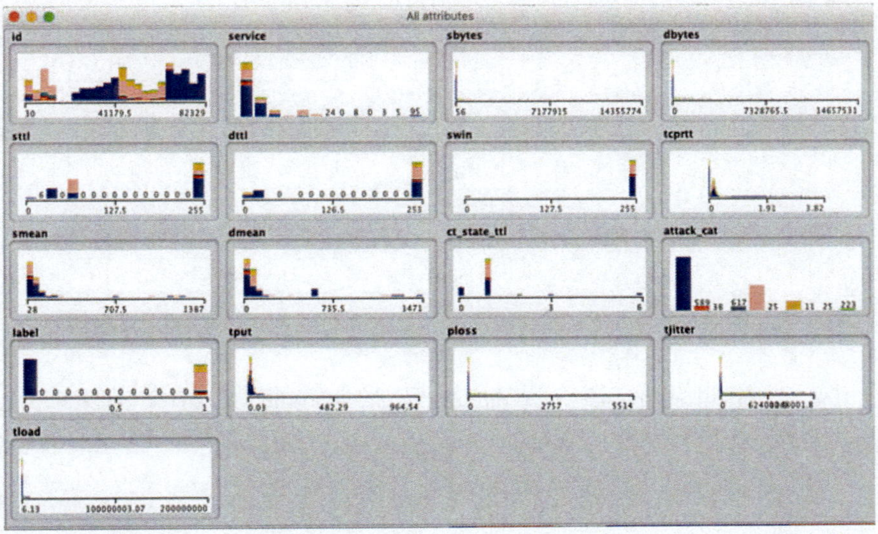

Fig. 1 Visualisation of modified dataset features

The selected six classification methods were used to classify the reduced training dataset, and the results are displayed in Table 2.

Seeing the results, it is clear that the classification algorithms LazyIBK and RandomTree are performing better in comparison with other classification algorithms.

3.3 *Testing Process*

Testing process is performed with the same features of the training dataset. The total instances in the testing dataset are 30,676. The testing dataset is loaded in the WEKA, and the results achieved from the classification algorithms are given in Table 3.

Table 2 Results of classifiers on training dataset

S. No.	Classifier algorithm	Testing time (in seconds)	Accuracy (%)
1	DecisionTable	0.04	89.16
2	Bagging	0.09	95.12
3	MLP	0.21	89.70
4	RepTree	0.03	93.59
5	LazyIBK	13.04	100
6	RandomTree	0.12	100
7	J48	0.05	95.95

Table 3 Results of classifiers on testing dataset

S. No	Classifier algorithm	Testing time (in seconds)	Accuracy (%)
1	DecisionTable	0.27	35.64
2	Bagging	0.43	83.25
3	MLP	0.64	70.49
4	RepTree	0.25	83.59
5	LazyIBK	30.13	71.46
6	RandomTree	3.31	78.81
7	J48	0.27	81.07

It is obvious from the accuracy rate of the classifiers that RepTree and Bagging produce better outcomes.

4 Analysis and Discussion

4.1 Feature Selection Method Based Results

Following feature selection techniques have been applied on the original dataset, and results were analysed.

- Correlation-based feature selection (CFS) feature selection
- Correlation attribute feature selection
- Proposed ensemble feature selection

The training and testing model constructed for the three reduced datasets produced distinct outcomes, which were analysed for the three dataset perspectives using the parameters below:

1. Accuracy: Figure 2 shows that the proposed EFS approach is more accurate than other methods in terms of accuracy.

$$\text{Accuracy} = \frac{\text{TP} + \text{TN}}{\text{TP} + \text{TN} + \text{FP} + \text{FN}}$$

where TP = True Positive Rate, TN = True Negative Rate, FP = False Positive Rate, and FN = False Negative Rate.

2. Recall: In comparison with the existing feature selection methods, the suggested method has a higher detection rate (as shown in Fig. 3).

$$\text{Recall} = \frac{\text{TP}}{\text{TP} + \text{TN}}$$

where TP = True Positive Rate and TN = True Negative Rate.

3. Testing Time: Our proposed method requires extremely little testing time (shown in Fig. 4).

As a result, it is a more practical and efficient approach for larger real-time datasets. The LazyIBK classifier has a testing time of 169 s; hence, it is not included in the graph.

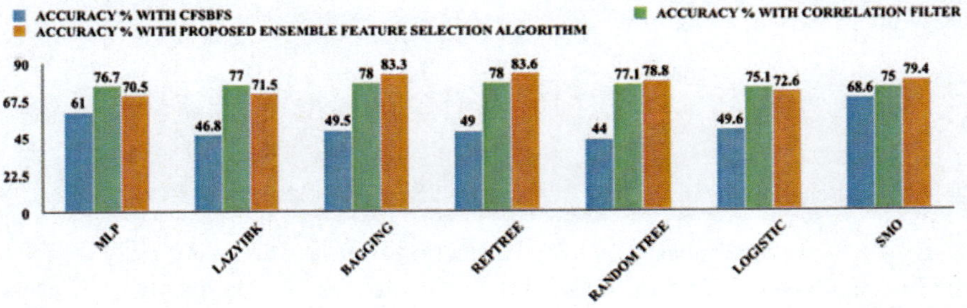

Fig. 2 Accuracy

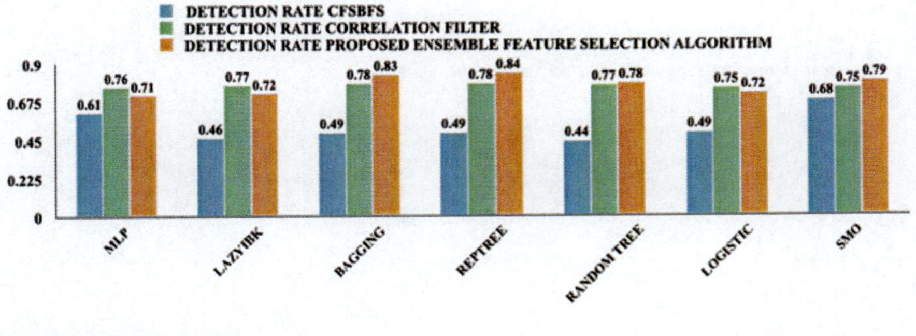

Fig. 3 Recall (detection rate)

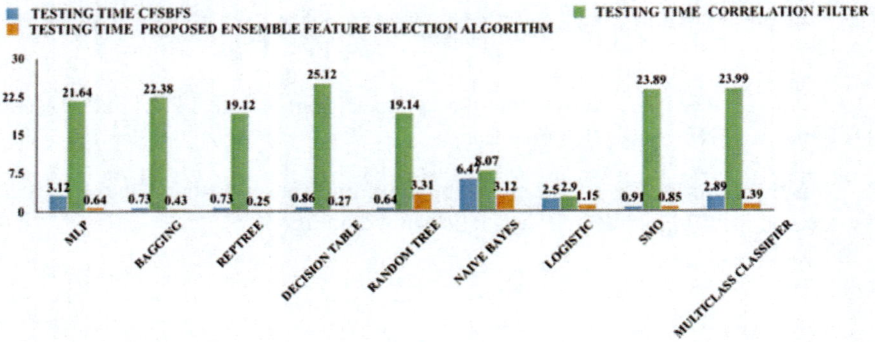

Fig. 4 Testing time

4.2 *Classifier-Based Results on EFS Applied Dataset*

The performance of the classification algorithms is shown on the basis of the following parameters.

1. True Positive Rate (TP Rate): It is the highest in RepTree and Bagging classifiers (shown in Fig. 5).

$$\text{TPR} = \frac{\text{TP}}{\text{TP} + \text{FN}}$$

2. ROC Area: It is the highest in Bagging and RepTree classifiers (shown in Fig. 6).

By seeing the results based on feature selection and classifier, it is clearly seen that the selected features have a great importance while training and testing of datasets. The size of the datasets is generally too large, and if smart selection of features is not done, then it may lead to low accuracy results. Thus, an efficient algorithm is required which can select the key features that have a major impact on the results of

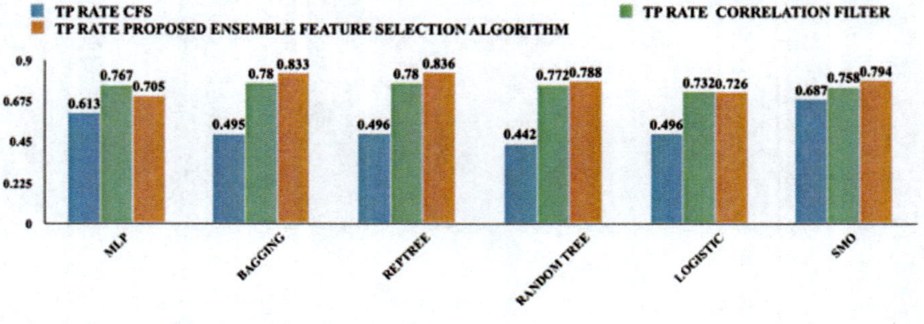

Fig. 5 True Positive Rate

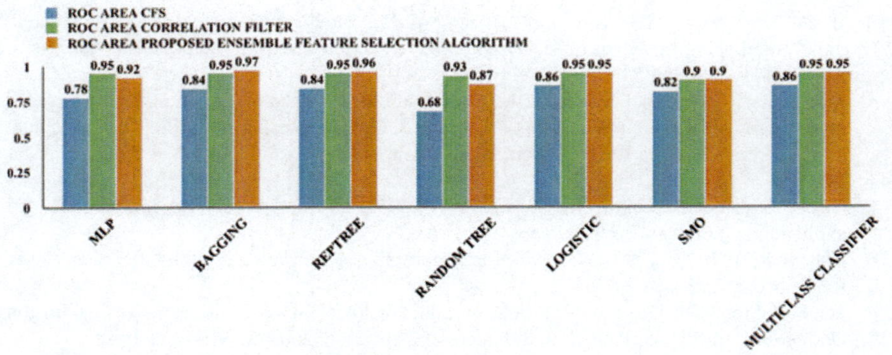

Fig. 6 ROC area

the dataset. Ensemble feature selection algorithm is working in the same manner so as to improve the overall results.

5 Conclusion

Malicious intrusion is an important field of research in network security. Past few researches had given good results, but the analysis of features of network traffic represents more clear picture of attacks. The proposed ensemble-based feature selection has utilised this concept, has transformed the large dataset and has given the prominent features responsible for the attacks. Along with the reduction of dataset, it was found that the classification algorithms are performing better with the reduced dataset obtained from ensemble feature selection approach. The classification of attacks is measured by using different performance parameters including accuracy rate, testing time, false detection rate, etc. The Bagging and RepTree classifiers are giving best results with respect to each parameter. The testing time has also been greatly reduced with the proposed EFS approach. The proposed approach can be used for more recent real-time attack datasets so as to identify the key features causing the attack. The future work involves the study of node behaviour according to the features, due to which the node becomes malicious in the network.

References

1. Rai AK, Tewari RR, Upadhyay SK (2010) Different types of attacks on integrated MANET-internet communication. Int J Comput Sci Secur 4(3):265–275
2. Franklin S, Graser A (1996) Is it an agent or just a program? In: ECAI '96 Proceedings of the workshop on intelligent agents III, agent theories, architectures, and languages. Springer, London

3. Jaisankar N, Yogesh SGP, Kannan A, Anand K (2012) Intelligent agent based intrusion detection system using fuzzy rough set based outlier detection. In: Soft computing techniques in vision science, SCI 395. Springer, pp 147–153
4. Magedanz T, Rothermel K, Krause S (1996) Intelligent agents: an emerging technology for next generation telecommunications. In: INFOCOM'96 Proceedings of the fifteenth annual joint conference of the IEEE Computer and Communications Societies, San Francisco, Mar 24–28
5. Guyon I, Gunn S, Nikravesh M, Zadeh L (2006) Feature extraction: foundations and applications. Springer, Berlin
6. Kohavi R, John G (1997) Wrappers for feature subset selection. Artif Intell J Spec Issue Relevance pp 273–324
7. Tan P-N, Steinbach M, Kumar V (2005) Introduction to data mining. Addison-Wesley, Boston
8. Duda RO, Hart PE, Stork DG (2000) Pattern classification, 2nd edn. Wiley, Hoboken
9. Stefano C, Sansone C, Vento M (2000) To reject or not to reject: that is the question: an answer in the case of neural classifiers. IEEE Trans Syst Manag Cyber 30(1):84–94
10. Sivatha Sindhu SS, Geetha S, Kannan A (2012) Decision tree based light weight intrusion detection using a wrapper approach. Expert Syst Appl 39:129–141
11. Ghadiri A, Ghadiri N (2011) An adaptive hybrid architecture for intrusion detection based on fuzzy clustering and RBF neural networks. In: Proceedings of the 2011 ninth IEEE conference on annual communication networks and services research conference, Otawa. IEEE Computer Society, Washington, pp 123–129
12. Chebrolu S, Abraham A, Thomas JP (2005) Feature deduction and ensemble design of intrusion detection systems. Comput Secur 24(4):295–307
13. Zhang W, Teng S, Zhu H, Du H, Li X (2010) Fuzzy multi-class support vector machines for cooperative network intrusion detection. In: Proceedings of 9th IEEE international conference on cognitive informatics (ICCI'10). IEEE, Piscataway, pp 811–818
14. Zadeh L (1998) Role of soft computing and fuzzy logic in the conception, design and development of information/intelligent systems. In: Proceedings of the NATO advanced study institute on soft computing and its applications held at Manavgat, Antalya, Turkey, 21–31 Aug 1996, vol 162. of NATO ASI Series. In: Kaynak O, Zadeh L, Turksen B, Rudas I (eds) Computational intelligence: soft computing and fuzzy-neuro integration with applications. Springer, Berlin, pp 1–9
15. UNSW-NB15 dataset. Available: http://www.unsw.adfa.edu.au/australian-centre-for-cyber-security/cybersecurity/ADFA-NB15-Datasets. Retrieved 15 Dec 2016
16. Moustafa N, Jill S (2015) UNSW-NB15: a comprehensive data set for network intrusion detection systems (UNSW-NB15 network data set). In: Military communications and information systems conference (MilCIS). IEEE
17. Moustafa N, Jill S (2016) The evaluation of network anomaly detection systems: statistical analysis of the UNSW-NB15 dataset and the comparison with the KDD99 dataset. Inf Secur J Glob Perspect, pp 1–14
18. Kalmegh SK (2015) Analysis of WEKA data mining algorithm REPTree, simple cart and RandomTree for classification of Indian news. IJISET—Int J Innov Sci Eng Technol 2(2): 438–446a, ISSN 2348-7968

Automatic Criminal Recidivism Risk Estimation in Recidivist Using Classification and Ensemble Techniques

Aman Singh and Subrajeet Mohapatra

Abstract Committing a crime and getting sanctioned for parole and bail is highly exposed to the risk of recidivism rate in the current world. Track the risk of every recidivist in real time is a huge assignment. As a result, a computer-assisted technique for early risk assessment among habitual offenders employing quantitative analysis is recommended, such as machine learning. Thus, the current study proposes multiple classification and ensemble machine learning techniques to assess the risk of recidivism in every individual recidivist. The proposed system helps in dividing the recidivist based on their risk vulnerability. The quantitative analysis is done based on the survey questionnaire, which consists of socio-economic and demographic factors with a well-known risk assessment tool, HCR-20. Stratified K-fold cross-validation is used to eliminate bias and create a more resilient system. The simulation results on the datasets show that the treebagger ensemble model outperforms with 79.24% accuracy then the traditional classification techniques in terms of accuracy, AUC, and F-measure.

Keywords Criminal recidivism · Recidivist · Classification · Ensemble learning · Risk assessment · Treebagger

1 Introduction

Criminal recidivism is a global problem that must be handled to maintain social unity and stability. Criminal recidivism is a traitor sickness that repeatedly causes criminals to perpetrate the same crime. A "recidivist" is a person or person who has been convicted of crimes multiple times under the Indian Penal Code (IPC). Criminals who have a high chance of re-offending must be identified and prohibited from

A. Singh (✉) · S. Mohapatra
Department of Computer Science Engineering, Birla Institute of Technology Mesra, Ranchi, Jharkhand, India
e-mail: amansingh281291@gmail.com; phdcs10068.17@bitmesra.ac.in

S. Mohapatra
e-mail: subrajeet@bitmesra.ac.in

M. Thakur et al. (eds.), *Soft Computing for Problem Solving*, Lecture Notes in Networks and Systems 547, https://doi.org/10.1007/978-981-19-6525-8_22

release. Screening and sorting convicted criminals into recidivist categories aid the government in reducing recidivism and lowering the rising crime rate. Quantitative criminology and quantitative psychology are valuable tools for investigating criminal behavior. Data are crucial in extracting hidden insights into the present world. Data mining, pattern recognition, the KDD process, and other methods for discovering knowledge from datasets exist. Many scholars are concentrating their efforts on figuring out how to extract information from the data that is accessible or how to use data to solve real-world problems. Analyzing such a large amount of information is challenging; hence specific methods and approaches for classifying a tremendous amount of information are essential. Recidivism behavior can be identified by various statistical and risk assessment tools, with a prediction accuracy of around (0.65–0.74) [1]. While using these instruments with state-of-the-art techniques, we find a better prediction accuracy and minimize the prediction error [2]. There are different categories of crime where researchers have implemented machine learning and classification techniques to get more fair results, e.g., predicting recidivism in homicide offenders using classification tree analysis, predictive utility of LR, and classification for inmate misconduct, decision tree approach for domestic violence as well as sexual violence, burglaries, homicide, theft, and violent recidivism [3–6]. Many studies have also been made on the fairness and effectiveness of the machine learning approach in predicting recidivism [7]. Apart from their predictive validity, these models focus on fewer attributes or common characteristics to deduce the conclusion.

The literature on how machine learning outperforms human judgment in predicting and assessing danger among convicted criminals is abundant. The following are some of the most recent researchers:

Fredrick David and Suruliandi [8] have covered all of the data mining approaches that have been used in crime research and prediction so far.

Mehta et al. [9] discussed how classification strategies are used to reduce criminal recidivism and describes the level of risk that can be split into three categories: low, moderate, and high.

Kirchebner et al. [10] find critical characteristics in schizophrenic offender patients assessed as recidivists for criminal recidivism.

Ghasemi et al. [11] have described how machine learning techniques can improve the predictive validity of a risk assessment tool.

Watts et al. [12] described how actuarial risk estimators and machine learning models could detect the risk and vulnerability among psychiatric disorder offenders.

Singh and Mohapatra [13] Using the ensemble learning technique, scholars in India highlighted the importance of psychological qualities and contextual elements that can affect first-time offenders in committing a crime.

Ngo [14] has attempted to assess the efficacy of parole and re-entry programs among recidivists in the United States, intending to lower the recidivism rate in the federal system.

Aziz et al. [15] describe a soft computing-based approach for crime data prediction that combines several regression techniques with machine learning models and concludes how these crime analysis techniques can assist reduce recidivism rates in India.

Following a thorough examination and analysis of the literature, it is concluded that using classifications and an ensemble learning-based method to reduce recidivism rates and assess the risk level of recidivists has sufficient potential. The conduct of a criminal is collected in the proposed study using a survey questionnaire that includes psychological information, the personality of individual recidivists, socio-economic factors, demographic characteristics, and historical parameters. It is further examined by an expert panel where scores for each attribute have been assigned; these attributes or features will be evaluated using machine learning techniques. The workflow of the proposed system has been introduced in Fig. 1, and here, we can see how the data was collected, standardized, and verified. We were able to determine how independent risk factors and socio-economic and demographic factors can help in determining the likelihood of recidivism in repeat offenders based on our extensive literature study. Many studies have utilized various classification algorithms to determine the level of risk, but none have coupled ensemble bagging and boosting techniques to obtain a reliable result.

The data and requisite methods are described in Sect. 2 of the article. The experimental results and discussion are presented in Sect. 3, and the conclusion is presented in Sect. 4.

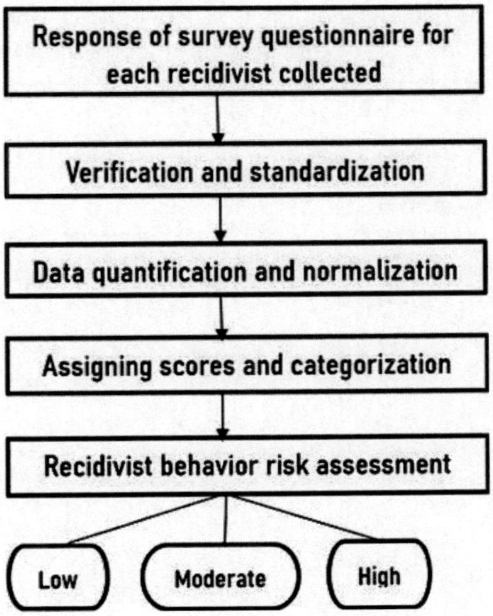

Fig. 1 Workflow of the automated recidivism detection

2 Data and Methods

In this section, we will go over the specifics of gathering behavioral data and the structure of the proposed method:

2.1 Study Subject Selection

In order to reintegrate repeat offenders back into society, risk assessments are required. The present recidivism rate in the Indian state of Jharkhand is roughly 6% [16], but the apprehension rate is substantially higher, owing to a lack of computer-assisted systems in the criminal justice system. Most of the culprits who were apprehended were granted bail or release due to the absence of evidence. Several Jharkhand prisons were surveyed for information on recidivist behavior. The present data are raw data acquired for the purpose of conducting an experiment [17]. Over the course of four months, trained correctional counselors and clinical psychologists examined the inmate's susceptibility. The experimental study included 104 male recidivist data, and the Jail Superintendent provided rigorous guidelines. The ages of the individuals ranged from 18 to 45, with a minimum of two convictions in the previous five years. A group of psychologists designed and approved the questionnaire used to collect the data. In addition, it included the H-10 subscale of the widely used risk indicator HCR-20 [18]. In addition, the present observation behavior was examined by the lead investigator and Jail Superintendent, and their criticism was taken into account to remove bias from the data (Fig. 2).

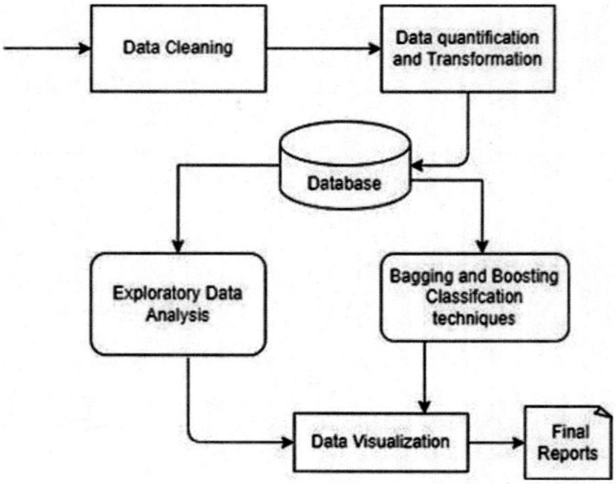

Fig. 2 Proposed architecture for automatic recidivism risk assessment

2.2 Data Acquisition

The information was gathered via a survey questionnaire that included personality, parental and family components, as well as environmental, demographic, socio-economic, and offense details, as well as a common risk assessment instrument called HCR-20 and cumulative jail behavior elements. Counselors conducted individual interviews with each participant, which took approximately 25–30 min to complete.

2.3 Data Preprocessing

This is the essential phase for the data, as all the irrelevant, unwanted, and redundant items are removed. In this phase, preprocessing the data is cleaned and removal of incomplete and inaccurate data is done manually. Out of the original data, the cases with missing values on predictors have been removed. Some of the factors in the data are noisy enough to be excluded from our preliminary experiments. After completion of this phase, 2 participants' responses were omitted due to a lack of sufficient values. Now, 102 participants' responses are used for processing to the recommended system. A random sampling approach is used to remove bias from the raw data. In addition, independent attributes must be normalized and feature scaling performed. The essential scaling and scoring methodologies have been carried out based on the recommendations from the expert panel.

2.4 Data Quantification and Transformation

A robust representation of behavioral traits is essential for computer-assisted risk assessment. The questionnaire has a total of 37 qualities, 36 of which were exclusively taken into account for this study. After converting all of the participants' responses to quantitative values, conventional psychological scaling was applied. The entire scoring and scale procedure were carried out under the supervision of an expert panel. HCR-20 is widely considered for measuring violent behavior and risk factor; in our study, the participants were interviewed based on HCR 20, where the scoring of each item was either 0 (absent), 1 (minor or moderately present), or 2 (definitely present) which total sums up to maximum 40 and minimum 20. These psychological scores were further verified and converted into numeric scores by the panel such that we can train the system. Since the HCR-20 is a well-known scale, and a subscale of it is employed in the experiment, no changes to the scoring of the H-10 subscale were made.

2.5 *Classification*

Classification [19] is a term used to describe any situation in which a specified type of class label must be predicted from a given field of data. There are four types of classification techniques, in which the multiclass classification technique is used in the proposed work. The datasets are divided into two categories: training and testing, which are used to train and test the system separately. The art of bringing together a different group of learners (individual models) to increase the model's stability and predictive performance is known as ensemble learning [20]. A generalized ensemble model structure is presented in Fig. 3. Here, in the proposed work, we are using two of three techniques of the ensemble; bagging and boosting [21].

1. Bagging is a term used to describe a group of independent learners who are all learning independently and averaging the results of each model.
2. The boosting ensemble technique is a sequential learning strategy, in which the outcomes of the base model or algorithms are influenced by the results of prior learners.

2.5.1 Classification Methods

a. Naïve Bayes classifier [22] is a classification technique based on the Bayes theorem, which uses probabilistic theory to classify data. It is explicitly designed for classification problems where it assumes the presence or absence of features independent of any other feature.
b. Multilayer perceptron [23] is the most widely used classifier based on an artificial feed-forward neural network developed to perform nonlinear mappings. MLP can

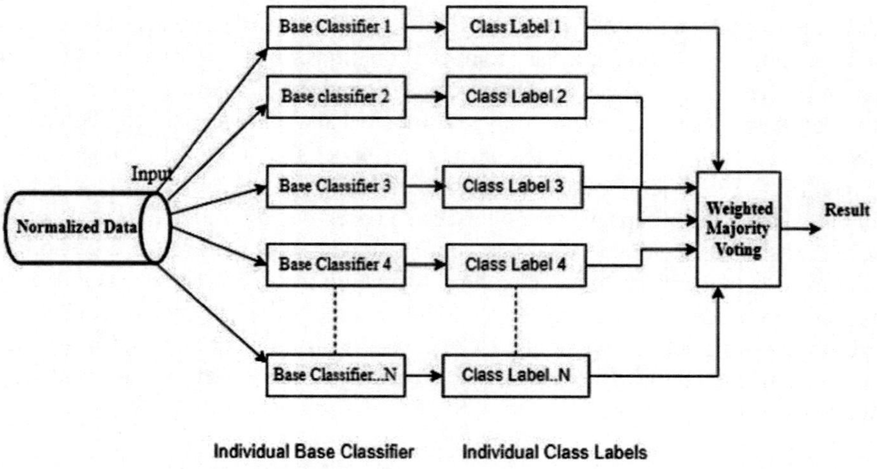

Fig. 3 Generalized ensemble classifier framework using majority voting

be represented as the neural network connecting with multiple layers of input nodes connected as a directed graph between the output layers.

c. Support Vector Machine [24]: It was developed by Vapnik in 1995, in which it separates the classes by creating a hyperplane. It is utilized for both linear and nonlinear datasets. It searches the optimal hyperplane, separating the classes, and acting as a decision boundary.
d. Random Forest [25]: It is a multiple-tree ensemble of DT classifiers. A random array of features is used to determine the division for each array, resulting in a specific decision tree.
e. Logit Boost Classifier [26]: A boosting technique in which logistic regression plays an additive weight with logistic regression. It is similar to AdaBoost, except it minimizes the logit loss.
f. Treebagger [27]: It compiles a set of decision trees for classification and regression. "Bagging" is the term for aggregating Bootstrap components. Every tree in the ensemble is built on an independently drawn bootstrap replica of the input data.

2.6 *Proposed Methodology*

Ensemble learning bagging and boosting approaches have been used in the proposed work. The proposed system entire workflow has been depicted in Fig. 1. Each participant's response was initially collected, and scores were allocated according to the expert's panel suggestions. After data cleaning and preprocessing evaluation, the recidivism risk has been divided into three groups: low, moderate, and high. Figure 2 depicts the proposed architecture, which includes data cleansing, classification, and normalization. These features were also stored in a database that can be used to test exploratory data analysis or ensemble bagging and boosting approaches. This experiment focuses on ensemble learning; hence, the developed feature sets will be input to ensemble algorithms that produce the desired outputs. In Table 1, the confusion matrix, precision, and recall metrics are utilized to evaluate and validate performance.

The confusion matrix for multiclass problem is described in Table 1 based on the identified class labels of the recidivist, where TP represents the true positive instances with respect to a (low), b (mid), and c (high), and all performance measures were derived using the above values:

Table 1 Confusion matrix evaluation for the models

Classifier output (actual)	Predicted		
	a (Low)	b (Mod)	c (High)
a (Low)	TP_{aa}	FP_{ba}	TN_{ca}
b (Mod)	FN_{ab}	TP_{bb}	FN_{cb}
c (High)	TN_{ac}	FP_{bc}	TP_{cc}

$$\text{Accuracy} = \frac{\text{TP} + \text{TN}}{\text{TP} + \text{FP} + \text{TN} + \text{FN}} * 100$$

$$\text{Precision} = \frac{\sum(\text{TP})}{\sum(\text{TP} + \text{FP})} * 100$$

$$\text{Recall} = \frac{\sum(\text{TP})}{\sum(\text{TP} + \text{FN})} \times 100$$

$$F\text{-measure} = 2 \times \frac{\text{Precision} * \text{Recall}}{\text{Precision} + \text{Recall}}$$

where

TP True positive,
TN True negative,
FP False positive and
FN False negative.

Here, we have used multiple ensembles approaches one of which is ensemble of 3 classifiers (EOC_3) which consists of NBC, SVM, and MLP classifiers. The result section contains all of the evaluation output for each ensemble approach.

3 Results

To evaluate the level of risk among repeat offenders, multiple ensemble bagging and boosting strategies were devised in the current study. The suggested system was constructed in MATLAB 2020a, and experimental simulations were run on an Intel Core i7 8th generation PC with 16 GB of RAM running Windows 10.

Recidivists are serial offenders who are judged on the basis of their vulnerability to crime and other risk factors. Recidivism risk was classified as low, moderate, or high based on the recidivist's assessment. Section 2 delves into the specifics of recidivist data gathering and analysis. On the basis of performance and validation, the traits or features that analyze risks were processed. Experiments were conducted using features as input from recidivist sample datasets, with personality and H-10 subscale features playing a crucial role in assessing recidivist risk. These characteristics were employed in the ensemble bagging and boosting techniques for assessing offenders' risks. In addition, to remove bias, a K-fold cross-validation approach is employed to train and test the datasets. Considering the fact of the collected data for this study is too small, the K-fold cross-validation resampling technique is enforced for training and testing the gathered re-offending risk features [28]. Taking the K value as 5, the complete dataset is randomly divided into five parts, in which each label is defined approximately in the same ratio of the original dataset. Four parts of the data are used for training the classifier and one part is used for testing. The method is carried out five times with a distinct combination of training and testing data.

Table 2 Average accuracy of all the ensemble bagging and boosting techniques over fivefold

Ensemble techniques	Accuracy (*K*-fold validation)					Average
	$k = 1$	$k = 2$	$k = 3$	$k = 4$	$k = 5$	
Random forest	71.56	75.49	77.54	80.00	73.52	75.62
Logit boost	66.34	62.74	78.36	63.72	82.52	70.73
EOC_3	75.00	79.52	74.33	83.33	74.33	77.30
Treebagger ensemble	74.54	76.67	86.67	80.00	78.33	79.24

Table 3 Performance measure for each ensemble techniques

Ensemble classifier	*F*-measure	Kappa value	Precision	Recall	FP-rate
Random forest	0.774	0.697	0.765	0.793	0.158
Logit boost	0.802	0.735	0.774	0.889	0.092
EOC_3	0.841	0.749	0.840	0.823	0.163
Treebagger ensemble	0.874	0.785	0.874	0.867	0.092

With the ensemble, random forest, ensemble of three classifiers, logit boost, and ensemble treebagger, a comparative recidivism risk assessment is analyzed. The accuracy, precision, recall, *F*-measure, FP-rate, and kappa values for each ensemble technique are determined. Table 2 gives the average prediction performance of each ensemble technique, with ensemble treebagger outperforming all ensemble methods as well as independent classifiers SVM, NBC, and MLP whose prediction accuracies are 71.66%, 72.91%, and 75% with **79.24%** accuracy, an ensemble of 3 independent classifiers coming in second with 77.30% accuracy, and Logit boost coming in third with 70.73%. To summarize ensemble treebagger, we found that this ensemble technique outperformed the other techniques on all performance measures. Table 3 summarizes the performance evaluations of all ensemble approaches. In this case, logit boost and treebagger have the lowest false positive (FP) rate, but ensemble treebagger and ensemble of three classifiers are superior to all other techniques utilized in terms of all other performance evaluations.

4 Conclusion

Repetitive offenses are a serious worry for the criminal justice system, and reintegrating these criminals into society without proper risk assessment is a significant risk. As India's population grows, so does its recidivism rate. Risk assessment and computer-assisted technologies for bail, parole, and judgment are needed in our country. Early detection of recidivists is critical, but behavior sampling using just criminologists is impossible; consequently, a machine learning technology paired with human insights can help. The level of risk among repeat offenders was assessed using a questionnaire based on personality, socioeconomics, demographics, environment, and the H-10 subscale. The panel of experts used human insights to ensure that the system would be free of prejudice. Data about recidivists were collected from various prisons and persons in Jharkhand. There were 36 features, with 10 of them being HCR-20 history items and the remaining 26 were personality, criminal details, present prison conduct, socio-economic, and demographic aspects. Compared to all other ensemble techniques, the ensemble treebagger has a prediction accuracy of **79.24%**, with total precision and recall levels over **86%**. Because there can be much data to process, the findings of this study motivate further work to use deep learning techniques to improve predictive accuracy.

References

1. Abbiati M et al (2019) Predicting physically violent misconduct in prison: a comparison of four risk assessment instruments. Behav Sci Law 37(1):61–77. https://doi.org/10.1002/bsl.2364
2. Liu YY et al (2011) A comparison of logistic regression, classification and regression tree, and neural networks models in predicting violent re-offending. J Quant Criminol 27(4):547–573. https://doi.org/10.1007/s10940-011-9137-7
3. Lussier P et al (2019) Using decision tree algorithms to screen individuals at risk of entry into sexual recidivism. J Crim Just 63:12–24. https://doi.org/10.1016/j.jcrimjus.2019.05.003
4. Neuilly MA et al (2011) Predicting recidivism in homicide offenders using classification tree analysis. Homicide Stud 15(2):154–176. https://doi.org/10.1177/1088767911406867
5. Ngo FT et al (2015) Assessing the predictive utility of logistic regression, classification and regression tree, chi-squared automatic interaction detection, and neural network models in predicting inmate misconduct. Am J Crim Justice 40(1):47–74. https://doi.org/10.1007/s12103-014-9246-6
6. Wijenayake S et al (2018) A decision tree approach to predicting recidivism in domestic violence
7. Karimi-Haghighi M, Castillo C (2021) Enhancing a recidivism prediction tool with machine learning: effectiveness and algorithmic fairness. In: Proceedings of the 18th international conference on artificial intelligence and law, ICAIL 2021. Association for Computing Machinery, Inc., pp 210–214. https://doi.org/10.1145/3462757.3466150
8. Fredrick David HB, Suruliandi A (2017) Survey on crime analysis and prediction using data mining techniques. ICTACT J Soft Comput 7(3):1459–1466. https://doi.org/10.21917/ijsc.2017.0202
9. Mehta H et al (2020) Classification of criminal recidivism using machine learning techniques. Int J Adv Sci Technol 29(4):5110–5122
10. Kirchebner J et al (2020) Identifying influential factors distinguishing recidivists among offender patients with a diagnosis of schizophrenia via machine learning algorithms. Forensic Sci Int 315:110435. https://doi.org/10.1016/j.forsciint.2020.110435
11. Ghasemi M et al (2021) The application of machine learning to a general risk-need assessment instrument in the prediction of criminal recidivism. Crim Justice Behav 48(4):518–538. https://doi.org/10.1177/0093854820969753
12. Watts D et al (2021) Predicting offenses among individuals with psychiatric disorders—a machine learning approach. J Psychiatr Res 138:146–154. https://doi.org/10.1016/j.jpsychires.2021.03.026
13. Singh A, Mohapatra S. Development of risk assessment framework for first time offenders using ensemble learning. https://doi.org/10.1109/ACCESS.2017.3116205

14. Ngo TT (2021) Recidivism and prisoner re-entry for firearm violations University of Central Oklahoma. Probation and parole re-entry education program: recidivism and prisoner re-entry for firearm violations
15. Aziz RM et al (2022) Machine learning-based soft computing regression analysis approach for crime data prediction. Karbala Int J Mod Sci 8(1):1–19. https://doi.org/10.33640/2405-609X.3197
16. National Crime Bureau, Govt. O.H.A.I. (2021) Crime in India 2020. Government of India
17. Singh A (2022) First time offender data. https://data.mendeley.com/datasets/8j3tf5zfd9/4. https://doi.org/10.17632/8J3TF5ZFD9.4
18. Douglas KS, Webster CD (1999) The HCR-20 violence risk assessment scheme. Crim Justice Behav 26(1):3–19. https://doi.org/10.1177/0093854899026001001
19. Kotsiantis SB et al (2006) Machine learning: a review of classification and combining techniques. Artif Intell Rev 26(3):159–190. https://doi.org/10.1007/s10462-007-9052-3
20. Polikar R (2006) Ensemble based systems in decision making. IEEE Circuits Syst Mag 6(3):21–45. https://doi.org/10.1109/MCAS.2006.1688199
21. Yaman E, Subasi A (2019) Comparison of bagging and boosting ensemble machine learning methods for automated EMG signal classification. Biomed Res Int 2019:1–13. https://doi.org/10.1155/2019/9152506
22. Duda RO, Hart PE, Stork DG (2006) Pattern classification. Wiley, Hoboken
23. Feng S et al (2003) Using MLP networks to design a production scheduling system. Comput Oper Res 30(6):821–832. https://doi.org/10.1016/S0305-0548(02)00044-8
24. Duwe G, Kim K (2017) Out with the old and in with the new? An empirical comparison of supervised learning algorithms to predict recidivism. Crim Justice Policy Rev 28(6):570–600. https://doi.org/10.1177/0887403415604899
25. Ani R et al (2016) Random forest ensemble classifier to predict the coronary heart disease using risk factors. https://doi.org/10.1007/978-81-322-2671-0_66
26. Kadkhodaei HR et al (2020) HBoost: a heterogeneous ensemble classifier based on the boosting method and entropy measurement. Expert Syst Appl 157:113482. https://doi.org/10.1016/j.eswa.2020.113482
27. Singh Y et al (2022) Betti-number based machine-learning classifier frame-work for predicting the hepatic decompensation in patients with primary sclerosing cholangitis. In: 2022 IEEE 12th Annual computing and communication workshop and conference (CCWC). IEEE, pp 0159–0162. https://doi.org/10.1109/CCWC54503.2022.9720887
28. Krogh A (2008) What are artificial neural networks? Nat Biotechnol 26(2):195–197. https://doi.org/10.1038/nbt1386

Assessing Imbalanced Datasets in Binary Classifiers

Pooja Singh and Rajeev Kumar

Abstract Despite continuous improvements in learning from imbalanced datasets, it remains a challenging research problem in machine learning. Classification models exhibit biases toward reducing the error rate on the majority samples by neglecting the minority samples. This paper aims to determine the impact of varying degrees of imbalances on a few selected classification models. Using comparative analysis of classification models on six binary imbalanced datasets with varying degrees of imbalances, we empirically analyze the effect of the degree of imbalancing on four classification models, namely decision tree (DT), random forest (RF), multilayer perceptron (MLP), and support vector machine (SVM). We show that imbalance distribution affects the performance of classification models, and the relation between the imbalance ratio and accuracy rate is convex.

Keywords Classification · Supervised · Degree of imbalance · Imbalance dataset · Abnormality detection

1 Introduction

Most machine learning algorithms are data-driven. In the machine learning domain, classification algorithms have significant importance. Classification models presume that the distribution of samples among the target classes is almost balanced, and they try to increase the predictive accuracy of the classification model on it [11]. If the dataset is imbalanced, then the distribution of samples points between the classes is unequal [4], increasing the overall accuracy of the model results in higher accuracy on the majority target and performing poorly on the minority target class, i.e., minority samples remain unknown, neglected, or treated to be noisy data as shown by, (e.g., Sun et al. [14]). The accuracy measure will no more be a valid evaluation criterion, and the classifiers could generate deceptive results, particularly for the minority class [18].

P. Singh (✉) · R. Kumar
Data to Knowledge (D2K) Lab School of Computer and Systems Sciences Jawaharlal Nehru University, New Delhi 110067, India
e-mail: poojasingh.ds7@gmail.com

M. Thakur et al. (eds.), *Soft Computing for Problem Solving*, Lecture Notes in Networks and Systems 547, https://doi.org/10.1007/978-981-19-6525-8_23

The class imbalance problem had several application areas, including corner detection by Kumar et al. [10], fraud detection by Kasasbeh et al. [8], medical diagnostics by Ahsan et al. [1], detection of anomalies by Kong et al. [9], detection of protein sequence by Ahsan et al. [2], and several others where the minority samples are more prominent than the majority samples. Our concern is to identify the minority target class, which is significant as in all the above applications. Here, we only work with the binary imbalanced datasets. Misclassifying the minority class can be costly, as failing to identify minority class instances might result in massive losses. As a result, the classifier must lower its classification error for both the minority and majority classes in an unbalanced situation.

While dealing with the problem of imbalanced class distribution, most popular classification learning methods seem inefficient (e.g., Sun et al. [14]). Napierala and Stefanowski [12] examined the neighborhood of each skewed class example and categorized them into one of four established groups, namely safe, rare, borderline, and anomalies. Their study provided a new edge to analyze the imbalanced dataset with their difficulties from minority class structure, but this attempt fails in real-world application domains.

In this work, we investigate the impact of varying imbalance factors on the classification model's performance using performance metrics driven by the confusion matrix, precision, recall, accuracy, geometric mean (G-mean), and receiver operating characteristics (ROC) curves.

The rest of the paper is organized as follows. Section 2 describes the related work on imbalance datasets. Section 3 presents the framed methodology to examine the performance of classifiers using evaluation metrics. Section 4 datasets and preprocessing steps. Section 5 includes the results and does empirical analysis. Finally, Sect. 6 concludes the paper.

2 Related Work

The issue of imbalanced class distributions exists in many real-world domains. Several researchers have studied this problem in the past, yet this remained an open problem. They have provided several solutions to handle class imbalance problems, such as the data-based techniques, algorithm-based techniques, and a hybrid approach, which incorporate the previous two techniques Wang et al. [16].

The data-based methods, also known as sampling methods, use a sampling approach to account for skewed distribution without altering a classification algorithm [6]. Chawla et al. [4] introduced the SMOTE algorithm that generates minority sample points randomly to increase the imbalance ratio; however, the marginalization generation and parameter selection limit its application. Wang et al. [17] introduced a new enhanced version of the SMOTE algorithm to address this issue based on Gaussian distribution. The algorithm augments minority samples by linear interpolation based on the Gaussian distribution trend between minority data points and the minority center. In order to avoid the expanded data being marginalized, the additional sample points are dispersed closer to the center of the minority sample with a higher probability.

Existing data-level methods become unsuitable for real-world skewed datasets with all categorical feature variables and mixed continuous and categorical variables. Park et al. [13] introduced a new resampling method consisting of ranking and relabeling strategies to produce balanced samples to address this issue. This approach generates more minority class samples from the majority class through ranking, resampling, and relabeling operations. The algorithm-based techniques change the traditional classification algorithm to deal with class imbalance [11]. Ali et al. [3] reviewed many research papers on imbalanced datasets and concluded that the hybrid sampling method suffers from overfitting problems. At the same time, ensemble learning methods address the problem of overfitting and improve the generalizability of unbalanced class problems.

Japkowicz et al. [7] proposed an approach to understand the relationship between imbalance levels of the classes, the size of the training set, and concept complexity. Also, they discussed various resampling or cost-modifying techniques to handle the class imbalance distribution problem. They concluded that the class imbalance factor impacts a classifier's generalization capacity when data complexity rises. Visa et al. [15] revealed that balance distribution between classes does not ensure improved classifier performance. Thus, imbalance distribution between samples is not only the reason that affects the classifier's performance, but other factors also play a significant role in better classification, such as training sample size, class complexity, and overlap between samples.

Sun et al. [14] suggested that learning from an imbalance dataset is challenging; they have shown that the within-class imbalance problem affects the performance of the classification model because it forces the classifier to learn the concept of the minority class. Another method to handle the skewed dataset classification problem is feature selection. Grobelnik and Marko [5] demonstrated that irrelevant features did not increase the performance of the classification model considerably, implying that adding more features might delay the induction process. Thus, by using feature selection, one may neglect irrelevant or noisy data that results in the issue of class complexity or overlap.

Although many researchers have proposed various methods to solve the problem of learning from imbalanced datasets and improving the accuracy of classification models, it is still a challenging task. The previous literature work done in imbalanced dataset classification highlighted the usefulness of data-level approaches, algorithmic approaches, ensemble-based techniques, intrinsic characteristics of imbalance data distribution such as small disjuncts, lack of training data, class overlap, feature selection methods by avoiding the effect of degree of imbalance on the performance of the classification models.

The novelty of the study done in this paper is to understand the impact of varying degrees of imbalance. We perform extensive experiments in this paper using eight different imbalance ratios of six imbalanced datasets, namely 15%:85%, 25%:75%, 35%:65%, 45%:55%, 55%:45%, 65%:45%, 75%:25%, 85%:15% and then compare the performance of four standards classification models to seek their behavior using

Table 1 Imbalanced dataset description

Dataset	# Instances	# Features	Imbalance ratio
Pima	768	8	0.57
Breast cancer wisconsin	569	32	0.59
Heart disease	304	13	0.78
Spambase	4601	58	0.65
Ionosphere	351	34	0.56
Monks-problems-2	601	8	0.52

performance measures that help researchers to understand the nature of the class imbalance problems and how the class imbalance hinders the performance of four classification models, namely decision tree (DT), random forest (RF), multilayer perceptron (MLP), and support vector machine (SVM).

3 The Methodology

We introduced a method to determine the impact of varying degrees of imbalance ratios on the performance of classification models. The method aims to perform experiments on six binary classification datasets with varying class ratios. Several samples with varying imbalance ratios are generated for each dataset, namely 15%:85%, 25%:75%, 35%:65%, 45%:55%, 55%:45%, 65%:45%, 75%:25%, and 85%:15%. Then, four standard classification models, such as decision tree (DT), multilayer perceptron (MLP), random forest (RF), and support vector machine (SVM), are applied to each sample whose performances can be analyzed using performance metrics such as precision, recall, accuracy, geometric mean (G-mean), and receiver operating characteristics (ROC) curves. The ten-fold stratified cross-validation method is applied in each experiment. These evaluation methods will provide us a clearer picture of the overall impact of class imbalance on these classification models.

4 Datasets and Preprocessing

Experiments are conducted with six binary class imbalance datasets from the UCI repository, as shown in Table 1. Preparing and transforming data into a suitable form for conducting experiments are one of the most critical tasks. We remove the missing values and insignificant variables from the model in the datasets and convert the categorical variables to numerical ones.

To examine the effect of different imbalance ratios on classifiers' performance, we first separate majority and minority classes for each dataset. Then to create eight different versions of each dataset with alter class ratios, namely 15%:85%, 25%:75%, 35%:65%, 45%:55%, 55%:45%, 65%:45%, 75%:25%, and 85%:15%, we begin to choose percentage of majority and minority sample accordingly. For example, to create a sample with a class ratio of 15%:85%, we selected 15% of the minority class and 85% of the majority class. Similarly, for a class ratio of 25%:75%, we selected 25% from the minority class and 75% from the majority class. Thus, all different samples of each dataset are created in this way. Experiments are performed using a 10-fold stratified cross-validation method on each dataset sample and then compare the performance of the classification model using evaluation metrics to derive the results. Here, we are interested in calculating the mean accuracy rate over 100 runs, precision rate, recall rate, G-mean, and ROC curves on the experimental datasets. We use comparative analysis of the classifier's performance to understand their behavior on different class ratios. We perform a train-test split of each dataset in an 80:20 ratio to form ROC curves using true positive rate (TPR) and false positive rate (FPR).

5 Experimental Results

This section compares and evaluates the performance of the four classification models using performance metrics. Here, we discussed the precision results achieved from classification models on the experimental datasets as shown in Fig. 1. The plot's x-axis represents the fraction of minority samples in respective datasets, and the y-axis shows the precision values. Results show that the precision rate is high when the class ratio is 15%:85% or 85%:15%. The behavior of machine learning models varies from one dataset to another. In Fig. 1a, c, and e when the imbalance ratio is 15%:85%, then MLP outperforms the other classifiers, but when the ratio begins to increase, its performance decreases, also shown in Table 2. At class ratio 25%:35%, the random forest gives the best precision results for all datasets except the ionosphere dataset.

In Fig. 1b–e, the curve of SVM classifiers shows a sharp increase in its slope at class ratio 35%:65% and begins to increase till it achieves a ratio of 85%:15%, except in Fig. 1d, where after obtaining the highest peak at ratio 45%:55%, it begins to decrease and then again increases till it achieves class ratio 85%:15%. Random forest shows the best results among all other classifiers and attains the highest value at a class ratio of 85%:15%.

Figure 2 shows the plot between the value of recall rate of classification models on the y-axis with respect to the fraction of minority class on the x-axis. The reported results conclude that class ratios affect the performance of the classification models, as shown in Table 3. In Fig. 2a, e, there is continuous improvement in the performance of all classifiers with an increasing fraction of the minority class. From Fig. 2b–d and f, we observed that when the imbalance ratio varies from 15%:85% to 45%:55%,, the

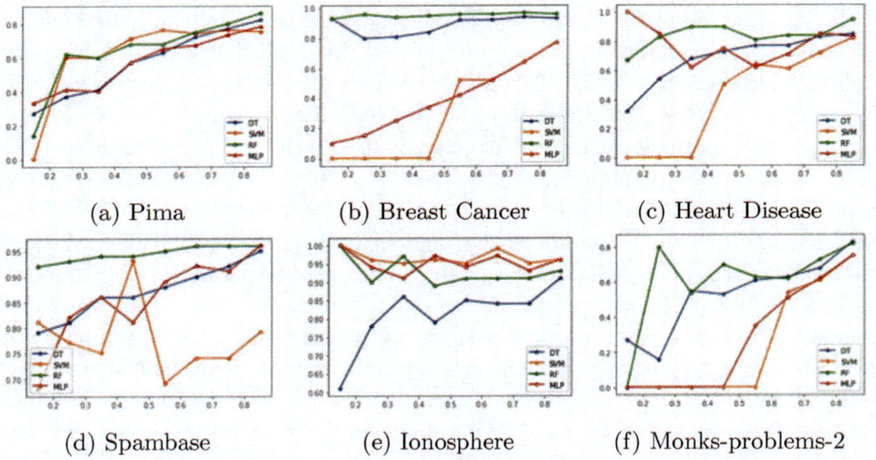

(a) Pima (b) Breast Cancer (c) Heart Disease

(d) Spambase (e) Ionosphere (f) Monks-problems-2

Fig. 1 Precision curves

Table 2 Precision table

Dataset	Degree of imbalance		Classification model			
	Percentage	Ratio	DT	RF	MLP	SVM
Pima	15%:85%	0.09	0.27	0.14	**0.33**	0.00
	25%:75%	0.18	0.37	**0.62**	0.41	0.60
	35%:65%	0.29	0.41	**0.60**	0.40	**0.60**
	45%:55%	0.44	0.57	0.68	0.57	**0.71**
Breast cancer	15%:85%	0.11	**0.93**	**0.93**	0.10	0.00
	25%:75%	0.19	0.80	**0.96**	0.15	0.00
	35%:65%	0.32	0.81	**0.94**	0.25	0.00
	45%:55%	0.48	0.84	**0.95**	0.34	0.00
Spambase	15%:85%	0.12	0.79	**0.92**	0.68	0.81
	25%:75%	0.22	0.81	**0.93**	0.82	0.77
	35%:65%	0.35	0.86	**0.94**	0.86	0.75
	45%:55%	0.53	0.86	**0.94**	0.81	0.93
Monks-problems-2	15%:85%	0.09	**0.27**	0.00	0.00	0.00
	25%:75%	0.18	0.16	**0.80**	0.00	0.00
	35%:65%	0.28	**0.55**	0.54	0.00	0.00
	45%:55%	0.43	0.53	0.70	**1.00**	0.00

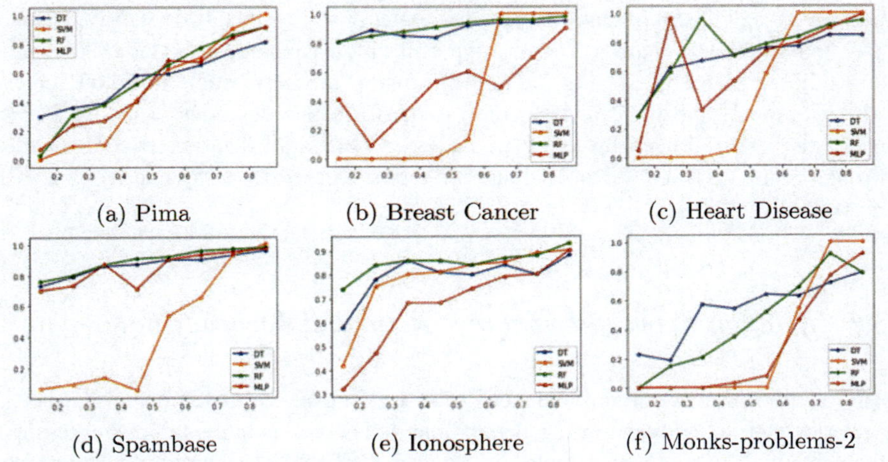

(a) Pima (b) Breast Cancer (c) Heart Disease

(d) Spambase (e) Ionosphere (f) Monks-problems-2

Fig. 2 Recall curves

Table 3 Recall table

Dataset	Degree of imbalance		Classification model			
	Percentage	Ratio	DT	RF	MLP	SVM
Pima	15%:85%	0.09	**0.30**	0.03	0.07	0.00
	25%:75%	0.18	**0.36**	0.31	0.24	0.09
	35%:65%	0.29	**0.39**	0.38	0.27	0.10
	45%:55%	0.44	**0.58**	0.52	0.40	0.41
Breast cancer	15%:85%	0.11	**0.81**	**0.81**	0.41	0.00
	25%:75%	0.19	**0.89**	0.85	0.09	0.00
	35%:65%	0.32	0.85	**0.88**	0.31	0.00
	45%:55%	0.48	0.84	**0.91**	0.53	0.00
Spambase	15%:85%	0.12	0.73	**0.76**	0.70	0.06
	25%:75%	0.22	0.79	**0.80**	0.73	0.09
	35%:65%	0.35	0.86	**0.87**	**0.87**	0.13
	45%:55%	0.53	0.87	**0.91**	0.71	0.05
Monks-problems-2	15%:85%	0.09	**0.23**	0.00	0.00	0.00
	25%:75%	0.18	**0.19**	0.15	0.00	0.00
	35%:65%	0.28	**0.57**	0.21	0.00	0.00
	45%:55%	0.43	**0.54**	0.35	0.03	0.00

performance of the SVM classifier is worst. After reaching the ratio of 45%:55%, the SVM curve shows a sharp increase, and it keeps increasing and gives the best results at a ratio of 85%:15%. In Fig. 2a, f, DT provides the best results in initial stages till it attains a ratio of 55%:45%, then its performance degrades. Random forest performs best, followed by a DT, MLP, and SVM on all datasets, except for the Monks-problems-2 dataset and the last two ratios, namely 75%:25%, and 85%:25%, respectively.

5.1 Relation Between Imbalance Ratio and Accuracy Rate

This section discusses the imbalanced ratio's effect on the experimental datasets' accuracy metric, as shown in Fig. 3 and Table 4. The x-axis in the figures represents the percentage of minority samples on experimental datasets. The y-axis represents the average accuracy rate of classification models, calculated using 10-fold stratified cross-validation over 100 runs.

The value of the accuracy rate is high when the class ratio is 15%:85% and then decreases until it reaches the ratio of 45%:55% to 65%:35%; then, the accuracy rate increases until the ratio reaches 85%:15%. There is a slight variation in the behavior of SVM on breast cancer and spambase dataset; it shows a sudden increase in accuracy rate at class ratio 45%:55%; then, it begins to decrease till the ratio reaches 65%:35% and then again begins to increase till obtaining the class ratio of 85%:15%. SVM shows unusual behavior on Pima and ionosphere datasets, achieving higher values of accuracy rate than MLP and DT compared to other datasets. Thus, we conclude that it has a higher accuracy rate when a dataset is highly imbalanced.

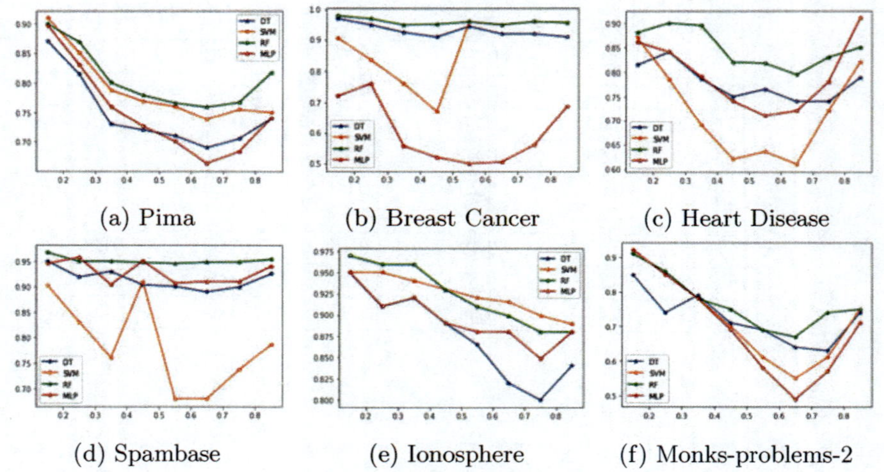

(a) Pima (b) Breast Cancer (c) Heart Disease

(d) Spambase (e) Ionosphere (f) Monks-problems-2

Fig. 3 Accuracy curves

Table 4 Accuracy table

Dataset	Degree of imbalance		Classification model			
	Percentage	Ratio	DT	RF	MLP	SVM
Pima	15%:85%	0.09	0.87	0.90	0.90	**0.91**
	25%:75%	0.18	0.82	**0.87**	0.83	0.85
	35%:65%	0.29	0.73	**0.80**	0.76	0.79
	45%:55%	0.44	0.72	**0.78**	0.73	0.77
Breast cancer	15%:85%	0.11	0.97	**0.98**	0.72	0.91
	25%:75%	0.19	0.95	**0.97**	0.76	0.84
	35%:65%	0.32	0.93	**0.95**	0.56	0.76
	45%:55%	0.48	0.91	**0.95**	0.52	0.67
Spambase	15%:85%	0.12	0.95	**0.97**	0.95	0.90
	25%:75%	0.22	0.92	0.95	**0.96**	0.83
	35%:65%	0.35	0.93	**0.95**	0.90	0.76
	45%:55%	0.53	0.90	**0.95**	**0.95**	0.91
Monks-problems-2	15%:85%	0.09	0.85	0.91	**0.92**	**0.92**
	25%:75%	0.18	0.74	**0.86**	0.85	0.85
	35%:65%	0.28	**0.79**	0.78	0.78	0.78
	45%:55%	0.43	0.71	**0.75**	0.69	0.70

This happens because the fraction of the majority class positively affects the overall accuracy rate, which is not necessarily true for the minority class. Thus from Fig. 3, we observed that curves of accuracy rate are formed in the form of convex curves; hence, the relation between accuracy rate and imbalance ratio is convex.

Figure 4 represents the plots of G-mean results of classification models. The x-axis represents the same as in the previous experimental datasets. The y-axis represents the G-mean of the classifiers. From the figures, we observe that when the dataset is highly imbalanced, having a class ratio of 15%:85% or 85%:15%, then classifiers have a lesser value of G-mean. When the dataset is slightly imbalanced, having a class ratio between 45%:55% to 65%:35%, then the high value of G-mean is obtained by classifiers. Random forest (RF) outperforms all other classifier's overall ratios, followed by a DT, MLP, and SVM, except in the Monks-problems-2 dataset where the decision tree outperforms random forest. SVM performs poorly on all datasets except for the Ionosphere dataset.

The geometric mean, or G-Mean, is a metric that combines sensitivity and specificity into a single value that balances both objectives [11]. The higher the sensitivity and specificity values, the better the model correctly identifies positive and negative cases. From Table 5, we observe that random forest behaves exceptionally on

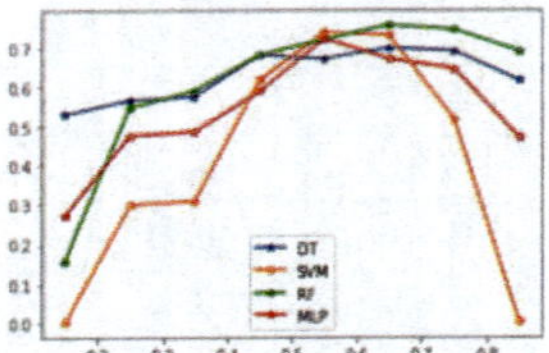

(a) Pima

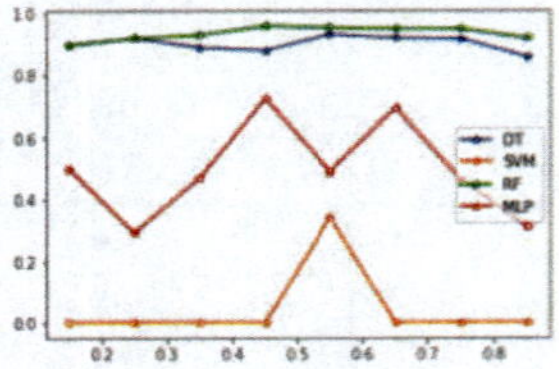

(b) Breast Cancer

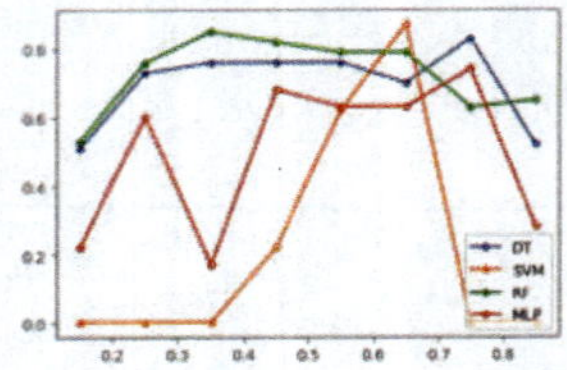

(c) Heart Disease

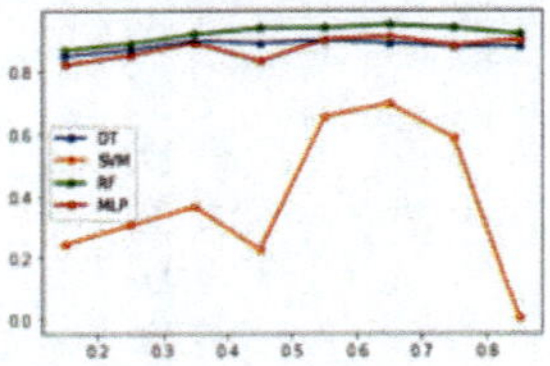

(d) Spambase

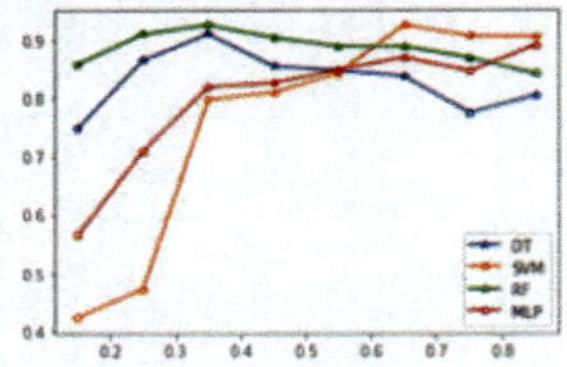

(e) Ionosphere

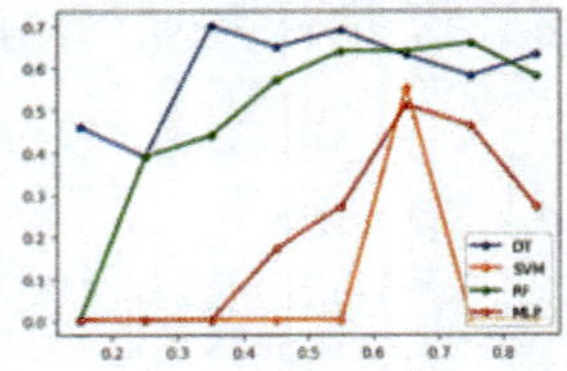

(f) Monks-problems-2

Fig. 4 G-mean curves

Table 5 Sensitivity and specificity table

Dataset	Imb. ratio	Sensitivity				Specificity			
		DT	RF	MLP	SVM	DT	RF	MLP	SVM
Heart disease	15%:85%	0.91	0.98	**1.00**	**1.00**	**0.29**	**0.29**	0.05	0.00
	25%:75%	0.85	0.97	0.96	**1.00**	**0.62**	0.59	0.38	0.00
	35%:65%	0.86	**0.96**	0.09	0.00	0.67	**0.75**	0.33	0.00
	45%:55%	0.82	0.95	0.88	**0.97**	**0.71**	**0.71**	0.53	0.05
	55%:45%	0.77	**0.81**	0.54	0.66	0.76	**0.79**	0.74	0.58
	65%:35%	0.64	0.74	0.50	**0.76**	0.77	0.84	0.80	**1.00**
	75%:25%	0.56	0.44	**0.61**	0.00	0.85	0.92	0.90	**1.00**
	85%:15%	0.32	**0.44**	0.08	0.00	0.85	0.95	0.99	**1.00**
Ionosphere	15%:85%	0.96	**1.00**	**1.00**	**1.00**	0.58	**0.74**	0.32	0. 42
	25%:75%	0.96	0.98	**1.00**	0.99	0.78	**0.84**	0.50	0.47
	35%:65%	0.96	**0.99**	0.98	**0.99**	**0.86**	**0.86**	0.68	0.80
	45%:55%	0.90	0.95	**0.99**	0.98	0.81	**0.86**	0.68	0.81
	55%:45%	0.90	0.94	**0.97**	**0.97**	0.80	**0.84**	0.74	**0.84**
	65%:35%	0.84	0.91	0.98	**0.99**	0.84	**0.87**	0.79	**0.87**
	75%:25%	0.75	0.86	0.89	**0.93**	0.79	**0.88**	0.80	**0.88**
	85%:15%	0.74	0.77	**0.88**	**0.88**	0.88	**0.93**	0.90	**0.93**
Monks-problems-2	15%:85%	**0.23**	0.00	0.00	0.00	0.94	**1.00**	**1.00**	**1.00**
	25%:75%	**0.19**	0.15	0.00	0.00	0.82	0.99	**1.00**	**1.00**
	35%:65%	**0.57**	0.21	0.00	0.00	0.87	0.95	**1.00**	**1.00**
	45%:55%	**0.54**	0.35	0.03	0.00	0.79	0.94	**1.00**	**1.00**
	55%:45%	**0.64**	0.52	0.08	0.00	0.74	0.80	0.90	**1.00**
	65%:35%	0.63	**0.69**	0.46	0.54	**0.64**	0.59	0.57	0.56
	75%:25%	0.72	0.91	0.77	**1.00**	0.46	**0.48**	0.27	0.00
	85%:15%	0.79	0.99	0.92	**1.00**	**0.50**	0.34	0.08	0.00

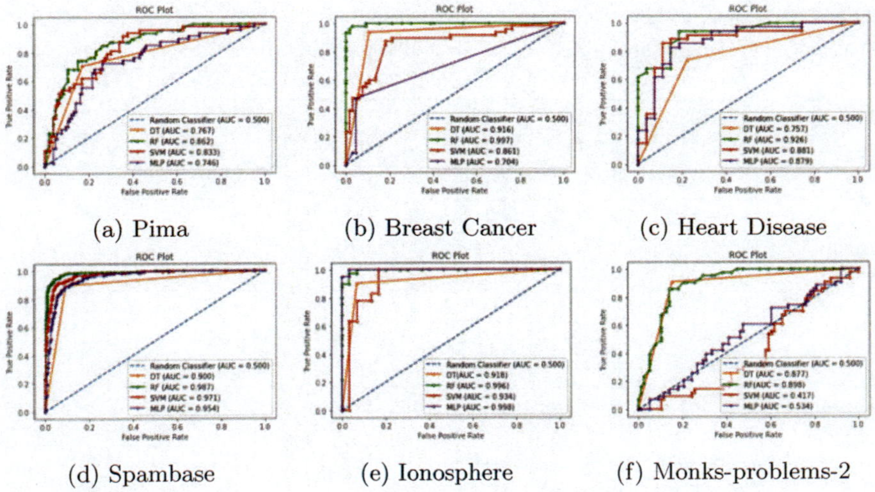

(a) Pima (b) Breast Cancer (c) Heart Disease

(d) Spambase (e) Ionosphere (f) Monks-problems-2

Fig. 5 ROC curves

Monks-problems-2 having low sensitivity and low specificity values. In contrast, random forest attains high sensitivity and specificity values on the heart disease and ionosphere datasets.

Receiver operating characteristics (ROCs) curves are an effective tool to visualize and assess the performance of classifiers. The area under curve (AUC) summates the ROC curve and measures a classifier's ability to differentiate among classes. The larger the region beneath the curve better is the classification model. Figure 5 shows that the random forest gives the best results. The behavior of other classifiers varies from dataset to dataset. Table 6 shows the performance of random forest for all datasets. Random forest outperforms all other classification models at all imbalance ratios.

6 Conclusion

Learning from imbalanced data distribution is challenging because minority samples remain undiscovered or ignored. They are treated as anomalies or supposed to be noise, resulting in more misclassification of the minority samples. This work attempted to assess the impact of the imbalance ratio on the classifier model's performance using evaluation metrics. We performed experiments on six imbalanced datasets using four classifiers: decision tree (DT), support vector machine (SVM), multilayer perceptron (MLP), and random forest (RF) with varying imbalance ratios. The results show that random forest performs best among other algorithms and SVM the worst. The performance of decision trees and MLP varies from dataset to dataset

Table 6 Performance of random forest

Dataset	Degree of imbalance		Evaluation metrics			
	Percentage	Ratio	Precision	Recall	Accuracy	G-mean
Pima	15%:85%	0.15	0.14	0.03	0.90	0.16
	25%:75%	0.28	0.62	0.31	0.87	0.55
	35%:65%	0.45	0.60	0.38	0.80	0.59
	45%:55%	0.68	0.68	0.52	0.78	0.68
Breast cancer	15%:85%	0.15	0.93	0.81	0.98	0.90
	25%:75%	.28	0.96	0.85	0.97	0.92
	35%:65%	0.45	0.94	0.88	0.95	0.93
	45%:55%	0.68	0.95	0.91	0.95	0.96
Heart disease	15%:85%	0.09	0.67	0.29	0.88	0.53
	25%:75%	0.19	0.83	0.59	0.90	0.76
	35%:65%	0.30	0.90	0.96	0.90	0.85
	45%:55%	0.46	0.90	0.71	0.82	0.82
Spambase	15%:85%	0.15	0.92	0.76	0.97	0.92
	25%:75%	0.28	0.93	0.80	0.95	0.94
	35%:65%	0.45	0.94	0.87	0.95	0.95
	45%:55%	0.68	0.94	0.91	0.95	0.94
Ionosphere	15%:85%	0.15	1.00	0.74	0.97	0.86
	25%:75%	0.28	0.90	0.84	0.96	0.91
	35%:65%	0.45	0.97	0.86	0.96	0.93
	45%:55%	0.68	0.89	0.86	0.93	0.90
Monks-problems-2	15%:85%	0.09	0.00	0.00	0.91	0.00
	25%:75%	0.18	0.80	0.15	0.86	0.39
	35%:65%	0.28	0.54	0.21	0.78	0.44
	45%:55%	0.43	0.70	0.35	0.75	0.57

and lies in between them. The results show that classifiers achieve high-performance measures when the dataset is highly imbalanced except for the SVM, which showed improved performance with balancing. This is an area of future research.

References

1. Ahsan MM, Siddique Z (2022) Machine learning-based heart disease diagnosis: a systematic literature review. Artif Intell Med 102289
2. Ahsan R, Ebrahimi F, Ebrahimi M (2022) Classification of imbalanced protein sequences with deep-learning approaches; application on influenza a imbalanced virus classes. Inf Med Unlocked 29:100860

3. Ali H, Salleh MNM, Saedudin R, Hussain K, Mushtaq MF (2019) Imbalance class problems in data mining: a review. Indonesian J Eng Comput Sci 14(3):1560–1571
4. Chawla NV, Bowyer KW, Hall LO, Kegelmeyer WP (2002) SMOTE: synthetic minority over-sampling technique. J Artif Intell Res 16:321–357
5. Grobelnik M (1999) Feature selection for unbalanced class distribution and naive bayes. In: Proceeding 16th international conference machine learning (ICML), Citeseer, pp 258–267
6. He H, Garcia EA (2009) Learning from imbalanced data. IEEE Trans Knowl Data Eng 21(9):1263–1284
7. Japkowicz N, Stephen S (2002) The class imbalance problem: a systematic study. Intell Data Anal 6(5):429–449
8. Kasasbeh B, Aldabaybah B, Ahmad H (2022) Multilayer perceptron artificial neural networks-based model for credit card fraud detection. Indonesian J Electr Eng Comput Sci 26(1):362–373
9. Kong J, Kowalczyk W, Menzel S, Back T (2020) Improving imbalanced classification by anomaly detection. In: Proceeding of international conference parallel problem solving from nature. Springer, pp 512–523
10. Kumar R, Chen WC, Rockett P (199) Bayesian labelling of image corner features using a grey-level corner model with a bootstrapped modular neural network. In: Proceeding of 5th international conference artificial neural networks (Conf. Publ. No. 440), pp 82–87
11. Lin WJ, Chen JJ (2013) Class-imbalanced classifiers for high-dimensional data. Briefings Bioinform 14(1):13–26
12. Napierala K, Stefanowski J (2016) Types of minority class examples and their influence on learning classifiers from imbalanced data. J Intell Inf Syst 46(3):563–597
13. Park S, Lee HW, Im J (2022) Raking and relabeling for imbalanced data. IEEE Trans Knowl Data Eng
14. Sun Y, Wong AK, Kamel MS (2009) Classification of imbalanced data: a review. Int J Pattern Recogn Artif Intell 23(04):687–719
15. Visa S, Ralescu A (2005) The effect of imbalanced data class distribution on fuzzy classifiers-experimental study. In: Proceeding 14th IEEE international conference, fuzzy systems. IEEE, pp 749–754
16. Wang L, Han M, Li X, Zhang N, Cheng H (2021) Review of classification methods on unbalanced data sets. IEEE Access 9:64606–64628
17. Wang S, Dai Y, Shen J, Xuan J (2021) Research on expansion and classification of imbalanced data based on smote algorithm. Sci Rep 11(1):1–11
18. Yang Q, Wu X (2006) 10 challenging problems in data mining research. Int J Inf Technol Decis Making 5(04):597–604

A Hybrid Machine Learning Approach for Multistep Ahead Future Price Forecasting

Jahanvi Rajput

Abstract In the financial sector, prediction of the stock market is one of the imperative working areas. The financial market index price is an important measure of financial development. The objective of this paper is to improve the forecasting accuracy of the closing price of different financial datasets. This work proposes a hybrid machine learning approach incorporating feature extraction methods with baseline learning algorithms to improve the forecasting ability of the baseline algorithm. Support vector regression (SVR) and two faster variants of SVR (least square SVR and proximal SVR) are taken as baseline algorithms. Kernel principal component analysis (KPCA) is introduced here for features extraction. A large set of technical indicators are taken as input features for index future price forecasting. Various performance measures are used to verify the forecasting performance of the hybrid algorithms. Experimental results over eight index future datasets suggest that hybrid prediction models obtained by incorporating KPCA with baseline algorithms reduce the time complexity and improve the forecasting performance of the baseline algorithms.

Keywords Support vector regression · Least square support vector regression · Proximal support vector regression · Kernel principal component analysis · Technical indicators

1 Introduction

Financial market is a market where trade and exchange of different type of financial instrument are based on the price taken place by people and companies. These prices are based on supply and demand. It is important for economy to profit by predicting the rise and fall of price of these instruments. According to the definition of Svirydzenka, financial markets index shows how developed a financial market is, and it also includes the depth, access, and efficiency of financial market. Hence, it is

J. Rajput (✉)
Institute of Technical Education and Research, Siksha 'O' Anusandhan, Bhubaneswar, Odisha, India
e-mail: jahanvirajput@soa.ac.in

M. Thakur et al. (eds.), *Soft Computing for Problem Solving*, Lecture Notes in Networks and Systems 547, https://doi.org/10.1007/978-981-19-6525-8_24

an important measure of financial development. That is why it contains information of broad coverage of key indicators, including stocks traded to GDP, stock market turnover ratio, total number of issuers of debt, etc.

Movements in stock market depend on so many factors that is why it is a critical work to do. Hence, the question arises: How much profit someone can make? The answer is, it all depends on the preciseness of the prediction of the movement of stocks. So when accuracy is high for prediction, then profit will be high. There are so many techniques which can be applied for the prediction, so everything depends on the accuracy of algorithms. In this paper, support vector regression (SVR) and its two variants least square support vector regression (LS-SVR) and proximal support vector regression (PSVR), with kernel principal component analysis, are used for price forecasting.

There are so many real-life applications of SVR and its variants. Some of them are financial time series forecasting, price prediction, population prediction, high-resolution temperature, and salinity model analysis [1]. In financial time series forecasting, the initial forecasts are closing and opening price prediction and trend forecast.

The main work of this paper is to work on financial datasets to predict the closing price of next five days using KPCA on SVR and its variants, namely LS-SVR and PSVR with different kernels for both KPCA and SVR, LS-SVR, and PSVR. Previously, work has been done on the same datasets without KPCA. There the prediction of 5 d ahead closing price has been made using algorithms SVR, LS-SVR, and PSVR without KPCA. With these algorithms, accuracy parameters are calculated and best method is found, but previously obtained results need to be enhanced along with computational time and also need to be reduced so that working with these algorithms on larger datasets will be economic in all respect. From overcoming these problems, KPCA is introduced with different kernels. In this paper, the accuracy of predicted closing price is found. How the accuracy and computational time are changing using different algorithms will be discussed. Both tabular results and graphical representation are given. This paper is divided into five sections. Section 2 gives the brief introduction of SVR and its variants, after that the mathematical background of KPCA is described. Section 3 describes the proposed method, and in the last two sections, results obtained are discussed and conclusion is given.

2 Methodology/Mathematical Background

2.1 *Support Vector Regression*

In SVR, we need to find a function that approximated the mapping from training samples. Our objective to find best fit hyperplane such that it has maximum number of points because in SVR basically, we need to consider the points which lie inside the decision boundary line, i.e., we need to find the optimum hard ε—band hyperplane. Two-dimensional graphical representation of ε-SVR is shown in Fig. 1.

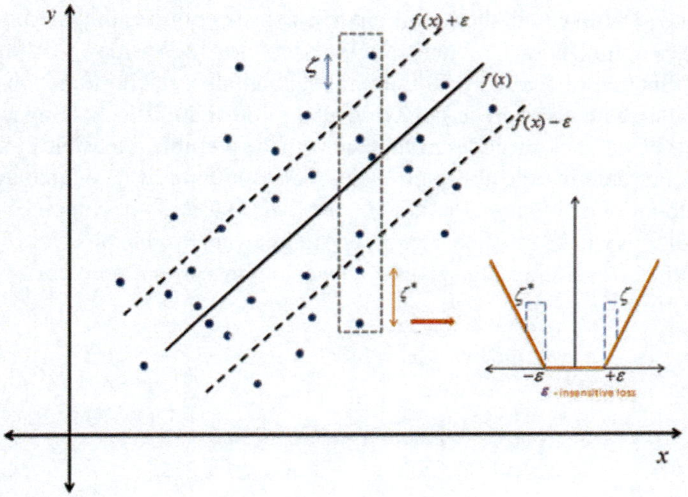

Fig. 1 Two-dimensional geometric interpretation of ε-SVR

In this section, theory based on formulation of Vapnik [2] will be discussed behind SVR equations. Considering training set A is given by

$$A = \{(x_1, y_1), (x_2, y_2), \ldots, (x_m, y_m)\} \tag{1}$$

where $x \in X \subset \mathbb{R}^n$ denotes training inputs and $y \in Y \subset \mathbb{R}$ represents the continuous values of training outputs.
Let us take there exists a nonlinear function, $f(x)$. Given by:

$$f(x) = w^T(x_i) + b \tag{2}$$

where w denotes the weight vector and b is called the bias. Now our aim is to fit training set A to function $f(x)$ such that the greatest deviation of $f(x)$ from all training points y_i is ϵ, at the same time it should be minimum.

We know that for the classification problem in $\mathbb{R}^{n+1}$ and given linear hyperplane is $w \cdot x + b$, then the corresponding optimization problem (it is also known as optimization problem of constructing a hard ε—band hyperplane) is given by:

$$\begin{aligned} &\text{minimize} \quad \frac{1}{2} w^T w \\ &\text{subject to:} \begin{cases} y_i - (w^T(x_i) + b) \le \epsilon \\ y_i - (w^T(x_i) + b) \ge \epsilon \end{cases} \end{aligned} \tag{3}$$

where ϵ (≥ 0) is the maximum acceptable deviation and it is user defined.

Equation (3) satisfies all the condition to be convex optimization problem. Hence, it is convex optimization problem of support vector regression. Now the aim of objective function of Eq. (3) is to minimize w and also making function as flat as possible along with it should be satisfying all the constraints. For solving Eq. (3), we need to introduce slack variables to cope up with the possible infeasible optimization problems. For defining the above problem, the assumption is taken that the convex optimization problem is feasible, i.e., $f(x)$ exists. But it is not compulsory that all these conditions will be satisfied. Hence, one might want to trade off errors by flatness of the estimate. Therefore, this gives us the following primal problem as stated by Vapnik [3]:

$$\text{minimize} \quad \frac{1}{2}w^T w + C\sum_{i=1}^{n}(\xi_i^+ + \xi_i^-)$$
$$\text{subject to:} \begin{cases} y_i - w^T(x_i) - b \le \epsilon + \xi_i^+ \\ w^T(x_i) + b - y_i \le \epsilon + \xi_i^- \\ \xi_i^+, \xi_i^- \ge 0 \end{cases} \tag{4}$$

where C (it is always positive) represents the weight of the loss function and it is known as regularization constant. To make function as "flat" as possible. In objective function, the term $(w^T w)$ is used, and hence, it is called regularized term and $C\sum_{i=1}^{n}(\xi_i^+ + \xi_i^-)$ is called empirical term, and it is used to measure the ϵ-insensitive loss function.

The dual of above primal problem is given by Eq. (5):

$$\text{maximize} \quad \frac{1}{2}\sum_{i,j=1}^{m} K(\mathbf{x}i, \mathbf{x}j)\left(\alpha_i^+ - \alpha_i^-\right)\left(\alpha_j^+ - \alpha_j^-\right)$$
$$+\epsilon\sum_{i=1}^{m}\left(\alpha_i^+ + \alpha_i^-\right) - \sum_{i=1}^{m} y_i\left(\alpha_i^+ - \alpha_i\right) \tag{5}$$
$$\text{subject to} \begin{cases} \sum_{i=1}^{m}\left(\alpha_i^+ - \alpha_i^-\right) = 0 \\ \alpha_i^+, \alpha_i^- \in [0, C] \end{cases}$$

where $K(x_i, x_j) = \psi(\mathbf{x}_i)^T \psi(\mathbf{x}_j) = (\mathbf{x}_i)^T(\mathbf{x}_j)$ is known as kernel function.

Nonlinear function approximation for SVR is possible due to kernel function. **Kernel functions** maintains the simplicity and efficiency same as linear SVR. For defined quadratic optimization problem for optimal solution, the kernel function must be positive semi-definite. Mainly, polynomial kernel of different degrees and Gaussian kernel are used. In next sections, the formulation of variants of SVR will be discussed.

2.2 Least Square Support Vector Regression (LS-SVR)

It is a variant of SVR which is very much similar to the standard SVR. In the standard SVR algorithm, if the inequality constraints are replaced by equality, then the least squares SVR algorithm is obtained. When inequality constraints in the standard SVR are replaced by equality, then the LS-SVR has convex linear system solving problem, which is easier to solve. Hence, it speeds up the training [4].

In this, the first need is to find the mapping from a sample set which consists of independent and also identically distributed (i.i.d) elements, where $x \in \mathbb{R}^d$ is input variable and $y \in \mathbb{R}$ is output variable. Let $\mathbf{y} = (y_1, y_2, y_3, \ldots, y_l)^T \in \mathbb{R}^l$. Using LS-SVR, the problem is solved. Here objective is to find $\mathbf{w} \in \mathbb{R}^{n_h}$ and $b \in \mathbb{R}$. This is obtained by solving the given equation:

$$\begin{aligned} \min_{\mathbf{w}\in\mathbb{R}^{n_h}, b\in\mathbb{R}} \quad & \mathcal{G}(\mathbf{w}, \zeta) = \frac{1}{2}\mathbf{w}^{\mathrm{T}}\mathbf{w} + \gamma\frac{1}{2}\zeta^{\mathrm{T}}\zeta \\ \text{subject to:} \quad & \mathbf{y} = \mathbf{X}^T\mathbf{w} + b\mathbf{1}_l + \zeta \end{aligned} \tag{6}$$

where $\mathrm{X} = (\psi(\mathbf{x}_1), \psi(\mathbf{x}_2), \ldots, \psi(\mathbf{x}_l)) \in \mathbb{R}^{n_h \times 1}$, $\psi : \mathbb{R}^d \rightarrow \mathbb{R}^{n_h}$ is mapping to some higher dimension with n_h dimensions, $\zeta = (\zeta_1, \zeta_2, \ldots, \zeta_1)^{\mathrm{T}} \in \mathbb{R}^l$ is a vector which consists of slack variables, and $\gamma \in \mathbb{R}^+$ denotes a regularized parameter which takes positive real values.

2.3 Proximal Support Vector Regression (PSVR)

It is quite similar to LS-SVR, but there are very slight changes in its formulation. Here two unparallel hyperplanes are obtained for data, whereas in LS-SVR, there is only one hyperplane [5].

Consider a regression problem with training dataset $\{(x_i, y_i)\}_{i=1}^n$, where the input vector $x_i \in \mathbb{R}^n$ and the corresponding target $y_i \in \mathbb{R}$. The main objective of regression is to find a function $f(x)$ which finds out the relationship between input vectors and their targets. PSVR is formulated as follows [6]:

$$\begin{aligned} & \min_{w,\zeta i} \frac{1}{2}\|w\|^2 + \frac{1}{2}b^2 + \frac{C}{2}\sum_{i=1}^{n}\zeta_i^2 \\ & \text{s.t. } w^T\psi(x_i) + b - y_i = \zeta_i, \quad i = 1, \ldots, n \end{aligned} \tag{7}$$

where ζ_i is the training error and $C > 0$ is given parameter.

LS-SVR and PSVR are two variants which are used for the prediction in this paper. Now how dimension will be reduced so that accuracy can be increased and mathematical formulation of KPCA need to be discussed.

2.4 Feature Dimensionality Reduction

Feature extraction is an important aspect while working with SVR for forecasting closing price of stock datasets because in stock data the trading data volume is large which contains some irrelevant indices. These will effect working with algorithms for the predictions such that irrelevant data need to be removed and only highly correlated data need to be used. Feature extraction can be done by **principal component analysis (PCA)** [7] method. PCA is used to transform data from higher dimension to lower dimension by removing components which are uncorrelated. PCA depends upon the eigen decomposition of positive semi-definite matrices and upon the singular value decomposition (SVD) of rectangular matrices. Eigenvalues and eigenvectors are calculated so that information about structure of matrix can be obtained. In this, data need to be converted into $m \times n$ matrix, and then mean is subtracted from each value and calculated SVD. After using PCA also, there are some problems with algorithms to overcome this KPCA [8] which is developed where the kernel method is introduced into PCA. In next section KPCA, the mathematical formulation of KPCA is given.

2.5 Kernel Principal Component Analysis

Nonlinear data need to be handled as it will affect the working of algorithms. Handling this will be done using **KPCA** where the original data, which are of lower dimensional, are projected into the higher dimensional. After this projection, PCA operation will be applied [9, 10].

Let $X = [x_1, x_2, \ldots, x_n]^T \in R^{n\times 1}$ represent input matrix where x_i denotes the observation vector at given time i. A nonlinear function $\zeta(.)$ maps data from input space to feature space:

$$R^m \xrightarrow{\zeta(.)} F^h \tag{8}$$

The input vector x_i transforms into $\zeta\,(x_i)$. Now covariance matrix will be calculated, and it is given by Eq. (9).

$$S^F = \frac{1}{n}\sum_{i=1}^{n}\zeta\,(x_i)\,\zeta\,(x_i)^T \tag{9}$$

where $\zeta\,(x_i)$ is scaled as zero mean. Doing this is not a simple task. Obtaining this need to find the eigenvalue decomposition. This decomposition will be in kernel space. It will be calculated by Eq. (10).

$$\lambda v = S^F v = \left(\frac{1}{n}\sum_{i=1}^{n}\zeta\,(x_i)\,\zeta\,(x_i)^T\right) v = \frac{1}{n}\sum_{i=1}^{n}\langle\zeta\,(x_i)\,,v\rangle\zeta\,(x_i) \tag{10}$$

where A and v are the eigenvalue and eigenvector, respectively, of S^F and $< \ldots >$ is the inner product of given values init. When $\lambda \neq 0$, then v lies in the span of training data into the kernel space. The coefficients $\alpha_{i \in \{1,2,3,\ldots\}}$ exist and satisfy (11).

$$v = \sum_{i=1}^{n} \alpha_i \zeta\left(x_i\right) \tag{11}$$

Multiplying both sides of Eqs. (3) and (4) with $\zeta\left(x_k\right)$, the equation becomes

$$\lambda \sum_{j=1}^{n} \alpha_j \left\langle \zeta\left(x_j\right), \zeta\left(x_k\right)\right\rangle = \frac{1}{n} \sum_{i=1}^{n} \sum_{j=1}^{n} \alpha_j \left\langle \zeta\left(x_j\right), \zeta\left(x_i\right)\right\rangle \left\langle \zeta\left(x_k\right), \zeta\left(x_i\right)\right) \tag{12}$$

Thus, the solution to decompose the covariance matrix is to obtain the inner product in nonlinear space. Kernel matrix K is given by Eq. (13), which is of size $n \times n$.

$$K_{ij} = \left\langle \zeta\left(x_i\right), \zeta\left(x_j\right)\right\rangle \tag{13}$$

The Gaussian kernel is used, and it is defined by Eq. (14).

$$K(x, y) = \exp\left(\frac{-\|x - y\|^2}{\sigma}\right) \tag{14}$$

where σ denotes a constant.

Kernel matrix needs to be centralized, and it is given by Eq. (15).

$$K - I_n K - K I_n + I_n K I_n \rightarrow K, I_n = \frac{1}{n} I_1 \in R^{n \times n} \tag{15}$$

Now, Eq. (12) can be rewritten, and it is given by Eq. (16).

$$\lambda \alpha = \frac{1}{n} K \alpha, \alpha = \left[\alpha_1, \alpha_2, \ldots, \alpha_n\right]^T \tag{16}$$

Now, the question is how to select principal components and answer is it is based on the value of λ. Principal components with large value of λ should be used into PC space, and other should be placed in residual space. Now the jth extracted PC will be calculated. It will be obtained by mapping training data $\zeta(x)$ in feature space onto eigenvector v_j and given by Eq. (17).

$$t_j = \left\langle v_j, \zeta(x)\right\rangle = \sum_{i=1}^{n} a_i^j \left\langle \zeta(x), \zeta\left(x_i\right)\right\rangle, j = 1, 2, \ldots, k \tag{17}$$

where k denotes the number of principal components extracted in principal component space.

3 Proposed Hybrid Approach

3.1 Input Feature

Forecasting for future price is an interesting and important work. This is studied in different fields, for example in trading, finance, statistics, and computer science. The main idea behind it is to enhance the profit by bought and sold of the stocks. Technical analysis is used by so many people so that they can have decisions regarding investments. Technical analysis is based on the study of previous price changes of stocks. Technical analysis [11] generates technical indicators using the price of the stocks, which are then used as an input measure by machine learning algorithms [12]. Some of the crucial and effective technical indicators are used for the prediction and listed below (Table 1).

3.2 Multistep Ahead Forecast Price

Multistep ahead time series forecasting [13] forecasts future time series using current or previous observations, i.e., forecasting ϕ_{N+h}, $(h = 1, 2, 3, \ldots, H)$, where H is an integer greater than one, using ϕ_t, $(t = 1, 2, \ldots, N)$. There are different strategies used for multistep forecasting, namely iterative strategy, direct strategy, and multiple input multiple output $(MIMO)$. In this paper, direct strategy is used for forecasting. This strategy is applied on different financial datasets to forecast closing price.

3.2.1 Direct Strategy

Direct strategy is first suggested by Cox [14]. Here a set of predicted models are constructed for each horizon where past observations are used. The associated squared multistep ahead errors are minimized [15]. Direct strategy estimates H different models between the inputs and the H outputs to predict ψ_{N+h}, $(h = 1, 2, 3, \ldots, H)$, respectively. The direct strategy first inserts the original series into H datasets

$$\begin{aligned} D_1 &= \{(\mathbf{x}_i, y_{t1}) \in (\mathbb{R}^m \times \mathbb{R})\}_{t=d}^{N}, \\ &\vdots \\ D_H &= \{(\mathbf{x}_t, y_{tH}) \in (\mathbb{R}^m \times \mathbb{R})\}_{t=d}^{N}. \end{aligned} \tag{18}$$

where $x_t \subset \{\psi_t, \ldots, \psi_{t-d+1}\}$, $y_{th} = \psi_{t+h}$. Then, the direct prediction strategy learns H direct models on $D_h \in [D_1 \cdots, D_H]$, respectively.

$$\psi_{t+H} = f_h(\mathbf{x}_t) + \omega_h, \quad h \in \{1, \ldots H\} \tag{19}$$

Table 1 Technical indicators

Indicator	Notation	Formula or definition
Open price	OP	Cost of a stock at the start of trading day
Closing price	CL	Cost of the stock at the end of a trading day
Low price	LP	Lowest cost of the day
High price	HP	Highest cost of the day
Trade volume	TV	Direction of a price trend
Stochastic %K	$\%K$	$\%K\,(m) = \frac{P_\alpha - LL_{\alpha-m}}{HH_{\alpha-m} - LL_{\alpha-m}}$
Larry William's	$\%R$	$\%R\,(m) = \frac{HH_{\alpha-m} - P_\alpha}{HH_{\alpha-m} - LL_{\alpha-m}}$
Moving average of %K	$\%D$	$\%D\,(m) = \frac{1}{m}\sum_{t=0}^{m-1} \%K_{\alpha-t}$
Bias	$BIAS$	$BAIS_\alpha\,(m) = \frac{P_\alpha - MA\,(m)}{m}$
Moving average	SMA	$SMA_{x_i,n} = \frac{x_i + x_{i-1} + x_{i-2} + \cdots + x_{i-n-1}}{n}$
Signal line	SL	$SL\,(m, n) = \frac{(MA\,(m) - MA\,(n)) + MA\,(n)}{10 * MA\,(n)}$
Exponential moving average	EMA	$EMA_\alpha\,(m) =$ $\beta\left(P_\alpha - EMA_{\alpha-1}\right) + EMA_{\alpha-1}$
Relative difference in %	RDP	$RDP_\alpha(m) = \frac{P_\alpha - P_{\alpha-m}}{P_{\alpha-m}} * 100$
Price rate of change	ROC	$ROC_\alpha = \frac{P_\alpha}{P_\alpha - m} * 100$
Momentum measures change in stock price	MTM	$MTM_\alpha = P_\alpha - P_{\alpha-m}$
MA convergence and divergence	$MACD$	$MACD_\alpha\,(m, n) = EMA\,(m) - EMA\,(n)$
Highest price	HH	$HH(m) = max\,(HI_{\alpha-m})$
Price oscillator	$OSCP$	$OSCP_\alpha\,(m, n) = \frac{MA\,(m) - MA\,(n)}{MA\,(m)}$
Commodity channel index	CCI	$CCI_\alpha\,(m) = \frac{P_\alpha - MA\,(m)}{0.015\sigma}$
Ultimate oscillator	UO	$UO\,(m, n, p) =$
		$\frac{100\,(4 * avg\,(m) + 2 * avg\,(n) + avg\,(p))}{4 + 2 + 1}$
Ulcer index	$Ulcer$	$Ulcer\,(m) = \sqrt{R_1^2 + R_2^2 + R_3^2 + \cdots + R_j^2}$
Average true range	ATR	$ATR_\alpha\,(m) = \frac{1}{m}\left(ATR_{\alpha-1} * (m - 1) + TR_\alpha\right)$
True strength index	TSI	$TSI\,(m) = \frac{EMA\,(EMA\,(MTM\,(1)\,, m))}{EMA\,(EMA\,(MTM\|\,(1)\,, m\|))}$

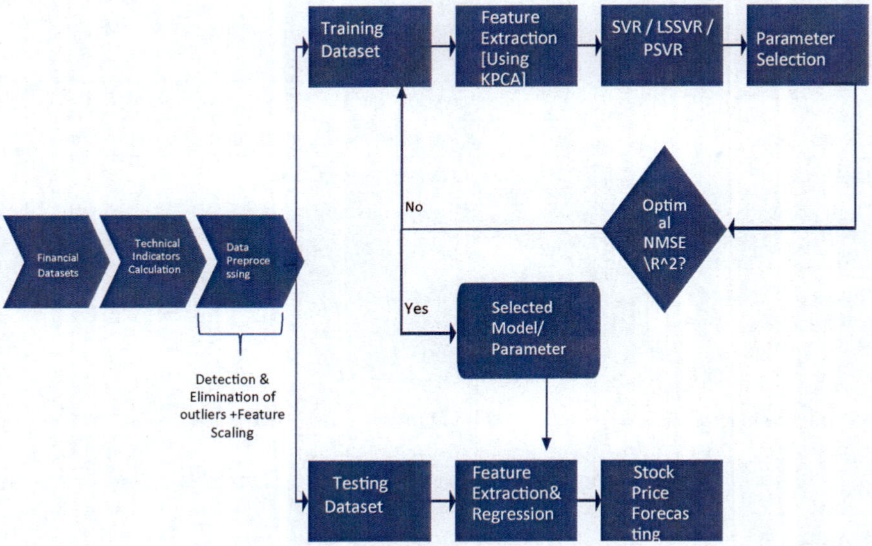

Fig. 2 Proposed model for stock price forecasting

where ω denotes the additive noise. After the learning process, the estimation of the H next values is returned by:

$$\hat{\psi}_{t+h} = \hat{f}_h\left(\psi_t, \psi_{t-1}, \ldots, \psi_{t-d-1}\right), \quad h \in \{1, \ldots, H\} \tag{20}$$

3.3 Proposed Hybrid Models

In this paper, the model used is support vector regression and its variants after applying KPCA techniques on different financial datasets. The input features are different technical indicators which are discussed in previous section. Initially, mainly thirty technical indicators which are listed above are used. But some of them are not that much useful, or if they are removed from the data, it will be giving approximately same accuracy with very less amount of time. Hence, principal component analysis(PCA) is used for reducing feature space so that time consumption can be reduced with same accuracy. Financial datasets are used in this paper, and these datasets are nonlinear datasets that is why kernel principal component analysis (KPCA) is used.

Using KPCA, dimension is reduced which will speed up the algorithm with same or more accuracy. Previously, 30 technical indicators are used for the prediction of 5 d ahead closing prices of given datasets. After using KPCA, 10, 15, and 20 indicators are taken and then prediction is done. The process is explained in Fig. 2.

The same can be explained by three steps. The following steps are followed for the predictions:

- **Stage 1:** KPCA is applied on the datasets, obtained using different technical indicators, for reducing the dimensions using **Gaussian kernel** and then **PCA** is applied for different number of components.
- **Stage 2:** Input the new dataset obtained by **KPCA** to **SVR, LSVR** and **PSVR** algorithms using different kernels. **Cross-validation** is used to the trained model for optimum parameters.
- **Stage 3:** The last stage is **multistep forecasting** of closing price. Here next five days closing price is predicted using obtained trained model and calculates different **Accuracy Parameters** for testing model.

4 Results and Discussion

4.1 Datasets

Financial datasets are used for predictions by discussed algorithms with optimal parameters so that accuracy can be achieved. Ten different datasets are used in this paper and will be discussed in this section. Datasets were taken from Yahoo finance. From 1985 to 2020, data are present in all the datasets. Data are divided into training and testing of ratio $80 - 20\%$, respectively.

Dow Jones index (DJI) consists of 30 prominent companies listed on stock exchanges in the USA. It is a price-weighted measurement index of stocks. The **NIFTY 50** is a benchmark Indian stock market index that represents the weighted average of 50 of the largest Indian companies listed on the National Stock Exchange. **Nifty Bank** represents the 12 most liquid and large capitalized stocks from the banking sector which trade on the National Stock Exchange (NSE) of India. The **Nasdaq** stock market is an American stock exchange based in New York City. The performance of 500 large companies of stock exchanges in the USA is represented by **The Standard and Poor's 500**, or simply the **S&P 500**. The **Korea Composite Stock Price Index** or **KOSPI** is the index of all common stocks traded on the Stock Market Division of the Korea Exchange. The **Hang Seng Index (HSI)** is a freefloat-adjusted market capitalization-weighted stock market index in Hong Kong. These 64 constituent companies represent about 58% of the capitalization of the Hong Kong Stock Exchange. The **Nikkei** is a stock market index for the Tokyo Stock Exchange. **Russellchicago** is the stock market index of Chicago. **TSECtaiwan** is the stock market index that measures the aggregate performance of listed stocks on TWSE; it is the most prominent and most frequently quoted index of stock performance of Taiwanese public companies.

4.2 Performance Evaluation Criteria

Some of the performance evaluation parameters for regression will be discussed in this section.

1. MSE (**Mean Square Estimation**): As the name suggests, MSE is an estimation of mean squared, i.e., it is a measurement of average-squared difference between the predicted value and the actual value of the data. It is always strictly positive. If it is close to zero, then our prediction of data is good.
   For m data point dataset if Z is the actual values vector and $\hat{Z}$ is the predicted values vector, then MSE (Mean Square Estimation) is given by:

$$\text{MSE} = \frac{1}{n} \sum_{i=1}^{m} \left(Z_i - \hat{Z}_i \right)^2 \tag{21}$$

2. RMSE (**Root Mean Square Estimation**): It is the standard deviation between predicted and actual values of a dataset. It is always larger or equal to MSE. If values of $RMSE$ and MSE are equal, then all the errors are of same magnitude. Lower value implies better result. Now for m data point dataset if Z is the actual values vector and $\hat{Z}$ is the predicted values vector, then $RMSE$ (Root Mean Square Estimation) is given by:

$$\text{RMSE} = \frac{1}{n} \sqrt{\sum_{i=1}^{m} \left(Z_i - \hat{Z}_i \right)^2} \tag{22}$$

3. NMSE (**Normalized Mean Square Error**): It is totally related to MSE. Smaller value of $NMSE$ indicates better model performance. When MSE is defined by equation (21), then $NMSE$ is given by the following expression:

$$\begin{aligned} \text{NMSE}(x, z) &= \frac{MSE(x, z)}{MSE(x, 0)} \\ &= \frac{\|x - z\|_2^2}{\|x\|_2^2} \end{aligned} \tag{23}$$

4. R^2 (**R Squared**): Here for m data points if z_is are the data points and $\hat{z}_i$s are predicted values, then their residual is defined as follows: $\epsilon_i = z_i - \hat{z}_i$.
   Let us take $\bar{z}$ to denote the mean of the original data than it is given by:

$$\bar{z} = \frac{1}{m} \sum_{i=1}^{m} z_i$$

Also the total sum of squares (proportional to the variance of the data) is given by:

$$SS_{\text{tot}} = \sum_i (z_i - \bar{z})^2$$

And the sum of squares of residuals, also called the residual sum of squares, is given by:

$$SS_{\text{res}} = \sum_i (z_i - \hat{z}_i)^2 = \sum_i e_i^2$$

Then R^2 is given by:

$$R^2 = 1 - \frac{SS_{\text{res}}}{SS_{\text{tot}}}$$

If it is close to 1, then prediction is good. If its value is exactly 1, then predicted value and true value are same, i.e., ideal condition. If it is negative, then prediction is bad.

5. MAE (**Mean Absolute Error**):
   It is an arithmetic mean of the absolute value of predicted and true value of the prediction. If $\hat{z}_i$ denotes the predicted value of z_i of n number of points, then MAE is given by:

$$\text{MAE} = \frac{\sum_{i=1}^{n} |\hat{z}_i - z_i|}{n} \quad (24)$$

4.3 *Result Analysis*

Results for financial datasets are listed in Table 2 using KPCA for ten components on SVR and its variants. ε-SVR gives good results, but LS-SVR and PSVR are much faster than ε-SVR with same results. If the accuracy of the algorithm is high, then

1. Value of NMSE, RMSE, and MSE will be close to zero.
2. Value of R^2 will be close to 1, i.e., 0.9999.
3. Value of MAE will be small.

Table 2 contains the values of different accuracy parameters, namely NMSE, RMSE, MSE, R^2, and MAE (up to 4–5 decimal places), for the prediction of 1 step ahead, 3 step head, and 5 step ahead days of the financial datasets, DJI, Nifty50, Nifty bank, Nasdaq, S&P500, KOSPI, HSI, Nikkeitokyo, Russellchicago, and TSECtaiwan using algorithms SVR, LS-SVR, and PSVR with KPCA.

The results for SVR are good because value of NMSE is quite small, value of R^2 is also close to 1, and MAE is smaller in comparison with other algorithms. The

Table 2 Results using KPCA for Gaussian Kernel

	Steps	NMSE	RMSE	MSE	R_Squared	MAE
DJI						
SVR	1 Ahead	0.0031	382.1757	146058.2328	0.9562	238.4757
	3 Ahead	0.0032	619.8292	384188.2327	0.8862	413.9457
	5 Ahead	0.0031	803.1574	645061.8742	0.8112	536.2112
LS-SVR	1 Ahead	1.8780	555.8205	308936.4600	0.9074	367.6138
	3 Ahead	1.8612	758.0469	574635.0590	0.8298	497.6496
	5 Ahead	1.8419	915.5906	838306.0902	0.7547	596.3718
PSVR	1 Ahead	1.9101	565.6011	319904.5686	0.9029	386.0746
	3 Ahead	1.9210	790.4810	624860.2722	0.8191	507.0197
	5 Ahead	1.9350	987.5863	975326.6473	0.7399	606.7695
NIFTY 50						
SVR	1 Ahead	0.0027	157.8440	24914.7204	0.9702	105.9706
	3 Ahead	0.0027	268.1496	71904.1857	0.9147	187.6582
	5 Ahead	0.0027	358.0582	128205.6850	0.8493	249.0224
LS-SVR	1 Ahead	1.3491	236.0821	55734.7728	0.9333	165.6623
	3 Ahead	1.3705	340.8833	116201.4310	0.8622	236.4805
	5 Ahead	1.3949	427.2407	182534.6468	0.7855	286.9269
PSVR	1 Ahead	1.4412	267.9904	71818.8460	0.9205	146.5600
	3 Ahead	1.4441	345.4834	119358.7826	0.8954	218.1407
	5 Ahead	1.4492	440.1835	193761.4891	0.7530	356.9275
NIFTY Bank						
SVR	1 Ahead	0.0027	157.8440	24914.7142	0.9702	105.9706
	3 Ahead	0.0027	268.1496	71904.1857	0.9147	187.6582
	5 Ahead	0.0027	358.0582	128205.6845	0.8493	249.0224
LS-SVR	1 Ahead	1.0795	828.1194	685781.7264	0.9591	602.7192
	3 Ahead	1.0805	1184.4007	1402805.1180	0.9163	849.1901
	5 Ahead	1.0812	1478.6955	2186540.4078	0.8696	1029.1876
PSVR	1 Ahead	1.0993	830.4204	689598.0373	0.9398	609.8939
	3 Ahead	1.1009	1199.0850	1437804.8840	0.9108	860.7943
	5 Ahead	1.1085	1498.0140	2244045.9208	0.8599	1036.8530
NASDAQ						
SVR	1 Ahead	0.0026	140.7362	19806.6664	0.9909	94.4722
	3 Ahead	0.0026	222.0228	49294.1284	0.9778	157.0509
	5 Ahead	0.0026	281.6245	79312.3464	0.9650	194.1328
LS-SVR	1 Ahead	2.1990	1238.7161	1534417.6625	0.2923	934.6352
	3 Ahead	2.1836	1259.6015	1586595.9419	0.2841	949.5474
	5 Ahead	2.1670	1278.1178	1633585.2036	0.2787	963.2212
PSVR	1 Ahead	2.3312	1240.9000	1539832.7068	0.2692	978.2589
	3 Ahead	2.3525	1268.6868	1609566.0990	0.2790	988.2438
	5 Ahead	2.3728	1299.2973	1688173.4712	0.2892	996.5100

(continued)

Table 2 (continued)

SP500						
SVR	1 Ahead	0.0031	42.3123	1790.3294	0.9772	26.2978
	3 Ahead	0.0031	67.1123	4504.0553	0.9437	44.2996
	5 Ahead	0.0031	86.2656	7441.7495	0.9085	57.0733
LS-SVR	1 Ahead	2.0287	60.4861	3658.5722	0.9535	40.7247
	3 Ahead	2.0171	82.2019	6757.1578	0.9155	55.3074
	5 Ahead	2.0033	99.0806	9816.9722	0.8793	66.2303
PSVR	1 Ahead	2.1579	60.8015	3696.8216	0.9500	48.1495
	3 Ahead	2.1995	83.2953	6938.1112	0.9019	57.2875
	5 Ahead	2.2421	99.6576	9931.6311	0.8580	68.7925
KOSPI						
SVR	1 Ahead	0.0029	28.4512	809.4714	0.9785	19.7631
	3 Ahead	0.0028	53.4205	2853.7532	0.9266	37.9100
	5 Ahead	0.0027	73.2340	5363.2252	0.8665	50.6539
LS-SVR	1 Ahead	1.4618	46.3846	2151.5301	0.9429	32.3471
	3 Ahead	1.4191	63.1473	3987.5836	0.8974	46.0826
	5 Ahead	1.3791	78.2886	6129.1024	0.8475	56.7479
PSVR	1 Ahead	1.5061	66.6394	4440.8050	0.9129	31.0435
	3 Ahead	1.6403	75.2228	5658.4651	0.8598	48.3695
	5 Ahead	1.7019	86.9636	7562.6700	0.8117	57.7032
Hangsenghsi						
SVR	1 Ahead	0.0021	326.6889	106725.6615	0.9628	241.4892
	3 Ahead	0.0021	554.5304	307503.9360	0.8926	440.1694
	5 Ahead	0.0020	725.1978	525911.8190	0.8158	574.9844
LS-SVR	1 Ahead	1.2260	389.3523	151595.2387	0.9471	301.3788
	3 Ahead	1.2104	608.4641	370228.6167	0.8707	485.4698
	5 Ahead	1.1986	769.7449	592507.1753	0.7925	605.8457
PSVR	1 Ahead	1.3513	415.5742	172701.9504	0.9107	321.1611
	3 Ahead	1.3695	655.9208	430232.1003	0.8430	542.8853
	5 Ahead	1.3922	782.1349	611734.9693	0.7598	650.4033
Nikkeitokyo						
SVR	1 Ahead	0.0054	261.1695	68209.5193	0.9767	183.1353
	3 Ahead	0.0054	491.6516	241721.3360	0.9191	345.6765
	5 Ahead	0.0054	649.6142	421998.6692	0.8618	449.3797
LS-SVR	1 Ahead	2.8676	433.7496	188138.7397	0.9358	289.7003
	3 Ahead	2.9372	601.0323	361239.8427	0.8790	402.8527
	5 Ahead	2.9219	731.8341	535581.1501	0.8246	487.2349
PSVR	1 Ahead	2.8889	446.1040	199008.7691	0.9211	300.6353
	3 Ahead	2.9210	613.1150	375909.9884	0.8980	401.5942
	5 Ahead	2.9959	796.6076	634583.6750	0.8322	500.3504

(continued)

Table 2 (continued)

Russellchicago						
SVR	1 Ahead	0.0016	24.9095	620.4852	0.9716	16.4859
	3 Ahead	0.0016	42.4925	1805.6151	0.9193	28.4713
	5 Ahead	0.0016	55.2070	3047.8095	0.8666	37.4297
LS-SVR	1 Ahead	1.0161	36.0814	1301.8659	0.9405	24.2396
	3 Ahead	1.0163	50.9233	2593.1820	0.8841	34.0682
	5 Ahead	1.0147	62.5917	3917.7203	0.8286	42.0243
PSVR	1 Ahead	1.0711	39.6727	1573.9261	0.9279	29.8674
	3 Ahead	1.0802	53.5454	2867.1046	0.8672	38.7494
	5 Ahead	1.0818	66.7037	4449.3882	0.8012	49.9280
TSECtaiwan						
SVR	1 Ahead	0.0018	117.1096	13714.6581	0.9892	81.2580
	3 Ahead	0.0018	212.2706	45058.7909	0.9656	150.0747
	5 Ahead	0.0018	287.4891	82649.9539	0.9384	197.8245
LS-SVR	1 Ahead	1.0965	181.2310	32844.6759	0.9743	125.4087
	3 Ahead	1.0948	259.7581	67474.2560	0.9485	178.3392
	5 Ahead	1.0982	322.7211	104148.9197	0.9224	218.3286
PSVR	1 Ahead	1.1058	201.0277	40412.1267	0.9703	133.9107
	3 Ahead	1.1072	285.6555	81599.0379	0.9317	204.0110
	5 Ahead	1.1095	374.6619	140371.5220	0.9115	270.9809

same can be verified using graphical representation also. Figure 3 is the graphical representation of closing price actual value, 1 step ahead, 3 step ahead, and 5 step ahead prediction of next five days for DJI dataset from year 2018 to 2020 using algorithm SVR (linear kernel) with KPCA (Gaussian kernel). Figure 4 is the graphical representation of closing price actual value, 1 step ahead, 3 step ahead, and 5 step ahead prediction of next five days for HSI dataset from year 2018 to 2020 using algorithm SVR (linear kernel) with KPCA (Gaussian kernel). Figure 5 is the graphical representation of closing price actual value, 1 step ahead, 3 step ahead, and 5 step ahead prediction of next five days for Russellchicago dataset from year 2018 to 2020 using algorithm SVR (linear kernel) with KPCA (Gaussian kernel). Figure 6 is the graphical representation of closing price actual value, 1 step ahead, 3 step ahead, and 5 step ahead prediction of next five days for SP500 dataset from year 2018 to 2020 using algorithm SVR (linear kernel) with KPCA (Gaussian kernel). Figure 7 is the graphical representation of closing price actual value, 1 step ahead, 3 step ahead, and 5 step ahead prediction of next five days for TSECtaiwan dataset from year 2018 to 2020 using algorithm SVR(linear kernel) with KPCA(Gaussian kernel).

Figures 3, 4, 5 and 7 have overlapping graphs which show good accuracy. Results of algorithms using KPCA are better than without KPCA. This comparison can take place by value of NMSE and R^2. The value of NMSE is smaller, and the value of R^2

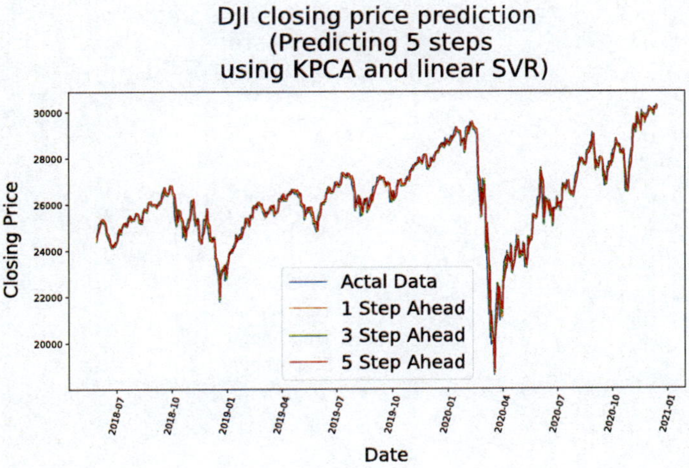

Fig. 3 DJI predictions using KPCA (Gaussian kernel) SVR (linear kernel)

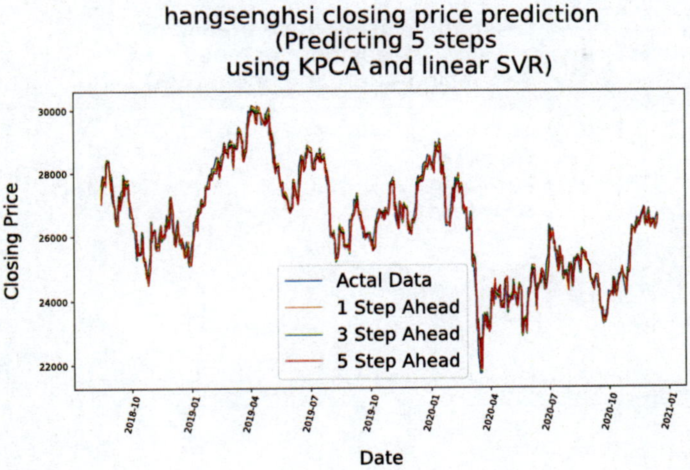

Fig. 4 Hangsenghsi predictions using KPCA (Gaussian kernel) SVR (linear kernel)

is near to 1 when KPCA is used. Also there is computational time which is also less in comparison with algorithms without KPCA.

5 Conclusion

In this paper, KPCA is applied on SVR and its variants for feature extractions. KPCA handles nonlinear data by projecting original data space into a high-dimensional

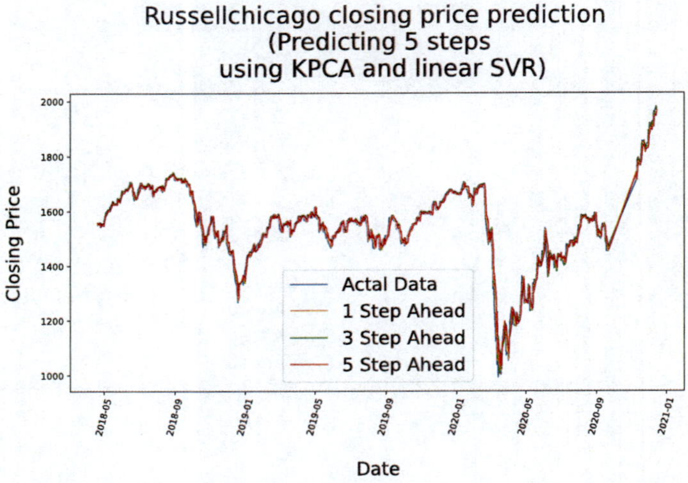

Fig. 5 Russellchicago predictions using KPCA (Gaussian kernel) SVR (linear kernel)

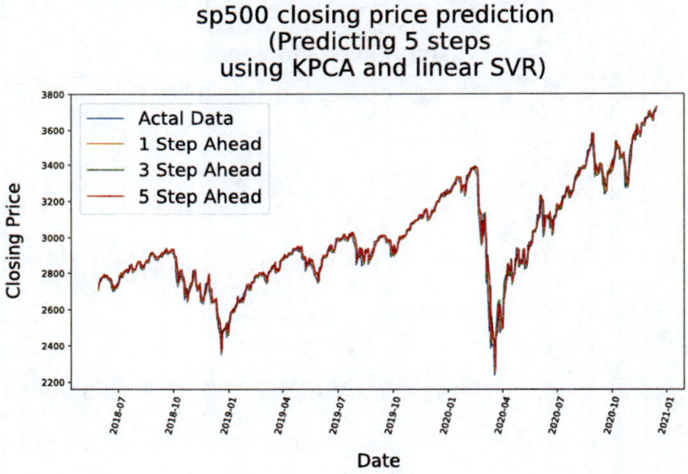

Fig. 6 SP500 predictions using KPCA (Gaussian kernel) SVR (linear kernel)

feature space before implementing PCA operation. Ten different financial datasets are used for the implementation. In the previous section, all results are listed for all the datasets along with their graphical representations. Results in comparison with SVR and its variants without KPCA are better when using KPCA as value of $NMSE$ is smaller and R^2 is close to 1 also time consumption decreased because of feature extraction. LS-SVR and PSVR are without KPCA also giving results fast in comparison with SVR, but using KPCA, the computational time of SVR is also reduced and LS-SVR & PSVR are faster than before. Hence, method PSVR with

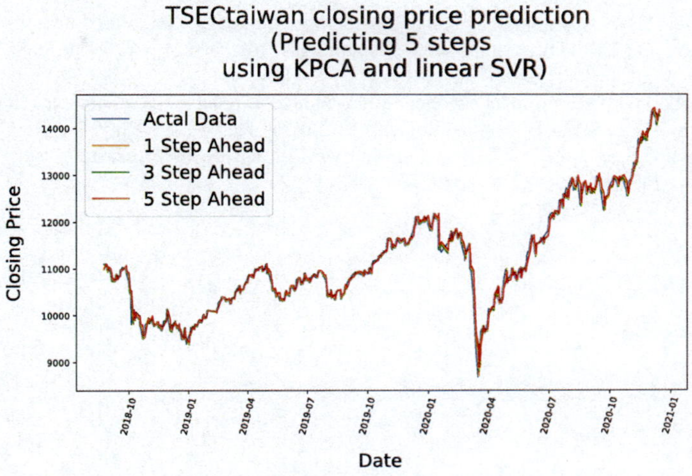

Fig. 7 TSECtaiwan predictions using KPCA (Gaussian kernel) SVR (linear kernel)

KPCA is giving better results with very less time period and implies that PSVR with KPCA is the best method among all.

Further work to explore methods for increasing accuracy with same or less time consumptions can be done. Other variants of SVR can be implemented using KPCA for increasing accuracy as well as reducing computational time.

References

1. Jiang Y, Zhang T, Gou Y, He L, Bai H, Hu C (2018) High-resolution temperature and salinity model analysis using support vector regression. J Ambient Intell Humanized Comput 1–9
2. Drucker H, Burges CJ, Kaufman L, Smola A, Vapnik V (1996) Support vector regression machines. Adv Neural Inf Proc Syst 9
3. Smola AJ, Schölkopf B (2004) A tutorial on support vector regression. Stat Comput 14(3):199–222
4. Vapnik V (1999) The nature of statistical learning theory. Springer Sci Bus Med
5. Kumar S, Mohri M, Talwalkar A (2012) Sampling methods for the Nyström method. J Mach Learn Res 13(1):981–1006
6. Mangasarian OL, Wild EW (2001) Proximal support vector machine classifiers. In: proceedings KDD-2001: knowledge discovery and data mining
7. Schölkopf B, Smola A, Müller KR (1998) Nonlinear component analysis as a kernel eigenvalue problem. Neural Comput 10(5):1299–1319
8. Alcala CF, Qin SJ (2010) Reconstruction-based contribution for process monitoring with kernel principal component analysis. Ind Eng Chem Res 49(17):7849–7857
9. Lee JM, Yoo C, Choi SW, Vanrolleghem PA, Lee IB (2004) Nonlinear process monitoring using kernel principal component analysis. Chem Eng Sci 59(1):223–234
10. Cheng CY, Hsu CC, Chen MC (2010) Adaptive kernel principal component analysis (KPCA) for monitoring small disturbances of nonlinear processes. Ind Eng Chem Res 49(5):2254–2262
11. Murphy JJ (1999) Study guide to technical analysis of the financial Markets: a comprehensive guide to trading methods and applications. Penguin

12. Turner T (2007) A beginner's guide to day trading online, 2nd edn. Simon and Schuster
13. Chevillon G (2007) Direct multi-step estimation and forecasting. J Econom Surv 21(4):746–785
14. Cox DR (1961) Prediction by exponentially weighted moving averages and related methods. J Royal Stat Soc: Series B (Methodological) 23(2):414–422
15. Franses PH, Legerstee R (2010) A unifying view on multi-step forecasting using an autoregression. J Econom Surv 24(3):389–401

Soft Computing Approach for Student Dropouts in Education System

Sumin Samuel Sybol, Shilpa Srivastava, and Hemlata Sharma

Abstract The education system has increased the number of dropouts in the coming years, decreasing the number of educated people. Education system refers to a group of institutions like ministries of education, local education bodies, teacher training institutes, universities, colleges, schools, and more whose primary purpose is to provide education to all the people, especially young people and children in educational settings. The research aims to improve the student dropout rate in the education system by focusing on students' performance and feedback. The students' dropout rate can be calculated based on complexity, credits, attendance, and different parameters. This study involves the extensive study that inculcates student dropout with their performance and other parameters with soft computing approaches. There are various soft computing approaches used in the education system. The approaches and techniques used are sequential pattern mining, sentimental analysis, text mining, outlier decision, correlation mining, density estimation, etc. The approaches and techniques will be beneficial to calculating and decreasing the rate of dropout of students in the education system. The research will make a unique contribution to improved education by calculating the dropout rate of students. In particular, we argue that the dropout rate is increasing, so soft computing techniques can be the solution to improvise/reduce the dropout rate.

Keywords Education · Soft computing · Educational technology

1 Introduction

The education system refers to a group of institutions whose primary purpose is to provide education to all the people in educational settings. The dropout rate of the students is increasing yearly because of many parameters such as academic

S. S. Sybol (✉) · S. Srivastava
CHRIST (Deemed to be University), NCR, Delhi, India
e-mail: suminsamuel.sybol@res.christuniversity.in

H. Sharma
Sheffield Hallam University, Sheffield, England

M. Thakur et al. (eds.), *Soft Computing for Problem Solving*, Lecture Notes in Networks and Systems 547, https://doi.org/10.1007/978-981-19-6525-8_25

and personal issues. The research aims to improve the student dropout rate in the education system by focusing on students' performance and feedback. The collected data is categorized into structured data such as grades, progression rate, performance, and marks. In addition to the structured data, the unstructured data are the opinions and the feedback received through forms, surveys, etc., can be used to calculate and reduce the rate of dropout in the education system. The student's drop and performance can be calculated on various aspects. The data can help improve Indian Higher Education. In addition, it can also upgrade the overall education system.

The data collected from structured and unstructured data are obtained from education system. The data collected from the students' academic progression report subject-wise and the feedback received from the students are considered. The data acquired will be processed as per the requirement, and then, data is categorized in different levels as per the parameters being considered. The data categorization is done as per the soft computing techniques to understand the student performance and grades [1]. Then, the testing is done on a model classified using soft computing techniques concerning various behaviors. The data can be subjected to multiple soft computing techniques like Fuzzy Logic, Artificial Neural Networks, Genetic Algorithms, Bayesian Networks, Swarm Intelligence, *K*-means Clustering, etc. [2].

This paper has eight sections. Section 2 describes the preliminary where the soft computing techniques/technology are described. Section 3 describes the related work to the topic, and Sect. 4 describes the methodology where the flow and method of the research work are being demonstrated. Then Sect. 5 describes the dataset. Section 6 describes the proposed model of the research work. Then, Sect. 7 describes the result and conclusion of the research work. Then, the study is concluded in Sect. 8 by giving some future directions related to the research.

2 Preliminaries

The research has adopted soft computing techniques/technologies such as SVM and the Naive Bayes Algorithm.

2.1 Support Vector Machine

Support vector machine (SVM) is one of the supervised machine learning algorithms used for both regression and classification. The main objective of the algorithm is to find a hyperplane in *n*-dimensional space that classifies the data points.

The dimensions will depend on the number of factors. If the number of factors is 2, the hyperplane will be just a line. If the number of factors is 3, the hyperplane will be a 2D plane. If the number of factors exceeds three, it is not very easy to imagine. It picks the extreme vectors that assist in developing hyperplanes. Thus, the severe

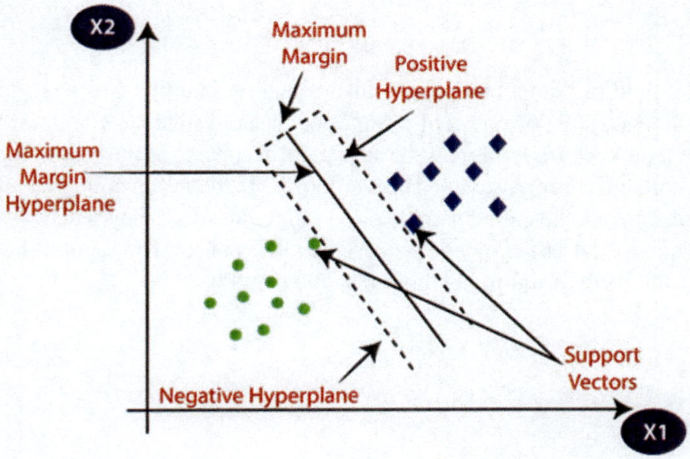

Fig. 1 Support vector machine graph

cases are known as support vectors, and so the algorithm is named a support vector algorithm. The graph for the support vector machine is explained in Fig. 1.

The SVM kernel converts the low-dimensional space into higher dimensional space. It can be primarily helpful in nonlinear separation problems. The kernels some extremely complex data transformations and then determine how to separate the data based on defined labels or outputs.

2.2 *Naïve Bayes*

Naïve Bayes Theorem is the classifiers, a collection of classification algorithms based on Bayes theorem. It is a family of algorithms where each algorithm shares a common principle, i.e., each pair of features is independent of each other. Bayes Theorem is stated mathematically as the following equations.

$$P\left(\frac{A}{B}\right) = \frac{P(B/A)P(A)}{P(B)}$$

Bayes Theorem finds the probability of the event occurring given the probability of the other event that has already happened.

2.3 N-*Gram*

N-gram is a field of computational linguistics and probability. It is a sequence of *n* number of items from a given set of samples of speech and texts. The items included phonemes, letters, words, syllables, or base pairs according to the requirements. The *n*-gram is collected from a text or speech corpus. If the items collected are words, then *n*-gram can also be called shingles. *N*-gram comes into play when we have the text with data in Natural Language processing (NLP) tasks. The *n*-gram is explained in Fig. 2 in the form of unigrams, bigrams, and trigrams.

3 Related Work

Numerous articles and papers were done on the problem of improving the student learning experience. The papers can be categorized to be divided with different techniques or the level of education used in the article to work on it. The new education 4.0 was discussed compared to the industry 4.0 era [3]. Education 4.0 is a purposeful approach along with industry 4.0, and it is about transforming education in the future using advanced technologies and automation. It explains the educational innovators who can design new educational models, learning methodologies, teaching tools, and infrastructure [4]. There was an evaluation of students' behavior in an E-learning environment where the work is done through an open university with 170,000 students in several modules. The final evaluation table was acquired using two different techniques: decision tree and apriori implemented on the information, and a combination of fuzzy with apriori was proposed [5].

The technologies like virtualization come into the role during the pandemic, and it offers improved performance for the teaching and learning process. The cost involved in building a cloud architecture for education is causing trouble [6, 7]. Different logics use clustering algorithms to illustrate the datasets and then demonstrate their potential use in education management. This application will explain the performance of the clustering algorithms [8]. Among the machine learning techniques, the most accessible model that can be introduced to the users is the decision tree, which

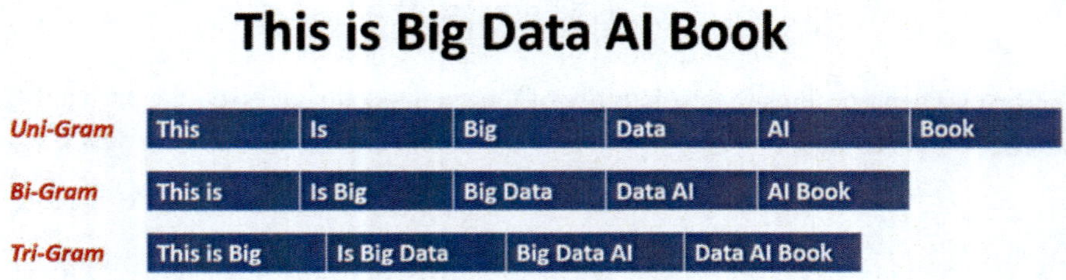

Fig. 2 Unigrams, bigrams, and trigrams

shows the essential factors are algorithm final result, calculus subjects, and discrete mathematics [9].

E-learning also assumes a fundamental part in the field of education [10]. E-learning using confidence-based online assessment was intended to dispose of the speculating strategies that the students use in standard assessment tests [11]. Various machine learning algorithms were executed to accomplish the same [12]. By this, it is learned that the student's behavior needs to be observed while working on courses to detect the sequential dimension correctly. The data size will be measured per the number of units, the examples, and the exercises [13].

The student's feedback is equally important to access the student's performance academically with their input so that the learning and teaching experience can be improved according to the input [14]. The feedback mining system is there to inquire about points and study the produced criticism [15]. This strategy helps enhance the student's knowledge and the educator's process [16]. Opinion mining was done with various techniques [17]. Naïve Bayes algorithms have achieved the highest precision and the *K*-Nearest Neighbor algorithm [18].

Various reviews have been done over the past ten years among 25 papers to conclude the feedback system for the students in the learning education system of the Indian Higher Education System. These reviews demonstrate different techniques and methods implemented in various higher education scenarios to improve student feedback [19]. Students' grading is also essential in resolving the competence-based learning approach [20]. There is firmer information, possibilities to customize the learning, inspiration to accomplish better grades, and a clearer picture of students about learner's present state, which was exceptionally esteemed [21]. The personalized, versatile E-learning system can gauge the effectiveness of online students and prescribe the proper activity to the learners [22].

4 Methodology

The collection of data has many benefits for students, and the lectures like it can help improve teaching and learning behavior. According to this, the dropout rate of the students will also decrease considering the parameters. It will enhance the communication between the students and the lecturer, allowing them to consider their individual opinions. The flow of the process is demonstrated graphically in Fig. 3 [23, 24].

4.1 Collection of Data

Firstly, the data is collected in structured data and unstructured data. The data obtained will be as per the real-world data, which can be noisy, inconsistent, and incomplete.

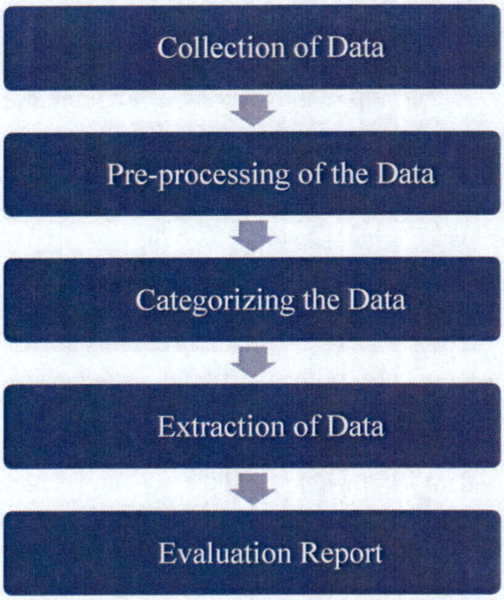

Fig. 3 Flowchart of the process or method

4.2 Preprocessing of the Data

Then, the data is preprocessed for the analysis purpose with the relevant data obtained. Preprocessing and cleaning the information are essential tasks for creating a dataset that can be used for extraction, as the real-world data can be noisy, inconsistent, and incomplete. It is needed to be cleaned before processing to the next step.

4.3 Categorizing the Data

Thirdly, the cleaned data is obtained from the preprocessing stage to get the subset of the data. There will be a sentimental analysis of the information preprocessed and cleaned. The data will be categorized in the form of positive and negative comments [25].

4.4 Extraction of Data

The evaluation of the data responses is data-driven and straightforward. The classification of questions is in the form of natural language text which involves sentimental analysis. Understanding the data pattern generated by the data obtained is crucial to improving the institution's effectiveness and creating plans to improve the teaching and learning experience.

4.5 Evaluation Report

Then the final report of the extracting and testing is obtained in the form of a report that can be used per the research. It will evaluate all the data and categorize the data's polarity. The evaluation will be data-driven and straightforward, while the text or data classification will involve sentimental analysis [26].

The soft computing techniques categorize the data to understand the teacher's and students' performance concerning various features such as teaching behavior, learning behavior, modules, pedagogy, and the structure of assessments. The preprocessed data can be subjected to multiple soft computing techniques like Fuzzy Logic, Artificial Neural networks, Genetic Algorithms, Bayesian networks, Swarm Intelligence, *K*-means Clustering, etc.

5 About Dataset

Dataset: 650 Rows and 33 Columns. Column Details: clg (College), gnd (Gender), age (Age), addr (Address), famsz (Family Size), parcost (parent's cohabitation status), medu (mother's education), fedu (father's education), mjob (mother's job), fjob (father's job), reasclg (reason to choose college), stugrd (student's guardian), htschtra (home to school travel time), wkstutm (weekly study time), pastcls (no of past class failures), edusup (extra educational support), famedu (family educational support), excls (extra paid classes), excur (extra-curricular activities), schl (attended school), highedu (higher education), intacc (Internet access), romrel (romantic relationship), qultyfam (quality of family relationships), tmschl (free time after school), gngfrndz (going out with friends), wrkalccons (workday alcohol cons), wkalccons (weekend alcohol cons), curhlthstat (current health status), noschlabse (no. of school absences), frstgrade (first-period grade), secgrade (second-period grade), fgrade (final grade).

6 Proposed Model

The previous research which was done on this is related to the

- Student performance
- Faculty performance
- Academic performance.

The calculation is done to improvise the teaching and learning process in the above.

The contribution of this research work is to

- Evaluate the dropout rate

- Decrease the dropout rate of the students yearly.

The above contribution will help to improve the educational methods and make them better for a better future.

The student's dropout and performance can be calculated on various aspects. The model can be divided into phases where the data collection till the result is evaluated to come with the outcome. The model is demonstrated in Fig. 4. The phases are

Phase 1: Collection of Data

The data can be categorized in structured data such as grades, enrollment, progression rate, performance, and marks. In addition to the structured data, the unstructured data are the student's opinions and the feedback expressed through forms, surveys, etc. In addition, unstructured data such as opinions or feedback through feedback forms, and surveys, can also upgrade the overall education system. The data obtained can help improve the education system of Indian Higher Education.

Feedback can feature various issues or aspects that the students face with academics. Feedback is usually collected at the end of the unit, yet it is more advantageous to take it continuously. Collecting feedback has various advantages for the lecturer and their students, like further developing instructing and understanding of students' learning conduct.

Phase 2: Pre-processing of the Data

The data which was collected is being used for analysis purposes. Before this, preprocessing and cleaning of the obtained data occur to create different datasets used for extraction. This is done as the data obtained or the real-world data will be noisy, consistent, and incomplete. The data is needed to be cleaned before moving it to the next phase of the model.

Phase 3: Analysis of Data

The data obtained from the previous phase is cleaned so that the subsets of the data can be obtained. On this, the sentimental analysis is done as per the data obtained. The data

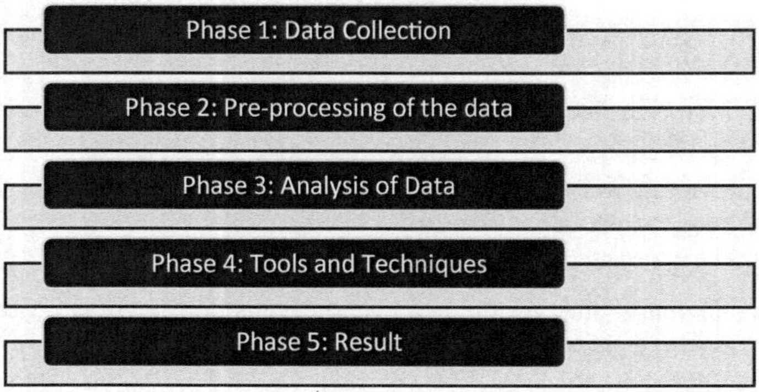

Fig. 4 Phases flow of the model

can be categorized as positive and negative comments, which will be helpful for the techniques and the tools where the final result could be obtained. The classification of questions is in the form of natural language text which involves sentimental analysis.

Phase 4: Tools and Techniques

The soft computing tools/techniques used here are SVM, Naïve Bayes, and *N*-gram in this research. The data testing depends on the supervised learning algorithm's training model. There are other tools to understand the student's performance concerning various features, i.e., Fuzzy Logic, Artificial Neural networks, Genetic Algorithms, Bayesian networks, Swarm Intelligence, *K*-means Clustering, etc. This research uses these techniques/tools to get a precise result as these have a high precision rate compared to other soft computing techniques/tools.

Phase 5: Result

After evaluating the data as per the techniques and tools, it will give the precise value of the dropout rate in the system, which helps the teachers or the management improve it accordingly and decrease the rate of dropouts.

7 Result and Discussion

The result obtained is beneficial for the students to understand their learning behavior. As per the study, the dropout rate of the students can be obtained with various parameters, which can be kept to determine which parameters the dropout rate of the students is increasing. By this, the institution or the college will be able to improve the various field where the extra efforts are to be put. According to the parameters, a primary example could be when the students are unaware of the syllabus, evaluation patterns, or course conduct methods, which might also increase the dropout rates. Secondly, sometimes the students may lose interest in the subject and create misconceptions due to their faculties having a very slow or fast teaching pace. Thirdly, sometimes the student might face issues according to the geographical issue, mainly language barriers, and psychological barriers.

Also, in the current pandemic scenario, when the world was shifted to the digital means for their day-to-day activities, either willingly or unwillingly, both students and the faculties faced several issues. The issues are mainly by using different platforms for the online classes where the interest of the student or the faculty in a particular session is get decreased. The concentration of the student and the faculties will get reduced. The parameters to evaluate or get the result for the dropout rate in Indian Higher Education will vary as per the students and the institution's geographical location, academic pace, and institution administration.

8 Conclusion and Future Scope

This research proposes a customized activity based on the proposed machine learning algorithms that can estimate the student's dropout rate and the learning performance and recommend the improvements that are being done in different domains such as curriculum, teaching, and learning pattern, which byways to minimize the rate of dropout in education system. Compared to the traditional methods, the education system 4.0 is being implemented during the pandemic compared to the industry 4.0. The proposed method demonstrates the various points to calculate the accuracy and precision of the obtained structured and unstructured data values. The future scope of the research field is to implement the model to decrease the student dropout rate and to improve the students' learning patterns and performance of each individual to in a better way in Indian Higher Education System.

References

1. Tanuar E et al (2019) Using machine learning techniques to earlier predict student's performance. In: 1st 2018 Indonesian association for pattern recognition international conference, INAPR 2018—proceedings, pp 85–89. https://doi.org/10.1109/INAPR.2018.8626856
2. Agius NM, Wilkinson A (2014) Students' and teachers' views of written feedback at undergraduate level: a literature review. Nurse Educ Today 34(4):552–559. https://doi.org/10.1016/j.nedt.2013.07.005
3. Miranda J et al (2021) The core components of education 4.0 in higher education: three case studies in engineering education. Comput Electr Eng 93(Feb). https://doi.org/10.1016/j.compeleceng.2021.107278
4. Chen JF, Hsieh HN, Do QH (2015) Evaluating teaching performance based on fuzzy AHP and comprehensive evaluation approach. Appl Soft Comput J 28:100–108. https://doi.org/10.1016/j.asoc.2014.11.050
5. Dhanalakshmi V, Bino D (2019) About 2019 4th MEC international conference on big data and smart city (ICBDSC). In: 2019 4th MEC International conference on big data and smart city, ICBDSC 2019, pp VI–VIII. https://doi.org/10.1109/ICBDSC.2019.8645612
6. García P et al (2007) Evaluating Bayesian networks' precision for detecting students' learning styles. Comput Educ 49(3):794–808. https://doi.org/10.1016/j.compedu.2005.11.017
7. Sobers Smiles David G, Anbuselvi R (2015) An architecture for cloud computing in higher education. In: Proceedings of the IEEE international conference on soft-computing and network security, ICSNS 2015. https://doi.org/10.1109/ICSNS.2015.7292432
8. Gogo KO, Nderu L, Mwangi RW (2018) Fuzzy logic based context aware recommender for smart e-learning content delivery. In: 5th International conference on soft computing and machine intelligence, ISCMI 2018, pp 114–118. https://doi.org/10.1109/ISCMI.2018.8703247
9. Hafidi M, Lamia M (2015) A personalized adaptive e-learning system based on learner's feedback and learner's multiple intelligences. In: 12th International symposium on programming and systems, ISPS 2015, vol 3, pp 74–79. https://doi.org/10.1109/ISPS.2015.7244969
10. Aderibigbe SA (2021) Can online discussions facilitate deep learning for students in general education? Heliyon 7(3):e06414. https://doi.org/10.1016/j.heliyon.2021.e06414
11. Shvets O, Murtazin K, Piho G (2020) Providing feedback for students in e-learning systems: a literature review, based on IEEE explore digital library. In: IEEE Global engineering education conference, EDUCON, 2020-Apr, pp 284–289. https://doi.org/10.1109/EDUCON45650.2020.9125344

12. Hardgrave BC, Wilson RL, Walstrom KA (1994) Predicting graduate student success: a comparison of neural networks and traditional techniques. Comput Oper Res 21(3):249–263. https://doi.org/10.1016/0305-0548(94)90088-4
13. Harwati H, Virdyanawaty RI, Mansur A (2016) Drop out estimation students based on the study period: comparison between Naïve Bayes and support vector machines algorithm methods. IOP Conf Ser Mater Sci Eng 105(1). https://doi.org/10.1088/1757-899X/105/1/012039
14. Aldowah H, Al-Samarraie H, Fauzy WM (2019) Educational data mining and learning analytics for 21st century higher education: a review and synthesis. Telematics Inform 37:13–49. https://doi.org/10.1016/j.tele.2019.01.007
15. Alemán JLF, Palmer-Brown D, Jayne C (2011) Effects of response-driven feedback in computer science learning. IEEE Trans Educ 54(3):501–508. https://doi.org/10.1109/TE.2010.2087761
16. Hu S et al (2019) A dual-stream recurrent neural network for student feedback prediction using Kinect. In: International conference on software, knowledge information, industrial management and applications, SKIMA, 2018-Dec, pp 1–8. https://doi.org/10.1109/SKIMA.2018.8631537
17. Seerat B (2016) Opinion mining: issues and challenges (a survey). Int J Comput Appl 49(Apr):42–51
18. Karunya K et al (2020) Analysis of student feedback and recommendation to tutors. In: Proceedings of the 2020 IEEE international conference on communication and signal processing, ICCSP 2020, pp 1579–1583. https://doi.org/10.1109/ICCSP48568.2020.9182270
19. Katragadda S et al (2020) Performance analysis on student feedback using machine learning algorithms. In: 2020 6th International conference on advanced computing and communication systems, ICACCS 2020, pp 1161–1163. https://doi.org/10.1109/ICACCS48705.2020.9074334
20. Sindhu I et al (2019) Aspect-based opinion mining on student's feedback for faculty teaching performance evaluation. IEEE Access 7:108729–108741. https://doi.org/10.1109/ACCESS.2019.2928872
21. Khan M et al (2018) Soft computing applications in education management—a review. In: 2018 IEEE International conference on innovative research and development, ICIRD 2018 (May), pp 1–4. https://doi.org/10.1109/ICIRD.2018.8376331
22. Ko M, Tiwari A, Mehnen J (2010) A review of soft computing applications in supply chain management. Appl Soft Comput J 10(3):661–674. https://doi.org/10.1016/j.asoc.2009.09.004
23. Ma J, Yang J, Howard SK (2019) A clustering algorithm based on fuzzy sets and its application in learning analytics. In: IEEE International conference on fuzzy systems, June 2019, pp 1–6. https://doi.org/10.1109/FUZZ-IEEE.2019.8858930
24. Ravikiran RK, Anil Kumar KR (2021) Experimental performance analysis of confidence-based online assessment portal in e-learning using data mining. Mater Today Proc 47(17):5912–5917. https://doi.org/10.1016/j.matpr.2021.04.456
25. Saeed EMH, Hammood BA (2021) Estimation and evaluation of students' behaviors in e-learning environment using adaptive computing. Mater Today Proc. https://doi.org/10.1016/j.matpr.2021.04.519
26. Tancock S, Dahnoun Y, Dahnoun N (2018) Real-time and non-digital feedback e-learning tool. In: Proceedings—2018 international symposium on educational technology, ISET 2018, pp 57–59. https://doi.org/10.1109/ISET.2018.00022

Machine Learning-Based Hybrid Models for Trend Forecasting in Financial Instruments

Arishi Orra, Kartik Sahoo, and Himanshu Choudhary

Abstract Forecasting trends in financial markets have always been an engaging task for traders and investors as they make a profit by accurately predicting the buying and selling points. This work proposes to develop hybrid predictive models that integrate feature selection methods with support vector machines to predict the trends of various financial instruments. Five variants of SVM, namely standard SVM, least squares support vector machine (LSSVM), proximal support vector machine (PSVM), multisurface PSVM via generalized eigenvalues (GEPSVM), and twin SVM (TWSVM), are used as baseline predictive algorithms for the hybrid models. Random forest and ReliefF algorithm are utilized for selecting an optimal input feature subset from a wide range of technical indicators. The proposed set of hybrid models along with baseline algorithms is tested over three principal financial instruments: Commodities, Cryptocurrency, and Foreign Exchange, for its applicability in trend forecasting. The empirical findings of the experiment demonstrated the superiority of hybrid models over the baseline algorithms.

Keywords Trend forecasting · Support vector machines · Feature selection · Technical indicators · Hybrid forecasting

1 Introduction

Financial market analysis has consistently grabbed the eye of numerous experts and researchers. Due to the growing technology and computational efficiency, buying and selling of financial instruments are much more swift and leisurely. The price of the financial instruments are highly influenced by a number of fundamental factors, such as business revenues and viability, and technical elements, such as chart patterns, momentum, and trader sentiments. However, predicting the future stock price is a non-viable task because of its dynamic, nonlinear, non-stationary, noisy, and chaotic behavior [1, 2]. But it is evident from the literature that the price's movement (rise

A. Orra (✉) · K. Sahoo · H. Choudhary
Indian Institute of Technology-Mandi, Mandi, Himachal Pradesh 175001, India
e-mail: arishi.orra98@gmail.com

M. Thakur et al. (eds.), *Soft Computing for Problem Solving*, Lecture Notes in Networks and Systems 547, https://doi.org/10.1007/978-981-19-6525-8_26

or fall) can be predicted instead of the actual price value [1]. Investors and traders are keen on the trend forecasting problem to profit from the financial market.

With the advent of machine learning techniques in the past decades, it is now feasible to handle trend forecasting problems in finance. Recently, artificial neural networks (ANNs), support vector machines (SVMs), decision trees, Bayesian networks, etc., are the techniques widely employed in directional prediction. Among them, ANNs are pretty prominent to researchers in trend forecasting problems [3, 4]. ANNs often tend to overfit the data due to learning of a large number of parameters during training. On the other hand, SVMs, based on the structural risk minimization (SRM) principle, provide an alternate approach for directional prediction. Due to the SRM principle, SVM offers better generalization capability than ANNs and prevents overfitting [5]. Various researches have demonstrated that SVM edge over ANN in financial trend prediction [6–8].

Market trends are shaped by various elements such as fundamental factors, technical analysis, expectations, and emotions. Therefore, selecting a suitable set of input attributes is crucial in trend forecasting. Many researchers have adopted to utilize technical indicators and oscillators as input to the SVM model for predicting market trends. Foremost, Kim [6] demonstrated the practicability of using technical indicators with SVM by forecasting the daily movements of the KOSPI index. The empirical findings suggested that the technical indicators and SVM-based approach produced better classification accuracy than the neural network. In [9], Kumar and Thenmozhi used the same 12 technical indicators as in [6] for predicting the day-ahead trend of the S&P CNX NIFTY index. The experimental outcomes evidenced that SVM outperformed random forest [10] and other classifiers. Thakur and Kumar [11] presented a hybrid approach for developing an automated trading framework integrating weighted multi-category generalized eigenvalue support vector machine [12] with random forest. They employed a collection of 55 frequently used technical indicators for predicting the trends of five different index futures.

Incorporating technical indicators raises the complexity of the input vectors, incurring high computational cost and menace of overfitting. Many studies tend to use feature selection methods to reduce the dimensionality of the input attributes by pruning a subset of the most relevant features to the problem. In recent years, researchers employed many hybrid SVM models with feature selection techniques such as principal component analysis (PCA), random forest, rank correlation, and genetic algorithm. Lee [13] introduced a hybrid feature selection approach that combined F-score and supported sequential forward search with SVM to forecast the NASDAQ index's daily trend. In another study, Ni et al. [14] applied a hybrid prediction model integrating a fractal feature selection approach with SVM for predicting the daily movements of the Shanghai Stock Exchange Index. Lin et al. [15] used a correlation-based SVM filter for ranking and evaluating the significance of input characteristics for stock market trend forecasting. The approach picks the features positively correlated with the trend signal and uncorrelated among themselves. Kumar et al. [16] proposed four hybrid models in a recent study by integrating PSVM with several feature selection approaches, including linear correlation, rank correlation, regression relief, and random forest. The efficacy of the presented models was evaluated by forecasting the daily movements of twelve distinct stock indices, and the experimental findings indicated that the union of PSVM and random forest outperformed the other models.

In recent studies, the random forest [16] and ReliefF algorithm [17] are the popularly employed methods for feature selection in financial trend forecasting problems. SVMs have shown better performance over financial datasets due to its high generalization capability in classification tasks. Many researchers have used several variants and hybrid models of SVM [11, 16, 18, 19] and reported that they produce superior performance than conventional SVM. This study proposes to generate hybrid models integrating random forest (RF) and ReliefF algorithm (RR) with improved variants of SVM for day-ahead trend forecasting of financial instruments. The proposed work considers five variants of SVM, viz., standard SVM, least squares support vector machine (LSSVM), proximal support vector machine (PSVM), multisurface PSVM via generalized eigenvalues (GEPSVM), and twin SVM (TWSVM) as baseline algorithms. The two feature selection methods, i.e., RF and RR, are incorporated with baseline algorithms to form hybrid predictive models: RF-SVM, RF-LSSVM, RF-PSVM, RF-GEPSVM, RF-TWSVM, RR-SVM, RR-LSSVM, RR-PSVM, RR-GEPSVM, and RR-TWSVM. In contrast to the bulk of research, which primarily concentrates on predicting stock market trends, this study focuses on trend forecasting of three extensively traded financial assets other than stocks. The presented hybrid models are assessed for their ability to forecast the following day's price movement over three crucial financial instruments, namely Commodities, Cryptocurrency, and Forex.

The rest of this study is organized as follows. Section 2 provides a brief overview of SVM variants and the feature selection methods utilized in this paper. The overall framework of the proposed hybrid models is discussed in Sect. 3. The experimental comparison and analysis of results are presented in Sect. 4. Section 5 concludes the findings of the work.

2 Methodology

This section goes through SVM variants and the feature selection approaches utilized for building the hybrid models.

2.1 Classification Models

2.1.1 SVM

Support vector machine (SVM) is a supervised machine learning algorithm used for both classification and regression [5]. The basis of SVM lies in creating maximum margin decision planes that define decision boundaries. Intuitively, a good separation means the hyperplane (constructed by the SVM technique) has the most significant distance from any class's nearest data points (support vectors) [20].

Consider a problem of classifying N points in n-dimensional space $\mathbb{R}^n$ into two classes $y = \{+1, -1\}$. For a data point $x \in \mathbb{R}^n$, SVM looks for the hyperplane $\omega^T \phi(\mathbf{x}) + \beta = 0$ and classifies a new data sample using the decision planes as

$$\begin{aligned} \omega^T \phi(\mathbf{x}) + \beta &\geq 1, \quad y = +1 \\ \omega^T \phi(\mathbf{x}) + \beta &\leq -1, \quad y = -1 \end{aligned} \tag{1}$$

where $\omega \in \mathbb{R}^n$ and $\beta \in \mathbb{R}$ define the orientation of the plane, and $\phi(\mathbf{x})$ is a mapping that maps the input points into some feature space [21]. The maximum margin hyperplane is determined by maximizing the margin $\frac{1}{||\omega||}$ between the points of two classes by solving the following optimization problem:

$$\begin{aligned} \min_{\omega,\beta,\xi} \quad & \frac{1}{2}\|\omega\|^2 + \nu \sum_{i=1}^{N} \xi_i \\ \text{subject to} \quad & y_i(\omega^T \phi(\mathbf{x}) + \beta) \geq 1 - \xi_i \\ & \xi_i \geq 0 \quad \forall i = 1, \ldots, N \end{aligned} \tag{2}$$

where $\nu > 0$ is a penalty parameter and $\xi_i (i = 1, \ldots, N)$ are the slack variables corresponding to the violation caused by the training samples. The optimization problem (2) is a convex quadratic programming problem and can be easily solved using KKT conditions [22].

2.1.2 LSSVM

The least squares support vector machine (LSSVM) [23] is obtained by a slight modification in the formation of the traditional SVM by replacing the inequality constraints with the equality ones. Due to the equality constraints, a system of linear equations will now be solved instead of solving an extensive system of QPP as earlier. Also, the two decision planes $\omega^T \phi(\mathbf{x}) + \beta = \pm 1$ are not bound to be together; rather, they can be as far as possible from each other. The minimization problem for LSSVM is represented as

$$\begin{aligned} \min_{\omega,\beta,\xi} \quad & \frac{1}{2}\|\omega\|^2 + \frac{\nu}{2}\sum_{i=1}^{N} \xi_i^2 \\ \text{subject to} \quad & y_i(\omega^T \phi(\mathbf{x}) + \beta) = 1 - \xi_i \quad \forall i = 1, \ldots, N. \end{aligned} \tag{3}$$

Also, no non-negativity constraint is needed on ξ because it is replaced with ξ^2 in the objective function, making $\xi > 0$ redundant. A new test sample $\mathbf{x}_j$ is classified by the decision rule defined in (1).

2.1.3 Proximal SVM

Proximal support vector machine (PSVM) [24] is very similar to LSSVM except has an additional objective function term. PSVM seeks an optimal margin hyperplane such that the two decision planes are proximal to the dataset of their respective classes. It classifies points depending on their proximity to one of the two parallel planes pushed apart as far as possible. The optimization problem for PSVM is formulated as

$$\begin{aligned} &\min_{\omega,\beta,\xi} \quad \frac{1}{2}\left(\|\omega\|^2 + \beta^2\right) + \frac{\nu}{2}\sum_{i=1}^{N}\xi_i^2 \\ &\text{subject to} \quad y_i(\omega^T\phi(\mathbf{x}) + \beta) = 1 - \xi_i \quad \forall i = 1,\ldots,N. \end{aligned} \tag{4}$$

The objective of (4) minimizes $\left(\|\omega\|^2 + \beta^2\right)$ term, which maximizes the margin between the decision planes with respect to the orientation (ω) of the hyperplane as well as its relative location to the origin (β). Furthermore, this formulation of PSVM leads to strong convexity of the objective function, which plays a vital role in speedy computational time.

2.1.4 GEPSVM

Multisurface PSVM via generalized eigenvalues (GEPSVM) [25] is a variant of SVM that aims at finding non-parallel planes for classification. In this algorithm, the parallelism condition on the decision planes is dropped, and each plane needs to be as close as conceivable to one of the data points and as far as possible from the other class. The non-parallel kernel generated surface $\mathcal{K}(\mathbf{x}, C)\mu_1 + \beta_1 = 0$ formed by GEPSVM is obtained by solving the following minimization problem:

$$\min_{\mu_1,\beta_1} \frac{\left\|\mathcal{K}(\mathcal{A}, \mathcal{C}^T)\mu_1 + e_1\beta_1\right\|^2 + \delta\left\|\begin{bmatrix}\mu_1\\ \beta_1\end{bmatrix}\right\|^2}{\left\|\mathcal{K}(\mathcal{B}, \mathcal{C}^T)\mu_1 + e_2\beta_1\right\|^2} \tag{5}$$

where $\mathcal{A}$ and $\mathcal{B}$ are the matrices containing the points of class $+1$ and -1, respectively, $\mathcal{C} = [\mathcal{A} \;\; \mathcal{B}]^T$, $\delta > 0$ is a regularization parameter, and e_1 and e_2 are the vector of ones. The numerator of (5) tries to make the hyperplane proximal to the dataset of class $+1$, while the denominator ensures that the hyperplane is farthest from the points of class -1. Similarly, the optimization problem for the other non-parallel kernel surface $\mathcal{K}(\mathbf{x}, C)\mu_2 + \beta_2 = 0$ is given as

$$\min_{\mu_2,\beta_2} \frac{\left\|\mathcal{K}(\mathcal{B}, \mathcal{C}^T)\mu_2 + e_2\beta_2\right\|^2 + \delta \left\|\begin{bmatrix}\mu_2 \\ \beta_2\end{bmatrix}\right\|^2}{\left\|\mathcal{K}(\mathcal{A}, \mathcal{C}^T)\mu_2 + e_1\beta_2\right\|^2} \tag{6}$$

After simplification, the optimization problems (5) and (6) get converted into a well-known Rayleigh quotient form. The global minimum to these forms is achieved at an eigenvector of the generalized eigenvalue problem corresponding to the least eigenvalue. Once both the hyperplanes are known, a fresh point is classified depending upon its proximity to either of the planes.

2.1.5 Twin SVM

Based on the idea of GEPSVM, twin SVM (TWSVM) also produces a pair of non-parallel hyperplanes, but its formulation is quite similar to standard SVM [26]. In TWSVM, a couple of QPPs are tackled, yet in SVM, a single QPP is solved. Additionally, in SVM, all data points constitute the constraints, while in TWSVM, points of one class constitute the constraints of the other QPP and the other way around. Solving two littler estimated QPPs make TWSVM operate faster than standard SVM. The TWSVM induced non-parallel planes are obtained by solving the following optimization problems:

$$\begin{aligned} \min_{\mu_1,\beta_1,\xi} \quad & \frac{1}{2}\left\|\mathcal{K}(\mathcal{A}, \mathcal{C}^T)\mu_1 + e_1\beta_1\right\|^2 + \nu_1 e_2^T\xi \\ \text{subject to} \quad & -(\mathcal{K}(\mathcal{B}, \mathcal{C}^T)\mu_1 + e_2\beta_1) + \xi \geq e_2 \\ & \xi \geq 0 \end{aligned} \tag{7}$$

and

$$\begin{aligned} \min_{\mu_2,\beta_2,\xi} \quad & \frac{1}{2}\left\|\mathcal{K}(\mathcal{B}, \mathcal{C}^T)\mu_2 + e_2\beta_2\right\|^2 + \nu_2 e_1^T\xi \\ \text{subject to} \quad & \mathcal{K}(\mathcal{A}, \mathcal{C}^T)\mu_2 + e_1\beta_2 + \xi \geq e_1 \\ & \xi \geq 0 \end{aligned} \tag{8}$$

where $\nu_1 > 0$ and $\nu_2 > 0$ are penalty parameters. Like GEPSVM, any unknown sample is assigned to a class if it lies closest to the corresponding plane.

2.2 Feature Selection Methods

2.2.1 Random Forest

Random forest is a supervised machine learning algorithm mainly used for classification but can be used for regression. It is an ensemble learning technique that builds multiple unpruned decision trees constructed using a random subset of data [10]. Every individual decision tree is trained using bootstrapped data, ensuring no two decision trees are correlated. Also, every decision tree is built from a randomly chosen subset of features.

For training purpose, just two-thirds of randomly split data are utilized, and the rest one-third data is called out of a bag (OOB) samples, which is utilized to obtain the predictive accuracy of the build decision tree and also for calculating feature significance scores. All constructed decision trees are evaluated using their separate OOB samples, and the error rate Q_i of each ith tree is reported. To obtain the significance score of feature F, disrupted OOB samples are created for each decision tree. This is achieved by arbitrarily permuting the feature among the samples. Once more, the error rate Q_i' for every tree is determined for the disrupted OOB samples. Now, the feature significance score Z_F is determined by the following formula:

$$Z_F = \frac{1}{K}\sum_i (Q_i - Q_i')$$

where K is the total number of decision trees constructed.

2.2.2 ReliefF Feature Selection

Kira and Rendell [27] proposed Relief algorithm that uses a filter-based technique for feature pruning tasks. Relief uses the notion of the nearest neighbor algorithm for calculating k-nearest *hits* and *misses* and thus determining the feature rankings for the input instances. Kononenko [28] introduced ReliefF as an extension of Relief to address the challenges of multi-class and incomplete data. In ReliefF, all the feature weights are set initially to zero. Then, any instance $\mathcal{X}_i$ is chosen randomly, and the algorithm finds its k nearest hits x^+ and misses x^-. In ith iteration, the algorithm updates the weight iteratively as

If x^+ and x^- are members of the same class,

$$Z_F^i = Z_F^{i-1} - \frac{d\left(x^+, x^-\right)}{N} \cdot \delta\{x^+, x^-\}$$

If x^+ and x^- are not in the same class,

$$Z_F^i = Z_F^{i-1} + \frac{P^+}{1 - P^-} \cdot \frac{d\left(x^+, x^-\right)}{N} \cdot \delta\{x^+, x^-\}$$

where

N	Total number of iterations
P^+	Prior probability of the class in which x^+ lies
P^-	Prior probability of the class in which x^- lies
$d\left(x^+, x^-\right)$	Difference between the class labels of x^+ and x^-
$\delta\{x^+, x^-\}$	Distance between the samples x^+ and x^-.

3 Proposed Hybrid Methods

The overall architecture of the proposed hybrid models for financial trend forecasting is presented in this section. Also, it discusses the details of the input characteristics, training, and parameter selection for the models.

3.1 Input

The foremost step in developing a forecasting algorithm is the selection of input variables. Other than OHLC prices, researchers choose to utilize a variety of technical indicators and oscillators as input [6]. Technical indicators are mathematical calculations based on the past price that are used to predict future price movement and market volatility [29]. In general, technical indicators overlay on value outline information to show where the market is heading, whether the stock is in an overbought state or oversold. While indicators indicate the market's trend, oscillators define the market's momentum, which is constrained by upper and lower bands. Traders tend to define many technical indicators and oscillators such as moving average (MA), exponential moving average (EMA), rate of change, price oscillator (OSCP), and true strength index (TSI) for getting better profits. In this study, a wide range of technical indicators that had been substantially used in earlier research and were well-known for their utility in technical analysis are utilized as input for various forecasting models. A thorough description of all the employed indicators and oscillators can be found in [11, 16].

3.2 Hybrid Models

This study proposed using ten hybrid prediction models to forecast the trend in financial instruments. Two feature selection methods, random forest and ReliefF, are coupled with five SVM variants to construct the hybrid models. These models are labeled as $\mathrm{RF} - F_i$ and $\mathrm{RR} - F_i$, where RF and RR stand for random forest and ReliefF, respectively, and F_i is the ith SVM variant.

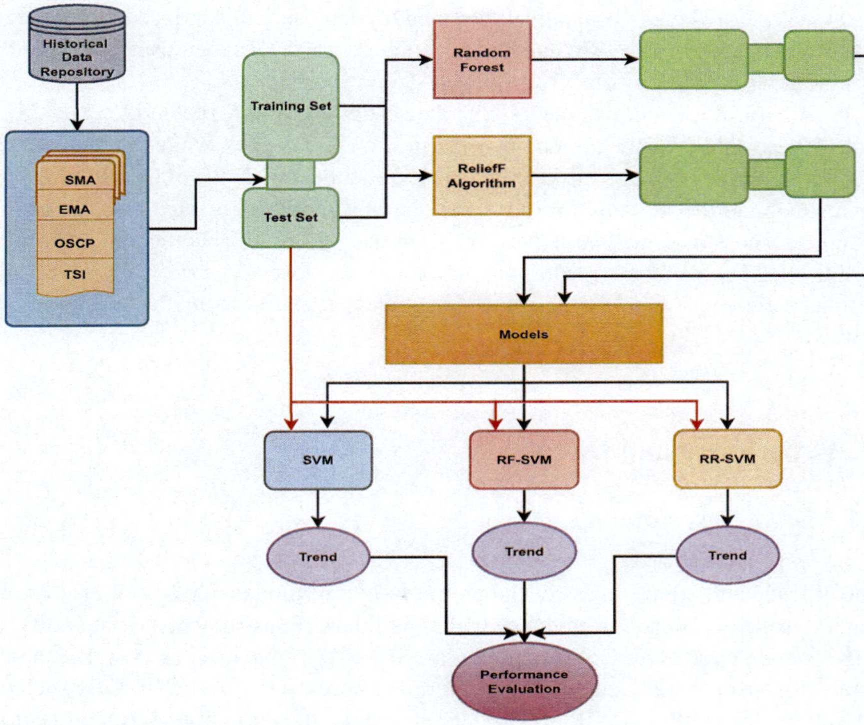

Fig. 1 Framework for the proposed hybrid models

Here, the two feature selection techniques are used for shedding the technical indicators with the least importance score and accelerating the algorithm's performance. Both methods assign an importance score to all input indicators, and all the features are ranked in descending order of their importance score, i.e., the component with maximum score is placed at the top. At each iteration, the model is trained using the top k features, and the accuracy for each subset is recorded. Finally, the subset having maximum accuracy is selected as an optimal input feature vector, and the model is trained using this subset. The overall framework for the proposed hybrid models is presented in Fig. 1.

3.3 Training and Parameter Selection

The proposed hybrid models are trained by partitioning the dataset into training and testing sets. Initial seventy percent data is utilized for training and feature selection tasks, while the remaining data is used to test the models' efficacy. Random forest and ReliefF utilize the training data for ranking the features and deciding the optimal feature subset. All the hybrid models are then trained using this reduced feature set. The optimal value of the SVM and kernel parameters is determined using the K-fold cross-validation (CV) methodology [30].

In the K-fold CV method, the training data is divided into K blocks of equal size. Out of these K blocks, the training is performed on K-1 blocks, while the remaining block is used for testing. This step is repeated K times, with each block used once for testing, and the validation results of all the blocks are aggregated to give single accuracy. The K-CV method is performed on the grid of all specific values of the parameters. The parameters with maximum accuracy over the grid are selected as optimal parameters. In this work, all the parameters are tuned using the five-fold CV method.

4 Experiment and Discussion

4.1 Data Description

The efficacy and effectiveness of the proposed hybrid models have been evaluated using five elements each from three significant financial instruments: Commodities (viz. Copper, Crude Oil, Gold, Natural Gas, Silver), Cryptocurrency (viz. Binance, Bitcoin, Dogecoin, Ethereum, LUNA coin), and Forex (viz. EUR/INR, GBP/AUD, GBP/JPY, USD/INR, USD/JPY). The dataset for this study is deliberately selected to cover the most widely traded financial instruments. The daily historical open, high, low, close (OHLC) price datasets considered for the experiment are collected from Yahoo Finance. The Commodity and Forex datasets cover the period from January 2011 to December 2020, while the Cryptocurrency data spans from January 2017 to December 2020. A large variety of most commonly used technical indicators and oscillators are calculated and fed as input to the various presented hybrid models. The entire dataset is divided into two parts: an initial 70% is utilized for training the models, while the remaining 30% is used to test the generalized performance of the models. A similar experimental setup is employed for all proposed hybrid models to generate a homogenous setting for comparison.

4.2 Performance Measures

Predicting the downtrend is equally crucial as an uptrend in the financial market. Therefore, a single measure cannot be relied upon to assess the algorithm's effectiveness. Different metrics derived from the confusion matrix are employed to evaluate the proposed hybrid models performance. The confusion matrix for a binary classification problem is given by

		Predicted	
		Positive	Negative
Actual	Positive	TP	FN
	Negative	FP	TN

4.2.1 Accuracy

Accuracy of any model is defined as the percentage of correctly classified samples and is given as

$$\text{Accuracy} = \frac{\text{TP} + \text{TN}}{\text{TP} + \text{FP} + \text{FN} + \text{TN}}.$$

4.2.2 Precision

The fraction of samples claimed by the model to be significant and are actually significant is referred to as precision. Precision for positive P^p and negative P^n class is defined by

$$P^p = \frac{\text{TP}}{\text{TP} + \text{FN}} \quad \text{and} \quad P^n = \frac{\text{TN}}{\text{TN} + \text{FP}}.$$

4.2.3 Recall

The capacity of the model to discover all relevant samples in data is called recall. Recall for positive R^p and negative R^n class is formulated as

$$R^p = \frac{TP}{TP + FP} \qquad \text{and} \qquad R^n = \frac{TN}{TN + FN}.$$

4.2.4 F_1-Score

F_1-score or f-score considers the knowledge of both precision and recall and is defined as their harmonic mean. F_1 score for positive F_1^p and negative F_1^n class is given by

$$F_1^p = \frac{2 \times P^p \times R^p}{P^p + R^p} \quad \text{and} \quad F_1^n = \frac{2 \times P^n \times R^n}{P^n + R^n}.$$

The value of the F_1-score ranges between 0 and 1, where 1 signifies a perfect classification, while 0 indicates the worst possible case.

4.3 Results and Discussion

This study proposes several hybrid models to predict the direction of daily change in the prices of various financial assets. Therefore, the performance of the proposed hybrid and baseline models has been assessed using classification accuracy and F_1-score for three widely used financial instruments. The detailed results in terms of classification accuracy and F_1-score are depicted in Tables 1, 2, and 3. The best results in the tables are highlighted, and the models with superior performance across both metrics are deemed preferable. The findings of the experiment are analyzed in two different instances. The first one compares the baseline algorithms and the hybrid models that integrate baseline algorithms with feature selection methods. And the second instance examines the superiority of the various hybrid models among themselves.

Table 1 shows a comparison of baseline and hybrid models over five commodities datasets. All models are compared based on classification accuracy and f-score measure. TWSVM, RF-LSSVM, RF-GEPSVM, RF-TWSVM, and RR-LSVM achieved the maximum classification accuracy over the five given datasets. Incorporating feature selection techniques into the baseline algorithms increases their performance. Four out of five times, the hybrid models perform better than the baseline algorithms, and random forest integrated models show superior performance among the hybrid models. In terms of f-score, PSVM (twice), TWSVM, RF-LSSVM (thrice), RF-PSVM, RR-LSSVM, RR-PSVM, and RR-TWSVM attained the highest f-scores. Again, the hybrid models outperform the baseline algorithms, with the random forest combined models exhibiting the best performance. RF-LSSVM and RR-LSSVM are the only models having superior results for both accuracy and f-score over the Copper and Silver datasets, respectively. With an accuracy of 54.95% and an f-score of 0.54, RF-LSSVM is the best performing model for the given datasets.

Similar comparative performance of the models over the five Cryptocurrency datasets, namely Binance, Bitcoin, Dogecoin, Ethereum, and Luna coin, is presented in Table 2. PSVM, RF-TWSVM, RR-SVM, and RR-GEPSVM (twice) are among the models having the highest accuracy for the five datasets. It is observed that among the baseline algorithms, PSVM attained maximum accuracy of 55.04% for the DOGE dataset, while the best performance of hybrid models is always higher than this. The results indicate that the hybrid models attained better accuracy four out of five times than the baseline models. Unlike the case of Commodity datasets, here, the hybrid models integrating the ReliefF algorithm have superior performance to the other hybrid models. For the other metric f-score, PSVM, TWSVM, RF-LSSVM, RF-TWSVM, RR-SVM (twice), RR-LSSVM, RR-PSVM, and RR-GEPSVM have shown the highest measures. It is clear that the hybrid models again outperform the baseline ones. Five out of the nine best performances have been observed for the ReliefF algorithm combined models. However, the best performance across both metrics is attained only for two models, PSVM and RR-SVM, over DOGE and

Table 1 Experimental results of baseline and hybrid models for the Commodity datasets

	Copper	Crude oil	Gold	Natural gas	Silver
Accuracy	%	%	%	%	%
F_1-score	0-1	0-1	0-1	0-1	0-1
SVM	52.31	53.17	50.86	53.43	52.58
	0.52	0.47	0.51	0.52	0.51
LSSVM	51.15	51.83	49.79	51.69	54.68
	0.51	0.5	0.49	0.5	0.52
PSVM	52.51	52.78	49.44	53.18	54.27
	0.52	**0.52**	0.5	0.51	**0.54**
GEPSVM	50.33	52.78	51.56	51.96	52.13
	0.49	0.36	0.5	0.42	0.51
TWSVM	50.86	53.04	53.37	**55.09**	51.85
	0.49	0.39	0.48	0.5	**0.54**
RF-SVM	48.59	49.33	49.24	52.14	48.26
	0.48	0.48	0.49	0.51	0.48
RF-LSSVM	51.42	52.23	52.91	50.88	**54.95**
	0.51	**0.52**	**0.53**	0.5	**0.54**
RF-PSVM	51.69	53.05	52.91	51.56	53.73
	0.51	0.48	**0.53**	0.51	0.53
RF-GEPSVM	50.39	**54.27**	52.13	52.91	52.23
	0.5	0.47	0.5	0.51	0.52
RF-TWSVM	50.75	52.91	**53.86**	51.69	53.59
	0.38	0.33	0.48	0.38	0.53
RR-SVM	52.65	52.91	51.56	52.37	52.37
	0.52	0.41	0.51	0.5	0.5
RR-LSSVM	**54.01**	51.42	50.47	51.11	53.46
	0.54	0.49	0.49	0.51	0.53
RR-PSVM	49.66	52.37	50.21	54.68	53.18
	0.49	0.42	0.49	**0.55**	0.53
RR-GEPSVM	49.66	53.05	49.25	51.69	50.88
	0.46	0.43	0.48	0.42	0.49
RR-TWSVM	52.83	52.91	53.18	51.96	52.13
	0.5	0.44	**0.53**	0.47	0.46

BNB datasets, respectively. Although RF-TWSVM achieves the highest accuracy of 57.18% for the ETH dataset, RR-SVM is considered the top-performing model for crypto data with an accuracy of 56.26% and an f-score of 0.56 for the BNB dataset.

The forecasting report of the hybrid models and the baseline algorithms over the five Forex datasets have been presented in Table 3. Across the given five datasets, RF-SVM (thrice) and RR-LSSVM (twice) attained the highest classification accuracy.

Table 2 Experimental results of baseline and hybrid models for the Cryptocurrency datasets

	BNB	BTC	DOGE	ETH	LUNA
Accuracy	%	%	%	%	%
F_1-score	0-1	0-1	0-1	0-1	0-1
SVM	50.28	54.55	51.73	47.98	50.86
	0.51	0.38	0.49	0.47	0.38
LSSVM	53.52	54.87	53.21	50.45	54.12
	0.53	0.49	0.52	0.49	0.46
PSVM	53.51	54.86	**55.04**	50.15	53.79
	0.53	0.46	**0.55**	0.5	0.48
GEPSVM	49.54	49.79	51.37	43.73	51.68
	0.49	0.46	0.54	0.33	0.38
TWSVM	55.09	53.64	52.61	50.57	53.46
	0.52	**0.53**	0.39	0.51	0.51
RF-SVM	54.59	54.48	50.51	53.57	47.44
	0.54	0.52	0.5	0.48	0.44
RF-LSSVM	53.51	54.86	48.92	52.91	56.27
	0.53	0.49	0.47	**0.53**	0.55
RF-PSVM	49.54	54.86	48.01	51.07	55.35
	0.48	0.5	0.47	0.51	0.49
RF-GEPSVM	55.65	54.86	50.76	44.67	55.96
	0.42	0.46	0.36	0.52	0.34
RF-TWSVM	55.65	54.86	50.76	**57.18**	54.12
	0.51	**0.53**	0.5	0.48	0.53
RR-SVM	**56.26**	54.87	54.74	49.23	53.51
	0.56	0.46	**0.55**	0.49	0.48
RR-LSSVM	51.98	54.86	50.15	49.84	55.35
	0.5	**0.53**	0.5	0.49	0.55
RR-PSVM	53.21	54.86	51.68	48.01	55.65
	0.51	0.51	0.51	0.47	**0.56**
RR-GEPSVM	46.78	**55.16**	51.68	45.56	**56.29**
	0.4	0.43	0.51	0.43	0.53
RR-TWSVM	55.66	54.86	53.51	56.57	55.96
	0.44	0.51	0.51	0.44	**0.56**

RF-SVM achieves the maximum accuracy of 81.61% among all models over the GBP/JPY dataset. In this case, hybrid models outperform the baseline algorithms for each dataset. Random forest integrated SVM showed superior results three out of five times among the feature selection hybrid models. PSVM, RF-SVM (twice), RR-LSSVM, RR-PSVM, and RR-TWSVM gained the highest f-scores over the given five datasets. Again the hybrid models beat the performance of the baseline algorithms on four out of five occasions. In contrast to the classification accuracy, the

Table 3 Experimental results of baseline and hybrid models for the Forex datasets

	EUR/INR	GBP/AUD	GBP/JPY	USD/INR	USD/JPY
Accuracy	%	%	%	%	%
F_1-score	0-1	0-1	0-1	0-1	0-1
SVM	77.56	77.41	79.93	75.79	78.79
	0.75	0.77	0.8	0.76	0.78
LSSVM	72.77	77.25	69.19	74.73	80.36
	0.72	0.77	0.69	0.75	0.8
PSVM	76.62	79.43	80.24	74.42	78.77
	0.77	0.79	0.79	0.74	0.78
GEPSVM	71.57	77.48	74.67	74.05	74.63
	0.37	0.77	0.74	0.73	0.75
TWSVM	69.63	70.42	72.51	71.72	74.08
	0.68	0.7	0.72	0.71	0.73
RF-SVM	**78.03**	80.45	**81.61**	76.09	**81.07**
	0.77	0.8	**0.81**	0.76	0.8
RF-LSSVM	76.43	79.18	79.71	74.73	78.64
	0.76	0.79	0.79	0.75	0.79
RF-PSVM	76.43	79.18	79.71	73.95	79.05
	0.76	0.79	0.8	0.74	0.79
RF-GEPSVM	72.34	75.52	75.68	72.47	75.65
	0.71	0.76	0.76	0.65	0.76
RF-TWSVM	75.39	78.14	77.87	74.08	74.34
	0.75	0.68	0.78	0.74	0.73
RR-SVM	73.29	78.01	71.98	76.96	76.83
	0.72	0.78	0.72	0.77	0.76
RR-LSSVM	76.04	**81.15**	73.43	**79.18**	78.92
	0.76	**0.81**	0.79	0.74	0.79
RR-PSVM	72.64	77.25	73.69	72.95	80.36
	0.72	0.77	0.74	**0.8**	0.8
RR-GEPSVM	70.26	72.36	68.56	72.47	78.43
	0.7	0.68	0.66	0.71	0.78
RR-TWSVM	73.56	79.13	77.65	73.27	74.42
	0.73	0.79	0.77	0.73	**0.81**

ReliefF algorithm combined models produce better f-scores than the random forest ones. RF-SVM and RR-LSSVM are the only two models that simultaneously have the highest accuracy and f-score. However, RF-SVM turns out to be the ideal model for Forex data with the highest accuracy of 81.61% with an f-score of 0.81.

5 Conclusion

This study proposes to use feature selection-based hybrid predictive models for financial trend forecasting. The two most commonly used feature selection methods have been incorporated with five improved variants of SVM leading to ten hybrid models. An extensive set of technical indicators has been utilized for feeding input to the models. The efficacy of the proposed hybrid models has been evaluated over three primary financial instruments: Commodity, Cryptocurrency, and Forex.

The presented approach is assessed for predicting the day-ahead trends of financial instruments by using classification accuracy and f-score measures. The numerical results demonstrated that the hybrid models outperformed the baseline algorithms in all three financial instruments. The empirical findings of the experiment suggest utilizing hybrid models comprising feature selection techniques for financial trend prediction tasks. It is observed that the random forest-based hybrid models showed superior performance over Commodity and Forex datasets, while ReliefF algorithm-based models performed better for Cryptocurrency data. Although two different feature selection approaches are applied, the hybrid models perform similarly. That indicates that the feature selection techniques also depend upon the underlying structure of the dataset. It implies that one should not rely on only one approach.

References

1. Abu-Mostafa YS, Atiya AF (1996) Introduction to financial forecasting. Appl Intell 6(3):205–213
2. Blank SC (1991) "Chaos" in futures markets? A nonlinear dynamical analysis. J Futures Markets (1986–1998) 11(6):711
3. Roh TH (2007) Forecasting the volatility of stock price index. Exp Syst Appl 33(4):916–922
4. De Faria EL, Albuquerque MP, Gonzalez JL, Cavalcante JTP, Albuquerque MP (2009) Predicting the Brazilian stock market through neural networks and adaptive exponential smoothing methods. Exp Syst Appl 36(10):12506–12509
5. Cortes C, Vapnik V (1995) Support-vector networks. Mach Learn 20(3):273–297
6. Kim KJ (2003) Financial time series forecasting using support vector machines. Neurocomputing 55(1–2):307–319
7. Cao LJ, Tay FEH (2003) Support vector machine with adaptive parameters in financial time series forecasting. IEEE Trans Neural Networks 14(6):1506–1518
8. Huang W, Nakamori Y, Wang SY (2005) Forecasting stock market movement direction with support vector machine. Comput Oper Res 32(10):2513–2522
9. Kumar M, Thenmozhi M (2006) Forecasting stock index movement: a comparison of support vector machines and random forest. In: Indian institute of capital markets 9th capital markets conference paper
10. Hartshorn S (2016) Machine learning with random forests and decision trees: a visual guide for beginners. Kindle edition
11. Thakur M, Kumar D (2018) A hybrid financial trading support system using multi-category classifiers and random forest. Appl Soft Comput 67:337–349
12. Kumar D, Thakur M (2016) Weighted multicategory nonparallel planes SVM classifiers. Neurocomputing 211:106–116

13. Lee MC (2009) Using support vector machine with a hybrid feature selection method to the stock trend prediction. Exp Syst Appl 36(8):10896–10904
14. Ni LP, Ni ZW, Gao YZ (2011) Stock trend prediction based on fractal feature selection and support vector machine. Exp Syst Appl 38(5):5569–5576
15. Lin Y, Guo H, Hu J (2013) An SVM-based approach for stock market trend prediction. In: The 2013 international joint conference on neural networks (IJCNN). IEEE, New York, pp 1–7
16. Kumar D, Meghwani SS, Thakur M (2016) Proximal support vector machine based hybrid prediction models for trend forecasting in financial markets. J Comput Sci 17:1–13
17. Huang CJ, Yang DX, Chuang YT (2008) Application of wrapper approach and composite classifier to the stock trend prediction. Exp Syst Appl 34(4):2870–2878
18. Hao PY, Kung CF, Chang CY, Ou JB (2021) Predicting stock price trends based on financial news articles and using a novel twin support vector machine with fuzzy hyperplane. Appl Soft Comput 98:106806
19. Abdollahi H (2020) A novel hybrid model for forecasting crude oil price based on time series decomposition. Appl Energy 267:115035
20. Hastie T, Tibshirani R, Friedman JH, Friedman JH (2009) The elements of statistical learning: data mining, inference, and prediction, vol 2. Springer, New York, pp 1–758
21. Schölkopf B, Smola AJ, Bach F (2002) Learning with kernels: support vector machines, regularization, optimization, and beyond. MIT Press
22. Mangasarian OL (1994) Nonlinear programming. Soc Ind Appl Math
23. Suykens JA, Vandewalle J (1999) Least squares support vector machine classifiers. Neural Process Lett 9(3):293–300
24. Mangasarian OL, Wild EW (2001) Proximal support vector machine classifiers. In: Proceedings KDD-2001: knowledge discovery and data mining
25. Mangasarian OL, Wild EW (2005) Multisurface proximal support vector machine classification via generalized eigenvalues. IEEE Trans Pattern Anal Mach Intell 28(1):69–74
26. Khemchandani R, Chandra S (2007) Twin support vector machines for pattern classification. IEEE Trans Pattern Anal Mach Intell 29(5):905–910
27. Kira K (1992) Rendell LA (1992, July) The feature selection problem: traditional methods and a new algorithm. AAAI 2:129–134
28. Kononenko I (1994) Estimating attributes: analysis and extensions of RELIEF. In: European conference on machine learning. Springer, Berlin, pp 171–182
29. Murphy JJ (1999) Technical analysis of the financial markets: a comprehensive guide to trading methods and applications. Penguin
30. Cao LJ, Chua KS, Chong WK, Lee HP, Gu QM (2003) A comparison of PCA, KPCA and ICA for dimensionality reduction in support vector machine. Neurocomputing 55(1–2):321–336

Support Vector Regression-Based Hybrid Models for Multi-day Ahead Forecasting of Cryptocurrency

Satnam Singh, Khriesavinyu Terhuja, and Tarun Kumar

Abstract After the introduction of Bitcoin in 2008, the cryptocurrency rose to popularity at an exponential rate, and currently, it is one of the most traded financial instruments worldwide. The nature of cryptocurrency is not only complicated but is also a bemusing financial instrument due to its high volatility. In this study, we present a novel hybrid machine learning model with a goal to perform multi-day ahead price forecasting of cryptocurrency. This study proposes four hybrid models that combine random forest (RF) with four variants of support vector regression (SVR): LSSVR, PSVR, ϵ-TSVR, and GEPSVR. The performance of these models is evaluated over six popular cryptocurrencies employed over a large scale of technical indicators based on two performance metrics, RMSE, and R^2 Score. The empirical results obtained over various cryptocurrency dataset show that hybrid models outperform the basic variants of SVR.

Keywords SVR · Cryptocurrency · Random Forest · Feature selection · Forecasting · Technical indicators

1 Introduction

In recent years, accurately predicting financial time series has become an essential issue in investment decision making. However, predicting the price of time series financial markets, which have non-stationary nature, is a challenging problem [1]. They are dynamic, nonlinear and chaotic [2] as they are influenced by the general economy, government policies and even psychology of investors. In the last decade, a new financial instrument called cryptocurrency rose to popularity after the introduction of Bitcoin in 2008. A cryptocurrency is a digital currency that is transferred between peers, without the intervention of a third party, like a bank. This principle of decentralized currency is the driving reason why cryptocurrency popularity is rising

S. Singh (✉) · K. Terhuja · T. Kumar
Indian Institute of Technology Mandi, Mandi, Himachal Pradesh 175001, India
e-mail: brarsatnam25299@gmail.com

M. Thakur et al. (eds.), *Soft Computing for Problem Solving*, Lecture Notes in Networks and Systems 547, https://doi.org/10.1007/978-981-19-6525-8_27

exponentially. Hence, it has also become an exciting research area, and now many researchers are looking for ways to analyze its features, and their impact on real-life.

Researchers have used various approaches to predict the price of cryptocurrency. Traditional, statistical and econometric are one of the approaches; some of the methods include ARIMA model and autoregressive model [3], moving average model [4], k-chainlets [5], more statistical and econometric models such as [5, 6] exits in the literature.

In the recent decade, the use of artificial intelligence to predict time series financial markets has gained lots of attention in the research community as these techniques could infer patterns and make predictions from the nonlinearity, chaotic, and randomness of the data. To predict the price of cryptocurrency various machine learning technique such as logistic regression [7], XGBoost [8], SVM [9], SVR [10] were used.

Among the various deep learning techniques RNN, LSTM [11], and GRU [12] have been widely used to predict trends of cryptocurrency, as these models are tailor-made to handle sequential data; moreover, these models are good in handling the complex and volatile nature of cryptocurrency.

Though these traditional machine learning/deep learning perform well, their performance can be enhanced by using hybrid models. An empirical study on financial time series by Kumar et al. [13] found that when a hybrid model of random forest and PSVM is used, it outperformed the original PSVM. Here, random forest was used to extract features from the technical indicators, hence reducing the computational complexity and improving the accuracy of the model by selecting the best features. Similar works on hybrid models to predict the price of cryptocurrency exits in the literature. Derbentsev et al. [14] used a hybrid of ARIMA and autoregressive binary tree to predict the price of cryptocurrency, and Livieris et al. [15] used a hybrid model of CNN and LSTM; here, CNN was used to extract features, and LSTM to make the prediction using the extracted features the overall study found that using this hybrid model reduced overfitting, and also the computational complexity.

This work proposes a novel hybrid framework incorporating random forest with SVR and its variants for multi-day ahead cryptocurrency price forecasting. In order to make multi-days ahead forecasting, two forecasting strategies have been used, direct strategy and MIMO strategy. The variants of SVRs are combined with random forest, which is used for feature selection. The use of random forest also reduces the computational complexity and interpretability of the model. The contribution of this paper can be summarized as follows:

- A novel hybrid model of random forest and SVR and its variants for multi-day cryptocurrency price forecasting.
- Used random forest for feature selection and LSSVR [16], PSVR [17], ϵ-TSVR [18], and GEPSVR [19] for price forecasting.
- Proposed models are RF-LSSVR, RF-PSVR, RF-ε-TSVR and RF-ReGEPSVR.
- Used five cryptocurrency datasets employed over a large set of technical indicators.
- Comparison of direct strategy and MIMO strategy for multi-days ahead forecasting.
- Model comparison of RF-LSSVR, RF-PSVR, RF-ε-TSVR, and RF-GEPSVR.

The rest of the paper is structures as follows: Sect. 2 gives a brief description of the various forecasting techniques. Architecture of the proposed model and two predicting strategy used in this study is discussed in Sect. 3. The dataset used in this study and the various measures of performance along with the results of the implemented models are discussed in detail in Sect. 4. Conclusion of the paper is drawn in Sect. 5.

2 Methodology

2.1 Forecasting Methods

This section briefly describes various forecasting techniques used in this paper.

Support Vector Regression: Support vector regression (SVR) is a supervised machine learning technique for regression problems that is based on the support vector machine (SVM). The standard formulation of SVR is described in this section. SVR algorithm aims to find the best approximation decision function to fit the training data points $\{(z_1, y_1), (z_2, y_2), \ldots, (z_l, y_l)\}$. Where, $z_i \in R^d$ is the input and $y_i \in R$ corresponds the output. Support Vector Regression aims to find the parameter $\omega \in R^d$ and $\omega_0 \in R$ by solving the following constraints:

$$\begin{aligned} &\min \tfrac{1}{2}\|\omega\|^2 + C\left(\sum_{i=1}^{l} \xi_i^+ + \sum_{i=1}^{l} \xi_i^-\right) \\ &\text{subject to} \begin{cases} (\boldsymbol{\omega} \cdot \mathbf{z}_i) + \omega_0 - y_i \le \epsilon + \xi_i^- \ \forall i \\ y_i - (\boldsymbol{\omega} \cdot \mathbf{z}_i) - \omega_0 \le \epsilon + \xi_i^+ \ \forall i \\ \xi_i^+, \xi_i^- \ge 0 \qquad \forall i \end{cases} \end{aligned} \tag{1}$$

where $C > 0$, $\varepsilon > 0$ are input parameters and ξ_i is the slack variable. The solution is obtained by introducing the Lagrangian multiplier as described in the paper [20]. Solving that convex quadratic programming, the two Lagrangian parameters β^+ and β^- are obtained. Then, ω and ω_0 becomes:

$$\omega = \sum_{j=1}^{l} (\beta_i^- - \beta_i^+) z_i$$

$$\omega_0 = y_i - \omega^{\top}.z_i + \varepsilon$$

Regression function for the input z becomes:

$$f(z) = \omega^\top . z + \omega_0$$

Least Square Support Vector Regression: The least square support vector regression (LSSVR) converts convex quadratic programming problems to a convex linear system by changing the inequality constraint's to equality [16]. LSSVR minimizes the following objective function with constraint:

$$\begin{aligned} &\min J(\omega, e) = \tfrac{1}{2}\omega^\top \omega + \gamma \tfrac{1}{2} e^\top e \\ &\text{s.t. } y = Z^\top \omega + \omega_0 1_l + e \end{aligned} \tag{2}$$

where $e = (e_1, e_2, \ldots, e_l)^\top \in R^l$ is a vector which consist of slack variables and $\gamma \in R^+$ denotes a regularized parameter which takes positive real value. By introducing the Lagrangian form for Eq. 2 and solving the Lagrangian form by using the KKT condition as described in paper [16], the required parameters for the regression function are obtained.

Proximal Support Vector Regression: Proximal support vector regression (PSVR) is formed by adding a term $(1/2)\omega_0^2$ called bias in the objective function of LSSVR, which converts it into a strongly convex objective function. PSVR can be taken as a particular case of regularized LSSVR, which results in the optimal solution and fast computational time.

Both LSSVR and PSVR minimize the mean square error (MSE) on the training dataset. For a regression problem, search a function $f(z)$ which results in the best relationship between the input vector and their corresponding output. PSVR minimizes the following objective function with constraints:

$$\begin{aligned} &\min\ J(\omega, \eta) = \tfrac{1}{2}\|\omega\|^2 + \tfrac{1}{2}\omega_0^2 + \tfrac{C}{2}\sum_{i=1}^{h} \eta_i^2 \\ &\text{s.t. } w^\top \phi(z_i) + \omega_0 - y_i = \eta_i, \quad i = 1, 2, \ldots, \end{aligned} \tag{3}$$

where η_i is the training error and $C > 0$ is a given parameter. By introducing the Lagrangian form for Eq. 3 and solving the Lagrangian form by using the KKT condition as described in paper [17], the required parameters for the regression function are obtained.

ε-Twin Support Vector Regression: ϵ-Twin Support Vector Regression (ϵ-TSVR) which is a new variant of TSVR is introduced by following the idea of TSVM and TSVR. ϵ-TSVR aims to find two ϵ-insensitive proximal linear functions by introducing the regularization terms $\frac{1}{2}(\omega_1^\top \omega_1 + \omega_0^2)$ and $\frac{1}{2}(\omega_2^\top \omega_2 + \omega_0^{\prime 2})$ in the objective function. The quadratic programming problem becomes:

$$\begin{aligned} &\min \tfrac{1}{2}c_3(\omega_1^\top \omega_1 + \omega_0^2) + \tfrac{1}{2}\eta_1^{*\top}\eta_1^* + c_1 e^\top \eta_1 \\ &\text{s.t: } \begin{cases} Y - (Z\omega_1 + e\omega_0) \geq -\varepsilon_1 e - \eta_1, \eta_1 \geq 0 \\ Y - (Z\omega_1 + e\omega_0) = \eta_1^* \end{cases} \end{aligned} \tag{4}$$

$$\begin{aligned} &\min \tfrac{1}{2}c_4(\omega_2^\top \omega_2 + \omega_0'^2) + \tfrac{1}{2}\eta_2^{*\top}\eta_2^* + c_2 e^\top \eta_2 \\ &\text{s.t:} \begin{cases} (Z\omega_2 + e\omega_1') - Y \geq \varepsilon_2 e - \eta_2, \eta_2 \geq 0 \\ (Z\omega_2 + e\omega_1') - Y = \eta_2^* \end{cases} \end{aligned} \tag{5}$$

where $c_1, c_2, c_3, c_4, \varepsilon_1$ and ε_2 are positive parameters, η_1 and η_2 are slack variables. By introducing the Lagrangian form for Eqs. 4 and 5 and solving the Lagrangian form by using KKT condition as described in paper [18], the required parameters for regression function are obtained for both ϵ-insensitive proximal linear functions. Then, the estimated regressor function is constructed by taking the average of both functions.

Generalized Eigen Value Proximal Support Vector Regression: Linear generalized eigen value proximal support vector regression (GEPSVR) algorithm aims to find two regression functions where each one determines the ε-insensitive bounding regressor [19]. The first regressor function forms the following optimization problem:

$$\min_{\omega_1,\omega_0 \neq 0} \frac{\|Z\omega_1 + e\omega_0 - (Y - e\epsilon)\|^2 / \left\| [\omega_1^\top \; \omega_0]^\top \right\|^2}{\|Z\omega_1 + e\omega_0 - (Y + e\epsilon)\|^2 / \left\| [\omega_1^\top \; \omega_0]^\top \right\|^2} \tag{6}$$

Also assumed that $Y \neq e\varepsilon$, $(\omega_1, \omega_0) \neq 0$ and $Z\omega_1 + e\omega_0 - (Y - e\epsilon) \neq 0$. The above problem can be regularized by introducing the Tinkhov regularization term and converting it into an eigenvalue problem by using the Rayleigh quotient as described in paper [19]. The solution for the first ε-insensitive bounding regressor can be obtained by finding the minimum value of the eigenvalue and normalizing it to the corresponding of its eigenvector, which yields ω_1 and ω_0.

The second regressor function forms the following optimization problem:

$$\min_{\omega_2,\omega_0'} \frac{\left\|Z\omega_2 + e\omega_0' - (Y + e\epsilon)\right\|^2 / \left\| [\omega_2^\top \; \omega_0']^\top \right\|^2}{\left\|Z\omega_2 + e\omega_0' - (Y - e\epsilon)\right\|^2 / \left\| [\omega_2^\top \; \omega_0']^\top \right\|^2} \tag{7}$$

Also assumed that $(\omega_2, \omega_0') \neq 0$, and $Z\omega_2 + e\omega_0\prime - (Y - e\epsilon) \neq 0$. Similarly, the above problem can be regularized and converted into an eigenvalue problem. The solution for the second ε-insensitive bounding regressor can be obtained by finding the maximum value of the eigenvalue and normalizing it to the corresponding eigenvector, which yields ω_2 and $\omega_0\prime$ as described in the paper [19]. Then the estimated regressor is constructed by taking the average of both the ε-insensitive bounding regressors.

2.2 Feature Selection

Selecting only those features that best represent the data can significantly reduce the computational complexity and improve the model's overall performance. Random forest is one of the techniques that can be used for feature selection.

Random forest (RF) [13] is a statistical method used for regression and classification problems; it is also used for correlation and provides feature importance. The idea of RF is to construct a number of decision trees by bootstrapping training data with replacements for each tree. In RF, only a subset of randomly selected features is considered instead of choosing the entire features from the training set.

Let a dataset contain p number of training data and q number of features, where m is the number of decision trees in random forest. For each m decision tree, p data are sampled with replacement. From q features, randomly $p_{\text{try}} \ll q$ features are selected to decide the best split to construct the decision trees. All m trees give the final regression output by aggregating it. To improve the generalization of RF, different bootstraping samples are used to decide the best split and construct different trees. Of the randomly split bootstrap data, two-third of the split is used for training, and the remaining one-third is used to quantify the feature score, also known as out-of-a-bag (OOB). The error rate E_{γ_i} is computed for all decision trees $\{\gamma_i, i = 1, 2, \ldots, m\}$ with respective to OOB samples. For each tree, perturbed OOB samples are produced by taking a random permutation of the features among the samples to compute the importance score f. Again for each tree, error rate $E_{\gamma'_i}$ recorded for the perturbed OOB samples. The formulation to compute the feature importance score is given as:

$$I_f^{\text{RF}} = \frac{1}{p} \sum_i (E_{\gamma_i} - E_{\gamma'_i})$$

To take down more informative feature, the features are sorted in the descending order according to I_f^{RF}.

3 Proposed Forecasting Model

Hybrid models of random forest and four variants of support vector regression are proposed for multi-step ahead forecasting for the closing price of cryptocurrencies. The proposed models are denoted by RF-LSSVR, RF-PSVR, RF-ε TSVR, and RF-GEPSVR.

3.1 Cryptocurrency

Cryptocurrency is a digital currency that allows peer-to-peer transactions without the intervention of central authority. Bitcoin, the first fully implemented cryptocurrency by Nakamoto [21], led to the boom of Cryptocurrency, after which several Cryptocurrencies came into existence; as of Dec 2019, there exist as many as 4950 cryptocurrencies with a net worth of approximately 190 billion dollars. Cryptocurrency leverages blockchain technology, which is a public ledger that maintains a record of all transactions so as to gain decentralization and transparency [22].

The mechanism of cryptocurrency is as follows:

- A user is assigned a wallet with an address, which acts as a public key.
- The wallet has a private key used to authenticate the transaction.
- A transaction between the payer and payee is documented using the payer's private key.
- Transaction is verified by the process of mining [23].

3.2 Input Features

To develop a financial trading system, the selection of input features plays a crucial role. A large set of technical indicators are used as input features to forecast the five steps ahead closing price of cryptocurrency. Some of the technical indicators are selected from previous studies done in [13] while the rest are collected on the basis of their popularity and application in technical analysis. Some previous researchers indicate that a certain subset of technical indicators is more suitable and effective to predict the future price of the financial market. Some of the popularly used technical indicators are relative strength index (RSI), William's oscillator percent R (WR), moving average (MA, EMA), and price oscillators (OSCP). In order to reduce the computational complexity, the technical indicator with the best information is selected by employing techniques such as RF; here, RF returns a subset of the technical indicators ranked in order of importance.

3.3 System Architect

Figure 1 illustrates the flowchart of the proposed hybrid model. Firstly, the dataset is split into training and testing in the ratio of 80:20, respectively. By using the importance score RF ranks the features by taking the training dataset as input. In SVR hybrid models, these features are added one by one according to their importance score, which is arranged in ascending order. The parameters of SVR play a crucial role in improving the model's accuracy. The parameters are optimized by using time

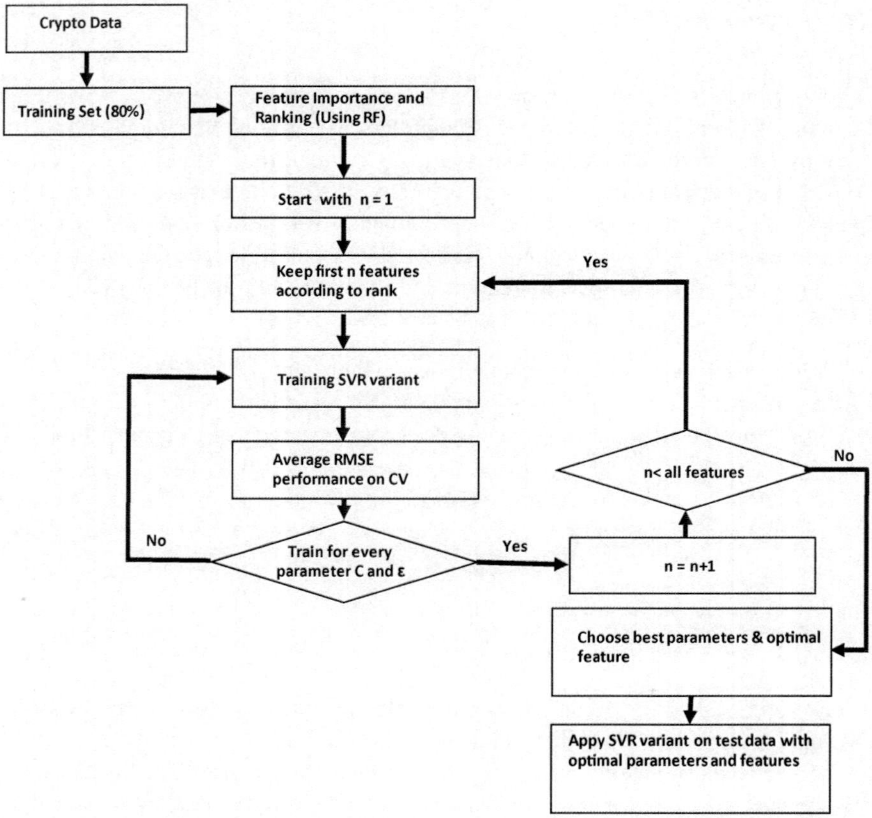

Fig. 1 Flowchart of the proposed hybrid forecasting models

series cross-validation (TSCV). In TSCV, the dataset is split into train and cross-validation (CV) according to their timestamp. In a particular iteration of time series data, the next instance can be treated as validation data. TSCV is performed for every value of the model's parameter. The optimum value is chosen by minimizing the average MSE. The model is further trained by using the optimized parameter from the previous iteration, and the training result is recorded. The same procedure is performed after every iteration when a new feature is added to the training dataset until all features are added. Finally, the feature subset is chosen as an optimal feature set for the model, which results in minimum MSE.

3.4 *Multi-step Ahead Forecasting Strategies*

In order to make N multi-day price forecasting of Cryptocurrency, where N is the projection period. The two different approaches are:

- **Direct strategy**: Direct strategy consists of multiple model methods to make a multi-step ahead forecasting [24]. In this method, N different models are used to forecast N steps. The input is the same for every model, but outputs are independent of each other. Hence, N models are trained for N steps ahead forecasting. However, every model of direct strategy has different architecture, hence making this method computationally expensive.
- **MIMO strategy**: MIMO is also a multi-step ahead forecasting method [24], but unlike direct strategy, it has only one model to produce all N predictions at once. Hence, it is computationally less expensive than the direct strategy.

4 Experiments and Discussion

4.1 *Dataset Description*

The daily closing price of the most popular six cryptocurrency datasets is used to evaluate the performance of the proposed models. The variables of each dataset are Date, Open, High, Low, Close, Adj Close, and Volume. The dataset was collected from https://finance.yahoo.com/, and the end date for each dataset is 21-Feb-2022. The cryptocurrency datasets used are as follows: Bitcoin (BTC), Ethereum (ETH), Crypto.com (CRO), Binance Coin (BNB), Cardano (ADA), and Tron (TRX).

4.2 *Performance Analysis*

The different performance measures used are

1. **Root Mean Square Error**: Root mean square error is square root of the average of summation sum of square of difference between actual value and predicted value.

$$\text{RMSE} = \sqrt{\frac{\sum_{i=1}^{n}(z_i - \hat{z}_i)^2}{n}}$$

2. R^2 **Score**: R^2 Score is also known as coefficient of determination, and it measures that provide information about the goodness of fit of a model. It is defined as:

$$R^2 = 1 - \frac{\text{SSR}}{\text{SST}}$$

where SSR is sum of squared of residual and SST is total sum of square. If the value of R^2 is close to 1, then prediction is good. If its value close to zero prediction is bad.

4.3 Parameter Selection

In order to get optimum results using SVR, choosing the right values for parameters C and ε is crucial. C can take any value from the set $\{2^{-22}, 2^{-20}, \ldots, 2^{20}, 2^{22}\}$ similarly ε can also take on any values from the set $\{2^{-22}, 2^{-20}, \ldots, 2^{20}, 2^{22}\}$, an optimum tuple of C and $\varepsilon(C, \varepsilon)$ is obtained by minimizing MSE over all possible combination of C and ε.

4.4 Implementation of Forecasting Model

The algorithm was implemented in Python 3.8.8 using Spyder 5.1.5, and the desktop machine had a configuration of 32 GB RAM and 2.10 GHz processor.

4.5 Results and Discussion

The performance measure root mean square error (RMSE) and R^2 Score have been used over six cryptocurrencies dataset to evaluate the performance of the proposed hybrid models.

The performance of the proposed models are reported from Tables 1, 2, 3, 4, 5, 6, 7, and 8, where the best results are stated in bold. It is clear from the tables that for all the cryptocurrency dataset the proposed models have better performance than the basic variant of SVR.

It is observed that for the Cardano dataset the proposed model RF-ϵ TSVR has the best performance for 1-day ahead, 3-days ahead, and 5-days ahead forecasting. From Table 2, it is observed that for 3-days ahead and 5-days ahead forecasting both the proposed models RF-LSSVR and RF-PSVR has the better performance. From Table 3, it is clear that for Bitcoin the proposed model RF-ϵ TSVR has the best performance

Table 1 Performance on ADA using direct strategy

Model	Cardano (ADA)					
	1 day ahead		3 days ahead		5 days ahead	
	RMSE	R^2 score	RMSE	R^2 score	RMSE	R^2 score
LSSVR	0.035572	0.975106	0.079997	0.846087	0.114607	0.627589
PSVR	0.035569	0.975111	0.079889	0.846594	0.113979	0.632803
GEPSVR	0.067734	0.916872	0.076984	0.892802	0.094261	0.794052
ε-TSVR	0.032459	0.979708	0.060982	0.914084	0.070182	0.882106
RF-LSSVR	0.031497	0.981157	0.073367	0.876527	0.102715	0.725509
RF-PSVR	0.031497	0.981156	0.072849	0.879312	0.102715	0.725509
RF-GEPSVR	0.055936	0.942725	0.057090	0.941170	0.071822	0.905478
RF-εTSVR	**0.031267**	**0.981536**	**0.051974**	**0.947907**	**0.064524**	**0.920921**

Table 2 Performance on BNB using direct strategy

Model	Binance coin (BNB)					
	1 day ahead		3 days ahead		5 days ahead	
	RMSE	R^2 score	RMSE	R^2 score	RMSE	R^2 score
LSSVR	0.035617	0.982894	0.062378	0.945158	0.086286	0.889778
PSVR	0.035613	0.982899	0.062351	0.945208	0.086060	0.890379
GEPSVR	0.075556	0.927437	0.090779	0.898050	0.106408	0.808248
ε-TSVR	0.049002	0.966227	0.064521	0.935875	0.079231	0.901126
RF-LSSVR	0.034934	0.983645	**0.058413**	**0.952795**	**0.078860**	**0.910880**
RF-PSVR	0.034928	0.983660	**0.058413**	**0.952795**	**0.078860**	**0.910880**
RF-GEPSVR	**0.034725**	**0.984156**	0.075596	0.927752	0.102156	0.867985
RF-εTSVR	0.044704	0.970697	0.066298	0.939836	0.082610	0.906485

Table 3 Performance on BTC using direct strategy

Model	Bitcoin (BTC)					
	1 day ahead		3 days ahead		5 days ahead	
	RMSE	R^2 score	RMSE	R^2 score	RMSE	R^2 score
LSSVR	0.024604	0.989703	0.042960	0.967721	0.056992	0.942011
PSVR	0.024604	0.989703	0.042960	0.967721	0.056991	0.942011
GEPSVR	0.054664	0.949875	0.065442	0.930010	0.074537	0.907496
ε-TSVR	0.025327	0.988857	0.041132	0.970235	0.054059	0.949392
RF-LSSVR	0.024524	0.989764	0.042371	0.968591	0.055995	0.944000
RF-PSVR	0.024508	0.989782	0.042234	0.968909	0.055995	0.944000
RF-GEPSVR	0.042795	0.970481	0.055589	0.948136	0.068900	0.928526
RF-εTSVR	0.024849	0.989399	0.040895	0.970866	0.053516	0.949320

Table 4 Performance on CRO using direct strategy

Model	Crypto.com (CRO)					
	1 day ahead		3 days ahead		5 days ahead	
	RMSE	R^2 score	RMSE	R^2 score	RMSE	R^2 score
LSSVR	0.028351	0.980061	0.048553	0.941073	0.065496	0.893933
PSVR	0.028351	0.980061	0.048553	0.941072	0.065497	0.893932
GEPSVR	0.063279	0.902611	0.112030	0.766321	0.099191	0.573871
ε-TSVR	0.030242	0.975497	0.047174	0.945517	0.065318	0.887996
RF-LSSVR	0.027928	0.980836	0.047307	0.945086	0.064279	0.897233
RF-PSVR	0.027779	0.981059	0.047307	0.945086	0.064870	0.894665
RF-GEPSVR	0.042971	0.955982	0.075901	0.853592	0.083746	0.826948
RF-εTSVR	0.028492	0.980894	0.046641	0.945950	0.064248	0.888191

Table 5 Performance on ETH using direct strategy

Model	Ethereum (ETH)					
	1 day ahead		3 days ahead		5 days ahead	
	RMSE	R^2 score	RMSE	R^2 score	RMSE	R^2 score
LSSVR	0.030708	0.983350	0.053463	0.947327	0.071831	0.899584
PSVR	0.030710	0.983348	0.053462	0.947328	0.071822	0.899615
GEPSVR	0.070667	0.916219	0.079101	0.862185	0.091178	0.867897
ε-TSVR	0.032115	0.981360	0.053496	0.948889	0.066488	0.915696
RF-LSSVR	0.030243	0.983877	0.052419	0.949309	0.071831	0.899584
RF-PSVR	0.030243	0.983877	0.052824	0.948312	0.071822	0.899615
RF-GEPSVR	0.040818	0.971582	0.056201	0.946190	0.069210	0.919748
RF-εTSVR	0.030524	0.983102	0.049519	0.957866	0.065147	0.922593

Table 6 Performance on TRX using direct strategy

Model	TRON (TRX)					
	1 day ahead		3 days ahead		5 days ahead	
	RMSE	R^2 score	RMSE	R^2 score	RMSE	R^2 score
LSSVR	0.027703	0.956337	0.050682	0.833082	0.071495	0.602763
PSVR	0.027709	0.956320	0.050677	0.833111	0.071493	0.602782
GEPSVR	0.046737	0.889430	0.057863	0.830470	0.065892	0.779048
ε-TSVR	0.025636	0.963548	0.042123	0.887251	0.056685	0.765833
RF-LSSVR	0.026766	0.959612	0.047358	0.860296	0.068492	0.643287
RF-PSVR	0.026765	0.959618	0.047299	0.860665	0.068772	0.644296
RF-GEPSVR	0.054072	0.871569	0.053069	0.855391	0.051981	0.843202
RF-εTSVR	0.025407	0.965975	0.041226	0.897228	0.054464	0.792326

Table 7 Performance on BTC, BNB, and TRX using MIMO strategy

Model	Bitcoin (BTC)		Binance coin (BNB)		Tron (TRX)	
	RMSE	R^2 score	RMSE	R^2 score	RMSE	R^2 score
LSSVR	0.043053	0.967745	0.062818	0.945433	0.054317	0.797742
PSVR	0.043052	0.967746	0.062932	0.945141	0.054314	0.797764
GEPSVR	0.066108	0.928840	0.093359	0.895686	0.057250	0.841588
ε-TSVR	0.041615	0.969786	0.067904	0.937632	0.044601	0.871550
RF-LSSVR	0.042044	0.969303	0.058493	0.953425	0.051779	0.821638
RF-PSVR	0.042053	0.969319	0.058491	0.953435	0.051615	0.823814
RF-GEPSVR	0.057353	0.947761	0.076127	0.924241	0.057048	0.841171
RF-εTSVR	0.040846	0.971015	0.064760	0.937582	0.043395	0.874439

Table 8 Performance on ADA, CRO, and ETH using MIMO strategy

SVR variant	Cardano (ADA)		Crypto.com (CRO)		Ethereum (ETH)	
	RMSE	R^2 score	RMSE	R^2 score	RMSE	R^2 score
LSSVR	0.082670	0.843471	0.053077	0.927291	0.054080	0.947528
PSVR	0.082370	0.844772	0.053078	0.927289	0.054076	0.947535
GEPSVR	0.0.68290	0.918929	0.091769	0.796419	0.067559	0.920522
ε-TSVR	0.057459	0.918929	0.053476	0.924809	0.050849	0.954241
RF-LSSVR	0.074828	0.881797	0.053071	0.927699	0.053782	0.948334
RF-PSVR	0.075103	0.881192	0.053142	0.927519	0.053944	0.948057
RF-GEPSVR	0.073946	0.899260	0.076622	0.847628	0.066619	0.928625
RF-εTSVR	0.052903	0.947689	0.052582	0.926011	0.049919	0.956847

for 3-days ahead and 5-days ahead forecasting. From Table 4, it is observed that for the dataset Crypto.com, for 1 day ahead forecasting the proposed model RF-PSVR has better performance, similarly for 3-days and 5-days ahead forecasting the proposed models RF-ϵ TSVR and RF-LSSVR have the best performance, respectively. From Table 4, it is observed that for Ethereum the proposed models RF- ϵ TSVR performs the best for 3-days and 5-days ahead forecasting. From Table 6, it is observed that for TRON RF-ϵ TSVR performs the best for 1-day and 3-days ahead forecasting.

For multi-day ahead two forecasting strategies have been used direct strategy and MIMO strategy. Tables 1, 2, 3, 4, 5, and 6 give the performance of the models over various cryptocurrency using direct strategy. Similarly, Tables 7 and 8 give the performance of the models over various cryptocurrency using MIMO strategy. For all the hybrid models, direct strategy and MIMO strategy have been used for multi-day ahead forecasting. It is observed that direct strategy has slightly better performance than MIMO strategy, but MIMO is computationally much faster than direct strategy. Hence, there exists a trade-off between the two strategies for better performance direct strategy is preferred and for faster computation MIMO is preferred.

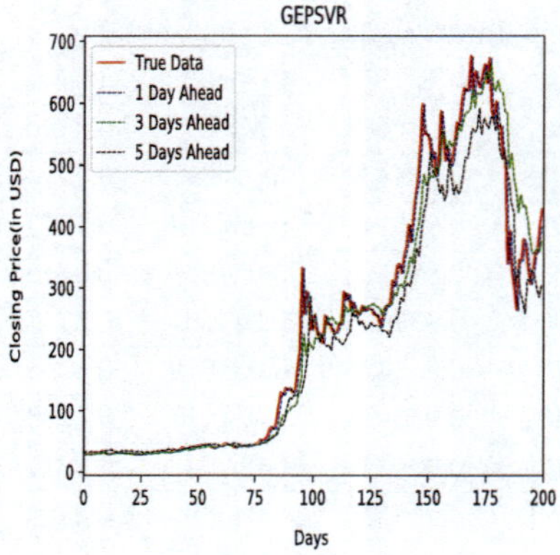

(a) Binance Coin

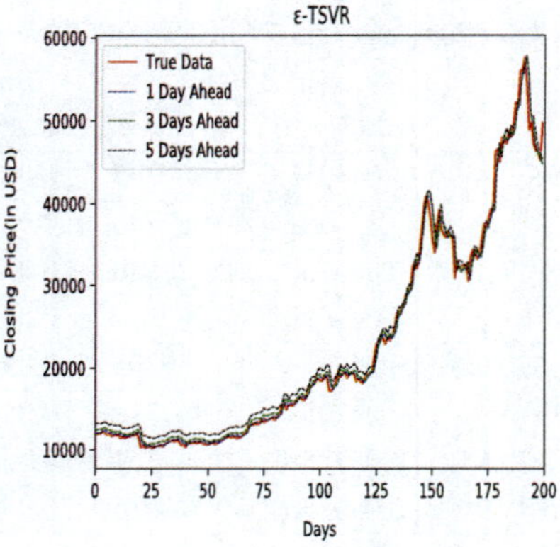

(b) Bitcoin

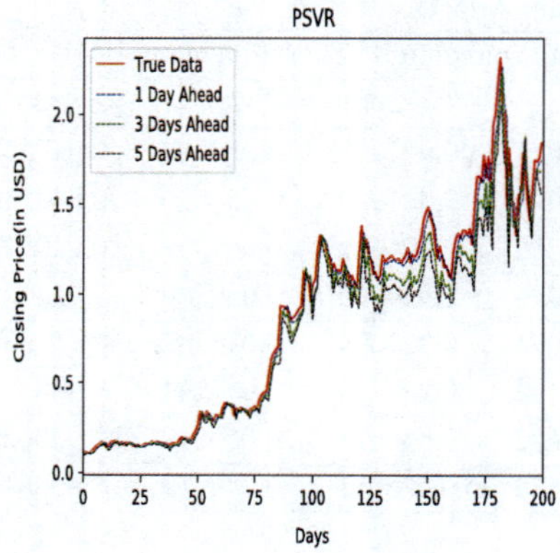

(c) Cardano

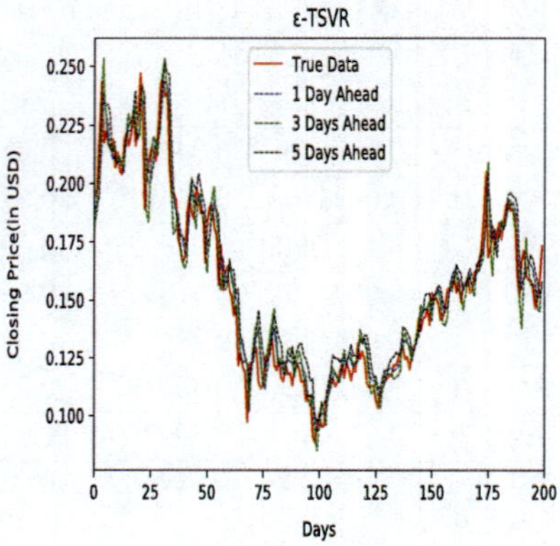

(d) Crypto.com

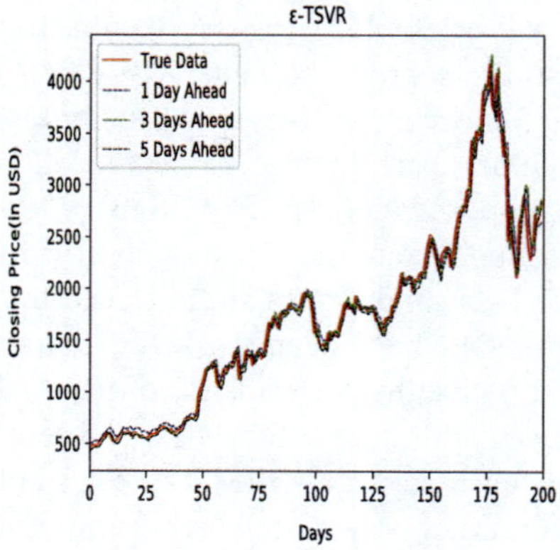

(e) Ethereum

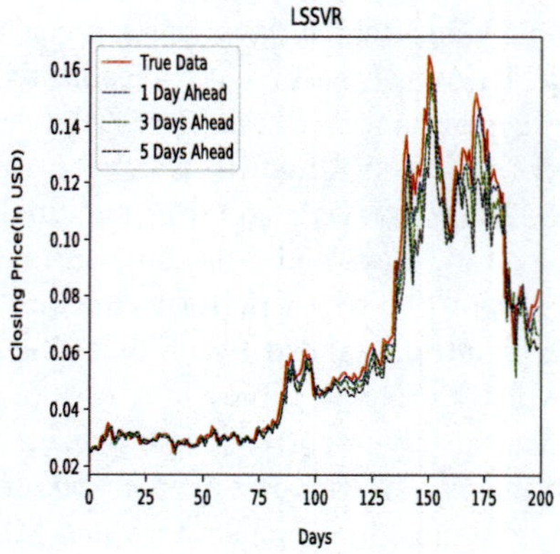

(f) Tron

Fig. 2 Direct strategy

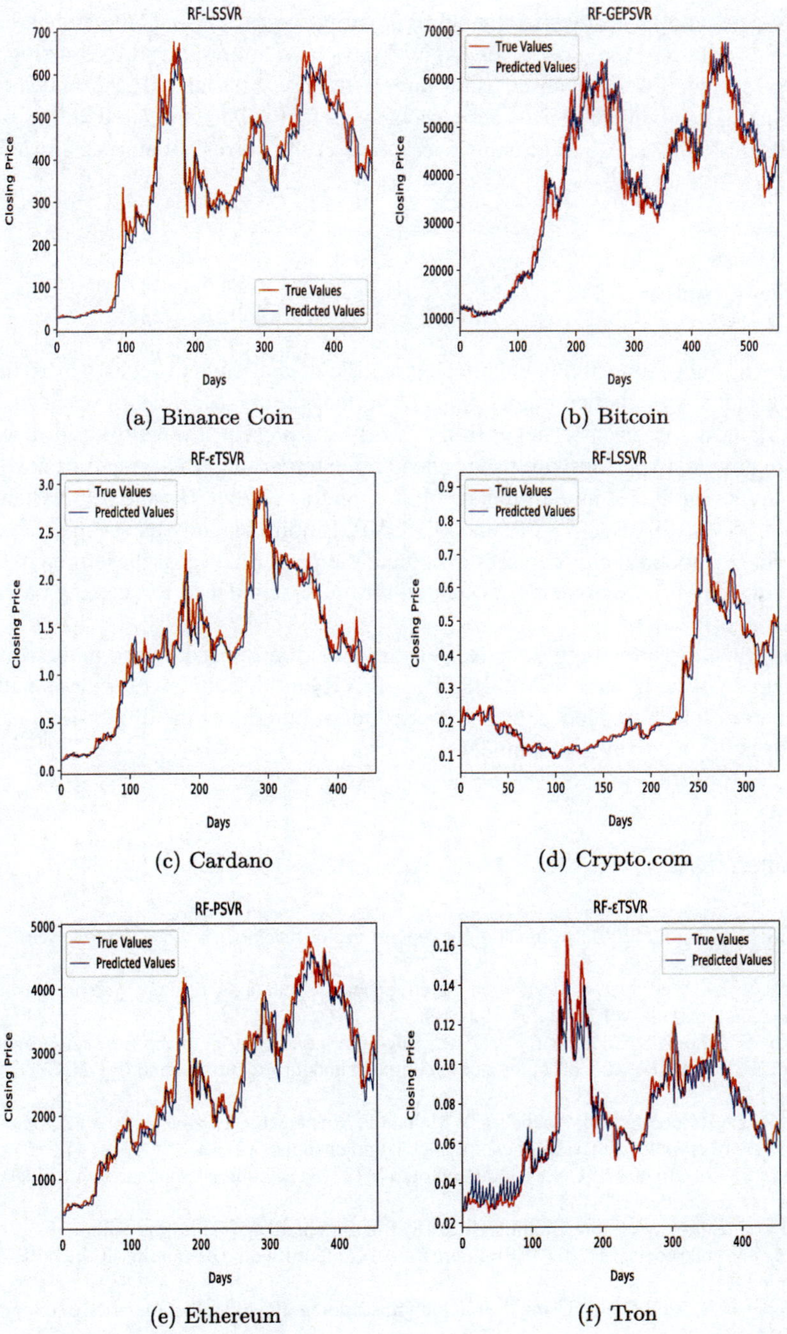

Fig. 3 MIMO strategy

Figure 2 and its corresponding subfigures Fig. 2a–f represents the actual versus predicted values of the models for 1-day, 3-days, and 5-days ahead forecasting over the six cryptocurrency dataset using direct strategy. Similarly, Fig. 3 and its corresponding subfigures Fig. 3a–f represents the actual versus predicted values of the models for 5-days ahead forecasting over the six cryptocurrency dataset using MIMO strategy.

5 Conclusion

Cryptocurrency being highly volatile, it is a difficult challenge to develop a predictive model, but this predictive model gives important inputs to investors when making a profitable transaction strategy. In this work, we present a novel hybrid machine learning model to perform multi-day ahead price forecasting of cryptocurrency. This study uses four hybrid models that combines random forest (RF) and four variants of SVR: LSSVR, PSVR, ϵ-TSVR, and GEPSVR. Empirical study has been performed over six cryptocurrencies dataset to evaluate the performance of these models. The performance of these models is evaluated for multi-step ahead forecasting based on RMSE and R^2 score.

Empirical findings suggest better performance of proposed hybrid models, when compared with original LSSVR, PSVR, ε-TSVR, and GEPSVR algorithms without any feature selection. Hence, hybrid model not only reduces the dimension but also improves the accuracy of the model.

References

1. Tay FEH, Cao L (2001) Application of support vector machines in financial time series forecasting. Omega 29(4):309–317
2. Deboeck GJ (ed) (1994) Trading on the edge: neural, genetic, and fuzzy systems for chaotic financial markets, vol 39. Wiley, Hoboken
3. Roy S, Nanjiba S, Chakrabarty A (2018) Bitcoin price forecasting using time series analysis. In: 2018 21st International conference of computer and information technology (ICCIT). IEEE, pp 1–5
4. Adeleke I, Zubairu UM, Abubakar B, Maitala F, Mustapha Y, Ediuku E (2019) A systematic review of cryptocurrency scholarship. Int J Commer Finan 5(2):63–75
5. Akcora CG, Dixon MF, Gel YR, Kantarcioglu M (2018) Bitcoin risk modeling with blockchain graphs. Econ Lett 173:138–142
6. Guo T, Bifet A, Antulov-Fantulin N (2018) Bitcoin volatility forecasting with a glimpse into buy and sell orders. In: 2018 IEEE International conference on data mining (ICDM). IEEE, pp 989–994
7. Greaves A, Au B (2015) Using the bitcoin transaction graph to predict the price of bitcoin. No data
8. Li TR, Chamrajnagar AS, Fong XR, Rizik NR, Fu F (2019) Sentiment-based prediction of alternative cryptocurrency price fluctuations using gradient boosting tree model. Front Phys 7:98

9. Poongodi M, Sharma A, Vijayakumar V, Bhardwaj V, Sharma AP, Iqbal R, Kumar R (2020) Prediction of the price of Ethereum blockchain cryptocurrency in an industrial finance system. Comput Electr Eng 81:106527
10. Peng Y, Albuquerque PHM, de Sá JMC, Padula AJA, Montenegro MR (2018) The best of two worlds: forecasting high frequency volatility for cryptocurrencies and traditional currencies with Support Vector Regression. Expert Syst Appl 97:177–192
11. McNally S, Roche J, Caton S (2018) Predicting the price of bitcoin using machine learning. In: 2018 26th Euromicro international conference on parallel, distributed and network-based processing (PDP). IEEE, pp 339–343
12. Phaladisailoed T, Numnonda T (2018) Machine learning models comparison for bitcoin price prediction. In: 2018 10th International conference on information technology and electrical engineering (ICITEE). IEEE, pp 506–511
13. Kumar D, Meghwani SS, Thakur M (2016) Proximal support vector machine based hybrid prediction models for trend forecasting in financial markets. J Comput Sci 17:1–13
14. Derbentsev V, Datsenko N, Stepanenko O, Bezkorovainyi V (2019) Forecasting cryptocurrency prices time series using machine learning approach. In: SHS Web of conferences, vol 65. EDP Sciences, p 02001
15. Livieris IE, Kiriakidou N, Stavroyiannis S, Pintelas P (2021) An advanced CNN-LSTM model for cryptocurrency forecasting. Electronics 10(3):287
16. Xu S, An X, Qiao X, Zhu L, Li L (2013) Multi-output least-squares support vector regression machines. Pattern Recogn Lett 34(9):1078–1084
17. Wang K, Pei H, Ding X, Zhong P (2019) Robust proximal support vector regression based on maximum correntropy criterion. Sci Program
18. Shao YH, Zhang CH, Yang ZM, Jing L, Deng NY (2013) An ε-twin support vector machine for regression. Neural Comput Appl 23(1):175–185
19. Khemchandani R, Karpatne A, Chandra S (2011) Generalized eigenvalue proximal support vector regressor. Expert Syst Appl 38(10):13136–13142
20. Deng N, Tian Y, Zhang C (2012) Support vector machines: optimization based theory, algorithms, and extensions. CRC Press
21. Nakamoto S (2008) Bitcoin: a peer-to-peer electronic cash system. Decentralized Bus Rev 21260
22. Fang F, Ventre C, Basios M, Kanthan L, Martinez-Rego D, Wu F, Li L (2022) Cryptocurrency trading: a comprehensive survey. Finan Innov 8(1):1–59
23. Mukhopadhyay U, Skjellum A, Hambolu O, Oakley J, Yu L, Brooks R (2016) A brief survey of cryptocurrency systems. In: 2016 14th Annual conference on privacy, security and trust (PST). IEEE, pp 745–752
24. Sahoo D, Sood N, Rani U, Abraham G, Dutt V, Dileep AD (2020) Comparative analysis of multi-step time-series forecasting for network load dataset. In: 2020 11th International conference on computing, communication and networking technologies (ICCCNT). IEEE, pp 1–7

Image Segmentation Using Structural SVM and Core Vector Machines

Varuun A. Deshpande and Khriesavinyu Terhuja

Abstract In this paper, we study the performance of two variants of support vector machines, namely structural support vector machines and core vector machines, on large-scale data. We have used image segmentation as a mechanism to test the classification abilities of these variants on large-scale data. The images are converted to numeric data using various filters, and the labels are generated using the available ground truth image segmentation mask.

Keywords Large-scale data · Structural SVM · Core vector machine · Image segmentation

1 Introduction

Image segmentation is the process of grouping pixels of an image based on the homogeneity of the features such as color or intensity to extract meaningful information. Image segmentation is an important area of computer vision as it has a wide range of applications ranging from traffic surveillance, satellite imaging, and automated driving cars to medical diagnosis.

Various approaches and techniques to segment images exist in the literature. Thresholding, histogram-based bundling, and region-growing algorithms are some of the classical methods that are used for image segmentation. These algorithms have the limitations that they cannot handle the real-life complexity that is tolerant of uncertainty and approximation.

This led to the computational approach that uses artificial intelligence, and supervised machine learning techniques such as SVM [27], random forest [24], and unsupervised machine learning technique such as k-means [12] are some machine learning-based image segmentation technique that has been researched heavily in the past decade [25].

V. A. Deshpande (✉) · K. Terhuja
Indian Institute of Technology, Mandi, HP, India
e-mail: deshpandevaruun@gmail.com

M. Thakur et al. (eds.), *Soft Computing for Problem Solving*, Lecture Notes in Networks and Systems 547, https://doi.org/10.1007/978-981-19-6525-8_28

Song et al. [27] used a hybrid pixel object approach to extract the roads in an image by using SVM to perform segmentation; Schroff et al. used a random forest for image segmentation along with textons, color, and HOG features to enhance the performance of the model [24]. The use and success of deep learning for computer vision rose to popularity after the ILSVRV 2012 image classification challenge for large image datasets using CNN. The use of CNN became successful in almost all computer vision problems, including image segmentation. One of the most successful CNN-based architectures, FCN by Shelhamer et al. [21], used fully convolution network to segment images. The FCN architecture used already pre-trained models such as AlexNet, VGGNet, and GoogleNet by replacing the classifier layers with a 1×1 convolution layer. This produces a low-resolution heat map of the image; this heat map is up-sampled by using bi-linear interpolation to segment the images. More work on CNN-based image segmentation frameworks such as Dialatednet, DeepLab, Deconvnet, and U-Net exists in the literature.

Structural SVM and CVM are two variants of SVMs that can handle large-scale datasets; considering the fact that images consist of a large number of datapoints, our work proposes an experiment to show the use of structure SVM and CVM to segment images. The contribution of this paper can be summarized as follows:

- Used structured SVM and CVM to segment images.
- Experimented over the dataset Weizmann horses dataset and CMU-Cornell iCoseg dataset.
- Used accuracy as a performance measure to compare SSVM and CVM.

This paper is divided as follows: Sect. 2 discusses the SVM variants structural support vector machines and core vector machines. Section 3 presents the results.

2 Methodology

Support vector machine [1, 23], abbreviated and commonly referred to as SVM, is a supervised learning algorithm for binary classification that constructs a hyperplane, the decision boundary, that separates classes by maximizing the margin between the sample points on either side of the hyperplane. The rationale behind maximizing the margin is to improve the confidence in the results achieved by the classifier.

2.1 Structural Support Vector Machines (SSVM)

Given training examples $(x_1, y_1), \ldots, (x_n, y_n)$, where x_is are the datapoints to be classified, $x_i \in X$ and $y_i \in Y$. Y is the set of possible outputs. The object is to obtain a classification map $f : X \rightarrow Y$, by constructing a classifier using the training examples of the form

$$h : X \times Y \to \mathbb{R} \tag{1}$$

such that

$$f(x) = \underset{y \in Y}{\text{argmax}}\ h(x, y). \tag{2}$$

Define

$$h(x, y) := a^T \Psi(x, y). \tag{3}$$

Here $\Psi : X \times Y \to \mathbb{R}^m$ is the joint feature map [3]. Joint feature map measures the compatibility of $x \in X$ with class $y \in Y$. The optimization problem for training the structural SVM is-

$$\begin{aligned} &\min_{a,\epsilon}\ \frac{1}{2}\|a\|^2 + c\sum_{i=1}^{n} \epsilon_i \\ &a^T\Psi(x_1, y_1) \geq l(y_1, y) - \epsilon_1 + a^T\Psi(x_1, y),\ y \neq y_1 \\ &\ldots\ldots\ldots\ldots\ldots\ldots\ldots\ldots\ldots\ldots \\ &\ldots\ldots\ldots\ldots\ldots\ldots\ldots\ldots\ldots\ldots \\ &a^T\Psi(x_n, y_n) \geq l(y_n, y) - \epsilon_n + a^T\Psi(x_n, y),\ y \neq y_n. \end{aligned} \tag{4}$$

This problem can be solved using the algorithm proposed by Joachims et al. [18] that employs the cutting plane method with polynomial time complexity. The algorithm starts with an empty set of constraints, say C. Out of (x_j, y_j), compute the most violated constraint. If the violation is more than the required precision, ζ, then the constraint is added to C. The optimization problem is then solved with constraints in C. Given any optimization problem, the aim is to find the set of optimal points S. Let x be some point in the domain of definition of the optimization problem; then, either $x \in S$, in which case we are done or there is some hyperplane that perfectly separates the set S and the point x, i.e., $a^T y \geq b, \forall y \in S$, and $a^T x \leq b$. This hyperplane is called a cut since it 'cuts' the domain into two parts such that S lies in one half and the other half is discarded [7]. The cutting plane method progressively strikes out the infeasible region by adding linear inequality constraints until it reaches within an acceptable distance of the optimum set. The following is taken from Sra et al. [28]. Consider the optimization problem -

$$\min\{f(x) | x \in X \subset \mathbb{R}^n, f : \mathbb{R}^n \to \mathbb{R}\} \tag{5}$$

where X is convex. The cutting plane method iteratively encloses the set of optimum points by constructing polyhedra by removing the infeasible region. When a point in the optimum set is attained within some tolerance, the iteration stops. The pseudocode of the cutting plane method is as follows:
The algorithm starts with a polyhedron that encloses the feasible set X, say X_0. It then finds the set of optimum points in X_0. If the intersection between the set

Algorithm 1

Initialize $t \leftarrow 0$ and $X \subset X_0$
repeat
  $x_t \in \underset{x \in X_t}{\operatorname{argmin}} f(x)$
  Find the hyperplane $a^T x \geq b$ that separates x_t from X
  $X_{t+1} \leftarrow \{x | a^T x \geq b\} \cap X_t$
  $t \leftarrow t + 1$
until $x_t \in X$

of optimum points and the feasible set X is non-empty, then we have found the point we were looking for. Otherwise, we find a hyperplane or the cutting plane that separates the optimum point, x_0, from the feasible set and discard the part which contains x_0. Repeat the process on the set left after the removal of the unwanted area. Iterations are carried out until we get an optimum point in the feasible set. One of the advantages of the cutting plane method is that at each iteration the evaluation of constraints is not required. Therefore, for problems with numerous constraints, like the structural SVMs, the cutting plane method is efficient [7]. The algorithm proposed by Tsochantaridis et al. [31] and Joachims et al. [18] leverages this characteristic of the cutting plane method to solve structural SVMs. Initially, C is empty. Constraints are then added to C that are violated more than the desired precision ζ under the current solution. These constraints, therefore, act as the cutting planes. The algorithm terminates when there is no violated constraint. The following algorithm proposed by Tsochantaridis et al. [31] for 4 is as follows:

Algorithm 2

Initialize: $S = (x_1, y_1), \ldots, (x_n, y_n), c, \zeta$
$C_i \leftarrow \emptyset, \epsilon_i \leftarrow 0, i = 1, \ldots, n$
repeat
  for $i = 1, \ldots, n$ **do**
    $y' \leftarrow \underset{y \in Y}{\operatorname{argmax}} \{l(y_i, y) - a^T[\Psi(x_i, y_i) - \Psi(x_i, y)]\}$
    if $l(y_i, y) - a^T[\Psi(x_i, y_i) - \Psi(x_i, y)] > \epsilon_i + \zeta$ **then**
      $C_i \leftarrow C_i \cup \{y'\}$
      $(a, \epsilon) \leftarrow$ Solution to 4 where $y \in C_i$
    end if
  end for
until no C_i changes
return (a, ϵ)

The above algorithm starts with an empty set $C = \cup_{i=1}^{n} C_i$ and then appends the most violated constraint. Although it solves the optimization problem for SVM in polynomial time but for large datasets, this can prove to be highly inefficient. The algorithm proposed by Joachims et al. [18] uses only one slack variable instead of

n and is called the 1-slack formulation. This reduces computational complexity and time complexity. The algorithm is as follows:

Algorithm 3

1: **Initialize**: $S = (x_1, y_1), \ldots, (x_n, y_n), c, \zeta$
2: $C \leftarrow \emptyset$
3: **repeat**
4: $(a, \epsilon) \leftarrow \underset{a,\epsilon \geq 0}{\text{argmin}} \frac{1}{2} a^T a + c\epsilon$
5: s.t $\frac{1}{n} a^T \sum_{i=1}^{n} [\Psi(x_i, y_i) - \Psi(x_i, y_i')] \geq \frac{1}{n} l(y_i, y_i') - \epsilon, \forall (y_1', \ldots, y_n') \in C$
6: **for** $i = 1, \ldots, n$ **do**
7: $y_i' \leftarrow \underset{y \in Y}{\text{argmax}} \{l(y_i, y) - a^T [\Psi(x_i, y_i) - \Psi(x_i, y)]\}$
8: **end for**
9: $C \leftarrow C \cup \{(y_1', \ldots, y_n')\}$
10: **until** $\frac{1}{n} l(y_i, y_i') - \frac{1}{n} a^T \sum_{i=1}^{n} [\Psi(x_i, y_i) - \Psi(x_i, y_i')] \leq \zeta + \epsilon$
11: return (a, ϵ)

The above algorithm is very similar to the algorithm for n-slack formulation. It constructs a set of constraints C and at each iteration appends the most violated constraint (Line 7 to Line 9). It iterates until no constraint is found to violate the given precision ζ.

2.2 *Core Vector Machines (CVM*

Core vector machines introduced by Tsang et al. [30] formulate the SVM with kernel as a minimum enclosing ball (MEB) problem. Given a set of points S, the minimum enclosing ball, MEB(S), is the ball with a minimum radius that contains S. This minimum ball is unique [15, 32]. Finding the minimum enclosing ball for high dimensional data is hard. Approximation algorithms are proposed by [2, 20], where a $(1 + \epsilon)$ approximate of the MEB is attained. The underpinning concept behind this is the idea of core sets.

Definition 1 (*Minimum Enclosing Ball*) Given a set of points, $S = \{x_1, \ldots, x_n\}$, $x_i \in \mathbb{R}^k$, the minimum enclosing ball centered at c and radius R, $B(c, R)$ is a ball with minimal radius such that $S \subset B(c, R)$. Let the minimal radius R be denoted by $R_{MEB(S)}$.

Definition 2 (*$(1 + \epsilon)$ -approximation of MEB(S)*) Given $\epsilon > 0$,the $(1 + \epsilon)$-approximation of MEB(S) is the ball $B(c, (1 + \epsilon)R)$, where $R \leq R_{MEB(S)}$ and $S \subset B(c, (1 + \epsilon)R)$.

Definition 3 (*Core Set*) Given $\epsilon > 0$, $S' \subseteq S$ is a core set of S if $MEB(S') = B(c, R_{S'})$ and $S \subset B(c, (1 + \epsilon)R_{S'})$.

Tsang et al. [30] conclude a relation between the MEB problem and SVM training, where they develop an approximation algorithm for the training.

Given training data $(x_1, y_1), \ldots, (x_n, y_n)$, where $x_i \in \mathbb{R}^k$ and $y_i \in -1, !$, the SVM training problem is

$$\begin{aligned} \min_{a,b,\rho,\epsilon} \quad & a^T a + b^2 - 2\rho + c \sum_{i,j=1}^{n} \epsilon_i^2 \\ \text{s.t.} \quad & y_i(a^T \phi(x_i) + b) \geq \rho - \epsilon_i, i = 1, \ldots, n \\ & \epsilon_i \geq 0, i = 1, \ldots, n \end{aligned} \tag{6}$$

The dual for this training problem then becomes

$$\begin{aligned} \max_{\lambda} \quad & -\sum_{i,j=1}^{n} \lambda_i \lambda_j \left(y_i y_j + y_i y_j k(x_i, x_j) + \frac{\delta_{i,j}}{c} \right) \\ \text{s.t.} \quad & \sum_{i=1}^{n} \lambda_i = 1 \\ & \lambda_i > 0, i = 1, \ldots, n, \end{aligned} \tag{7}$$

$\delta_{i,j}$ is the Kronecker delta.

Given the training points, the MEB problem is to find a ball of minimum radius such that the training points are contained inside this ball. The corresponding optimization problem then is

$$\begin{aligned} \min_{c,R} \quad & R^2 \\ \text{s.t.} \quad & \|\phi(x_i) - c\|^2 \leq R^2, i = 1, \ldots, n. \end{aligned} \tag{8}$$

The dual for this problem is

$$\begin{aligned} \max_{\lambda} \quad & \sum_{i=1}^{n} \lambda_i k(x_i, x_j) - \sum_{i,j=1}^{n} \lambda_i \lambda_j k(x_, x_j) \\ \text{s.t.} \quad & \sum_{i=1}^{n} \lambda_i = 1, \quad \lambda_i > 0, i = 1, \ldots, n. \end{aligned} \tag{9}$$

Consider kernels such that

$$k(x, x) = k_0. \tag{10}$$

Any kernel k can be normalized to satisfy 10. Gaussian kernels in particular satisfy the above condition. Kernels that satisfy the above property map datapoints to a higher dimension such that the image of these points is of the same length. In other words, these points are mapped to a sphere of radius $\sqrt{k_0}$.

Now the dual of the MEB problem can be modified by dropping the constant term $\sum_{i=1}^{n} \lambda_i k(x_i, x_i) = k_0 \sum_{i=1}^{n} \lambda_i = k_0$. The MEB problem then becomes

$$\begin{aligned} \max_{\lambda} \quad & -\sum_{i,j=1}^{n} \lambda_i \lambda_j k(x_i, x_j) \\ \text{s.t.} \quad & \sum_{i=1}^{n} \lambda_i = 1, \quad \lambda_i > 0, i = 1, \ldots, n. \end{aligned} \tag{11}$$

Define

$$\overline{k}(x_i, x_j) := \left(y_i y_j + y_i y_j k(x_i, x_j) + \frac{\delta_{i,j}}{c} \right). \tag{12}$$

So the modified SVM training problem is

$$\begin{aligned} \max_{\lambda} \quad & -\sum_{i,j=1}^{n} \lambda_i \lambda_j \overline{k}(x_i, x_j) \\ \text{s.t.} \quad & \sum_{i=1}^{n} \lambda_i = 1, \quad \lambda_i > 0, i = 1, \ldots, n. \end{aligned} \tag{13}$$

Therefore, the MEB problem and the SVM training are the same optimization problem. Solving one will solve the other. The MEB problem can be solved using approximation algorithms.

The CVM algorithm starts by initializing the set of core vectors S_0 and ball centered at c_0 with radius R_0. Starting with an arbitrary point in the training set, and find a point that is farthest away from it in the feature space, say x_1, which goes into S_0. Again, find a point farthest away from x_1, x_2 and append this to S_0. R_0 is initialized as

$$R_0 = \frac{1}{2}\sqrt{2\overline{k}(x_1, x_1) - \overline{k}(x_1, x_2)}. \tag{14}$$

Second, find a point x such that x is farthest away from c_0 in the feature space. Update the set S_0 as, $S_1 = S_0 \cup \{x\}$. Distance between the center in iteration t, c_t, and an arbitrary point x_i in the feature space is

$$\|c_t - \overline{\phi}(x)\|^2 = \sum_{x_i, x_j \in S_t} \lambda_i \lambda_j \overline{k}(x_i, x_j) - 2 \sum_{x_i \in S_t} \lambda_i \overline{k}(x_i, x) + \overline{k}(x, x), \tag{15}$$

where λ is the solution to 9. To find the farthest point, probabilistic speedup method is used where a set of size 59 is sampled from the training set. The point that is farthest away from the center is added to S_{t+1}. Point obtained by this method is in the 5% of points in the training set farthest away from the center with 95% probability. The radius is also updated

$$R_{t+1} = \sqrt{\sum_{i=1}^{n} \lambda_i k(x_i, x_j) - \sum_{i,j=1}^{n} \lambda_i \lambda_j k(x, x_j)}. \tag{16}$$

This process is repeated until all training points lie inside $B(c_t, (1+\epsilon)R_T)$.

We use the RBF kernel to conduct our experiments. The RBF kernel is given by

$$k(x, x') := e^{\frac{-\|x-x'\|^2}{2\sigma^2}} \tag{17}$$

3 Results and Discussion

3.1 Segmentation Using SSVM and CVM

The underpinning concept behind segmenting images using SSVM and CVM used in this paper was simple classification. Images were converted to numerical data, and the corresponding labels were generated using the ground truth image segmentation mask. The humongous size of the data thus generated makes it prohibitive to use SVM, which scales to $O(n^3)$ in training time, where n is the number of training points. However, the advantages ensured by margin maximization, the concept on which SVM relies on, cannot be ignored. SSVM and CVM approximate the solution of the QP within a required error. Our results indicate that SSVM and CVM can be implemented on large datasets and still preserve and maintain the accuracy for which it is known for when working with larger datasets. The segmentation problem here is, thus, treated as a classification problem.

We use the various filters like Gabor filters, Canny edge, Scharr edge, Prewitt edge, Robert Cross edge, Sobel edge, and various Gaussian filters for feature selection. Each image is passed through these filters to generate an image endemic data with each filter corresponding to one feature. We use a total of 41 filters to generate the data. The labels are created using the available ground truth image segmentation mask.

A Gabor filter is obtained by the modulation of Gaussian with the sinusoidal [16, 22]. These filters are particularly adroit at dealing with textured and complicated images and provide optimal resolution in both space and spatial frequency domains and are close to mimicking the the mammalian visual perception [6, 14, 29]. The Canny edge filter is a popular tool used to detect edges in an image where a pixel is classified as an edge if in the direction of maximum intensity change, the magnitude of the gradient of that pixel is greater than its adjacent pixels [8, 13], The Scharr edge, Prewitt edge, Robert Cross edge, Sobel edge [9, 19, 26] further augment to the edge detection. The Gaussian filters, with varying variance, are used for noise reduction [11].

3.2 Experimental Dataset Description

We conduct the experiment on two benchmark datasets—Weizmann horses[1] and the CMU-Cornell iCoseg dataset [4, 5][2] to test the two algorithms. The Weizmann horses contain 327 images of horses with their respective ground truth image segmentation masks. The iCoseg datasets contain 38 groups with 643 images in each group. The segmentation was carried out on four images from the Weizmann horses dataset and four images from the iCoseg dataset (two images from two different groups).

3.3 Performance Measure

In order to measure the performance of the two models, the performance measure *accuracy* is used which is defined as follows:

$$\text{Accuracy} = \frac{\text{TP} + \text{TN}}{\text{TP} + \text{FP} + \text{TN} + \text{FN}} \times 100$$

where TP, TN, FP, and FN represent True Positive, True Negative, False Positive, and False Negative, respectively.

3.4 Implementation of Prediction Model

The datasets are divided into training and testing sets in 20% and 80% ratio, respectively. We then conduct hyperparameter tuning using grid search. The parameters to be tuned for SSVM are C. The parameters to be tuned for CVM are C and the RBF kernel parameter σ. For every value of C in a certain range, the SSVM algorithm is run and the accuracy is noted. The value of C at which the accuracy is maximum is taken as the final value for C. The SSVM is then again trained using the optimal value of C. Similarly, for every value of C and σ, CVM is trained. The combination of C and σ that produce the maximum accuracy is taken, and the CVM is trained again using these parameters. We search for the optimal C in $\{2^{-4}, \ldots, 2^{24}\}$ and σ in $\{10^{-3}, \ldots, 10^{5}\}$. SSVM requires precision within which the solution is required. The precision is kept at 0.1 for all datasets. Similarly, CVM requires ϵ. We set $\epsilon = 10^{-5}$ for all datasets. The number of iterations in CVM is solely dependent on the value of ϵ. SSVM also requires a feature map that measures the compatibility between the feature vector and the output vector. We take

[1] https://www.msri.org/people/members/eranb/.

[2] http://chenlab.ece.cornell.edu/projects/touch-coseg/.

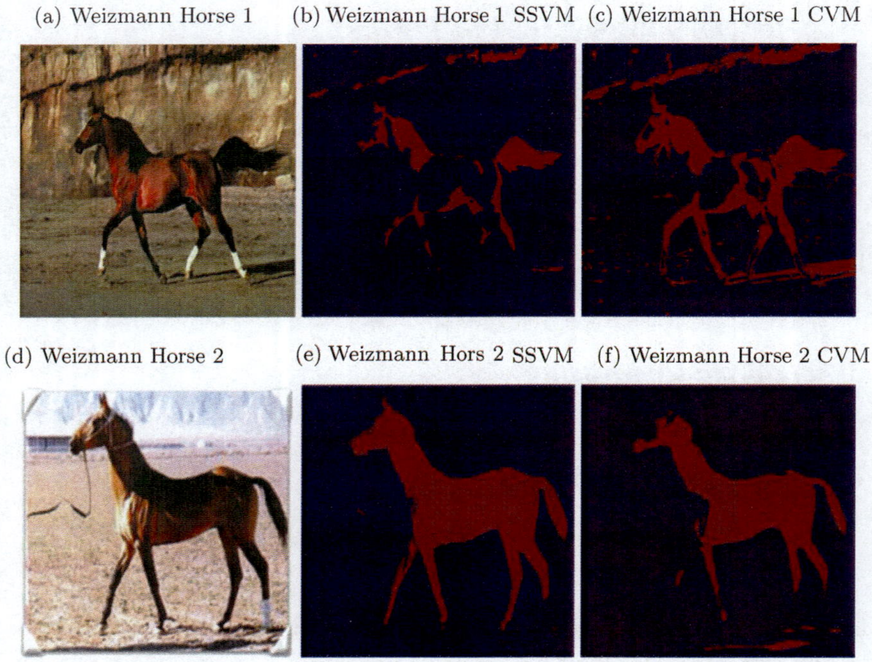

Fig. 1 Weizmann horse dataset

$$\psi(x, y) = \begin{bmatrix} 0 \\ \vdots \\ x \\ \vdots \\ 0 \end{bmatrix} \tag{18}$$

where x is placed at the yth position [18]. This formulation makes the SVM problem equivalent to the variant proposed by Crammer and Singer [10]. All the experiments were done on 3GHz Intel Core i3 with 8 GB RAM. The optimization problems were solved using [17] (Figs. 1 and 2).

3.5 Experimental Results

We conducted experiments on a total of eight images: four from the Weizmann horse dataset and four from the iCoseg dataset. Two images from the pyramids group, two from the statue of liberty group, and four from the Weizmann horse dataset were randomly selected. The complexity of training an SVM classifier comes

(a) Weizmann Horse 3 (b) Weizmann Horse 3 SSVM (c) Weizmann Horse 3 CVM

Fig. 2 Weizmann horse dataset

with an increase in the number of datapoints. But, SSVM and CVM deftly handle this complexity. SSVM employs the cutting plane method and has a training time complexity of $O(n)$. CVM solves the MEB problem to ascertain the solution for SVM training and has $O(n)$ asymptotic training time. Both variants significantly reduce the time complexity from $O(n^3)$ to $O(n)$. This, therefore, makes SVM a viable algorithm for classifying large datasets. The segmented images are presented below

(a) Pyramid 1 (b) Pyramid 1 SSVM (c) Pyramid 1 CVM

Fig. 3 iCoseg dataset

Fig. 4 iCoseg dataset

in three columns. The first column contains the original image, the second column contains the image segmented using SSVM, and the third column contains the image segmented by CVM. The target objects are separated from the background; the target object is colored in red, and the background is colored in blue (Figs. 3 and 4).

Table 1 summarizes the accuracy attained on the training set of each image along with the number of datapoints in the image data.

These results indicate that when there is a clear separation between the object and the background, these variants work well and segment the image with high accuracy. But, when the image contains additional objects in the foreground, like Pyramid 2

Table 1 Accuracy result on testing data and number of datapoints in datasets

Dataset	SSVM (%)	CVM(%)	# of Datapoints
Pyramid 1	98.54	91.48	166,500
Pyramid 2	78.20	80.05	166,500
Statue of liberty 1	97.46	94.36	187,500
Statue of liberty 2	96.69	97.15	187,500
Weizmann horse 1	86.22	88.82	472,000
Weizmann horse 2	96.46	92.07	57,028
Weizmann horse 3	98.63	93.39	376,425
Weizmann horse 4	80.07	86.53	307,200

and Weizmann Horse 2, the accuracy starts falling and segmentation becomes more erratic. Yet, when it comes to images where there is a clear distinction between the target object in the foreground and the background, the accuracy is high.

4 Conclusion and Scope

In this paper, we experimented with various images and segmented them via the classification mechanism using SSVM and CVM. The testing accuracy ranged between 78 and 99%. These algorithms give state-of-the-art accuracy by employing the maximum margin and also reducing the time complexity associated with training SVMs. Larger and more complex datasets can be applied to these variants of SVM to test their limits.

References

1. Abe S (2005) Support vector machines for pattern classification, vol 2. Springer
2. Bādoiu M, Har-Peled S, Indyk P (2002) Approximate clustering via core-sets. In: Proceedings of the thiry-fourth annual ACM symposium on Theory of computing, pp 250–257
3. BakIr G, Hofmann T, Smola AJ, Schölkopf B, Taskar B (2007) Predicting structured data. MIT press
4. Batra D, Kowdle A, Parikh D, Luo J, Chen T (2010) ICOSEG: Interactive co-segmentation with intelligent scribble guidance. In: 2010 IEEE computer society conference on computer vision and pattern recognition. IEEE, pp 3169–3176
5. Batra D, Kowdle A, Parikh D, Luo J, Chen T (2011) Interactively co-segmentation topically related images with intelligent scribble guidance. Int J Comput Vis 93(3):273–292
6. Bovik AC (1991) Analysis of multichannel narrow-band filters for image texture segmentation. IEEE Trans Sig Proc 39(9):2025–2043
7. Boyd S, Vandenberghe L (2007) Localization and cutting-plane methods. From Stanford EE 364b lecture notes

8. Canny J (1986) A computational approach to edge detection. IEEE Trans Pattern Analysis and Machine Intell 6:679–698
9. Chaple GN, Daruwala R, Gofane MS (2015) Comparisons of robert, prewitt, sobel operator based edge detection methods for real time uses on fpga. In: 2015 International conference on technologies for sustainable development (ICTSD). IEEE, pp 1–4
10. Crammer K, Singer Y (2001) On the algorithmic implementation of multiclass kernel-based vector machines. J Mach Learn Res 2(Dec):265–292
11. Deng G, Cahill L (1993) An adaptive gaussian filter for noise reduction and edge detection. In: 1993 IEEE conference record nuclear science symposium and medical imaging conference. IEEE, pp 1615–1619
12. Dhanachandra N, Manglem K, Chanu YJ (2015) Image segmentation using k-means clustering algorithm and subtractive clustering algorithm. Proced Comput Sci 54:764–771
13. Ding L, Goshtasby A (2001) On the canny edge detector. Pattern Recogn 34(3):721–725
14. Dunn D, Higgins WE (1995) Optimal gabor filters for texture segmentation. IEEE Trans Image Proc 4(7):947–964
15. Fischer K, Gärtner B (2004) The smallest enclosing ball of balls: combinatorial structure and algorithms. Int J Comput Geom Appl 14(04n05):341–378
16. Gabor D (1946) Theory of communication. part 1: the analysis of information. J Inst Electr Eng-Part III: Radio Commun Eng 93(26):429–441
17. Gurobi Optimization, LLC (2022) Gurobi optimizer reference manual
18. Joachims T, Finley T, Yu C-NJ (2009) Cutting-plane training of structural SVMS. Mach Learn 77(1):27–59
19. Kumar M, Saxena R et al (2013) Algorithm and technique on various edge detection: a survey. Sig Image Proc 4(3):65
20. Kumar P, Mitchell JS, Yildirim EA (2003) Approximate minimum enclosing balls in high dimensions using core-sets. J Exper Algor (JEA) 8:1
21. Long J, Shelhamer E, Darrell T (2015) Fully convolutional networks for semantic segmentation. In: Proceedings of the IEEE conference on computer vision and pattern recognition, pp 3431–3440
22. Prasad VSN, Domke J (2005) Gabor filter visualization. J Atmos Sci 13:2005
23. Schölkopf B, Smola AJ, Bach F et al (2002) Learning with kernels: support vector machines, regularization, optimization, and beyond. MIT press
24. Schroff F, Criminisi A, Zisserman A (2008) Object class segmentation using random forests. In: BMVC, pp 1–10
25. Seo H, Badiei Khuzani M, Vasudevan V, Huang C, Ren H, Xiao R, Jia X, Xing L (2020) Machine learning techniques for biomedical image segmentation: an overview of technical aspects and introduction to state-of-art applications. Med Phys 47(5):e148–e167
26. Shrivakshan G, Chandrasekar C (2012) A comparison of various edge detection techniques used in image processing. Int J Comput Sci Issues (IJCSI) 9(5):269
27. Song M, Civco D (2004) Road extraction using SVM and image segmentation. Photogrammetric Eng Remote Sens 70(12):1365–1371
28. Sra S, Nowozin S, Wright SJ (2012) Optimization for machine learning. Mit Press
29. Tsai D-M, Wu S-K, Chen M-C (2001) Optimal gabor filter design for texture segmentation using stochastic optimization. Image Vis Comput 19(5):299–316
30. Tsang IW, Kwok JT, Cheung P-M, Cristianini N (2005) Core vector machines: fast svm training on very large data sets. J Mach Learn Res 6(4)
31. Tsochantaridis I, Joachims T, Hofmann T, Altun Y, Singer Y (2005) Large margin methods for structured and interdependent output variables. J Mach Learn Res 6(9)
32. Zhou G, Sun J, Toh K-C (2003) Efficient algorithms for the smallest enclosing ball problem in high dimensional space. Novel Approaches to Hard Discrete Optim 37:173

Identification of Performance Contributing Features of Technology-Based Startups Using a Hybrid Framework

Ajit Kumar Pasayat and Bhaskar Bhowmick

Abstract In the present scenario, a country's economy is positively affected with the growth of the technology-based startups. It is important to identify and understand the relevant features that contributes to the performance of these startup in terms of success or failure. Existing work focuses on predictive behavior and has analyzed the features subjectively. True to our knowledge, none of the existing studies discusses feature identification schemes to highlight the crucial elements contributing to the technology-based startup's performance in terms of success. A framework based on feature correlation, an evolutionary algorithm, and chi-square test is proposed to identify the performance contributing features of the technology-based startup. A publicly available dataset is used for the evaluation of the proposed framework. The identified features match closely with the subjective results of the startup feature analysis studies. Furthermore, on training with the features obtained by the proposed framework with popular machine learning classification techniques, results in remarkable improvement in classification accuracy.

Keywords Start-up · Feature selection · Feature correlation chi-square test algorithm

1 Introduction

Technology-based startups are popular as a new development engine that generates employment and added value to the economy. These startups are also promoting national competitiveness among themselves. As a result, this brings a lot of variations in the products and services. At the present moment, nations all over the globe consider the growth of technology-based startups as a critical policy concern that aims to make policy measures to resuscitate startups and boost companies' innova-

A. K. Pasayat (✉) · B. Bhowmick
IIT Kharagpur, Kharagpur, West Bengal, India
e-mail: akpasayat@iitkgp.ac.in

B. Bhowmick
e-mail: bhaskar@see.iitkgp.ac.in

M. Thakur et al. (eds.), *Soft Computing for Problem Solving*, Lecture Notes in Networks and Systems 547, https://doi.org/10.1007/978-981-19-6525-8_29

tive skills. It is becoming increasingly crucial for the growth and advancement of technology-based startup enterprises that pioneer new markets and invigorate the economy based on original and innovative technologies to ensure dominance in the harsh market rivalry. As a result, it is critical to do systematic research and prepare for the survival and growth of technological startups.

Technology-based startup (TBS) is a youthful, dynamic, adaptive, high-risk organization that delivers unique products or services. These startups are defined upon common consensus than the conventional definition of startups. These ventures are small-scale organizations in the initial phase which are dynamic, adaptive, and capable of taking risks when required. Furthermore, TBS is a firm whose mission is to deliver technological goods or services to the market. These firms provide new technology-oriented goods or services to solve society's fundamental problems. Technology entrepreneurship, according to [1], is an investment in a business that gathers and employs specialist individuals and diverse scientific and technology knowledge-based assets to develop and generate value for the company [1]. As entrepreneurs worldwide try to transform their inventions into new goods and services, many technology-based firms have been developed. According to researchers, TBS's have received much attention in the past two decades as mentioned in [2]. The authors in [3] were the first to investigate the critical success features of TBS. In [4], the authors identified 21 critical features for startup success. Furthermore, the rate of failure of these firms is substantial, implying the presence of other features.

There are several measures of success for TBS, and there is no specific agreement on this part. Entrepreneurs, for example, describe success as the capacity to create new employment and gain personal fulfillment, while investors measure success as the capability to make a profit as mentioned in [2]. A vast amount of literature on the features determining TBS's success has been established in [5]. However, there is a lack of agreement on what these features are selected?. As a result, it is vital to define the critical success features of the TBS to reduce the risks of failure and boost their success. TBS defines firm-level competitiveness as being driven by three components: entrepreneur specific attributes, firm-specific characteristics, and external entrepreneurial environment (ecosystem) related characteristics [6, 7]. Moreover, in recent years, characteristics such as past startup experience and the entrepreneur's entrepreneurial mentality have received much attention. Another essential feature determining the competitiveness of TBS is the entrepreneur's age [7, 8]. Researchers have characterized the benefits of human resources as enabling startups to manage and alleviate finance, marketing, network, and R&D issues [9, 10]. According to research, geographical locations and socioeconomic and cultural considerations impact entrepreneurial activity in developed and developing nations as mentioned in [6, 11].

The above literature indicates that few work has been performed on the identification of the features responsible for the TBS's success. Furthermore, these works are subjective in nature. To the best of our knowledge, there had been little effort to establish a feature identification framework to determine these critical features. This prompted us to establish a feature identification framework for determining the features deciding the TBS's success. The fundamental contribution of this research is the development of a hybrid framework based on feature correlation, an evolutionary algorithm, and the chi-square test to discover critical aspects of the TBS.

The paper is structured into four sections. In Sect. 2, we have described the proposed framework. The results have been discussed in the Sect. 3. The paper has been concluded in the Sect. 4.

2 Proposed Framework

The proposed framework is divided into four phases. In these phases the best possible features are identified. The identified features are then used for finding the performance of the TBS based in terms of success or failure. The phases are defined as: Data Pre-processing, Feature Correlation Analysis- Particle Swarm optimization-based feature identification, chi-square-based feature identification, and intersection.

Phase 1: Data Pre-processing: The first part of the analysis is the pre-processing of the datasets, which is an essential step in developing machine learning models. Pre-processing includes transforming the obtained datasets into an understandable and consistent format. The raw dataset may consist of missing values, error values, or even a few outliers that may generate wrong results without pre-processing. When a model is trained, ambiguity emerges due to unimportant and redundant data; therefore, it is essential to remove the redundant data before model training. Here missing numeric variables were computed using the median of the corresponding data field. The categorical variables were converted to binary values by one-hot encoded. After pre-processing, data is passed to the subsequent phases.

Phase 2: Feature Correlation-PSO-based Feature Identification: In this phase, the Feature Correlation (FC) analysis and Particle Swarm Optimization (PSO) are used together to identify essential features. In Feature Correlation (FC) analysis, the ranking of the features is performed concerning the label provided from the dataset. High correlation among features and low correlation between labels have a detrimental impact on classification performance. Features having lower linear connections with respect to other features and strong linear relations to labels perform better in terms of accuracy [12]. The most important task is to select these features up to a particular rank. This rank is an unknown quantity that requires empirical analysis for its determination. This process involves bias. To solve this problem, the FC is used along with PSO to determine the correct rank for which the classification error is less. PSO is a popular and straightforward evolutionary method [13]. PSO algorithm is selected at this stage for its simplicity in functionality and its ability to converge faster to the optimal solution. Usually, PSO-based feature identification schemes are developed with the aim of identifying a combination of features for which the classification error is minimum [14, 15]. The objective function has two parts. The first part is a feature ranking-based selection scheme. In this experiment, the features are selected using FC analysis. The second part is a classifier that accounts for relevant and accurate predictions by using the selected features. In this experiment, kNN is used as a classifier. The primary goal is to decrease the fitness function as defined in Eq. (1).

$$\text{error} = 1 - \frac{\text{CP}}{\text{VP}}, \tag{1}$$

where CP and VP signify relevant and accurate predictions, respectively. The solution space N comprises particle where each particle comprises of two values: the rank up to the which the features need to be selected and the value of k in kNN classifier. These values are integer values. PSO defines each particle in a swarm with periodic updates as one that acts autonomously while socially contributing to the best of the swarm. Each particle advances in the direction of its best prior position (pbest) as well as the swarm's overall best location (gbest) [13, 15], represented by Eqs. (2–3), respectively.

$$\text{pbest}(j, m) = \arg_k \min[f(P_j(k))],\ j \epsilon \{1, 2, \ldots N\} \tag{2}$$

$$\text{gbest}(m) = \arg_{k,j} \min[f(P_i(k))],\ k = \{1, 2, \ldots N\} \tag{3}$$

where i, j corresponds to ith, jth particle, m, f, V, and P represents the iteration, fitness function, velocity and position. The velocity and position of the particle is updated using Eqs. (4–5), respectively.

$$V_j(m+1) = \omega V_j(m) + c_1 s_1(\text{pbest}(j, m) - P_i(m)) + c_2 s_2 (\text{gbest}(m) - P_j(m))) \tag{4}$$

$$P_j(m+1) = P_j(m) + V_j(m+1) \tag{5}$$

where inertial weight, random variables and the real integer-based accelerating coefficients are represented by W, s_1, s_2, c_1 and c_2, respectively.

Phase 3: Chi–square (X^2) based feature selection: This feature selection measures the dependency strength between independent categorical features and dependent categorical value [16]. The smaller the Chi-square value is, the more independent both features are. So, for a strong relationship between two categorical variables, we need the score comparatively larger as defined by Eq. (6).

$$X_c{}^2 = \frac{\sum(O_i - E_i)}{E_i}, \tag{6}$$

where

$X_c{}^2$ Chi-square score
O Observed value
E Expected value
C degrees of freedom

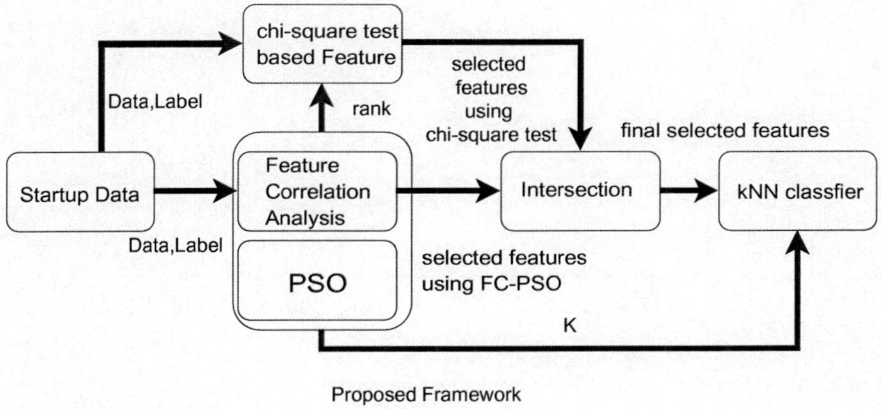

Fig. 1 Block Diagram of the proposed feature identification framework for Technological-based startup's performance

Phase 4: Intersection: The intersection of the features obtained from Chi-square and FC-PSO gives the final selected features. The selected features can be provided to any classifier for performance analysis.The complete workflow of the framework is illustrated in Fig. 1.

3 Results

The dataset used in this research is collected from [17]. The data comprises 472 TBSs and 116 features describing the TBS's characteristics. These features were present in both numerical and category forms. As mentioned in the above section, the data is pre-processed to avoid ambiguity. The categorical data is then transformed into numerical data using the one-hot coding approach. The data have been normalized in the range between $[-1, 1]$ to improve scalability.

The PSO algorithm's population size (N), the cognitive factor, social factor, and the inertial weight are set to 100, 2, 2, and 1, respectively. The FC analysis gives the ranks of the features in the data with respect to the labels. k-Nearest Neighbors (kNN) classifier is used in the objective function to classify the features. The FC-PSO analysis gives a rank of 41 and k as 24. The value of k is set constant throughout the experiment. It means the features ranked up to 41 are selected and then used for further analysis. This rank is also used to retain and select features from the data using the chi-square test-based approach. The intersection of these 41 features provides 14 features that are used for classification. The suggested framework is built upon MATLAB 2020b on an INTEL Core i7 CPU and 32 GB of RAM. Figure 2 shows the convergence curve and minimal fitness values obtained during Phase 2.

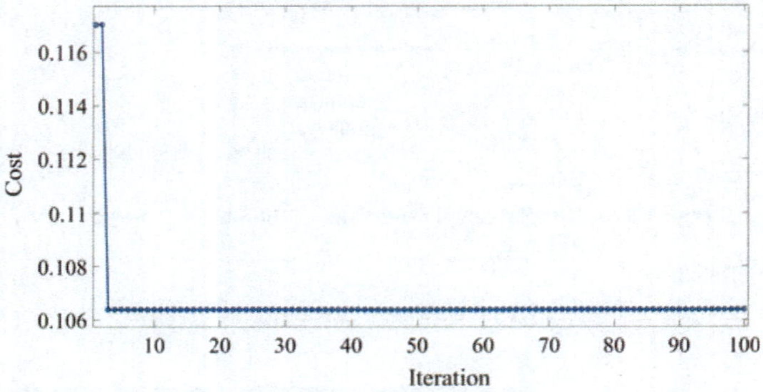

Fig. 2 Convergence curve for FC-PSO feature identification phase

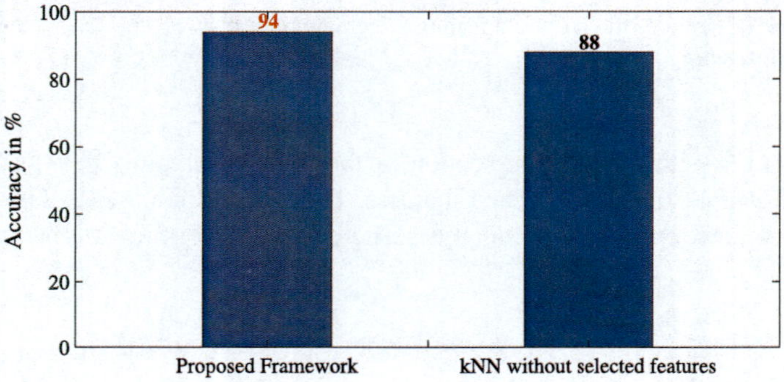

Fig. 3 Comparison in terms of accuracy for the proposed framework and kNN-based classifier implemented on dataset without feature selection

The proposed framework identifies 14 essential features out of 116 features. This obtained sub-dataset is utilized to evaluate TBS's performance in terms of success. Popular machine learning classifiers such as Logistic Regression (LR), Support Vector Machine (SVM), and Decision Tree (DT) have been used for this experiment. In terms of classifier metrics, the proposed framework outperforms machine learning frameworks such as Support Vector Machine (SVM), Decision Tree (DT), and Logistic Regression (LR). The classification results using kNN with and without the selected features are displayed. in Fig. 3. The figure shows that kNN with the given features outperforms KNN without the selected features. In another experiment, the suggested framework's output is compared to the SelectKBest method. The top 14 features were chosen and utilized for classification in SelectKBest, and the outputs were analyzed using LR, SVM, and DT algorithms. It is observed that the features identified using the proposed framework, results in an improvement of around 2–5% in accuracy. The results are presented in Table 1 and Fig. 4, respectively.

It can be observed that 14 highly ranked features that are obtained from the proposed framework match with the ground truth features derived from the stud-

Table 1 Comparison table showing improvement in classification accuracy using different classifiers using the proposed framework and SelectKBest feature selection scheme

Framework	SVM	DT	LR
Proposed	**93.8**	**90.3**	**91.2**
SelectKBest	90.8	89.1	90.1

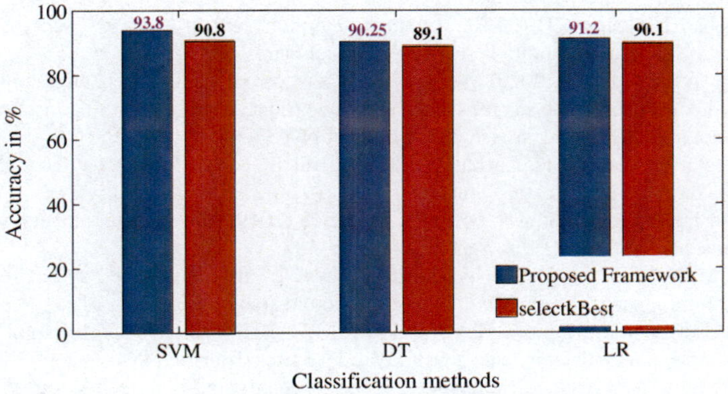

Fig. 4 Comparison in terms of accuracy obtained after classification of selected features obtained using proposed framework and SelectKBest feature selection algorithm for various classifiers

ies [18, 19]. The obtained 14 features *Team Size, No of partner, R&D, Funding Rounds, M&A/IPO, Last Funding at, Entrepreneur background, Service/Product, Social Media Presence, Location, Business Plan, Funding Amount, Percent skill Domain ,Category*, were found to be consistent with the understanding of the features contributing to the TBS success.

4 Conclusion

The proposed hybrid framework for feature identification identifies contributing features for TBS's success. The framework finds 14 out of 116 features, which are validated by literature that defines these aspects subjectively. The classification using the selected features improves the classification using existing feature identification techniques. The classification accuracy by taking these identified features is also higher than the conventional features classification accuracy. The features obtained from this hybrid framework can assist TBS owners and investors in financial-related decision-making and further data analysis.

References

1. Satyanarayana K, Chandrashekar D, Mungila Hillemane BS (2021) An assessment of competitiveness of technology-based startups in India. Int J Glob Bus Competitiveness 16(1):28–38
2. Kim B, Kim H, Jeon Y (2018) Critical success factors of a design startup business. Sustainability 10(9):2981
3. Reynolds P, Miller B (1992) New firm gestation: conception, birth, and implications for research. J Bus Ventur 7(5):405–417
4. Santisteban J, Mauricio D (2017) Systematic literature review of critical success factors of information technology startups. Acad Entrepreneurship J 23(2):1–23
5. Roy S, Modak N, Dan P (2020) Managerial support to control entrepreneurial culture in integrating environmental impacts for sustainable new product development. In: Sustainable waste management: policies and case studies. Springer, pp 637–646
6. Santisteban J, Mauricio D, Cachay O (2021) Critical success factors for technology-based startups. Int J Entrepreneurship Small Bus 42(4):397–421
7. Wiklund J, Nikolaev B, Shir N, Foo MD, Bradley S (2019) Entrepreneurship and well-being: past, present, and future. J Bus Ventur 34(4):579–588
8. Furdas M, Kohn K (2011) Why is start-up survival lower among necessity entrepreneurs? a decomposition approach. In: Workshop on entrepreneurship research, p 24
9. Criaco G, Minola T, Migliorini P, Serarols-Tarrés C (2014) To have and have not: founders' human capital and university start-up survival. J Technol Transfer 39(4):567–593
10. Adler P, Florida R, King K, Mellander C (2019) The city and high-tech startups: the spatial organization of Schumpeterian entrepreneurship. Cities 87:121–130
11. Vuong QH (2016) Impacts of geographical locations and sociocultural traits on the Vietnamese entrepreneurship. SpringerPlus 5(1):1–19
12. Kim JK, Kang S (2017) Neural network-based coronary heart disease risk prediction using feature correlation analysis. J Healthcare Eng 2017
13. De La Iglesia B (2013) Evolutionary computation for feature selection in classification problems. Wiley Interdisc Rev Data Mining Know Discov 3(6):381–407
14. Zhang C, Chan E, Abdulhamid A (2015) Link prediction in bipartite venture capital investment networks. CS224-w report, Stanford
15. Pasayat AK, Bhowmick B (2021) An evolutionary algorithm-based framework for determining crucial features contributing to the success of a start-up. In: 2021 IEEE technology and engineering management conference-Europe (TEMSCON-EUR). IEEE, pp 1–6
16. McHugh ML (2013) The chi-square test of independence. Biochemia medica 23(2):143–149
17. Github-dmacjam/startups-success-analysis: which startups are successful? data analysis. https://github.com/dmacjam/startups-success-analysis. Accessed on 27 April 2022
18. Pasayat AK, Bhowmick B, Roy R (2020) Factors responsible for the success of a start-up: a meta-analytic approach. IEEE Trans Eng Manage
19. Song M, Podoynitsyna K, Van Der Bij H, Halman JI (2008) Success factors in new ventures: a meta-analysis. J Product Innovat Manage 25(1):7–27

Fraud Detection Model Using Semi-supervised Learning

Priya and Kumuda Sharma

Abstract Everything is moving to online platforms in this digital age. The frauds connected to this are likewise rising quickly. After COVID, the amount of fraudulent transactions increased, making this a very essential area of research. This study intends to develop a fraud detection model using machine learning's semi-supervised approach. It combines supervised and unsupervised learning methods and is far more practical than the other two. A bank fraud detection model utilizing the Laplacian model of semi-supervised learning is created. To determine the optimal model, the parameters were adjusted over a wide range of values. This model's strength is that it can handle a big volume of unlabeled data with ease.

Keywords Optimization · Manifold · Laplacian SVM

1 Introduction

Undoubtedly, the banking techniques have evolved over the past few years but it has some cons too. According to a survey by ToI, India loses almost 100 crores to bank frauds everyday which is a serious matter of concern. If an algorithm can follow the activities of a fraudulent transaction, it can be incorporated in financial systems to prevent the user from withdrawing any money until it is approved by a trusted person. The banking sector can then significantly improve economic circumstances and avoid losses caused by fraudulent actions. But it is not easy. Due to the dynamic nature of frauds and also the fact that they have no defined pattern, frauds cannot be easily identified [1]. Several models have been created already harnessing different machine learning methods. Neural networks can be created using the past records of the user and then detecting the possible outlier in it [2]. Unsupervised learning can also be applied on the same for an outlier detection. The problem of fraud detection

Priya (✉) · K. Sharma
ITER College, SOA University, Bhubaneshwar, Odisha, India
e-mail: priya@soa.ac.in

K. Sharma
e-mail: kumudasharma@soa.ac.in

M. Thakur et al. (eds.), *Soft Computing for Problem Solving*, Lecture Notes in Networks and Systems 547, https://doi.org/10.1007/978-981-19-6525-8_30

has a big drawback that it deals with a humongous data set. A solution to that is proposed using the outlier detection algorithm where in order to imitate the real situation, a large number of objects are generated in to simulate the distribution of the original data set, and the distance D is defined to limit the number of outliers to within a few percent of all objects [3]. Other techniques such as Bayesian networks have also been implemented [4]. A feedforward multilayer perceptron approach also proves to be a good one as mentioned in the same. The approach of KNN is also used by Khodabakhshi and Fartash [5] where two data sources are used in the proposed method to separate transactions. Transactions with a fraud suspicion are one of them, while transactions with regular patterns are another.

The major challenge that the fraud card detection algorithms face is the scarcity of data and the fact that real-time data is not available readily [6]. Moreover, since there is no defined pattern in fraud detection, a slight change in the pattern can hamper the adaptability of the model [7].

Given the enormous imbalance in the data, this research suggests a novel method for turning the fraud detection problem into a semi-supervised one. Every fraud detection model has been studied using supervised and unsupervised learning, and the main aim of this approach is to take the middle ground and convert this into a semi-supervised model using Laplacian SVM for putting the correct label. The data has been taken from the Kaggle repository. In each run, it is made sure that some ratio of the labeled data is present, i.e., Class 0(authenticate transaction) and Class 1(fake transaction). This has been done in order to ensure that the unlabeled data gets the most accurate label and that every run gets at least some labeled data since the instances are taken at random. The model is executed several times with each of the kernel to get even more accurate results. The strength of this model is that its outputs can be interpreted as applying to other relevant data sets because it was trained on broad data. In particular, the outcomes can be relied upon to function effectively when applied to new bank transaction data.

The manifold regularization and Laplacian SVM, as well as their mathematical formulations, are briefly discussed in the section that follows. The subsequent section provides explanations of the experimental setup and the model's operation with various parameters. Finally, technical analysis has been provided, and some suggestions for the shortcomings have been made in order to improve this model even more.

2 Methodology

2.1 Working of Laplacian Models

To classify the data more distinctively, the techniques used here consist of Laplacian RLS and Laplacian SVM [8, 9]. To understand what such techniques will do, or what void will these techniques fill, there is a need to get an insight on the kind of data set that is dealt when it comes to semi-supervised learning.

2.2 *Importance of Unlabeled Data in SSL*

Since some amount of data is labeled in a typical SSL setting one might wonder about the significance of unlabeled data. It can be shown with a very trivial example that unlabeled data has the potential to completely change the prediction boundary [10]. Assume that the analyst has access to only the labeled data in an SSL setting and it looks somewhat as shown in Fig. 1.

Using any standard supervised algorithm such as SVM and KNN, the boundary between the data points can be made. Suppose the hyperplane separating the two classes looks as shown in Fig. 2.

Whereas if the unlabeled data is also put into the picture, the whole scenario will be changed. And it can be seen as shown in Fig. 3.

So, it is evident that only the labeled data can produce haphazard results which are nowhere near to the instinct of the data. In this case, the whole geometry of the hyperplane has been changed. To figure out the actual underlying geometry or the distribution of the data, unlabeled data is important. In the subsequent sections, the Laplacian models which incorporate unlabeled data to a large extent have been explained.

Fig. 1 Labeled data

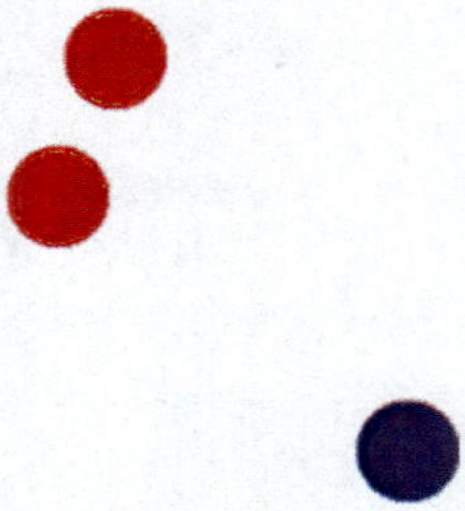

Fig. 2 Separating hyperplane

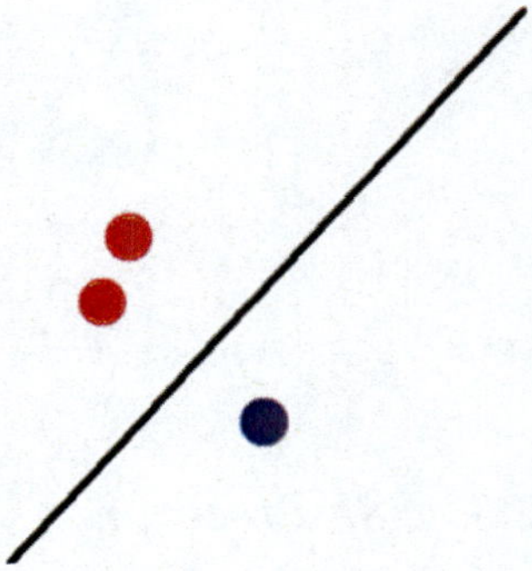

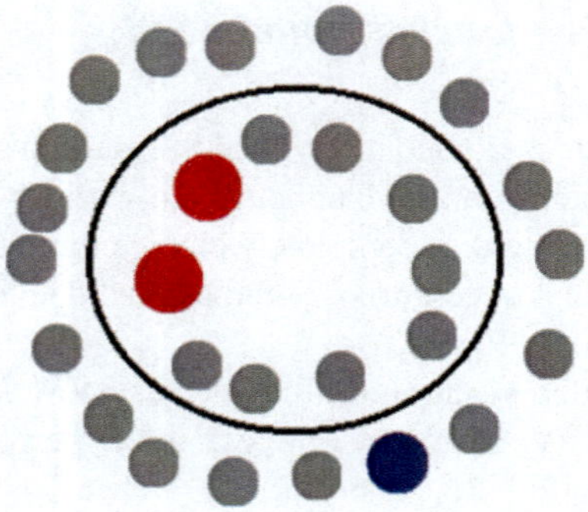

Fig. 3 Actual data

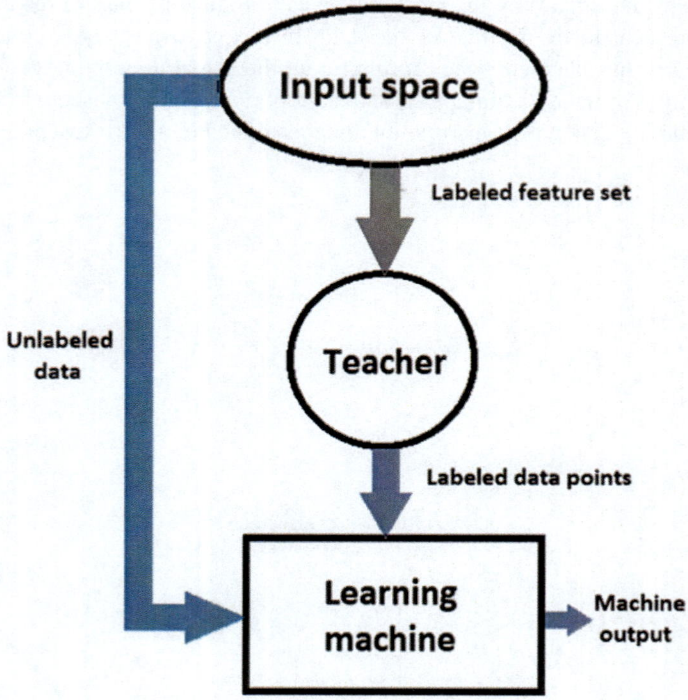

Fig. 4 Model of semi-supervised learning process

2.3 SSL Procedure

SVM is a well-known algorithm used heavily in supervised learning. In semi-supervised, an intrinsic regularization term is associated with SVM to construct an optimisation problem. A general framework of the working of a regularization algorithm is given below in Fig. 4 [11].

Mathematical explanation of the figure is as follows:

- Suppose the data set has n points. This input space consists of two sets of data points, viz., $\{x_i\}_{i=1}^{l}$ (labeled) and $\{x_i\}_{i=l+1}^{n}$ (unlabeled). Both of these are directed by a fixed distribution $p_X(x)$.
- The *teacher* assigns a label d_i to all points of $\{x_i\}_i^l$.
- The *learning machine* produces an output using the combination of both labeled and unlabeled data.

2.4 Assumptions in SSL

For a good classifier and to reach an optimum solution, the following two assumptions are made in mostly all the semi-supervised learning algorithms [12]:

- *Manifold assumption*: It states that a significantly large subset of the data comes from a manifold, which in layman terms is a topological space with some useful properties.
  A **manifold** is a topological space that locally resembles a Euclidean space. To be exact, a manifold is a structure that has the property that each point has a neighborhood homeomorphic to Euclidean space [13].
- *Cluster assumption*: It states that data with dissimilar labels is likely to be apart [14]. Consequently, data points with similar label are supposed to be close to each other. And this most certainly implies that the target function must not change quickly in places where density of data points is high.
  In other words, the function learned from SSL should be *smooth*. This assumption is of great help while constructing the labeling function, because the unlabeled data could instruct the function where to change quickly and where to not.

2.5 Why Manifolds?

Suppose, we have unlabeled data points as $\{x_1, x_2, ...\}$. Say, each data point has dimensionality n [11]. These sample points are represented by points of an n-dimensional Euclidean space. In unsupervised learning, only the ambient space constructed by these unlabeled data points is used.

However, if we are able to construct a manifold with dimension lower than n, such that the real data lies on or around that manifold, then by using the properties of that manifold, we might be able to make a better classifier.

The advantage of this classifier would be that it will have properties of the ambient space as well as the underlying geometric properties of the manifold so obtained. The target function thus obtained will be more successful than the one that would have been constructed using the ambient space only.

2.6 Manifold Regularization

Manifold regularization mainly reduces over-fitting. Moreover, it tries to achieve a well posed target function, and it does so by giving penalty parameters to complex solutions.

Over-fitting refers to constructing a labeling function such that it performs astonishingly well with the training data but does very poor if subjected to some new data set. An over-fitted model does not produce generalized results and hence is of no use when it comes to predicting something completely unseen. Manifold regularization is a modified version of what is called Tikhonov regularization. Under manifold regularization, Tikhonov regularization is applied to reproducing kernel Hilbert space.

A standard Tikhonov problem tries to choose the best fitting function from a hypothesis space of candidate functions $\mathcal{H}$ which is a reproducing kernel Hilbert space [15]. Since, it is RKHS, kernel is attached with each function, so each function has a norm $||f||_N$ which intuitively represents the complexity of the functions present in $\mathcal{H}$. Since, a well posed problem is needed, there is a need to penalize complex functions. And so, here, the complexity parameter is assigned to every function on basis of norm.

A typical Tikhonov problem has a loss function L associated with penalized norm. So, if a the labeled data $(x_i, y_i)_{i=1}^{l}$ is considered, then the problem appears to be as following:

$$f^* = \arg\min_{f \in \mathcal{H}} \frac{1}{l} \sum_{i=1}^{l} L(f(x_i), y_i) + \gamma ||f||_K^2$$

Here, γ is the hyperparameter that administers the use of simpler functions for best fitting curve. Extending this further, to accommodate the unlabeled data as well, an *intrinsic regularization* term is added, which converts this into a manifold regularization problem.

Note: The construction of a *Reproducing kernel Hilbert space* from a Hilbert space is a complex one. In simpler terms, it can be said that if two functions f and g are in RKHS [16], and their norm $||f - g||$ is less in magnitude then the point wise difference, i.e., $|f(x) - g(x)|$ is also less $\forall x$. And for the sake of intuition, RKHS establishes a linear relationship in a Hilbert space of different functions.

2.7 Laplacian SVM

In supervised learning, SVM is the most used machine learning algorithm. SVM, in simple terms tries to maximize the distance between the boundary vectors, majorly known as the *support vectors*, i.e., SVM tries to create a separating line such that the margin between the classes is maximum.

2.8 Mathematical Formulation

The formulation of LapSVM is on the similar grounds of LapRLS where only the distance function has to be changed [17]. While in LapRLS, the minimum squared distance is used, in LapSVM, the hinge loss function is used, i.e., in Tikhonov regularization algorithm, SVM can be incorporated by taking the loss function L as the *Hinge loss function* which is $L(x, y) = \max(0, 1 - yf(x))$. The optimization function for LapSVM can hence be written as

$$f^* = \arg\min_{f \in \mathcal{H}} \frac{1}{l} \sum_{i=1}^{l} \max(0, 1 - y_i f(x_i)) + \gamma ||f||_K^2$$

After adding the *intrinsic regularization* term, the problem statement for LapSVM is obtained

$$f^* = \arg\min_{f \in \mathcal{H}} \frac{1}{l} \sum_{i=1}^{l} \max(0, 1 - y_i f(x_i)) + \gamma ||f||_K^2 + \frac{\gamma_I}{(l+u)^2} f^T L f$$

Now applying the Representer theorem, solution can again be expressed in terms of kernel evaluated at sample points:

$$f^*(x) = \sum_{i=1}^{l} \alpha_i^* K(x_i, x)$$

Now, α can be evaluated by converting the problem into linear, and further, the dual problem will be solved to obtain the following solution:

$$\alpha = (2\gamma_A I + 2\frac{\gamma_I}{(l+u)^2} L K)^{-1} J^T Y \beta^*$$

where, Y: Vector of labels,

K: Kernel matrix (which is calculated as $K_{ij} = K(x_i, x_j)$ for any two data points x_i and x_j,

J: (l+u)×(l+u) block matrix with 0 for unlabeled samples and 1 for labeled samples, i.e., $J = \begin{bmatrix} I_l & 0 \\ 0 & 0_u \end{bmatrix}$,

β^* is the solution of the dual problem:

$$\begin{aligned} &\max_{b \in \mathbb{R}} \textstyle\sum_{i=1}^{l} \beta_i - \frac{1}{2} \beta^T Q \beta \\ &\text{s.t.} \quad \textstyle\sum_{i=1}^{l} \beta_i y_i = 0 \\ &\quad 0 \le \beta_i \le \frac{1}{l} \quad i = 1, 2, \ldots, l \end{aligned} \tag{1}$$

Q is given by

$$Q = YJK(2\gamma_A I + 2\frac{\gamma_I}{(l+u)^2}LK)^{-1}J^T Y$$

The solution can hence be obtained.

3 Proposed Fraud Detection Model

The main aim of a fraud detection model is to first identify the data and then create a model that decides whether a particular transaction is fraudulent or not. So, a fraud detection model can be thought of as a classification problem with two labels 0, if the transaction is fake and 1, if the transaction is authenticated by the user. And there is a Nan value for every unlabeled instance.

3.1 Experimental Setup

The data set used here is taken from the Kaggle repository. The data set contains all credit card transactions performed in September 2013 by the cardholders of Europe. The first few instances of the data set have been shown below in Fig. 5:

Due to confidentiality issues, every single column cannot be explained but some important features such as time taken for each transaction or the amount of money withdrawn in each transaction. Some of the columns are as follows:

- **Number of times the password has been entered.** If an impostor is trying to perform a fake transaction, then the password might have to be entered multiple times because the password is not known to the third party. So, if there are numerous attempts at entering the password, then the transaction must be stopped immediately.
- **Time taken during each transaction** If there is a drastic mismatch in the time taken for every transaction by any specific user, then an issue can be raised.

	Time	V1	V2	V3	...	V27	V28	Amount	Class
0	0.0	-1.359807	-0.072781	2.536347	...	0.133558	-0.021053	149.62	0
1	0.0	1.191857	0.266151	0.166480	...	-0.008983	0.014724	2.69	0
2	1.0	-1.358354	-1.340163	1.773209	...	-0.055353	-0.059752	378.66	0
3	1.0	-0.966272	-0.185226	1.792993	...	0.062723	0.061458	123.50	0
4	2.0	-1.158233	0.877737	1.548718	...	0.219422	0.215153	69.99	0

Fig. 5 First few instances of the data set

- **Amount processed in the transaction.** Looking at the previous records of a user the amount at the time of a new transaction can be compared. If there is a significant difference between both the amounts, then most likely, it is not an authenticated transaction.
- **Failed transaction.** Suppose a user has processed a transaction that involves sending money to another user. If the transaction has been processed at one end, and the money has been withdrawn from that user's account, but the money has not reached the user intended, then it can be put under a fake transaction.The last column is the class to distinguish between the fraudulent and authenticate transactions. Rows belonging to Class 1 represent that the transaction is fake, whereas Class 0 represents an authenticate transaction.

The 28 columns, viz., V1-V28 in the middle have been transformed using the PCA technique [18] of dimensionality reduction. This has been done to ensure that the data is not leaked and misused.

One thing to notice is that there are only 492 Class 1 cases out of the total 284,807 transactions which also includes a large amount of unlabeled ones as well. So it can be seen that there is a huge imbalance between the number of instances of the two classes.

3.2 *Results*

Taking advantage of the fact that the amount of data set is quite large, the data set has been trained on lesser number of points and tested on noticeably more number of data points. The variation of the results produced by choosing different kernels has been shown ahead.

The data set has been trained using the **Polynomial kernel** and by varying degrees. Each one has been run ten times, and then the mean and standard deviation have been calculated. The results are shown below in Table 1.

The results produced are not that well, so a different kernel should be chosen.

The data set has now been trained using the **Linear kernel** and by varying the constant term c. The result on linear kernel with various values of c has been shown below in Table 2.

Table 1 Fraud detection model with polynomial kernel

Kernel degree	Accuracy (%)	Deviation (%)
2	64.22	±4.06
3	57.8	±1.7
4	76	±4.53
5	73.5	±1.22

Table 2 Fraud detection model with linear kernel with different values of c, $\lambda_k = 0.1$, $\lambda_u = 1.7$

c	Accuracy (%)	Deviation (%)
30	64.3	±2.3
50	71.06	±2.01
60	71.05	±2.25
90	60.24	±2.06
110	78.68	±2.39

Table 3 Fraud detection model with RBF kernel with different values of σ, $\lambda_k = 0.1$, $\lambda_u = 1.5$

σ	Accuracy (%)	Deviation (%)
0.01	69.55	±1.57
0.05	72.12	±3.23
0.1	80.9	±5.12
0.2	91.56	±3.81
1	72.18	±1.45

Now, the data set has been trained using the **RBF kernel** and by varying degrees. Each one has been run ten times again, and the mean and standard deviation have been calculated. Here, the variance hyperparameter, viz., σ has also been varied.

The result on RBF kernel with different values of σ has been shown below in Table 3.

4 Conclusion

We have applied the Laplacian model to the fraud detection by converting it into a semi-supervised problem. In the same, we have used different kernels to check the performance of each kernel. The labeled and unlabeled data have been taken randomly in each run, and it is also made sure that a fixed amount of unlabeled data is available at each run.

The program has then been executed 10 times, and the average of all the values of the accuracy thus obtained has been calculated along with the deviations.

It can be seen that the *RBF* kernel has produced the best results. The value of σ which gives the highest accuracy is 0.2, with a standard deviation of almost 4%. Note here that a deviation as large as 5% is also detected here. Also note that in some cases, the accuracy is as high as 91.56%, whereas in some cases, it is as low as 69.55%.

These disparities are bound to happen due to several reasons. The model proposed can perform even better if the following fixes to the limitations are followed:

- **Imbalanced data**: It was told earlier that out of almost 3 lakh cases including the unlabeled data, there are merely 500 frauds which makes the data prone to haphazard results.
  Fix: To overcome this, there are certain techniques which involve normalizing [19] the data without hampering any of the instances. And also, some other metrics such as F1 score can be used to conclude how well the model is performing.
- **Shuffling of the data** The data, being large, has been trained on lesser number of data points, and in order to create promising results, the code should be run for a certain number of times and the mean should be calculated. To include all the variations the model could handle with a certain set of parameters, the data has been shuffled. And shuffling in an already imbalanced data can lead to such results.
  Fix: One solution to this is that some of the data instances can be fixed in the same ratio of positive and negative classes as is presented in the original data set. In the case of semi-supervised, this becomes a bit harder because the unlabeled data has to be included as well but this can be sorted by giving the unlabeled data a dummy label.

The problem of fraud detection should be further scrutinized in order to reduce the losses that occur due to it. This model can be used to save a common man's hard earned money. To produce more promising results, the techniques explained above can be applied to create a better model.

References

1. Raghavan P, El Gayar N (2019) Fraud detection using machine learning and deep learning. In: 2019 international conference on computational intelligence and knowledge economy (ICCIKE). IEEE, pp 334–339
2. Aleskerov E, Freisleben B, Rao B (1997) Cardwatch: a neural network based database mining system for credit card fraud detection. In: Proceedings of the IEEE/IAFE 1997 computational intelligence for financial engineering (CIFEr). IEEE, pp 220–226
3. Hung E, Cheung DW (2002) Parallel mining of outliers in large database. Distrib Parallel Datab 12(1):5–26
4. Maes S, Tuyls K, Vanschoenwinkel B, Manderick B (2002) Credit card fraud detection using Bayesian and neural networks. In: Proceedings of the 1st international naiso congress on neuro fuzzy technologies, vol 261, p 270
5. Khodabakhshi M, Fartash M (2016) Fraud detection in banking using knn (k-nearest neighbor) algorithm. In: International conference on research in science and technology
6. Rafalo M (2017) Real-time fraud detection in credit card transactions. Data Sci Warsaw
7. West J, Bhattacharya M (2016) An investigation on experimental issues in financial fraud mining. In: 2016 IEEE 11th conference on industrial electronics and applications (ICIEA). IEEE, pp 1796–1801
8. Sindhwani V, Niyogi P, Belkin M (2005) A co-regularization approach to semi-supervised learning with multiple views. In: Proceedings of ICML workshop on learning with multiple views, vol 2005. Citeseer , pp 74–79
9. Wu J, Diao YB, Li ML, Fang YP, Ma DC (2009) A semi-supervised learning based method: Laplacian support vector machine used in diabetes disease diagnosis. Interdisc Sci Comput Life Sci 1(2):151–155

10. Ren Z, Yeh R, Schwing A (2020) Not all unlabeled data are equal: learning to weight data in semi-supervised learning. Adv Neural Inf Proc Syst 33:21786–21797
11. Haykin S (2010) Neural networks: a comprehensive foundation. 1999. Mc Millan, New Jersey, pp 1–24
12. Melacci S, Belkin M (2011) Laplacian support vector machines trained in the Primal. J Mach Learn Res 12(3)
13. Loeff N, Forsyth D, Ramachandran D (2008) ManifoldBoost: stagewise function approximation for fully-, semi-and un-supervised learning. In: Proceedings of the 25th international conference on machine learning, pp 600–607
14. Chapelle O, Zien A (2005) Semi-supervised classification by low density separation. In: International workshop on artificial intelligence and statistics. PMLR, pp 57–64
15. Sindhwani V, Rosenberg DS (2008) An RKHS for multi-view learning and manifold co-regularization. In: Proceedings of the 25th international conference on machine learning, pp 976–983
16. Nadler B, Srebro N, Zhou X (2009) Semi-supervised learning with the graph Laplacian: the limit of infinite unlabeled data. Adv Neural Inf Proc Syst 22:1330–1338
17. Gómez-Chova L, Camps-Valls G, Munoz-Mari J, Calpe J (2008) Semi supervised image classification with Laplacian support vector machines. IEEE Geosc Remote Sens Lett 5(3):336–340
18. Sanguansat P (ed) (2012) Principal component analysis: engineering applications. BoD-Books on Demand
19. Blagus R, Lusa L (2010) Class prediction for high-dimensional class-imbalanced data. BMC Bioinf 11(1):1–17

A Modified Lévy Flight Grey Wolf Optimizer Feature Selection Approach to Breast Cancer Dataset

Preeti and Kusum Deep

Abstract Breast cancer is one of the most common and deadly diseases, accounting for 25% of all cancer in women worldwide. Early detection plays a vital role in preventing the disease by providing an appropriate cure. This study uses an enhanced grey wolf optimizer as a wrapper feature selection method to diagnose breast cancer. Grey wolf optimizer (GWO) is a population-based swarm-inspired algorithm well known for its superior performance than other well-established nature-inspired algorithms (NIAs). However, it is prone to local stagnation and low convergence. To boost the local search, a class of random walks called the Lévy flights taken from the Lévy distribution is integrated into the wolf hunting process. Also, a dynamic scaling factor is introduced in the wolf updating position for a high exploration phase in the early stages and high exploitation in the latter stage. To investigate the best identification between the classes, the simulation of Lévy Walk grey wolf-based feature selection (LW-GWO) is recorded for five different machine learning classifiers. The finding shows that LW-GWO has better classification results on breast cancer data than GWO. The results indicate that LW-GWO obtains the best accuracy using Logistic Regression (LR) classifier.

Keywords Grey wolf optimizer · Feature selection problem · Wisconsin breast cancer data · Classification algorithm

1 Introduction

Breast cancer is the most common cancer globally and one of the most frequent reasons for death among women. According to a survey done by World Health Organisation (WHO), nearly 2.3 million women were diagnosed with breast cancer, and there were 685,000 deaths recently. As of year 2020, around 7.8 million women

Preeti (✉) · K. Deep
Indian Institute of Technology Roorkee, Roorkee, Uttrakhand, India
e-mail: preeti@ma.iitr.ac.in

K. Deep
e-mail: kusum.deep@ma.iitr.ac.in

M. Thakur et al. (eds.), *Soft Computing for Problem Solving*, Lecture Notes in Networks and Systems 547, https://doi.org/10.1007/978-981-19-6525-8_31

have been alive and diagnosed with breast cancer in the last five decades. The prediction of breast cancer disease is more prominent in high-income countries and is also rising in countries like Asia, Africa, and Latin America [12]. The estimated age of the at-risk population for breast cancer has been seen in the age around 43–46 years in India. Women aged 53–57 years are likely susceptible to breast cancer [2]. The increase in cancer is due to many factors such as lifestyle and diet habits developed due to genetic mutations, which trigger an uncontrolled growth of cells within breast tissues. A primary tumor is defined as the center where the cancer cell starts to develop. The excessive growth forms a secondary tumor that may spread to other body parts like the lymphatic and immune systems, the blood circulation system, and the hormone regularity [5]. The major symptoms of breast cancer include abnormal bleeding, prolonged cough, unexplained weight loss, and change in bowel movements. Although the disease is unlikely to show any symptoms in the initial stages, an early diagnosis plays an essential step in high survival rates. Many new technologies have been used for the early detection of breast cancer. However, since different patients show different symptoms, it is necessary to characterize the distinct features of the different patients for patient-specific treatment. The accurate diagnosis results must differentiate between the two classes of breast cancer, i.e. benign and malignant tumors.

Several data mining techniques have been used, such as classification, clustering, feature selection, and regression, to understand the hidden patterns and improve classification accuracy correctly. In this regard, K Nearest Neighbor (KNN), Support Vector Machine (SVM), Naive Bayes (NB), Classification Tree (CT), and Logistic Regression (LR) are the most common classification algorithms used to predict breast cancer data. A good classification model provides a low false positive and negative rate. However, using all the features for classification has limitations like computational complexity and long runtime. Choosing the efficient features in the data reduces the complexity of the model and the overfitting problem. Feature selection influences the classification models by eliminating redundant and unnecessary data. For n features, there are $2^n - 1$ possible feature subsets to search in the feature selection method. This problem is considered an NP-hard problem for a large n.

The objective of the proposed work is to search for optimum features using an enhanced Lévy fight-based grey wolf optimizer in feature space such that classification error is minimized. Grey wolf optimizer is a swarm-based nature-inspired algorithm known for its superior performance in NIAs. However, it suffers from an imbalance between the exploitation and exploration search process and premature convergence. Lévy flights are introduced to the nature-inspired optimizer [7] to boost the optimal search. They are defined as the random walks whose step length is drawn from the Lévy distribution. To enrich the local and global optimum search ability, we introduce the Lévy steps and scaling factor in grey wolf optimizer for breast cancer classification problem. The proposed LW-GWO has improved the classification results over five different classifiers.

2 Literature Review

The grey wolf optimizer for features selection was first introduced in [3], which uses a sigmoid function to change the continuous space into binary space. Many improvements have been made in GWO and applied to the breast cancer diagnosis problem. An opposition learning-based GWO was applied to breast cancer classification in [4], but the average accuracy seems to be much less. A hybrid of grey wolf and dragonfly is presented in [10] for breast cancer and heart disease. A mammogram image analysis is performed using the GWO-FS problem in [14]. In [17], a hybrid grey wolf and whale optimizer is used for breast cancer using SVM. An enhanced version of GWO-SVM is proposed for breast cancer diagnosis [8]. A New Sequential and Parallel Support Vector Machine with grey wolf optimizer for breast cancer diagnosis is proposed in [1]. In [15], the breast cancer classification is utilized using LR, SVM and GWO, where the outcomes are assessed with the exactness, accuracy, explicitness, and false positive rate boundaries. A GWO-based neural network is used in [13] for the classification problem. Lévy fights have been shown very effective in enhancing GWO, and there is no experiment seen in the literature for analyzing the Lévy flight-based GWO for breast cancer feature selection problem. Hence, an experiment is carried out on breast cancer data Lévy flight-based GWO and taking five different classifiers into account.

The proposed method is presented in Sect. 3. The discussion on the experimental results is done in Sect. 4. The conclusion and the future research scope are drawn in Sect. 5.

3 Materials and Methods

3.1 Details on Dataset

In this study, the breast cancer data is downloaded from the UCI machine learning repository for evaluating the classification results. The data is named Breast Cancer Wisconsin (Diagnostic) Data Set and are obtained from the records of 569 patient samples. The normal group includes 212 patient cases and is classified as benign, denoted as 'B'. The other class, called malignant, consists of 357 cases of samples out of the total sample and is denoted as 'M'. Each sample has thirty-two features which include patient ID, diagnosis results, and thirty real-valued input features. The ten features out of the total features are computed from a digitized image of a fine needle aspirate (FNA) of a breast mass. This breast mass represents the characteristic of each cell nuclei. The characteristics are described as follows:

(F1) The first feature is the radius of the cell nuclei, which is calculated as the mean of distances from the center to points on the perimeter.

(F2) The second feature taken from the cell nuclei is the texture, defined as the standard deviation of greyscale values.

(F3) The third feature represents the perimeter of the cell nuclei.
(F4) The fourth feature is the area of the cell nuclei.
(F5) The fifth feature is termed smoothness which is the local variation in radius lengths.
(F6) The sixth feature is compactness calculated as $\frac{\text{perimeter}^2}{\text{area}} - 1.0$.
(F7) The seventh feature represents the concavity which is the severity of concave portions of the contour.
(F8) The eighth feature represents the concave points which is the number of concave portions of the contour.
(F9) The ninth feature is symmetry.
(F10) The tenth feature is the fractal dimension which is calculated as the coastline approximation -1.

The rest of the features are found using the mean, the standard error, and the worst or the largest of these features for each image. For example, F3 is the mean perimeter, F13 is the perimeter standard error, and F23 is the worst perimeter. All feature values are recoded with four significant digits with no missing attribute.

3.2 *Grey Wolf Optimization (GWO)*

Grey wolf optimization (GWO) is the swarm intelligence optimization technique which was first introduced in [9]. It is inspired by the leadership hierarchy and hunting process of the grey wolf in nature. The simple mechanism of GWO makes it easy to implement over other NIAs. Also, it has fewer decision variables, less storage required, and does not possess any rigorous mathematical equations of the optimization problem. Thus, applying GWO in selecting features can ease the classification problem. Muro et al. [11] explained the hunting behavior of wolf into three stages as:

1. **Social Hierarchy**: The social hierarchy of grey wolf has four levels: alpha α, beta β, delta δ, and omega ω. The leaders are responsible for decision-making and is denoted as the alpha wolf. The second level called beta wolf works as a helping hand to the alpha for any activity. In the third level, the delta wolf is placed, which plays the role of scapegoat in grey-wolf packing. The rest of the wolves are categorized as omega wolf and are dominated by all other wolves.
2. **Encircling the Prey**: The encircling process is given by the following mathematical equation:

$$D = \overrightarrow{C} \cdot \overrightarrow{X_{p,t}} - \overrightarrow{X_t} \tag{1}$$

$$\overrightarrow{X_{t+1}} = \overrightarrow{X_{p,t}} - \overrightarrow{A} \cdot D \tag{2}$$

where $\overrightarrow{A}$ and $\overrightarrow{C}$ are coefficient vectors, $\overrightarrow{X}_{p,t}$ denotes the position vector of the prey at current iteration t, and $\overrightarrow{X_{t+1}}$ denotes the position vector of a grey wolf at next iteration.
The vectors are determined as:

$$\overrightarrow{A} = 2a \cdot r_1 - a \tag{3}$$

$$\overrightarrow{C} = 2r_2 \tag{4}$$

The vector a is a linearly decreasing function from 2 to 0 and r_1 and r_2 are random vectors in [0,1].

3. **Hunting**: To encircle the position of prey, and the wolf position is approximated by the average of alpha, beta, and gamma wolf positions, the below equations are used:

$$D_\alpha = C_1 \cdot X_\alpha^d - X_t^d \tag{5}$$

$$D_\beta = C_2 \cdot X_\beta^d - X_t^d \tag{6}$$

$$D_\delta = C_3 \cdot X_\delta^d - X_t^d \tag{7}$$

$$X_1^d = X_\alpha^d - A_1 \cdot D_\alpha \tag{8}$$

$$X_2^d = X_\beta^d - A_2 \cdot D_\beta \tag{9}$$

$$X_3^d = X_\delta^d - A_3 \cdot D_\delta \tag{10}$$

The position of prey is estimated as:

$$X_{t+1} = \frac{X_1 + X_2 + X_3}{3} \tag{11}$$

However, the conventional GWO has an insufficient diversification of the wolf, which can lead to pre-mature convergence. To improve search ability, a Lévy flight along with a scaling factor is introduced in the hunting mechanism of GWO.

3.3 *Grey Wolf Based on Lévy Flight Feature Selection Method*

In GWO, the search agents update their position according to alpha, beta, and gamma wolf according to Eq. (11). This inclines the search process to converge into a local optima. To elevate the local stagnation problem, the wolf position is updated using Lévy based hunting pattern. Lévy flight is scale-free walks which are randomly drawn from a heavy-tailed distribution. The concept of Lévy flight was introduced in [6, 16], which describes the foraging behavior of the animals can be interpreted based on Lévy flight patterns. In order to maintain a balance between the exploitation and exploration in the algorithm, a scaling factor S is employed. The new position prey is updated as:

$$X_{t+1} = X_{t+1} + S * \mathrm{LF}(R) * \left(\frac{X_1 + X_2 + X_3}{3} - X_{t+1} \right) \tag{12}$$

The term in the bracket corresponds to the difference between the position of the current wolf and the position obtained by the best three wolves. This tells the dissimilarity between the solution quality and helps in discovering the new solutions around the best solutions. The scaling factor S maintains the search process balanced and is given as,

$$S = \frac{T - (t - 1)}{T} \tag{13}$$

and LF is Lévy flight jumps controlling the newly created solutions taken from Lévy distribution with a large step as:

$$L(s) \sim |s|^{-1-\beta}, \quad 0 < \beta \leq 2, \tag{14}$$

where s is the variable and β is the Lévy index for controlling the stability. This has an infinite variance with an infinite mean. Obviously, these step sizes are not trivial and need to be reformed according to the given space. In [18, 19], a simple scheme is described for Lévy flight as:

$$L(s) = 0.01 * \frac{u}{v^{1/\beta}} \tag{15}$$

u and v are taken from normal distribution as:

$$u \sim N(0, \sigma_u^2), \quad v \sim N(0, \sigma_v^2) \tag{16}$$

with,

$$\sigma_u = \frac{\tau(1+\beta)(\sin(\beta/2)}{\tau[(1+\beta)/2]\beta 2^{(\beta-1)/2}}^{\frac{1}{\beta}}, \quad \sigma_v = 1, \tag{17}$$

The LW-GWO starts with an initial population of size ($N * D$), N is the number of wolf and D is the dimension. Each vector in the population represents feature indices. A sequential step of the method is presented in Fig. 2. The fitness function is defined as:

$$f = 0.99 * E_s + 0.01 * \frac{|\mathrm{SF}|}{|\mathrm{TF}|} \tag{18}$$

where E_s is the classification error of the considered classifier, SF is the total number of features selected, and TF is the total number of feature. If the fitness of the current iteration is found to be better than the previous iteration, the wolf update the position, and it is set to be 1 or 0 using the following threshold:

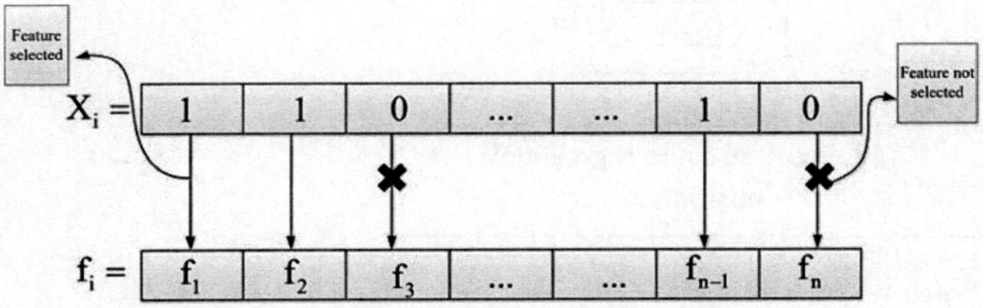

Fig. 1 A solution representation of selecting feature

$$f(X) = \begin{cases} 1, & \text{if } \Theta \geq 0.5 \\ 0, & \text{if } \Theta \leq 0.5 \end{cases} \tag{19}$$

A feature ith is selection if the corresponding ith coordinate of X_α is 1 else not selected as shown in Fig. 1.

4 Experimental Results

All the numerical experiments are performed using MATLAB 2021b. The values of the parameters are given in Table 1. The training data consist of 80% data, while 20% is used for the test with fivefold cross-validation. Due to the proposed method's stochastic nature, the simulation is done for twenty runs and average of obtained results is recorded.

4.1 Performance Evaluation

In order to investigate the efficacy of LW-GWO, the following evaluation criteria are taken: accuracy, precision, fitness, specificity and sensitivity, feature size, time, and area under the ROC curve (AUC). A set of five different classifiers is used to determine which one provides the best results in identifying the benign and malignant

Table 1 Parameter settings

Data	Points
Population size	10
Maximum iteration	200
Runs	20

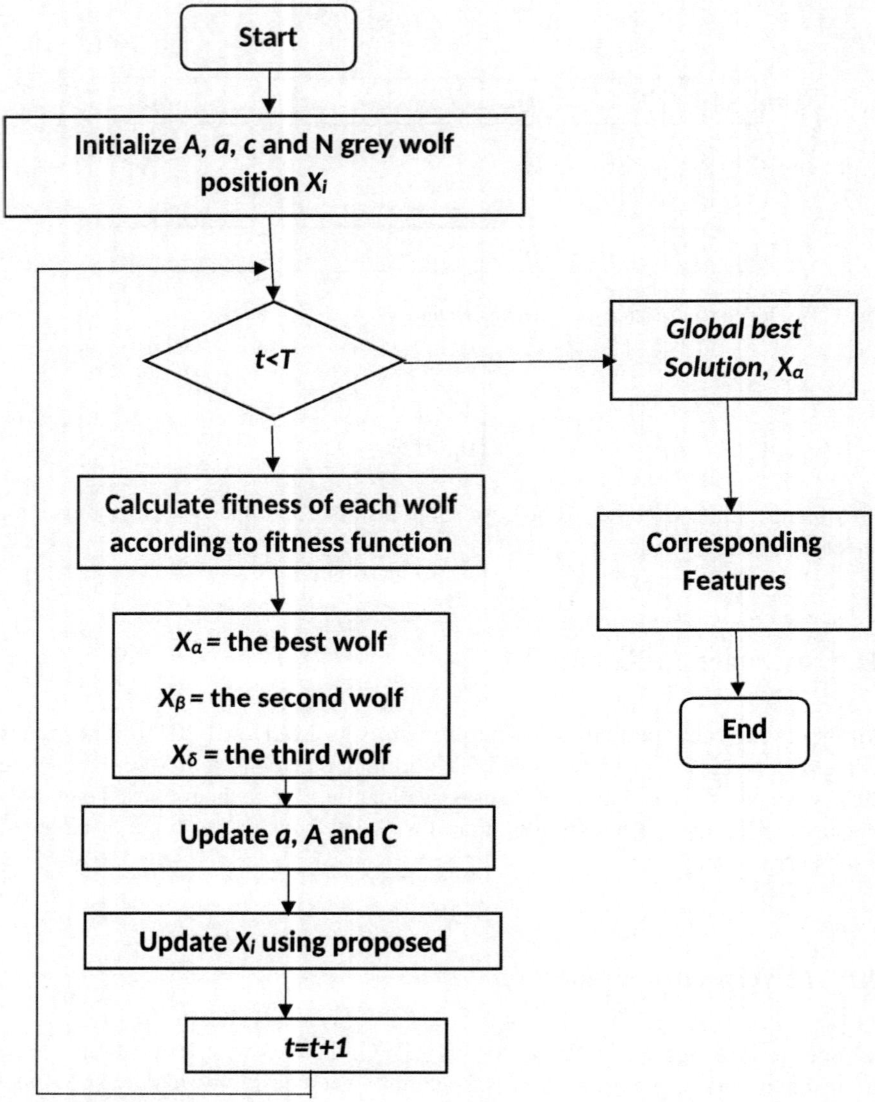

Fig. 2 Flow chart of LW-GWO

Table 2 Accuracy, fitness, and precision

Method	Accuracy		Fitness		Precision	
	GWO	LW-GWO	GWO	LW-GWO	GWO	LW-GWO
KNN	93.90	**94.65**	0.0496	**0.0415**	0.9395	**0.9602**
SVM	**93.81**	93.72	**0.0429**	0.0461	0.9466	**0.9510**
NB	93.01	**93.15**	0.0478	**0.0370**	**0.9435**	0.9236
CT	92.22	**93.50**	0.0512	**0.0324**	0.9058	**0.9162**
LR	94.70	**95.49**	0.0482	**0.0415**	0.9443	**0.9594**

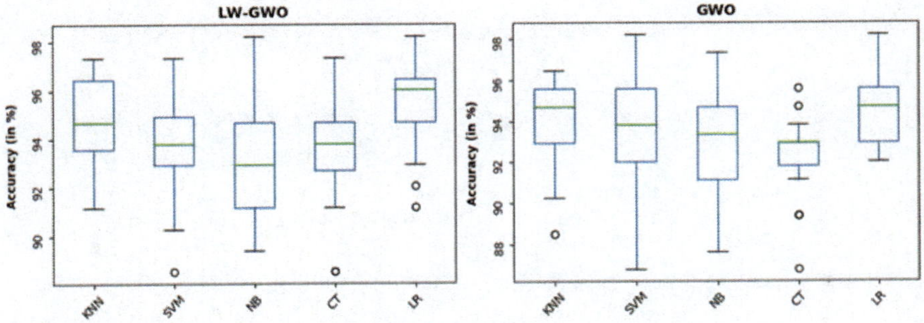

Fig. 3 Boxplot of accuracy results on each classifier between LW-GWO versus GWO

classes using the selected features. The value of classification accuracy, fitness value, and precision of LW-GWO in comparison with GWO is shown in Table 2 for each classifier. The best classification accuracy is obtained by LW-GWO-LR with 95.49% following by LW-GWO-KNN with 94.65%, GWO-SVM with 93.81%, LW-GWO-CT with 93.5%, and LW-GWO-NB with 93.15%. A distribution of classification accuracy over 20 runs for each classifier is compared between GWO and LWGWO in Fig. 3. The observation indicates that classifier CT and LR obtained few of the extreme accuracy values for the method LW-GWO and GWO. The lower the fitness value, the better is the classification model. It is observed that out of five classifiers , four have much lower fitness value when the LW-GWO approach is applied to the feature selection method. Precision is an important measure to predict the total true positive class from the overall estimated positive class for an imbalance classification which lies between 0 and 1, with zero as the worst value and one as the best value. As can be seen, feature selection using the LW-GWO method has a much closer value to 1 than GWO when tested on BCD. To test the number of cases with malignancy and the number of cases without any malignant tumor, sensitivity and specificity values are recorded in Table 3. The values obtained are better for LW-GWO feature selection problem. The area under the curve (AUC) is a metric between 0 and 1 used to calculate the overall performance of a classification model. A high value indicates the better performance value which LW-GWO better yields. Figure 4 shows the comparison of

Table 3 Sensitivity, specificity, and AUC

Method	Sensitivity		Specificity		AUC	
	GWO	LW-GWO	GWO	LW-GWO	GWO	LW-GWO
KNN	**0.8953**	0.8941	0.9648	**0.9775**	0.9301	**0.9358**
SVM	**0.8846**	0.8774	0.9698	**0.9726**	**0.9272**	0.9250
NB	0.8667	**0.8905**	**0.9677**	0.9557	0.9172	**0.9231**
CT	0.8858	**0.9108**	0.9437	**0.9493**	0.9147	**0.9301**
LR	0.9120	**0.9179**	0.9677	**0.9768**	0.9398	**0.9474**

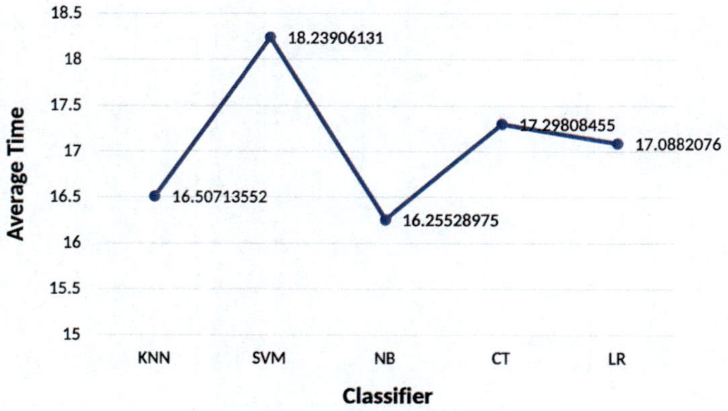

Fig. 4 Average time taken by LW-GWO

average time taken by LW-GWO by different machine learning classifier, in which NB takes the minimum time (Fig. 2).

To analyze the outcome of LW-GWO feature selection method, there are four possible cases:

True Positives (TP) Output of predicted positives that is positive.
False Positives (FP) Output of predicted positives that is negative.
False Negatives (FN) Output of predicted negatives that is positive.
True Negatives (TN) Output of predicted negatives that is negative.

The average number of TP, FP, TN, and FN is calculated in Table 4. The results show that the number of predicted outcome for each class is achieved better by LW-GWO than GWO.

4.2 *Relevant Feature Selected*

On the other hand, the most important features obtained using the GWO and LW-GWO are noted in Table 5. These subset features are generated by taking the top 70

Table 4 Average true positive, true negative, false positive, and false negative

Classifier	LW-GWO				GWO			
	TP	FP	TN	FN	TP	FP	TN	FN
KNN	69.4	37.55	1.6	4.45	68.5	37.6	2.5	4.4
SVM	69.05	36.85	1.95	5.15	68.85	37.15	2.15	4.85
NB	67.85	37.4	3.15	4.6	68.7	36.4	2.3	5.6
CT	67.4	38.25	3.6	3.75	67	37.2	4	4.8
LR	69.35	38.55	1.65	3.45	68.7	38.3	2.3	3.7

Table 5 Selected features

Classifier	Method	Selected feature
KNN	GWO	F1, F2, F11, F12, F21, F26, F27
	LW-GWO	F1, F11, F20, F21, F27, F28, F29
SVM	GWO	F1, F21, F26, F27
	LW-GWO	F1, F10, F11, F12, F16, F19, F21, F25, F26, F27, F28, F29
NB	GWO	F1, F11, F12, F21, F26, F27, F29
	LW-GWO	F1, F5, F7, F11, F13, F21, F26, F27, F29
CT	GWO	F1, F3, F11, F12, F21, F26, F27
	LW-GWO	F1, F8, F10, F11, F12, F13, F15, F17, F21, F25, F26, F27
LR	GWO	F1, F11, F12, F21, F23, F26, F27
	LW-GWO	F1, F5, F8, F9, F12, F20, F21, F26, F27

% features that occurred in 20 runs by FS algorithm. The final subset of features selected using GWO comprises F1, F2, F3, F11, F12, F21, F23, F26, F27, and F29, whereas the subset of features obtained using LW-GWO includes F1, F5, F7, F8, F9, F10, F11, F12, F13, F15, F16, F17, F19, F20, F21, F25, F26, F27, F28, and F29. It can be observed that F1-'radius', F21-'worst radius', and F27-'worst concavity' are the unique feature attained by all the considered classifiers for both the methods. It implies that the features of the mean radius, the largest radius, and the largest concavity play an important role as the features for accurately differentiating the two classes. Although the average feature size of LW-GWO is more than GWO as shown in Fig. 5, the evaluation results show LW-GWO outperforms GWO.

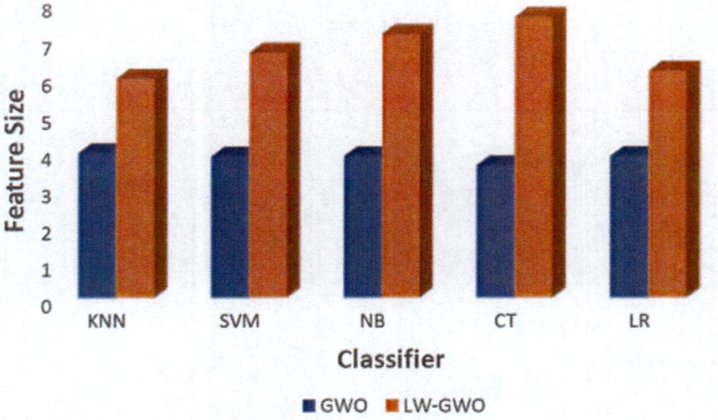

Fig. 5 Average feature size

5 Conclusion

Identifying the benign and malignant classes more accurately helps in early treatment. In this paper, an improvised GWO termed as LW-GWO is applied to the feature selection method to increase breast cancer classification accuracy. The optimal features were selected using the five machine learning classifiers: KNN, SVM, CT, LR, and NB. The best results were obtained from the experiment done so far by LW-GWO using LR classifier to find an efficient feature subset. The proposed method not only uplifts the local stagnation but also balances the exploration and exploitation throughout the search process. In addition, different performance metric like true positive and false positive is calculated for each class, showing the much better results attained by LW-GWO.

In the future, an experiment can be performed by taking a different number of training and testing data samples and comparing which one obtained the best results. Further, investigation can also be expanded on predicting another type of disease like heart disease or kidney disease.

References

1. Badr E, Almotairi S, Salam MA, Ahmed H (2022) New sequential and parallel support vector machine with grey wolf optimizer for breast cancer diagnosis. Alexandria Eng J 61(3):2520–2534
2. Chaurasia V, Pal S (2017) A novel approach for breast cancer detection using data mining techniques. Int J Innov Res Comput Commun Eng 2
3. Emary E, Zawbaa HM, Hassanien AE (2016) Binary grey wolf optimization approaches for feature selection. Neurocomputing 172:371–381
4. Hans R, Kaur H (2020) Opposition-based enhanced grey wolf optimization algorithm for feature selection in breast density classification. Int J Mach Learn Comput 10(3):458–464

5. Howlader KC, Das U, Rahman M (2019) Breast cancer prediction using different classification algorithm and their performance analysis. Int J Recent Eng Res Dev (IJRERD) 4(2):2455–8761
6. Humphries NE, Queiroz N, Dyer JR, Pade NG, Musyl MK, Schaefer KM, Fuller DW, Brunnschweiler JM, Doyle TK, Houghton JD (2010) Hays: environmental context explains Lévy and Brownian movement patterns of marine predators. Nature 465(7301):1066–1069
7. Kamaruzaman AF, Zain AM, Yusuf SM, Udin A (2013) Lévy flight algorithm for optimization problems—a literature review. In: Applied mechanics and materials, vol 421. Trans Tech Publications Ltd., pp 496–501
8. Kumar S, Singh M (2021) Breast cancer detection based on feature selection using enhanced Grey Wolf optimizer and support vector machine algorithms. Vietnam J Comput Sci 8(02):177–197
9. Mirjalili S, Mirjalili SM, Lewis A (2014) Grey wolf optimizer. Adv Eng Softw 69:46–61
10. Moturi S, Rao ST, Vemuru S (2021) Grey wolf assisted dragonfly-based weighted rule generation for predicting heart disease and breast cancer. Comput Med Imaging Graph 91:101936
11. Muro C, Escobedo R, Spector L, Coppinger RP (2011) Wolf-pack (*Canis lupus*) hunting strategies emerge from simple rules in computational simulations. Behav Proc 88(3):192–197
12. Osman AH (2017) An enhanced breast cancer diagnosis scheme based on two-step-SVM technique. Int J Comput Sci Appl 8(4):158–165
13. Pal SS (2018) Grey wolf optimization trained feed foreword neural network for breast cancer classification. Int J Appl Ind Eng 5(2):21–29
14. Sathiyabhama B, Kumar SU, Jayanthi J, Sathiya T, Ilavarasi AK, Yuvarajan V, Gopikrishna K (2021) A novel feature selection framework based on grey wolf optimizer for mammogram image analysis. Neural Comput Appl 33(21):14583–14602
15. Senthil T, Deepika J, Nithya R (2021) Detection and classification of breast cancer using improved Grey Wolf algorithm. IOP Conf Ser Mater Sci Eng 1084(1):012023
16. Sims DW, Southall EJ, Humphries NE, Hays GC, Bradshaw CJ, Pitchford JW, James A, Ahmed MZ, Brierley AS, Hindell MA, Morritt D (2008) Scaling laws of marine predator search behaviour. Nature 451(7182):1098–1102
17. Singh I, Bansal R, Gupta A, Singh A (2020) A hybrid grey wolf-whale optimization algorithm for optimizing SVM in breast cancer diagnosis. In: 2020 Sixth international conference on parallel, distributed and grid computing. IEEE, pp 286–290
18. Yang XS, Deb S (2010) Engineering optimisation by cuckoo search. Int J Math Model Numer Optimisation 1(4):330–343
19. Yang XS (2010) Engineering Optimization: an introduction with metaheuristic applications. Wiley

Feature Selection Using Hybrid Black Hole Genetic Algorithm in Multi-label Datasets

Hitesh Khandelwal and Jayaraman Valadi

Abstract In multi-label classification, each instance is assigned to a group of labels. Due to its expanding use in applications across domains, multi-label classification has gained prominence in recent years. Feature selection is the most common and important preprocessing step in all machine learning and data mining tasks. Removal of highly correlated, irrelevant, and noisy features increases the performance of an algorithm and reduces the computational time. Recently, a Black Hole metaheuristic algorithm, inspired by the phenomenon of Black Holes, has been developed. In this study, we present a modified standalone Black Hole algorithm by hybridizing the standard Black Hole algorithm with two genetic algorithm operators, namely crossover and mutation operators. We carried out experiments on several benchmarking datasets across domains. The modified feature selection algorithm was also employed on a multi-label epitope dataset. The results were compared with the standalone Black Hole and other most common evolutionary algorithms such as Ant Colony, Advance Ant Colony, genetic algorithms, and Binary Gravitational Search Algorithm. The results showed that our improved hybrid algorithm outperforms the existing evolutionary algorithms on most datasets. The synergistic combination of Black Hole and Genetic Algorithms can be used to solve multi-label classification problems in different domains.

Keywords Black hole · Genetic algorithms · Feature selection

1 Introduction

Feature selection is the process of selecting a subset of the most relevant feature set. Feature selection approaches are broadly classified into three classes of algorithms: filter, wrapper, and embedded algorithms [1]. Filter algorithms work independently of the learning algorithm and rank all the features using a specific criterion (usually a

H. Khandelwal · J. Valadi (✉)
Vidyashilp University, Bangalore, India

H. Khandelwal
e-mail: hitesh.khandelwal@vidyashilp.edu.in

M. Thakur et al. (eds.), *Soft Computing for Problem Solving*, Lecture Notes in Networks and Systems 547, https://doi.org/10.1007/978-981-19-6525-8_32

statistic criterion). The top-ranked features are considered most relevant, while all the bottom-ranked attributes are removed. Filtering algorithms are generally employed in datasets with high dimensions due to their relatively higher speed. Mutual information, Chi-square statistics, Pearson correlation, maximum relevance, and minimum redundancy are among the most popular filter ranking algorithms available in the literature. Unlike the filtering algorithms, wrapper methods are learning algorithm-dependent methods wherein an exhaustive search is performed to select the features with the best predictive power. Therefore, the classification algorithm has to be employed repeatedly in these methods. The performance of wrapper methods is usually better than filter methods, but this comes at the cost of higher computational time and resources. Some of the conventional wrapper algorithms include forward and backward elimination algorithms. Several stochastic and nature-inspired algorithms have recently become very popular for selecting the most informative features. These algorithms include simulated annealing, genetic algorithms, Ant Colony Optimization, Particle Swarm Optimization, Gravitational Search, and Black Hole algorithm. The feature selection process is embedded within the classification algorithm during the learning phase in the embedded feature selection methods. These methods are relatively less expensive. Examples of embedded feature selection algorithms include support vector machines (SVM), recursive feature elimination, and random forest mean decrease in GINI.

1.1 Multi-label Classification

In single-label classification tasks, each input instance is mapped to only one label from all possible labels in the data. Datasets in which each instance is mapped to a group of labels simultaneously are called multi-label datasets. Elimination of redundant, correlated, and noisy features is an important preprocessing step in both single-label and multi-label classification tasks. Multi-label classification problems are critical in different domains of science and engineering [2]. Some real-life label classification examples include hate speech classification [3], multi-label image classification [4], multi-label classification of music into emotions, etc. [5]. Multi-label classification problems are usually solved by employing two approaches: problem transformation and algorithm adaptation. In problem transformation approach, the multi-label classification problems are transformed into multiple single-label classification problems. Binary relevance and label powerset are common examples of problem transformation methods. In algorithm adaptation strategy, the classification algorithms are modified to adapt to the multi-label scenarios. Examples of this category of algorithms include multi-label K nearest neighbors (MLKNN) and Bayesian classifier.

Evaluating all feature subsets to find their relevance or predictive power is an expensive operation. As data scales in terms of the number of features, it becomes infeasible to perform feature selection using the brute force methods. Employing heuristics for feature selection represents an attractive alternative. These methods use only a subset of all possible feature combinations to obtain near-global optimal solutions. This study has employed an improved Black Hole algorithm hybridized with genetic algorithms for multi-label feature selection.

2 Related Works

Several metaheuristic algorithms inspired by the real-world phenomenon have been employed in recent literature to solve feature selection problems in single-label and multi-label datasets. We compared our work with some popular heuristic algorithms, namely Advanced Binary Ant Colony Optimization (ABACO) [6], Binary Gravitational Search algorithm (BGSA) [7], and Ant Colony Optimization (ACO) [8, 9], and genetic algorithms (GA) [10].

ACO: Ant Colony Optimization is inspired by the real-world behavior of ants. Ants deposit pheromones and are attracted to pheromones while moving from the nest to the food source and vice versa. Due to this indirect communication facility, they can optimize their movements. Artificial ant algorithms mimic the real-life behavior of ants to solve optimization problems, including feature selection.

ABACO: Advanced Binary Ant Colony Optimization is an improvement of the ACO algorithm. In this algorithm, the attributes are considered graph nodes. All the features are fully connected and there are two sub-nodes for every node. One sub-node is for selecting an attribute, while the other sub-node carries out the task of deselecting an attribute. A software ant visits all the nodes and completes the tour. The results are compiled as a binary vector with selected/deselected feature information.

GA: Genetic algorithms are inspired by the phenomenon of natural evolution and natural selection. In GA, we start with a population of trial solutions, and in each iteration, the GA operator's natural selection, crossover, and mutation are applied. The process is stopped when a suitably selected termination criterion is satisfied.

BGSA: Binary Gravitational Search algorithm derives its inspiration from the laws of gravitation and mass interactions. Here, each subset is represented as the mass and the interaction between these subsets are simulated based on Newton's laws of gravitation. In each iteration, the position of the solution, its velocity, and mass are updated based on the universal law of gravitation. Thus, the solutions are evolved until a termination criterion is met.

3 Proposed Method

Recently, a novel metaheuristic algorithm inspired by the real-life behavior of the Black Hole and stars has been proposed for selecting the most informative attributes [11]. This algorithm is simple but very effective for solving various optimization algorithms [12, 13]. The algorithm is based on the principle that stars are attracted toward the Black Hole due to the massive gravitational pull and they are eventually sucked into it if they are inside a radius known as Schwarzschild radius, determined by laws of motion.

3.1 Standalone Binary Black Hole Algorithm (SBH)

The Black Hole feature selection heuristics mimic the real-life behavior of stars and Black Holes [13]. The basic binary Black Hole algorithm was initiated by creating a set of a population of N stars. Each star represented a distinct subset of the original dataset and consists of a vector with a size equivalent to the number of features in a given dataset. The elements of each star were randomly filled with zeros and ones. Zeros indicate that the corresponding features are not selected and the ones indicate the corresponding features are selected. The subset of selected features corresponding to each star was assessed by a multi-label classification algorithm and the fitness was evaluated. The star with the highest fitness was made the Black Hole. In each iteration, the stars were randomly moved toward the Black Hole with some probability and the fitness was reevaluated. In this work, we have kept the probability as 0.5.

The star with the best fitness was redesignated as Black Hole, and the process was repeated until convergence. A star falling within a neighborhood known as the Schwarzschild radius was destroyed and a random star was again generated. The Schwarzschild radius was calculated using Eq. (1)

$$\text{Schwarzschild radius} = \text{Fitness of Black Hole}/\text{Sum of fitness of all stars} \tag{1}$$

3.2 Improved Hybrid Black Hole Genetic Algorithm for Multi-label Feature Selection

We modified the standalone binary Black Hole algorithm for improving the algorithm performance. First, a preprocessing filter step was performed in which the features with the lowest rankings were removed by using Chi-square (χ^2) heuristic. In the improved algorithm, we have hybridized genetic algorithm, crossover, and mutation steps in every 5th iteration. In the 5th iteration, the one point crossover and mutation steps of the genetic algorithm were incorporated after the movement of the stars toward the Black Hole.

The crossover probability was kept at 0.5 and the mutation probability was kept as 1. We found that incorporation of genetic algorithms reduces the rapid convergence to local minima and diversifies feature subsets of the star population. We also

Table 1 Single-point crossover illustration

Star 1	1 1 1 | 0 0 0 1 0
Star 2	0 0 0 | 1 1 0 1 0
NewStar 1	1 1 1 | 1 1 0 1 0
NewStar 2	0 0 0 | 0 0 0 1 0

incorporated a term in the fitness equation to control the tradeoff between performance and the number of features selected. The modified algorithm was initiated by randomly generating stars and the modified fitness was evaluated. Subsequently, the star with the best fitness was assigned as Black Hole and the stars were moved randomly toward the Black Hole. After this step, the genetic algorithm operators were incorporated after every 5 iterations. In the crossover step, two stars were randomly selected with a crossover probability, and a crossover point was randomly selected. The process is illustrated in Table 1.

This process was repeated until the new population size was same as the original population. Next, the standard flip bit mutation operator was applied in which the selected features of a star were deselected with the mutation probability and vice versa. This mutation step prevented rapid convergence to local minima and diversified the feature subsets. The modified fitness was again calculated, and the iterations were repeated until convergence.

3.3 Datasets

Two different types of multi-label datasets were employed in this work: dataset-I and dataset-II. In dataset-I, the standard benchmarking datasets which were employed by the researchers for multi-label classification in the past were used [2]. These included the Scene, Yeast, Emotions, and Medical datasets. Scene is a benchmarking dataset from image domain while Yeast is from biology. Emotions dataset is from music, whereas medical is from text domain. The details of input features, output label sizes, and number of instances in each dataset are illustrated in Table 2.

Feature selection was also carried out on dataset-II, which denotes a very important multi-label dataset in the bioinformatics domain. This dataset was used to develop an algorithm that deals with prediction of antibody classes (types) to which an epitope can simultaneously bind [14]. This multi-label classification problem is important in

Table 2 Benchmarking data (Dataset-I)

Dataset	Instances	Features	Labels
Scene	2407	294	6
Yeast	2417	103	14
Emotions	593	72	6
Medical	978	1449	45

Table 3 Epitope dataset (Dataset-II)

Dataset features	Instances	Features	Labels
Amino acid	10,744	20	4
Dipeptide	10,744	400	4

the design of novel vaccines and diagnostic strategies. The dataset was compiled from the Immune Epitope Database (IEDB) [14] by taking into account linear (sequential) B-cell epitopes of length 5–50 amino acids from only positive B-cell assays. The dataset consists of 10,744 epitope sequences. The authors extracted 20 amino acid composition features and 400 dipeptide composition features [14] from these sequences and were used as inputs in our work. The output comprised of four labels, viz. the four antibody classes: IgG, IgE, IgA, and IgM. In our work, both the stand alone and the hybrid Black Hole genetic algorithm feature selection methods were employed on these datasets. The details of the dataset-II are given in Table 3.

3.4 Simulation Setup

Dataset-I: In all our simulations for dataset I, we randomly divided it into 60:40 training and test splits following the earlier work [2]. In order to facilitate easy comparison, the MLKNN multi-label classifier was employed with the parameter $K = 10$ (number of nearest neighbors), as per the previous work [15]. The number of stars/population size used for the experiment was 20. The training sets were used to build the models, and their performance was evaluated on the corresponding test sets.

For the standalone Black Hole algorithm, the most relevant feature sets were determined by using only the Hamming Loss (HL) which denotes one of the most common performance metrics in multi-label classification problems. In multi-label classification, Hamming Loss is defined as the Hamming distance between ground truth labels and the predicted label set. It is the ratio of wrongly predicted labels relative to the total number of labels in the dataset. The lesser the Hamming Loss, the better is the performance of an algorithm. The feature set with the least Hamming Loss value was observed to be the most relevant. Defining 'M' as number of examples in the dataset, Y_i the number of correct labels, Z_i is the number of predicted labels, and L is the label set the Hamming Loss can be defined as:

$$\frac{1}{|M|}\sum_{i=1}^{|M|}\frac{Y_i \Delta Z_i}{|L|} \tag{2}$$

The equation for fitness for the standalone Black Hole algorithm can be written as:

$$\text{Fitness} = (1 - \text{Hamming's loss}) \tag{3}$$

We also calculated corresponding subset accuracies. Subset accuracy is the number of exact matches between predicted labels and true labels. The equation for subset accuracy can be given as:

$$\frac{1}{|M|}\sum_{i=1}^{M}\frac{|Y_i \cap z_i|}{|Y_i \cup z_i|} \tag{4}$$

For the improved hybrid algorithm, a modified fitness equation was employed with the inclusion of a Lambda (λ) parameter, which characterizes the tradeoff between Hamming Loss and number of features. This equation can be written as:

$$\text{Fitness} = (1 - \text{Hamming's loss})/\text{Lambda} * \text{Number of features selected} \tag{5}$$

The maximum number of iterations were set to 50. The experiments were repeated 20 times and the performance metrics were then averaged.

Dataset-II: We randomly split the dataset into 60:40 training-test splits, respectively. We also used random forest classifier and binary relevance multi-label classification methodology. Hamming Loss was maintained as the performance metric in the fitness equation.

Parameters Used in the Algorithm: Following are the parameters and their corresponding values which were used during the simulations.

1. Population size = $M = 20$
2. Number of features = N = Number of attributes present in each dataset
   Dataset-I: (N(Scene) = 294, N(Yeast) = 103, N(Emotions) = 72, N(Medical) = 1449).
   Dataset-II: (N(Amino) = 20, N(Dipeptide) = 400).
3. Maximum iteration = 50
4. Crossover probability and movement probability = $P = 0.5$
5. Mutation probability = $1/M$
6. MLKNN—Number of nearest neighbors $K = 10$
7. Lambda = for each dataset lambda is tuned. Range = (2.e−10, 0).

The details of the **Algorithm 1**: Simple Binary Black Hole algorithm are as mentioned in Fig. 1.

The details of the **Algorithm 2**: Hybrid Binary Black Hole algorithm are as mentioned in Fig. 2.

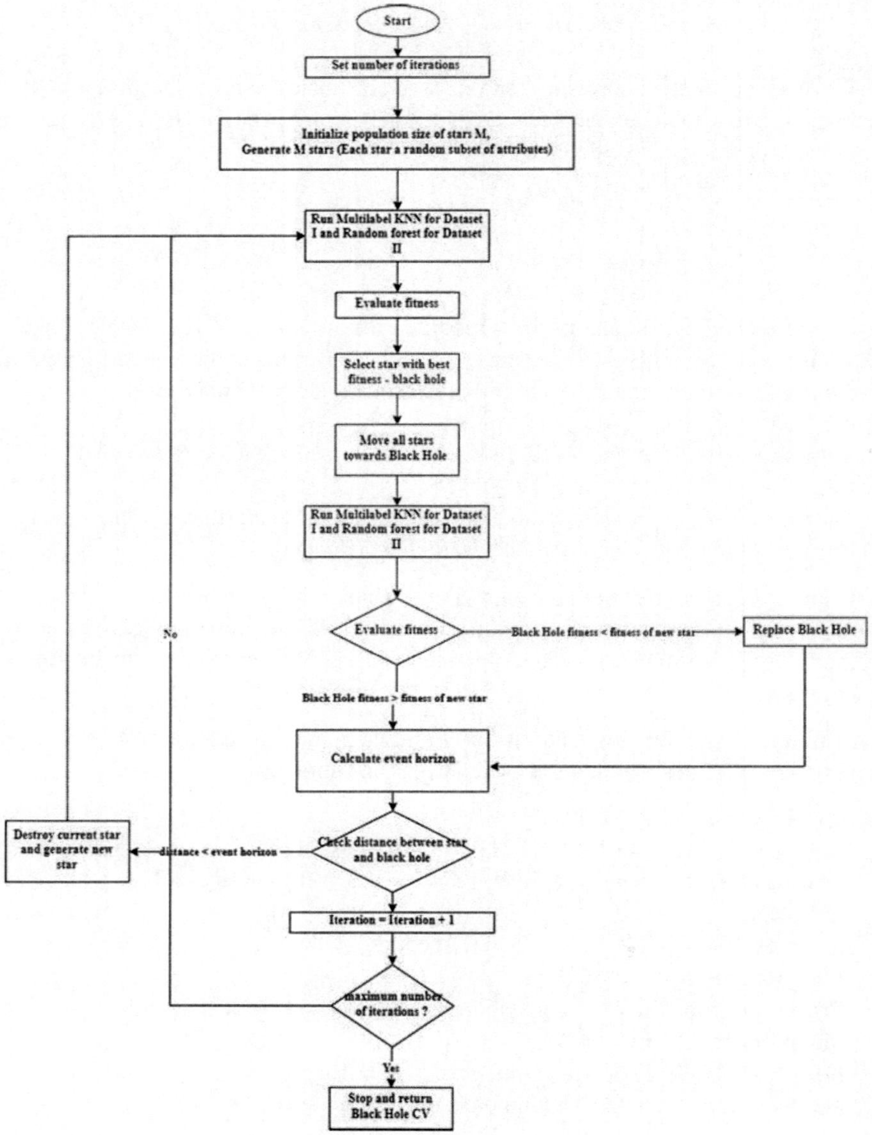

Fig. 1 Flowchart of simple Binary Black Hole algorithm

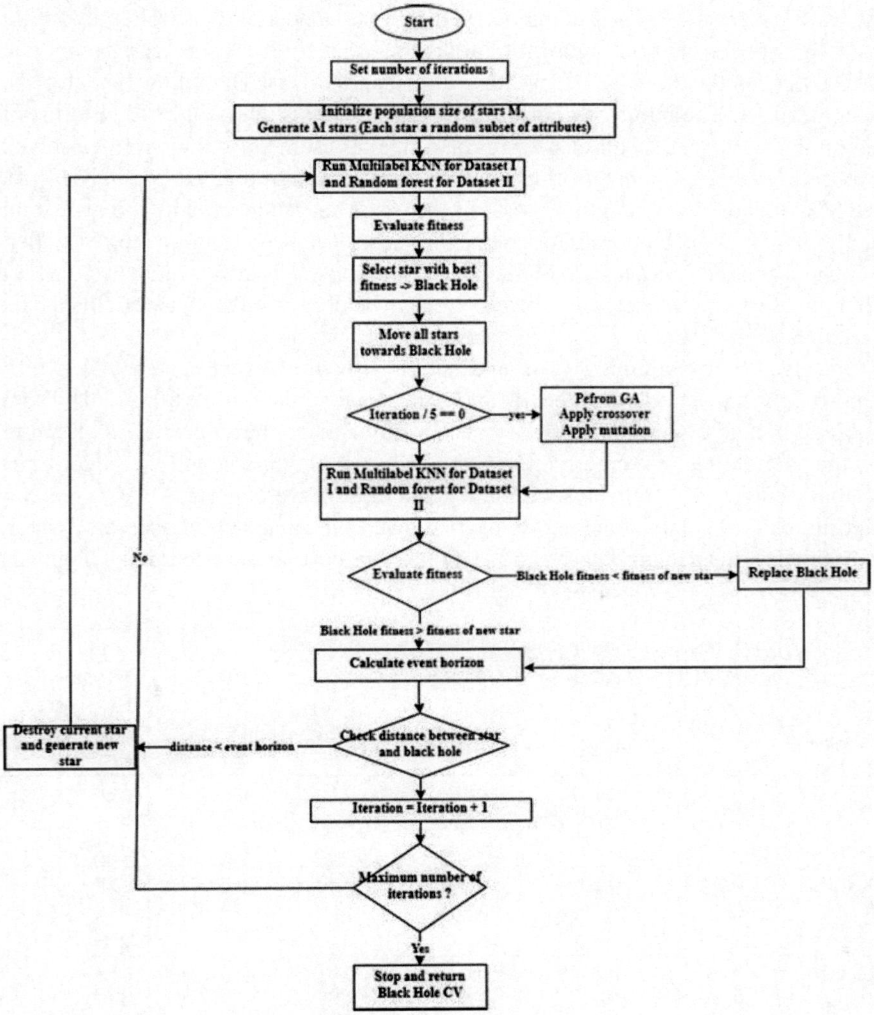

Fig. 2 Hybrid Binary Black Hole algorithm

4 Experimental Results

4.1 Dataset-I

Based on the simulation settings explained earlier, the models were built and results were compiled for the four datasets. These results are given in Tables 4 and 5. In Table 4, we have illustrated the average Hamming Loss (HL) and the average subset size (SS) obtained with the standalone Black Hole algorithm and the improved hybrid algorithm along with the results obtained by earlier authors employing ABACO,

ACO, BGSA, and GA. The Lambda (λ) value for the standalone Black Hole algorithm and the improved hybrid algorithm are also mentioned. It can be seen from Table 4 that the standalone Black Hole algorithm has the least Hamming Loss for the Scene dataset. The hybrid algorithm performance was slightly inferior in terms of Hamming Loss but the subset size was also smaller. Both algorithms were observed to outperform the earlier algorithms, both in terms of Hamming Loss and subset size. ABACO had the least Hamming loss for the Yeast dataset and the hybrid algorithm had slightly higher Hamming Loss while the subset size was also found to be smaller. For the rest of the datasets, i.e., Medical and Emotions, hybrid algorithm had the best Hamming loss and subset size. Therefore, it can be observed that our work displayed better Hamming Loss in 3 out of 4 datasets.

In addition to Hamming Loss and subset size, the subset accuracy was also reported for the datasets-I (Table 4) using the same simulation settings [3.4]. From Table 5, it can be seen that both of our algorithms have the same values of subset accuracy (Bold entries represent the best results). As compared to other algorithms on the Scene Dataset, our subset accuracy and the size were better. ABACO and our algorithms (both standalone and hybrid) showed the same subset accuracy, which is optimal. For medical dataset, ABACO had the best subset accuracy which was

Table 4 Comparison of Hamming Loss on Dataset-I

Data	ABACO [HL] [SS]	ACO [HL] [SS]	BGSA [HL] [SS]	GA [HL] [SS]	BH [HL] [SS] (λ)	Hybrid BH [HL] [SS] (λ)
Scene	0.104 154	0.107 157	0.106 158	0.103 156	0.092 146 0	0.093 143 0.00005
Yeast	0.20075 63	0.2019 59	0.2013 61	0.2018 58	0.203 62 0	0.206 49 2.e−10
Medical	0.1355 927	0.0148 817	0.0137 874	0.0138 837	0.0136 726 0	0.0133 714 2.e−10
Emotions	0.226 37	0.225218	0.241236	0.23 35	0.2022 37 0	0.200 28 0

Table 5 Comparing subset accuracy on Dataset-I

Data	ABACO	ACO	BGSA	GA	BH	Hybrid BH
Scene	0.64	0.61	0.61	0.61	**Acc = 67**	**Acc = 67**
Yeast	**0.50**	0.49	0.49	0.50	Acc = 50	Acc = 50
Medical	**0.674201**	0.5964	0.6348	0.6387	Acc = 64	Acc = 67
Emotions	0.50481	0.496	0.45	0.48	**Acc = 53**	**Acc = 54**

Table 6 Comparing Hamming Loss on dataset-II

Dataset	AbCPE	BH (Hamming loss and subset size)	Hybrid BH (Ham loss and subset size)
Amino	0.1433	0.141, 16	0.140, 16
Dipeptide	0.1370	0.135, 231	0.134, 199

slightly better than our algorithm. Lastly, the hybrid algorithm had the best subset accuracy for emotions dataset as well. Thus, our work provided the best subset accuracy for 3 out of 4 datasets.

4.2 Dataset-II

Standalone Black Hole and Hybrid Black Hole algorithm were used with the setup discussed in Sect. 4.2, and the results were compared with AbCPE. Referring to Table 6. The results were compared by using only Hamming Loss metric on random forest binary classifier. It is observed that the feature selection was not performed during AbCPE algorithm development. In our work, both the algorithms provided better results than those reported in AbCPE.

4.3 Computational Complexity

In each iteration, the complexity of the original algorithm is $O(M * N)$. The revised algorithm has two components, i.e., standalone algorithm for four iterations out of every 5 iterations and genetic algorithms, crossover and mutations steps incorporated in one out 5 iterations. For each iterations, the complexity of the original algorithm is $O(M * N)$ and complexity of genetic algorithm for each iteration is $O(2 * M * N)$. So the complexity for G number of iterations is $O(((4/5) * G * M * N) + (1/5 * 2 * G * M))$ which reduces to $O(G * M * N)$.

5 Conclusion

In this work, we have developed an improved algorithm for multi-label feature selection. This algorithm hybridizes the standalone binary Black Hole algorithm with two genetic algorithm operators, viz. one point crossover and mutation. The hybridized algorithm performs very well on majority of the benchmarking datasets and fares better than the earlier algorithms. We also conducted simulations on real-life bioinformatics dataset dealing with multi-label epitopes. Our algorithm is able to find the most informative subset for this dataset. This work will be useful for multi-label feature selection tasks in different domains.

References

1. Talavera L (2005) An evaluation of filter and wrapper methods for feature selection in categorical clustering. In: International symposium on intelligent data analysis. Springer, Berlin, pp 440–451
2. Kashef S, Nezamabadi-pour H (2017) An effective method of multi-label feature selection employing evolutionary algorithms. In: 2nd Conference on swarm intelligence and evolutionary computation 2017. IEEE, pp 21–25
3. Ibrohim MO, Budi I (2019) Multi-label hate speech and abusive language detection in Indonesian twitter. In: Proceedings of the third workshop on abusive language online, pp 46–57
4. Wang J, Yang Y, Mao J, Huang Z, Huang C, Xu W (2016) CNN-RNN: a unified framework for multi-label image classification. In: Proceedings of the IEEE conference on computer vision and pattern recognition, pp 2285–2294
5. Spolaôr N, Cherman EA, Monard MC, Lee HD (2013) A comparison of multi-label feature selection methods using the problem transformation approach. Electron Notes Theor Comput Sci
6. Kashef S, Nezamabadi-pour H (2015) An advanced ACO algorithm for feature subset selection. Neurocomputing 147:271–279
7. Rashedi E, Nezamabadi-Pour H, Saryazdi S (2010) BGSA: binary gravitational search algorithm. Nat Comput 9(3):727–745
8. Al Salami NM (2009) Ant colony optimization algorithm. UbiCC J 4(3):823–826
9. Touhidi H, Nezamabadi-pour H, Saryazdi S (2007) Feature selection using binary ant algorithm. In: Frist joint congress on fuzzy and intelligent systems
10. Kumar M, Husain DM, Upreti N, Gupta D (2010) Genetic algorithm: review and application. [Online] papers.ssrn.com
11. Deeb H, Sarangi A, Mishra D, Sarangi SK (2020) Improved black hole optimization algorithm for data clustering. J King Saud Univ Comput Inf Sci
12. Kumar S, Datta D, Singh SK (2015) Black hole algorithm and its applications. In: Computational intelligence applications in modeling and control. Springer, Cham, pp 147–170
13. Hatamlou A (2013) Black hole: a new heuristic optimization approach for data clustering. Inf Sci 222:175–184
14. Kadam K, Peerzada N, Karbhal R, Sawant S, Valadi J, Kulkarni-Kale U (2021) Antibody Class(es) predictor for epitopes (AbCPE): a multi-label classification algorithm. Front Bioinf 37
15. Zhang ML, Zhou ZH (2007) ML-KNN: a lazy learning approach to multi-label learning. Pattern Recogn 40(7):2038–2048

Design and Analysis of Composite Leaf Spring Suspension System by Using Particle Swarm Optimization Technique

Amartya Gunjan, Pankaj Sharma, Asmita Ajay Rathod, Surender Reddy Salkuti, M. Rajesh Kumar, Rani Chinnappa Naidu, and Mohammad Kaleem Khodabux

Abstract Recently, weight optimization has been popping into a vital production and mechanical design tool. In this paper, particle swarm optimization (PSO) is introduced and applied to the design along with the optimization of composite leaf spring suspension systems. Since leaf springs constitute long and slender panels linked to one another with the frame of a trailer, they can improve the vehicle's suspension system on the road. This paper is centered on replacing steel metal springs with leaf springs to reduce weight when the load is applied. The main design limitation is the spring stiffness. Because of its high strength-to-weight ratio as well as superior corrosion resistance, mono-composite materials such as E-glass/epoxy are utilized to construct a typical mono-composite leaf spring plate for light-duty commercial vehicles. Furthermore, a constant cross-sectional design is adopted because of its mass-production operations as well as the potential to develop continuous fiber reinforcement. Bending deformation and deflection are the fundamental design limitations in the production of springs. An experimental test has depicted that the deformation and deflection of composite springs are substantially lower than steel springs. Furthermore, results have shown that PSO has overtaken SA between the two approaches.

Keywords Particle swarm optimization · Composite leaf spring · Weight optimization

---

A. Gunjan
School of Electronics Engineering, Vellore Institute of Technology, Vellore, India

P. Sharma · A. A. Rathod
School of Electrical Engineering, Vellore Institute of Technology, Vellore, India

S. R. Salkuti
Department of Railroad and Electrical Engineering, Woosong University, Daejeon, South Korea

M. Rajesh Kumar · R. C. Naidu (✉) · M. K. Khodabux
Faculty of Sustainable Development and Engineering, Université Des Mascareignes, Beau Bassin-Rose Hill, Mauritius
e-mail: rnaidu@udm.ac.mu

M. Thakur et al. (eds.), *Soft Computing for Problem Solving*, Lecture Notes in Networks and Systems 547, https://doi.org/10.1007/978-981-19-6525-8_33

1 Introduction

The current combination of lightweight composites in automobiles and other transportation systems has reduced vehicles' loads in recent years. This decrease in weight has led to a reduction in fuel consumption for automobiles, and hence, it has justified an increase in speed it can attain. The leaf spring suspension represents an essential feature for weight decrease in automobiles as it constitutes almost 10% to 20% of components whose weight is not subjected to the suspension spring of the vehicles. Multi-leaf steel springs are being substituted by mono-leaf composite springs due to composite materials having the highest tensile strength, lowest modulus of elasticity, as well as lowest mass density than steel [1, 2]. As a result, composite materials can attain an excellent strength-to-weight ratio while reducing significant weight [3]. The weight reduction that comes from composite materials results in significant fuel savings [4, 5]. Composite materials are also highly fatigue resistant and long lasting [6].

This paper discusses the application of PSO and SA for leaf spring design optimization. The PSO algorithm is based on a population chaotic optimization method that mimics the social behavior of a group of animals, for instance, a flock of birds or a school of fish, to solve optimization problems [7]. Particles, which are real solutions in PSO, adopt the new best possible particles through the actual problem area [8]. In comparison, SA is a process that requires regulated high temperature and low temperature of a material to increase the performance and reduce errors related to the chemical change of hardening of metal and glass [9, 10]. The heat causes the atoms to become unstuck from their starting positions as well as move randomly through via an excited state; the slow cooling increases their change [11, 12]. In this research topic, a steel leaf spring is substituted with a mono-composite leaf spring, made up of E-glass epoxy composites in the automobile. Composite leaf springs constitute the same sizes and as many leaves as steel leaf springs. The essential purpose of the composite leaf spring is to reduce weight.

This paper is structured as follows: Section 2 introduces the literature survey; Sect. 3 introduces the problem statement; Sect. 4 presents the conventional leaf spring; Sect. 5 presents the composite leaf spring; Sect. 6 discusses PSO; Sect. 7 presents the algorithm; Sect. 8 presents the results, and the final section concludes this paper.

2 Literature Review

Few experimental tests on the composite spring have conducted in the early 1960s. However, due to stress and fatigue, it has later shown that composite springs fail to provide manufacturing resources. Therefore, several researchers have looked into the design of composite springs to find scope for better springs. In this context, this paper focuses on analyzing composite leaf springs while applying the PSO approach. First, static as well as dynamic analyses of composite leaf springs for big vehicles are discussed in [13]. The primary aim was to evaluate the load-bearing strength, material stiffness, and weight imparted to the composite spring before actually conducting tests on the steel spring [14]. In contrast, [15] discusses the planning, manufacturing weight assessment, along with testing of a composite-integrated back suspension. Moreover, the spring has been utilized to develop complex composite layout techniques. Besides, composite springs have not been subjected to component testing since the early 1960s. Thus, their unpredictable fatigue outcome has not been assessed. Furthermore, the critical necessity for mass reduction has been overlooked.

3 Problem Statement

A leaf spring is a simple kind of spring that is repeatedly utilized to enhance driving stability in cars. A leaf spring is a thin, rectangular-sectioned arc-shaped segment of spring steel. A leaf spring is made up of several leaves arranged on multiple levels on top of one another, mostly with selectively shorter leaves for huge vehicles. Leaf springs are mainly constructed from simple carbon steel of about 0.90 to 1.0% of carbon.

Multi-leaf steel springs are being changed with mono-leaf composite springs since composite materials provide higher strength, lower modulus of elasticity, and relatively low-mass density than steel. A leaf spring can be designed via one of three design criteria: constant width with variable thickness, constant thickness with variable width, or constant cross-section design, in which both thickness as well as width are changed across the leaf spring, while the cross-section area is kept constant along the length of the leaf spring. As a result, the middle width and thickness are assessed for optimization. The heat treatment of metallic spring components is sometimes very strong, leading to high-load capacity springs. They also have good conductivity and are fatigue resistant. The swept ratio is used to calculate the final thickness and breadth. The size of an actual traditional leaf spring for a light commercial automobile is presented in Table 1.

Table 1 Parameters and dimensions

S. No.	Parameters	Dimensions
1	Design load (W)	4500 N
2	Spring length under (L)	1220 mm (straight condition)
3	Maximum allowable vertical deflection	160 mm
4	Spring rate (K)	28–32 N/mm

4 Conventional Leaf Spring

Leaf springs are traditionally composed of good-quality steel. These are among the earliest types of springs, dating back to the mid-seventeenth century. They are also known as carriage springs or elliptical springs. Leaf springs were widely employed in automobiles until the 1970s in Europe as well as Japan and in the late 1970s in the United States [16]. The transition to front-wheel drive, as well as more complex suspension systems, drove car manufacturers to coil springs [17]. This is because the weight of powerful cars is more evenly distributed throughout the chassis, while the coil springs keep it stable to an unconnected point. Leaf springs, unlike coil springs, also placed the back axle, removing the requirement for trailing hands as well as a track bar, resulting in a lower cost and weight for an efficient stay axle rear suspension [18, 19].

A leaf spring is composed of several plates of varying lengths that are connected with each other by clamps and bolts. This spring is commonly seen in carriages, as well as in cars, lorries, and railway trains [20]. To begin, all of the plates are bent to the same radius and loosened so that they may be slipped one over the other. The master leaf is the tallest and longest of the three leaves. The eye is a slot that may be used to connect the spring to any other machine. Camber refers to the amount of bend imparted to the spring from the principle line that runs across the eyes.

The camber is set such that the extendable spring's maximum limit of deflection in length is reached, and the deflected spring does not come into contact with the machine to which it is connected. The primary clamp is essential to retain the spring's leaves properly. On the other hand, the bolt holes must interact with the bolts to clamp the leaves and slightly weaken the spring. Rebound clips support the weight distribution from the main leaf to the graded leaf.

Plain carbon steel is the most common material used to make conventional leaf springs, containing 0.90 to 1% carbon [21]. All steel plates used in springs are heat treated after the manufacturing process. When these steel plates are heat treated, they provide material with increased strength, load capacity, fatigue characteristics, and deflection range. Leaf springs are made of chromium-nickel-molybdenum steel, chromium-vanadium steel, and silicon-manganese steel, among other materials [22]. These materials are more prone to corrosion and can be replaced with composite materials to make them more efficient. Some current materials have qualities that are superior to those mentioned above. These composite materials provide several benefits.

5 Composite Leaf Spring

A composite material is a substance consisting of two or more elements that have been physically or chemically combined on a macroscopic scale. The materials' properties are preserved in the composite. The additives can typically be identified completely, as well as a distinction between them. Serval composite materials have a composite of conductivity as well as rigidity comparable to or better than traditional steel materials. Certain composite substances' strength weight ratios, as well as modulus weight ratios, are significantly better than those of metallic substances due to the relatively low precise gravities [23].

Furthermore, severe fatigue weight ratios harm many composite laminates' tolerances. As a result, fiber composites have evolved into a major class of structural fabric that is utilized or contemplated as a metal replacement in various applications, including aerospace, automotive, and other sectors. The significant internal damping effect of fiber-reinforced composites is another distinctive characteristic. This results in higher vibration power absorption inside the material, as well as a reduction in noise and vibration transmission to nearby systems.

Composite materials with high damping potential may be useful in automotive applications where noise, vibration, and durability are essential for passenger comfort [24]. However, a few of the environmental factors that might cause degradation within the mechanical components for a few polymeric matrix composites include temperature expansion, corrosive fluids, and UV rays. At high temperatures, oxidation of the matrix generates a harmful chemical interaction between nanocomposites in many metal matrix composites.

5.1 *Objective Function*

The purpose is to make the composite leaf spring as light as possible.

$$f(w) = \rho L b t \tag{1}$$

L stands for the length of the leaf spring, *b* for the breadth at the center, *t* for the thickness at the middle, and ρ for the density of the composite leaf spring material.

5.2 *Design Variables*

The range for the design variables is defined as follows:

- $b_{max} = 0.050$ m & $b_{min} = 0.020$ m, $t_{max} = 0.050$ m & $t_{min} = 0.01$ m.

where the center width, b and center thickness, t.

5.3 Design Parameters

The design variables are generally not dependent on the design parameters. They are the design load (W), leaf spring length (L), and composite material parameters, such as density (ρ), Young's modulus (E), and maximum allowable stress ($S_{\max}$).

5.4 Design Constraints

This section discusses the composite leaf spring's design parameters and resource application. The limitations are defined by formulae and pertain to the bending stress, Sb, and vertical deflection, d.

$$Sb = \frac{1.5WL}{bt^2} \tag{2}$$

$$d = \frac{WL^3}{4Ebt^3} \tag{3}$$

The upper and lower limits for the constraints are provided as:

- $\mathrm{Sb}_{\max} = 550$ MPa & $\mathrm{Sb}_{\min} = 400$ MPa, $d_{\max} = 0.160$ m & $d_{\min} = 0.120$ m

6 Particle Swarm Optimization

Dr. Eberhart and Dr. Kennedy presented the PSO based on the global optimization evolutionary method. PSO is a social influence and social learning-based optimization technique that is inspired by the social and psychological characteristics of flocking birds or schools of fish. A PSO algorithm generates a swarm of particles, leading to a sensible solution. Compared to evolutionary concepts, a swarm is the same as a population, and a particle is the same as an individual. It is based on the nature or behavioral patterns of groups of animals, where they have identified that as information sharing within groups improves the survival advantage increases. For example, a bird seeking food on an erratic basis can improve its results by fitting in with the crowd. Working as a group can simply share the best-found knowledge, allowing them to choose the finest spot to eat quickly [7, 25].

In basic words, the particles are scattered throughout a multidimensional sampling area, with each particle's location adjusted according to its own and neighbors' occurrences. PSO initiates with a collection of random particles (solutions) as well as then improves fresh generations to track for optima (the best-updated fitness). In every iteration, the two "best" values pbest (particle best) and gbest (global best) are used to update each particle. Each particle tracks its location in the concerned space, which is linked to the best solution (fitness) found so far. pbest is the name given to this value. The alternative best value that the PSO looks for is the best value captured so far by somewhat particle in the particle's neighborhood. gbest is the name given to this value (global best). At each time step, the PSO approach enhances each particle by adjusting its velocity (or accelerating) toward its pbest (particle best) and gbest (global best) positions [26]. Each particle seeks to change its real position as well as velocity according to the distance between its present location along with the pbest and the distance between its original position and gbest [27, 28].

6.1 Velocity Clamping

It was identified in basic PSO that the velocity shortly reaches huge values, particularly for particles far from the neighborhood's as well as personal optimal locales. As a result, particles have high-velocity feedback, allowing them to move beyond the search space's limits. As a result, the velocities are required to stay within the limits of the boundary constraints. If a particle's velocity surpasses the given velocity, the particle's velocity is lowered to the maximum velocity. This leads to the need to find an appropriate value for each maximum velocity while considering whether to go too rapidly or too slowly. The maximum values are chosen to be a percentage of the domain for each dimension of the selected area as stated by the equation:

$$V_{\max} = \frac{kX(X_{\max} - X_{\min})}{2} \tag{4}$$

where $X_{\max}$ as well as $X_{\min}$ denotes the domain of x maximum and minimum values, respectively, and k denotes a user-defined velocity fixing factor, $0.1 \leq k \leq 1$.

7 Algorithm

The PSO method involves filling numerous particle solutions in the study region one at a time. Every particle solution is selected by the objective function for optimization and evaluating the fitness of that particular solution throughout each iteration of the algorithm. Each particle solution may be considered a particle "flying" across the fitness approach, searching for the objective function's maximum or minimum value. The steps of the PSO algorithm are described as follows:

- Step 1: Initiate a population of particles that are evenly dispersed over X.
- Step 2: Using the objective function, calculate the fitness of each particle.
- Step 3: If the particle's current position is superior to its prior best position, it should be modified.
- Update velocity:

$$V_i^{t+1} = W \cdot V_i^t + c_1 U_1^t \left(P_{b1}^t - P_i^t\right) + c_2 U_2^t \left(g_b^t - P_i^t\right) \quad (5)$$

- Update position:

$$P_i^{t+1} = P_i^t + V_i^{t+1} \quad (6)$$

- Step 2 should be repeated once more. Until the end criteria are met.

This process continues until a stop condition is met. Finally, a computer program Matlab based is given to carry out the optimization process as well as achieve the best possible structure. The flow graph shows in Fig. 1 how to optimize the composite leaf spring using PSO step by step.

8 Result

By allocating width and thickness to the mono-leaf composite spring, we may lower the weight of the weighted spring. Next, choose the spring's length compared to a standard leaf spring that would be used on the same car. Used the input parameter listed in Table 2 as our preliminary step. The breadth increases as the thickness reduces in this problem, while the length remains constant. Composite leaf spring input parameters.

As the algorithm's number of iterations rises, the mass converges toward its minimum value from its maximum value and becomes optimized. High strength, corrosiveness, and low specific gravity are all characteristics of the composite leaf spring. Figure 2 depicts the variations in mass as a function of the number of PSO iterations. As a result, the spring load is 85% lower when using PSO and 79% lower when applying SA. According to results from the analysis, the composite leaf spring is lighter as well as highly cost-effective compared to the typical steel spring with corresponding design specifications. PSO reduces the weight of the mono-leaf spring from 6.40742 kg to 6.34400 kg. PSO saves 9.89% of its weight as a result.

9 Conclusion

PSO has one of the best practical mono-composite leaf spring optimization approaches. In this paper, PSO, as well as SA techniques, is introduced and applied to the design, along with the optimization of composite leaf spring suspension systems. The suspension system of any automobile plays an essential role in proper functioning. Leaf spring is used in heavy automobiles. In this paper, it is observed that the addition of inertia weight and velocity to particle swarm optimization is necessary to assure convergent behavior. PSO reduces the weight of the mono-leaf spring from

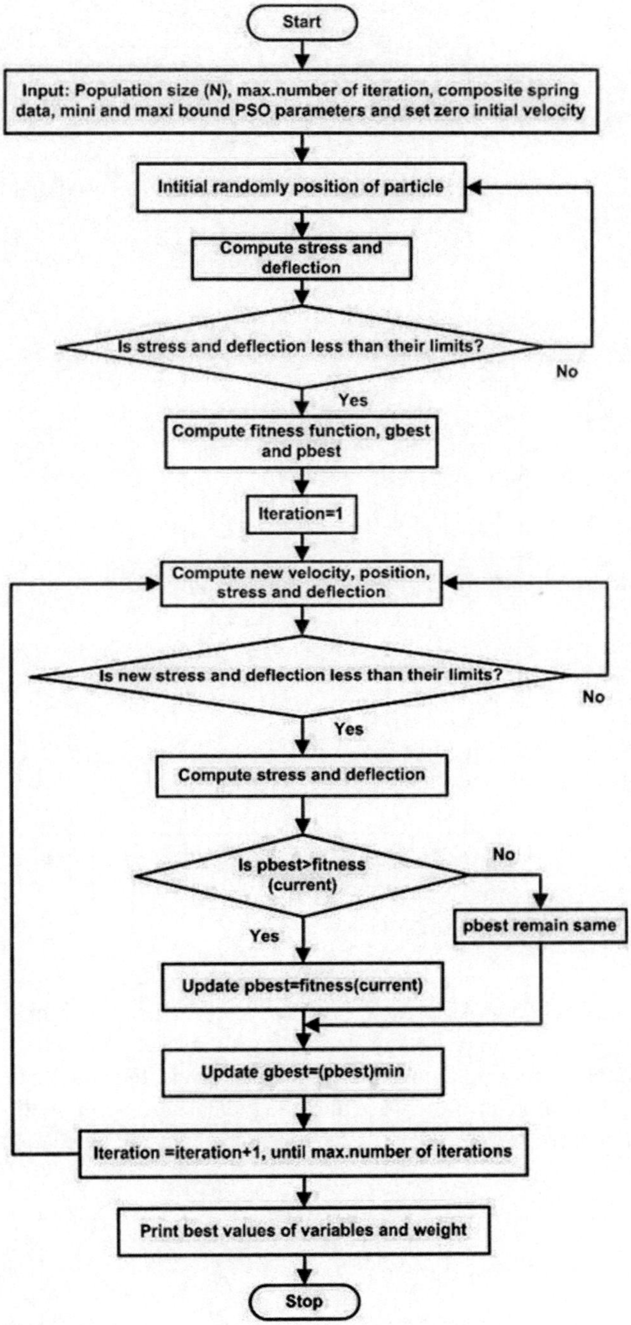

Fig. 1 Flow graph

Table 2 Input parameters

S. No.	Parameters	Composite spring
1	Spring length	1220 mm (stretched condition)
2	Spring length	32.5 mm (un-stretched condition)
3	Material density	2600 kg/m^3
4	Load	4500 N
5	Maximum allowable stress	550 MPa

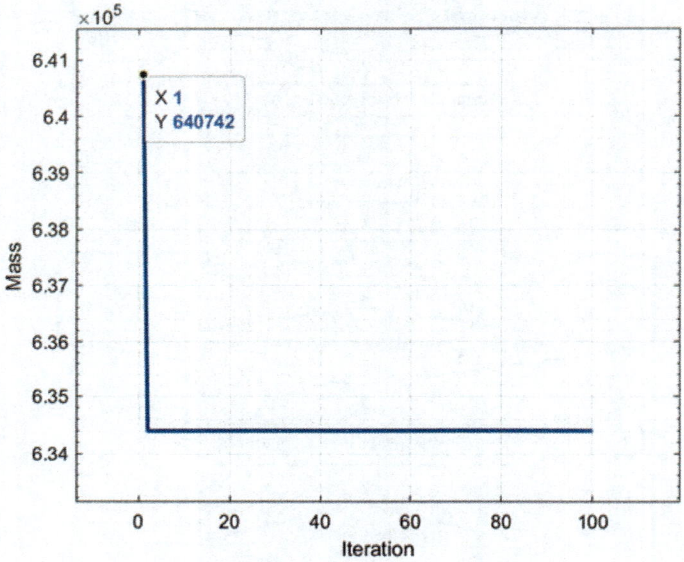

Fig. 2 Mass variation with number of iteration

6.40742 kg to 6.34400 kg. As a result, PSO saves 9.89% of its weight. This demonstrates that PSO is a reliable strategy for solving such issues and aids in improving physical products. According to results from the analysis, the composite leaf spring is lighter as well as more cost-effective than the typical steel spring with comparable design specifications.

References

1. Mani A, Adhikari U, Risal S (2021) Finite element analysis of Epoxy/E-Glass composite material based mono leaf spring for light weight vehicle 4:138–145
2. Patil DK, Hv P, Kumar A (2021) Numerical investigation of composite mono leaf spring. Gradiva Rev J 7:275–285

3. Naik V, Kumar M, Kaup V (2021) A review on natural fiber composite material in automotive applications. Eng Sci 18:1–10
4. Muhammad A, Rahman MR, Baini R, Bin Bakri MK (2021) Applications of sustainable polymer composites in automobile and aerospace industry. In: Rahman MR (ed) Advances in sustainable polymer composites. Elsevier, pp 185–207
5. Kumar Sharma A, Bhandari R, Sharma C, Krishna Dhakad S, Pinca-Bretotean C (2022) Polymer matrix composites: a state of art review. Mater Today: Proc 57:2330–2333
6. Rajak DK, Pagar DD, Behera A, Menezes PL (2022) Role of composite materials in automotive sector: potential applications. In: Advances in engine tribology. Springer, Heidelberg, pp 193–217
7. Kennedy J, Eberhart R (1995) Particle swarm optimization. In: Proceedings of ICNN'95—international conference on neural networks. IEEE, pp 1942–1948
8. Abd Elaziz M, Elsheikh AH, Oliva D, Abualigah L, Lu S, Ewees AA (2022) Advanced metaheuristic techniques for mechanical design problems: review. Arch Comput Methods Eng 29:695–716
9. Vamsi Krishna Reddy AK, Venkata Lakshmi Narayana K (2022) Meta-heuristics optimization in electric vehicles—an extensive review. Renew Sustain Energy Rev 160:112285
10. Ingber L (1993) Simulated annealing: practice versus theory. Math Comput Model 18:29–57
11. Hilleke KP, Bi T, Zurek E (2022) Materials under high pressure: a chemical perspective. Appl Phys A 128:441
12. Bruns M, Varnik F (2022) Enhanced dynamics in deep thermal cycling of a model glass. J Chem Phys
13. Hwang WKSH (1986) Fatigue of composites—fatigue modulus concept and life prediction 2:154–165
14. Tanabe K, Seino T, Kajio Y (1986) Characteristics of carbon/glass fiber reinforced plastic leaf spring. In: {SAE} Technical paper series. SAE International, 400 Commonwealth Drive, Warrendale, PA, United States 2, pp 154–165
15. Lakshmi BV, Satyanarayana I (2012) Static and dynamic analysis on composite leaf spring in heavy vehicle. Int J Adv Eng Res Stud 2:80–84
16. Waghmare S, Nimbalkar DH, Mane VV, Patel A (2021) Design & analysis of (Mahindra-Bolreo) leaf spring by using composite material 2:511–519
17. Hilgers M, Achenbach W (2021) Chassis and axles. Springer, Heidelberg
18. Pütz R, Serné T (2022) Springs. In: Race car handling optimization. Springer Fachmedien Wiesbaden, Wiesbaden, pp 163–204
19. Kumar T, Stephen R, Zaeimi M, Wheatley G (2020) Formula SAE rear suspension design. Mobility Veh Mech 46:1–18
20. Sajan S, Philip Selvaraj D (2021) A review on polymer matrix composite materials and their applications. Mater Today: Proc 47:5493–5498
21. Nallusamy S, Rekha RS, Saravanan S (2018) Study on mechanical properties of mono composite steel plate cart spring using pro engineer and ANSYS R16.0. Int J Eng Res Afr 37:13–22
22. Baravkar PS (2022) Experimental and fea investigation of V shape spring with materials. SSRN Electronic J 1:43–47
23. Ghosh AK, Dwivedi M (2020) Advantages and applications of polymeric composites. In: Processability of polymeric composites. Springer India, New Delhi, pp 29–57
24. Mathew J, Joy J, George SC (2019) Potential applications of nanotechnology in transportation: a review. J King Saud Univ—Sci 31:586–594
25. Begambre O, Laier JE (2009) A hybrid particle swarm optimization—simplex algorithm (PSOS) for structural damage identification. Adv Eng Softw 40:883–891
26. Gaing Z-L (2004) A particle swarm optimization approach for optimum design of PID controller in AVR system. Energy conversion. IEEE Trans Energy Conversion 19:384–391
27. Sahab MG, Toropov VV, Gandomi AH (2013) A review on traditional and modern structural optimization. In: Gandomi AH, Yang X-S, Talatahari S, Alavi AH (eds) Metaheuristic applications in structures and infrastructures. Elsevier, Oxford, pp 25–47

28. Yang X-S (2021) Particle swarm optimization. In: Yang X-S (ed) Nature-inspired optimization algorithms. Elsevier, pp 111–121

Superpixel Image Clustering Using Particle Swarm Optimizer for Nucleus Segmentation

Swarnajit Ray, Krishna Gopal Dhal, and Prabir Kumar Naskar

Abstract In this decade, various superpixel-based techniques have been proposed by various authors for image segmentation where few of them are for medical images. These algorithms are evaluated using several evaluation measures and datasets, resulting in inconsistency in algorithm comparison. We noticed that some superpixel-based algorithms are performing better than other algorithms, for example, clustering-based superpixel methods are more efficient than graph-based superpixel techniques. In this paper, we choose simple linear iterative clustering (SLIC) for its less computation time and great performance on pathology images. In this paper, particle swarm optimization (PSO) and k-means clustering (KM) methods are used both with superpixel preprocessing for kidney renal cell carcinoma images. Clustered images are compared with ground truth images, and SLIC with PSO outperformed all other methods.

Keywords Image segmentation · Superpixel · Pathology image · PSO

1 Introduction

Superpixel-based segmentation has been widely utilized in computer vision as an efficient technique to minimize computation time. Using superpixels, redundancy in the image can be eliminated, and image processing efficiency can be increased. As a

S. Ray (✉)
Department of Computer Science and Engineering, Maulana Abul Kalam Azad University of Technology, Nadia, West Bengal 700064, India
e-mail: swarnajit32@gmail.com

K. G. Dhal
Department of Computer Science and Application, Midnapore College (Autonomous), Paschim Medinipur, West Bengal 721101, India
e-mail: krishnagopal.dhal@midnaporecollege.ac.in

P. K. Naskar
Department of Computer Science and Engineering, Government College of Engineering and Textile Technology, Serampore, West Bengal 712201, India
e-mail: cse.prabir@gmail.com

M. Thakur et al. (eds.), *Soft Computing for Problem Solving*, Lecture Notes in Networks and Systems 547, https://doi.org/10.1007/978-981-19-6525-8_34

result, various superpixel segmentation methods have been developed, each with its own set of advantages and disadvantages. There have already been several superpixel generation algorithms proposed. SLIC algorithm is the best superpixel generation algorithm out of all of them. Depending on the application, a different superpixels algorithm may be more appropriate. The SEEDS [1] algorithm may be preferable if speed is a concern. The normalized cuts (NCs) [2] algorithm produces more regular superpixels with a more attractive appearance when using superpixels to form a graph. If the superpixel technique is used as preprocessing steps in the segmentation algorithm, then a technique like SLIC [3] vastly improves the performance of the segmentation algorithm. While determining the best method for every situation is difficult, the following features are always desirable: (a) To recall as many pixels as possible from the object boundaries, the boundaries of the superpixels should closely resemble those of the object. (b) In a preprocessing phase, superpixels should be easy to use and fast to compute, with low memory needs. (c) When used for segmentation, superpixels should speed up the process as well as improve the quality of the results.

The approach of superpixel-based segmentation is frequently employed for a variety of image segmentation tasks. Numerous authors applied the SLIC preprocessing approach to different types of images like natural-color images, MRI brain tumor images, ultrasounds images, and histopathological images. Elkhateeb et al. [4] used superpixels for optical remote sensing images (RSIs) to segment out sea-land of the coastal region. In this paper, the author introduces a novel superpixel fuzzy c-means (SPFCM) method to reduce the amount of redundant information and make use of spectral and spatial data. A Jaccard similarity coefficient (JSC) of 97.1 has been calculated for the proposed technique, with an average accuracy of 98.9%. Liu et al. [5] implemented SLIC with K-means for the segmentation of color images. K-means is employed in this case to partition the superpixel image into smooth and complex sections. Detection of tampering in complex regions is accomplished using scale-invariant feature transforms (SIFTs). To identify tampering in smooth sections, sector mask and RGB color features are proposed. Siyuan and Xinying [6] also proposed a SLIC with K-means method for image segmentation. Ilesanmi et al. [7] developed a SLIC with a graph cut segmentation method for segmentation of boundary region of tumor from breast ultrasounds images. The proposed method outperformed other methods in terms of segmentation average accuracy of 94% and the speckle noise. Gaussian noise can also be removed with this technique. Ghaffari et al. [8] employed a two-step superpixel generation method to develop a fast, weighted conditional random fields (FWCRFs) algorithm for synthetic-aperture radar (SAR) image segmentation. A fast robust fuzzy c-means clustering technique (FRFCM) is used to divide the SAR image into homogeneous and heterogeneous regions, respectively. Simultaneously, the image is segmented into superpixels using the simple linear iterative clustering (SLIC) technique. As a result of using DBSCAN, all of the superpixels that belong to the same class are combined into one large image. An improvement over traditional FCM was achieved using the new method. One fast and robust fuzzy c-means (FRFCM) was proposed by Kim et al. [9] where SLIC was used as preprocessing of gastric endoscopy images. It was found that using the SLIC superpixel algorithm, performance could be improved, while segmentation computation time was decreased. The FRFCM algorithm reduces artifacts like noise and out-of-place clustering. Kumar et al. [10] developed a superpixel-based FCM (SPOFCM) method for the detection of suspicious lesions in the brain, mammogram, and breast magnetic resonance images. The suggested technique can improve performance for organ or suspicious lesion segmentation in CAD-based clinical applications. Giraud and Berthoumieu [11] proposed a new nearest neighbor-based superpixel clustering (NNSC) for standard composite texture image segmentation. Patch-based nearest neighbor matching is used by the author to introduce a new clustering framework in place of the more traditional K-means clustering. The patch space allows us to directly group pixels, which allows us to capture texture information. Mohamed et al. [12] employed statistical pixel-level (SPL) method to build an automatic system for diagnosing eye disease severity, which is very helpful to detect glaucoma. Here, the author uses SLIC for image preprocessing.

The SLIC method was used to perform image preprocessing for the superpixel-based image segmentation technique described above. Apart from SLIC method, other superpixel methods are also used for image segmentation. Chakraborty and Mali [13] proposed a fuzzy image segmentation method based on superpixels and meta-heuristics to explain COVID-19 radiological images. Here, author used the watershed-based superpixel method, and in terms of rapid and accurate diagnosis, the proposed method significantly reduces the overhead of processing a large amount of geographic data. An image is often made up of many small overlapping portions that are difficult to separate, so this is an important consideration. For MRI brain tumor image segmentation, Khosravanian et al. [14] used enhanced lattice Boltzmann method (LBM) and fuzzy c-means (En-FCM). Multiscale morphological gradient reconstruction and watershed transform (MMGR-WT) operations are used to generate superpixel regions. In this method, there is no need to worry about noise, initialization, or non-uniform intensity. With an average running time of 3.25 s, it outperformed other cutting-edge segmentation algorithms in terms of speed and accuracy. Rela et al. [15] proposed CT abdominal image segmentation by employing a fast fuzzy c-means clustering method where superpixels are used to reduce computation time and eliminate the manual interface. In this algorithm, a multiscale morphological gradient reconstruction process was used to make a superpixel image with perfect contours.

The author explored different superpixel-based approaches and their use in the medical image domain in this section. In next section, we will discuss on superpixel generation technique and clustering methodologies. Section 3 is for result and discussion, while the final section has the conclusion.

2 Methodology

In image segmentation, clustering has been proven to be a useful way to do it. An image dataset is divided into several disjoint groups or clusters where there are a lot of intra-cluster similarities and not much inter-cluster similarity. One of the primary goals of the clustering-based technique is to increase internal cluster similarity while decreasing external cluster similarity. Objective functions are defined as a result of this concept [16–18]. Using the strategy of minimizing or maximizing, one or more objective functions can result in optimal data partitioning.

2.1 Superpixel Generating Techniques

Superpixel segmentation is a technique for preprocessing images that divides the image into small areas. A common definition of a superpixel region in an image is an area that is both homogeneous and uniform to the human eye. Image segmentation can be more efficient and effective when using superpixel. Using local spatial information from the images, superpixel can perform image pre-segmentation.

Superpixel segmentation methods have proliferated in recent years. There are numerous subcategories of superpixel segmentation techniques, such as watershed based, graph based, path based, density based, and clustering based; each has a different set of categorization criteria. Simple linear iterative clustering (SLIC) [3] comes under the clustering-based method. Seed pixels are used to initialize the superpixel algorithm [19], which is based on clustering algorithms like k-means and uses color, spatial, and additional information like depth to do so. Intuitively, the number of created superpixels and their compactness can be adjusted. The brief discussion on the SLIC algorithm has been presented as follows.

2.2 SLIC (Simple Linear Iterative Clustering) Algorithm for Making Superpixels

SLIC is easy to learn and use. Because of this, the algorithm's only input is the desired number of approximately equal-sized superpixels, which is set K by default. Suppose the image has N number of pixels and K amount of superpixels. Then, the average area of superpixels is N/K. The distance between centers is $S = \sqrt{N/K}$. A single superpixel is a group of five values (Eq. 1). One can say it can be represented by five-dimensions [labxy] space.

$$C_k = [l_k, a_k, b_k, x_k, y_k,]^T \tag{1}$$

Here, [lab] denotes the pixel color vector in CIELAB color space, and x, y represent the pixel position. Here, author introduced a new distance measurement technique in 5D space considering the superpixel size because the maximum possible distance in CIELAB color space between two colors is limited. So, distance measurement technique needs to add special distance in x, y plane along with the distance between color spaces which is calculated by Euclidean distance. The details distance measurement technique is discussed below.

$$d_{\text{lab}} = \sqrt{(l_k - l_i)^2 + (a_k - a_i)^2 + (b_k - b_i)^2} \tag{2}$$

$$d_{xy} = \sqrt{(x_k - x_i)^2 + (y_k - y_i)^2} \tag{3}$$

$$D_s = d_{\text{lab}} + \frac{m}{s} d_{xy} \tag{4}$$

Here, S is the distance between two superpixel centers. The m parameter will determine the influence of the spatial distance between pixels when computing the final distance. The compact of the cluster is proportionate to the value of m, and the value lies between 1–20.

Algorithm 1. Superpixel Generation Technique

Step-1: *Equation (1) is used to set up the cluster centers so that pixels are sampled at regular grid steps S.*

Step-2: *Set the lowest gradient cluster centers in an N-by-N neighborhood.*

Step-3: ***While*** *($E \leq threshold$) does not meet.*

Step-4: ***for*** *each cluster center C_k do.*

Step-5: *Compute the distance around the cluster center from a $2S \times 2S$ square according Eq. (4) and assigned it.*

Step-6: ***end for***

Step-7: *Recompute new cluster centers and measure residual error E (distance between new and old).*

Step-8: ***end***

Step-9: *Enforce connectivity*

2.3 Objective Function of Superpixel Image-Based Segmentation

Using local spatial information from the images, superpixel can perform image pre-segmentation. For color image segmentation using superpixel-based clustering, the objective function is taken as:

$$J_{\text{SC}} = \operatorname{argmin} \sum_{l=1}^{ns} \sum_{k=1}^{K} S_l u_{kl}^{e} \left\| \left(\frac{1}{S_l} \sum_{p \in R_l} z_p \right) - m_k \right\|^2 \tag{5}$$

where l is color level and $1 \leq l \leq ns$, ns is the superpixel image's region count. S_l is the number of pixels in the l th region R_l, and z_p is the color pixel in the l th region of the superpixel that is obtained by any superpixel generating method.

For hard clustering like K-means, the membership U and centers have been computed as follows:

$$\boldsymbol{m}_k = \frac{\sum_{l=1}^{ns} u_{kl}^{e} \sum_{p \in R_l} z_p}{\sum_{l=1}^{ns} S_l u_{kl}^{e}} \tag{6}$$

$$u_{kl} = \begin{cases} 1 \text{ if } k = \underset{K}{\operatorname{argmin}} \left\| \left(\frac{1}{S_l} \sum\limits_{p \in R_l} z_p \right) - \boldsymbol{m}_k \right\|^2 \\ 0 \qquad \text{otherwise} \end{cases} \tag{7}$$

Finding global optima by minimizing Eq. (5) is computationally expensive and regarded as an NP-hard problem. Therefore, the researchers are also concentrating to introduce the efficient nature-inspired optimization algorithms (NIOAs) to perform clustering to find global or nearly global optima of this non-convex optimization problem. Here, well-known NIOA called particle swarm optimization (PSO) algorithm has been used. The following subsection provides a brief mathematical implementation of the PSO algorithm.

2.4 *Particle Swarm Optimization (PSO)*

In 1995, Kennedy and Eberhart created particle swarm optimization (PSO) in response to the way some animals, like flocking birds and schooling fish, search for food together [20]. This is one of the most well regarded algorithms in the literature of optimization, and it is widely used in image processing. Here, each solution candidate is described as "particles." Each particle represents a d-dimensional point in the search space when there are d variables in the optimization problem. It is represented by the below solution.

$$\overrightarrow{X_i^t} = (x_{i1}, x_{i2}, \ldots, x_{id})^T \tag{8}$$

$\overrightarrow{X_i^t}$ stored the position vector of the ith particle in d-dimension space $(x_{i1}, x_{i2}, \ldots, x_{id})$. t denotes the iteration. To update the position vector, we need to define the movement direction and speed. This is done by a vector called velocity. Each particle has its velocity vector $\overrightarrow{V_i^t} = (v_{i1}, v_{i2}, \ldots, v_{id})^T$. The initial position and velocity of the particle are generated randomly within the search space.

Algorithm 2. Particle Swarm Optimization (PSO) Algorithm

Step-1: *Initialize the parameters* $(n, c_1, c_2, w_{\min}, w_{\max}, MaxIter)$.
Step-2: *Initialize the population of n particles.*
Step-3: ***do***
Step-4: ***For** each particle*
Step-5: *Calculate the objective of the particles*
Step-6: *Update* P_{best} *if required.*
Step-7: *Update* g_{best} *if required.*
Step-8: ***End for***
Step-9: *Update the inertia weight using Linear Decreasing rule* [21].
Step-10: ***For** each PSO algorithm's distance rule.*
Step-11: *Update the velocity* $\left(\vec{V}\right)$ *using* Eq. (9).
Step-12: *Update the position* $\left(\vec{X}\right)$ *using* Eq. (10).
Step-13: ***End for***
Step-14: ***While** the termination condition is unsatisfied*
Step-15: ***Return*** g_{best} *as the best estimation of the global optimum.*

In each iteration, the velocity vector is updated as follows.

$$\overrightarrow{V_i^{t+1}} = w\vec{V_i^t} + c_1 \cdot r_1^t\left(\overrightarrow{P_{\text{best},i}^t} - \vec{X_i^t}\right) + c_2 \cdot r_2^t\left(\overrightarrow{g_{\text{best}}^t} - \vec{X_i^t}\right) \text{ for } i = 1, 2, \ldots, d \quad (9)$$

Here, w is the inertia of the particles, normally $w \in 0.9$ to 0.4, which tunes exploitation and exploration. Exploration is the lowest level when $w = 0$. This value decreases linearly when the number of iterations increases. c_1 and c_2 are the acceleration coefficients. When $c_1 = 0$, then exploration is the lowest level, and exploitation is the highest level, and when $c_2 = 0$, then it is reverse of previous. r_1 and r_2 are the random numbers which are drawn from a uniform distribution within the interval [0, 1]. $P_{\text{best},i}$ is personal best of ith particle, and g_{best} is the global best. The position of a particle is modified as follows.

$$\overrightarrow{X_i^{t+1}} = \vec{X_i^t} + \vec{V_i^t}; \text{ for } i = 1, 2, \ldots, d \quad (10)$$

Based on the above discussion, the algorithm of PSO has been presented as **Algorithm 2**. Whereas, the summarization of the complete PSO-based superpixel image clustering process is presented as **Algorithm 3**.

Algorithm 3. Procedure of Superpixel Image Clustering Using PSO

Step-1: *Use SLIC superpixel generation technique to compute the superpixel image.*
Step-2: *Consider the objective functions specified in* Eq. (5).
Step-3: *Each solution should start with randomly chosen K cluster centers.*

Step-4: *Each solution fitness calculated by assigning each superpixel to the closest cluster by computing the distance from centers.*

Step-5: ***While*** *(the halting condition) is not satisfied*

Step-6: *Update the solution by utilizing PSO's operators.*

Step-7: ***For*** *each PSO's solution*

Step-8: ***For*** *each pixel contained inside the image*

Step-9: *Calculate the distance from the cluster center to each superpixel.*

Step-10: *Assign the superpixel to the closest cluster.*

Step-11: *Recalculate the fitness of the solutions.*

Step-12: *Choose the optimal global solution.*

Step-13: ***end for***

Step-14: ***end for***

Step-15: ***end While***

3 Results and Dıscussıon

Using Matlab R2018b and ×64-based Windows 11 OS, Ryzen 5 CPU with 8 GB DR-4 RAM, the experiment was carried out on 100 kidney renal cell images. The data for this research came from the Beck Laboratory at Harvard University. Histopathological images of renal cell carcinoma at high resolution were obtained from the portal of The Cancer Genome Atlas (TCGA) and made publicly available for research purposes. At a resolution of 400 × 400 pixels, there are 810 histological images of ten different types of kidney renal cell carcinoma images. The National Cancer Institute and The National Human Genome Research Institute in the United States together fund the Translational Cancer Genome Alliance (TCGA), a major cancer research organization. To create superpixels for preprocessing tasks, the simple linear iterative clustering (SLIC) method is used. KM and particle swarm optimization (PSO) have been used for kidney renal cell segmentation. Number of superpixel used to segment the input image is 4000. For PSO, the parameter setting has been performed empirically. The acceleration coefficients are set to 2; population size (n) is set to 100, $w_{\min} = 0.4$, $w_{\max} = 0.9$; termination condition has been taken as the maximum iteration ($MaxIter$) which is set to 150. KM terminates the process if the change in consecutive centers values is smaller than $\eta = 10^{-5}$. The number of clusters required to perform nuclei segmentation is empirically investigated for different values ranging from 3 to 10 for the images under consideration. It has been discovered that SLIC-PSO produces the best results when the cluster number is 4. As a result, it is reasonable to use a cluster number of 4 for the experiments conducted. For measuring the segmentation efficiency of the proposed and other tested clustering algorithms, four well-known ground truth-based image segmentation quality parameters were used. Quality parameters include segmentation accuracy (SA), dice index (DI), Jaccard index (JI), and Matthews correlation coefficient (MCC) [18]. These parameters with higher values indicate better segmentation.

3.1 *Results and Discussion of Kidney Renal Cell Images*

KM and PSO algorithms have been utilized to segment the kidney renal cell images. Simple linear iterative clustering is used to create superpixels (SLIC). In this section, the author used a simple clustering method and clustering with superpixel preprocessing. Additionally, the KM and PSO clustering methods were compared to K-means clustering and PSO based on superpixels. Figures 1 and 2 represent the original color kidney renal cell images, the ground truth of segmented images provided by experts, and binary segmented kidney renal cell images by employing four clustering methods, namely KM, SLIC with KM (SLIC-KM), PSO, and SLIC with PSO (SLIC-PSO). The visual analysis of both Figs. 1 and 2 demonstrates that the SLIC-PSO produces the best segmentation results. SLIC-PSO-based method segment out the nucleus from the images is almost near to ground truth which is also given in Figs. 1 and 2. In comparison to the ground truth image, the segmentation accuracy of the first image is near 93.22% and for the second image is 93.40%. In addition to visual analysis, four well-known segmentation quality metrics, namely SA, DI, JI, and MCC, were used to evaluate the segmentation performance. Segmentation quality metrics for two samples of kidney renal cells are shown in Table 1 for reference. In the same Table 1, the average values for the segmentation quality parameter for 100 images are listed. The bolded values in Table 1 represent the best numerical values. The majority of the quality parameters clearly indicate that SLIC-PSO outperforms the other clustering techniques. Figure 3 illustrates the relationship between iteration and fitness value at each iteration graphically. The use of superpixel as a preprocessing task clearly reduces overall computational time. For example, the computational time of PSO for original kidney renal cell images is 46.28 s, but superpixel-based PSO has a lower computational time of 42.12 s. Due to the fact that the number of superpixels used to segment the input image (K) value varies according to the user, the time complexity associated with superpixel generation may vary.

4 Conclusion

In this work, various clustering approaches are compared, as well as their performance with and without the usage of a superpixel as a preprocessing method. K-means (KM) and particle swarm optimization (PSO) are two of the most commonly used clustering techniques which are used to segment kidney renal cell images. Simple linear iterative clustering (SLIC) is one of the most common ways to produce superpixels. Experiments show that for kidney renal cell images, the SLIC-PSO is best at the segmentation of nucleus from the background. The segmentation accuracy of the SLIC-PSO is 87.69% whereas KM, PSO, SLIC-KM are associated with 87.20%, 87.07%, 87.65%. According to segmentation accuracy and other quality parameter, the SLIC-PSO algorithm outperformed the others. In the future, we will

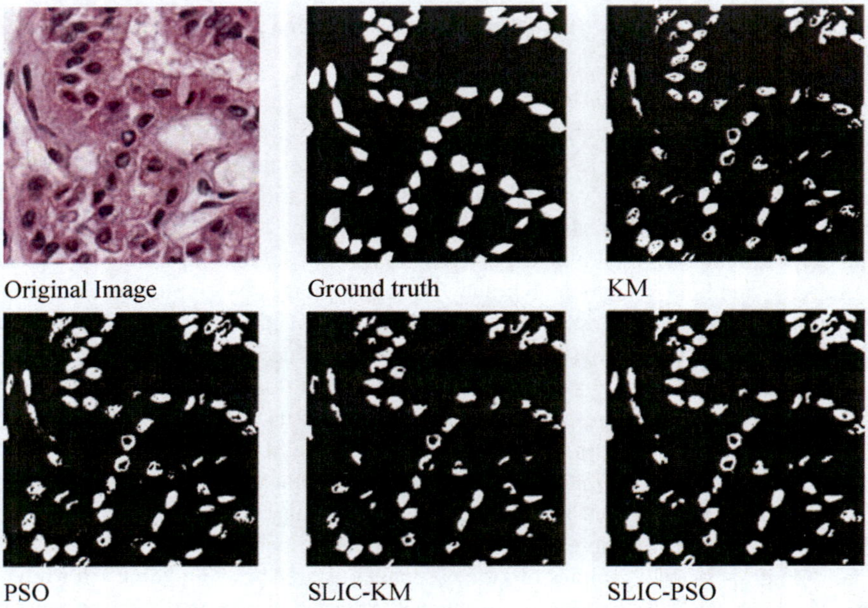

Fig. 1 Nucleus segmented results for first pathology ımage

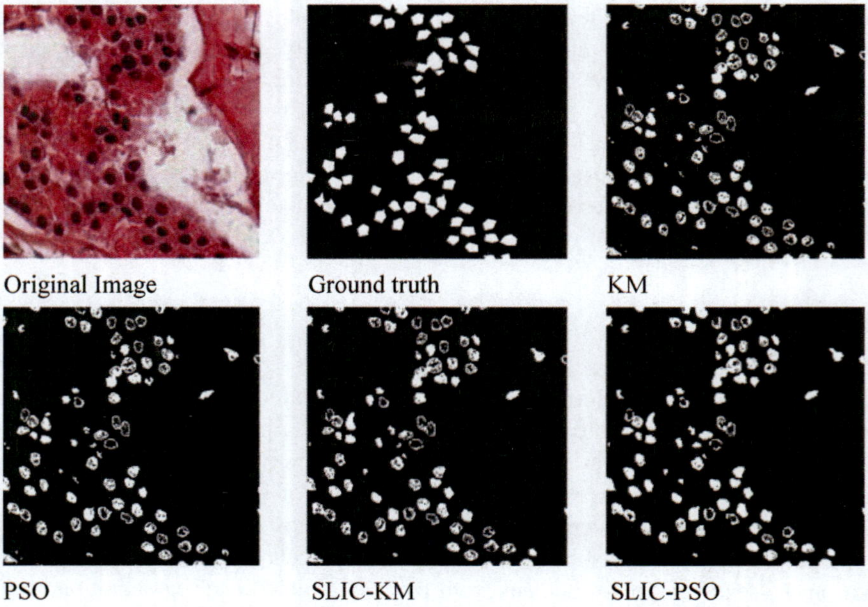

Fig. 2 Nucleus segmented results for second pathology ımage

Table 1 Quality parameters of two kidney renal cell sample images and the average of fifty images are given

Sample No	Method	Accuracy	MCC	DICE	JACCARD
Figure 1	KM	0.9219	0.6580	0.6798	0.5149
	PSO	0.9308	0.7089	0.7438	0.5921
	SLIC-KM	0.9206	0.6513	0.6723	0.5063
	SLIC-PSO	**0.9322**	**0.7147**	**0.7485**	**0.5981**
Figure 2	KM	0.9318	0.5767	0.6134	0.4424
	PSO	0.9307	0.5747	0.6124	0.4414
	SLIC-KM	0.9280	0.5677	0.6073	0.4360
	SLIC-PSO	**0.9340**	**0.6201**	**0.6558**	**0.4879**
Average	KM	0.8720	0.5146	0.5590	0.3912
	PSO	0.8707	0.5179	0.5677	0.4028
	SLIC-KM	0.8765	0.5398	0.5813	0.4129
	SLIC-PSO	**0.8769**	**0.5533**	**0.6017**	**0.4352**

(Results are shown in bold for the best ones)

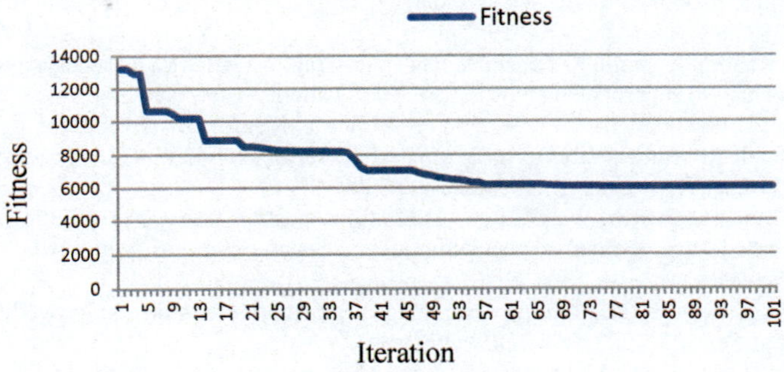

(a)

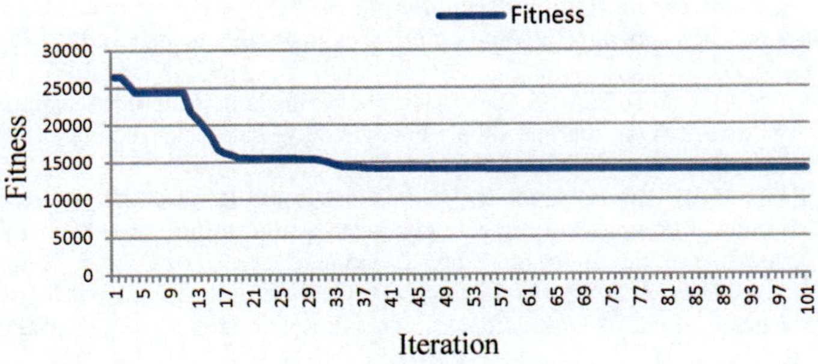

(b)

Fig. 3 Graphs for iteration versus fitness values: **a** for Fig. 1; **b** for Fig. 2

apply this technique to a variety of medical images and study several emerging technologies, like color segmentation, image fusion, and deep learning.

Funding: This work has been partially supported by the grant received in the research project under RUSA 2.0 component 8, Govt. of India, New Delhi.

Conflict of Interest: On behalf of all authors, the corresponding author states that there is no conflict of interest. The authors declare that they have no conflict of interest.

Ethical Approval: This article does not contain any studies with human participants or animals performed by any of the authors.

References

1. Van den Bergh M, Boix X, Roig G, de Capitani B, Van Gool L (2012) Seeds: superpixels extracted via energy-driven sampling. In: European conference on computer vision. Springer, Heidelberg, pp 13–26
2. Shi J, Malik J (2000) Normalized cuts and image segmentation. IEEE Trans Pattern Anal Mach Intell 22(8):888–905
3. Achanta R, Shaji A, Smith K, Lucchi A, Fua P, Süsstrunk S (2012) SLIC superpixels compared to state-of-the-art superpixel methods. IEEE Trans Pattern Anal Mach Intell 34(11):2274–2282
4. Elkhateeb E, Soliman H, Atwan A, Elmogy M, Kwak KS, Mekky N (2021) A novel coarse-to-fine sea-land segmentation technique based on superpixel fuzzy C-means clustering and modified Chan-Vese model. IEEE Access 9:53902–53919
5. Liu Y, Wang H, Chen Y, Wu H, Wang H (2020) A passive forensic scheme for copy-move forgery based on superpixel segmentation and K-means clustering. Multimedia Tools Appl 79(1):477–500
6. Siyuan R, Xinying L (2020) Superpixel ımage segmentation based on ımproved K-means. J Phys: Conf Se 1533(3):032067 (IOP Publishing)
7. Ilesanmi AE, Idowu OP, Makhanov SS (2020) Multiscale superpixel method for segmentation of breast ultrasound. Comput Biol Med 125:103879
8. Ghaffari R, Golpardaz M, Helfroush MS, Danyali H (2020) A fast, weighted CRF algorithm based on a two-step superpixel generation for SAR image segmentation. Int J Remote Sens 41(9):3535–3557
9. Kim DH, Cho H, Cho HC (2019) Gastric lesion classification using deep learning based on fast and robust fuzzy C-means and simple linear iterative clustering superpixel algorithms. J Electrical Eng Technol 14(6):2549–2556
10. Kumar SN, Fred AL, Varghese PS (2019) Suspicious lesion segmentation on brain, mammograms and breast MR images using new optimized spatial feature based super-pixel fuzzy c-means clustering. J Digit Imaging 32(2):322–335
11. Giraud R, Berthoumieu Y (2019) Texture superpixel clustering from patch-based nearest neighbor matching. In: 2019 27th European Signal Processing Conference (EUSIPCO). IEEE, pp 1–5
12. Mohamed NA, Zulkifley MA, Zaki WMDW, Hussain A (2019) An automated glaucoma screening system using cup-to-disc ratio via simple linear iterative clustering superpixel approach. Biomed Signal Process Control 53:101454
13. Chakraborty S, Mali K (2021) SuFMoFPA: a superpixel and meta-heuristic based fuzzy image segmentation approach to explicate COVID-19 radiological images. Expert Syst Appl 167:114142

14. Khosravanian A, Rahmanimanesh M, Keshavarzi P, Mozaffari S (2021) Fast level set method for glioma brain tumor segmentation based on superpixel fuzzy clustering and lattice boltzmann method. Comput Methods Programs Biomed 198:105809
15. Rela M, Rao SN, Reddy PR (2020) Liver tumor segmentation using superpixel based fast fuzzy C means clustering. Int J Adv Comput Sci Appl (IJACSA) 11(11)
16. Dhal KG, Gálvez J, Ray S, Das A, Das S (2020) Acute lymphoblastic leukemia image segmentation driven by stochastic fractal search. Multimedia Tools Appl 1–29
17. Dhal KG, Fister Jr I, Das A, Ray S, Das S (2018) Breast histopathology image clustering using cuckoo search algorithm. In: Proceedings of the 5th student computer science research conference, pp 47–54
18. Dhal KG, Das A, Ray S, Gálvez J (2021) Randomly attracted rough firefly algorithm for histogram based fuzzy image clustering. Knowl-Based Syst 216:106814
19. Stutz D, Hermans A, Leibe B (2018) Superpixels: an evaluation of the state-of-the-art. Comput Vis Image Underst 166:1–27
20. Eberhart R, Kennedy J (1995) A new optimizer using particle swarm theory. In; Micro Machine and Human Science, MHS'95, IEEE, pp 39–43
21. Bansal JC, Singh PK, Saraswat M, Verma A, Jadon SS, Abraham A (2011) Inertia weight strategies in particle swarm optimization. In: 2011 third world congress on nature and biologically inspired computing. IEEE, pp 633–640

Whale Optimization-Based Task Offloading Technique in Integrated Cloud-Fog Environment

Haresh Shingare and Mohit Kumar

Abstract The adoption of Internet of Things (IoT) applications is growing rapidly in various sectors to create a smart atmosphere like automobiles, medicine, smart cities, agriculture, etc. Each IoT device produces enormous data that requires the resources for processing, but these devices have resource constraints. However, the cloud platform executes such types of tasks, but it cannot process the real-time latency-sensitive applications due to communication delay, scalability, and high power consumption. Task offloading becomes a challenging issue, where tasks should be allocated to an efficient resource that can process them without delay and save energy. Hence, fog computing is a promising solution to address the mentioned issue, and it provides a computation-oriented service, is closer to data sources, and improves the QoS metrics including delay and energy. Several task offloading approaches have been proposed to allocate the resource for IoT applications, but are unable to take the appropriate offloading decision and best resource allocation which degrades the QoS parameters. In this paper, a whale optimization approach is proposed that fills up research gaps by taking the offloading decision at run time. The simulation results proved that the proposed approach performs better as compared with the state-of-the-art approach.

Keywords Cloud computing · Fog computing · Internet of Things · Task offloading · Whale Optimization(WOA)

1 Introduction

In today's era, mobile devices are using different applications such as 3D gaming, virtual reality (VR), smart homes, and many more. The data generated by these applications is vast and difficult to implement on limited resource available devices like

H. Shingare (✉)
Department of CSE, Dr. B.R. Ambedkar National Institute of Technology, Jalandhar, India
e-mail: haresh.sith@gmail.com

M. Kumar
Department of IT, Dr. B.R. Ambedkar National Institute of Technology, Jalandhar, India
e-mail: kumarmohit@nitj.ac.in

M. Thakur et al. (eds.), *Soft Computing for Problem Solving*, Lecture Notes in Networks and Systems 547, https://doi.org/10.1007/978-981-19-6525-8_35

personal laptops or mobile devices. It takes a large amount of time to process and sometimes fails to satisfy the requirements of the application. The cloud paradigm provides the services to IoT applications because it has infinite resources in virtualization mode and offers the services as per the demand. There are several characteristics of the cloud paradigm like resource pooling, elasticity, on-demand services, etc., which confirms that it is the most momentous and well-known evolution in the computing paradigm, but it fails to provide the services to latency-sensitive applications. Hence, a new paradigm has been introduced named fog computing to process the real-time applications without delay and accomplish the demands of IoT applications at local nodes [1]. The main goal of IoT-based applications is to provide end-to-end service without the involvement of the user. The fog device provides the services to real-time IoT applications such as healthcare systems, video surveillance, parking systems, and many more. Fog computing develops its architecture using different routers. There are many advantages of fog computing like it creates very low latency in between data generation devices and processing or analytics devices, and it also reduces bandwidth. It also has more user privacy because it processes data locally [2]. Uncountable transmitting or processing delay from IoT device to cloud device becomes a highly impactful factor while processing data with maximum network bandwidth, transmitting data from IoT applications to the cloud [3]. The main goal of fog computing is to create an intermediate platform for processing the data. Such systems significantly reduce the transmitting, communicating, and receiving delay, thus reducing the overall propagation delay. There are many more applications of IoT available that are facing the issue of resource constraints in fog environments because the fog paradigm has limited computation and storage capability and cannot execute a large amount of data in a limited period of time affecting other QoS parameters as well. On the contrary, the centralized cloud infrastructure can process a large amount of data but fails to provide accurate results to real-time IoT applications, whereas the fog paradigm is a decentralized architecture in which many devices are connected. Hence, to overcome the issue of cloud and fog, we need the integrated cloud-fog architecture that can process every type of application (latency sensitives or others) without violating the constraints. The goal of integrated architecture is to allocate the resource for IoT applications where resources may be fog oriented or cloud oriented. Hence, identifying the applications and offloading them to suitable platforms (cloud or fog) as per their demands become a challenging issue in an integrated cloud-fog environment [4].

1.1 Task Offloading

It is a mechanism to allocate the appropriate resources for IoT applications as per their requirement. The required resource can belong to the cloud or fog paradigm to process the diverse IoT applications. Hence, it becomes a challenging issue to search and allocate the best possible resources for delay and resource concerning applications. After the adoption of IoT-based applications in various domains, it has

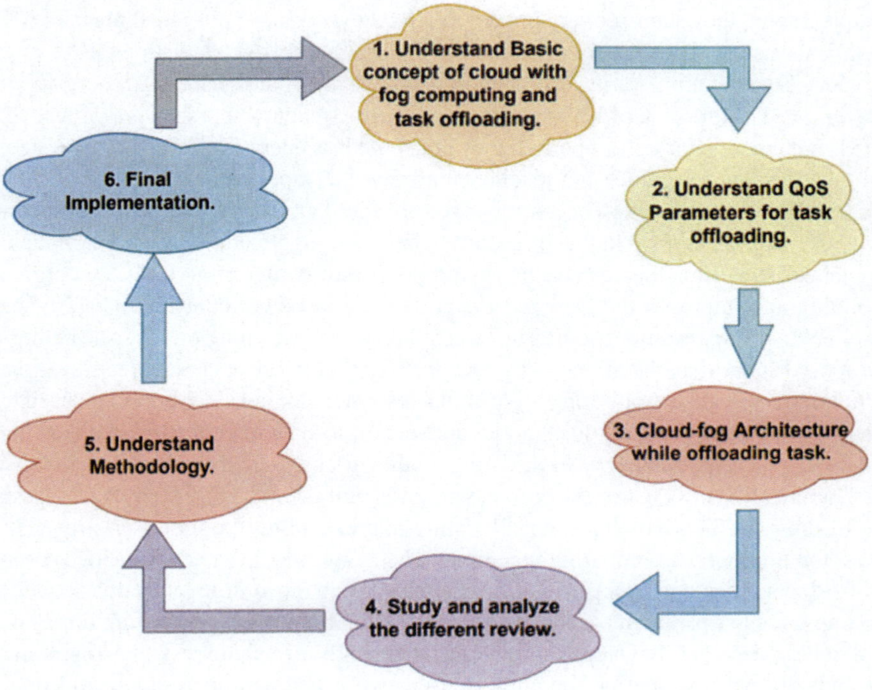

Fig. 1 Research approach

received the attention of various authors, scientists, and researchers. The IoT devices should offload the applications to an appropriate device based on QoS metrics like latency, bandwidth, power consumption, scalability, and many more.
The rest of the paper is organized as follows. Section 2 depicts the literature review, and Sect. 3 discusses the whale optimization algorithm. We introduce the architecture of an integrated cloud-fog environment for the task offloading as described in Sect. 4. The performance of the proposed algorithm with results is discussed in Sects. 5 and 6 followed by a conclusion.

2 Literature Review

Several techniques have been proposed by several researchers to offload the task efficiently in an integrated cloud-fog environment. Some of the recent techniques are discussed below:
The authors proposed the LOTEC algorithm to study energy-efficient processing of data for managing the trade-off between the time it takes to analyze data and the cost of running the system. To evaluate the algorithm performance, the authors have

implemented the round-robin approach to allocate the cloud or fog resources to IoT applications [1]. The authors implemented the PSO and BAT algorithms. BAT outperforms the other in a variety of scenarios. Makespan reduction leads to energy conservation. The idle time of VMs is reduced when the system make span is diminished. VMs are inactive for a short period; they consume less energy [5]. The fog nodes are not appropriate for processing resource-intensive IoT applications because of their limited resource capacity. The authors have proposed an integrated cloud-fog infrastructure to process the delay and resource-concerned applications using an efficient metaheuristic offloading approach. The proposed firefly algorithm finds the optimal solution and improves the QoS parameter's time and power consumption [2]. The authors focus on resource monitoring and the service allocation process. The authors proposed support and confidence-based approach that improves resource utilization. Traditional resource monitoring algorithms, resource monitoring models, are used to monitor resources, and the push model delivers up-to-date resource information and improves service efficiency [6]. Existing studies focus on real-time delay-sensitive IoT applications. Researchers propose a new delay-dependent priority aware task offloading (DPTO) technique for allocating and executing the end user's requests over the appropriate computing machines. The suggested approach prioritizes each work depending on its completion time and places it in the multilevel feedback queue to improve QoS parameters such as communication time, average queuing time, and offloading time [7]. IoT applications must efficiently distribute tasks on fog nodes to increase service quality, the authors used two schedulers, particle swarm optimization (PSO), and ant colony optimization (ACO) for successfully balancing IoT jobs among fog nodes while keeping transmission costs and response time in mind. The proposed ACO-based scheduler improves IoT application response time and efficiently balances the load [8]. Mobile devices can offload their heavy duties to fog nodes. The authors describe a module placement strategy based on classification and regression tree algorithms in this research. In cloud and fog computing platforms, simulated studies suggest that MPCA outperforms FF and local mobile processing approaches. MPMCP has a faster response time and higher performance [3]. This article presents a dependable scheduling strategy for assigning customer requests to cloud-fog environments' resources. The load balanced service scheduling approach (LBSSA) addresses load balancing across resources when assigning requests. The quicker the method for scheduling requests, the greater the value, particularly for real-time queries [9]. Load balancing is another important aspect of cloud technology since it allows for a balanced distribution of load over several servers to meet customers' increasing demands. Particle swarm optimization (PSO), cat swarm optimization (CSO), BAT, cuckoo search algorithm (CSA) optimization algorithm, and whale optimization algorithm (WOA) were used in this study to balance the load, improve energy efficiency, and better resource scheduling [10]. Emerging vehicular applications require a real-time connection between the data center and end users. Three-layer service architecture with the resources of vehicular fog (VF), fog server (FS), and central cloud are used cooperatively. Computational results demonstrate the probabilistic task offloading (PTO) issue to reduce the execution time and energy consumption. The task offloading approach combines the alternating direction method of multipliers with particle swarm optimization to tackle the PTO issue [11]. In VFC, several task offloading issues are necessary to be solved. Authors propose the energy-efficient scheduling approach to schedule the task on fog nodes to satisfy the task deadline and resource allocation using a reinforcement-based algorithm. With a fuzzy logic heuristic reinforcement approach, it will outperform other scheduling approaches like rate monotonic FCFS [12].

Motivation and Goal: Motivation and Goal: Task offloading becomes a significant aspect of the integrated cloud-fog environment for IoT networks and directly influences the mentioned QoS parameters. Hence, an efficient task offloading technique is required to accomplish the objectives without compromising any constraints. The goal of the paper is to understand the concept of task offloading, get an overview the existing state-of-the-art strategy, and analyze the different QoS parameters that are considered for task offloading to ensure that the system needs to link all tasks with available resources efficiently. The research methodology figure is shown in Fig. 1 [13].

3 Whale Optimization Algorithm

WOA is a nature-inspired algorithm inspired by humpback whales' bubble-net hunting behavior and searches for the optimal solution to the hard computational problems in the field of computer science [14]. Bubble-net feeding is a term used to describe the foraging behavior of humpback whales. Humpback whales tend to hunt krill or small fish near the water's surface. We have mapped the WOA algorithm with our task offloading problem where the algorithm finds the optimal resource over the fog server or cloud data center for n number of requests. Each WOA represents the solution after mapping with a whale. It is represented with 1 × n where n is a task. Each element of the mapping whale has 1 and m numbers, where m is in total VMs.

Exploration phase: The search agent finds the best solution randomly based of current position of search agent.

$$\Delta = |\nu * \sigma - \chi| \tag{1}$$

$$\chi(t+1) = \sigma - \alpha * \Delta \tag{2}$$

σ = is random position vector from population

Encircling Phase: During hunting, humpback whales surround their prey. The current best candidate solution is considered the best and closest to the optimal option. Following is the encircling behavior model that is used to update the position of the other whales concerning the best search agent.

$$\Delta = |\nu * \Psi(t) - \chi(t)| \tag{3}$$

$$\chi(t+1) = \Psi(t) - \alpha * \Delta \tag{4}$$

αandν = are coefficient vector
Ψ = is the best solution position
χ = is position vector
The value of αandν calculated as following:

$$\alpha = 2 * \mu * \omega - \mu \tag{5}$$

$$\nu = 2 * \omega \tag{6}$$

Bubble-net Hunting: This bubble-net hunting strategy uses two approaches: shrinking encircling mechanism and a spiral updating position. This approach begins with calculating the distance between the hunt (X', Y') and the whale (X, Y).

$$\chi(t+1) = \begin{cases} |\Psi(t) - \chi(t)| * e^{xy} * \cos(2 * \pi * l) + \Psi(t) & \text{if } x \geq 0.5 \\ \Psi(t) - \alpha * \Delta & \text{otherwise} \end{cases} \tag{7}$$

$|\Psi(t) - \chi(t)|$= distance between whale and hunt.
x is the constant value, y is the random number$[-1,1]$.

4 Architecture of Integrated Cloud-Fog Environment

We have proposed an integrated cloud-fog framework that offers the services to upcoming demand in an efficient way and improves the QoS metrics. This framework has been divided into three parts: IoT layer, fog layer and cloud layer as shown in Fig. 2.

4.1 IoT Layer

Components for IoT Device Communication [15]

1. **IoT Device**: Any IoT device, from the tiniest temperature sensor to a massive industrial robot, is an IoT device.
2. **Local Communications**: The process through which the gadget communicates with nearby devices.
3. **Application Protocol**: The architecture that determines how information material is transmitted is known as the application protocol.
4. **Gateways**: Often used to connect local device networks to the Internet, gateways translate and re-transmit data.

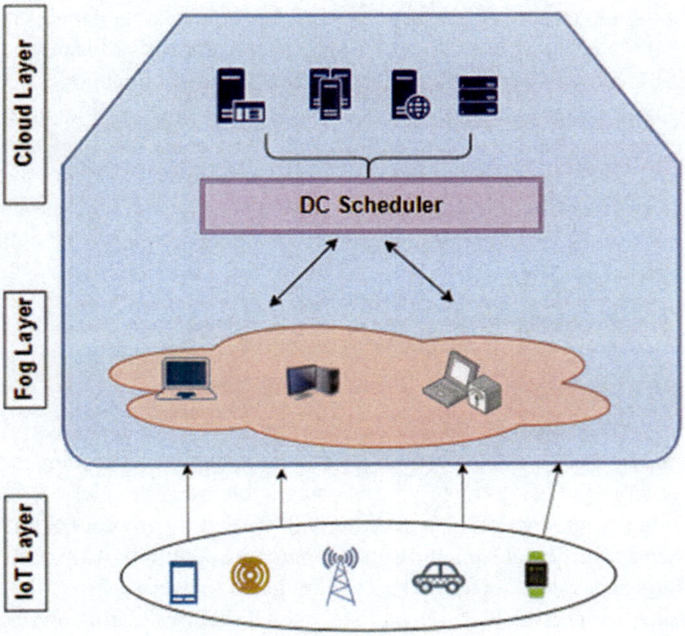

Fig. 2 Cloud-fog architecture

5. **Network Servers**: Systems that control the receipt and transmission of IoT data, which are usually housed in data centers.
6. **Cloud Applications**: Transform IoT data into actionable information for users.
7. **User Application**: This is where users can view IoT data, alter it, and send commands to IoT devices.

Sensor devices may connect in a variety of ways and with a variety of protocols. That is because how they communicate is determined by what they are doing, where they are doing it, what other devices and systems they need to communicate with, and what they have to say. There is not a single optimum protocol, which is effectively a "language" for routing data from one IoT device to another. The best option is always determined by the application's requirements.

4.2 Fog Layer

Fog computing allows IoT devices to do computing, decision-making, and action-taking while only sending relevant data to the cloud. Cisco invented the term "Fog computing", and it works as a computing paradigm. The fog brings the cloud closer to the devices that generate and act on IoT data. Fog nodes, as they are known, may be placed everywhere there is a network connection [6]. In this section, we will describe the fog model. In a fog environment, there is one fog node that is the master and other nodes work as slaves (leader). Fog master's aim is to allocate and manage the resources in a fog environment. It accepts requests from client nodes and assigns them to fog slave nodes to process. Fog slave's nodes accept the client node request, create different small chunks of operation, and send requests to sensors and fog devices to process them. Each fog device collects processing data from sensors and executes the request on source data. The fog device node receives the service request from the fog slave node and creates a number of container requests that collects data from different IoT sensors and return the result to fog slave nodes [6].

4.3 Cloud Layer

This is the top layer in the architecture. It has a large number of resources available to process or store the data remotely as it is accessible over the Internet. The task is offloaded to the cloud environment when it realizes that it can process the data at a fog level due to insufficient availability of resources. Sometimes tasks are offloaded to fog by cloud because they require less communication overhead over the network. Also, fog nodes offload the task to the cloud when it requires storing and processing the task for a long period of time.

5 Result and Discussion

In this section, we will analyze the performance of the proposed whale optimization algorithm in cloud and fog environments based upon some QoS metrics. We have also considered some test cases for comparative analysis between cloud and fog services for latency-sensitive IoT applications or CPU intensive applications.

5.1 Experimental Setup

The authors have experimented to implement the proposed whale optimization algorithm over a system with configuration 8GB RAM, AMD RYZEN processor, and Windows 11 Operating System with Java as programming language. Authors have evaluated the performance by changing the simulation parameters like task length, task deadline, and resource processing speed in different scenarios. The authors have considered two QoS parameters named makespan time and execution time for this experiment. To calculate the mentioned parameters, we follow the problem formulation from the article [16].

5.2 *Experimental Analysis*

In this section, we have done comparative analysis of whale optimization algorithm with two test cases.

5.2.1 Test Case 1 with Task Size 1000

In the first discussion, we have taken the 25 resources initially and increased up to 75 for this experiment, where all the resources are heterogeneous and their processing speed or CPU cycle computation varies from 1500 to 2000. For cloud environment processing, transmission delay is considered 0.25 s. We run the proposed whale optimization algorithm and found the makespan time for fog is 0.28, 0.24, 0.13 and cloud is 0.32, 0.26, and 0.14. Figure 3 shows the comparative analysis of whale optimization with respect to diverse tasks and heterogeneous VMs.

5.2.2 Test Case 2 with Task Size 1500

In this test case, we have taken 25 resources initially and increased up to 75 for this experiment, where all the resources are heterogenous and their processing speed or CPU cycle computation varies from 1500 to 2000. For cloud environment processing, transmission delay is considered 0.25 s. The makespan time for fog is 0.39, 0.25, and 0.19 and cloud is 0.43, 0.29, and 0.21. Figure 4 shows the comparative analysis of whale optimization concerning diverse tasks and heterogeneous VMs.

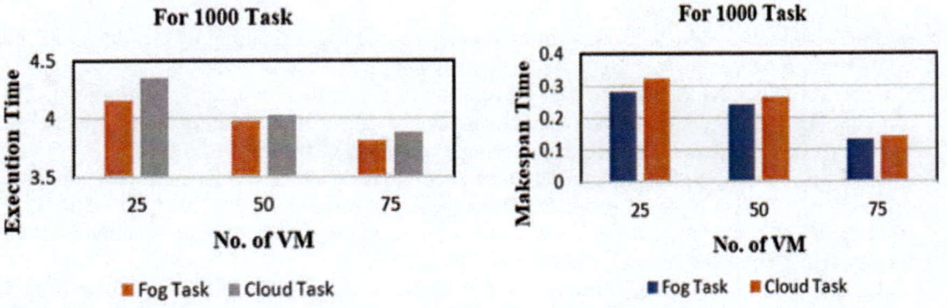

Fig. 3 Analysis for test case 1: execution and makespan time

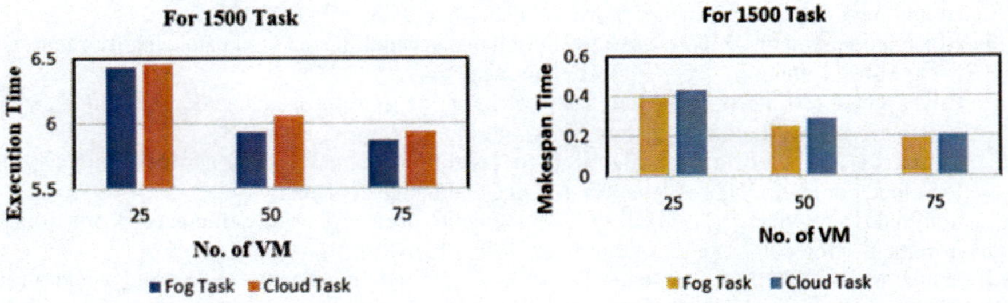

Fig. 4 Analysis for test case 2: execution and makespan time

6 Conclusion

In this paper, firstly we discussed the cloud and fog paradigm and their applications. The research methodology approach and motivation of the article are depicted in the introduction section. Further, we rigorously reviewed some state-of-the-art approaches for task offloading and resource allocation in an integrated cloud-fog environment. An integrated cloud-fog framework has been proposed for efficient task offloading to improve different QoS parameters. In addition, the structure of cloud fog with different layers along with their working principle is discussed. In the end, we have implemented a metaheuristic algorithm named whale optimization algorithm to offload the task in an integrated cloud-fog environment and allocate the best resource for the processing of IoT applications. We have evaluated the performance of the proposed algorithm over the simulation environment and found that the cloud paradigm is unable to process the latency-sensitive IoT applications due to its high delay, while such types of applications can process over the fog node due to less delay. The cloud paradigm can execute the high workload that can afford the latency. For future work, we plan to explore this approach to improve other QoS parameters.

References

1. Nan Y et al (2017) Adaptive energy-aware computation offloading for cloud of things systems. IEEE Access 5: 23947–23957
2. Adhikari M, Gianey H (2019) Energy efficient offloading strategy in a fog-cloud environment for IoT applications. Internet of Things 6:100053
3. Rahbari D, Nickray M (2020) Task offloading in mobile fog computing by classification and regression tree. Peer-to-Peer Netw Appl 13(1):104–122
4. Kumar M, Dubey K, Pandey R (2021) Evolution of emerging computing paradigm cloud to fog: applications, limitations and research challenges. In: 2021 11th international conference on cloud computing, data science engineering (confluence), IEEE
5. Mishra SK et al (2018) Sustainable service allocation using a metaheuristic technique in a fog server for industrial applications. IEEE Trans Industrial Informatics 14(10): 4497–4506
6. Battula SK et al (2019) An efficient resource monitoring service for fog computing environments. IEEE Trans Services Comput 13(4): 709–722
7. Adhikari M, Mukherjee M, Srirama SN (2019) DPTO: a deadline and priority-aware task offloading in a fog computing framework leveraging multilevel feedback queueing. IEEE Internet Things J 7(7): 5773–5782
8. Hussein MK, Mousa MH (2020) Efficient task offloading for IoT-based applications in fog computing using ant colony optimization. IEEE Access 8: 37191–37201
9. Alqahtani F, Amoon M, Nasr AA (2021) Reliable scheduling and load balancing for requests in cloud-fog computing. Peer-to-Peer Netw Appl: 1–12
10. Goyal S et al (2021) An optimized framework for energy-resource allocation in a cloud environment based on the whale optimization algorithm. Sensors 21(5): 1583
11. Liu Z et al (2021) A distributed algorithm for task offloading in vehicular networks with hybrid fog/cloud computing. IEEE Trans Syst Man Cybernetics: Systems
12. Vemireddy S, Rout RR (2021) Fuzzy reinforcement learning for energy efficient task offloading in vehicular fog computing. Comput Netw 199: 108463
13. Kumar M et al (2019) A comprehensive survey for scheduling techniques in cloud computing. J Netw Comput Appl 143: 1–33

14. Mirjalili S, Lewis A (2016) The whale optimization algorithm. Adv Eng Softw 95:51–67
15. https://data-flair.training/blogs/how-iot-works/
16. Xu J, Hao Z, Sun X (2019) Optimal offloading decision strategies and their influence analysis of mobile edge computing. Sensors 19(14):3231

Solution to the Unconstrained Portfolio Optimisation Problem Using a Genetic Algorithm

Het Shah and Millie Pant

Abstract A critical aspect of the mechanism of financial decision-making is the evaluation of promising assets and the distribution of resources between them. This aspect was formulated by modern portfolio theory as a quadratic optimisation problem of optimising expected returns and reducing portfolio risk. The model proposed by Markowitz is a perfect example of a multi-objective optimisation problem, in which the aim is to find the best trade-off solutions. Here, the trade-offs are between the portfolio return and its variance. In this paper, the multi-objective version of the unconstrained portfolio optimisation problem is presented. Henceforth, it is proved that there is no difference in the solutions obtained by solving the unconstrained multi-objective portfolio optimisation problem and the single-objective weighted sum approach of the unconstrained portfolio optimisation problem. A genetic algorithm is then suggested to solve the unconstrained problem of portfolio optimisation. Genetic algorithm is used on four existing data sets in the literature to find the efficient frontiers. The obtained unconstrained efficient frontiers are analysed using six different performance metrics in comparison with other state-of-the-art meta-heuristics.

Keywords Portfolio optimization · Multi-objective optimisation · Genetic algorithm · Evolutionary algorithm · Evolutionary computation

1 Introduction

In 1952, Harry Markowitz [1] provided a breakthrough in the field of portfolio allocation. The Markowitz mean–variance model provided the answer to a very relevant question: How does an investor distribute funds from the investment options that are

H. Shah (✉) · M. Pant
Indian Institute of Technology, Roorkee, UK 247667, India
e-mail: het0314@gmail.com

M. Pant
e-mail: pant.milli@as.iitr.ac.in

M. Thakur et al. (eds.), *Soft Computing for Problem Solving*, Lecture Notes in Networks and Systems 547, https://doi.org/10.1007/978-981-19-6525-8_36

available? Essentially, the model does a trade-off between the return and risk associated with a portfolio. The model constructed is a quadratic optimisation problem. Since then, many researchers have developed meta-heuristic algorithms based on the principles of evolution [2] to solve the Markowitz mean-var portfolio optimization problem.

The first paper to use a genetic algorithm to solve the portfolio optimization problem was published in 1993 [3]. After that, evolutionary algorithms have been extensively used to solve the Markowitz mean–variance model [4]. Many researchers have modified the portfolio optimization model to solve a multi-period portfolio optimization problem [5–7] as well as a fuzzy portfolio optimization problem [8–11].

Many researchers have also included real-life constraints in Markowitz's original model [12, 13]. Some of the modifications include adding cardinality constraints, using different risk measures, and adding constraints related to brokerage [14].

In this article [15], researchers have used artificial bee colony (ABC) heuristic to solve the cardinality-constrained portfolio optimization problem with feasibility enforcement. Whilst in another research paper [16], an improved ABC has been implemented to solve the portfolio optimization problem. Portfolio optimization problem was tackled by using differential evolution (DE) in this research articles [17, 18]. Particle swarm optimization (PSO) has been the most frequently used swarm intelligence algorithm in solving the portfolio optimization problem [19–21]. Simulated annealing is a single-solution-based heuristic which has been utilised in solving the portfolio optimization problem by many researchers [22, 23].

In this paper, the focus is on two main aspects. The first is to prove that the Pareto-optimal set is obtained successfully with the help of the weighted sum strategy for the unconstrained multi-objective portfolio optimisation problem. And the second is to propose a GA to solve the problem of unconstrained portfolio optimisation.

The presented theory of mean–variance by Markowitz is generally called modern portfolio theory (MPT) [24]. At the same time, the model itself is referred to as mean–variance optimisation (MVO), investment portfolio optimization (IPO), or portfolio selection problem (PSP). Many modifications are made to the initial portfolio optimisation model by different researchers by adding additional constraints or changing the measures of risks [25]. Whilst the model dealt with in this paper is unconstrained, it is only appropriate to coin the term unconstrained portfolio optimization (UPO) [23]. Throughout this paper, UPO will refer to the original Markowitz mean–variance model, as proposed in 1952.

The majority of the article is broken up as follows:

Basic concepts of multi-objective optimisation problems are discussed. Following this section is the discussion of the problem of portfolio optimisation. This section focuses on the paper's first goal, out of the two main goals, which is to show that the weighted sum approach is sufficient to find the entire Pareto-optimal set. Next, after a brief introduction to genetic algorithms, we propose a GA specifically designed to solve the portfolio optimization problem. Lastly, after defining the performance metrics used, we present the results and conclusion.

2 Multi-objective Optimisation Problems

In this section, a brief introduction to multi-objective optimisation problems (MOOPs) is presented. Also, it is discussed here what is the final aim of solving such problems.

The general form of a multi-objective optimisation problem (MOOP) is [26] as follows:

$$\text{minimise } f_m(x) \; m = 1, 2, \ldots, M; \tag{1}$$

$$\text{subject to } g_j(x) \geq 0, \; j = 1, 2, \ldots, J; \tag{2}$$

$$h_k(x) = 0, \; k = 1, 2, \ldots, K; \tag{3}$$

$$x_i^{(L)} \leq x_i \leq x_i^{(U)}, \; i = 1, 2, \ldots, n. \tag{4}$$

Here,

$x = (x_1, x_2, \ldots, x_n)$ is the vector of n decision variables,
f_m's are the objective functions,
g_j's are the inequality constraints,
h_k's are the equality constraints, and
$x_i^{(U)}$ is the upper bound and $x_i^{(L)}$ the lower bound of x_i.

The constraint in Eq. (4) constitutes a *decision variable space* D.

A solution x is called a *feasible solution* that meets all the constraints and variable limits. *The feasible region,* S, is the set of all the feasible solutions. It is frequently referred to as simply the *search space.*

Next, a few definitions are presented to discuss the concept of optimality.

Definition 1 Let $x^{(1)}$ and $x^{(2)}$ be two solutions such that:

1. $f_m\left(x^{(1)}\right) \leq f_m\left(x^{(2)}\right), \forall m$, i.e. $x^{(1)}$ is no worse than $x^{(2)}$ with respect to all the objectives.
2. $\exists m \in \{1, 2, \ldots, M\} \ni f_m\left(x^{(1)}\right) < f_m\left(x^{(2)}\right)$, i.e. $x^{(1)}$ is strictly better than $x^{(2)}$ in at least one objective.

Then, it is said that $x^{(1)}$ dominates $x^{(2)}$.

When $x^{(1)}$ dominates $x^{(2)}$ mathematically, we write it as $x^{(1)} \preccurlyeq x^{(2)}$.

Definition 2 Let $x \in S$ be such that no member of S dominates x. Let P be the set of all such solutions, i.e. $P = \{x \in S | \nexists y \in S \text{ such that } y \preccurlyeq x\}$. Such a set P is called the *Pareto-Optimal set*. The image of the Pareto-optimal set P in the objective function space is called the Pareto front, also referred to as the efficient frontier.

The main aim of the MOOP is to find an efficient frontier. It is typically represented graphically to visualise the trade-offs between the objectives. Hence, the concept of optimality in a MOOP is to find a set of Pareto-optimal solutions and not just a single optimal solution as in the case of single-objective optimisation problems. All the Pareto-optimal set members are equally good, and a member can be further chosen depending on the criteria set by the user.

In the next section, we discuss the unconstrained portfolio optimization (UPO) as a multi-objective optimisation problem (MOOP).

3 Portfolio Optimization Problem

The basic concepts associated with UPO are presented in this section [27]. Finally, the UPO model is formulated.

Let,

N be the number of assets to choose from,
μ_i be the expected return of asset i,
σ_{ij} be the covariance between assets i and j,
w_i be the proportions of the asset i.

Clearly, w_i's are the decision variables and $0 \leq w_i \leq 1$. Also, the proportions of the assets have to add up to one. Hence, the constraint $\sum_{i=1}^{N} w_i = 1$ is involved whilst choosing the values for the decision variables. The total variance of the assets is given by

$$\sum_{i=1}^{N}\sum_{j=1}^{N} w_i w_j \sigma_{ij} \tag{5}$$

And the portfolio has an expected return of

$$\sum_{i=1}^{N} w_i \mu_i \tag{6}$$

The total variance is the risk measure. Further, let ρ_{ij} be the correlation between assets i and j and s_i, s_j be the standard deviations in the returns for these assets. Then, σ_{ij} can be replaced by $\rho_{ij} s_i s_j$ where $-1 \leq \rho_{ij} \leq 1$.

The unconstrained portfolio optimization (UPO) problem is as below:

$$\text{minimise} \sum_{i=1}^{N}\sum_{j=1}^{N} w_i w_j \rho_{ij} s_i s_j \tag{7}$$

$$\text{maximise} \sum_{i=1}^{N} w_i \mu_i \tag{8}$$

subject to

$$\sum_{i=1}^{N} w_i = 1 \tag{9}$$

$$0 \leq w_i \leq 1,\ i = 1, 2, \ldots, n \tag{10}$$

The UPO is a typical example of a MOOP with two conflicting objectives: the risk measure and the total return of the portfolio. There are no inequality type constraints. There is one equality type constraint, and we have variable bounds.

Next, we claim that the above UPO problem is a convex MOOP.

Definition 3 When all of a MOOP's objective functions are convex, its inequality constraints are nonconvex, and its equality constraints are linear; we say that the MOOP is convex [26].

In the above definition, the objective functions are assumed to be of minimisation form. Hence, we need to show that the risk measure and negative of total return are convex.

As negative of total return is a linear function, it is both convex and concave.

Also, the risk measure can be written as.

$f(w) = wQw^T$, where $Q = \left(\sigma_{ij}\right)_{N\times N}$ is the covariance matrix and $w = (w_1, w_2, \ldots, w_N)$ is the weight vector. The covariance matrix is always symmetric and positive semidefinite. Hence, the risk measure is a quadratic form. A quadratic form is always convex, and hence, the risk measure is convex.

As there are no inequality constraints, the second condition is trivially true. Also, our equality constraint is linear.

Hence, it is shown that the UPO is a convex MOOP. This result is used in further discussions.

The Pareto-optimal front obtained by solving the UPO is conveniently called the unconstrained efficient Frontier (UCEF) throughout the paper. The UCEF gives the best possible trade-off of risk against return.

By assigning a weighting parameter of λ $(0 \leq \lambda \leq 1)$, the UPO can be represented as a single-objective optimisation problem as below:

$$\text{minimise } \lambda\left[\sum_{i=1}^{N}\sum_{j=1}^{N} w_i w_j \rho_{ij} s_i s_j\right] - (1-\lambda)\left[\sum_{i=1}^{N} w_i \mu_i\right] \tag{11}$$

subject to

$$\sum_{i=1}^{N} w_i = 1 \tag{12}$$

$$0 \leq w_i \leq 1, i = 1, 2, \ldots, N \tag{13}$$

For each value of λ, a unique solution is obtained for the above problem. We put the following claim:

By varying the values of λ in the interval (0, 1), the entire true Pareto-optimal set can be obtained except for at most two points. To prove the claim, the following theorems are used [28]:

Theorem 1 *Consider,*

$$\text{minimise } F(x) = \sum_{m=1}^{M} w_m f_m(x),\ m = 1, 2, \ldots, M; \tag{14}$$

$$\text{subject to } g_j(x) \geq 0,\ j = 1, 2, \ldots, J; \tag{15}$$

$$h_k(x) = 0,\ k = 1, 2, \ldots, K \tag{16}$$

$$x_i^{(L)} \leq x_i \leq x_i^{(U)},\ i = 1, 2, \ldots, n \tag{17}$$

Here, $w_m \in [0, 1]$ and $\sum_{m=1}^{M} w_m = 1$.

If $w_m > 0$ for all $m = 1, 2, \ldots, M$, then the solution to the problem presented in the above equations is Pareto-optimal.

Theorem 2 *Given any Pareto-optimal solution, y, of a convex MOOP, there exists a weight vector w* $\left(w_m \geq 0, m = 1, 2, \ldots, M, \sum_{m=1}^{M} w_m = 1\right)$ *such that y is a solution to the optimisation problem presented in the above equations.*

It can be deduced from Theorem 1 that any solution of the above equations for $\lambda \in (0, 1)$ is a Pareto-optimal solution. As when $\lambda \in (0, 1)$, λ, $(1 - \lambda) > 0$.

From Theorem 2, it can be concluded that there is no Pareto-optimal solution of the UPO, which is not a solution of the above equations for $\lambda \in [0, 1]$. This holds true as UPO is, in fact, a convex MOOP, as discussed previously.

Hence, the following procedure is proposed to find the UCEF.

1. Let $\lambda_x = x/L$, $x = 0, 1, \ldots, L$ for a sufficiently large L.
2. Solve the single-objective problem for each value of λ_x. For each value of λ_x, solve the optimisation problem presented in Eqs. (11)–(13).
3. Check whether the solutions obtained for λ_0 and λ_L are not dominated by other obtained solutions. If any of these two solutions are dominated, discard it/them.
4. The remaining set of solutions is the accurate representation of the Pareto-optimal set.
5. Plot the image of the Pareto-optimal set to obtain the UCEF.

4 Genetic Algorithms

Genetic algorithms are search and optimisation tools based on the principles of natural selection. John Holland originally conceived the concept and theoretical foundations of GAs.

GAs have been applied to solve the optimisation problems arising in various fields, including finance. The reason behind their popularity is their ability to explore enormous search spaces, together with their flexibility with respect to the problem to be solved [29]. In this section, a GA variant is proposed specifically to solve the UPO.

Following are the building blocks of the proposed GA.

4.1 Encoding

In our genetic algorithm, the encoding is done in such a way that the linear equality weight constraint $\sum_{i=1}^{N} w_i = 1$ is easily satisfied. Each population member contains N randomly generated numbers r_i $(0 \leq r_i \leq 1)$. Now, the weights of the assets are obtained as below:

$$w_i = \frac{r_i}{\sum r_i} \tag{18}$$

In this way, the linear constraint from Eq. (12) is satisfied by default, as the sum of the weights assigned using Eq. (18) is always going to be 1.

4.2 Fitness Evaluation

The fitness in our GA is nothing but the function value calculated from the objective function in Eq. (11). The fitness of each population is evaluated.

4.3 Selection

Once the fitness of each individual is known, the parents are chosen using the binary tournament selection. In this selection, two pools are made, each pool containing two randomly selected individuals. From each pool, one individual is selected to be the parent based on the individual's fitness values in the pool. Hence, we have selected two parents.

4.4 Crossover

To form the child from the selected parents, the asset i in the child is randomly allotted weight w_i either from parent 1 or from parent 2. Hence, the child obtained after the crossover operator will have a randomly chosen weight w_i, for asset i, from either parent.

4.5 Mutation

An asset i is randomly chosen in the child obtained after crossover. The weight w_i of this asset is either increased or decreased by 10% with equal probability.

Note that the sum of the weights may not be equal to 1 in the child obtained after crossover and mutation operations. But, this will be handled whilst calculating the fitness value of the child. The new weights of the assets in the child can be calculated using the same trick as in Eq. (18):

$$w_i^{'} = \frac{w_i}{\sum w_i} \tag{19}$$

In our GA, we use a steady-state population replacement strategy. Each new child is immediately placed in the population, replacing the worst member of the population. In Figs. 1 and 2, the complete GA heuristics can be seen in pseudo-code.

5 Performance Metrics

Some measures of performance to compare the UCEF obtained using the proposed GA to the true efficient frontier are required. Based on the two different priorities in MOOP, the performance of the efficient frontier can be measured. They are as follows:

1. Discover solutions as near as possible to the true efficient frontier
2. Find solutions as widespread as achievable in the UCEF obtained

Diversity can be further classified into two categories: *extent* and *distribution*. How good is the spread of extreme solutions is measured by *extent*, whilst *distribution* is a measure of the relative distance between the solutions [30].

Towards these goals, the following six-performance metrics are defined. In each definition, it is assumed that P^* is the true Pareto front, whilst Q is the Pareto front obtained using the GA.

```
// This program finds the best portfolio associated with each value of λ
using a Genetic Algorithm.
P = Set of population members.
E = The total of λ values to be used
t* = Total number of iterations
N = Total number of assets

begin
  for e = 1 to E
    λ = (e − 1)/(E − 1) // Generating E number of λ values equally
  spaced in [0,1].
    Initialize the population P of size S such that each member of P is a
  N-tuple of the form (r_s1, r_s2, ..., r_sN) where r_sj ∈ (0,1) is randomly
  generated.
    Evaluate the fitness of each p_s ∈ P:
    evaluate(p_s) s = 1,2, ..., S
    for t = 1 to t*
      Select two population members p* and p** from P using Binary
    Selection Operator
      The child C is produced by crossover between p* and p**.
      Next, to perform mutation an asset j is randomly selected from the
    child C and its value c_j is multiplied by 0.9 or 1.1 with equal proba-
    bility.
      evaluate(C)
      Substitute the weakest member of the population with the child C.
    end for
  end for
end
```

Fig. 1 Pseudo-code for the proposed GA

5.1 Set Coverage Metric

The metric $C(A, B)$ of the set coverage measures the proportion of solutions in B dominated by solutions in A [31]:

$$C(A, B) = \frac{|\{b \in B | \exists a \in A : a \preccurlyeq b\}|}{|B|} \tag{20}$$

```
// This program block defines the function evaluate.

evaluate(p)
begin
    Calculate the weights of the population member p using equation
(18).
    Calculate the objective function value f(p).
    if f(p) < V
        V = f(p)
end
```

Fig. 2 Pseudo-code for the fitness evaluation function in GA

We expect $C(Q, P^*) = 0$. Also, the worst case is $C(P^*, Q) = 1$, which tells us that all members in Q are dominated by members of P^*. Whilst the best case is $C(P^*, Q) = 0$. Hence, we expect the value to be as close to zero as possible.

5.2 *Generational Distance*

The average distance of the members of Q from P^* is found in this metric, as follows [32]:

$$GD = \frac{\left(\sum_{i=1}^{|Q|} d_i^p\right)^{1/p}}{|Q|} \tag{21}$$

The parameter d_i, for $p = 2$, is the Euclidean distance (in the objective space) between the solution $i \in Q$ and the nearest member of P^*:

$$d_i = \min_{k=1}^{|P^*|} \sqrt{\sum_{m=1}^{M} \left(f_m^{(i)} - f_m^{*(k)}\right)^2} \tag{22}$$

where the m-th objective function value of the k-th member of P^* is $f_m^{*(k)}$. It is quite intuitive that the smaller the value of GD closer are the solutions to the true Pareto front. Generational distance is a measure of closeness.

5.3 Maximum Pareto-Optimal Front Error

The worst distance d_i (d_i's are as computed in GD) amongst all members of Q is evaluated by this metric. The metric is exactly what its name suggests, i.e. maximum distance of a member of Q from the nearest member of P^*. The smaller the value of MFE, the closer the solutions are to the true Pareto front. Again, this is a measure of closeness.

5.4 Spacing

Spacing [33], as the name suggests, is a diversity measure. The formula to evaluate this performance metric is represented in Eq. (23).

$$S = \sqrt{\frac{1}{|Q|}\sum_{i=1}^{|Q|}\left(d_i - \overline{d}\right)^2} \tag{23}$$

where $d_i = \min\limits_{k \in Q \wedge k \neq i} \sum\limits_{m=1}^{M} \left| f_m^i - f_m^k \right|$ and $\overline{d} = \sum\limits_{i=1}^{|Q|} d_i / |Q|$ which is the mean value of d_i. In this case, d_i is determined so that it is the smallest possible sum of the absolute differences between the values of the objective function for the i-th solution and those of all other solutions in the resulting non-dominated set. The smaller the spacing value more uniformly the solutions are distributed.

5.5 Spread

In addition to the uniformity in the solutions, this metric also considers the extent of spread in the solutions. The metric is given as [34]:

$$\Delta = \frac{\sum_{m=1}^{M} d_m^e + \sum_{i=1}^{|Q|} \left| d_i - \overline{d} \right|}{\sum_{m=1}^{M} d_m^e + |Q| d} \tag{24}$$

Here, the distances between the neighbouring solutions are d_i's, which can be calculated by any standard distance measure. The average of these distances d_i's is $\overline{d}$. By calculating the difference between extreme solutions of P^* and Q corresponding to the m-th objective function, the parameter d_m^e is measured. If in the obtained solution, the true extreme Pareto-solutions exist, and if all the solutions are uniformly distributed, then $\Delta = 0$, and this is the ideal distribution.

5.6 Maximum Spread

As the name suggests, this metric measures the extent of the spread of the obtained solutions. It is not a measure of the exact distribution of intermediate solutions. The metric is defined as below:

$$\overline{D} = \sqrt{\frac{1}{M}\sum_{m=1}^{M}\left(\frac{\max_{i=1}^{|Q|} f_m^i - \min_{i=1}^{|Q|} f_m^i}{F_m^{\max} - F_m^{\min}}\right)^2} \tag{25}$$

Here, $F_m^{\max}$ and $F_m^{\min}$ are the maximum and minimum values of the m-th objective in the chosen set of Pareto-optimal solutions, P^*. A widely dispersed collection of solutions is produced if the metric's value is one.

6 Result and Analysis

The proposed GA was used on the four data sets available on OR-Library [35]. The sizes of the data sets range from $N = 31$ to $N = 98$. The available data in the OR-Library are the correlation coefficients ρ_{ij}, and s_i, s_j the standard deviations in the means of the assets i and j, respectively. Also, the true efficient frontiers are available for all the four data sets. The obtained UCEFs are compared with the true efficient frontiers using the six-performance metrics, and the results are presented. UCEFs are also obtained for these four different data sets with different state-of-the-art meta-heuristic algorithms. The techniques used are artificial bee colony, differential evolution, particle swarm optimisation, and simulated annealing. The primary reason for selecting these four meta-heuristics is there frequent use in the literature [36] concerned with the field of portfolio optimisation.

All the algorithms are performed with the parameters set as below:

$E = 1000$, the number of equally spaced λ values in [0, 1]
$t^* = 1000 * N$, the total number of iterations/generations for each value of λ
$S = 100$, size of initial population

Let $Q_{i.j}$ be the obtained set of solutions for data set i, where $i = 1, 2, \ldots, 4$ and $j = 1, 2, \ldots, 5$. Here, the algorithms are represented in the sequence GA, ABC, DE, PSO, and SA. That is, $Q_{2,4}$ is the set of solutions obtained for the 2nd data set obtained by PSO. Then, $\left|Q_{i,j}\right| = 1000$. Kung et al.'s efficient method to find a non-dominated set was used for the obtained sets $Q_{i,j}$'s. Let $Q^{'}_{i,j}$ be the non-dominated sets obtained from $Q_{i,j}$. Then, it was observed that.

It can be inferred from Table 1, that out of the 1000 solutions obtained, the non-dominated solutions obtained by GA are the highest. At the same time, the second-best non-dominated solutions were obtained by DE. On average, only, 218 out of

Table 1 Number of members in the non-dominated solution sets

Heuristic	Data set-1	Data set-2	Data set-3	Data set-4
GA	679	855	954	929
ABC	98	56	56	89
DE	214	174	212	274
PSO	22	16	15	33
SA	30	22	21	40

Table 2 Time taken (in seconds) for respective data sets

Heuristic	t_1	t_2	t_3	t_4
GA	874	6175	6881	9640
ABC	1249	4152	4277	8918
DE	1454	4687	4825	9363
PSO	1300	4289	4367	8710
SA	4421	13,584	14,360	26,140

1000 solutions obtained by DE are non-dominated. This empirical result verifies the efficiency of GA.

The time taken t_i (in seconds), for each data set, i is as in Table 2. It is to be noted that the data sets are in increasing order of the number of assets. The first data set has the least assets, and the last has the maximum number of assets. The best time for each asset is highlighted in the table.

In Figs. 3 and 4, UCEFs obtained using the five meta-heuristics are plotted. Only, GA and DE are able to cover the entire Pareto front. Also, they perform better in terms of closeness and diversity of the solution. This can be further verified with different numerical values of the performance metrics. To better understand the obtained solutions, performance metrics are used. The following tables show the values of the performance metrics:

6.1 Set Coverage Metric

Observe from Table 3 that for all other heuristics, except GA, the value of $\mathcal{C}(P^*, Q_{i,j}) = 1, \forall i, j \neq 1$. This is also evident from Fig. 3. This suggests that there are no solutions obtained by the other heuristics that are not dominated by the true optimal solutions. Whilst the metric values for GA suggests that GA has obtained a good proportion of the Pareto front, which is not dominated by the solutions of the true Pareto front. Also, it was verified that $\mathcal{C}(Q_{i,j}, P^*) = 0, \forall i, j$.

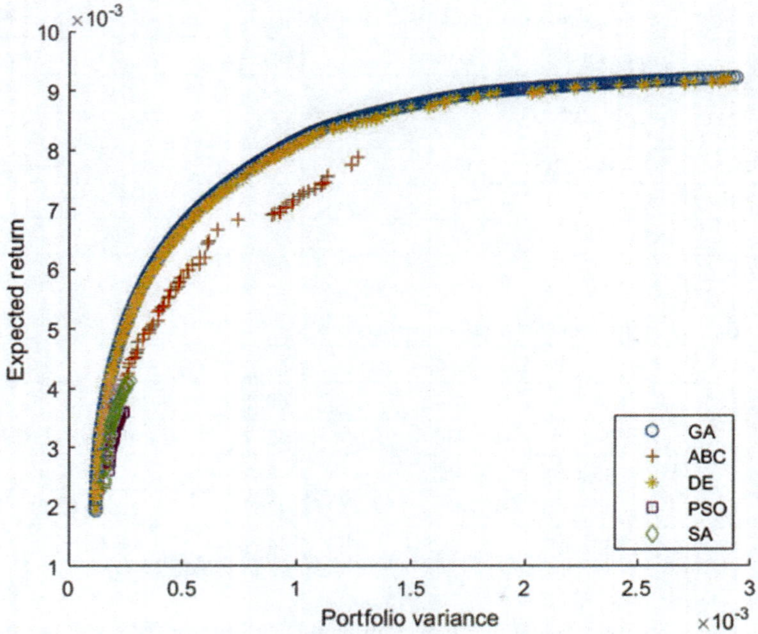

Fig. 3 UCEF's for the largest data set

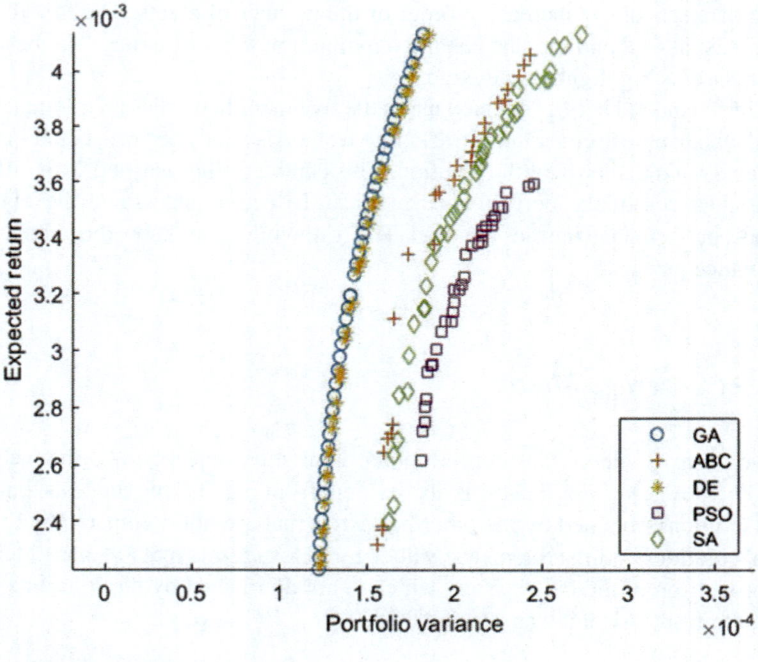

Fig. 4 Close-up of UCEFs from Fig. 3

Table 3 Values of the set coverage metric $\mathcal{C}(P^*, Q_{i,j})$

Heuristic	$\mathcal{C}(P^*, Q_{1,j})$	$\mathcal{C}(P^*, Q_{2,j})$	$\mathcal{C}(P^*, Q_{3,j})$	$\mathcal{C}(P^*, Q_{4,j})$
GA	0.676	0.3357	0.1614	0.211
ABC	1	1	1	1
DE	1	1	1	1
PSO	1	1	1	1
SA	1	1	1	1

6.2 Generational Distance and MFE

In Table 4, the best values of the generational distance for each data set are highlighted in light grey, whilst the second-best values in dark grey. GA outperforms all other heuristics for all four data sets.

The obtained values of the maximum Pareto-optimal front error, as shown in Table 5, are on the same lines as the generation distance values. A similar conclusion can be inferred from MFE values.

Table 4 Values of generational distance for different data sets

Heuristic	GD-1	GD-2	GD -4	GD -4
GA	7.39E-08	7.74E-08	2.55E-08	8.51E-08
ABC	2.32E-05	1.67E-05	1.74E-05	2.10E-05
DE	4.37E-06	3.05E-06	1.84E-06	1.73E-06
PSO	6.40E-05	3.09E-05	2.20E-05	1.13E-05
SA	3.87E-05	2.53E-05	1.68E-05	9.18E-06

Table 5 Values of MFE for different data sets

Heuristic	MFE-1	MFE-2	MFE-3	MFE-4
GA	3.83E-06	1.04E-05	2.28E-06	1.40E-05
ABC	6.19E-04	2.91E-04	2.58E-04	3.98E-04
DE	1.83E-04	1.28E-04	8.02E-05	8.14E-05
PSO	5.10E-04	2.02E-04	1.16E-04	9.09E-05
SA	4.86E-04	3.59E-04	1.18E-04	9.15E-05

Table 6 Values of spacing metric for different data sets

Heuristic	S_1	S_2	S_3	S_4
GA	8.47E-06	1.55E-05	9.65E-06	1.05E-05
ABC	4.72E-05	5.56E-05	4.69E-05	4.30E-05
DE	2.70E-05	3.09E-05	1.58E-05	1.70E-05
PSO	4.50E-05	1.48E-05	4.50E-05	2.04E-05
SA	5.56E-05	5.98E-05	2.29E-05	2.47E-05

6.3 Spacing

The smaller the spacing values more uniformly the solutions are spread. Except for the second data set as shown in Table 6, the result is similar to the ones obtained from GD and MFE values.

6.4 Spread and Maximum Spread

The spread metric measures the uniformity of the solutions as well as the extent of spread in the obtained UCEFs. From the obtained values, we can infer that the uniformity of the solutions obtained via GA has a scope for improvement, as we can observe from Table 8 that GA has the values of maximum spread closer to 1. Hence, GA has the best extent of spread amongst all the algorithms. From Table 7, we conclude that the solutions obtained via DE have the best uniformity in comparison with the other algorithms.

To further analyse the values of spread metric Δ, the metric is evaluated for the true Pareto-optimal set, and its values for the four different data sets are as below:

Here, from Eq. (24), it is clear that the term $\sum_{m=1}^{M} d_m^e = 0$. Hence, the Δ values indicate just the uniformity of solutions in the true efficient frontier. Again, the values of maximum spread are very close to one for the UCEFs obtained via GA and DE. Hence, comparing the values of spread Δ for the UCEFs from Table 7 and the true efficient frontier from Table 9, it gets clear that the uniformity of the solutions can be significantly improved in the obtained UCEFs.

Table 7 Values of spread Δ for different data sets

Heuristic	Δ_1	Δ_2	Δ_3	Δ_4
GA	0.496	1.05	1.025	0.8209
ABC	0.6968	0.8347	0.7681	0.7753
DE	0.5766	0.5756	0.6118	0.6214
PSO	0.8958	0.9784	0.959	0.9539
SA	0.8631	0.9405	0.9471	0.9279

Table 8 Values of maximum spread $\overline{D}$ for different data sets

Heuristic	$\overline{D}_1$	$\overline{D}_2$	$\overline{D}_3$	$\overline{D}_4$
GA	0.9987	0.9999	0.9997	1
ABC	0.6172	0.5277	0.4993	0.6081
DE	0.9573	0.9825	0.9823	0.993
PSO	0.1589	0.055	0.09981	0.09625
SA	0.3009	0.1317	0.1501	0.1703

Table 9 Values of spread Δ corresponding to the true Pareto-optimal set

Δ_1	Δ_2	Δ_3	Δ_4
0.2905	0.3395	0.1703	0.3097

7 Conclusion

In this paper, it is shown analytically and empirically that the solutions to the weighted sum approach of the UPO, for varying values of λ, are, in fact, non-dominated solutions (except for the values of $\lambda = 0 \text{or} 1$). Additionally, it is demonstrated that they are the only Pareto-optimal solutions, meaning that the discovered solutions cover the complete Pareto-optimal set.

Later, the results obtained for four different asset sets were compared with the exact solutions using six different performance metrics.

GA outperforms all the algorithms with respect to all but one performance metric. There is a scope for improvement in the uniformity of the solutions in the UCEF obtained via GA. The cardinality of non-dominated solutions obtained by GA is the best as compared to the other four algorithms. Also, it is evident from Table 3 that with respect to closeness and extent of spread, GA is better than the other algorithms, whilst DE is placing second.

7.1 *Future Scope*

Most of the time, investors are only interested in considering the decrease in a stock price as a risk and are not interested in considering the increase in the asset prices as a risk. Hence, one can consider using the downside risk measure, also known as semi-variance, as a risk measure to improve presented UPO.

Again, investors are not always interested in a large number of assets in their portfolios. So, considering the constraint on the number of assets, known as cardinality constraint, can be modified in UPO.

As DE and ABC were able to produce more uniformly distributed solutions, whilst GA has the best closeness to the true Pareto front, as a further extension of this work, one can combine GA and DE to expect better results.

References

1. Markowitz H (1952) Portfolio selection. J Finance 7(1):77–91
2. Metaxiotis K, Liagkouras K (2012) Multiobjective evolutionary algorithms for portfolio management: a comprehensive literature review. Expert Syst Appl 39(14):11685–11698
3. Arnone S, Loraschi A, Tettamanzi A (1993) A genetic approach to portfolio selection. Neural Netw World 3(6):597–604
4. Ponsich A, Jaimes AL, Coello CAC (2013) A survey on multiobjective evolutionary algorithms for the solution of the portfolio optimization problem and other finance and economics applications. IEEE Trans Evol Comput 17(3):321–344
5. Mei X, DeMiguel V, Nogales FJ (2016) Multiperiod portfolio optimization with multiple risky assets and general transaction costs. J Bank Finance 69:108–120
6. Nesaz HH, Jasemi M, Monplaisir L (2020) A new methodology for multi-period portfolio selection based on the risk measure of lower partial moments. Expert Syst Appl 144:113032
7. Yang X, Liu W, Chen S, Zhang Y (2021) A multi-period fuzzy mean-minimax risk portfolio model with investor's risk attitude. Soft Comput 25(4):2949–2963
8. Chen W, Xu W (2019) A hybrid multiobjective bat algorithm for fuzzy portfolio optimization with real-world constraints. Int J Fuzzy Syst 21(1):291–307
9. Fang Y, Lai KK, Wang S (2008) Fuzzy portfolio optimization: theory and methods. Springer, Heidelberg
10. Gupta P (2014) Fuzzy portfolio optimization Advances in Hybrid Multi-criteria Methodologies. Springer, Heidelberg
11. Yue W, Wang Y, Xuan H (2019) Fuzzy multi-objective portfolio model based on semi-variance–semi-absolute deviation risk measures. Soft Comput 23(17):8159–8179
12. Rubinstein M (2002) Markowitz's "portfolio selection": a fifty-year retrospective. J Financ 57(3):1041–1045
13. Tapia MGC, Coello CAC (2007) Applications of multi-objective evolutionary algorithms in economics and finance: a survey. In: 2007 IEEE congress on evolutionary computation. IEEE, Piscataway, NJ, pp 532–539
14. Kolm PN, Tütüncü R, Fabozzi FJ (2014) 60 years of portfolio optimization: practical challenges and current trends. Eur J Oper Res 234(2):356–371
15. Kalayci CB, Ertenlice O, Akyer H (2017) Aygoren, H: An artificial bee colony algorithm with feasibility enforcement and infeasibility toleration procedures for cardinality constrained portfolio optimization. Expert Syst Appl 85:61–75
16. Suthiwong D, Sodanil M (2016) Cardinality-constrained portfolio optimization using an improved quick artificial bee colony algorithm. In: 2016 International Computer Science and Engineering Conference (ICSEC). IEEE, pp 1–4
17. Zaheer H, Pant M (2016) Solving portfolio optimization problem through differential evolution. In: 2016 International Conference on Electrical, Electronics, and Optimization Techniques (ICEEOT). IEEE, pp 3982–3987
18. Lwin K, Qu R (2013) A hybrid algorithm for constrained portfolio selection problems. Appl Intell 39(2):251–266
19. Yin X, Ni Q, Zhai Y (2015) A novel PSO for portfolio optimization based on heterogeneous multiple population strategy. In: 2015 IEEE Congress on Evolutionary Computation (CEC). IEEE, pp 1196–1203
20. Abbas A, Haider S (2009) Comparison of AIS and PSO for constrained portfolio optimization. In: 2009 international conference on information and financial engineering. IEEE, pp 50–54
21. Farzi S, Shavazi AR, Pandari A, Graduated MA (2013) Using quantum-behaved particle swarm optimization for portfolio selection problem. Int Arab J Inf Technol 10(2):111–119
22. Woodside-Oriakhi M, Lucas C, Beasley JE (2011) Heuristic algorithms for the cardinality constrained efficient frontier. Eur J Oper Res 213(3):538–550
23. Chang TJ, Meade N, Beasley JE, Sharaiha YM (2000) Heuristics for cardinality constrained portfolio optimisation. Comput Oper Res 27(13):1271–1302

24. Elton EJ, Gruber MJ, Brown SJ, Goetzmann WN (2017) Modern portfolio theory and investment analysis. Wiley, Hoboken
25. Rom BM, Ferguson KW (1994) Post-modern portfolio theory comes of age. J Invest 3(3):11–17
26. Deb K (2001) Multi-objective optimization using evolutionary algorithms. Wiley India, New Delhi
27. Cornuejols G, Tütüncü R (2018) Optimization methods in finance. Cambridge University Press, Cambridge
28. Miettinen KM (1999) Nonlinear multiobjective optimization. Kluwer Academic Publishers, Boston
29. Goldberg DE (1989) Genetic algorithms in search, optimization, and machine learning. Addison-Wesley, Boston
30. Zitzler E, Deb K, Thiele L (2000) Comparison of multiobjective evolutionary algorithms: empirical results. Evol Comput 8(2):173–195
31. Zitzler E (1999) Evolutionary algorithms for multiobjective optimization: methods and applications. Eidgenössische Technische Hochschule, Zürich
32. Van Veldhuizen DA, Lamont GB (2000) Multiobjective evolutionary algorithms: analyzing the state-of-the-art. Evol Comput 8(2):125–147
33. Scott JR (1995) Fault tolerant design using single and multi-criteria genetic algorithms. Master's thesis, Department of Aeronautics and Astronautics, Massachusetts Institute of Technology
34. Deb K, Pratap A, Agarwal S, Meyarivan TAMT (2002) A fast and elitist multiobjective genetic algorithm: NSGA-II. IEEE Trans Evol Comput 6(2):182–197
35. OR-Library Homepage. http://people.brunel.ac.uk/~mastjjb/jeb/info.html. Last accessed 2022/06/11
36. Kalayci CB, Ertenlice O, Akbay MA (2019) A comprehensive review of deterministic models and applications for mean-variance portfolio optimization. Expert Syst Appl 125:345–368

Task Scheduling and Energy-Aware Workflow in the Cloud Through Hybrid Optimization Techniques

Arti Yadav, Samta Jain Goyal, Rakesh Singh Jadon, and Rajeev Goyal

Abstract Task Scheduling is very complex in terms of resource utilization for the workload distributed on Virtual Machines. It is a very typical issue for the task of scheduling distributed on virtual machines, which is the most specific problem in cloud computing. Without scheduling, load balancing is not proper for balancing distributed on a virtual machine because maximum makes the unbalancing. So, it will be indispensable for which the reduced response time, makespan time with scheduling. The first experiment aims to compare traditional Ant Colony Optimization algorithms (ACO), the second experiment to solve the maximum global optimization problems by Genetic Algorithms (GA), and the third experimental series the performance of Particle Swarm Optimization algorithms (PSO) for increasing the resources' utilization for synthetically and actual trace data, respectively. Furthermore, the result shows simulation by clouds in this paper and problems identified by NIA algorithms.

Keywords Task scheduling · Virtual machine · Energy consumption · Makespan · Cloud computing · Meta-heuristics algorithm · Heuristics algorithms

1 Introduction

Cloud infrastructure is a big platform of cloud environment, load balancing, and task scheduling maximum use of the resources with all virtual machines, and maximum

A. Yadav (✉) · S. J. Goyal
Department of CSE, Amity University Madhya Pradesh, Gwalior, India
e-mail: aaruyn8@gmail.com

S. J. Goyal
e-mail: sgoyal@awa.amity.edu

R. S. Jadon
Department of Computer Applications, MITS, Gwalior, India

R. Goyal
Department of CSE, Vellore Institute of Technology, Bhopal, India
e-mail: goyalrajeev@gmail.com

M. Thakur et al. (eds.), *Soft Computing for Problem Solving*, Lecture Notes in Networks and Systems 547, https://doi.org/10.1007/978-981-19-6525-8_37

resources utilization with all virtual machines used by the physical machine. Moreover, all resources are being utilized in cloud computing and cloud service providers [14]. These resources can be used dynamically for tasks to achieve an optimal solution [15]. This resource has typically offered to users by a pay-per-use model, in which all agreements between user and service for Quality of Services and Service Level Agreements (SLA) [17].

The shown in Fig. 1 represents cloud infrastructure. The cloud infrastructure is a big platform of cloud computing and these are maximum resource utilization by the virtual machine and according to cloud services provided in form of SAAS, PAAS, IAAS, and internet are connected the between in services analysis and then PAYG plan by QoS and SLA. The virtual machine is available techniques GA, ACO and PSO are calculates and finding the energy consumption so that all data centers are easily reaching on the all datacenter to distribute the among the virtual machine by at least Physical machine.

The biggest challenge for the cloud computing process is assigning tasks to potential resources to reduce response time and find a better performance by scheduling algorithms [2]. So, this process avoids the utilization of resources overloaded and solves the complex problem [12].

Load balancing algorithms such as Round Robin (RR), First Come First Served (FCFS), Shortest Job First (SJF), and Weighted Round Robin [17] have been proposed for traditional tasks. Intelligent algorithms are used to initialize tasks.

The shown in Fig. 2 is represented by Scheduling Algorithms. Task scheduling is very usable for cloud computing and it has three levels such as task level, scheduling level, and virtual machine level scheduling algorithms are performed in a maximum virtual machine, and scheduling algorithms are used for strategy.

The optimal solutions are based on services and applications-based for cloud computing [15]. The main focus is load balancing, which reduces the response time and makespan [6]. Task scheduling will be found the necessary resources problem by task scheduling algorithms [15], which has compared with NIA algorithms for cloud data centers so that they are widely used to virtualization the resources. The cloud computing process is an opportunity for simultaneously running resources multiple times [19]. So maximum Virtual Machines (VMs) work according to the least Physical Machine (PM) in cloud computing. The cloud data center has become unsustainable due to load balancing [12]. The cloud services provided have two types, so there are inter-related such as Service Level Agreements (SLAs) and Quality of services (QoS) [12]. Moreover, several technical applications such as migration time, makespan time, throughput, response time, and request time accessible works have focused on the optimization approach related to Nature-inspired algorithms [11]. The research for load balancing with the help of task scheduling and cloud services providers is used for performing Quality of services and Services level agreement. It will be a critical task schedule but better for cloud computing [16]. Requirement of the optimal solution for load balancing of utilization resources and with the schedule the performances by task scheduling is used for scheduling all resources optimizes and making effective performances enhance algorithms in cloud computing just because of the optimization of approaches [20]. The task scheduling procedure runs

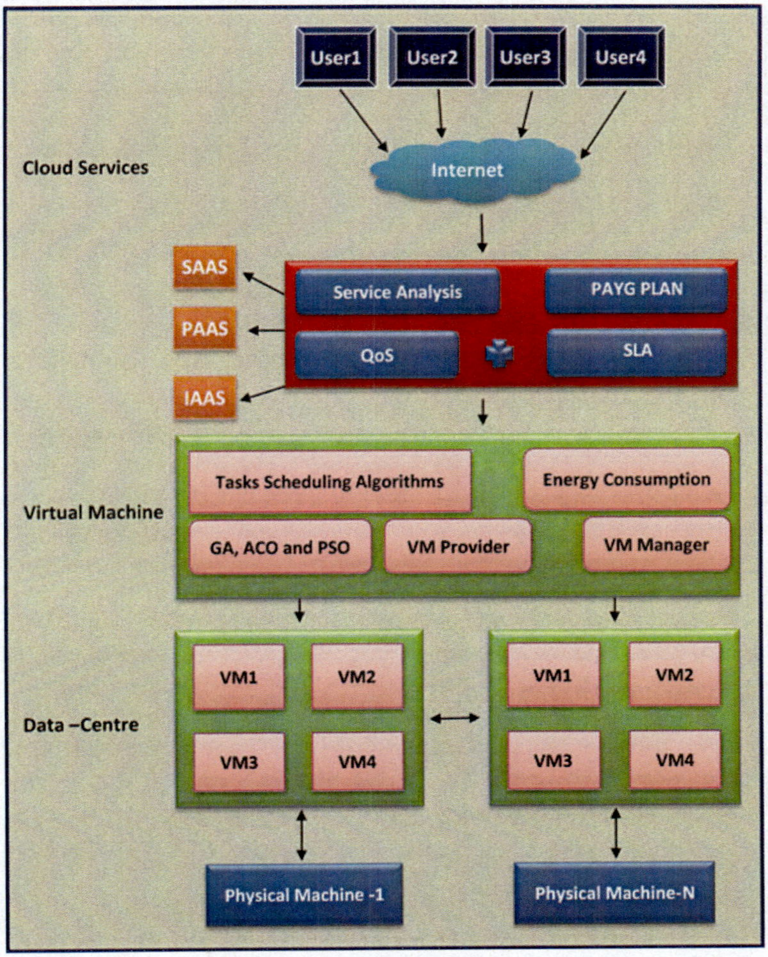

Fig. 1 Cloud infrastructure

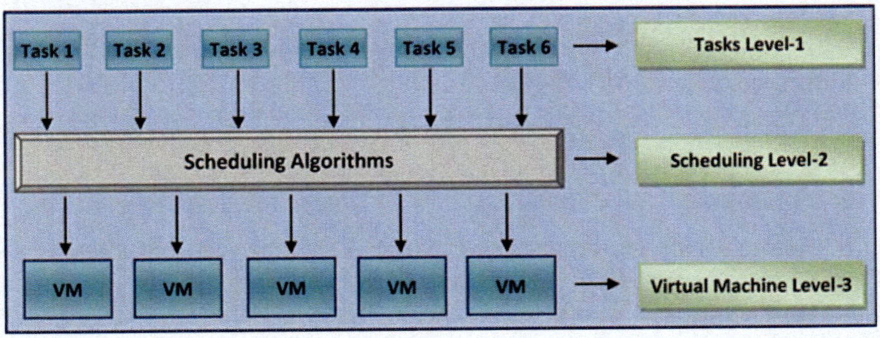

Fig. 2 Scheduling algorithms

process by NP-complete problem and NP-Hard problem by scheduling [8]. The task scheduling process is efficient in cloud computing, which effectively performs load balancing metrics—scheduling procedures such as reducing makespan and response time [21]. The load balancing and task scheduling are heuristic algorithms, and hybrid techniques are Meta-heuristic algorithms. NIA proposed the hybrid algorithms types, and the previous features rearrange by new algorithms so that more are performed by representing the exploration of cloud computing [16].

2 Related Work

(a) Dhirendra kumarsukla, et al., 2015. "Task scheduling to reduce energy consumption and makespan of cloud computing using NSGA-I". This paper discusses NSGA and NSGA-2 work for sorting concepts. However, this research is about reducing makespan and response time [9]. It is more resource utilization.
(b) Negar et al., 2017. "A hybrid particle swarm optimization and hill-climbing algorithm for task scheduling in cloud environments". The task scheduling algorithm first solves the same issues [15]. However, the research in this paper related to more parameters in cloud computing with NIA algorithms [6]. Its effective dynamic load balancing Task scheduling.
(c) Mohomed et al., 2020. "Task scheduling in cloud computing based on hybrid moth search algorithm and differential evolution". This paper says about moth search algorithms just like work fly behavior [15]. Two concepts Photos axis levy flight areas known as explorations exploitation [8]. Furthermore, in this research, cloud computing environments for task scheduling strategies depend on the work of heuristics and meta-heuristics experiments, so the issue is optimization for a solution [3].
(d) Rambabu-Media et al., 2021. "Energy-aware workflow task scheduling in clouds with virtual machine consolidation using discrete water wave optimization". The research work for load balancing with task scheduling of multiple resources utilizes one to more parameters applied to cloud computing [6]. We face more phases in this paper for solving the situation according to the paper.

- **Task Scheduling Algorithms**

Task scheduling with nature-inspired algorithms and task optimization should be given priority according to the best datacenters hosts. Following are the queries that have been solved.

Q.1. How to find the priority according to the best given the data center host?
Q.2. How to find the Passing the task node to the other machines?
Q.3. How to select the priority to give one to another machine?
Q.4. How to find more effective task scheduling work?
Q.5. How will it compare to maximum algorithms by task scheduling method?

The task scheduling algorithms support the three levels of scheduling in cloud computing. Levels are working on the performance of virtualization in cloud computing. So that all are maximized resources utilized and all parameters overall in cloud infrastructure. However, performance is needed to reduce makespan time and response time for all tasks in cloud computing. Task scheduling algorithms are designed for maximum resource utilization in cloud computing [15]. These features are related to the task community for unbalancing situations in cloud computing [9]. So we will perform on NIA algorithms in the cloud computing environment [23].

This research work focuses on maximum resource utilization and parameters distributed virtual machine and the main factor of criteria performance of tasks. These are working on reducing response time and makespan time and many parameters utilized for all tasks [10]. To manage the unbalancing load problem by task scheduling algorithms and increasing the number of users for giving the tasks according to the real-time system [8]. So that improving the new tasks for arriving by scheduling can include maximum virtual machines. Static and dynamic scheduling is considered for comparison in cloud computing. The current systems of VMS are then divided into all traffic areas, so research will have to need to define areas of balancing so that they are distributed among all the VMs [5]. However, we are optimized for dynamic scheduling and consider the current systems of VMs. Dynamic scheduling algorithms perform on FCFS, SJF, RR, and WRR, and according to the capacity of all available VMs, Pre-emptive and Non-Preemptive scheduling, each task is interrupted during execution [1].

This is shown in Fig. 3 represented by Task Scheduling Performance. The real-time task scheduling algorithms (RTTA) is having in cloud computing usable for computing. RTTA has performed into two parts dynamic and statics so dynamic scheduling decides the pre-emotive and Non-Pre-sensitive. When do we services provide to the user according to QoS and SLA?

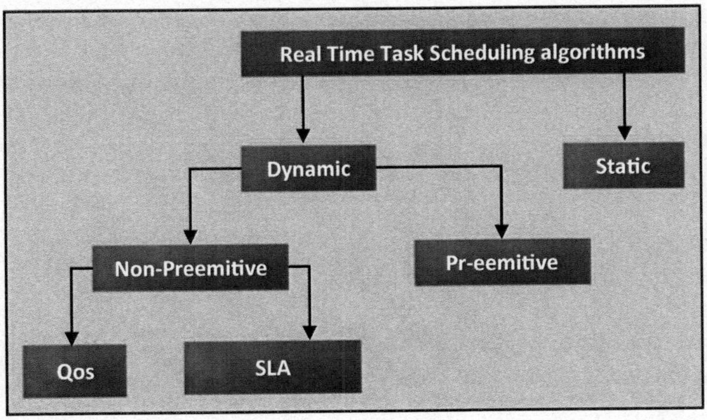

Fig. 3 Task scheduling performance

Algorithms of Task scheduling

The classification of the task scheduling is included in the task scheduling module consisting of two algorithms. Beginning algorithms and End-algorithms for association rule the generated tasks and work between in VM of given tasks. So, initializes the based on the beginning and End algorithm and design the fundamental solution of the approaches is to schedule tasks by the first one to prioritize the set of completion times of which tasks. So, performance is given priority tasks according to which one data center hosts and Crossover the mutation to calculate the which value best is fit value for resolution. Because it is necessary to know about the completion time for tasks, and task scheduling algorithms initialize all values.

Algorithms

Step 1: Input: The infrastructure function applies for task scheduling for minimum response time and makspane time.
Step 2: Define the NIA algorithms and initialization of random value $K < n = T$;
Step 3: Updates the random value $K < n = T$ (initial stage)
Step 4: Calculates the random value (Fitness function) and then $T = 1$(Response time) &&Crossover function with Mutation time (Fitness value).
Step 5: While il = 10 k/3do Multiple resources utilization with virtualization.
Step 6: Calculates the random with more function and updation While $T <= 0$, $K = 2/0:n$; Step 4 and Step 7:
Step 7: Update value (0);
Step 8: End do

End.

Flowchart of Task Scheduling Algorithms

This is shown in Fig. 4 representing task Scheduling Performance. The Flowchart of task scheduling NIA with deadline and time based and sort queuing compare with ascending and descending form and both are solutions apply weighted fair queuing for selecting the tasks and then for each VM calculate turnaround time and assigns the tasks to VM with minimum and other solutions for each VM calculates the time for response time both are tasks available in solution yes or no and combination of both algorithms and execute the tasks and ending the process of flowchart. The Flowcharts work according to the NIA algorithms which use task scheduling and load balancing. The result of task schedule compared with the activity is shown in Table 1 and comparison chart is shown in Fig. 5.

Result: Task schedule compared with the activity

See Table 1.

Chart

See Fig. 5.

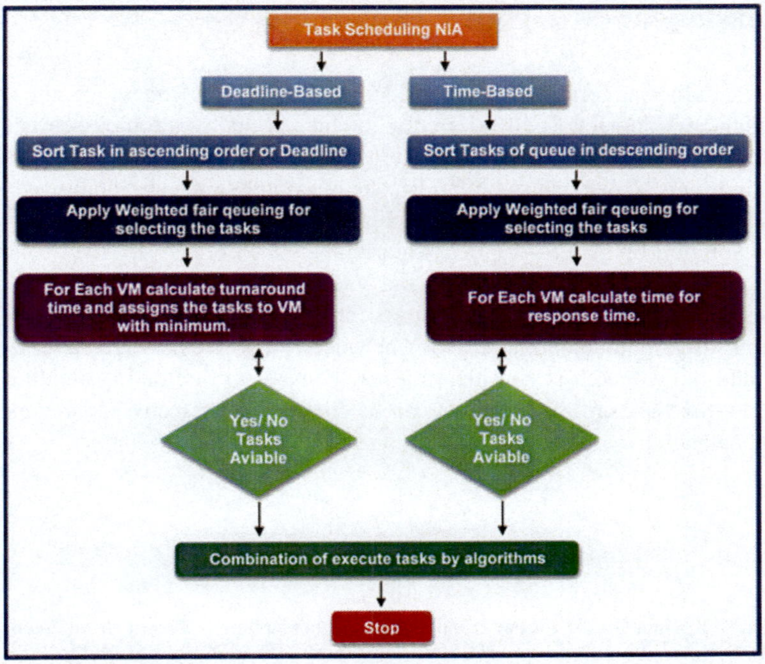

Fig. 4 Task scheduling algorithms

Table 1 Table of task scheduling comparing

S. No	Tasks	Virtual machine	TVM	Max value
1	Task 1	VM2	TIVM2	1.02
2	Task 2	VM4	T2VM4	2.04
3	Task 3	VM6	T2VM6	3.06
4	Task 4	VM8	T2VM8	4.08

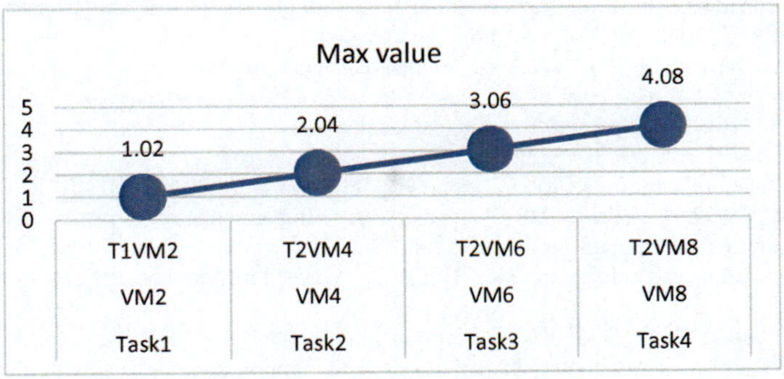

Fig. 5 Charts of comparison value

3 Conclusion

Task scheduling algorithms are considered critical challenges for the cloud computing environment. So, these are finding the optimal solution by heuristic and meta-heuristics algorithms in task scheduling strategies that have been used in cloud computing, and NIA performs hybrid techniques and so can be optimized more approaches according to the situation such as PSO, ACO, and GA. We can find the optimal solution by Nature-Inspired Algorithms, which effectively load balancing with a schedule to tasks distributed to the VM. The result simulation by cloudsim because cloudsim is a framework for the task and essential work for NIA algorithms.

The novelty of cloud computing for a new define a problem by NIA hybrid algorithms and our research is considered a new convenient in cloud computing load balancing. So these are methods that can be considered for many parameters used for load balancing.

References

1. Afzal S, Kavitha G (2019) Load balancing in cloud computing—a hierarchical taxonomical classification. J Cloud Comput 8(1). https://doi.org/10.1186/s13677-019-0146-7
2. Ahmad MO, Khan RZ (2019) PSO-based task scheduling algorithm using adaptive load balancing approach for the cloud computing environment. Int J Sci Technol Res 8(11):457–462
3. Almutairi J, Aldossary M (2021) A novel approach for IoT tasks offloading in edge-cloud environments. J Cloud Comput 10(1). https://doi.org/10.1186/s13677-021-00243-9
4. Behal V, Kumar A (2014) Cloud computing: performance analysis of load balancing algorithms in cloud heterogeneous environment. In: Proceedings of the 5th international conference on confluence 2014: the next generation information technology summit, pp 200–205. https://doi.org/10.1109/CONFLUENCE.2014.6949291
5. Daryapurkar A, Deshmukh M (2013a) Efficient load balancing algorithm in cloud environment. Int J Comput Sci Appl 6(2):308–312. http://researchpublications.org/IJCSA/NCAICN-13/205.pdf
6. Daryapurkar A, Deshmukh M (2013b) Efficient load balancing algorithm in cloud environment. Efficient load balancing algorithm in cloud environment 6(2):308–312. http://researchpublications.org/IJCSA/NCAICN-13/205.pdf.
7. Dubey U, Songare LS (2019) Analysis of load balancing in cloud computing. Int J Sci Technol Res 8(12):3912–3914
8. Garg S (2016) Algorithm for virtual machines in cloud
9. Hu Y, Wang H, Ma W (2020) Intelligent cloud workflow management and scheduling method for big data applications. J Cloud Comput 9(1). https://doi.org/10.1186/s13677-020-00177-8
10. Joshi N, Kotecha K, Choksi DB, Pandya S (2018) Implementation of novel load balancing technique in cloud computing environment. In: 2018 international conference on computer communication and informatics, ICCCI 2018, pp 1–5. https://doi.org/10.1109/ICCCI.2018.8441212
11. Liu Z, Qiao B, Fang K (2020) Joint optimization strategy for QoE-aware encrypted video caching and content distributing in a multi-edge collaborative computing environment. J Cloud Comput 9(1). https://doi.org/10.1186/s13677-020-00204-8
12. Mapetu JPB, Chen Z, Kong L (2019) Low-time complexity and low-cost binary particle swarm optimization algorithm for task scheduling and load balancing in cloud computing. Appl Intell 49(9):3308–3330. https://doi.org/10.1007/s10489-019-01448-x

13. Mesbahi M, Masoud Rahmani A (2016) Load balancing in cloud computing: a state of the art survey. Int J Modern Educ Comput Sci 8(3):64–78. https://doi.org/10.5815/ijmecs.2016.03.08
14. Mittal S, Katal A (2016) An optimized task scheduling algorithm in cloud computing. In: Proceedings of 6th international advanced computing conference, IACC 2016, vol 5(22), pp 197–202. https://doi.org/10.1109/IACC.2016.45
15. Shafiq DA, Jhanjhi NZ, Abdullah A (2021) Load balancing techniques in cloud computing environment: a review. J King Saud Univ Comput Inform Sci. https://doi.org/10.1016/j.jksuci.2021.02.007
16. Sreelakshmi, Sindhu S (2019) Multi-objective PSO based task scheduling—a load balancing approach in cloud. In: Proceedings of 1st international conference on innovations in information and communication technology, ICT (ICT 2019), pp 1–5. https://doi.org/10.1109/ICIICT1.2019.8741463
17. Srichandan S, Ashok T, Bibhudatta S (2018) Task scheduling for cloud computing using multi-objective hybrid bacteria foraging algorithm. Future Comput Inform J 3(2):210–230. https://doi.org/10.1016/j.fcij.2018.03.004
18. Student MT (2015) A review of load balancing in cloud computing using swarm intelligence techniques. 3(1):9–14
19. Surjeet K, Sabyasachi P, Ranjan A (2021) A particle swarm and ant colony optimization based load balancing and virtual machine scheduling algorithm for cloud computing environment. Turkish J Comput Math Educ 12(11):3885–3898
20. Tang L, Tang B (2021) Joint optimization of network selection and task offloading for vehicular edge computing
21. Usman Sana M, Li Z (2021) Efficiency aware scheduling techniques in cloud computing: a descriptive literature review. PeerJ Comput Sci 7:e509. https://doi.org/10.7717/peerj-cs.509
22. Wang Y, Ren Z, Zhang H, Hou X, Xiao Y (2018) "Combat Cloud-Fog" network architecture for internet of battlefield things and load balancing technology. In: 2018 IEEE international conference on smart internet of things (SmartIoT), pp 263–268.https://doi.org/10.1109/smartiot.2018.00054
23. Yu L, Chen L, Cai Z, Shen H, Liang Y, Pan Y (2016) Stochastic load balancing for virtual resource management in datacenters. 7161(November 2014). https://doi.org/10.1109/TCC.2016.2525984.
24. Chiang ML, Hsieh HC, Tsai WC, Ke MC (2017) An improved task scheduling and load balancing algorithm under the heterogeneous cloud computing network. In: Proceedings of 2017 IEEE 8th international conference on awareness science and technology, ICAST 2017, 2018-Janua (iCAST), pp 290–295. https://doi.org/10.1109/ICAwST.2017.8256465

A Hyper-Heuristic Method for the Traveling Repairman Problem with Profits

K. V. Dasari and A. Singh

Abstract This paper addresses a recently introduced problem called traveling repairman problem with profits (TRPP). The problem can be regarded as a variation of the well-known traveling salesman problem, where a repairman does not need to visit all the nodes. In TRPP, each node is linked with a profit, which is time-dependent. The objective of TRPP is to maximize the total revenue obtained. Our approach uses a greedy randomized construction method to generate the initial solution. A hyper-heuristic approach with a greedy selection mechanism (HH-GREEDY) is used for iteratively improving the solution. TRPP instances available in literature are utilized to evaluate the effectiveness of the proposed approach. Our approach obtained solutions of quality comparable to the state-of-the-art approaches at a fraction of time needed by these approaches.

Keywords Heuristic · Traveling repairman problem with profits · Hyper-heuristic · Traveling salesman problem

1 Introduction

A variation of the well-known traveling salesman problem (TSP) is traveling repairman problem (TRP) in which a repairman is expected to visit all the locations exactly once and make repairs at each of these locations. The goal is to minimize the sum of arrival times at the locations. In the literature, TRP is also defined as minimum latency problem or traveling deliveryman problem [1]. Real-world TRP applications include routing problems in service systems where the performance criterion is customer satisfaction, home delivery and machine scheduling problem [2]. We will use location, node and vertex interchangeably in this paper.

K. V. Dasari · A. Singh (✉)
School of Computer and Information Sciences, University of Hyderabad, Hyderabad, Telangana 500046, India
e-mail: alokcs@uohyd.ernet.in

K. V. Dasari
e-mail: dasarivisu@gmail.com

M. Thakur et al. (eds.), *Soft Computing for Problem Solving*, Lecture Notes in Networks and Systems 547, https://doi.org/10.1007/978-981-19-6525-8_38

Traveling repairman problem with profits (TRPP) is a recently introduced extension of TRP [3]. The TRPP can be described as, let $G = (N, E, w)$ be a complete weighted undirected graph, where $N = \{0, 1, 2, \ldots, n\}$ represents a set of locations that can be visited starting from vertex 0 which is called the depot. Each vertex $i \neq 0$ has an associated positive integer profit p_r. The depot is assumed to have zero profit, i.e., $p_0 = 0$. $E = \{(l, m) : l, m \in N, l \neq m\}$ is the set of edges. Each $(l, m) \in E$ has its corresponding weight w_{lm} that denotes the travel time between vertex l and vertex m, which is symmetric ($w_{lm} = w_{ml}$). A repairman is supposed to visit some locations starting from the depot (vertex 0) at time 0 and collects a revenue $p_r - t_r$ for each location r that it visits, where t_r is the time of arrival of the repairman at location r. Repairman can visit each location at most once and need not return to depot. There is no need to visit all the locations. In fact, repairman cannot visit any unvisited location r for which t_r becomes larger than p_r as visiting such a location yields a negative revenue. The goal of TRPP is to visit a subset of vertices, which maximizes the total collected revenue. The TRPP is classified as $\mathcal{NP}$-hard, as it can be considered as a generalization of TSP which is $\mathcal{NP}$-hard [3]. The TRPP finds real-world applications in transportation planning related to humanitarian aid operations and emergency relief logistics [3].

In the class of routing problems with profits, orienteering problem (OP) [4] and profitable tour problem (PTP) [5, 6] are closely related to TRPP. The goal of PTP is to determine a route so as to get the difference between total collected profit and total travel time maximized. The key distinction between PTP and TRPP is that the travel time is used in the objective function of the former and the arrival time is used in the objective function of the latter. An example TRPP instance with 11 nodes numbered 0 to 10 is specified in Table 1. For this instance, the optimal solution is $0 \rightarrow 10 \rightarrow 2 \rightarrow 9 \rightarrow 1 \rightarrow 4 \rightarrow 8 \rightarrow 7 \rightarrow 5$. Revenue calculation for this optimal solution is given in Table 2. From this table, we can see that for locations 10, 2, 9, 1, 4, 8, 7, and 5, the profit is greater than the arrival time. Hence, in each of these locations, the repairman will earn a revenue equivalent to the difference between the profit and the arrival time. Therefore, the total revenue collected is $395 + 473 + 419 + 299 + 27 + 314 + 128 + 341 = 2396$. Please note that unvisited locations 3 and 6 cannot be added after 5 as doing so will yield a negative revenue.

TRPP is formulated by Dewilde et al. [3]. The authors also introduced a mathematical model, and a Tabu Search (T_S) approach with multiple neighborhoods for addressing the TRPP. In the mathematical model, the number of vertices to be visited is decided in advance. For the tabu search approach, authors proposed ten heuristics, each of which creates a neighborhood of their own. These heuristics were divided into two groups based on subset selection characteristics and permutation characteristic. The subset selection group includes deletion, insertion and replacement. The permutation group includes remove-insert, a generalized Or-opt, swap-adjacent, swap, 2-opt, move-down and move-up. Each step of the tabu search algorithm utilizes a variable neighborhood descent (VND) search on the neighborhoods created by these 10 heuristics. VND explores the permutation group except Or-opt first and then subset selection group followed by Or-opt in the end. The heuristic in a group is explored in the order, in which they are listed here except for Or-opt which is explored at the end.

Table 1 Travel time between locations and profits

	0	1	2	3	4	5	6	7	8	9	10	profits
0	0	54	14	47	47	64	36	64	54	49	9	0
1	54	0	40	61	29	57	49	40	34	7	55	366
2	14	40	0	47	37	59	30	55	44	35	16	498
3	47	61	47	0	34	24	77	38	33	62	55	133
4	47	29	37	34	0	28	62	18	8	32	52	123
5	64	57	59	24	28	0	87	20	22	60	72	475
6	36	49	30	77	62	87	0	79	70	42	28	112
7	64	40	55	38	18	20	79	0	10	45	70	242
8	54	34	44	33	8	22	70	10	0	38	59	418
9	49	7	35	62	32	60	42	45	38	0	49	479
10	9	55	16	55	52	72	28	70	59	49	0	404

Table 2 Revenue calculation for route $0 \rightarrow 10 \rightarrow 2 \rightarrow 9 \rightarrow 1 \rightarrow 4 \rightarrow 8 \rightarrow 7 \rightarrow 5$

Locations r	0	10	2	9	1	4	8	7	5
Arc travel time	–	9	16	35	7	29	8	10	20
Arrival time (δ_r)	0	9	25	60	67	96	104	114	134
Profit p_r	0	404	498	479	366	123	418	242	475
Revenue ($p_r - \delta_r$)	0	395	473	419	299	27	314	128	341

Performance of these approaches was evaluated on a set of 120 TRPP benchmark instances having $n \in \{10, 20, 50, 100, 200, 500\}$.

Avci and Avci [2] introduced a new mixed integer linear programming formulation for TRPP. Unlike the model of Dewilde et al. [3], which required the number of vertices to be visited pre-specified, the new model has no such restriction. The authors proposed a powerful hybrid algorithm, a greedy randomized adaptive search procedure (GRASP) with iterated local search (ILS) to address the TRPP. Specifically, an initial solution is constructed for each iteration by a GRASP mechanism. An ILS procedure then enhances this initial solution that integrates a tabu-enhanced VND search with a flexible perturbation procedure. Detailed computational results presented in [2] show that GRASP-ILS approach outperformed T_S approach presented in [3].

Lu et al. [7] proposed the first population-based metaheuristic called hybrid evolutionary search algorithm (HESA) to tackle the TRPP, which integrates a GRASP for the initial solution generation, a variable neighborhood search (VNS) approach to enhance the initial solution. This solution is further enhanced by employing a crossover operator and the VNS process. The detailed computational results in [7] show the superiority of HESA over GRASP-ILS [2] and T_S [3]. Pei et al. [8] introduced general variable neighborhood search (GVNS) to find the solution of TRPP. The GVNS uses VND as a local search which explores various neighborhoods in a new random sequential fashion. The experimental results on benchmark instances demonstrate that GVNS outperforms T_S and GRASP-ILS [8].

To address the traveling repairman problem with profits (TRPP), we have developed a simple yet effective method based on hyper-heuristic (HH) approach with the greedy selection mechanism and variable degree of perturbation [9]. The computational results demonstrate the superiority of our approach in comparison with the state-of-the-art as our approach obtained comparable quality solutions in much shorter times.

Remainder of this paper is organized in the following manner: Sect. 2 provides an overview of hyper-heuristics. Section 3 describes our hyper-heuristic-based approach for TRPP. Section 4 provides the computational results and their analysis. Finally, Sect. 5 outlines the contributions of this paper and some possible directions for future work.

2 Overview of Hyper-Heuristic

Numerous metaheuristic algorithms such as genetic algorithms (GA), tabu search (TS), ant colony optimization (ACO) and simulated annealing (SA) have been designed and effectively implemented to solve several difficult combinatorial optimization problems. However, knowledge and experience are needed to develop an effective metaheuristic method to solve a particular problem. Hyper-heuristics provide a new approach to avoid the problem of such dependence on metaheuristics. The inspiration behind hyper-heuristics is to increase the degree of generality at which search techniques can perform, and thereby, delivering good quality solutions [10]. The term *hyper-heuristics* defined by Cowling et al. [11] as "a heuristic to select heuristics in the perspective of combinatorial optimization." Recently, the interpretation of hyper-heuristics has been extended to relate to a high-level automated exploration technique that explores the low-level heuristics search space to solve computationally challenging problems [12]. Hyper-heuristic techniques are segregated into two types :

- *Heuristic selection*: Techniques for selecting heuristics from available low-level heuristics.
- *Heuristic generation*: Techniques for creating new heuristics using the available low-level heuristics elements.

Heuristic selection-based hyper-heuristics have two main elements: a move acceptance criteria and a heuristic selection approach. In these hyper-heuristics, a low-level heuristic is chosen and applied at each step on the current solution. Then, a decision is taken whether or not to approve the newly created solution. Several selection mechanisms such as random, choice function-based, and greedy have been proposed in the literature. One of the heuristics is chosen at random in the random selection criteria. In contrast, all the available heuristics in greedy criteria are applied to the solution at hand. The best solution produced by all the heuristics is taken as the newly created solution to be subjected to acceptance criteria. In choice function-based selection method, a heuristic is chosen by employing a function whose preference changes dynamically over the iterations of the algorithm. Acceptance criteria determines whether the newly created solution can update the existing solution. In all acceptance (AA) criteria, newly created solution always replaces the current solution. In only improvement move acceptance (OI) criteria, newly created solution can replace the existing solution only if it is better than the existing solution. The readers may refer to [12, 13] for an elaborate discussion on hyper-heuristics and their applications.

3 Proposed Hyper-Heuristic Method

In the current section, we outline TRPP method. We will call our method hyper-heuristic with greedy selection mechanism (HH-GREEDY). The prominent elements of our method are described in the following sub-sections.

3.1 Generation of the Initial Solution

To get an initial solution S, we have proposed a greedy randomized construction method based on the nearest neighbor heuristic of TSP which operates as follows:

- Initialize S with the predefined depot as start vertex and a randomly chosen vertex as the second vertex. The vertices that are part of S will be termed as visited vertices and remaining as unvisited vertices.
- Select an unvisited vertex with the highest positive revenue out of 3 nearest adjacent unvisited vertices with respect to the last vertex of the route S, and it has to be added at the end of the route S. This procedure has to be iterated till no unvisited vertex with positive revenue remains.

Clearly, the complexity of generating an initial solution is $O(n^2)$. This initial solution is improved further by *hyper-heuristic*, which uses low-level heuristic for generating new solutions. Eight low-level heuristics are utilized by our hyper-heuristic approach, viz. *insert* (LH_1), *swap*(LH_2), *exchange* (LH_3), *remove* (LH_4), *2-Opt* (LH_5), *relocate* (LH_6), *MRR* (LH_7) and MRA (LH_8). The heuristics from LH_2 to LH_5 follow the first improvement strategy to produce a new solution, whereas the other heuristics follow the best improvement strategy.

3.2 Low-Level Heuristics

- *Insert* (LH_1): This heuristic attempts to add an unvisited location with the highest revenue in the tour at the best possible position, which increases the objective value. The best possible position means the position that produces the greatest rise in the objective value. An unvisited location needs to be checked in m positions of the tour, where m is current length of tour including the depot. Clearly, the worst-case complexity of LH_1 is $O(n)$.

- *Swap* (LH_2): The positions of two visited nodes are swapped within the tour to get an improved solution. This is possible by swapping each of the $m - 1$ visited nodes with the remaining $m - 2$ nodes of the tour, excluding the depot node. The complexity is $O(n^2)$.
- *Exchange* (LH_3): Here, a visited node can be attempted for exchange with an unvisited one. It tries to exchange $m - 1$ visited nodes with $n - m$ unvisited nodes. So the complexity of LH_3 is $O(n^2)$.
- *Remove*(LH_4): This heuristic deletes a visited node from the tour if doing so yields a higher objective value. Thus, the complexity of LH_4 is $O(n)$.
- *2-Opt* (LH_5): This heuristic aims to enhance the solution by deleting two edges from the tour and then inserting two new edges to create a legitimate tour. The complexity of 2-Opt heuristic is $O(n^2)$.
- *Relocate* (LH_6): This heuristic randomly removes five consecutive visited nodes and reinserts them in the order opposite to the order in which they are removed at their best possible positions. The complexity of LH_6 heuristic is $O(n)$.
- *MRR* (LH_7): This heuristic removes each visited node from the tour with an expectation (Deg_p) that decreases over iterations. We call this concept *variable degree of perturbation* which is described below. Then, in some random order, these removed nodes are inserted one by one back into the tour at their best possible positions. The complexity of MRR heuristic is $O(n^2)$. MRR heuristic complexity is $O(n^2)$.
- *MRA* (LH_8): It tries to remove each visited node from the tour by following the variable degree of perturbation principle mentioned in LH_7. Such removed nodes are added to the set of unvisited nodes. Then, all unvisited nodes are investigated for insertion into the tour in an iterative process. Initially, all unvisited nodes are specified as untested. Then, an untested unvisited node is chosen randomly and investigated for insertion into the tour at the best possible position during each iteration. If the trial indicates that inserting the node is beneficial, it is inserted. Otherwise, it is marked as tested. This process is repeated till no untested unvisited node remains. The complexity of MRA heuristic is $O(n^2)$.

A variable degree of perturbation is used in HH-GREEDY method. The degree of perturbation in the starting iterations must be high, and it must be decreased as the approach advances to find good quality solutions [9, 14, 15]. The degree of perturbation Deg_p decreases from max_{Deg_p} to min_{Deg_p} over first $iter_{max}$ iterations. Beyond $iter_{max}$ iterations, it remains fixed at min_{Deg_p}. For an iteration $iter$, the value of Deg_p is determined according to:

$$Deg_p = \begin{cases} \left(\frac{max_{Deg_p} - min_{Deg_p}}{iter_{max}}\right)(iter_{max} - iter) + min_{Deg_p} & \text{if } iter \leq iter_{max} \\ min_{Deg_p} & \text{otherwise} \end{cases} \tag{1}$$

Here, max_{Deg_p} is maximum degree of perturbation, min_{Deg_p} is minimum degree of perturbation, $iter$ is current iteration and $iter_{max}$ is maximum number of iterations.

3.3 Proposed Algorithm

The hyper-heuristic performance relies on the heuristic selection and acceptance criteria. To solve TRPP, we have used the greedy selection and the only improvement move acceptance criteria. To generate a new solution S' from S, HH-GREEDY applies all the available low-level heuristics on the current solution to generate new solutions. From these new solutions, HH-GREEDY uses the solution with the largest objective value for comparison with the current solution. As per only improvement move acceptance criteria, if the solution with largest objective value is better than the current solution then it replaces the current solution, otherwise current solution remains unchanged.

If there is no improving move for any of LH_1, LH_2, LH_3, LH_4 and LH_5, then there is no reason in trying that heuristic again till the structure of the tour alters. Hence, the respective heuristic is not utilized as long as there is no change in structure of the tour. By structure of the tour, we mean the set of visited locations and their positions in the tour. So structure can change either when the set of visited locations changes or their positions in the tour changes.

The pseudocode for the HH-GREEDY is given in Algorithm 1, where N_{run} is number of times the approach makes a fresh start, LH is the set of afore-mentioned eight low-level heuristics. *Create_New_Solution*(S, LH) is a function that constructs a new solution S' from the current solution S by utilizing the low-level heuristics belonging to LH as per greedy selection mechanism. Another function *Create_Initial_Solution()* constructs an initial solution according to the description provided in Sect. 3.1.

Algorithm 1: HH-GREEDY pseudocode

```
Input: HH-GREEDY parameters and a TRPP instance
Output: Best solution obtained

for k ← 1 to N_run do
    S_k ← Create_Initial_Solution();
    while (stopping criterion not met) do
        S' ← Create_New_Solution(S_k, LH);
        if (S' > S_k) then
            S_k ← S';
        end
    end
end
best ← best solution among S_1, S_2, ..., S_N_run;
return best;
```

3.4 Complexity Analysis of HH-GREEDY

To generate a new solution, HH-GREEDY uses all existing low-level heuristics one by one on the current solution. As a result, the complexity of each iteration of HH-GREEDY is governed by the highest complexity low-level heuristic. Hence, the overall complexity of each iteration of HH-GREEDY approach is $O(n^2)$.

4 Computational Results

The program for the proposed method HH-GREEDY has been written in C programming language, and this program has been executed on an Ubuntu 18.04 system with 2.20GHz Core i3-2330M CPU and 4 GB RAM, which is slower than the machines used in [7, 8]. The performance of the HH-GREEDY is evaluated using 80 TRPP benchmark instances proposed in [3]. The number of nodes $n \in \{10, 20, 50, 100\}$ in these instances. There are 20 instances for each value of n, thereby leading to a total of 80 instances. The values of different parameters used in determining the degree of perturbation in HH-GREEDY are as follows: the maximum degree of perturbation $max_{deg_p} = 0.9$, the minimum degree of perturbation $min_{deg_p} = 0.5$, the maximum number of iteration $iter_{max} = 2n$, i.e., it is set to twice the number of nodes in an instance. The HH-GREEDY terminates when best solution fails to improve over 250 consecutive iterations. The HH-GREEDY has been executed 30 times independently on each test instance.

We have compared the results of our proposed method with those of T_S [3], GRASP-ILS [2], HESA [7] and GVNS [8]. We have reported the instance-by-instance results for various approaches in Tables 3, 4 and 5. In all these tables, column 1 provides the instance name. Table 3 presents the TRPP results for the instances having vertices 10 and 20. In this table, columns 2 and 7 report the known-optimal values computed in [2], columns 3 & 8, 4 & 9 and 5 & 10 report the best objective values of T_S, GRASP-ILS, and HESA respectively, whereas columns 6 & 11 report the best objective values obtained by our approach. Execution time is not presented for these instances as they are negligible (generally under a second). For instances 10, 20 and 50, the results of GVNS were not reported in [8], because no significant difference in outcomes with the state-of-art approaches (GRASP-ILS and T_S) was observed.

Table 4 reports the TRPP results for the instances having 50 vertices. Column 2 reports the tabu search(T_S) algorithm's best objective value, whereas its average computation time is reported in column 3. Columns 4 5 and 6 7 reports the same quantities for GRASP-ILS and HESA. Column 8 gives the best objective value of HH-GREEDY, whereas column 9 provides its average computation time. The proposed method's solution quality is very similar to the current state of the art (T_S, GRASP-ILS and HESA). Still, the computation time for HH-GREEDY is significantly shorter. The average executions time in seconds are: 14.2 for T_S, 9.4 for GRASP-ILS, 9.4 for HESA and 0.1 for HH-GREEDY.

Table 3 TRPP results for the instances having 10 and 20 vertices

Instance	n=10					n=20				
	MIP-Exact	T_S	GRASP-ILS	HESA	HH-GREEDY	MIP-Exact	T_S	GRASP-ILS	HESA	HH-GREEDY
1	2520	2520	2520	2520	2520	8772	8772	8772	8772	8772
2	1770	1770	1770	1770	1770	10,174	10,174	10,174	10,174	10,174
3	1737	1737	1737	1737	1737	7917	7917	7917	7917	7917
4	2247	2247	2247	2247	2247	7967	7967	7967	7967	7967
5	2396	2396	2396	2396	2396	7985	7985	7985	7985	7985
6	1872	1872	1872	1872	1872	7500	7500	7500	7500	7500
7	1360	1360	1360	1360	1360	9439	9439	9439	9439	9439
8	1696	1696	1696	1696	1696	7999	7999	7999	7999	7999
9	1465	1465	1465	1465	1465	6952	6952	6952	6952	6952
10	1014	1014	1014	1014	1014	8582	8582	8582	8582	8582
11	1355	1355	1355	1355	1355	7257	7257	7257	7257	7257
12	1817	1817	1817	1817	1817	6857	6857	6857	6857	6857
13	1585	1585	1585	1585	1585	7043	7043	7043	7043	7043
14	2122	2122	2122	2122	2122	6964	6964	6964	6964	6964
15	1747	1747	1747	1747	1747	6270	6270	6270	6270	6270
16	1635	1635	1635	1635	1635	8143	8143	8143	8143	8143
17	2025	2025	2025	2025	2025	10,226	10,226	10,226	10,226	10226
18	1783	1783	1783	1783	1783	7625	7625	7625	7625	7625
19	1797	1797	1797	1797	1797	7982	7982	7982	7982	7982
20	1771	1771	1771	1771	1771	7662	7662	7662	7662	7662

Table 5 reports the TRPP results for the instances having 100 vertices. Column 2 reports the best objective value of the T_S algorithm, column 3 provides its average computation time. Column 4 5, 6 7 and 8 9 reports the same for GRASP-ILS, HESA and GVNS. Column 10 gives best objective value of HH-GREEDY, while column 11 provides its average computation time. The solution quality of the proposed method is similar to the current state of the art (T_S, GRASP-ILS, HESA and GVNS). Still, the computation time for HH-GREEDY is significantly shorter. The average executions time in seconds for different approaches are: 144.9 for T_S, 106.6 for GRASP-ILS, 106.7 for HESA, 8.82 for GVNS and 1.16 for HH-GREEDY.

Table 6 presents the summary of results for the four sets of benchmark instances containing 10, 20, 50 and 100 vertices. For each instance and approach, we describe the average of the best objective value obtained by various approaches under column AA_B, the average gap (%) between the best-known and best-obtained results under column AA_G(%) and the average execution time in seconds under column AA_T(s). The average gap(%) for an approach is computed as $100 \times \frac{(L_2-L_1)}{L_2}$, where L_1 is the average of the best value obtained by that approach and L_2 is the best-known value [7]. This summary table shows that the solution quality of the proposed approach is similar to current state-of-the-art approaches (T_S, GRASP-ILS, HESA and GVNS), but at the expense of execution times that are only a small fraction of execution

Table 4 TRPP results for the instances having 50 vertices

Instance	T_S		GRASP-ILS		HESA		HH-GREEDY		
	Sol_b	Avg_t	Sol_b	Avg_t	Sol_b	Avg_t	Sol_b	Avg_s	Avg_t
1	50,921	14.10	50,921	11.50	50,921	11.50	50,921	50,843.04	0.11
2	52,594	12.60	52,594	8.40	52,594	8.40	52,594	52,377.67	0.10
3	52,144	12.50	52,144	9.40	52,144	9.40	52,144	52,032.00	0.11
4	45,465	12.40	45,465	7.30	45,465	7.30	45,465	45,204.07	0.09
5	45,489	11.20	45,489	8.40	45,489	8.40	45,489	45,417.57	0.10
6	55,630	12.80	55,630	7.30	55,630	7.30	55,630	55,539.23	0.11
7	44,300	9.20	44,302	9.40	44,302	9.40	44,302	44,247.77	0.08
8	55,753	12.30	55,801	10.50	55,801	10.50	55801	55,758.46	0.10
9	44,964	11.30	44,964	10.50	44,964	10.50	44,964	44,868.83	0.09
10	47,071	9.80	47,071	9.40	47,071	9.40	47,071	46,987.17	0.09
11	51,912	10.30	51,912	9.40	51,912	9.40	51,912	51,886.93	0.09
12	53,567	14.50	53,567	8.40	53,567	8.40	53,567	53,462.10	0.11
13	46,830	12.60	46,830	9.40	46,830	9.40	46,830	46,662.70	0.11
14	52,665	53.10	52,665	8.40	52,665	8.40	52,665	52,541.70	0.10
15	58,856	14.00	58,856	10.50	58,856	10.50	58,856	58,550.83	0.13
16	49,754	12.10	49,754	12.60	49,754	12.60	49,754	49,616.67	0.11
17	42,525	9.60	42,525	8.40	42,525	8.40	42,525	42,503.20	0.10
18	40,536	11.90	40,536	9.40	40,536	9.40	40,536	40,479.40	0.09
19	55,346	15.00	55,346	10.50	55,346	10.50	55,346	55,214.23	0.10
20	61,286	13.40	61,286	9.40	61,286	9.40	61,286	61,213.67	0.09

Table 5 TRPP results for the instances having 100 vertices

Instance	T_S		GRASP-ILS		HESA		GVNS		HH-GREEDY		
	Sol_b	Avg_t	Sol_b	Avg_t	Sol_b	Avg_t	Sol_b	Avg_t	Sol_b	Avg_s	Avg_t
1	209,952	119.7	209,952	92.2	209,952	92.2	209,952	1.89	209,952	209,154.17	1.21
2	196,318	101.3	196,318	135.1	196,318	135.2	196,318	22.92	196,318	195,810.41	0.93
3	211,937	126.7	211,937	107.9	211,937	108.0	211,937	5.15	211,937	211,397.41	1.05
4	217,685	112.1	217,685	89.0	217,685	89.1	217,685	2.14	217,614	217,032.83	1.22
5	215,119	169.2	215,119	131.0	215,119	131.0	215,119	5.52	215,119	214,348.94	1.12
6	228,687	144.0	228,687	102.7	228,687	102.8	228,687	2.23	228,431	227,722.80	1.23
7	200,060	347.3	200,064	128.9	200,064	129.0	200,064	6.36	200,032	199,444.94	1.26
8	205,760	117.8	205,760	119.4	205,760	119.5	205,760	8.93	205,760	205,276.56	1.16
9	226,240	93.7	226,240	112.1	226,240	112.1	226,240	0.79	226,240	225,655.36	0.90
10	218,202	162.5	218,202	88.0	218,202	88.0	218,202	1.19	218,202	217,601.80	1.23
11	212,503	126.8	212,503	125.7	212,503	125.8	212,503	5.17	212,503	211,809.77	1.15
12	222,249	148.0	222,249	98.5	222,249	98.6	222,249	2.25	222,249	221,880.91	1.08
13	206,878	145.2	206,957	89.0	206,957	89.1	206,957	0.99	206,957	206,418.33	1.17
14	215,690	126.8	215,690	85.9	215,690	86.0	215,690	2.41	215,690	215,337.70	1.22
15	213,758	157.7	213,811	89.0	214,041	89.1	214,041	16.40	213,988	213,163.06	1.05
16	214,036	152.0	214,036	89.0	214,036	89.0	214,036	13.89	214,036	213,385.20	1.17
17	223,636	136.6	223,636	118.4	223,636	118.5	223,636	25.70	223,636	222,879.06	1.31
18	192,849	122.5	192,849	97.4	192,849	97.4	192,849	4.47	192,849	192,268.70	1.14
19	206,755	174.9	206,755	128.9	206,755	128.9	206,755	20.62	206,755	206,008.83	1.34
20	198,842	213.4	198,908	104.8	198,908	104.8	198,908	27.42	198,908	198,300.91	1.36

Table 6 Summary Table

	T_S			GRASP-ILS			HESA			GVNS			HH-GREEDY		
n	AA_B	AA_G(%)	AA_T(s)	AA_B	AA_G(%)	AA_T(s)	AA_B	AA_G(%)	AA_T(s)	AA_B	AA_G(%)	AA_T(s)	AA_B	AA_G(%)	AA_T(s)
10	1785.7	0.00	< 1	1785.7	0.00	< 1	1785.7	0.00	< 1	–	–	–	1785.7	0.00	<1
20	7965.8	0.00	< 1	7965.8	0.00	<1	7965.8	0.00	<1	–	–	–	7965.8	0.00	<1
50	50,380.4	0.00	14.2	50,382.9	0.00	9.4	50,382.9	0.00	9.4	–	–	–	50382.9	0.00	0.1
100	211,857.8	0.01	149.9	211,867.9	0.01	106.6	211,879.4	0.00	106.7	211,879.4	0.00	8.82	211,858.8	0.01	1.16

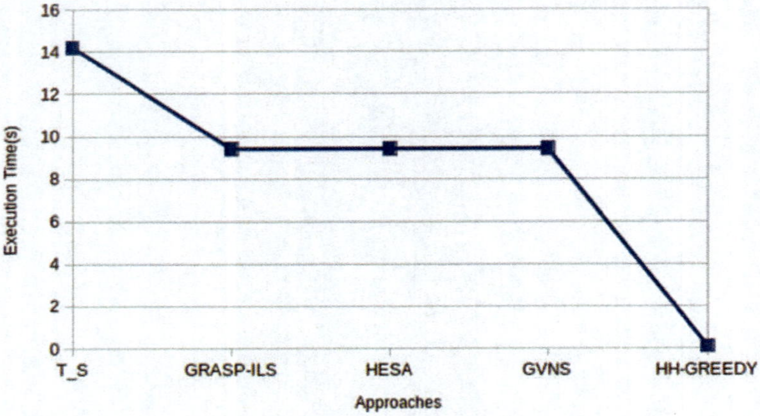

Fig. 1 Execution times of different approaches for instances with 50 vertices

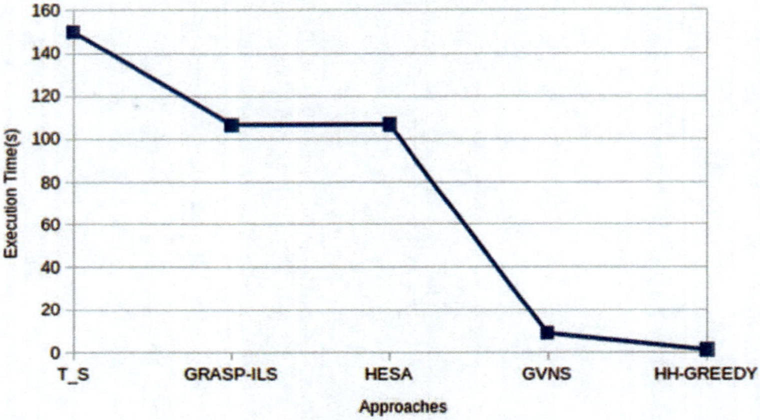

Fig. 2 Execution times of different approaches for instances with 100 vertices

times required by other approaches. HH-GREEDY is significantly faster than other approaches while providing the solutions of same quality. Figures 1 and 2 show the average execution times of different approaches on instances with 50 and 100 vertices, respectively. These figures also assert the clear superiority of our approaches in terms of computation time.

5 Conclusions

In this paper, the traveling repairman problem with profits (TRPP) is studied as an essential variant of the traveling salesman problem (TSP). To address this problem, we have presented a hyper-heuristic approach with a greedy selection strategy. A variable degree of perturbation over the iterations of our approach is used to direct the search process toward high quality solutions. The computational results demonstrate that the proposed approach is significantly faster while performing similar to state-of-the-art approaches (T_S, GRASP-ILS, HESA and GVNS) in terms of solution quality.

We will explore other hyper-heuristic selection mechanisms and new low-level heuristic methods in the future. We intend to incorporate local search procedures to enhance the performance of HH-GREEDY and test its performance on larger instances. Another potential future work is to use reinforcement learning or any other machine learning technique to boost the performance of the proposed approach.

Acknowledgements The first author is obliged to the University Grants Commission, Government of India, for supporting his research via a Senior Research Fellowship.

References

1. Goemans M, Kleinberg J (1998) An improved approximation ratio for the minimum latency problem. Math Programming 82:111–124
2. Avci M, Avci MG (2017) A grasp with iterated local search for the traveling repairman problem with profits. Comput Industrial Eng 113:323–332
3. Dewilde T, Cattrysse D, Coene S, Spieksma FC, Vansteenwegen P (2013) Heuristics for the traveling repairman problem with profits. Comput Oper Res 40:1700–1707
4. Vansteenwegen P, Souffriau W, Van Oudheusden D (2011) The orienteering problem: a survey. Eur J Oper Res 209(1):1–10
5. Dasari KV, Pandiri V, Singh A (2021) Multi-start heuristics for the profitable tour problem. Swarm Evol Comput 64:100897
6. Feillet D, Dejax P, Gendreau M (2005) Traveling salesman problems with profits. Transp Sci 39(2):188–205
7. Lu Y, Hao J-K, Wu Q (2019) Hybrid evolutionary search for the traveling repairman problem with profits. Inf Sci 502:91–108
8. Pei J, Mladenović N, Urošević D, Brimberg J, Liu X (2020) Solving the traveling repairman problem with profits: a novel variable neighborhood search approach. Inf Sci 507:108–123
9. Venkatesh P, Singh A (2015) Two metaheuristic approaches for the multiple traveling salesperson problem. Appl Soft Comput 26:74–89
10. Burke E, Kendall G, Newall J, Hart E, Ross P, Schulenburg S (2003) Hyper-heuristics: an emerging direction in modern search technology. In: Handbook of metaheuristics. Springer, Heidelberg, pp 457–474
11. Cowling P, Kendall G, Soubeiga E (2000) A hyperheuristic approach to scheduling a sales summit. In: International conference on the practice and theory of automated timetabling. Springer, Heidelberg, pp 176–190
12. Burke EK, Gendreau M, Hyde M, Kendall G, Ochoa G, Özcan E, Qu R (2013) Hyper-heuristics: a survey of the state of the art. J Oper Res Soc 64(12):1695–1724
13. Burke EK, Gendreau M, Hyde M, Kendall G, Ochoa G, Özcan E, Qu R (2020) Recent advances in selection hyper-heuristics. Eur J Oper Res 285(2):405–428
14. Venkatesh P, Srivastava G, Singh A (2019) A multi-start iterated local search algorithm with variable degree of perturbation for the covering salesman problem. In Harmony search and nature inspired optimization algorithms, pp 279–292

15. Venkatesh P, Singh A (2019) An artificial bee colony algorithm with variable degree of perturbation for the generalized covering traveling salesman problem. Appl Soft Comput 78:481–495

Economic Dispatch Using Adapted Particle Swarm Optimization

Raghav Prasad Parouha

Abstract In power system process Economic Dispatch (ED) foremost typical task. Minimize the total fuel cost of the generating units is the key objective of this problem subjected to satisfying inequality and equality constraints. Also, it converted into nonconvex and nonlinear complicated optimization problem, due to some practical constraints like prohibited operating zones, transmission losses and ramp rate limits. Recently, particle swarm optimization (PSO) is one of the well-known algorithm for solving nonconvex optimization problems. But, particularly for multimodal problem it may suffer to trap at local minima. To overcome the issues of PSO, this paper offering an adapted particle swarm optimization (viz. aPSO) for solving ED problems. In aPSO, new factors for inertia, cognitive and social components (i.e. acceleration coefficients and inertia weight) are familiarized. As per the concerts these factors are dynamically changed and properly maintained the diversity of the PSO. The projected aPSO is implemented to solve three altered unit test structures (3, 6 and 15) of ED problem. The experimental results of aPSO compared with related algorithms. It shows that the performance of aPSO are enhanced than the other algorithms in terms of searching ability and rate of convergence.

Keywords Economic dispatch · Particle Swarm Optimization · Position update · Velocity

1 Introduction

In electric power process Economic dispatch (ED) is the major problems. Fuel cost reduction of the generators with satiating numerous inequality and equality constraints is the main objective of ED problem. In the past decades, various salient mathematical approaches [1] have been recommended to solve ED problems. As an alternative to the conventional mathematical approaches, evolutionary optimization

R. P. Parouha (✉)
Indira Gandhi National Tribal University, Amarkantak, Madhya Pradesh 484887, India
e-mail: rparouha@gmail.com

M. Thakur et al. (eds.), *Soft Computing for Problem Solving*, Lecture Notes in Networks and Systems 547, https://doi.org/10.1007/978-981-19-6525-8_39

techniques such as Differential Evolution (DE) [2] and Particle Swarm Optimization (PSO) [3] and many more have been implemented on the ED problems. These methods are given that quality results with a reduced amount of time, few numbers of function evaluations etc. Though, by reason of premature convergence, these methods are not sure for optimum solution. Among successful swarm/evolutionary optimization techniques, PSO (particle swarm optimization) is broadly used for complex real world problems due to its advantages like global examining capability and simple application. Many modifications and hybrids of PSO methods were proposed for solving the complex optimization problem [4–18].

For instance, Lee and Chen [19] introduced iteration best index on of velocity equation of PSO (named as IPSO) and successfully applied to solve ED problem. But after some iteration IPSO still face the trap at local minima. Based on the information of bad experience of particles, Selvakumar and Thanushkodi [20] proposed a new PSO (NPSO) and applied on nonconvex ED problems. Due to the suitable parameters, NPSO produce better results compared to others. In [21], Chaturvedi et al., proposed novel PSO using time varying acceleration coefficients (TVAC) (named as PSO_TVAC). In [22], Khokhar et al., proposed a new PSO (named as TVAC_PSO) using time varying acceleration coefficients and effectively applied on nonconvex ED problems. Further, Abdullah et al. [23] proposed a modified PSO with time varying acceleration coefficients (named as MPSO-TVAC) for solving ED problems. As a result, by proper tuning of coefficients of PSO, the particles are guided towards optimum solution. But, most of these methods do not yield reliable results for many optimization problems. Moreover, the PSO still need further improvement to offer a good solution [16, 17].

In this research paper, an adapted PSO (viz. aPSO) presented to answer ED problems. It is based on new factors for, cognitive, social and inertia components (i.e. acceleration coefficients and inertia weight). As per the need of swarm optimization techniques the presented factors of aPSO produce the value in different stages as follows (i) beginning-small social and large cognitive factors, (ii) latter-large social and small cognitive and (iii) in beginning and latter inertia weight reduced accordingly.

The performance of the offered aPSO is tested on different ED problems (3, 6 and 15-unit test problems). In terms of effecting times and optimum cost, acquired results of aPSO related with classical PSO and DE. It shows that aPSO is better methods, due to the quality solution and faster convergence.

The rest article is ordered as Sect. 2 briefed ED problems; Sect. 3 explained proposed aPSO; Sect. 4 presents application and results; lastly, conclusion and future works given in Sect. 5.

2 Economic Dispatch (ED) Problem

In power system, Economic dispatch (ED) is one of the major task. Fuel cost (Fc) reduction/minimize of the generators with satiating numerous inequality and equality constraints is the main objective of ED problem. Mathematically, it can be presented as follows.

$$\text{Minimize } F_c = \sum_{i=1}^{N} F_i(P_i) \tag{1}$$

where, N is number of power generation units and F_i (cost function) is defined as follows.

$$F_i(P_i) = a_i P_i^2 + b_i P_i + c_i + \alpha \left| e_i \sin\left(f_i \left(P_i^{\min} - P_i \right) \right) \right| \tag{2}$$

where, P_i—ith generator power; a_i, b_i, c_i, e_i, f_i—factors of the cost function.

If $\alpha = 0$, ED problems is convex/smooth else $\alpha = 1$ then ED problems non-convex/non-smooth i.e. valve-point loading effect included in ED problem.

Constraints

Power balance

$$P_L(\text{total transmission loss}) = \sum_{i=1}^{N} P_i - P_D(\text{total power demand}) \tag{3}$$

using B matrix coefficients, the transmission loss can be calculated as

$$\sum_{i=1}^{N} \sum_{j=1}^{N} P_i B_{i,j} P_j + \sum_{i=1}^{N} B_{0i} + B_{00} = P_L \tag{4}$$

where, $B_{i,j}$ is the i, jth and B_{0i} is the ith element of B symmetric matrix and B_{00}—constant loss coefficient.

Generator

$$P_i^{\min}(\text{maximum limits}) \le P_i \le P_i^{\max}(\text{minimum limits}) \tag{5}$$

Prohibited operating zones (POZs):

$$AZ_i(\text{operation zones of unit } i) = \begin{cases} P_i^{\min} \le P_i \le P_{i,1}^{l} \\ P_{i,m-1}^{u} \le P_i \le P_{i,m}^{l} \\ P_{i,M_i}^{u} \le P_i \le P_i^{\max} \end{cases} \tag{6}$$

where, $i = 1, 2, \ldots, N$ and $m = 2, 3, \ldots, M_i$ (number of permitted POZ of unit i), $P_{i,m}^{u}$ and $P_{i,m}^{l}$—upper and lower bounds of mth POZ.

Ramp rate limits

$$\text{maximum}\left(P_i^0 - DR_i, P_i^{\min}\right) \le P_i \le \text{minimum}\left(P_i^0 - UR_i, P_i^{\max}\right) \tag{7}$$

where, P_i^0 and P_i—previous and current real output power of ith generator respectively, DR_i and UR_i—down and up ramp limit ramp limit of ith generator, $P_i \in AZ_i$.

3 Proposed Adapted Particle Swarm Optimization (aPSO)

In this section, presented standard PSO then explained the proposed adapted PSO (aPSO) algorithm.

3.1 Standard PSO

PSO was established by Kennedy and Eberhart in 1995 [3]. It stimulated by collective activities of beasts for instance birds and fish. The main advantages of PSO algorithm are simplicity, easy implementation, computational efficiency, robustness and ability to find a global optimum [4, 5]. So, currently PSO broadly used to solve numerous real-life applications. In D-dimensional search space, PSO initiates with a random swarm of particles. At 'tth' iteration 'ith' particle has an position $x_i^t = \left(x_{i,1}^t, x_{i,2}^t, \ldots, x_{i,D}^t\right)$ and a velocity $v_i^t = \left(v_{i,1}^t, v_{i,2}^t, \ldots, v_{i,D}^t\right)$. The personal best solution attained by ith particle is distinct as $p_{best_i}^t = \left(p_{best_{i,1}}^t, p_{best_{i,2}}^t, \ldots, p_{best_{i,d}}^t\right)$ and among of best $p_{best_i}^k$ is g_{best}^t called global best. The v_i^t and x_i^t of Dth particle will be updated as follows.

$$v_{i,D}^{t+1} = wv_{i,D}^t + c_1r_1\left(p_{best_{i,D}}^t - x_{i,D}^t\right) + c_2r_2\left(g_{best_D}^t - x_{i,D}^t\right) \tag{8}$$

$$x_{i,D}^{t+1} = v_{i,D}^{t+1} + x_{i,D}^t \tag{9}$$

where, $x_{i,D}^t$ and $v_{i,D}^t$—tth present position and velocity of the ith particle; $t(1 \le t \le t_{\max})$—present iteration, and $t_{\max}$—maximum iteration, i—ith particle of the population/swarm $(n_P - 1 \le i \le n_P)$.

Factor	Impact
c_1 and c_2—cognitive and social acceleration coefficient	• Switches particle direction to best position • Large c_1 and small c_2 permitted to move everywhere in search space at beginning [18] • Small c_1 and large c_2 allowed to converge particle to the global optima in latter part of optimization process [19]
r_1 and r_2—random numbers	• Uniformly distributed within the range [0, 1] to control the component of PSO [5]
w: inertia weight	• It was set to 1, [0, 1.4] and [0.9, 1.2] [19]. Their results directed that taking different values of w contributes a faster convergence • A large and small value of w enables exploitation and exploration ability [18] correspondingly. It improves search performance and convergence capability • It trade-off between global exploitation and local exploration abilities [19]

Observation and motivation

- Usually, in population based methods two things are desirable in different phases

(i) Early—it needed to inspire particle to move whole search space deprived of gathering the local solution

(ii) Latter—it needed vital enhancement in convergence for global solution

- PSO is simplest and efficient method for global optimization
- Particles trapped in local optima which indicate premature convergence of PSO
- Recently, many researches have been carried out to improve its performance of PSO through novel parameters [20]
- c_1 and c_2 factor helps PSO to fulfill above necessities of both phases
- Suitable w settings can advance presentation of PSO method effectively
- An adapted PSO (aPSO) is presented to advance performance of basic PSO in the next section

3.2 *Proposed aPSO*

Inspired by above explanations, an adapted particle swarm optimization (aPSO) is presented. It consists of new acceleration factor (c_1 and c_2). In aPSO, large c_1 and small c_2 are castoff at start phases. While, small c_1 and large c_2 are castoff at latter phases. Similarly, a new dynamically changing inertia weight (w) is castoff in aPSO. The c_1, c_2 and w specified as below.

$$c_1 = c_{1\text{initial}} \times \left(\frac{c_{1\text{final}}}{c_{1\text{initial}}}\right)^{\left(\frac{t}{t_{\max}}\right)^2} \text{ and } c_2 = c_{2\text{initial}} \times \left(\frac{c_{2\text{final}}}{c_{2\text{initial}}}\right)^{\left(\frac{t}{t_{\max}}\right)} \tag{10}$$

$$w = \begin{cases} w_{\text{maximum}} \times \left(\dfrac{\frac{w_{\text{minimum}}+w_{\text{maximum}}}{2}}{w_{\text{maximum}}}\right)^{\left(\frac{t}{t_{\max}/2}\right)^2}; & t \le \frac{t_{\max}}{2} \\ \dfrac{w_{\text{minimum}}+w_{\text{maximum}}}{2} \times \left(\dfrac{w_{\text{minimum}}}{\frac{w_{\text{minimum}}+w_{\text{maximum}}}{2}}\right)^{\left(\frac{t-t_{\max}/2}{t_{\max}/2}\right)^2}; & t > \frac{t_{\max}}{2} \end{cases} \tag{11}$$

Hence, in aPSO $v_{i,j}^{t+1}$ of the ith particle are updated by the following Eq. (12).

$$v_{i,D}^{t+1} = \left\{ \begin{array}{ll} w_{\text{maximum}} \times \left(\frac{\frac{w_{\text{minimum}} + w_{\text{maximum}}}{2}}{w_{\text{maximum}}} \right)^{\left(\frac{t}{t_{\max}/2}\right)^2} ; & t \leq \frac{t_{\max}}{2} \\ \frac{w_{\text{minimum}} + w_{\text{maximum}}}{2} \times \left(\frac{w_{\text{minimum}}}{\frac{w_{\text{minimum}} + w_{\text{maximum}}}{2}} \right)^{\left(\frac{t - t_{\max}/2}{t_{\max}/2}\right)^2} & ; t > \frac{t_{\max}}{2} \end{array} \right\} v_{i,D}^{t}$$
$$+ \left(c_{1\text{initial}} \times \left(\frac{c_{1\text{final}}}{c_{1\text{initial}}} \right)^{\left(\frac{t}{t_{\max}}\right)^2} \right) r_1 \left(p_{best_{i,D}}^{t} - x_{i,D}^{t} \right)$$
$$+ \left(c_{2\text{initial}} \times \left(\frac{c_{2\text{final}}}{c_{2\text{initial}}} \right)^{\left(\frac{t}{t_{\max}}\right)} \right) r_2 \left(g_{best_d}^{k} - x_{id}^{k} \right) \quad (12)$$

After extensive following values are set to use—$c_{1\text{initial}} = c_{2\text{initial}} = 2.5$; $c_{1\text{final}} = c_{2\text{initial}} = 0.5$, $w_{\text{maximum}} = 0.9$ and $w_{\text{minimum}} = 0.4$). The performance of c_1 and c_2) and w during evolution of aPSO are represented in Figs. 1 and 2 respectively. c_1 and c_2 vary as per the iteration number in aPSO, which maintain exploration and exploitation and result near optimum solution. In early iteration, large c_1 and small c_2, shoves the particles to change in whole search space. In later iteration, decrease c_1 and increase c_2 twitch the particles to the global solution. Also, small w enhance aPSO local search ability and large w improve aPSO global search capability. Accordingly, proposed aPSO will regularly provide good quality global solution.

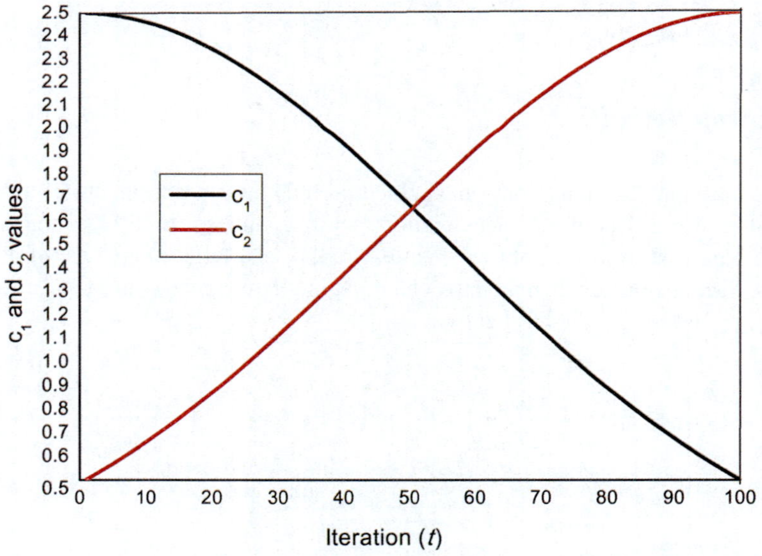

Fig. 1 Deviation of c_1 and c_2

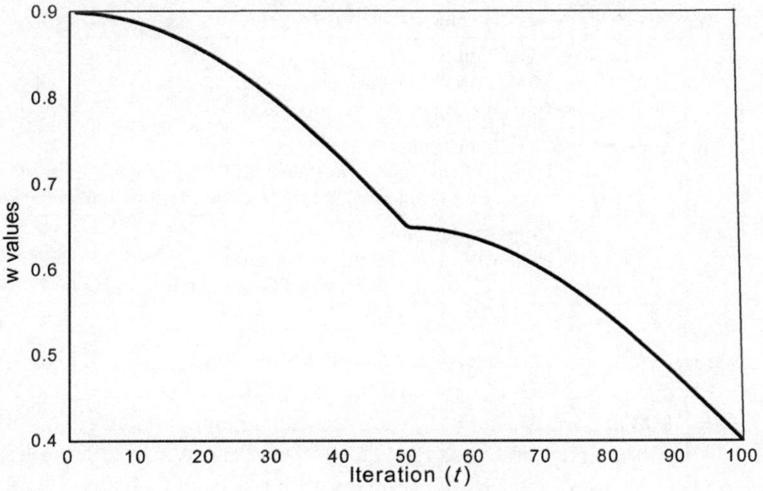

Fig. 2 Deviation of w

Steps of aPSO for ED problems

1-step	Initialize population $p_{i,D} = p_{i,D}^{\min} + rand(0, 1) \times \left(p_{i,D}^{\max} - p_{i,D}^{\min}\right)$
2-step	Objective function calculation $\sum_{i=1}^{N} F_i(P_i) + \beta(\text{penalty factor})\left\|\sum_{i=1}^{N} P_i - P_D - P_L\right\| = f(P_i)$
3-step	Set and update p_{best} and g_{best}
4-step	Update velocity by (12) and position using (9)
5-step	Checking and replacing $p_{i,D}^{t+1} = \begin{cases} p_{i,D}^{t} + v_{i,D}^{t+1} & \text{if } p_{i,D}^{\min} \leq p_{i,D}^{k+1} \leq p_{i,D}^{\max} \\ p_{i,D}^{\min} & \text{if } p_{i,D}^{t+1} \leq p_{i,D}^{\min} \\ p_{i,D}^{\max} & \text{if } p_{i,D}^{t+1} \geq p_{i,D}^{\max} \end{cases}$
6-step	Report obtained solution as per stopping criteria or repeat steps 2–6

4 Application and Results

To check the effectiveness of the presented aPSO, it used to solve ED problems having 3 typical unit generation test structures (TSt_s). These TSt_s are reported below.

Test structures (TSt_s)	Generation units
$TStm_s$-1 [24]	• 3-generating units • 850 MW (total real power demand) • Transmission losses are not considered
$TStm_s$-2 [25]	• 6-generating units • 1263 MW (total real power demand) • Considered-ramp rate limit, POZs, and transmission loss
$TStm_s$-3 [25]	• 15-generating units • 2630 MW (total real power demand) • Considered-ramp rate limits, POZs and transmission loss

Following parameters have been used for all $TStm_s$—$c_{1\text{initial}} = 2.5$, $c_{1\text{final}} = 0.5$, $c_{2\text{initial}} = 0.5$, $c_{2\text{final}} = 2.5$, $w_{\text{maximum}} = 0.9$ and $w_{\text{minimum}} = 0.4$, $\beta = 5 \times 10^3$, $n_P = 30$, $t_{\max} = 1000$ and trails/run = 50, for a reasonable comparison. All simulation experiments are conducted on Core (TM) i7-10750H GHz, Win10 OS, Intel(R) and 16 GB RAM. The achieved results of presented aPSO equated with DE [2], PSO [3], IPSO [19], NPSO [20], TVAC_PSO [22] and MPSO-TVAC [23]. Note that n_P, $t_{\max}$ and independent run of the proposed aPSO are same equated algorithms and the best results signified as bold letters in each tables. The results of Test Structures (TSts) are summarized as follows.

Test structures (TSts)	Results
TSts-1	The outcomes of this structure are demonstrated in Table 1 and the results of aPSO compared with DE [2], PSO [3], NPSO [20] and TVAC_PSO [22] in term of computation time and generation cost. From Table 1, all algorithms achieve an optimum result, due to the simple structures. But, the process time of aPSO is less than others
TSts-2	The results of this structure generated by aPSO recorded in Table 2 with comparison of DE [2], PSO [3], IPSO [19], MPSO-TVAC [23]. From this Table, it is observed that aPSO can yield lower generation cost with less Execution time than others
TSts-3	The results formed by aPSO for this structure is noted in Table 3 with the comparison of DE [2], PSO [3], IPSO [19], MPSO-TVAC [23]. It provided that aPSO produces lower generation cost with less execution time with others

Apart from this, convergence curves of aPSO with DE and PSO are depicted in revealed in Figs. 3, 4 and 5 for TSts-1, TSts-2 and TSts-3 separately. From these figures, it can be perceived that aPSO converge faster than others. From numerical and graphical comparisons, it can say that the presented aPSO is a well-organized method in terms of lowest cost. Its frequency to converging the better solution is always higher than other algorithms. Thus, it may be concluded that the aPSO is the most robust, consistent and free from premature convergence, as it attaining the lowest cost in all the runs. Comparatively, aPSO yields the lowest cost at lesser time than others. Hence, the all results reveal that the aPSO may be used as a good optimizer for the solution power systems optimization problems.

Table 1 Results for TSts-1

Output (MW)	DE [2]	PSO [3]	NPSO [20]	TVAC_PSO [22]	aPSO
u-1	300.27	300.27	300.27	300.27	300.27
u-2	400.00	400.00	400.00	400.00	400.00
u-3	149.73	149.73	149.73	149.73	149.73
Total power	850	850	850	850	850
Cost	**8234.07**	**8234.07**	**8234.07**	**8234.07**	**8234.07**
Time (s)	2.06	4.37	4.52	3.64	**1.03**

Table 2 Results for TSts-2

Output (MW)	DE [2]	PSO [3]	IPSO [19]	MPSO-TVAC [23]	aPSO
u-1	474.81	447.47	449.802	448.170	451.5204
u-2	178.64	173.10	171.042	173.291	172.1750
u-3	262.21	262.68	250.865	263.145	258.4186
u-4	134.28	139.42	150.000	138.714	140.6441
u-5	151.90	165.30	159.347	165.960	162.0797
u-6	74.180	87.980	94.633	86.691	90.34150
Total power	1276.03	1275.95	1275.69	1275.97	**1275.179**
Generation loss	13.03	12.95	12.69	12.97	**12.179**
Cost	15,459.0	15,450.0	15,453.50	15,449.92	**15,441.108**
Time (s)	26.59	49.00	14.89	12.68	**8.61**

5 Conclusion and Future Work

This article offers an adapted particle swarm optimization (viz. aPSO) for solving economic dispatch (ED) problem. In aPSO, new factors for cognitive, social and inertia components (i.e. acceleration coefficients and inertia weight) are familiarized. As per the concerts these factors are dynamically changed and properly maintained the exploration and exploitation.

The presentation of aPSO has been validated on 3-unit test structures (3, 6 and 15) of ED problems. The outcomes achieved by aPSO equated with DE, PSO as well as effective variants of PSO (IPSO, NPSO, TVAC_PSO and MPSO-TVAC) in terms of convergence capability and searching ability. For all test systems, presented aPSO is capable to find the best cost efficiently. Likewise, this algorithm is capable to grasp an optimum result within fewer calculation times, due to the effectiveness of its novel parameters c_1, c_2 and w.

Finally, presented aPSO can be an alternate methods or PSO variants in finding best solution for power system optimization problems. In future, aPSO can be applied for real-life multi-objective optimization problems.

Table 3 Results for TSts-3

Output (MW)	DE [2]	PSO [3]	IPSO [19]	MPSO-TVAC [23]	aPSO
u-1	439.12	415.31	455.00	455.00	454.9812
u-2	407.97	359.72	380.00	380.00	455
u-3	119.63	104.43	129.83	130.00	130
u-4	129.99	74.99	130.00	130.00	130
u-5	151.07	380.28	168.25	170.00	235.5844
u-6	460.00	426.79	459.93	459.99	460
u-7	425.56	341.32	430.00	430.00	465
u-8	98.57	124.79	98.26	72.60	60
u-9	113.49	133.14	37.59	58.32	25
u-10	101.11	89.26	160.00	159.73	28.9867
u-11	33.91	60.06	73.39	80.00	76.8357
u-12	79.96	50.00	80.00	80.00	80
u-13	25.00	38.77	25.00	25.01	25
u-14	41.41	41.94	18.49	15.00	15
u-15	35.61	22.64	15.66	15.00	15
Total power	2662.4	2668.4	2661.40	2660.66	**2656.3882**
Generation loss	32.43	38.28	31.40	30.66	**26.3882**
Cost	32,858	33,113	32,721.49	32,704.47	**32,542.7320**
Time (s)	26.59	49.3	19.15	12.78	**10.99**

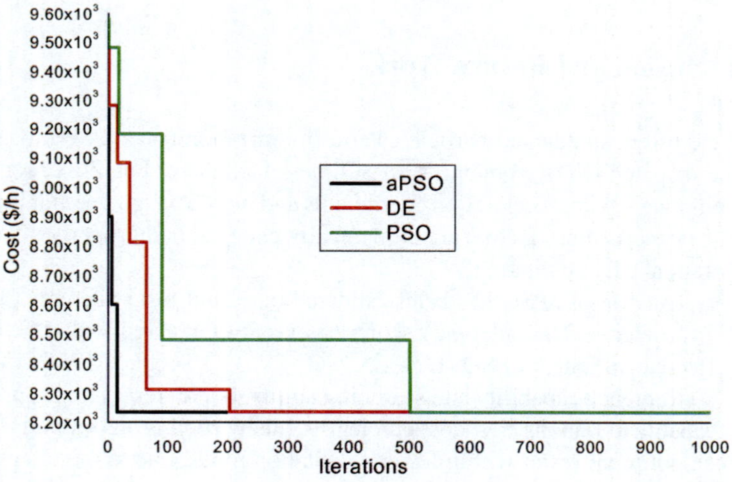

Fig. 3 Convergence for TSts-1

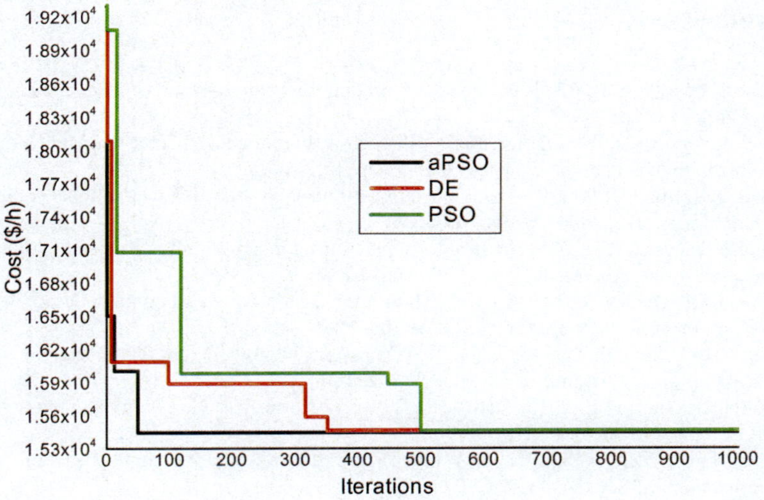

Fig. 4 Convergence for TSts-2

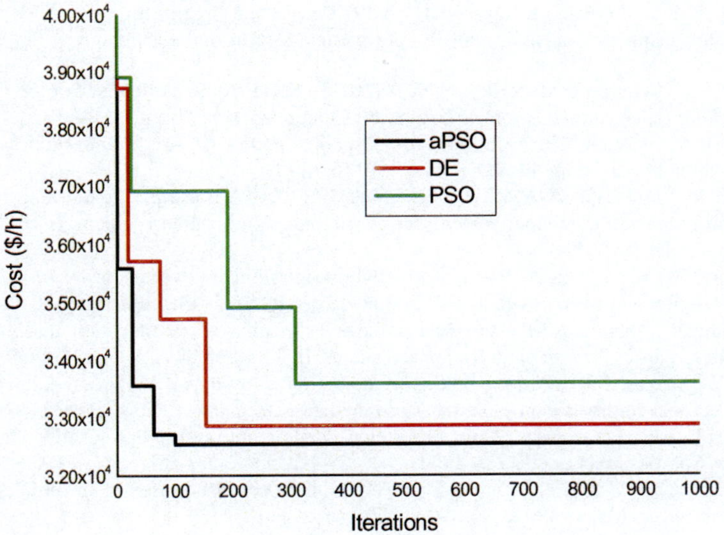

Fig. 5 Convergence for TSts -3

Acknowledgements This research supported by DST-SERB Govt. of India; grant number CRG/2020/000817.

References

1. Wood AJ, Wollenberg BF (1996) Power generation. In: Operation and control, 2nd ed. Wiley, New York
2. Storn R, Price K (1997) Differential evolution—a simple and efficient heuristic for global optimization over continuous spaces. J Glob Opt 11:341–359
3. Kennedy J, Eberhart R (1995) Particle swarm optimization. In: IEEE international conference on neural networks, vol 4, pp 1942–1948
4. Das KN, Parouha RP (2015) An ideal tri-population approach for unconstrained optimization and applications. App Math Comput 256:666–701
5. Parouha RP, Das KN (2016) A novel hybrid optimizer for solving Economic Load Dispatch problem. Int J Electr Power Energ Syst 78:108–126
6. Chen Y, Li L, Peng H, Xiao J, Wu Q (2018) Dynamic multi-swarm differential learning particle swarm optimizer. Swarm Evol Comput 39:209–221
7. Espitia HE, Sofrony JI (2018) Statistical analysis for vortex particle swarm optimization. App Soft Comput 67:370–386
8. Yu H, Tan Y, Zeng J, Sun C, Jin Y (2018) Surrogate-assisted hierarchical particle swarm optimization. Inf Sci 454–455:59–72
9. Chen Y, Li L, Xiao J, Yang Y, Liang J, Li T (2018) Particle swarm optimizer with crossover operation. Eng App Art Intel 70:59–169
10. Parouha RP (2019) Nonconvex/nonsmooth economic load dispatch using modified time-varying particle swarm optimization. Comput Intel 1–28. https://doi.org/10.1111/coin.12210
11. Hosseini SA, Hajipour A, Tavakoli H (2019) Design and optimization of a CMOS power amplifier using innovative fractional-order particle swarm optimization. App Soft Comput 85:1–10
12. Kohler M, Vellasco MMBR, Tanscheit R (2019) PSO+: a new particle swarm optimization algorithm for constrained problems. App Soft Comput 85:1–26
13. Khajeh A, Ghasemi MR, Arab HG (2019) Modified particle swarm optimization with novel population initialization. J Inf Opt Sci 40(6):1167–1179
14. Ang KM, Lim WH, Isa NAM, Tiang SS, Wong CH (2020) A constrained multi-swarm particle swarm optimization without velocity for constrained optimization problems. Exp Syst App 140:1–23
15. Parouha RP, Verma P (2020) An innovative hybrid algorithm to solve nonconvex economic load dispatch problem with or without valve point effects. Int Trans Elect Energ Syst 34(1):1–67
16. Huynh DC, Dunnigan MW (2010) Parameter estimation of an induction machine using advanced particle swarm optimization algorithms. IET J Elec Power App 4:748–760
17. Shi Y, Eberhart R (1998) A modified particle swarm optimizer. In: Proceedings of IEEE international conference on evolutionary computation, Piscataway, New Jersey, pp 69–73
18. Alatas B, Akin E, Ozer AB (2009) Chaos embedded particle swarm optimization algorithms. Chaos Solitons Fractals 40:1715–1734
19. Lee T-Y, Chen C-L (2007) Unit commitment with probabilistic reserve: an IPSO approach. Energ Convers Manag 48(6):486–493
20. Selvakumar AI, Thanushkodi K (2007) A new particle swarm optimization solution to nonconvex economic dispatch problems. IEEE Trans Power Syst 22:42–51
21. Chaturvedi KT, Pandit M, Srivastava L (2009) Particle swarm optimization with time varying acceleration coefficients for non-convex economic power dispatch. Int J Electr Power Energ Syst 31:249–257
22. Khokhar B, Parmar KPS, Dahiya S (2012) An efficient particle swarm optimization with time varying acceleration coefficients to solve economic dispatch problem with valve point loading. Energ Power 2(4):74–80
23. Abdullah MN, Bakar AH, Rahim NA, Mokhlis H, Illias HA, Jamian JJ (2014) Modified particle swarm optimization with time varying acceleration coefficients for economic load dispatch with generator constraints. J Electr Eng Technol 9(1):15–26

24. Sinha N, Chakrabarti R, Chattopadhyay PK (2003) Evolutionary programming techniques for economic load dispatch. IEEE Trans Evol Comput 7:83–94
25. Zwe-Lee G (2003) Particle swarm optimization to solving the economic dispatch considering the generator constraints. IEEE Trans Power Syst 18:1187–1195

A Mathematical Model to Minimize the Total Cultivation Cost of Sugarcane

Sumit Kumar and Millie Pant

Abstract In the present study, a non-linear mathematical model is proposed to minimize the total expenditure made on sugarcane cultivation. The objective here is to optimize the costs involved in transportation, cultivation, and management of the deterioration of sugarcane. The optimization tools used are Differential Evolution (DE) and Particle Swarm Optimization (PSO). Further, the model is validated by comparing the actual expenditure of sugarcane cultivation with results obtained by PSO and DE, which indicates that the application of optimization methods may help in reducing the expenditure made on sugarcane cultivation.

Keywords Sugarcane cultivation · Mathematical model · Optimization · PSO · DE

1 Introduction

Agriculture provides food for survival and raw materials to the manufacturing industries. It is the primary source of livelihood and helps economically. In agriculture, sugarcane is the most important and valuable crop globally. Sugarcane growing countries worldwide lay between the latitude 36.7° north and 31.0° south of the equator extending from tropical to subtropical zones. It is the main agricultural crop in tropical and sub-tropical regions in India and has been cultivated since the Vedic period. In the period 1400 to 1000 B.C., the earliest mention of sugarcane cultivation can be found in Indian literature [1]. Sugarcane is produced in over 110 countries globally,

S. Kumar (✉) · M. Pant
Department of Applied Mathematics and Scientific Computing, Indian Institute of Technology Roorkee, Roorkee, Uttarakhand, India
e-mail: skumar1@as.iitr.ac.in

M. Pant
e-mail: pant.milli@as.iitr.ac.in

M. Pant
Mehta Family School for Data Science and Artificial Intelligence, Indian Institute of Technology Roorkee, Roorkee, Uttarakhand, India

M. Thakur et al. (eds.), *Soft Computing for Problem Solving*, Lecture Notes in Networks and Systems 547, https://doi.org/10.1007/978-981-19-6525-8_40

mostly produced in Brazil, Thailand, India, Australia, China, Pakistan, the United States, and South Africa [2].

India is the second-largest producer of sugarcane after Brazil. It is the primary source of fodder for cattle. Because of the many uses of this crop in various industries and cattle forms, it is known as "Wonder cane." Like many other agricultural products, sugarcane is a seasonal crop influenced by weather and climatic conditions [3]. Sugarcane is a cash crop with a cultivation period of 8–10 months which is a very long and complicated process.

Uttar Pradesh is India's highest sugarcane-producing state in the subtropical zone. In this state, sugarcane is cultivated in about 22.7 lakh ha of area with a production of 135.64 million tons [1]. The primary region of sugarcane cultivation in Uttar Pradesh is western UP. There are two main types of sugarcane fields in Uttar Pradesh: Plant and Ratoon. After the plantation of sugarcane, the initial growing crop is called plant, and after one-time harvesting, the next growing crop is called ratoon.

1.1 A Sugarcane Supply Chain

The sugarcane supply chain is defined as an agro-industrial system for growing, cultivating, harvesting, transporting, and milling sugarcane from the field to the mill [4]. There are four main stakeholders involved in the system: farmers, field and harvesting management unit, transportation mode management, and sugar mill [5]. Because of weather, agricultural process, harvesting time effects, transportation challenges, cane bulk density influences, and various logistical, social, economic, and physiological links across the chain, the sugarcane supply chain has a lot of complexity [2]. The entire sugar supply chain is highly integrated, and it contains: (a) cane growing, (b) harvesting, (c) cane transport to the mill, (d) mill processing and refining, (e) sugar transported to the port or market, (f) storage, and (g) retailing to customers [6].

The primary processes in the complete cultivation of sugarcane in the UP are given below:

- ***Ploughing Process***: First, a field where sugarcane is to be planted is selected. This is followed by the irrigation and tilling process, which is done at least three times. This process requires a tractor, harrow, tiller, rotavator, etc.
- ***Planting Process***: In India, sugarcane is planted thrice a year during the months: February–March (spring), October–November (autumn), and July–August (adsali). In northern India, spring and autumn plantings are more common, while adsali planting is common in Maharashtra. Under the climate conditions of Uttar Pradesh, a large portion of sugarcane is planted in spring because it requires about 25–30 °C temperature for best germination, which is met twice a year in north India, i.e., in October and also in the months of February–March.
- ***Sowing***: The sowing methods of sugarcane are Flat Bed Planting, Furrow Planting, Ring Pit Planting, Trench/Jawa planting, Paired row planting, etc. After plantation,

fields are leveled by heavy planking. The previous year's sugarcane plants are used as seeds for growing sugarcane.

- ***Cultivation process***: After three weeks on the plantation, sugarcane begins to germinate. All germination completes within a month. After that weeding process is conducted manually or by the weeding machine, and then irrigating the fields using any irrigation methods; these procedures take almost 60–70 days. After 7–10 days of irrigation, some fertilizers like urea are utilized. The above process (weeding, irrigation, fertilization) repeats at least two times in the cultivation process. After 120 days, the star fruit sugarcane begins to be ready, called the grant growth phase. From this time, the growth of sugarcane will increase day by day. Fertilization and irrigation should be completed from time to time. The maturity phase begins from 270 days onwards. Germination, weeding, grand growth, and maturity are the critical stages for irrigation.
- ***Harvesting process***: The harvesting process must be completed in the maturity phase for the best recovery of sugarcane juice. In Uttar Pradesh, most of the sugarcane harvesting is done manually. However, sugarcane harvesting machines are provided by sugar mills. With the help of the machine, harvesting can be done quickly. The top green part of sugarcane is mainly used to feed domestic animals.
- ***Transportation process***: After harvesting, the sugarcane is transported to the sugar mills or collection centers from fields. It is the farmer's responsibility to transport the sugarcane. Some farmer uses a personal vehicle, and others take help from a third-party vehicle provider.
- ***Milling Process***: After transporting the sugarcane to the sugar mill, the milling process begins. In this process, the pure sugarcane is first weighed and then crushed to get the two major products, namely sugarcane juice, and bagasse. Sugarcane juice is used to make sugar, molasses, syrup, etc., while bagasse is used to make tableware items, straws, etc.

A large amount of the farmer's expenditure is made on the transportation and cultivation of sugarcane. Sufficient investment is to be made by the farmer to prevent the deterioration of the plant, which may occur due to a mismanaged time between harvesting and transportation [7].

Considering the significance of transportation, prevention of deterioration, and Cultivation (TDC), an optimization model is formulated in the paper through which the abovementioned three (TDC) are optimized.

The remainder of this paper is organized as follows. In Sect. 2, a brief literature review is provided. A mathematical model is presented in Sect. 3. The methodology, including data collection, is shown in Sect. 4. In Sect. 5, results are given along with a discussion and comparison between PSO, DE, and actual expenditure on sugarcane cultivation. Finally, the paper concludes with Sect. 6.

2 Literature Review

The sugar industry is the second largest industry in India, after the textile industry, and is significantly important for rural development and the growth of the Indian economy [8]. Sugarcane, one of the most important cash crops, can be used in several different ways like in the production of jaggery, molasses, and sugar; as a raw material for bioenergy; bagasse is used in energy cogeneration and can also use to make cement concrete [9–11].

India produces 370 million tonnes of sugarcane and holds the second position in the world in sugarcane production after Brazil. Throughout India, there is a constant growth in the production of sugarcane, though it is still less than production in Peru (127.81 t/ha), Ethiopia (9120.59 t/ha), and Egypt (114.98 t/ha) [12].

To improve the sugarcane production in India, a proper mechanism for sugarcane cultivation is not just desirable but is also necessary [13]. Harvesting can be the most crucial task in the lifetime of sugarcane, as proper harvesting can increase the yield and reduce the weight loss in sugarcane, thereby increasing the profitability [14, 15]. A multi-objective sugarcane harvesting model to optimize the harvesting schedule, optimize harvesting time-slot distribution, and maximize the supply chain efficiency was suggested in [18], and an accurate estimation of crop production before harvesting is important for deciding the prices, transportation services, and various food procurement policies was discussed in [16].

There have been limited studies on supply chain solutions for the sugar industry compared to the manufacturing and automobile sectors. This is mainly because agricultural industries, including sugar, are generally more complex [17, 18]. This paper presented a mathematical model for minimizing the transportation, deterioration, and cultivation costs of sugarcane together.

3 Problem Formulation

Maximizing earns or minimizing losses has always been a concern in every sector. So the farmers' priority is to reduce their expenditure on sugarcane cultivation and get more profit. Farmers have to choose the right harvesting time when the sugarcane is in its full mature age. After harvesting, sugarcane weight decreases day by day. The authors construct a non-linear sugarcane TDC cost optimization problem in this paper.

The flow chart of the sugarcane transporting processor is given in Fig. 1.

3.1 Mathematical Model

The notations used in the presented mathematical model are given in Table 1.

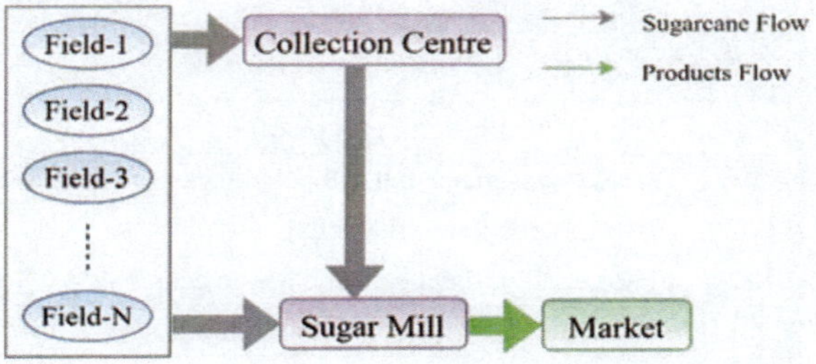

Fig. 1 Sugarcane transportation process

3.1.1 Set and Parameters

The variables in the model are quantity, days, tilling cost, plantation cost, and cultivation cost.

3.1.2 Objective Function

The proposed objective is made up of the sum of three objectives, Z1, Z2, and Z3, as follows:

$$Z = Z_1 + Z_2 + Z_3 \tag{1}$$

where;

Z_1 is the transportation cost of sugarcane from fields to sugar mill or collection center, defined as:

$$Z_1 = \sum_j \sum_k (Q_{jk}.u_{jk}) + \sum_k (Q_{kM}.u_{kM}) + \sum_j (Q_{jM}.u_{jM}) \tag{2}$$

Z_2 is the deterioration cost of sugarcane between harvesting and milling time, defined as:

$$Z_2 = \{\sum_j \sum_k (Q_{jk}.d_{jk}) + \sum_k (Q_{kM}.d_{kM}) + \sum_j (Q_{jM}.d_{jM})\}.\text{h} \tag{3}$$

Z_3 is the total cultivation cost of sugarcane which includes tilling cost, plantation cost, and IFW (irrigation, fertilizer, weeding) cost, and is defined as:

$$Z_3 = \sum_i \sum_j \sum_l a_{ij}(C_t + C_p + C_c) \tag{4}$$

Table 1 Notations and their description

Notations	Description
$i \in I$	The set of farmers or group of farmers
$j \in J(i)$	A set of sugarcane fields that belongs to the farmer $i \in I$
P_j	Total production of the field j
$k \in K$	The set of the sugarcane collection center
W_k	The capacity of the collection center k
M	Represent a sugarcane mill
W_M	The capacity of mill M
a_{ij}	The area of sugarcane field $j \in J$, related to sugarcane farmer $i \in I$
w_{ij}	Denotes the estimated yield of sugarcane field $j \in J(i)$
Q_{jk}	The quantity of sugarcane transported from sugarcane field $j \in J(i)$ to sugarcane collection center $k \in K$
Q_{kM}	The quantity of sugarcane transported from sugarcane collection center $k \in K$ to sugarcane mill M
Q_{jM}	The quantity of sugarcane transported from sugarcane field $j \in J(i)$ to sugarcane mill M
u_{jk}	The per-unit transportation cost of sugarcane from sugarcane field $j \in J(i)$ to sugarcane collection center $k \in K$
u_{kM}	The per-unit transportation cost of sugarcane from sugarcane collection center $k \in K$ to sugarcane mill M
u_{jM}	The per-unit transportation cost of sugarcane from sugarcane field $j \in J(i)$ to sugarcane mill M
d_{jk}	The number of days in which sugarcane quantity Q_{jk} transported from field $j \in J(i)$ to sugarcane collection center $k \in K$
d_{jk}^U	Maximum number of days in which sugarcane quantity Q_{jk} transported from field $j \in J(i)$ to sugarcane collection center $k \in K$
d_{kM}	Number of days in which sugarcane quantity Q_{kM} transported from the sugarcane collection center $k \in K$ to the sugarcane mill M
d_{kM}^U	Maximum number of days in which sugarcane quantity Q_{kM} transported from the sugarcane collection center $k \in K$ to the sugarcane mill M
d_{jM}	Number of days in which sugarcane quantity Q_{jk} transported from field $j \in J(i)$ to sugarcane mill M
d_{jM}^U	Maximum number of days in which sugarcane quantity Q_{jk} transported from field $j \in J(i)$ to sugarcane mill M
h	Deterioration cost per unit quantity per day
C_t	Tilling cost of sugarcane field $j \in J(i)$ per unit^2
C_{jt}^U	The maximum amount for tilling sugarcane fields $j \in J(i)$
C_p	Plantation cost of sugarcane field $j \in J(i)$ per unit^2
C_{jp}^U	The maximum amount for plantation in sugarcane field $j \in J(i)$
C_c	Total cultivation cost of sugarcane field $j \in J(i)$ per unit^2

(continued)

Table 1 (continued)

Notations	Description
C_{jc}^{U}	The maximum amount for total cultivation of sugarcane field j ∈ J(i)

Table 2 Constraints

Equation	Description
$\sum_k Q_{jk} + Q_{jM} \leq P_j, \quad J \in J(i)$	Total production of the field, j
$\sum_j Q_{jM} + \sum_k Q_{kM} \leq W_M$	Capacity of mill
$\sum_j Q_{jk} \leq W_k, \quad k \in K$	The capacity of the collection center k
$0 < d_{jk} \leq d_{jk}^{U}$ $0 < d_{kM} \leq d_{kM}^{U}$ $0 < d_{jM} \leq d_{jM}^{U}$	Constraints on number of the days between harvesting and transporting mill or collection center
$0 < C_{jt} \leq C_{jt}^{U}, \quad j \in J(i)$ $0 < C_{jp} \leq C_{jp}^{U}, \quad j \in J(i)$ $0 < C_{jc} \leq C_{jc}^{U}, \quad j \in J(i)$	Tilling, production, and cultivation cost constraints
$Q_{jk} > 0$ $Q_{kM} > 0$ $Q_{jM} > 0$	Non-negative restriction on quantity

3.1.3 Constraints

Constraints equations and their description are defined in Table 2.

4 Methodology

A non-linear mathematical model is developed in the paper to optimize the transportation cost, deterioration cost, and cultivation cost, collectively termed the TDC cost of sugarcane in the paper. The metaheuristic techniques PSO and DE are used to solve this non-linear sugarcane TDC cost optimization problem. The process used by PSO and DE to produce new solutions is based on floating-point arithmetic, which means that the density of solutions within the solution space can be considerably larger. The velocity and position update equations in PSO are used to build a new swarm of particles. As a result, all new particles will be very different from the old ones. By merging numerous solutions with the candidate solution, DE creates a new solution. These algorithms are simple to put into practice.

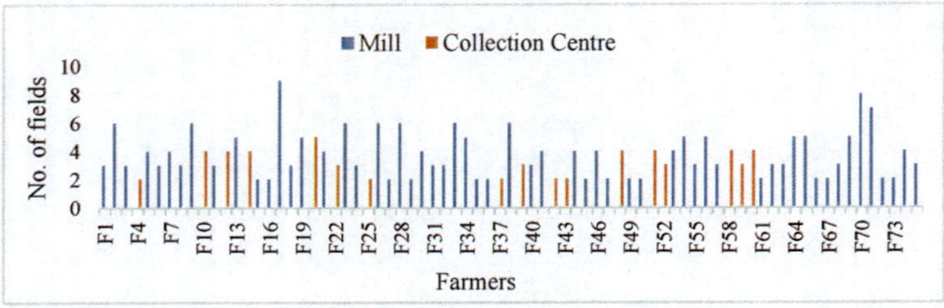

Fig. 2 Distribution of farmer's fields

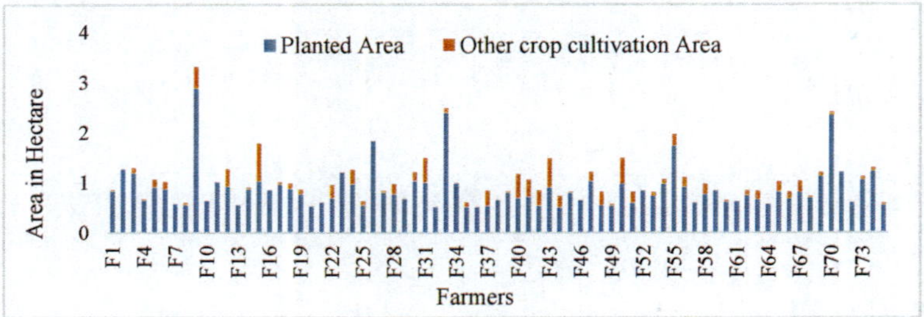

Fig. 3 Area distribution of farmers

4.1 Data Collection

The data collected is for village Najarpur, Amroha district, Uttar Pradesh [19]. The sugarcane of this village is transported to either Dhanaura mill or its collection center, Najarpur. The data collected is for 75 farmers with 273 fields.

Figure 2 shows the number of fields associated with each farmer. Figure 3 shows the total cultivation area of each farmer, which is separated into sugarcane cultivation and other crops.

4.2 Differential Evolution

DE was first introduced by Storn and Price [20]. It is a population-based search algorithm such as Evolution Strategies, Genetic Algorithms, etc. [21, 22]. Its working process is divided into two sections: Initialization and evolution. In the first stage, the population is randomly generated. In the second stage, the randomly generated population goes through three genetic operators' mutation, crossover, and selection processes which are repeated until a termination criterion is satisfied.

(a) **Initialization**: In this phase, a set of the uniformly distributed population is randomly generated, and calculated the fitness of all randomly generated populations. Let $P^t = \{X_i^t : i = 1, 2, 3, 4, \ldots, N\}$ be the population at any generation t, N denotes the size of the population. $X_i^t = \{x_{i,1}^t, x_{i,2}^t, \ldots, x_{i,n}^t\}$ be an n-dimensional vector.

(b) **Evolution**: when the initial population is generated, the second stage of evolution is performed. In evolution, the genetic operators are performed.

(1) ***Mutation***: In this operator, a mutant vector is generated by adding the weighted difference between two randomly chosen vectors to a third vector from the population. Mutant vector V_i^t for every target vector X_i^t at any generation t as follows;

$$V_i^t = X_{r1}^t + F * (X_{r2}^t - X_{r3}^t)$$

where; 'F' lies between 0 and 1 and is called a scaling factor. r_1, r_2, r_3 are mutually distinct randomly chosen vectors from 1 to N.

(2) ***Crossover***: During crossover, a new vector called the trial vector is generated by combining the variables of the target vector $X_i^t = \{x_{i,1}^t, x_{i,2}^t, \ldots, x_{i,n}^t\}$ and mutant vector $V_i^t = \{v_{i,1}^t, v_{i,2}^t, \ldots, v_{i,n}^t\}$, with the help of crossover probability P_c If it is denoted as $U_i^t = \{u_{i,1}^t, u_{i,2}^t, \ldots, u_{i,n}^t\}$, its component is defined as:

$$u_{i,n}^t = \begin{cases} v_{i,n}^t, \ if\ rand_{i,n}^t \le P_c \\ x_{i,n}^t, \ Otherwise \end{cases}$$

where; $rand_{i,n}^t$ are generated randomly between 0 and 1.

(3) ***Selection***: In this section, a comparison is made between the target vector and trail vector on the basis of their fitness value. After comparison, the vector which has better fitness is selected for the next generation. This operation is defined as

$$X_i^{t+1} = \begin{cases} U_i^t, \ if\ f(U_i^t) \le f(X_i^t) \\ X_i^t, \ Otherwise \end{cases}$$

The fitness value is the same as the function value.

The evolution process is repeated till the termination condition is not satisfied.

4.3 *Particle Swarm Optimization*

The idea behind PSO is to simulate the bird's behaviour in finding food through social learning, which means by observing the behaviour of nearby birds who appear to be near the food source [23]. It is a bio-inspired metaheuristic algorithm based on bird flocking or fish schooling and swarming. It is easy to implement and initialize by a group of random particles (each particle is a solution) [24].

Consider a population with P particles with a position vector $X_i^t = \left(x_{i,1}^t, x_{i,2}^t, \ldots, x_{i,n}^t\right)$ and a velocity vector $V_i^t = \left(v_{i,1}^t, v_{i,2}^t, \ldots, v_{i,n}^t\right)$ at a 't' iteration for each one of the ith particle in n-dimensional space. In the basic process of PSO, there are two equations, namely velocity and position update equations. Each particle's speed and position change as a result of updating equations such as:

$$v_{i,n}^{t+1} = w_i * v_i^t + c_1 * r_1^t * \left(pbest_{in}^t - x_{in}^t\right) + c_2 * r_2^t * \left(gbest_{in}^t - x_{in}^t\right),$$
$$x_{in}^{t+1} = x_{in}^t + v_{in}^{t+1}.$$

where; i = 1, 2, 3, P, t denotes the iteration, and n ∈ N denotes the dimension of the particles. To balance the global exploration and local exploitation, used inertia weight w. c_1, c_2 called "acceleration coefficients" are positive constant parameters. r_1, r_2 are uniformly distributed random variables which are generated between 0 and 1. To find the optimal solution in PSO, each particle moves in the direction of its previous best position (pbest means personal best position) and the global best position (gbest) in the swarm. $pbest_{in}^t$ is the n-dimensional quantity of the individual "i" at its best position in the iteration "t." $gbest_{in}^t$ is the n-dimensional quantity of the swarm(population) at its best position.

The working processor of PSO is given below:

Step 1. Initialization

- Initialize Parameters
- Initialize Population
  1. Initialize position (X_i^t) randomly for each particle
  2. Initialize Velocity ($V_i^t = 0.1 * X_i^t$) for each particle

Step 2. Evaluate Fitness $f\left(X_i^t\right)$

- Calculate the fitness value for each particle
- Choose a particle with the best fitness value as gbest

Step 3. For each particle, calculate velocity and position.

- Calculate particle position by: $x_{in}^{t+1} = x_{in}^t + v_{in}^{t+1}$
- Calculate particle velocity by: $v_{i,n}^{t+1} = w_i * v_i^t + c_1 * r_1^t * \left(pbest_{in}^t - x_{in}^t\right) + c_2 * r_2^t * \left(gbest_{in}^t - x_{in}^t\right)$

Step 4. Again Evaluate Fitness $f\left(X_i^{t+1}\right)$

- Again find the best fitness value gbest

Step 5. Update generation t = t + 1
Step 6. Evaluate gbest and X_i^t

Repeat the above from step 2 to step 6.

Table 3 Parameter for DE and PSO

Algorithm	Population size	Max iterations	Other parameters
DE	100	100	Scaling factor (F) = 0.5, Crossover probability $P_c = 0.7$
PSO	100	100	Inertia weight (w) = 0.9, acceleration constant $c_1 = c_2 = 2$

4.4 Parameter Setting

The PSO is influenced by a number of control parameters, namely the dimension of the problem, number of particles, acceleration coefficients, inertia weight, and number of iterations. Population size, differential weight, crossover, and generation number are the parameters for differential evolution algorithms. The parameters used in PSO and DE are defined in Table 3.

4.5 System Configuration

The experiments were executed on Spyder 4.1.5 Integrated Development Environment (IDE) with Python 3.8 through an Anaconda distribution on an Intel Xeon W-1290P CPU @ 3.70 GHz dual-processor system with 64 GB RAM, an Nvidia Quadro P1000 GPU, and a 64-bit Windows 10 Pro Operating System.

5 Results and Discussion

This mathematical model takes the three major problems in the cultivation of sugarcane. These are the transportation of sugarcane, deterioration of sugarcane, and total cultivation costs of sugarcane. To solve this non-linear sugarcane TDC cost optimization problem, using the PSO algorithm and DE Algorithm. Using the real data of sugarcane farmers, we run the code on spyder with python of this mathematical model with defined parameters by PSO and DE. The code of both algorithms on the model is executed 100 times, and the *best, mean, median,* and *standard deviation* of the objective function is calculated for each algorithm and shown in Table 4.

Table 4 Results by PSO and DE

Algorithm	Objective value (in INR)				
	Best	Worst	Mean	Median	Standard deviation
DE	10,963,276	11,055,173	11,002,181	11,000,543	15,636.85
PSO	12,054,215.1	12,299,215.1	12,161,797.2	12,153,194.5	52,189.946

Table 5 Results by Friedman statistical test

Algorithms	Mean rank
PSO	2
DE	1
N	100
Chi-square	100.000
df	1
Asymp. significance	0.000

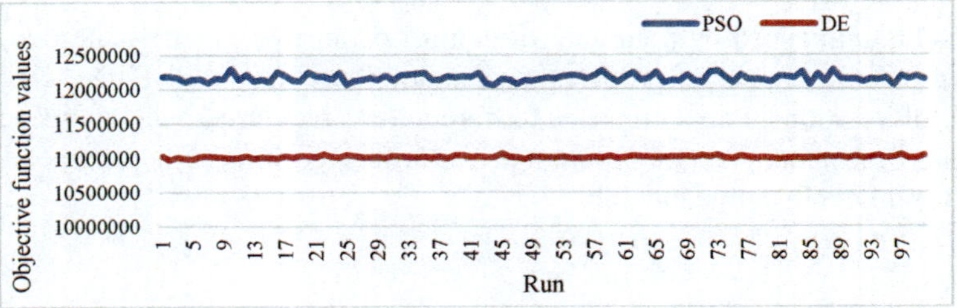

Fig. 4 Run wise comparison between PSO and DE

To confirm that the differences are significant, we use a non-parametric Friedman statistical test. It is a mean rank-based test. Its ranks table and statistics table is given in Table 5.

The above table provides the test statistic (χ^2) value ("Chi-square"), degrees of freedom ("df"), and the significance level ("Asymp. Sig.").

Executed suggested mathematical model using mentioned algorithms with real data of sugarcane farmers and their fields successfully. The code of PSO and DE for the mathematical model with the given real data runs 100 times. Run-wise comparison of objective function values using PSO and DE are shown in Fig. 4.

Out of 100 runs, we take the best run of each algorithm. Corresponding to these best runs, pick up the objective function values of each 100 generations of both algorithms. A comparison of all iterated objective function values of the best run of the algorithms is shown in Fig. 5.

5.1 Comparison of Expenditure (Actual v/s PSO and DE)

Table 6 shows the comparison of expenditure of actual versus PSO versus DE.

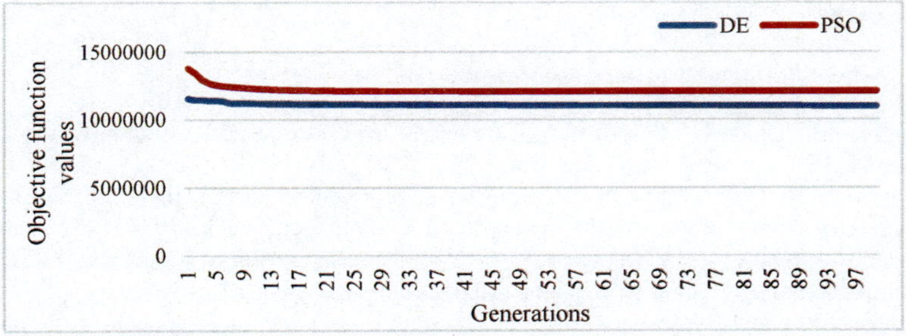

Fig. 5 Comparison of all iterative values for the best run of DE and PSO

Table 6 Comparison of the total expenditure

Total farmers	Total area (in Hec)	Actual expenditure (in INR)	Expenditure by PSO	% Reduction in expenditure by PSO	Expenditure by DE	% Reduction in expenditure by DE
75	65.919	13,896,150.06	12,059,215	13.219%	10,963,276	21.105%

6 Conclusion and Future Directions

The sugarcane cultivation process is a laborious and time-consuming process that takes 8–10 months to complete. A large amount of the total revenue of farmers by sugarcane goes through its cultivation process. Every farmer wants to reduce their expenditure and attain the maximum profit. For this, farmers could have to cultivate sugarcane in an organized way, make a schedule for one year of cultivation, and arrange all the logistics according to the schedule. In this paper, a non-linear TDC cost minimization model has been proposed to minimize the total expenditure of the farmers in the cultivation of sugarcane. Minimizing the sugarcane expenditure is an important contribution of this paper. We have used two metaheuristics to solve this optimization problem: PSO and DE. We implement these algorithms to the data of 75 sugarcane farmers with 273 fields in Uttar Pradesh. PSO reduces the farmers' expenditure in the collected data by 13%, and DE reduces by 21%. This paper shows that the DE algorithm works better than the PSO algorithm in all the iterations of this mathematical model with this data. This model can be extended individually in the transportation and harvesting models and make the mathematical plan for the whole cultivation period. In the future, we will work on selecting the best plantation and irrigation methods. On the other side, future research will focus on the possible environmental concerns in the process of transportation and milling of sugarcane.

References

1. https://farmer.gov.in/cropstaticssugarcane.aspx. Last accessed 2022/04/05
2. Higgins AJ, Laredo LA (2006) Improving harvesting and transport planning within a sugar value chain. J Oper Res Soc 57:367–376
3. Gbetibouo GA, Hassan RM (2005) Measuring the economic impact of climate change on major South African field crops: a Ricardian approach. Glob Planet Change 47:143–152
4. Gigler JK, Hendrix EMT, Heesen RA, Van Den Hazelkamp VGW, Meerdink G (2002) On optimisation of agri chains by dynamic programming. Eur J Oper Res 139:613–625
5. Gaucher S, Le Gal P-Y, Soler L-G (2003) Modelling supply chain management in the sugar industry. Sugar Cane Int 22:8–16
6. Higgins AJ, Muchow RC (2003) Assessing the potential benefits of alternative cane supply arrangements in the Australian sugar industry. Agric Syst 76:623–638
7. Higgins AJ, Muchow RC, Rudd AV, Ford AW (1998) Optimising harvest date in sugar production: a case study for the Mossman mill region in Australia I. Development of operations research model and solution. Field Crops Res 57:153–162
8. Solomon S (2016) Sugarcane production and development of sugar industry in India. Sugar Tech 18:588–602
9. Hiloidhari M, Banerjee R, Rao AB (2021) Life cycle assessment of sugar and electricity production under different sugarcane cultivation and cogeneration scenarios in India. J Clean Prod 290:125170
10. Bahurudeen A, Kanraj D, Gokul Dev V, Santhanam M (2015) Performance evaluation of sugarcane bagasse ash blended cement in concrete. Cem Concr Compos 59:77–88
11. Gardas BB, Raut RD, Narkhede B (2018) Evaluating critical causal factors for post-harvest losses (PHL) in the fruit and vegetables supply chain in India using the DEMATEL approach. J Clean Prod 199:47–61
12. Vishawajith KP, Sahu PK, Dhekale BS, Mishra P (2016) Modelling and forecasting sugarcane and sugar production in India. Indian J Econ Dev 12:71
13. Singh J, Singh AK, Sharma MP, Singh PR, Srivastava AC (2011) Mechanization of sugarcane cultivation in India. Sugar Tech 13:310–314
14. Piewthongngam K, Pathumnakul S, Setthanan K (2009) Application of crop growth simulation and mathematical modeling to supply chain management in the Thai sugar industry. Agric Syst 102:58–66
15. Stray BJ, van Vuuren JH, Bezuidenhout CN (2012) An optimisation-based seasonal sugarcane harvest scheduling decision support system for commercial growers in South Africa. Comput Electron Agric 83:21–31
16. Verma AK, Garg PK, Hari Prasad KS, Dadhwal VK, Dubey SK, Kumar A (2021) Sugarcane yield forecasting model based on weather parameters. Sugar Tech 23:158–166
17. Higgins A, Thorburn P, Archer A, Jakku E, Higgins A, Thorburn P, Archer A, Jakku E (2007) Opportunities for value chain research in sugar industries. Agric Syst 94:611–621
18. Bezuidenhout CN, Bodhanya S, Sanjika T, Sibomana M, Boote GLN (2012) Network-analysis approaches to deal with causal complexity in a supply network. 50:1840–1849
19. Cane UP Enquiry. Sugar Cane and cane development. https://enquiry.caneup.in/. Last accessed 2022/04/05
20. Storn R, Price K (1997) Differential evolution—a simple and efficient heuristic for global optimization over continuous spaces. J Glob Optim 11:341–359
21. Choi TJ, Lee Y (2018) Asynchronous differential evolution with strategy adaptation for global numerical optimization. In: ACM international conference proceeding series, pp 15–18 (2018).
22. Bilal, Pant M, Zaheer H, Garcia-Hernandez L, Abraham A (2020) Differential evolution: a review of more than two decades of research. Eng Appl Artif Intell 90:103479
23. Kennedy J, Eberhart R, Gov B (1995) Particle swarm optimization. IEEE, p 1942
24. Wang D, Tan D, Liu L (2018) Particle swarm optimization algorithm: an overview. Soft Comput 22:387–408

Genetically Optimized PID Controller for a Novel Corn Dryer

Raptadu Abhigna, Akshat Sharma, Kudumuu Kavya, Pankaj Sharma, Surender Reddy Salkuti, M. Rajesh Kumar, Rani Chinnappa Naidu, and Bhamini Sreekeessoon

Abstract Corn drying is a necessary step in processing corn for different food products. However, few solutions are present in the industry to automate this process. In this paper, a system is designed to use genetic optimization techniques. Controlling the corn dryer's plant may be a difficult task, given its sophisticated drying instrument. This work intends to investigate new control strategies for a configured grain dryer plant in order to smooth control execution, such as overshoot, accuracy, and impact limits disruption. Typically, regular control has fruitfully applied in locals where the structure is obvious. Still, it encounters difficulties in attaining complicated control objectives, particularly when the control objectives have been influenced by several components, such as dielectric limits, climatic conditions, material properties, as well as the drying framework. As a result, it is difficult to develop precise mathematical models that employ conventional standard control to address the challenges.

Keywords Corn drying · Agricultural products · Optimization

1 Introduction

The requirement for food items security is one of the worldwide difficulties within recent memory. Aside from a lack of required farming items, one of the significant issues with food items security is insufficient post-harvest handling as well as inadequate safeguarding procedures. This is even though the majority of agrarian food crops are transient as well as seasonal. Assuming that these agrarian items are not protected enough as well as put away, they will ruin and become undesirable for

R. Abhigna · A. Sharma · K. Kavya · P. Sharma
School of Electrical Engineering, Vellore Institute of Technology, Vellore, India

S. R. Salkuti
Department of Railroad and Electrical Engineering, Woosong University, Daejeon, South Korea

M. Rajesh Kumar · R. C. Naidu (✉) · B. Sreekeessoon
Faculty of Sustainable Development and Engineering, Université Des Mascareignes, Beau Bassin-Rose Hill, Mauritius
e-mail: rnaidu@udm.ac.mu

M. Thakur et al. (eds.), *Soft Computing for Problem Solving*, Lecture Notes in Networks and Systems 547, https://doi.org/10.1007/978-981-19-6525-8_41

human utilization. Food safeguarding is vital in light of the fact that it upgrades protection, stockpiling, and more straightforward transportation of such agrarian items while moderating the surprising ascent in costs of horticultural products. The spoilage of foods arises primarily from the activities of numerous microorganisms and compounds in agricultural products. The development of microorganisms frequently leads to a rapid generation of waste. The presence of water as dampness content is the central point that aids microorganism development and exercise because it builds their metabolic exercises. In the absence of water or the presence of a negligible amount of water, horticultural items will become unfriendly to microorganisms and hinder their development and exercise, thereby protecting the item for a more extended period of time. Corn is one of Turkey's most important agricultural products (creation of 2,300,000 tons in 2000). Corn is an essential modern natural substance in the starch industry. Natural products, vegetables, and oats such as red pepper, grape, corn, and apricot are typically dried in an open environment by exposing them to the sun. During this time, the item may become polluted as a result of the drying climate's sand particles, soil, and other matter [1]. In addition, corn has been contaminated with aflatoxins due to shape development during the sun-drying process, which is a harmful mycotoxin for both human and animal health. In this manner, hot air drying occurs quickly as well as provides consistency to the item. As a result, it has resulted in the development of sterile and high-quality products.

Furthermore, drying time is expected to increase the dryer's drying limit while decreasing energy consumption [2]. Drying is one of the post-collect dealing processes that include the expulsion of dampness from items because of synchronous hotness and mass exchange [3]. Drying is a Thermo-Physical cycle that can lower moisture levels to a safe level, preventing the growth as well as activity of microorganisms in agricultural derivatives [4, 5]. Dampness content evacuation should be done so that the item's ideal quality isn't jeopardized for an extended period of time [6, 7]. In writing, the flimsy layer drying of corn bits has been done utilizing various medicines and drying techniques. [8] proposed that encompassing air corn drying outperforms high-temperature drying in terms of corn quality.

Nonetheless, the drying process takes a much longer time to complete. As a result, a convincing control framework for corn drying should be created to get the drying control needs. When in doubt, the speed of the corn delivering motor is controlled to comprehend the drying objective by recognizing the power source corn dampness according to the looking at control computation in the real control cooperation of a corn dryer.

This paper is structured as follows: Sect. 2 presents the exhaustive literature survey. Section 3 presents the problem statement. Section 4 presents the proposed techniques and briefly describes the algorithm in Sect. 5. Section 6 provides the result analysis, and the final section concludes this paper.

2 Literature Survey

In [9], the numerical model of the grain drying process is a significant reason for the grain drying gear plan, programmed control of drying activity, and enhancement of the drying process have been presented. Ceaseless drying is one of the most broadly utilized types of grain drying. In the drying system, many elements influence the water content of the machine, like food temperature, food stickiness, hot air wind speed, and the underlying dampness in the food. Understanding the grain dryer's programmed control is critical for ensuring the consistency of food dampness and dry grain quality, reducing the administrator's work power and maximizing the dryer's production limit. In this paper, the gathered temperature control model of corn is advanced, and the pace of grain release dryer is controlled by the collected temperature worth to control the water content of grain. The exploratory outcomes show that the numerical model of the aggregated temperature can be utilized for the programmed control of the constant dryer. The gathered temperature can viably control the drying system and meet the activity's necessities. It has the attributes of straightforward calculation, quick reaction, and conquering the impact of a similar slack on the estimation and control. The drying conditions that influence the nature of dried soybean seeds have been presented in [10]. In this case, a precise depiction of the drying rate is required. This review offered the restricted component plan and arrangement of diffusive dampness move conditions to improve seed drying simulation of axis-symmetric structures. Fick's diffusive model has been established. 'Villiamz' assortment of slim layer soybean seed has dried in three reproductions at a constant drying airspeed of 1 ms^{-1} for the drying air temperatures of 30, 40, 50, 60 °C, as well as 70 °C. There is a high agreement level when the model's output has been compared to trial data. For contrasting re-enacted and exploratory data, the mean, relative deviation, modulus (*P*), as well as demonstrating proficiency (EF) have been used. Drying conditions influence the nature of dried soybean seeds. As a result, a precise depiction of the drying rate is required. This review has introduced the limited component definition and arrangement of diffusive dampness move conditions to ameliorate the reproduction of seed drying in axis-symmetric bodies. The diffusive model of Fick has been established. A clear picture emerged when the model's output has compared to that of exploratory data. In [11], the grain drying research at China Agricultural University (CAU) has been examined. CAU is heavily emphasized by numerical demonstration and programmatic experience. Since around 1988, researchers have been cultivating research on the re-enactment of simultaneous stream, counter-flow, cross-stream, and blended stream grain dryers. The created mixed stream grain drying program has been utilized in drier performance evaluation, planning, and development of new grain dryers, as well as the maintenance of existing ones. In addition, the influence of the size of the air channel, characteristics, number of lines, and system procedure on the exhibition of a blended stream grain dryer has been considered.

In [12], the work depicts an experiment conducted at Universidad de Los Llanos. It consists of the design of a PID regulator for temperature as well as security examination utilizing several procedures of control hypothesis utilized for grain corn drying process to work on the nature of the grain of corn, to avert the development of form and rot of wet seed, other than capacity bugs like weevils, which are more likely to duplicate quicker in the warm and sodden grain leading to the development of mushrooms. The paper is divided into three strategic phases: first, the attributes of the framework have been set up, then the PID regulator has scheduled, as well as third, results have been validated by tests recreated in MATLAB, getting the check of the boundaries essential for the framework's actual execution to address seed harm.

In this paper [13], the authors presented a checking arrangement of relativity stickiness, carbon monoxide (CO), carbon dioxide (CO_2), alkali (NH_3), and temperature control for a little shed, proposing an answer to the issue of openness to hurtful gases for the strength of creatures in poultry creation processes. The framework estimates the degrees of temperature, relative mugginess, carbon dioxide (CO_2), carbon monoxide (CO), and alkali (NH_3) utilizing the sensors MQ-7, MQ-135, and DHT11. The framework gets information from every sensor and settles the temperature of the climate at the ideal level. In [14], the authors examined the heterogeneous display of deep bed grain dryers based on a two-stage model that considers coupled hotness as well as mass exchange inside grains. This model also considers pivotal mass and hotness scattering in the liquid stage. The powerful two-stage conditions are handled mathematically by restricted distinction with rotating course-specific strategy computation and then employed to reproduce the stickiness along with the temperature profile of drying gas across dryers, as well as dampness content along with grain temperature. The capabilities of these models have been compared to exploratory information obtained from accessible writings under drying conditions such as temperature along with the plain stickiness of drying gas, as well as grain dampness content. The simulation results show that the two-stage model accurately predicts the dynamic of maize drying inside the bed.

Due to the incredible contrast in material and dampness restricting the energy of rural items in various drying stages, many problems, such as colossal energy utilization and low quality of drying products, have been presented in [15]. The variable temperature drying innovation is one of the best ways to address this problem. It cannot just save energy and improve color, nutrition, and the drying product's actual properties, but it also would not build the expense of drying equipment. This article has determined the definition and mechanical attributes of variable temperature drying. Meanwhile, this study discusses, analyzes, and assesses the new advances in factor temperature drying research with energy productivity and item quality as a standpoint. Finally, the temperature changing method is absent of level-headed choice, and the demonstration of the drying process is deficiently called attention to during the variable temperature drying. Therefore, the course of additional examination is proposed. In [16], the drying of five types of rice in a continuous temperature and impact broiler with an underlying dampness content of 21.1–24.4% wet premise along with the temperature range of 45–70 °C is used to explore the strong diffusivity, as well as glass progress temperature (Tg) of paddy during drying, which have been presented. The drying temperature successfully increased the desorption rate and dampness, but the drying span decreased. The bend of the desorption rate changed gradually at a drying temperature of 45 °C, along with a longer drying period. Most of the five paddy assortments had similar desorption rates for drying temperatures of 45–70 °C, along with their dampness and strong diffusivities in the range of 2.638×10^{-9} to 2.514×10^{-8} m 2 min^{-1} with functioning energy of 6.547 to 36.913 kJmol^{-1}. Tg upsides of long-grain assortment 'Zhunliang you' as well as two medium-grain assortments 'Zhongjia zao' not really settled utilizing the differential checking calorimeter technique augmented from range 38.7 to 51.2 °C when their dampness substance diminished from 20 to 13%. Differences in Tg upsides of the three paddy types have been seen under equivalent settings. When the moisture content of the 3 paddy types drops from 20 to 13% at drying temperatures of 50 to 70 °C, the drying time for portions with less than 10% damage should be 60–80 min.

Investigation of the consequence of drying air temperature on corn piece mechanical qualities has been carried out in [17]. The corn has dried in a convective drier operating in the range of 40–70 °C with a wind stream pace of 1.8 kg min^{-1}. The shards piled uniaxially at a 3 mm min^{-1} rate in a material testing machine until they shattered. An escalation of the drying temperature from 40–70 °C amplified part twisting at the break-point on average by 12%. Furthermore, the advantages of corn's power, stress, durability, and modulus of adaptability decreased by 21, 26, 36, as well as 38%, respectively, on average. Drying air temperature on maize piece mechanical properties has been investigated in [18]. The drying studies have been conducted at 45, 50, and 550 degrees Celsius, with air relative stickiness of 10, 20, as well as 30%, and at a constant airspeed of 1.0 ms^{-1}. To generate the drying bends for each test data, the dampness content data at various drying cool has converted to dampness percentage as well as plotted against time. The bends have fitted eight different delicate layer drying models to determine which is best for drying the kenaf core. The fit nature of the designs has been assessed dependent on the assurance coefficient ($R2$), decreased chi-square ($\chi 2$), as well as the root mean square error (RMSE). The two term model is the finest of the eight models for depicting the drying conduct of the kenaf center. Under the test conditions considered, the drying air temperature has a greater impact on the drying motor of the kenaf center than the drying air relative mugginess.

Many multi-reason drying frameworks employing the piece-wise drying method have been developed in [19]. To enhance the handling qualities of corn ears along with the utilization of drying equipment, as well as to decrease energy loss. It is distributed into two categories based on the distinct moisture content of maize seed: dry stockpiling and wet stockpiling. The wet stockpiling used minor-cycle drying, whereas the dry distribution center used a massive roundabout drying. When the moisture content of the corn ear reaches 18%, it has discharged and sent to the sifting region for filtering duties, and it has subsequently dried to safe hoarding dampness. Tests demonstrate that the framework can dry maize ears in a short period of time that the piece-by-piece dehydrating cycle may lessen energy utilization and production expenses by 20 to 30%, and that the normal rate of precipitation can be increased by 4.2% in comparison with the conventional drying process. In the meantime, the rate of seed squashing has been reduced by 1 to 1.5%. In [20], the author provides and characterizes a model for regulating paddy drying by combining collected temperatures. Drying tests have been conducted on newly gathered paddy at various degrees of drying air limits, comprising of temperatures = 27, 31, 35, 39, as well as 43 °C, with relative stickiness of RH varying from 45 to 65%, introductory dampness content, M_0 in the range of 17–25% and velocity values from 0.4 to 0.8 ms^{-1}). When the temperature is 31 °C, RH value of 60%, M_0 corresponding to 19%, and V indicating 0.5 ms^{-1}, the extra break rate arrived at the very least worth 0.508%, with the usual precipitation rate at 0.719% h. This publication also included a compiled temperature and quality outline. The x-arrangement in this diagram represents temperature, along with the underlying dampness content addresses the y-facilitate. It adequately covers the drying stages of the actual dryer and has a wide range of materialistic for climate certification. The model established in this study gives a conceptual reference to accurate drying as well as careful paddy management along with directs the actual drying activity.

A symmetrical investigation of hot air drying with variable temperature along with the stickiness has been planned to focus on the temperature's impact and stickiness on corn's drying and quality files and improves corn's dehydrating process [21]. The drying process of a genuine dryer has been duplicated by employing trial elements such as the angle of hot air temperature (x) as well as the inclination of evident dampness (y). The findings showed that the drying files (Drying time, standard drying rate, as well as amassed temperature) have perfect under-cooling and dehumidifying drying states, with 10 °C being the ideal value of x ideal worth of y amounted to 8 gm^{-3}. The quality records (degree of damage, unsaturated fat worth, and germination rate) arrived at the optimal qualities beneath warming and humidifying drying conditions; among them, the ideal worth of y for the three quality markers has 8 gm^{-3}. But the ideal worth of x has distinct, the ideal worth of x has 2.65 °C, 2.24 °C as well as 6.33 °C. The multi-objective capability has been combined into a single file using a hereditary calculation. The findings show that: when x equaled 2.26 °C, y has 4.84 gm^{-3}, the extensive qualities of the dehydrating system have the potential to be enhanced. In [22], it actively uses genetic optimization to tune a PID controller. Fuzzy immune is more accurate but takes significantly more time. To use MATLAB to make an SVR optimizer for the PID controller by genetic optimization support vector machine in [23]. Significantly more complex than normal genetic optimization but doesn't increase accuracy a lot.

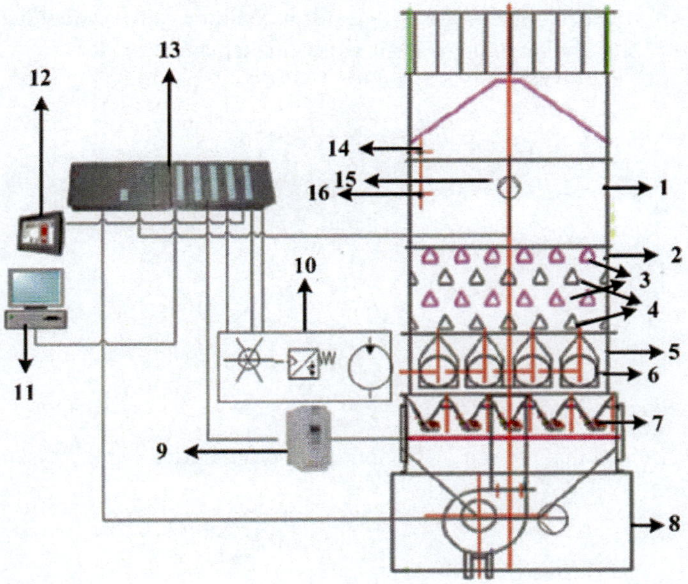

Fig. 1 Structure of the dryer [22]

3 Problem Statement

This paper set out to design and simulate a PID controller in MATLAB configured for a corn dryer, then use a genetic optimization to tune that PID controller according to the output of the grain dryer's output moisture content.

Figure 1 presented the structure of the dryer. The numbers shown in the figure represent the different parts of the dryer such as 1 represents storage, 2 for conventions, 3 shows wind angle inlet box, 4 presents exit wind angle box, 5 illustrates radiations, 6 demonstrates combustion compartment, 7 for corn discharging wheel, 8 corresponds to corn outlet temperature and humidity sensor, 9 denotes frequency converter, 10 implies other sensors assessing the drying parameters, 11 indicates computer, 12 for touch screen, 13 corresponds programmable logic controller, 14 expresses grain upper limit level, 15 for inlet corn temperature and humidity sensor, and last 16 shows corn's lower limit level.

4 Proposed Technique

The first step has to decide on the reference model. After some research, it ended up referring to this model. The next step has to research and make the dryer's mathematical model. This paper also needed the mathematical model of the output moisture content. Unfortunately, due to limited means, it couldn't gather actual data, so this paper used [22] as the reference for this paper mathematical model.

If

$$\begin{aligned} G(s) =& L(f(t)) = L(25.63Xe^{-0.094t}) \\ =& - 1s + 0.0940\infty 25.63Xe^{-(s+0.094)\text{td}-(s+0.094)t} \\ =& \int_0^\infty 25.63\,\text{Xe}^{-0.094t}e^{-\text{st}}\text{d}t \\ =& -\frac{25.63}{s+0.094}Xe^{-(s+0.094)t}\Big|_0^\infty \\ =& \frac{25.63}{s+0.094} = \frac{25.63}{1+10.65s} \end{aligned} \tag{1}$$

If it is assumed that the samples' input and output have a lag of 10 samples, get this equation

$$G(s) = \frac{273e^{-10s}}{10.65s+1} \tag{2}$$

where $T = 10.65$, $L = 10$, $K = 273$, $\text{ts} = 1$

This would be the transfer equation of this function. Dried corn has close to 15% moisture content in the industry, which is the model's target. Next, create a PID controller in MATLAB using the loop command to create a simple PID controller loop, giving it a step input of 0.15. The output of the PID controller loop is used in a cost function to guide the optimizer. Then, use a genetic optimizer that will take the cost function as its input and adjust the PID factors of the controller. The flowchart of the controller signal flow is presented in Fig. 2.

The algorithm used in this paper is given below:

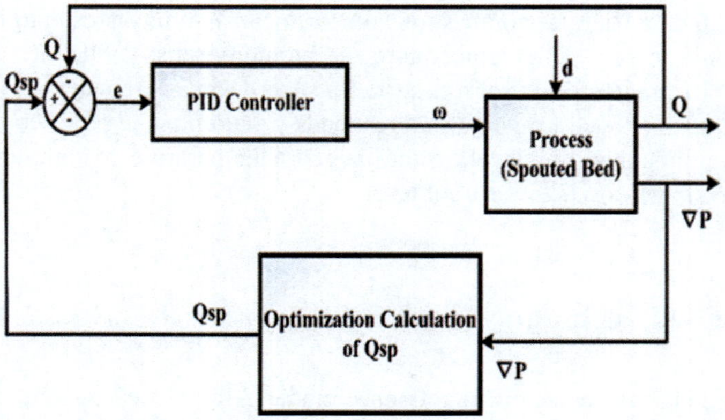

Fig. 2 Flowchart of controller signal flow

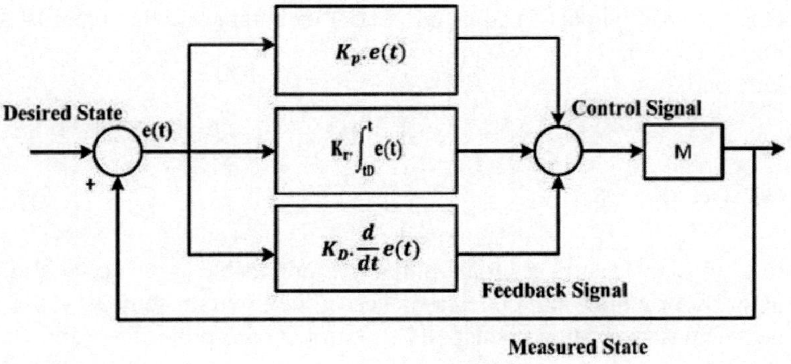

Fig. 3 PID signal loop

4.1 *PID Function: Part 1*

- Take inputs for the PID factors: The function starts by taking the proportional, integrative, and derivative factors of the PID controller as input.
- Create a PID controller signal loop using the values: Using the loop () function present in MATLAB, create a PID signal loop similar to a standard PID controller loop. PID signal loop is presented in Fig. 3.
- Input the step signal into the signal loop to generate the output: Next, use the step () signal function that is input in the signal loop and generate the output.
- Input the signal loop output in the cost function to generate a cost factor for the input PID factors: The output of the signal loop is then processed through a cost function to give a cost output. The function returns this cost output.

4.2 *Part 2*

- Set the factors for the GO algorithm: Start by initializing the factors for the genetic optimizer. The following values to set

  1. Population size
  2. Max generations
  3. Constraint equation factors

- Set the additional options for the optimizer: This paper uses the initialized factors and wraps them in an optimum-options () function. It also uses this function to decide the selection, crossover, and mutation method and factor to be used by the optimizer. These give the optimizer speed, efficiency, and process occupation.
- Configure the GO function to adjust 3 values according to minimize the cost function of part 1: Finally, use the g_a () function to start the genetic optimizer. Inside the function, call the function from part 1 and set the optimizer to minimize

the cost function output via adjusting the 3 PID parameters that input back into the part 1 function.

- Execute part 2

5 Result Analysis

The results are significantly accurate. Initial runs are unable to run successfully due to the algorithm running into edge cases due to the program settings. After some adjustment, and started getting results. The controller can optimize its values to 10% of the idea values in 5 secs and 1% in 9.3 s. The optimized values showed a rise time of 0.992 s and a settling time of 5 s, which is well within the expected parameters. Figure 4 represents the output step graph. Table 1 presents the result analysis. Table1 illustrates the expected and actual values of run time, rise time, settling time, and overshoot. The expected run time is the 30 s (Seconds) but the actual run time is 27.5 s. Similarly, expected rise time is 3 s and the actual rise time is 0.9 s. Settling time's expected and actual values are 10 and 5.94 s. 10 and 5.8% are the expected and actual values of overshoot.

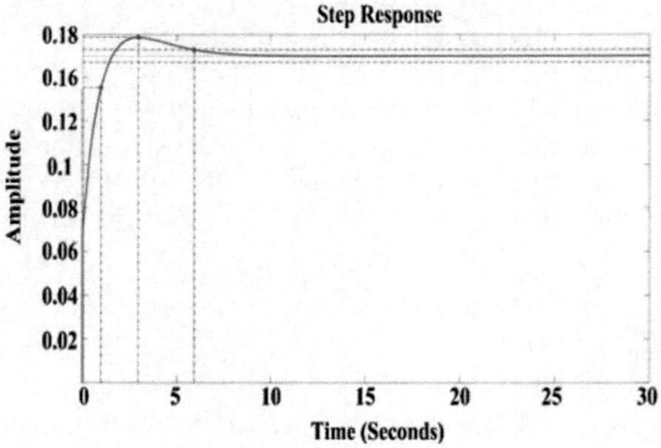

Fig. 4 Output step graph

Table 1 Result analysis

Parameter	Expected	Actual
Runtime	30 s	25.7 s
Rise time	3 s	0.9 s
Settling time	10 s	5.94 s
Overshoot	10%	5.8%

6 Conclusion

The constants to the PID regulator are appropriately adjusted for the drying of the corn grain in this endeavor tuning approach to limit the hazards of grain destruction. Furthermore, the data revealed by the regulator's method of soundness examination in time reaction has established that this framework can be steady in a brief time frame, and there is no error in the consistent state, allowing assumed execution for the next execution. Therefore, according to the data gathered in the regulator's plan via the recurrence (frequency) reaction, one might conclude that the corn grain drying system will produce good results in the final execution, permitting the commercialization of an acceptable grade item.

References

1. Amir Syariffuddeen MA, Yahya S, Ruwaida AW, Zainun MS, Shahrir A, Azman H, Shafie A, Zaimi ZAM, Hafiz MATM, Amir Redzuan S, Aliq J, Shukri J, Faewati AK, Mohsin Y, Shanmugevelu S (2020) Evaluation on drying temperature of grain corn and its quality using flat-bed dryer. ASM Sci J 13:78–83
2. Doymaz I, Pala M (2002) Hot-air drying characteristics of red pepper. J Food Eng 55:331–335
3. Onu CE, Igbokwe PK, Nwabanne JT, Nwajinka CO, Ohale PE (2020) Evaluation of optimization techniques in predicting optimum moisture content reduction in drying potato slices. Artifi Intell Agric 4:39–47
4. Sajith KG, Muraleedharan C (2014) Economic analysis of a hybrid photovoltaic/thermal solar dryer for drying amla. Int J Eng Res Technol (IJERT) 3:907–910
5. Wankhade PK, Sapkal RS, Sapkal VS (2013) Drying characteristics of okra slices on drying in hot air dryer. Procedia Eng 51:371–374
6. Correia A, Loro A, Zanatta S, Spoto M, Vieira T (2015) Effect of temperature, time, and material thickness on the dehydration process of tomato. Int J Food Sci 2015:1–7
7. B MS, El-Hana NH, G AE (2008) Effects of tomato geometries and air temperature on the drying behavior of plum tomato. Am J Appl Sci 5:1369–1375
8. Lynch BE, Morey RV (1989) Control strategies for ambient air corn drying. Trans ASABE 32:1727–1736
9. Wu Y, Wu W, Han F, Zhang Y, Xu Y (2017) Intelligent monitoring and control of grain continuous drying process based on multi-parameter corn accumulated temperature model. In: 2017 international conference on smart grid and electrical automation (ICSGEA). pp 77–80
10. Rafiee S, Keyhani A, Mohammadi A (2008) Soybean Seeds Mass Transfer Simulation During Drying Using Finite Element Method 4:284–288
11. Cao CW, Yang DY, Liu Q (2007) Research on modeling and simulation of mixed flow grain dryer. Drying Technol 25:681–687
12. Jittanit W (2007) Modelling of seed drying using a two-stage drying concept. A thesis submitted for the degree of Doctor of Philosophy
13. Piedrafita Moreno R (2004) Ingeniería de la automatización industrial. Ra-ma 2
14. Sitompul JP, Istadi Widiasa IN (2001) Modeling and simulation of deep-bed grain dryers. Drying Technol 19:269–280
15. Yao B, Zhang XK, Wen XD, Zou J (2015) Developments of variable temperature drying technology for agricultural products at home and abroad. Food Sci Technol 40:94–98
16. Li X, Wang X, Li Y, Jiang P, Lu H (2016) Changes in moisture effective diffusivity and glass transition temperature of paddy during drying. Comput Electron Agric 128:112–119

17. Abasi S, Minaei S (2014) Effect of drying temperature on mechanical properties of dried corn. Drying Technol 32:774–780
18. Misha S, Sohif M, Ruslan MH, Sopian K, Salleh E (2013) The effect of drying air temperature and humidity on the drying kinetic of kenaf core. Appl Mech Mater 315:710–714
19. Zhao WY, Shi ZL, Li YJ, Yang Z, Wei LJ, Jiang WY (2013) Study based on piecewise drying process of corn seed multi-purpose drying system. In: Mechanical science and engineering III. Trans Tech Publications Ltd, pp 93–99
20. Jin Y, Wong KW, Wu Z, Qi D, Wang R, Han F, Wu W (2019) Relationship between accumulated temperature and quality of paddy. Int J Food Prop 22:19–33
21. Chen J, Wu W, Cheng R, Jin Y, Liu Z (2020) Optimization of hot air drying process of corn using genetic algorithm and response surface methodology. Int J Food Prop 23:753–764
22. Dai A, Zhou X, Liu X (2017) Design and simulation of a genetically optimized fuzzy immune PID controller for a novel grain dryer. IEEE Access 5:14981–14990
23. Dai A, Zhou X, Wu Z (2020) Design of an intelligent controller for a grain dryer: a support vector machines for regression inverse model proportional–integral–derivative controller. Food Sci Nutr 8:805–819

Minimization of Molecular Potential Energy Function Using Laplacian Salp Swarm Algorithm (LX-SSA)

Prince, Kusum Deep, and Atulya K. Nagar

Abstract Optimization of molecular potential energy problem is a computationally expensive problem. The number of local minimizers shows exponential growth with the size of the molecules, showing that solving molecular potential energy function is a challenging task. The most stable conformation of a molecule corresponds to finding its global minimum potential energy, which dominates most of its properties. In this paper, a metaheuristic algorithm named Laplacian Salp Swarm Algorithm (LX-SSA) has been applied on the molecular potential energy function to find its global optimal solution, and the results have been compared with several other existing metaheuristic algorithms. The results display the efficiency of LX-SSA over other algorithms. On an independent set of coordinates which solely involves torsion angles, the minimum value of molecular potential energy is obtained. This study presents computational findings with 100 degrees of freedom efficiently.

Keywords Molecular potential energy function · Molecular conformation · Laplacian Salp Swarm Algorithm · Swarm

1 Introduction

Determining the stable structure of a molecular system is an alluring problem, due to its high complexity. The intricacy of the problem grows as the molecule size grows. The lowest energy structure of a molecule is very important as it dominates the majority of its attributes. According to Troyer et al. [1], the most stable configuration of a molecule corresponds to its optimal potential energy, therefore, it

Prince (✉) · K. Deep
Indian Institute of Technology Roorkee, Roorkee, Uttrakhand 247667, India
e-mail: prince@ma.iitr.ac.in

K. Deep
e-mail: kusum.deep@ma.iitr.ac.in

A. K. Nagar
Liverpool Hope University, Liverpool L16 9JD, UK
e-mail: nagara@hope.ac.uk

M. Thakur et al. (eds.), *Soft Computing for Problem Solving*, Lecture Notes in Networks and Systems 547, https://doi.org/10.1007/978-981-19-6525-8_42

can be modelled as an optimization problem. The energy function is a non-convex function. It is a computationally expensive problem because the count of local minimizers grows exponentially with the size of molecule [2]. The global minimizer represents the molecule's energetically optimal conformation, whereas these local minimizers represent the molecule's metastable state. This problem has been solved using a variety of optimization techniques, including the branch and bound approach, simulated annealing, genetic algorithms, diffusion equation method and smoothing techniques. A summary of various approaches for solving the molecular conformation problems may be found in [2–5]. Lavor et al. [5] developed a function to test the approaches used on this energy problem. Many researchers [6–14] have applied the metaheuristic methods to minimize the molecular potential energy function.

Mirjalili et al. [15] proposed the Salp Swarm Algorithm (SSA) in 2017. It mimics the swarming and foraging behaviour of salps. SSA has been extensively implemented for solving many real-life problems. Ibrahim et al. [16] suggested a novel SSA-based fish image segmentation technique. They optimized the parameters using SSA, which appeared in the segmentation of their model. Barik et al. [17] employed SSA to optimize load frequency management of the renewable microgrid. The study included the components such as rooftop solar arrays, fuel cells, biomass-fired combined heat and power and aqua-electrolyzer.

In this paper, a recently designed algorithm named Laplacian Salp Swarm Algorithm (LX-SSA) [18] has been applied on the molecular potential energy problem to test its efficacy over other existing metaheuristic algorithms. It is shown that LX-SSA solves this energy problem up to 100 degrees of freedom. This paper is organized as follows: Sect. 2 defines the molecular potential energy problem analytically. Section 3 is consisting of the details of the Laplacian Salp Swarm Algorithm (LX-SSA), whereas performance evaluation criteria of the algorithm is described in Sect. 4. We present the numerical results and analysis of the proposed algorithm in Sect. 5. Finally, in Sect. 6, the paper is concluded with some remarks.

2 Molecular Potential Energy Problem

In a three-dimensional search space, the simple molecular model contains a linear chain of n atoms centred at $x_1, x_2, \ldots, x_n$. Let $r_{i,i+1}$, the Euclidean distance between two successive atoms x_i and x_{i+1}, be the bond length. Let $\theta_{i,i+1}$ represents the bond angle corresponds to the third atom's relative position to the line joining the previous two. Let $\omega_{i,i+3}$ represents the angle of torsion between the normal of the planes determined by the atoms x_i, x_{i+1}, x_{i+2} and $x_{i+1}, x_{i+2}, x_{i+3}$. Figure 1 illustrates the above-mentioned atoms, bond lengths and the bond angles.

Force field potentials plays an essential role in explaining the concept of potential energy among the atoms of a system. Mathematically, it is defined as a function of system's conformation, which gives output in the form of energy of the system. Here, conformation of the system includes its bond angle, bond length, non-bonded interactions, etc.

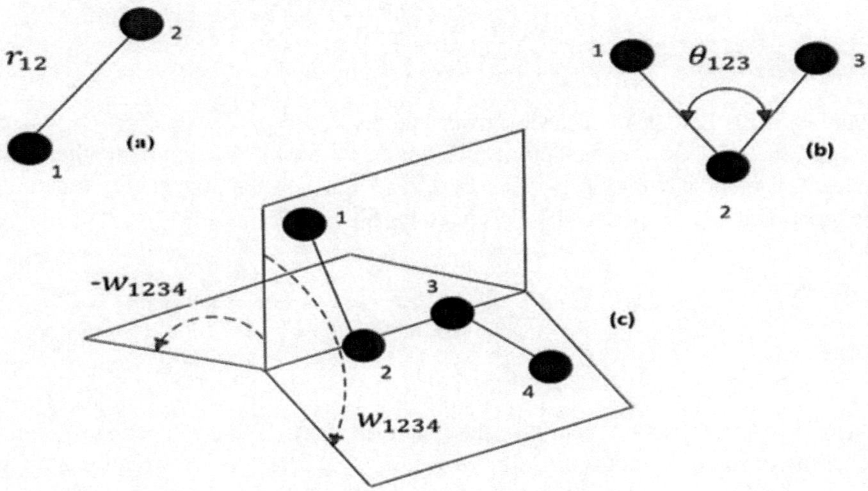

Fig. 1 (**a**) Euclidean distance (**b**) Bond angle (**c**) Torsion angle of the atomic chain

The mathematical formulation of the force field potentials is defined as:

(i) Bond length

$$E_1 = \sum_{(i,j)\in M_1} c_{ij}^1 \left(r_{ij} - r_{ij}^0\right)^2 \tag{1}$$

(ii) Bond angle

$$E_2 = \sum_{(i,j)\in M_2} c_{ij}^2 \left(\theta_{ij} - \theta_{ij}^0\right)^2 \tag{2}$$

(iii) Torsion angle

$$E_3 = \sum_{(i,j)\in M_3} c_{ij}^3 \left(1 + \cos\left(3\omega_{ij} - \omega_{ij}^0\right)\right) \tag{3}$$

where c_{ij}^k represent bond stretching, angle bending and torsion force constant for $k = 1, 2, 3$, respectively. The constants r_{ij}^0 and θ_{ij}^0 are the bond length and bond angle, respectively, and ω_{ij}^0 is the phase angle defining the position of minima. M_k, represents the set consisting of atoms separated by k covalent bonds, where $k = 1, 2, 3$. Additionally, a potential E_4 defines the two-body interatomic pair potential parted by more than two covalent bonds in the chain. Therefore, E_4 can be written as:

$$E_4 = \sum_{(i,j)\in M_3} \frac{(-1)^i}{r_{ij}} \tag{4}$$

where r_{ij} is the Euclidean distance between atoms x_i and x_j.

Thus, the aim is to find the optimal position of the atoms by minimizing the total molecular potential energy $E = E_1 + E_2 + E_3 + E_4$. The potential energy function is expressed using the parameters described in [5].

$$E = \sum_i \left(1 + \cos\left(3\omega_{i,i+3}\right)\right) + \sum_i \frac{(-1)^i}{\sqrt{(10.60099896 - 4.141720682\left(\cos \omega_{i,i+3}\right)}} \tag{5}$$

where $i = 1, 2, \ldots n - 3$ and n is the count of atoms in the molecular system. Thus, the problem is reduced to find $\omega_{i,i+3}$, $i = 1, 2, \ldots n - 3$. Because E is a non-convex function, even for small values of n, it has a significant count of local minima. These local minima correspond to a nearly stationary state of the molecule called metastable state, which is not truly stationary. According to Lavour et al. [5], the count of local minima of the function (5) is 2^N where $N = n - 3$, n is the count of atoms. The global minima of the energy function E is an alternate sequence of torsion angles $\langle p, q, p, q, p, q \ldots \rangle$ which is independent of other variables, where $p = 1.039195303$ and $q = 3.141592654$. The constraint $\omega_{i,j}$, $0 \leq \omega_{i,j} \leq 5$ assures only one global minimum. Barbosa et al. [6] demonstrated that the difference between the optimal solution E^* and second-best solution of (5) meets the following relation for all values of n,

$$\left|E^* - E_2\right| = 0.0816608225 \tag{6}$$

Because of the vast diversity of local minimizers, the problem still remains very challenging even after so many simplifications in E. According to Lavour et al. [5], the count of local minimizers corresponding to 20 atoms are $2^{17} = 131072$.

3 Laplacian Salp Swarm Algorithm (LX-SSA)

Mirjalili et al. [15] introduced the Salp Swarm Algorithm (SSA) in field of swarm intelligence techniques. The inspiration for designing of SSA was their swarming behaviour. Salps start foraging by forming a swarm known as a salp chain. The exploitation of SSA is better than its exploration. Because of less exploration, SSA suffers from premature convergence and slow convergence rate. To solve this difficulty, Prince et al. [18] proposed a new variant of SSA using Laplace Crossover Operator [19] called Laplacian Salp Swarm Algorithm (LX-SSA). After forming a salp chain, the salp population is bifurcated into leaders and followers. As the name implies, the leader salp is at the head of the chain and directs the swarm, while the others are followers and follow one another in succession. The position of the salps is defined in an n-dimensional search space, where n denotes the count of input variable. According to the food source H, the leader salp updates its position. The salp chain moves in the direction of food source if it has been replaced with a global optimal solution. However, if the global optimal solution of the optimization problem is unknown (as it is in real-life optimization problems), the best solution obtained by the algorithm is used as a food source. The following equations were used to update the position of the leader salp:

$$x_j^1 = \begin{cases} H_j + c_1\big((\mathrm{ub}_j - \mathrm{lb}_j)c_2 + \mathrm{lb}_j\big)c_3 \geq 0.5 \\ H_j - c_1\big((\mathrm{ub}_j - \mathrm{lb}_j)c_2 + \mathrm{lb}_j\big) \quad c_3 < 0.5 \end{cases} \tag{7}$$

where x_j^1 represents the position of leader salp, H_j denotes the food source's position, ub_j and lb_j denote the upper bound and lower bound in the jth dimension, c_1 is a controlling parameter and c_2, c_3 are uniformly generated random numbers between 0 and 1.

In addition, to sustain the balance between exploitation and exploration, the most essential parameter in SSA is c_1 and it is defined as:

$$c_1 = 2e^{-\left(\frac{4t}{T}\right)^2} \tag{8}$$

where $t(T)$ represents the current (maximum count of) iteration.

The follower salps update their position by following equation:

$$x_j^i = \frac{1}{2}\left(x_j^i + x_j^{i-1}\right) \tag{9}$$

where $i \geq 2$ and x_j^i denotes the ith salp's position in jth dimension.

After updating the leader and follower salps by using Eqs. (7) and (9), respectively, the situation occurs where SSA suffers from premature convergence due to less exploration. To overcome this problem, a new position update mechanism for the follower salps based on Laplace Crossover Operator [19] is used. For this, a random number β following the Laplace distribution [20] is generated:

$$\beta = \begin{cases} a - b * \log u, \; u \leq 0.5 \\ a + b * \log u, \; u > 0.5 \end{cases} \tag{10}$$

where u denotes uniformly distributed random number in [0, 1] and $a \in (-\infty, \infty)$ denotes location parameter and $b > 0$ is scale parameter.

After generating β, following new position update equation is utilized for the follower salps which enhances the exploration of LX-SSA and prevents the algorithm from premature convergence.

$$\text{currentsalp} = \text{currentsalp} + \beta(u, a, b) * (\text{foodposition} - \text{currentsalp}) \tag{11}$$

where $\beta(u, a, b)$ is a random number generated using Eq. (10), currentsalp denotes the current salp's position and foodposition represents the position of the food source H. After updating the positions of the follower salps using Eq. (11), a greedy search method is applied for updating the position of the current salp (follower salp) between Eqs. (9) and (11).

3.1 Computational Steps

The various computational steps involved in the algorithm are listed as follows:

Step I. Initialize a random population within the domain considered by the bounds.
Step II. Check for the stopping criteria, Halt, if satisfied; else proceed to III.
Step III. Calculate the position of initial population.
Step IV. Update position of leading salp and follower salps of initial population to update the population.
Step V. Increase the generation; update the existing population by new population; perform step II on the new updated population.

4 Performance Evaluation Criteria

To assess the effectiveness of the algorithm, the metrics are listed below to confirm the validity of the given method and to evaluate its performance:

1. Mean count of function evaluations (AFE).
2. Minimum count of function evaluations (MNFE).
3. Maximum count of function evaluations (MXFE).
4. Standard deviation of function evaluations (SDFE).
5. Computational time of a successful run.
6. Success Rate (SR) $= \frac{\text{number of successful runs}}{\text{Total number of runs}} \times 100$

Here, the parameters mentioned above are evaluated for the 100 runs for each problem size. Therefore, the parameters, MNFE and MXFE, i.e., the minimum and the maximum count of function evaluations, respectively, required to attain the threshold value for each problem size in the 100 runs. It should also be noted that out of the 100 runs that are conducted, AFE and SDFE represent the average count and standard deviation count of function evaluations of the successful runs, respectively.

A run is said to be successful, if the algorithm obtains a solution which satisfies $\left|\frac{f_{\min} - f_{\text{opt}}}{f_{\text{opt}}} \leq 0.01\right|$, where $f_{\min}$ is the best solution obtained by stated algorithm, while f_{opt} is the known global optimal solution. For computation of results, the population size for the algorithm is considered double the number of variables. The algorithm

Table 1 Global minima for various chain sizes (for $n = 20$ to 100)

n	20	40	60	80	100
E	−0.822366	−1.644732	−2.467098	−3.289464	−4.111830

terminates when one of the stopping criteria is achieved, i.e., either the maximum count of iterations (5000), or a known optimal solution is reached.

5 Numerical Results

This section deals with the numerical computation of proposed LX-SSA for the molecular potential energy. Shashi et al. [7] applied several real-coded genetic algorithms (RCGAs) which includes real-coded crossover operator LX [19] and mutation operator PM [21] to solve this energy problem. The results obtained by the LX-SSA have been compared with these real-coded genetic algorithms (RCGAs) developed. This problem is coded in MATLAB 2021 and runs on Intel i5 processor with a clock speed of 1.66 GHz and 8 GB of RAM. The minimum value of the energy function E is obtained in the search space $[0, 5]^n$, where n denotes the total count of atoms in the given system. After evaluating, the global minimum values of the considered function E for various chain sizes, i.e., $n = 20$, 40, 60, 80 and 100, we summarize results in Table 1.

The computational results in terms of function evaluations and computational time are given in Table 2. The results consist of minimum, maximum, average, standard deviation of successful runs and the time invested for the computation of these results for the different algorithms. The success rate (SR) of the Laplacian Salp Swarm Algorithm (LX-SSA) on SMPE investigated in this work is compared with other algorithms in Table 3. The method's reliability is directly related to the method's success rate.

6 Conclusion

In this study, a newly developed algorithm called Laplacian Salp Swarm Algorithm (LX-SSA) has been applied to a simplified molecular potential energy problem. Even though the energy problem is simplified; however, the major difficulty remains because the count of local minimizers increases exponentially with the problem size, which makes it hard to obtain the global minima. The degrees of freedom used in computational tests range from 20 to 100. The results show that the success rate of LX-SSA is quite low, but it finds the approximate global solution in very few counts of function evaluations and the least computational time with respect to

Table 2 Numerical results of the molecular model for $n = 20$ to 100 in terms of count of function evaluations and computational time

n	Algorithm	AFE	MNFE	MXFE	SDFE	Time
20	LX -PM	23,257	29,313	37,096	5523	0.53
	WX -PM	15,574	8412	24,051	3675	0.45
	LX -LLM	14,586	10,351	21,532	3021	0.34
	WX -LLM	28,969	17,643	44,867	7117	0.59
	DDX -LLM	14,255	8624	21,491	3273	0.33
	DDX -PM	13,523	8352	25,874	3849	0.32
	LX -SSA	**5468**	**2680**	**5980**	**1079**	**0.31**
40	LX -PM	71,336	45,544	99,370	15,184	2.57
	WX -PM	59,999	37,863	95,335	13,015	2.46
	LX -LLM	39,366	34,202	53,370	4342	1.47
	WX -LLM	89,478	61,765	120,042	18,059	3.34
	DDX -LLM	39,066	111,345	54,341	8889	1.43
	DDX -PM	38,095	24,918	57,693	8491	1.39
	LX -SSA	**5600**	**11,960**	**7689**	**2442**	**0.88**
60	LX -PM	280,131	203,618	329,337	35,492	15.65
	WX -PM	175,835	136,429	195,119	14,635	10.39
	LX -LLM	105,892	83,272	155,530	14,865	5.54
	WX -LLM	225,008	195,462	300,197	25,165	12.15
	DDX -LLM	75,086	42,146	99,813	9722	3.47
	DDX -PM	67,071	50,310	89,814	9083	3.21
	LX -SSA	**16,315**	**9300**	**17,940**	**3119**	**1.25**
80	LX -PM	326,287	281,876	369,451	27,841	25.12
	WX -PM	302,011	281,755	323,296	21,684	20.71
	LX -LLM	237,621	263,097	291,265	19,032	18.03
	WX -LLM	372,836	322,095	398,969	20,432	31.28
	DDX -LLM	146,043	111,345	180,127	14,265	8.73
	DDX -PM	108,909	46,586	152,025	24,155	7.16
	LX -SSA	**22,245**	**13,360**	**23,920**	**3412**	**2.1**
100	LX -PM	379,998	310,432	467,427	36,848	31.98
	WX -PM	369,376	324,621	412,876	30,276	30.62
	LX -LLM	320,146	268,764	356,729	32,412	26.48
	WX -LLM	443,786	399,601	498,352	28,659	39.91
	DDX -LLM	237,313	168,425	319,458	46,598	37.49
	DDX -PM	192,534	145,930	231,421	20,471	31.54
	LX -SSA	**28,466**	**18,800**	**29,900**	**3141**	**3.16**

Table 3 Comparison of LX-SSA for simplified molecular model based on success rate (SR)

n	LX-PM	WX-PM	LX-LLM	WX-LLM	DDX-LLM	DDX-PM	LX-SSA
20	100	100	100	100	100	100	7
40	100	100	100	98	100	100	11
60	91	93	95	89	100	100	31
80	79	71	82	78	92	97	29
100	88	89	86	82	98	100	26

other algorithms. LX-SSA performs better for large chain size ($n = 60, 80\&100$) as compared to small chain size ($n = 20\&40$).

For future work, LX-SSA can be incorporated with random walk or opposition-based learning techniques to improve its performance and can be applied to solve more energy problems.

References

1. Troyer JM, Cohen FE (1991) Simplified models for understanding and predicting protein structure. Rev Comput Chem (Wiley-VCH) 2:57–80
2. Wales DJ, Scheraga HA (1999) Global optimization of clusters, crystals, and biomolecules. Science 285(5432):1368–1372
3. Floudas CA, Klepeis JL, Pardalos (1999) Global optimization approaches in protein folding and peptide docking. In: Farach-Colton M, Roberts FS, Vingron M, Waterman M (eds) Mathematical support for molecular biology. DIMACS series, vol 47. American Mathematical Society, pp 141–171
4. Pardalos PM, Shalloway D, Xue GL (1994) Optimization methods for computing global minima of nonconvex potential energy functions. J Global Optim 4(2):117–133
5. Lavor C, Maculan N (2004) A function to test methods applied to global minimization of potential energy of molecules. Num Algor 35(2–4):287–300
6. Barbosa H, Lavor C, Raupp F (2005) A GA simplex hybrid algorithm for global minimization of molecular potential energy functions. Ann Oper Res 138(1):189–202
7. Shashi DK, Nagar AK, Katiyar VK (2012) Minimization of molecular potential energy function using newly developed real coded genetic algorithm. IJOCTA 2(1):51–58
8. Bansal JC, Shashi DK, Katiyar VK (2010) Minimization of molecular potential energy function using particle swarm optimization. Int J Appl Math Mech 6(9):1–9
9. Hedar AR, Ali AF, Hamid THA (2011) Genetic algorithm and tabu search based methods for molecular 3d-structure prediction. Num Algebra Control Optim 1(1):191–209
10. Ali AF, Hassanien AE (2013) Minimizing molecular potential energy function using genetic Nelder-Mead algorithm. In: 2013 8th international conference on computer engineering and systems (ICCES), pp 177–183
11. Agrawal S, Silakari S (2014) Fletcher-reeves based particle swarm optimization for prediction of molecular structure. J Mol Graph Model 49:11–17
12. Cheung NJ, Shen HB (2014) Hierarchical particle swarm optimizer for minimizing the non-convex potential energy of molecular structure. J Mol Graph Model 54:114–122
13. Tawhid MA, Ali AF (2017) A hybrid social spider optimization and genetic algorithm for minimizing molecular potential energy function. Soft Computi 21:6499–6514
14. Tawhid MA, Ali AF (2017) A hybrid grey wolf optimizer and genetic algorithm for minimizing potential energy function. Memetic Comp 9:347–359

15. Mirjalili S, Gandomi AH, Mirjalili SZ, Saremi S, Faris H, Mirjalili SM (2017) Salp swarm algorithm: a bio-inspired optimizer for engineering design problems. Adv Eng Softw 114:163–191
16. Ibrahim A, Ahmed A, Hussein S, Hassanien AE (2018) Fish image segmentation using salp swarm algorithm. In: International conference on advanced machine learning technologies and applications. Springer, Berlin, pp 42–51
17. Barik AK, Das DC (2018) Active power management of isolated renewable microgrid generating power from rooftop solar arrays, sewage waters and solid urban wastes of a smart city using salp swarm algorithm. In: 2018 technologies for smart-city energy security and power (ICSESP). IEEE, New York, pp 1–6
18. Prince, Deep K (2021) Laplacian salp swarm algorithm on CEC 2014 continuous optimization benchmarks. Int J Syst Assurance Eng Manag (Submitted)
19. Deep K, Thakur M (2007) A new crossover operator for real coded genetic algorithms. Appl Math Comput 188(1):895–912
20. Distribution L (2008) The concise encyclopaedia of statistics. Springer, New York, NY
21. Deep K, Thakur M (2007) A new mutation operator for real coded genetic algorithms. Appl Math Comput 193(1):211–230

Performance Evaluation by SBM DEA Model Under Fuzzy Environments Using Expected Credits

Deepak Mahla, Shivi Agarwal, and Trilok Mathur

Abstract Performance evaluation is always the interest of the manager in every sector like banks, supply chain management, etc. Data envelopment analysis (DEA) is an operations research methodology based on linear programming, which is used to evaluate the performance of decision-making units (DMUs). The available data in real-life scenarios may be present in the vague form. Therefore, the crisp DEA model cannot be used for efficiency evaluation. In the present study, the slacks based model (SBM) DEA model is integrated with the fuzzy set theory to overcome the problem. The expected credits approach is used to solve the fuzzy SBM DEA model in the proposed study. The relative efficiency of the Indian oil refineries is calculated using this methodology, and the rank of the companies is computed from their relative efficiency.

Keywords Data envelopment analysis · Uncertainty · SBM model · Credibility measure · Decision-making units · Expected credits

1 Introduction

Data envelopment analysis (DEA) has been used by various researchers from its notion by Charnes et al. [1] in various applications [2–4]. It is a nonparametric linear programming-based technique that uses multiple attributes to evaluates the relative efficiencies of decision-making units (DMUs). Various DEA models are existed in the literature like Charnes Cooper Rhodes (CCR) DEA model [1], Banker Charnes Cooper (BCC) DEA model [5], Additive model [6], slacks based measure (SBM) model [7], new slack model (NSM) [8], etc. The present study uses the non-radial SBM DEA model proposed by Tone [7] which can directly deal with the input excess and output shortfalls. Many times, available real-world data is in uncertain or qual-

D. Mahla · S. Agarwal (✉) · T. Mathur
Department of Mathematics, Birla Institute of Technology and Science Pilani, Pilani, India
e-mail: shivi@pilani.bits-pilani.ac.in

D. Mahla
e-mail: deepakmahlabits@gmail.com

M. Thakur et al. (eds.), *Soft Computing for Problem Solving*, Lecture Notes in Networks and Systems 547, https://doi.org/10.1007/978-981-19-6525-8_43

itative form. DEA models do not require any prior information related to data and use frontier to calculate the relative efficiencies and minimal uncertainty in data can change the frontier significantly, which leads to absurdity in the results. Therefore, the crisp DEA models are not the true representation under vague data. To handle vague data, various techniques have been used by researcher [9, 10]. Fuzzy set theory proposed by Zadeh [10] can handle this problem very efficiently. Many researchers proposed integrated fuzzy DEA models. Emrouznejad and Tavana [11] reviewed fuzzy DEA methods and classified them into six primary categories, namely tolerance approach [12], α-level-based approach [13], fuzzy ranking approach, possibility approach [14], fuzzy arithmetic, and fuzzy random/Type-2 fuzzy sets. Fuzzy DEA has a very vast literature history, and some important literature important to the proposed study is discussed below. Puri and Yadav [15] used fuzzy SBM DEA models to handle the imprecise data, and calculate the mix-efficiencies of State Bank of Patiala. Hekmatnia et al. [16] computed the relative efficiency of the six production factories of Iran using fuzzy SBM DEA model. Saini and Alirezaee [17] developed the fuzzy versions of trade-off DEA models by using ranking method based on the comparison of α-cuts. Wanke et al. [18] presented an analysis of the efficiency of Angolan banks using fuzzy DEA and stochastic DEA models based on the α-level approach and different tail dependence structure, respectively. Recently, Bakhtavar and Yousefi [19] used a special risk prioritization algorithm by failure mode using SBM DEA model under fuzzy conditions. As per the Zadeh examination of the fuzzy theory, the possibility distribution of the fuzzy variable replicates the probability distribution of the random variable in probability theory. Fuzzy linear models can be considered the evolution of conventional linear programming models, in which fuzzy variables play the role of fuzzy coefficients, and fuzzy events construct the constraints which control the model. The possibility theory determines the possibility of fuzzy events. Lertworasirikul et al. [14, 20] studied the fuzzy DEA models built by Guo et al. [21], which took the possibility criterion and the necessity criterion as a measure and solved the ranking problem with two distinct approaches, namely the possibility approach and the credibility approach. In [8], Agarwal used the possibility measure approach for fuzzy SBM DEA model and transforms it into the crisp linear model. The possibility measure is used extensively, but it has no self-dual property, which is undoubtedly needed for practice. The credibility theory which managed personal conviction degree numerically was proposed by Liu et al. [22–24]. Recently, Mahla and Agarwal [25] used credibility approach to solve fuzzy SBM DEA model and also performed sensitivity and stability analysis of this model in [26]. Arana-Jiménez et al. [27] proposed a fuzzy extension of the measure of inefficiency proportions. All the above-mentioned techniques required extra effort to rank the DMUs after computing the relative efficiency because the computed efficiency value is different at every level of significance. We further either need some index to rank the DMUs or need to specify the level of significance, which took more interactive data and time from the decision-maker. To overcome this problem, our methodology is based on the expected credits derived from the credibility measure which will give a single value of relative efficiency. It can be directly used to rank the DMUs by their relative efficiency score without the help of an expert or any significant level. Thus, it saves time and cost for the decision-maker. Lertworisikul et al. [20] proposed the expected credits approach for the fuzzy CCR DEA model, which is similar to the expected value approach in stochastic programming. In the proposed study, fuzzy SBM DEA model has been solved using expected credits and both the efficiency as well as input/output slacks are calculated. The paper upholds as follow: introduction of the SBM DEA and fuzzy SBM DEA model is described in the next section. Expected credits from credibility measure has been presented in Sect. 3. In Sect. 4, data of Indian oil companies [28] have been collected and the relative efficiencies of the companies are calculated using the proposed methodology. The obtained result and conclusion about the proposed model is discussed in Sect. 5.

2 Preliminaries

In this section, SBM and fuzzy SBM DEA models with required definitions for the study are discussed.

2.1 *Slacks Based Measure DEA Model*

Tone [7] proposed an unit invariant and non-radial DEA model that measures the relative efficiency of DMUs by minimizing the ratio of the reduction rate input slacks to the expansion rate of output slacks. Mathematically, the SBM DEA model for DMU$_k$ with p-inputs, q-outputs, and n-number of DMUs is formulated as,

$$\begin{aligned} \min\ \rho_k &= \frac{1 - \frac{1}{m}\sum_{i=1}^{p} s_{ik}^{-}/x_{ik}}{1 + \frac{1}{q}\sum_{j=1}^{q} s_{jk}^{+}/y_{jk}} \\ \text{subject to } & \sum_{o=1}^{n} \lambda_{ok} x_{io} + s_{ik}^{-} = x_{ik} \quad \forall i = 1, \ldots, p \\ & \sum_{o=1}^{n} \lambda_{ok} y_{jo} - s_{jk}^{+} = y_{jk} \quad \forall j = 1, \ldots, q \\ & \lambda_{ok} \geq 0, s_{ik}^{-} \geq 0, s_{jk}^{+} \geq 0, \quad \forall o = 1, \ldots, n. \end{aligned} \tag{1}$$

where x_{ik}= amount of ith input consumed by kth DMU, y_{jk}= amount of jth output produced by kth DMU, s_{ik}^{-}= slack in the ith input of the kth DMU, s_{jk}^{+}= slack in the jth output of the kth DMU, and λ_{ok} are intensity variables. The definition of the fuzzy number and fuzzy SBM DEA model has been described in the next subsections.

2.2 Fuzzy Numbers

A fuzzy set $\tilde{T}$ defined on real numbers $\mathbb{R}$ is said to be fuzzy number [29] if it satisfies the following properties:

1. $\tilde{T}$ is normal fuzzy set,
2. $(\tilde{T})_\alpha$ is a closed interval for every $\alpha \in (0, 1]$,
3. The support of $\tilde{T}$ is bounded.

There are many fuzzy numbers, depending on the nature of the user and application. But most commonly used and user-friendly membership functions are triangular, Gaussian, bell-shaped, and trapezoidal. Triangular and Trapezoidal fuzzy numbers are discussed below.

Definition 1 (*Triangular fuzzy numbers (TFN)*) A TFN $\tilde{T}$ is a particular type of fuzzy number and is considered as fuzzy variable determined by triplet (p, q, r) such that $(p < q < r)$, with membership function as follows,

$$\mu_{\tilde{T}}(x) = \begin{cases} \frac{x-p}{q-p}, & \text{if } p \leq x \leq q \\ \frac{r-x}{r-q}, & \text{if } q \leq x \leq r \\ 0, & \text{otherwise.} \end{cases} \tag{2}$$

Definition 2 (*Trapezoidal fuzzy numbers (TrFN)*) Another widely used fuzzy number is the TrFN. A trapezoidal fuzzy number $\tilde{T}$ membership function for (p, q, r, s) such that $(p < q < r < s)$, is defined as,

$$\mu_{\tilde{T}}(x) = \begin{cases} 0, & \text{if } x \leq p \\ \frac{x-p}{q-p}, & \text{if } p \leq x \leq q \\ 1, & \text{if } q \leq x \leq r \\ \frac{s-x}{s-r}, & \text{if } r \leq x \leq s \\ 0, & \text{if } x \geq r. \end{cases} \tag{3}$$

2.3 Fuzzy SBM DEA Model

Considering the inputs and outputs as fuzzy number, SBM DEA model has been changed into fuzzy SBM DEA model; it is described for DMU_k as

$$\begin{aligned}
\min\ \rho_k &= \frac{t - \frac{1}{p}\sum_{i=1}^{p} S_{ik}^{-}/\tilde{x}_{ik}}{t + \frac{1}{q}\sum_{j=1}^{q} S_{jk}^{+}/\tilde{y}_{jk}} \\
\text{subject to } & \sum_{o=1}^{n} \Lambda_{ok}\tilde{x}_{io} + S_{ik}^{-} = t\tilde{x}_{ik} \quad \forall i = 1, \ldots, p \\
& \sum_{o=1}^{n} \Lambda_{ok}\tilde{y}_{jo} - S_{jk}^{+} = t\tilde{y}_{jk} \quad \forall j = 1, \ldots, q \\
& \Lambda_{ok} \geq 0, S_{ik}^{-} \geq 0, S_{jk}^{+} \geq 0, t > 0 \quad \forall o = 1, \ldots, n.
\end{aligned} \tag{4}$$

where $\tilde{x}_{ik}$ is the ith fuzzy input consumed by the kth DMU and $\tilde{y}_{jk}$ is the jth fuzzy output produced by the kth DMU. It is not possible to solve the fractional objective function. Therefore, the objective function is transformed into linear by normalizing the denominator.

$$\begin{aligned}
\min\ \rho_k &= t - \frac{1}{p}\sum_{i=1}^{p} S_{ik}^{-}/\tilde{x}_{ik} \\
\text{subject to } & t + \frac{1}{q}\sum_{j=1}^{q} S_{jk}^{+}/\tilde{y}_{jk} = 1 \\
& \sum_{o=1}^{n} \Lambda_{ok}\tilde{x}_{io} + S_{ik}^{-} = t\tilde{x}_{ik} \quad \forall i = 1, \ldots, p \\
& \sum_{o=1}^{n} \Lambda_{ok}\tilde{y}_{jo} - S_{jk}^{+} = t\tilde{y}_{jk} \quad \forall j = 1, \ldots, q \\
& \Lambda_{ok} \geq 0, S_{ik}^{-} \geq 0, S_{jk}^{+} \geq 0, t > 0 \quad \forall o = 1, \ldots, n.
\end{aligned} \tag{5}$$

The model (5) is a fuzzy SBM DEA model, which cannot be solved directly. The expected credits from credibility measures are used to solve fuzzy SBM model which are described in the next section.

3 Expected Credits

The expected credits from credibility measure approach will be used to solve the fuzzy SBM DEA model. The credibility measure is self-dual in nature. Self-duality is, in fact, a generalization of the contradiction law and the law of the excluded middle. In other words, a mathematical system without self-duality assumption would be incompatible with laws [30]. The credibility measure is defines as,

Definition 3 Consider $\tilde{\Psi}$ be a nonempty set with the power set of $\tilde{\Psi}$ is $P\{\tilde{\Psi}\}$. Credibility set function Cr{.} is defined by Liu and Liu [22] as credibility measure if it satisfies the following conditions:

1. $\mathrm{Cr}\{\tilde{\Psi}\} = 1$,
2. $\mathrm{Cr}\{P\} \leq \mathrm{Cr}\{Q\}$ whenever $P \subset Q$,
3. $\mathrm{Cr}\{P\} + \mathrm{Cr}\{P\}^C = 1$ for any event P,
4. $\mathrm{Cr}\{\cup_i P_i\} = \mathrm{Sup}_i \mathrm{Cr}\{P_i\}$ for any events P_i with $\mathrm{Sup}_i \mathrm{Cr}\{P_i\} < 0.5$.

The triplet $(\tilde{\Psi}, P(\tilde{\Psi}), \mathrm{Cr})$ are called as the credibility space.

The expected credits value operator of a fuzzy variable $\tilde{\Psi}$ on credibility space $(\tilde{\Psi}, P(\tilde{\Psi}), \mathrm{Cr})$ is defined by Liu and Liu [22] as

$$E(\tilde{\Psi}) = \int_{0}^{+\infty} \mathrm{Cr}(\tilde{\Psi} \geq x)\mathrm{d}x - \int_{-\infty}^{0} \mathrm{Cr}(\tilde{\Psi} \leq x)\mathrm{d}x. \tag{6}$$

Some important definitions related to the study are discussed below.

Definition 4 (*Credibilistically efficient*) If the value $E(\rho_k)$ is equal to one, the DMU is a credibilistically efficient DMU. Otherwise, it is a credibilistically inefficient DMU.

Definition 5 (*Input slacks*) The excesses in the actual input known as input slacks are used to calculate the target input, which is required for the inefficient units to become the efficient units.

Definition 6 (*Output slacks*) The shortfalls in the actual output known as output slacks are used to calculate the target output, which is required for the inefficient units to become the efficient units.

Expected credits of the fuzzy variable replace the fuzzy objective function and fuzzy constraints.

$$\begin{cases} \min \rho_k = E\left(t - \frac{1}{p}\sum_{i=1}^{p} S_{ik}^-/\tilde{x}_{ik}\right) \\ \text{subject to } E\left(t + \frac{1}{q}\sum_{j=1}^{q} S_{jk}^+/\tilde{y}_{jk}\right) = 1 \\ E\left(\sum_{o=1}^{n} \Lambda_{ok}\tilde{x}_{io} + S_{ik}^-\right) = tE\left(\tilde{x}_{ik}\right) \qquad \forall i = 1, \ldots, p \\ E\left(\sum_{o=1}^{n} \Lambda_{ok}\tilde{y}_{jo} - S_{jk}^+\right) = tE\left(\tilde{y}_{jk}\right) \qquad \forall j = 1, \ldots, q \\ \Lambda_{ok} \geq 0, S_{ik}^- \geq 0, S_{jk}^+ \geq 0, t > 0 \qquad \forall o = 1, \ldots, n. \end{cases} \tag{7}$$

Lemma 1 [20] has been used to solve the model (7) for normal, convex trapezoidal fuzzy number.

Lemma 1 *If $\tilde{\Psi}$ is normal, convex trapezoidal fuzzy variable, the expected credits for the $\tilde{\Psi}$ is,*

$$E(\tilde{\Psi}) = \frac{1}{4}[(\tilde{\Psi})_1^U + (\tilde{\Psi})_1^L + (\tilde{\Psi})_0^U + (\tilde{\Psi})_0^L] \tag{8}$$

1. $(\Psi)_1^U$ is the upper bound of Ψ at $\alpha = 1$.
2. $(\Psi)_0^U$ is the upper bound of Ψ at $\alpha = 0$.
3. $(\Psi)_1^L$ is the lower bound of Ψ at $\alpha = 1$.
4. $(\Psi)_0^L$ is the lower bound of Ψ at $\alpha = 0$.

Model (7) is transformed into the model (9) when membership functions are normal, convex trapezoidal number, as follows,

$$\begin{cases} \min\ \rho_k = \frac{1}{4}\left[\left(t - \frac{1}{p}\sum_{i=1}^{p} S_{ik}^-/\tilde{x}_{ik}\right)_1^U + \left(t - \frac{1}{p}\sum_{i=1}^{p} S_{ik}^-/\tilde{x}_{ik}\right)_0^U \right. \\ \qquad \left. + \left(t - \frac{1}{p}\sum_{i=1}^{p} S_{ik}^-/\tilde{x}_{ik}\right)_1^L + \left(t - \frac{1}{k}\sum_{i=1}^{p} S_{ik}^-/\tilde{x}_{ik}\right)_0^L \right] \\ \text{subject to} \\ \frac{1}{4}\left[\left(t + \frac{1}{q}\sum_{j=1}^{q} S_{jk}^+/\tilde{y}_{jk}\right)_1^U + \left(t + \frac{1}{q}\sum_{j=1}^{q} S_{jk}^+\tilde{y}_{jk}\right)_0^U + \left(t + \frac{1}{q}\sum_{j=1}^{q} S_{jk}^+/\tilde{y}_{jk}\right)_1^L \right. \\ \left. + \left(t + \frac{1}{q}\sum_{j=1}^{q} S_{jk}^+/\tilde{y}_{jk}\right)_0^L \right] = 1, \\ \left[\left(\sum_{o=1}^{n} \Lambda_{ok}\tilde{x}_{io} + S_{ik}^-\right)_1^U + \left(\sum_{o=1}^{n} \Lambda_{ok}\tilde{x}_{io} + S_{ik}^-\right)_0^U + \left(\sum_{o=1}^{n} \Lambda_{ok}\tilde{x}_{io} + S_{ik}^-\right)_1^L + \right. \\ \left. \left(\sum_{o=1}^{n} \Lambda_{ok}\tilde{x}_{io} + S_{ik}^-\right)_0^L \right] = t\left[(\tilde{x}_{ik})_1^U + (\tilde{x}_{ik})_0^U + (\tilde{x}_{ik})_1^L + (\tilde{x}_{ik})_0^L\right] \quad \forall i = 1, \ldots, p \\ \left(\sum_{o=1}^{n} \Lambda_{ok}\tilde{y}_{jo} - S_{jk}^+\right)_1^U + \left(\sum_{o=1}^{k} \Lambda_{ok}\tilde{y}_{jo} - S_{jk}^+\right)_0^U + \left(\sum_{o=1}^{n} \Lambda_{ok}\tilde{y}_{jo} - S_{jk}^+\right)_1^L \\ + \left(\sum_{o=1}^{n} \Lambda_{ok}\tilde{y}_{jo} - S_{jk}^+\right)_0^L = t\left[(\tilde{y}_{jk})_1^U + (\tilde{y}_{jk})_0^U + (\tilde{y}_{jk})_1^L + (\tilde{y}_{jk})_0^L\right] \quad \forall j = 1, \ldots, q \\ \Lambda_{ok} \geq 0,\ S_{ik}^- \geq 0,\ S_{jk}^+ \geq 0,\ t > 0, \quad \forall o = 1, \ldots, n. \end{cases} \tag{9}$$

The model (9) is converted into the linear programming model by finding the lower and upper bounds at 0 and 1 using α-cut. Then, these linear programming problems have been solved using MATLAB 2019b.

4 Numerical Illustration

The proposed methodology is illustrated on the numerical data. Data of Indian oil refineries are collected for the financial year 2017–18 from their annual reports, and the relative efficiency, as well as the slacks, are calculated by using the proposed methodology. Indian oil sector is one of the largest and core areas which influences the other core sectors effectively. India is one of the biggest importers of oil, and the amount of imports is increasing year by year. Compared to 4.56 million barrels of oil per day consumption by India in 2016, the use increased to 4.69 million barrels of oil per day in 2017 [28]. The data of the 9 Indian oil companies in which 5 are PSUs, 2 are joint venture (enterprises which are run by two or more parties), and two are private limited are collected. A total of 249.4 million metric tons (MMT) oil are installed in their refineries, making it the second-largest refiner in Asia. The relative efficiency of nine companies is calculated using the proposed methodology by taking three inputs as capital expenditure, oil throughput, and the number of employees and one output as gross refining margin (GRM).

4.1 Inputs and Output

1. **Capital Expenditure (CaExp)**: The CaExp is the funds utilized by the companies to acquire all the physical resources needed for the new projects, the increment of the number of the employees, repairing the company's equipment, celebrating function for companies, etc. CaExp is taken as the input for our study.
2. **Throughput Utilization of Oil**: In oil refinery companies, the utilization of oil has been considered as the performance of the company. The usage of the oil depends on different factors like oil-seed crushing parity, etc. Hence, this input is considered as triangular fuzzy number for the proposed methodology.
3. **Employees**: The total strength of workers working in a company is the total employees. The employees are the input of the companies, and the number of the total employees is precisely known from annual reports of the oil companies. The value of the total number of employees is used as crisp input for the present study.
4. **Gross Refining Margins (GRM)**: The GRM is the difference between the revenue generated by all oil-based products manufactured by an oil refinery and the cost of the crude material, which is raw petroleum. The value of the GRM depends on many factors like:

    (a) Cost of sourcing crude petroleum
    (b) Manufacturing reliability and efficiency
    (c) Demand–supply jumbling
    (d) Higher custom duty would result in higher GRM.

    Some factors of GRM are uncertain, and therefore, it should be taken as fuzzy output. The companies performance is in the direct proportion of the value of GRM. GRM is one of the significant attributes affecting the economy of the company, and hence, it is in great interest of any company to get higher GRM.

The available data contains capital expenditure, number of employees, oil utilization, and GRM for the companies BPCL, ONGC, HPCL, IOCL, CPCL, NRL, BORL, RIL, and NEL. The relative efficiency of small companies and large companies calculated

Table 1 Data of Indian oil companies

Company size	Company	Capital expenditure	Crude oil throughput	GRM	Employees
Large	BPCL	1.89	4.20	(4,5,6)	3.41
	ONGC+MRPL	9.00	2.85	(6,7,8)	9.00
	HPCL	1.73	4.26	(7,8,9)	3.09
	IOCL	3.21	8.83	(6,7,8)	5.36
	RIL	5.34	9.00	(8,9,9)	5.56
Small	CPCL	3.57	5.17	(6,7,8)	5.00
	NRL	2.03	2.08	(7,8,9)	2.44
	BORL	9.00	3.59	(6,7,8)	9.00
	NEL	3.56	9.00	(8,9,9)	8.84

separately as the comparison of the comparative effectiveness of small companies with large companies is not reasonable. So, the data are divided into two types, large companies (employees $>$ 10,000) and small companies (employees $<$ 10,000). The large companies are BPCL, ONGC, HPCL, IOCL, and RIL, and the small companies are CPCL, NRL, BORL, and NEL. The values of GRM have some uncertainty; therefore, in the proposed work, the GRM is converted into triangular fuzzy number by fuzzification process. Table 1 contains the data of all inputs and outputs.

Further, the efficiencies of all companies are calculated by the proposed methodology using MATLAB 2019b. The relative efficiencies of the companies using α-cut, possibility measure, and credibility measure approach on SBM DEA are also computed. Table 2 shows the efficiencies of small and large companies using expected credits as well as the ranking of the companies based on relative efficiency. It also contains the input and output slacks for the inefficient DMUs. The results presented

Table 2 Relative efficiency of companies using expected credits approach

	Efficiency	s_1^-	s_2^-	s_3^-	s_1^+	Ranking
Small company						
CPCL	0.2178	1	745.6	9.2	1209.7	3
NRL	1	0	0	0	0	1
BORL	0.4804	1	2603.9	3.8	115.6	2
NEL	0.1614	1	658.6	18.5	9323.5	4
Large company						
BPCL	0.8807	1	1565.7	0	2000.8	3
ONGC	1	0	0	0	0	2
HPCL	1	0	0	0	0	1
IOCL	0.4041	0.1	1263.2	3.6	2125	5
RIL	0.5377	0.1	2946.2	2.6	652.3	4

Table 3 Relative efficiency of companies using credibility measure approach

Credibility level	0.5	0.6	0.7	0.8	0.9	1	Ranking
Large company							
BPCL	0.8571	0.8484	0.8387	0.8275	0.8148	0.8	3
ONGC + MRPL	1	1	1	1	1	1	1
HPCL	1	1	1	1	1	1	1
IOCL	0.7595	1	1	1	1	1	2
RIL	0.6655	0.6937	0.701	0.7014	0.7025	0.7029	4
Small company							
CPCL	0.5748	0.6836	0.6854	0.6872	0.6391	0.6309	2
NRL	1	1	1	1	1	1	1
BORL	1	1	1	1	1	1	1
NEL	1	1	1	1	1	1	1

Table 4 Comparison in ranking between all four methods

Approaches	α-cut	Possibility	Expected credits	Credibility
Large company				
BPCL	3	3	3	3
ONGC + MRPL	1	1	2	1
HPCL	1	1	1	1
IOCL	2	2	5	2
RIL	4	4	4	4
Small company				
CPCL	2	2	3	2
NRL	1	1	1	1
BORL	1	1	2	1
NEL	1	1	4	1

in Table 2 can be interpreted as follows: NRL is efficient company with relative efficiency 1, followed by BORL, CPCL, and NEL, respectively. It is also evident from Table 2 that HPCL is efficient among large companies using expected credits. All other companies are inefficient, and hierarchy of the ranking is as ONGC, BPCL, RIL, IOCL, respectively.

Further, the efficiency of these companies are also computed using the credibility measure approach, α-cut approach, and possibility measure approach (Table 4).

The efficiencies obtained from credibility measure approach, α-cut approach, and possibility measure approach are comparable with the proposed expected credits approach. Table 4 shows the comparison between all four methods based on the ranking. Also, there is a minimal difference in ranking order which demonstrates the robustness and validation of the proposed technique.

5 Conclusion

The expected credits from the credibility measure, which combines both the possibility and its dual necessity measure, are used to deal with the fuzzy environment in the fuzzy SBM DEA model. This approach changes the fuzzy model into credibility programming which is similar to the linear programming model. Expected credits approach handles fuzzy data significantly and a single value of relative efficiency is obtained, and therefore, there is no additional requirement for any expert advice or any level of significance for ranking purposes. The relative efficiency of the Indian oil refineries for the year 2017–18 has been computed using the proposed methodology, and also input slacks and output slacks are obtained simultaneously without any further calculations. We compared our obtained results with the results obtained from possibility measure, α-cut approach, and credibility approach. The results are comparable which shows the robustness of the model.

References

1. Charnes A, Cooper WW, Rhodes E (1978) Measuring the efficiency of decision making units. Eur J Oper Res 2(6):429–444
2. Chen W, Gai Y, Gupta P (2018) Efficiency evaluation of fuzzy portfolio in different risk measures via DEA. Ann Oper Res 269(1–2):103–127
3. Emrouznejad A, Parker BR, Tavares G (2008) Evaluation of research in efficiency and productivity: a survey and analysis of the first 30 years of scholarly literature in DEA. Soc-Econ Plan Sci 42(3):151–157
4. Liang L, Yang F, Cook WD, Zhu J (2006) DEA models for supply chain efficiency evaluation. Ann Oper Res 145(1):35–49
5. Banker RD, Charnes A, Cooper WW (1984) Some models for estimating technical and scale inefficiencies in data envelopment analysis. Manag Sci 30(9):1078–1092
6. Charnes A, Cooper WW, Golany B, Seiford L, Stutz J (1985) Foundations of data envelopment analysis for Pareto-Koopmans efficient empirical productions functions. J Econ 30(1):91–107
7. Tone K (2001) A slacks-based measure of efficiency in data envelopment analysis. Eur J Oper Res 130(3):498–509
8. Agarwal S, Yadav SP, Singh SP (2011) A new slack DEA model to estimate the impact of slacks on the efficiencies. Int J Oper Res 12(3):241–256
9. Banker RD, Chang H, Cooper WW (1996) Simulation studies of efficiency, returns to scale and misspecification with nonlinear functions in DEA. Ann Oper Res 66(4):231–253
10. Zadeh LA (1996) Fuzzy sets. In: Zadeh LA (ed) Fuzzy sets, fuzzy logic, and fuzzy systems: selected papers, pp 394–432
11. Emrouznejad A, Tavana M (eds) (2013) Performance measurement with fuzzy data envelopment analysis, vol 309. Springer, Berlin
12. Wendell RE (1985) The tolerance approach to sensitivity analysis in linear programming. Manag Sci 31(5):564–578
13. Kao C, Liu ST (2000) Fuzzy efficiency measures in data envelopment analysis. Fuzzy Sets Syst 113(3):427–437
14. Lertworasirikul S, Fang SC, Joines JA, Nuttle HL (2003) Fuzzy data envelopment analysis (DEA): a possibility approach. Fuzzy Sets Syst 139(2):379–394
15. Puri J, Yadav SP (2013) A concept of fuzzy input mix-efficiency in fuzzy DEA and its application in banking sector. Expert Syst Appl 40(5):1437–1450

16. Hekmatnia M, Allahdadi M, Payan A (2016) Solving fuzzy slack-based measure of efficiency model by possibilistic programming approach. Int J Oper Res 27(3):502–512
17. Sani MR, Alirezaee M (2017) Fuzzy trade-offs in data envelopment analysis. Int J Oper Res 30(4):540–553
18. Wanke P, Barros CP, Emrouznejad A (2018) A comparison between stochastic DEA and fuzzy DEA approaches: revisiting efficiency in Angolan banks. RAIRO-Oper Res 52(1):285–303
19. Bakhtavar E, Yousefi S (2019) Analysis of ground vibration risk on mine infrastructures: integrating fuzzy slack-based measure model and failure effects analysis. Int J Environ Sci Technol 16(10):6065–6076
20. Lertworasirikul S, Fang SC, Joines J, Nuttle H (2003) Fuzzy data envelopment analysis: a credibility approach. Fuzzy Sets Based Heuristics Optim 126:141–158
21. Guo P, Tanaka H, Inuiguchi M (2000) Self-organizing fuzzy aggregation models to rank the objects with multiple attributes. IEEE Trans Syst Man Cybern Part A Syst Hum 30(5):573–580
22. Liu B, Liu YK (2002) Expected value of fuzzy variable and fuzzy expected value models. IEEE Trans Fuzzy Syst 10(4):445–450
23. Liu B (2007) Uncertainty theory. Springer, Berlin, pp 205–234
24. Liu ST (2008) A fuzzy DEA/AR approach to the selection of flexible manufacturing systems. Comput Ind Eng 54(1):66–76
25. Mahla D, Agarwal S (2021) A credibility approach on fuzzy slacks based measure (SBM) DEA model. Iran J Fuzzy Syst 18(3):39–49
26. Mahla D, Agarwal S, Mathur T (2021) A novel fuzzy non-radial data envelopment analysis: an application in transportation. RAIRO-Oper Res 55(4):2189–2202
27. Arana-Jiménez M, Sánchez-Gil MC, Lozano S (2022) A fuzzy DEA slacks-based approach. J Comput Appl Math 404:113180
28. Ministry of Commerce and Industry, Government of India, India Brand Equity Foundation, viewed 30 Apr 2019. https://www.ibef.org/industry/oil-gas-india.aspx
29. Klir G, Yuan B (1995) Fuzzy sets and fuzzy logic, vol 4. Prentice Hall, New Jersey, pp 1–12
30. Liu B (2009) Some research problems in uncertainty theory. J Uncertain Syst 3(1):3–10

Measuring Efficiency of Hotels and Restaurants Using Recyclable Input and Outputs

Neha Sharma and Sandeep Kumar Mogha

Abstract During the time of COVID-19, the growth of the Hotels and Restaurants (H&Rs) sector fall down throughout the world. The overall performance of this sector falls during this challenging time. Therefore, this sector has to adopt new strategies to make the best use of the opportunities and counter the challenges in this crucial time. One of these challenges is how to assess the performance of H&Rs based on multiple criteria. Keeping this in view, this paper attempts to evaluate the efficiency of 45 large-scale H&R companies operating in India using the data envelopment analysis (DEA) technique. DEA-based CCR and BCC models are employed to evaluate the efficiency on the bases of recyclable input and output variables. The study is carried out with four input and two output variables. On an average, 78.30% H&Rs are found to be technically efficient (TE) which suggests a possible improvement of 21.70%. The stability of results is also carried out using recyclable input and output that suggest the importance of the used variables for the overall improvement of the efficiency of the deployed H&Rs.

Keywords Efficiency measurement · Data envelopment analysis · Hotel and restaurants

1 Introduction

The hospitality business in India has been growing at a moderate rate for the past few years, and it has a lot of scopes to grow even more in the future. Tourism is a significant source of foreign exchange in India like in many other countries. From the year 2016 to 2019, foreign exchange earnings have increased at a CAGR of 7% but fell in 2020 because of the COVID-19 epidemic. The tourist industry in India has employed 39 million people in FY20, accounting for 8.0% of the country's total workforce. By 2029, it is expected that 53 million jobs will be generated in this sector. In 2019, India ranked 10th out of 185 nations in terms of travel and

N. Sharma (✉) · S. K. Mogha
Department of Mathematics, Chandigarh University, Mohali, Punjab, India
e-mail: nsm250890@gmail.com

M. Thakur et al. (eds.), *Soft Computing for Problem Solving*, Lecture Notes in Networks and Systems 547, https://doi.org/10.1007/978-981-19-6525-8_44

tourism's total contribution to GDP, according to the World Travel and Tourism and Travel Council (WTTC). The contribution of travel and tourism to GDP in 2019 was 6.8% of total GDP or Rs. 13,68,1000 million. The government has made significant attempts to increase investment in the hospitality sector. The automatic route allows 100% Foreign Direct Investment (FDI) in the hotel and tourism sectors. To improve India's competitiveness as a tourism destination, the government decreased GST on hotel rooms with rates between Rs 1,001 (US$ 14.3) and Rs 7,500 (US$ 107.3) per night to 12% and those over Rs 7,501 (US$ 107.3) to 18% in 2019. The Hospitality Development and Promotion Board was established by the Ministry to monitor and assist hotel project clearances and approvals.

The competition in the H&R industry is a result of two aspects: changing customer profiles and technological adaptation. H&R who uses the resources in an organized manner are more successful. Such H&R sets the benchmarks for the other competitors. To identify these efficient and inefficient H&Rs, this paper applies the CCR and BCC model of DEA. CCR model deals with constant return of scale (CRS), whereas BCC Model deals with the variable return of scale (VRS). Analysis has been done on the basis of efficiency scores attained by different H&R under these two models. The result of efficiency has been analyzed with respect to each input and output variable.

The paper unfolds as follows: Sect. 2 includes the literature review followed by a brief description of DEA in Sect. 3. Section 4 presents the research design along with statics of data, and an explanation of the data and computation with overall performance score is given in Sect. 5. Under Sect. 6, recyclable input–output analysis has been done, and in the last section, a conclusion is given.

2 Literature Review

In the tourism sector, several studies examined the performance of the hotel industry in different countries in terms of efficiency by using different techniques which are given in the articles [1–3]. The studies relevant to the H&R sector have been reviewed here. Wu [4] used the non-radical DEA model to measure the efficiency of H&R companies in Taipei. The total number of employees, total number of the guest room, the total area of the food and beverages department, and total operating cost were considered as inputs, whereas room revenues, food and beverages revenue, and other revenues are considered as outputs. Avkiran [5] demonstrated data envelopment analysis of cross-sectional data to check the efficiency of hotels by keeping full-time staff members, permanent, part-time staff members, and the number of beds as inputs and revenue, and room rent as output variable. The results indicate that to become efficient, the number of beds and part-time staff can be reduced and revenue can be increased. Assaf and Agbola [6] adopted the DEA double bootstrap approach to examine the performance of Australian Hotels. This study suggests that to improve the efficiency, government should provide tax breaks for small and medium-sized hotels and invest in technology. Barros [7] investigated the technical progress of the Portuguese hotel industry by using the stochastic cost frontier method which indicates that to increase the level of efficiency hotels should focus on productivity and increase foreign exchange. Lado-Sestayo and Fernandez-castro [8] examine the impact of location on hotel efficiency by applying a four-stage DEA model over a sample of 4000 hotels with the novel feature of addressing location at a tourist destination level. This research helps in hotel management and policy making. Gossling [9] examined the influence of COVID-19 on the hospitality industry. They found that various types of restrictions during this pandemic had put a negative effect on the hotel and restaurant industry. Gursoy and Chi [10] showed the negative effect of COVID-19 on the hospitality industry. According to the study, a very small percentage of consumers choose to eat at a sit-down restaurant, travel to a destination, and stay in a hotel because of the COVID-19.

There is no sector that is untouched by the bad effect of the Covid-19 pandemic. Indian Hotel and restaurant industry have seen a downfall during this time but there are few types of research on the effectiveness of Indian hotels and restaurants. In this study, performance of H&R companies in India has been analyzed using the DEA-based models. Recyclable input–output approach has been adapted for the identification of improvement area of different H&R. DEA have applications in different sectors also except H&Rs and can be seen in Agarwal et al. [11, 12], Mogha et al. [13, 14], Tyagi et al. [15], Das [16], Ramaiah and Jayasankar [17]

3 Data Envelopment Analysis

DEA is a data-oriented, multi-factor performance measurement technique that is based on the linear programming problem of operations research. It originated in the decade of 1950s with the definition by Pareto Koopman extended by Farrell [18] with the idea of a frontier analysis approach to evaluate the efficiency of different DMU. However, the frontier approach is bound only for two inputs and one output case and vice-versa. Later on, a vast number of studies have been done on DEA literature, causing significant growth in its methodology and application in the real world. The first DEA model was developed by Charnes et al. [19] known as the CCR model with constant returns to scale (CRS) which was extended by Banker et al. [20] and Cooper to variable returns (VRS) to scale in 1984, known as the BCC model. The description of CCR and BCC model is described below.

3.1 DEA Model

Assume that there are n DMUs, each of which uses m inputs and generates s outputs. The input variables are denoted by x_{ik}, outputs denoted as y_{rk}, and corresponding weights of input and output by u_{ik} and v_{rk}. The amount of ith input used by kth DMU is denoted as x_{ik}, y_{rk} indicates the amount of rth output used by the kth DMU, u_{ik} is the weight assigned to the ith input of the kth DMU, v_{rk} is the weight assigned to the rth output of the kth DMU. The efficiency of the kth DMU is defined as the ratio of virtual output to virtual input. Mathematically, it is given by:

$$E_k = \frac{\sum_{r=1}^{s} v_{\text{rk}} y_{\text{rk}}}{\sum_{i=1}^{m} u_{\text{ik}} x_{\text{ik}}}, k = 1, 2, 3, \ldots n, \text{ where } E_k \in [0, 1]$$

Fractional programming can be written as

$$\max\ E_k = \frac{\sum_{r=1}^{s} v_{\text{rk}} y_{\text{rk}}}{\sum_{i=1}^{m} u_{\text{ik}} x_{\text{ik}}}$$

$$\text{s.t.} = \frac{\sum_{r=1}^{s} v_{\text{rk}} y_{\text{rj}}}{\sum_{i=1}^{m} u_{\text{ik}} x_{\text{ij}}} \le 1$$

$$u_{\text{ik}}, v_{\text{rk}} \ge 0, \forall i = 1, 2, \ldots m, r = 1, 2, \ldots s, j = 1, 2, \ldots n$$

After normalizing the numerator, we get the input minimization CCR model as follows:

$$\min\ E_k = \sum_{i=1}^{m} u_{\text{ik}} x_{\text{ik}}$$

$$s.t.\ \sum_{i=1}^{m} v_{\text{rk}} y_{\text{rk}} = 1$$

$$\sum_{i=1}^{m} u_{\text{ik}} x_{\text{ij}} - \sum_{r=1}^{s} v_{\text{rk}} y_{\text{rj}} \ge 0 \forall j = 1, 2, 3, \ldots n$$

$$u_{\text{ik}}, v_{\text{rk}} \ge \varepsilon \forall i = 1, 2, 3, \ldots m; r = 1, 2, 3, \ldots s$$

The dual of this problem is called output-oriented CCR Model, given by

$$\max\ E_k = \phi_k + \varepsilon \left(\sum_{r=1}^{s} s_{\text{rk}}^{+} + \sum_{i=1}^{m} s_{\text{ik}}^{-} \right)$$

$$\text{s.t.}\ \sum_{j=1}^{n} \lambda_{\text{jk}} y_{\text{rj}} - s_{\text{rk}}^{+} = \phi_k y_{\text{rk}} \forall r = 1, 2, 3, \ldots, s,$$

$$\sum_{j=1}^{n} \lambda_{\text{jk}} x_{\text{ij}} - s_{\text{ik}}^{-} = x_{\text{ik}} \forall i = 1, 2, 3, \ldots, m,$$

$$\lambda_{\text{jk}}, s_{\text{ik}}^{-}, s_{\text{rk}}^{+} \ge 0 \quad \forall i = 1, 2, 3, \ldots, m;\ j = 1, 2, 3 \ldots, n; r = 1, 2, 3, \ldots, s$$

And ϕ_k is unrestricted in sign, ε denotes non-Archimedean constant, λ_{jk} is a dual variable corresponding to the jth constraint and is called as intensity variable, s_{rk}^{+} is the slack in the rth output of kth DMU, s_{ik}^{-} is the slack in the ith input of kth DMU, along with these conditions, we take one more condition $\sum_{j=1}^{n} \lambda_{jk} = 1, \forall j = 1, 2, 3, \ldots, n$ then this model become output-oriented BCC model.

In this study, performance of the H&R industry has been analyzed with the output-oriented CCR model as main motive of this industry is to maximize the output with the given number of inputs. To check the impact of scale and size on H&Rs, the BCC model is also applied.

4 Research Design

4.1 Selection of DMUs

The first step of DEA analysis is to identify the set of DMUs, based on their homogeneity. A set of DMU can be considered homogeneous if the objective, working process, and aim are similar. Secondly, the number of inputs and outputs and DMUs are selected with the assumption 'number of DMU should be more than the product of the number of input and number of outputs' for better results. In the present study, a total of 45 large-scale H&Rs with four inputs and two outputs have been considered.

4.2 Data and Variables

Based on variables used in earlier studies and as per the availability of data, four inputs, namely capital employed, gross fixed assets, current assets, and operating costs are considered in the study, whereas two outputs operating income and profit before depreciation, interest, and tax (PBDIT) are considered. The data for the study has been collected from the Prowess database of the Centre for Monitoring Indian Economy (CMIE). The descriptive statistics of data and variables are given in Table1.

The analysis is carried out with the help of DEA solver software. The variation in resources used and output generation demonstrates that these H&Rs have used their resources to varying degrees. Along with this maximum and minimum values are also there in this table for more clarity.

Table 1 Statistic on input and output variables

	Total capital	Net fixed assets	Current assets	Total expenses	Total income	PBDITA
Max	13,926.5	26,915.2	3187.2	13,935.8	14,152.4	1341.6
Min	7.8	0.1	41	17.1	80.2	30.1
Average	1562.88	4214.96	698.57	1908.02	1832.80	367.70
SD	2811.86	4499.49	746.00	2562.54	2572.28	370.59

5 Results and Discussions

The efficiency score (OTE, PTE, and SE) of 45 private ltd. H&Rs have been estimated for the year 2019–20. Table 2 represents the efficiency score along with rank, reference set, peer count, and peer weights of the H&Rs. The DEA analysis also evaluates the reference set of H&R which construct the efficiency frontier. The H&R achieving an efficiency score equal to one constitute the efficiency frontier and the remaining are considered inefficient.

5.1 Overall Performance of H&R

For the overall performance of the H&R, four inputs and two outputs are considered under main model. CRS and VRS assumptions are taken for the analysis with an output-oriented approach. This analysis indicates that out of 45 H&R only nine DMUs ($H3, H9, H13, H15, H17, H20, H26, H28$, and $H34$) is technically efficient, so these 9 H&R are relatively efficient than the other 36 H&R and are forming efficiency frontier. Average efficiency scores come out as 0.783% which indicates on average inefficient H&R can limit their resources and expand output by 22%. From the ranks, it is clear that $H24$ is the most inefficient H&R out of 36 inefficient. $H33, H8, H12, H30, H16, H31, H1, H10, H44, H40$, and $H19$ are having efficiency scores more than average efficiency, whereas $H24$, $H5$, $H38$, $H25$, $H7$, $H29$, $H35$, $H14$, $H32$, $H4$, $H22$, $H23$, $H41$, $H18$, $H21$, $H2$, $H11$, $H43$, $H27$, $H42$, $H36$, $H39$, $H6$, $H37$, and $H45$ are having less efficiency score then the average efficiency. $H15$ is having maximum peer count 35 so it is considered as the most technically efficient.

Under VRS, 15 H&R is having efficiency score of 1 so these DMUs are considered the relatively efficient as compared to the remaining 30 H&R. Under VRS also $H15$ is having a maximum peer count of 26, so this is considered the most efficient H&R, whereas $H28$, $H39$, and $H45$ are not the part of reference set for any of the inefficient H&R as peer count for these H&R is zero. Only 12 H&R are forming the VRS efficient frontier. One interesting fact is that there are 6 such H&R those were inefficient under CRS but under VRS assumption they come out as efficient ones. For Example, $H39$ has a relatively low CRS score (0.7627) but the VRS score is one. This indicates that this H&R can convert its inputs into output efficiently, but due to the disadvantageous size, its technical efficiency is low.

A total of 9 H&R's ($H3$, $H9$, $H13$, $H15$, $H17$, $H20$, $H26$, $H28$, and $H34$) are scale efficient as they have scale efficiency one which means these H&R are working on optimal scale and size of these H&R 's will not put any negative impact on their performance. A total of 36 H&R is scale inefficient which indicates that either these H&R are too small or too big. H&R '$H9$' has the lowest scale efficiency among all H&R.

This study with the help of slacks suggests the average reduction in input variables and augmentation in output variables. The average reduction in total capital, net

Table 2 Efficiency score, reference set, peer weight, peer count, and rank calculated by CCR and BCC models

	CCR results					BCC results					
Code	OTE (CRS)	Reference set	Peer weights	Peer count	Rank	PTE (VRS)	Reference set	Peer weights	Peer count	Rank	S.E
*H*1	0.829	*H*9, *H*13, *H*15	0.22, 1.049, 0.372		16	0.909	*H*9, *H*13, *H*15	0.24, 0.09, 0.671		20	0.912
*H*2	0.723	*H*9, *H*15	0.119, 1.289		30	0.770	*H*9, *H*15	0.179, 0.821		28	0.940
*H*3	1.000	*H*3	1	3	1	1.000	*H*3	1	1	1	1.000
*H*4	0.668	*H*15, *H*20	0.644, 2.152		36	0.710	*H*15, *H*20	0.661, 0.339		36	0.940
*H*5	0.435	*H*9, *H*15	0.077, 0.431		44	0.437	*H*9, *H*15, *H*20	0.085, 0.362, 0.554		45	0.996
*H*6	0.765	*H*15, *H*26, *H*34	1.614, 0.518, 0.125		23	0.870	*H*3, *H*9, *H*15	0.103, 0.177, 0.72		21	0.879
*H*7	0.599	*H*15, *H*20	1.499, 2.705,		41	0.695	*H*9, *H*15	0.076, 0.924		38	0.863
*H*8	0.933	*H*9, *H*13, *H*15	0.023, 0.594, 0.400		11	0.938	*H*9, *H*13, *H*15	0.024, 0.57, 0.406		18	0.995
*H*9	1.000	*H*9	1	15	1	1.000	*H*9	1	15	1	1.000
*H*10	0.827	*H*15, *H*20, *H*26	1.336, 4.428, 0.066		17	0.970	*H*9, *H*15	0.061, 0.939		17	0.852
*H*11	0.742	*H*3, *H*15	0.033, 0.409		29	0.913	*H*12, *H*15, *H*16	0.343, 0.24, 0.416		19	0.812
*H*12	0.899	*H*15, *H*17	0.13, 0.154		12	1.000	*H*12	1	2	1	0.899
*H*13	1.000	*H*13	1	7	1	1.000	*H*13	1	7	1	1.000
*H*14	0.651	*H*15, *H*20	0.123,0.651		38	0.658	*H*9, *H*15, *H*20	0.004, 0.092, 0.905		40	0.989

(continued)

Table 2 (continued)

	CCR results					BCC results					
Code	OTE (CRS)	Reference set	Peer weights	Peer count	Rank	PTE (VRS)	Reference set	Peer weights	Peer count	Rank	S.E
*H*15	1.000	*H*15	1	35	1	1.000	*H*15	1	26	1	1.000
*H*16	0.887	*H*3, *H*15	0.043,0.086		14	1.000	*H*16	1	1	1	0.887
*H*17	1.000	*H*17	1	4	1	1.000	*H*17	1	1	1	1.000
*H*18	0.709	*H*15, *H*20, *H*26	0.052, 0.845, 0.152		32	0.711	*H*15, *H*20, *H*26	0.06, 0.8, 0.14		35	0.997
*H*19	0.791	*H*9, *H*13, *H*15	0.038, 0.307, 0.042		20	0.824	*H*9, *H*13, *H*20	0.02, 0.481, 0.499		23	0.960
*H*20	1.000	*H*20	1	15	1	1.000	*H*20	1	23	1	1.000
*H*21	0.723	*H*15, *H*20	0.410, 2.06		31	0.780	*H*15, *H*20	0.425, 0.575		26	0.927
*H*22	0.671	*H*9, *H*13, *H*15	0.027,0.134,0.048		35	0.737	*H*13, *H*20	0.383, 0.617		34	0.911
*H*23	0.674	*H*15, *H*20	0.287, 2.429		34	0.758	*H*15, *H*20	0.303, 0.697		32	0.889
*H*24	0.421	*H*15, *H*20	0.123, 2.292		45	0.561	*H*15, *H*20	0.137, 0.863		43	0.751
*H*25	0.563	*H*9, *H*15	0.017, 0.194		42	0.573	*H*9, *H*15, *H*20	0.03, 0.082, 0.888		42	0.981
*H*26	1.000	*H*26	1	7	1	1.000	*H*26	1	5	1	1.000
*H*27	0.745	*H*9, *H*15	0.04, 0.389		27	0.750	*H*9, *H*15, *H*20	0.049, 0.309, 0.642		33	0.994
*H*28	1.000	*H*28	1	0	1	1.000	*H*28	1	0	1	1.000

(continued)

Table 2 (continued)

	CCR results					BCC results					
Code	OTE (CRS)	Reference set	Peer weights	Peer count	Rank	PTE (VRS)	Reference set	Peer weights	Peer count	Rank	S.E
*H*29	0.635	*H*15, *H*20, *H*26	0.162, 1.155, 0.119		40	0.647	*H*15, *H*20, *H*26	0.234, 0.759, 0.007		41	0.982
*H*30	0.890	*H*9, *H*13	0.033, 0.372		13	0.989	*H*9, *H*13, *H*20	0.004, 0.55, 0.447		16	0.900
*H*31	0.841	*H*15, *H*26, *H*34	0.247, 0.602, 0.099		15	0.842	*H*15, *H*20, *H*26, *H*34	0.237, 0.062, 0.6,0.101		22	0.999
*H*32	0.655	*H*9, *H*13, *H*15	0.015, 0.029, 0.143		37	0.690	*H*9, *H*13, *H*15, *H*20, *H*33	0.014, 0.09, 0.04, 0.79, 0.05		39	0.949
*H*33	0.992	*H*9, *H*13, *H*15	0.01, 0.256, 0.263		10	1.000	*H*33	1	2	1	0.992
*H*34	1.000	*H*34	1	2	1	1.000	*H*34	1	2	1	1.000
*H*35	0.638	*H*15, *H*20	0.307, 0.431		39	0.764	*H*12, *H*15, *H*20, *H*44	0.285, 0.102, 0.372, 0.241		30	0.835
*H*36	0.749	*H*15, *H*20	0.28, 1.622		25	0.801	*H*15, *H*20	0.289, 0.711		24	0.936
*H*37	0.765	*H*15, *H*20, *H*26	0.735, 0.029, 0.022		22	0.770	*H*15, *H*20, *H*26, *H*34	0.695, 0.282, 0.015, 0.008		27	0.993
*H*38	0.489	*H*15, *H*20	0.303, 1.623		43	0.522	*H*15, *H*20	0.312, 0.688		44	0.938
*H*39	0.763	*H*3, *H*15, *H*17	0.002, 0.074, 0.281		24	1.000	*H*39	1	0	1	0.763

(continued)

Table 2 (continued)

	CCR results					BCC results					
Code	OTE (CRS)	Reference set	Peer weights	Peer count	Rank	PTE (VRS)	Reference set	Peer weights	Peer count	Rank	S.E
*H*40	0.792	*H*15, *H*20, *H*26	0.386, 0.601, 0.349		19	0.797	*H*15, *H*20, *H*26	0.441, 0.296, 0.263		25	0.994
*H*41	0.684	*H*15, *H*20	0.467, 0.875		33	0.696	*H*15, *H*20	0.471, 0.529		37	0.984
*H*42	0.745	*H*9, *H*15	0.015, 0.376		26	0.759	*H*9, *H*15, *H*17, *H*20	0.015, 0.276, 0.063, 0.646		31	0.983
*H*43	0.744	*H*9, *H*15	0.054, 0.111		28	0.767	*H*9, *H*13, *H*20, *H*33	0.053, 0.064, 0.811, 0.071		29	0.970
*H*44	0.827	*H*9, *H*15, *H*17	0.014, 0.251, 0.038		17	1.000	*H*44	1	1	1	0.827
*H*45	0.770	*H*9, *H*15, *H*17	0.024, 0.068, 0.037		21	1.000	*H*45	1	0	1	0.770
Mean	0.783					0.836					0.937

Table 3 Average reduction in input and increment in output for improving overall efficiency

Total capital	Net fixed assets	Current assets	Total expenses	Total income	PBDITA
1075.051	1893.227	176.441	28.067	0.851	118.514

fixed assets, current assets, and total expenses is 1075.051, 1893.227, 176.441, and 28.067, respectively, and the average augmentation in output variables total income and PBDITA is 0.851 and 118.514, respectively. Table 3 gives this reduction and increment in different inputs and outputs.

6 Post-DEA Analysis

Under DEA, there is a possibility for a DMU to be efficient if that is operating exceptionally well in one area and performing below average in others, however, such DMUs cannot be the part of any reference set. Therefore, if there is any DMU that was identified efficient previously, additional analysis has to be conducted termed as sensitivity analysis. If any DMU is having high peer count then that DMU is considered genuinely efficient but if some DMU has a low peer count, then further analysis is required.

6.1 Recyclable Input–output Analysis

The post-DEA analysis is carried out by considering recyclable inputs and outputs. In the first case, we have removed input variable total expenses, in the second, current assets, in the third, net fixed assets, and in the fourth case, we have removed total capital. Also, we have checked the variation in efficiencies by recycling the output variables. In the 5th case, we have dropped total income and, in the 6th, PBDITA. The average efficiency scores in different recyclable input–output cases are given in Table 4 and the comparison of these cases with main model is shown in Fig. 1.If we compare the overall average efficiency score of these models with the performance score of main model, we can observe that all these models are having less average efficiency scores than main model. There is a large difference between the average efficiency scores of models 2a, 2b, 2e, and main model, which indicates that the H&R is doing better in the given amount of total expenses, current assets, and the given level of output 'total income'. However, in the case of models 2c, 2d, and 2f, the difference in average efficiency score in comparison with main model is not much which shows the level of total capital, net fixed assets used, and amount of PBDITA generated are not much affecting the average efficiency score. This indicates that there is a need to work on the levels of these input and output variables to improve the overall efficiency score.

Table 4 Average efficiency score of recyclable input output models

Model 2a	Model 2b	Model 2c	Model 2d	Model 2e	Model 2f
0. 609	0.671	0.756	0.779	0.489	0.772

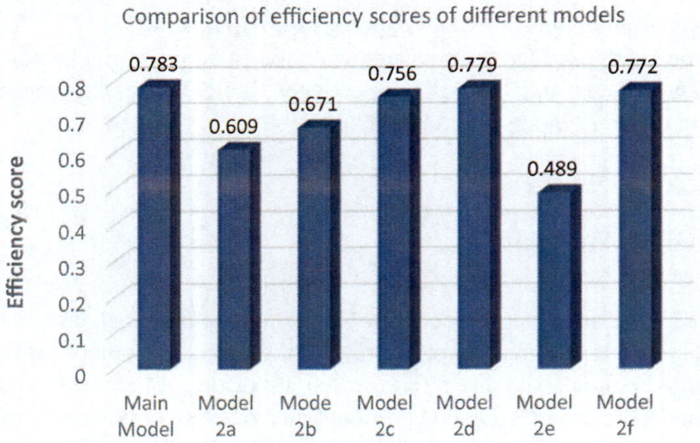

Fig. 1 Comparison of efficiency scores of different models

In this part by omitting all input and output one by one, robustness of the DEA result has been checked. Under this, six models are formed.

7 Conclusion

This paper examines the performance of selected large-scale H&R companies operating in India using DEA-based CCR and BCC models. Recyclable input/output criteria are used with different combinations of inputs and outputs. The study indicates that out of 45 H&Rs, only 9 (20%) are efficient and there is a huge scope for improvement for the rest of the inefficient H&Rs, an average OTE of 78.3% indicates that 21.7% of the potential of these H&R has not been used which shows that these H&R can raise the level of output by 21.7% with the existing level of inputs. The result of the BCC model shows that among 45 H&R only 15 (33.3%) is efficient, and the study suggests that on average remaining inefficient H&R may be able to augment their output level by 16.4% relative to the best H&R. Recyclable input–output approach indicates that there is an immense potential to reduce the level of current and net fixed assets and also to raise the level of PBDITA. Though there is a scope of improvement in the level of total expenses, total capital, and the total income also but that is not as much as the other remaining inputs and outputs.

References

1. Hsieh LF, Lin LH (2010) A performance evaluation model for international tourist hotels in Taiwan-an application of the relational network DEA. Int J Hosp Manag 29(1):14–24
2. Barros CP (2005) Measuring efficiency in the hotel sector. Ann Tour Res 32(2):456–477
3. Morey RC, Dittman DA (1997) An aid in selecting the brand, size and other strategic choices for a hotel. J Hosp Tour Res 21(1):71–99
4. Wu J, Tsai H, Zhou Z (2011) Improving efficiency in international tourist hotels in Taipei using a non-radial DEA model. Int J Contemp Hosp Manag 23(1):66–83
5. Avkiran NK (2002) Monitoring hotel performance. J Asia-Pacific Bus 4(1):51–66
6. Assaf AG, Agbola FW (2011) Modelling the performance of Australian hotels: a DEA double bootstrap approach. Tour Econ 17(1):73–89
7. Barros CP (2006) Analysing the rate of technical change in the Portuguese hotel industry. Tour Econ 12(3):325–346
8. Lado-Sestayo R, Fernández-Castro ÁS (2019) The impact of tourist destination on hotel efficiency: a data envelopment analysis approach. Eur J Oper Res 272(2):674–686
9. Gössling S, Scott D, Hall CM (2020) Pandemics, tourism and global change: a rapid assessment of COVID-19. J Sustain Tour 29(1):1–20
10. Gursoy D, Chi CG (2020) Effects of COVID-19 pandemic on hospitality industry: review of the current situations and a research agenda. J Hosp Mark Manag 29(5):527–529
11. Agarwal S, Yadav SP, Singh SP (2011) A new slack DEA model to estimate the impact of slacks on the efficiencies. Int J Oper Res 12(3):241–256
12. Agarwal S, Yadav SP, Singh SP (2014) Sensitivity analysis in data envelopment analysis. Int J Oper Res 19(2):174–185
13. Mogha SK, Yadav SP, Singh SP (2014) New slack model based efficiency assessment of public sector hospitals of Uttarakhand: State of India. Int J Syst Assur Eng Manag 5(1):32–42
14. Mogha SK, Yadav SP, Singh SP (2016) Estimating technical efficiency of public sector hospitals of Uttarakhand (India). Int J Oper Res 25(3):371–399
15. Tyagi P, Yadav SP, Singh SP (2009) Relative performance of academic departments using DEA with sensitivity analysis. Eval Program Plann 32(2):168–177
16. Das T (2021) Rating of State Co-operative Banks in India: a DEA approach. Bus Stud J 35(1):24–37
17. Ramaiah V, Jayasankar V (2022) Performance assessment of Indian electric distribution utilities using data envelopment analysis. DEA 11(3):192–202
18. Farrell MJ (1957) The measurement of productive efficiency http://www.jstor.org/stab. J R Stat Soc Ser A 120(3):253–290
19. Charnes A, Cooper WW, Rhodes E (1978) Measuring the efficiency of decision-making units. Eur J Oper Res 2(6):429–444
20. Banker RD, Charnes A, Cooper WW (1984) Some models for estimating technical and scale inefficiencies in data envelopment analysis. Manage Sci 30(9):1078–1092

Efficiency Assessment of an Institute Through Parallel Network Data Envelopment Analysis

Atul Kumar, Ankita Panwar, and Millie Pant

Abstract Conventional data envelopment analysis (DEA) is a non-parametric approach to examine the efficiency of similar decision-making units (DMUs) as a whole system without considering its internal structure, i.e., the system is considered as a black box. Therefore, a network DEA (NDEA) is needed to study the internal structure of a system. By allowing for the measurement of individual components, an NDEA model can reveal inefficiencies that a traditional DEA ignores. In this study, we use the parallel network DEA to calculate the efficiency of a higher education institute with 19 decision-making units (DMUs) using two parallel processes (teaching and research) and compare it with the conventional DEA, CCR model through a numerical example. The main advantages of the parallel NDEA model are (i) to identify which DMUs are inefficient and make necessary adjustments, and (ii) the parallel DEA model has a lower efficiency score than the traditional DEA model.

Keywords DEA · Parallel NDEA · CCR · Efficiency

1 Introduction

The linear programming approach of DEA is used to assess the performance of peer DMUs [1] and introduced by Charnes, Cooper, and Rhodes in 1978. The inputs (i/ps) and outputs (o/ps) are usually multiple but are common for every DMU in

A. Kumar · A. Panwar (✉) · M. Pant
Department of Applied Mathematics and Scientific Computing, Indian Institute of Technology Roorkee, Roorkee 247667, India
e-mail: apanwar@as.iitr.ac.in

A. Kumar
e-mail: atul_k@as.iitr.ac.in

M. Pant
e-mail: pant.milli@as.iitr.ac.in

M. Pant
Mehta Family School of Data Science and Artificial Intelligence, Indian Institute of Technology Roorkee, Roorkee 247667, India

M. Thakur et al. (eds.), *Soft Computing for Problem Solving*, Lecture Notes in Networks and Systems 547, https://doi.org/10.1007/978-981-19-6525-8_45

the DEA technique. The conventional DEA focuses on the initial input and final output information of the DMUs and considers each as a black box (ignoring the internal mechanism of DMUs) [2]. As the internal structure of DMUs is ignored, this approach may tend to inefficient scores or mislead the results. An NDEA approach is appropriate to study the internal structure of the system and examine the DMUs by making a network structure [3]. So, we can define that the DEA model to assess the performance of DMUs with a network structure is called NDEA. The NDEA approach proposed by Charnes et al. in 1986 when inputs and outputs of DMUs form a network structure [4]. To assess the efficiency of the DMUs, several NDEA models have been developed. Series and parallel network structures are two fundamental structures for NDEA models. Series NDEA is a generalization of a general two-stage network structure in which the system is divided into two stages, and the overall efficiency can be measured by taking the weighted average of their efficiencies [5]. In parallel NDEA, each DMUs is measured by two or more inter-dependent processes that run parallel to obtain the final result.

In this study, a parallel NDEA model is proposed to assess the efficiency of the DMUs (faculties) by considering two inter-dependent processes teaching and research in a parallel network structure [6]. A literature review of research employing the DEA approach has been conducted in this section of the study. Many studies in which DEA is used in various fields such as industry [7], automobiles [8], health [9], and agriculture [10] can be found in the review.

A literature review of research employing the network DEA approach has been conducted in this section of the study. Many authors used various NDEA model in the education sector. In this section, we have discussed how to apply NDEA models in higher education institutions.

Singh and Ranjan [11] analyzed the non-homogenous parallel sub-unit systems efficiency of the higher education institutions. Arya and Singh [12] developed the two-stage parallel-series model for the fuzzy dataset. Yang et al. [13] measured the 64 Chinese universities performance with respect to the research. They used the two-stage NDEA model. Kumar and Thakur [14] examined the performance of the various management institutes in India. They used the dynamic NDEA model in their study. Koronakos et al. [15] determined the research performance of the department of computer science in the UK through the network DEA model. Tavares et al. [16] used a multi-stage approach, network DEA model, and evaluated the efficiency of the higher education institutions of Brazil. Ding et al. [17] used a two-stage NDEA approach and assessed the research performance of Chinese universities.

The present study focuses on the performance of the teaching staff of the HEI through the parallel NDEA model. Through this model, we assess the efficiency of the DMUs and compare the results with the classical DEA model. The assessment is done concerning the teaching and research program. The aim of the parallel NDEA model is to identify the causes of inefficiency in process units and helping the decision maker to do the further improvement in the process units.

Teaching and research are the two essential parts of any higher education institution. The development and growth of any country are mainly dependent on its education system. So, improving the performance of higher education institutions is the primary concern of every country nowadays [18].

The study is summarized into five parts, including the introduction. Section 2 mainly focused on the methods. The methodology is described in the third section. The performance of the academic staff is done by the DEA and parallel NDEA is mentioned in the fourth section. The conclusion of the study is summarized in Sect. 5.

2 Methodology

This section discusses the different models used in this study. The models are the classical DEA model, CCR, and the parallel NDEA model under the output orientation.

2.1 CCR Model

The CCR model is the first DEA model to maximize the relative efficiency of DMUs. Sometimes it is also called the 'constant returns to scale (CRS)' [19] and was introduced by Charnes, Cooper, and Rhodes [1]. The mathematical structure of the CCR model is defined as follows.

$$
\begin{aligned}
E_i &= \max \sum_{s=1}^{S} u_s \text{Output}_{s,i} \\
&\text{s.t.} \sum_{r=1}^{m} v_r \text{Input}_{r,i} = 1 \\
&\sum_{s=1}^{S} u_s \text{Output}_{s,j} - \sum_{r=1}^{m} v_r \text{Input}_{r,j} \leq 0, \forall j = 1, \ldots, n \\
&u_s, v_r \geq 0; \forall s, r
\end{aligned} \tag{1}
$$

where E_i = efficiency of ith DMU, $\text{Output}_{s,i}$ = sth output of ith DMU, $\text{Input}_{r,i}$ = rth input of ith DMU, u_s = sth output multiplier, v_r = rth input multiplier, $i = 1, \ldots, n(\text{DMUs})$, $r = 1, \ldots, m(\text{inputs})$, $s = 1, \ldots, S(\text{outputs})$

If the score of the model is 100% or fully efficient, it is operating at the most productive scale size.

Figure 1 shows the structure of teaching staff (DMUs) considered in this study for assessment of the efficiency.

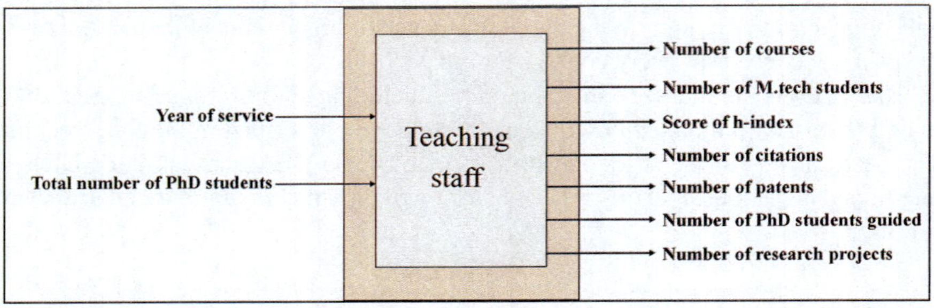

Fig. 1 Overall structure of the teaching staff

2.2 *Parallel Network DEA*

Teaching and research are the two essential components of a higher education institution. Every faculty contributes to teaching and research in the Institution. So the functioning of a faculty can be defined as a parallel network structure containing two processes teaching and research [20].

The input and output variables are identified for each process for the available data. We observed one input and two outputs in the first process unit (teaching) and one input and five outputs in the second process unit (research). Several indicators play an essential role in teaching and research in a higher education institute (HEIs).

Many authors studied the HEIs and used various input and output variables in their study [21–24].

According to the above literature review, we have selected the i/p and o/p variables for the present study. The total number of the i/p variables is two (Years of service, Total number of PhD. Students), and the number of the o/p variables is seven (Number of courses, Number of M.tech students taught, Score of *h*-index', 'Number of citations', 'Number of patents', 'Number of PhD students guided', and 'Number of research projects'). The details of the variables are discussed as follows.

Number of PhD students: This variable is representing the enrollments in PhD program(intake) under the guidance of a particular faculty (DMU) which is, one of the essential contributor to the development of research within the university.

Number of PhD students guided: This variable represents the number of Ph.D. degrees awarded by a Higher Education Institute (HEI) under the guidance of a particular faculty. We understand that a large amount of the scientific production developed in a HEI goes through research carried out by Ph.D. students which plays an important role to measure the efficiency of a faculty in research. So, this variable is taken as an output of a faculty.

Year of Service: Year of service is considered as a proxy variable with the availability of the data. M.tech students which are completed dissertations under the supervision of the faculty is taken as an output of that faculty. So, these variables play an essential role to measure the efficiency of the faculties.

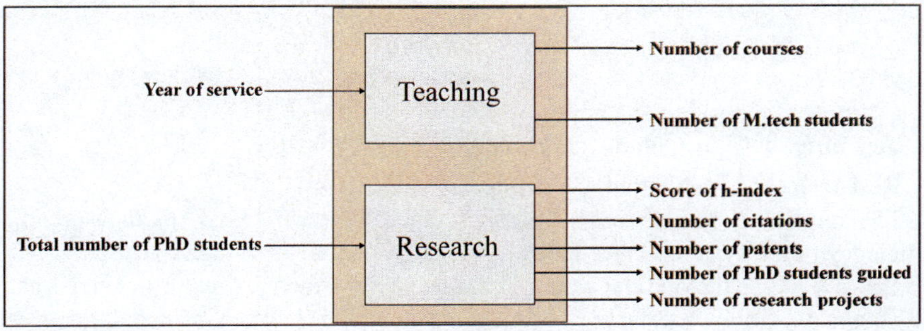

Fig. 2 Parallel structure of the teaching and research

Patents: This variable is an essential feature to assess the effectiveness of research work of a faculty that transform their theoretical aspect into innovation. This is basically an ownership of the researcher for their invention or a model. So it has significance in the measurement of the efficiency of the faculty.

***H*-index**: This variable permits the revealing of efficient researchers. It stimulates generation of quality publications that tend to have great impact on research, and analyze publication activity.

Citations: Research publications are important part to establish scientific discovery which is researcher's main objective. Therefore, higher education institute prefers to assess the performance of its researchers using publication and according citations. Thus citation is one of the most essential output variables to measure the efficiency of the researcher.

After that, the efficiency of each faculty is calculated by taking the weighted average of the efficiencies of teaching and research. So, a parallel network structure that involves two main processes (teaching and research) of a higher institution is formed in Fig. 2.

2.3 Mathematical Model of Parallel NDEA

To calculate the efficiency of the whole system as well as of process units, a mathematical model of parallel NDEA is developed. From Fig. 2, a parallel structure of two processes (teaching and research) contains rth input for kth process is $\text{Input}_{r,k}$ and sth output for kth process is $\text{Output}_{s,k,}$ where $r = 1, 2$ and $k = 1, 2$, for each DMU $i = 1, 2, \ldots, 19$. Let u_s and v_s be the output and input weights, respectively, then the efficiency of ith DMU using the parallel NDEA model is defined as follows.

$$E_i = \max \sum_{s=1}^{7} u_s \text{Output}_{s,i}$$

$$s.t. \sum_{r=1}^{2} v_r \text{Input}_{r,i} = 1 \tag{2}$$

Teaching: $\sum_{s=1}^{2} u_s \text{Output}_{s,j} - v_1 \text{Input}_{r,j} \leq 0, \quad j = 1, 2, \ldots, 19.$
Research: $\sum_{s=3}^{7} u_s \text{Output}_{s,j} - v_2 \text{Input}_{r,j} \leq 0, \quad j = 1, 2, \ldots, 19.$
The model (2) runs through 19 times, once for each DMU, to measure the efficiency of the systems as well as their processes units.

Let (u^*, v^*) be the optimal weights obtained by (2), then process efficiencies can be obtained as follows.

$$E_i^{(\text{Teaching})} = \frac{\sum_{s=1}^{2} u_s^* \text{Output}_{s,i}}{v_1^* \text{Input}_{1,i}}$$

$$E_i^{(\text{Research})} = \frac{\sum_{s=3}^{7} u_s^* \text{Output}_{s,i}}{v_2^* \text{Input}_{2,i}} \tag{3}$$

The overall aggregated input used and the overall aggregated output produced by the teaching staff are $\sum_{r=1}^{2} v_r^* \text{Input}_{r,i}$ and $\sum_{s=1}^{7} u_s^* \text{Output}_{s,i}$, respectively. Therefore, the faculty's efficiency is $\frac{\sum_{s=1}^{7} u_s^* \text{Output}_{s,i}}{\sum_{r=1}^{2} v_r^* \text{Input}_{r,i}}$ that is also the weighted average of teaching and research efficiencies which is defined as.

$$E_i^{(\text{overall})} = w_i^{(\text{Teaching})} * E_i^{(\text{Teaching})} + w_i^{(\text{Research})} * E_i^{(\text{Research})} \tag{4}$$

where

$$w_i^{(\text{Teaching})} = \frac{v_1^* * \text{Input}_{1,i}}{v_1^* * \text{Input}_{1,i} + v_2^* * \text{Input}_{2,i}},$$

$$w_i^{(\text{Research})} = \frac{v_2^* * \text{Input}_{2,i}}{v_1^* * \text{Input}_{1,i} + v_2^* * \text{Input}_{2,i}}$$

$$w_i^{(\text{Teaching})} = \text{Teaching weight for } i\text{th DMU}$$

$$w_i^{(\text{Research})} = \text{Research weight for } i\text{th DMU}$$

These (teaching and research) weights are input proportions [20].

The basic difference among parallel NDEA and the classical DEA is that the constraint for each unit has been replaced by those associated with its processes units. In model (2), the sum of the constraints (Teaching and Research) is equal to the constraint of model 1 (CCR). Also, the constraints in model (2) are stronger than in model (1) [25].

3 Problem Structure

The framework of the proposed methodology is shown in Fig. 3 and the structure for calculating the efficiency of the faculties through the basic DEA and parallel NDEA techniques is defined in the following steps.

Step 1: Decision-making units (DMUs): The total number of DMUs is 19. The teaching staff is considered as a DMUs.
Step 2: Input (i/p) and Output (o/p) variables: The input and output variables under the teaching and research management system are described in Tables 1 and 2.
Step 3: Calculation of relative efficiency: After selecting input and output variables according to the above step, calculate the relative efficiency of each DMUs using Eqs. 1–4.

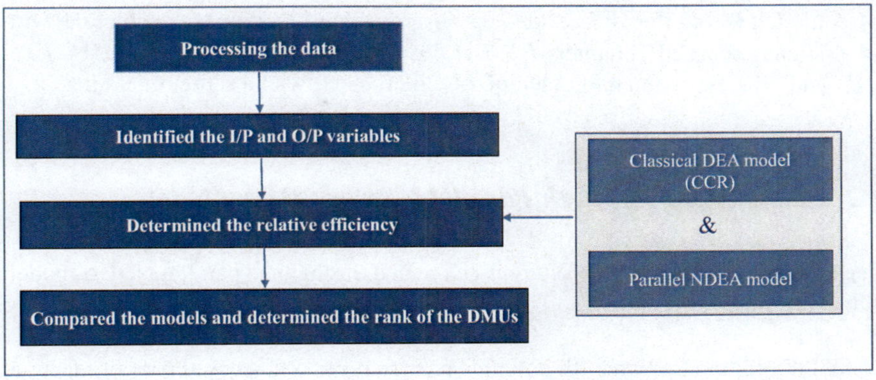

Fig. 3 The framework of the methodology

Table 1 i/p and o/p variables under teaching management

i/p variables	o/p variables
Years of service	Number of courses
	Number of M.Tech students

Table 2 i/p and o/p variables under research management

i/p variables	o/p variables
Total number of PhD. Students	Score of *h*-index
	Number of citations Number of patents Number of PhD students guided Number of research projects

Step 4: Analysis of results: The obtained results are analyzed and discussed in this step.

4 Results and Discussion

This section discusses the performance of the DMUs(faculties) measured by the conventional DEA model and parallel NDEA model with respect to the output orientation.

4.1 Assessment Through the Conventional DEA Model

The academic staff (faculty) measurement is done by the CCR model, a classical DEA model. Out of the 19 faculties, we observed that 13 faculties are efficient, with an efficiency score of 1, namely $F1$, $F2$, $F3$, $F4$, $F5$, $F6$, $F8$, $F11$, $F12$, $F14$, $F15$, $F16$, and $F19$. The remaining faculties are inefficient, with a score of less than 1.

4.2 Assessment Through Parallel Network DEA Model

The efficiency score of each faculty is calculated through the parallel NDEA model, which is defined by (2), (3), and (4). The efficiency of each faculty is given in Table 3.

In Table 3, the 2nd and 4th columns represent the efficiency scores of teaching and research, respectively, for each faculty. The overall efficiency measured by the weighted average of teaching and research efficiency scores is shown in column 6th. The 7th column represents the efficiency score of each faculty calculated by the conventional DEA model (CCR model) without considering teaching and research individually. In parallel NDEA technique, a faculty is overall efficient if it's both processes (teaching and research) are simultaneously efficient.

Five faculties are efficient in research with an efficiency score of 1, namely $F2$, $F4$, $F5$, $F6$, and $F8$. Also, the same faculties have higher overall efficiency scores, which are given in Table 3. On the other side, none of the faculty is efficient in teaching. Only faculty $F12$ has the highest teaching efficiency score (0.9561).

The difference between efficiency scores, in columns 6th and 7th in Table 3 and graphically shown in Fig. 4, represents that the parallel NDEA model is stronger than the basic DEA model because it is more capable of discerning between efficient faculties. This shows that a parallel NDEA model can enhance the discerning capacity of the basic DEA model with relatively fewer efficiency scores.

Therefore, the parallel NDEA model is more relevant to calculate the efficiency of faculties because it can better identify the place of inefficiency for each faculty.

Table 3 Efficiency measurement of the DMUs

DMUs	Teaching efficiency	Teaching weights	Research efficiency	Research weights	Overall efficiency	Basic DEA
*F*1	0.2458	0.0022	0.7712	0.9978	0.7700	1.0000
*F*2	0.2924	0.0027	1.0000	0.9974	0.9981	1.0000
*F*3	0.5364	0.0072	0.8695	0.9929	0.8700	1.0000
*F*4	0.3314	0.0012	1.0000	0.9969	0.9972	1.0000
*F*5	0.2974	0.0014	1.0000	0.9986	0.9990	1.0000
*F*6	0.1275	0.0024	1.0000	0.9976	0.9979	1.0000
*F*7	0.1912	0.0013	0.7410	0.9987	0.7403	0.8400
*F*8	0.2458	0.0011	1.0000	0.9989	0.9890	1.0000
*F*9	0.0682	0.0022	0.7038	0.9978	0.7024	0.7500
*F*10	0.0882	0.0021	0.8155	0.9979	0.8140	0.8700
*F*11	0.6605	0.0037	0.8157	0.9963	0.8200	1.0000
*F*12	0.9561	0.0010	0.8389	0.9990	0.8390	1.0000
*F*13	0.1912	0.0038	0.9211	0.9962	0.9183	0.9200
*F*14	0.1912	0.0027	0.6212	0.9973	0.6200	1.0000
*F*15	0.8621	0.9933	0.6137	0.0067	0.8600	1.0000
*F*16	0.3824	0.0032	0.8789	0.9997	0.8800	1.0000
*F*17	0.1730	0.0033	0.7830	0.9967	0.7810	0.8200
*F*18	0.1738	0.0035	0.8223	0.9965	0.8200	0.9200
*F*19	0.1478	0.0035	0.9428	0.9965	0.9400	1.0000

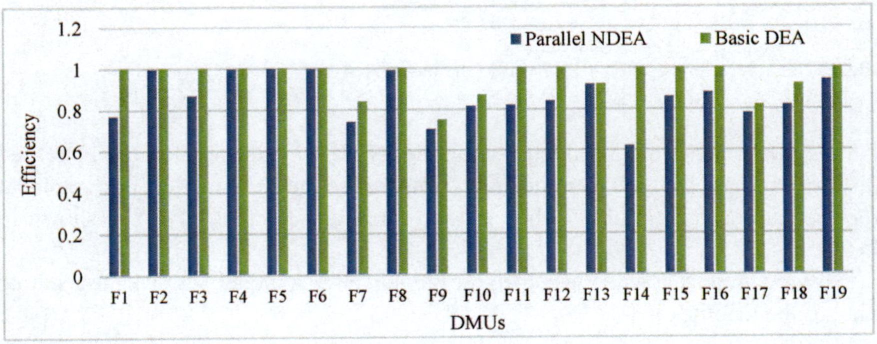

Fig. 4 Efficiency between parallel NDEA versus basic DEA

So, with this information, each faculty can distinguish which process requires more effort to improve the overall efficiency.

Table 4 represents the rank of the parallel DEA model and conventional DEA model (CCR model). Here we observe that the F5 has the best performance according to the parallel NDEA model, but under the conventional DEA model, total 13

Table 4 Rank of the DMUs

DMUs	Rank (parallel NDEA)	Rank (basic DEA)	DMUs	Rank (parallel NDEA)	Rank (basic DEA)
*F*1	16	1	*F*11	12	1
*F*2	2	1	*F*12	11	1
*F*3	9	1	*F*13	7	14
*F*4	4	1	*F*14	19	1
*F*5	1	1	*F*15	10	1
*F*6	3	1	*F*16	8	1
*F*7	17	17	*F*17	15	18
*F*8	5	1	*F*18	12	14
*F*9	18	19	*F*19	6	1
*F*10	14	16			

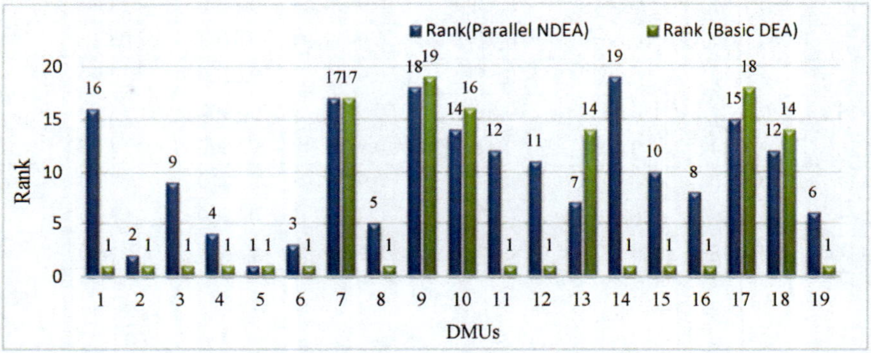

Fig. 5 Rank between parallel NDEA versus the basic DEA

DMUs performed well. Through the parallel NDEA model, we see the increased discriminating power between the units compared to the CCR model. The graphical representation of the ranks of the classical DEA and parallel NDEA is shown in Fig. 5.

So, we can say that with the help of the parallel NDEA model, we give the accurate rank of the faculties.

5 Conclusion

In this paper, we have used a parallel NDEA model to assess the efficiency of each faculty of a higher academic institute and compared it with the DEA-CCR model. In parallel NDEA, the system is called efficient if both the processes (teaching and research) are efficient. The overall efficiency of the system is calculated from the weighted average of teaching and research efficiencies of each DMU. This model shows the interrelation between system and processes efficiencies. That is why this encourages us to use the parallel NDEA model. We have also used the basic DEA model in this study and show the difference between the parallel NDEA and basic DEA model. Some essential remarks of the study are defined as follows.

- In the teaching program, we notice that the DMU F12 has the highest efficiency with a score of 0.9561, but DMU F9 has the least efficiency score with a value of 0.0682.
- In the research program, 5 DMUs are efficient ($F2$, $F4$, $F5$, $F6$, and $F8$) with efficiency value 1. The DMU $F15$ is the least efficient, with a score of 0.6137.
- DMU $F2$ has received the highest efficiency score (0.9981) under the overall NDEA model, and $F14$ is the least efficient DMU with a score of 0.6200.
- In the classical DEA model, out of the 19 DMUs, 13 DMUs are efficient (with an efficient score 1). But DMU $F9$ gives the least efficient with a score of 0.75.

In this study, the example interprets the usefulness of the parallel NDEA model for 19 DMUs (faculties) of an institution. The results demonstrate the efficiency scores of both processes for each faculty and identify the inefficiency in the processes, which helps for further improvement.

So, this method can be used as an alternative to assess the efficiency of an Institution or University.

References

1. Charnes A, Cooper WW, Rhodes E (1978) Measuring the efficiency of decision making units. Eur J Oper Res 2:429–444
2. Chen Y, Cook WD, Li N, Zhu J (2009) Additive efficiency decomposition in two-stage DEA. Eur J Oper Res 196:1170–1176
3. Ghafari Someh N, Pishvaee MS, Sadjadi SJ, Soltani R (2020) Performance assessment of medical diagnostic laboratories: a network DEA approach. J Eval Clin Pract 26:1504–1511
4. Charnes A, Cooper WW, Golany B, Halek R, Klopp G, Schmitz E, Thomas D (1986) Two-phase data envelopment analysis approaches to policy evaluation and management of army recruiting activities: tradeoffs between joint services and army advertising. Cent. Cybern. Stud. Univ. Tex.-Austin Austin Tex, USA
5. Kao C (2014) Network data envelopment analysis: a review. Eur J Oper Res 239:1–16
6. Park J, Lee BK, Low JMW (2020) A two-stage parallel network DEA model for analyzing the operational capability of container terminals. Marit Policy Manag 1–22
7. Tan Y, Despotis D (2021) Investigation of efficiency in the UK hotel industry: a network data envelopment analysis approach. Int J Contemp Hosp Manag
8. Jahangoshai Rezaee M, Yousefi S, Baghery M, Chakrabortty KR (2021) An intelligent strategy map to evaluate improvement projects of auto industry using fuzzy cognitive map and fuzzy slack-based efficiency model. Comput Ind Eng 151:106920
9. Tapia JA, Salvador B (2022) Data envelopment analysis efficiency in the public sector using provider and customer opinion: an application to the Spanish health system. Health Care Manag Sci

10. Pan W-T, Zhuang M-E, Zhou Y-Y, Yang J-J (2021) Research on sustainable development and efficiency of China's E-agriculture based on a data envelopment analysis-Malmquist model. Technol Forecast Soc Change 162:120298
11. Singh S, Ranjan P (2018) Efficiency analysis of non-homogeneous parallel sub-unit systems for the performance measurement of higher education. Ann Oper Res 269:641–666
12. Arya A, Singh S (2021) Development of two-stage parallel-series system with fuzzy data: a fuzzy DEA approach. Soft Comput 25:3225–3245
13. Yang G, Fukuyama H, Song Y (2018) Measuring the inefficiency of Chinese research universities based on a two-stage network DEA model. J Informetr 12:10–30
14. Kumar A, Thakur RR (2019) Objectivity in performance ranking of higher education institutions using dynamic data envelopment analysis. Int J Prod Perform Manag 68:774–796
15. Koronakos G, Chytilova L, Sotiros D (2019) Measuring the research performance of UK computer science departments via network DEA. In: 2019 10th international conference on information, intelligence, systems and applications (IISA), pp 1–7
16. Tavares RS, Angulo-Meza L, Sant'Anna AP (2021) A proposed multistage evaluation approach for higher education institutions based on network data envelopment analysis: a brazilian experience. Eval Program Plann 89:101984
17. Ding T, Yang J, Wu H, Wen Y, Tan C, Liang L (2021) Research performance evaluation of Chinese university: a non-homogeneous network DEA approach. J Manag Sci Eng 6:467–481
18. Lee BL, Johnes J (2021) Using network DEA to inform policy: the case of the teaching quality of higher education in England. High Educ Q
19. Panwar A, Tin M, Pant M (2021) Assessment of the basic education system of Myanmar through the data envelopment analysis. In: Singh D, Awasthi AK, Zelinka I, Deep K (eds) Proceedings of international conference on scientific and natural computing, vol 232. Springer, Singapore, pp 221–232
20. Kao C (2012) Efficiency decomposition for parallel production systems. J Oper Res Soc 63:64–71
21. Tang TL-P, Chamberlain M (2003) Effects of rank, tenure, length of service, and institution on faculty attitudes toward research and teaching: the case of regional state universities. J Educ Bus 79:103–110
22. Mammadov R, Aypay A (2020) Efficiency analysis of research universities in Turkey. Int J Educ Dev 75:102176
23. Popova O, Romanov D, Popov B, Karandey V, Kobzeva S, Evseeva M (2015) New methods and evaluation criteria of research efficiency. Mediterr J Soc Sci 6:212
24. Shamohammadi M, Oh D (2019) Measuring the efficiency changes of private universities of Korea: a two-stage network data envelopment analysis. Technol Forecast Soc Change 148:119730
25. Kao C (2009) Efficiency measurement for parallel production systems. Eur J Oper Res 196:1107–1112

Efficiency Measurement at Major Ports of India During the Years 2013–14 to 2018–19: A Comparison of Results Obtained from DEA Model and DEA with Shannon Entropy Technique

R. K. Pavan Kumar Pannala, N. Bhanu Prakash, and Sandeep Kumar Mogha

Abstract This paper measured the efficiency at selected container terminals in major ports of India during the period 2013–14 to 2018–19. Ports have proven to be a key node in growth of foreign trade and economic development. With growing containerization, shippers across the world are looking to container terminals that perform at better levels of efficiency and ensure greater cargo throughput. Terminals that operate with proven and consistent efficiency are preferred destinations for shippers. With growing dynamics in the maritime sector, numerous efficiency measurement techniques are evolving that are used by researchers. DEA has been a popular technique to measure efficiency of economic entities and numerous studies using these techniques have measured efficiencies of world ports. However, DEA technique is constrained by the possible number of variables to be selected against the number of decision making units (DMUs). Shannon entropy technique has proved to overcome this constraint in efficiency measurement. Therefore, this study has applied both DEA and Shannon entropy techniques and measured efficiencies of selected container terminals and found that the latter technique can overcome the constraint posed by DEA.

Keywords Efficiency · Data envelopment analysis · Shannon entropy method · Major ports of India

---

R. K. P. K. Pannala
Department of Mathematics, Sharda University, Greater Noida, India

N. Bhanu Prakash (✉)
School of Maritime Management, Indian Maritime University, Visakhapatnam, India
e-mail: bhanuprakashn@imu.ac.in

S. K. Mogha
Department of Mathematics, Chandigarh University, Punjab, India

M. Thakur et al. (eds.), *Soft Computing for Problem Solving*, Lecture Notes in Networks and Systems 547, https://doi.org/10.1007/978-981-19-6525-8_46

1 Introduction

Efficiency measurement of economic entities helps in devising plans for improvement. In the competitive business environment, performance evaluation among competing organizations helps to understand business practices followed at efficient entities in comparison with others. Globalization has bestowed an opportunity to business entities to expand their operations and converge with global trade. Initiatives of policy-makers that are aimed to boost foreign trade coupled with development of physical infrastructure had given impetus to global trade. Progression of global trade mandated development of strong transportation sector, especially the ports across the world.

Ports have, from time immemorial, acted as the nodal points of foreign trade and economic development. With shipping being a cost-effective transportation mode for bulk cargo, the demand for existence of robust ports can alone ensure smooth international cargo movement. Both shippers and carriers of the modern day look for ports that can efficiently handle their cargo safely and swiftly. Efficiency in operations is a major factor to improve outlook of a port and build confidence among the port users. Being nodal points for international trade, performance of ports had been focal area for various stakeholders including the international agencies, policy-makers, industry, port authorities, and researchers. Economic reforms paved way for private participation in the sector and had given stimulus for efficiency improvement at the port sector across the world.

Port sector in India was dominated by the public sector till reforms at major ports of India initiated in year 1995 [1] with an aim to accelerate their operational and financial efficiency. Port sector reforms backed by overall economic reforms have resulted in steady growth during the period 1995 to 2014 [2]. To enhance the efficiency levels, government of India privatized various key operations in a phased manner which included berth operations, yard operations and maintenance, stevedoring, crane operations, etc., at the major ports through PPP mode [1]. Privatization and supportive regulatory framework in the sector have heightened performance expectations [3] from the major ports of India. With a buoyant economic outlook and progressive scope for foreign trade, efficient performance of major ports that account for over 55% of India maritime trade would alone supplement the demands of growing Indian industry [4]. Therefore, assessment of efficiency improvements recorded by these ports especially after privatization would help the government in framing policies and strategies for impending future demands.

Measurement of operational and financial efficiency at seaports using a wide variety of modals including mathematical, econometric is attempted by numerous researchers. Research works aimed to measure efficiency improvements with the backdrop of reforms in the sector using both parametric and nonparametric techniques had proved positive results of privatization.

This paper measures efficiency levels reported at container terminals of selected major ports of India by integrating DEA with Shannon's entropy model. It is interesting to note that there exist limited number of studies on efficiency measurement at container terminals in Indian context. Traditionally, major ports of India maintained multi-purpose berths and terminals. The need for exclusive terminals to handle containers is still evolving, and therefore, studies on efficiency measurement of container terminals are virtually non-existent. Studies on Indian ports for long time have considered the overall performance indicators of ports to measure efficiency. Thus, this research paper and its results would fill the gap in dearth of studies covering efficiency measurement at container terminals of India. The paper is divided into five heads with the first part giving an introduction to the paper, the second part covering literature review. The third part will discuss the research methodology applied, the fourth giving analysis, and the fifth part detailing about the results, conclusions, and scope for further studies.

2 Literature Review

To have a clear picture on the existing literature on efficiency measurement at ports, this study reviewed seminal works both at International and Indian context and are presented as under.

2.1 Studies Covering International Ports

Studies on performance of seaports have been initiated lately with limited number of studies available vis-à-vis the other sectors of infrastructure [5]. Few of the authors [6–10] proposed inclusion of performance parameters for better assessment of port efficiencies.

Park and Prabir [11] reviewed existing literature on port efficiency measurement across the world. Their study divided overall efficiency into several stages by transforming inputs and outputs at each stage reflecting efficiencies according to production process and stage-wise role of inputs and outputs. Bichou and Gray [12] studied performance in logistics and supply chain management with a special reference to port sector. Their study traced that the concept of efficiency is vague and its measurement to be complex in port sector. It further concluded that port activities that extend into production, trading, and service industries make assessment more complicated.

Empirical studies covering container ports by Barros et al. [13–25] found that improvement in efficiency is possible with better strategies, financial discipline, growth of economy, planned reforms, infrastructural and technological investments, port location and accessibility, and improved service standards.

Few authors [26–28] linked efficiency at container ports to reform program and suggested for existence of strong support from government in the form of regulatory environment, reforms covering overall economy, and authors such as [23, 29] found that privatization without existence of strong regulatory environment cannot be a complete solution for improving port efficiency. Authors like [24, 30] compared the techniques of efficiency evaluation and suggested data envelopment analysis (DEA) to be a robust technique to trace efficiency.

Kaisar [31] assessed efficiency levels at selected ports of US and discovered a set of best practices for inefficient ports. The study found that sources and extent of inefficiencies that a port should concentrate for improving operations. Estache [16] studied the sources of efficiency improvement at Mexican terminals for the years 1996 to 1999. The study found constrained efficiency improvements at ports even after investments in improved technologies with a large gap between actual and expected performance levels. Weille [32] studied on capacity of port operations that can result in optimal net benefits to both the port authority and ship owners. They concluded that specialized berths would enhance the performance for ports. The study also suggested that reduction in service time can enhance efficiency levels.

Although studies measuring port efficiency have initiated during the 1970s, the number and type of variables taken by the researchers are mostly influenced by data availability and context of work done. While technical indicators of efficiency were considered by most of the researchers, some of the researchers had considered the financial aspects of port operations. By and large, researchers have traced efficiency improvements at the ports considered for evaluation. Most studies had traced positive correlation between privatization and efficiency at ports across the world. Some studies had also proved efficiency gains due to containerization.

2.2 Studies Covering Indian Ports

Studies covering Indian Ports, by Gaur et al. [33–37] found that efficiency of a port is not influenced by size of port but due to factors such as infrastructure setup, technological upgradation, investments into capacity buildings, business and operational acumen, investor-friendly policies, and ability to utilize resources. Rajasekar et al. [37] traced efficiency improvements at smaller ports due to long-term strategies and that size does not matter in improving port performance.

Jim Wu and Lin [34] found that India, bestowed with relative comparative advantage over many other industrialized nations, needs to overhaul its port infrastructure to accommodate its ever growing imports and exports.

Analysis of the above literature proves efficiency improvements at ports across the world, including India, due to growth in world trade, liberal policies of government, private participation, and strategic formulations of the port authorities. It can also be observed that while most of the studies at international level are focusing on container ports, studies in Indian context attempted to assess overall port efficiency improvements. Studies covering container terminals at Indian ports are almost non-existent. With growing containerization across the world, India needs to focus on assessing and improving efficiency at container terminals. This is more so important with a growing varieties of cargos being moved through containers and dedicated containers terminals being developed in India. Therefore, this study is aimed to fill the gap and an attempt to trace efficiency improvements at selected container terminals in India.

3 Research Methodology

While measuring efficiency levels of business entities, often taken as decision making units (DMUs) of sectors including banks, hospitals, seaports, airports, universities, etc., that experience multiple inputs and outputs, where the relative efficiencies cannot be determined using the simple output–input ratio methods. Data envelopment analysis (DEA), a popular nonparametric technique proposed by Charnes et al. [38] which was enhanced by Banker et al. [39], has been a popular technique used to measure efficiency levels for business entities that work on several input and output variables. However, the technique with its inherent restriction on the maximum possible number of entities that can be considered for evaluation posed a problem to the researchers. There exist several general rules regarding the total number of DMUs vis-à-vis the sum of inputs and outputs variables [40, 41]. When the number of DMUs does not satisfy this condition, most of the considered DMUs report to be efficient, even if they are really inefficient or vice versa. In these situations, identifying the most efficient DMU among the selected DMUs becomes a tedious task.

To overcome this anomaly, DEA is integrated with Shannon's entropy to rank the DMUs effectively [4, 42–44]. The present study had used [4] model to obtain the comprehensive efficiency scores (CES) and rank the DMUs.

3.1 Data Envelopment Analysis

DEA is associated with the concept of linear algebra, fractional linear programming, linear programming along with duality relations in linear programming. DEA's mathematical formulation is constructed as fractional linear programming problem by assigning weights to all the input and output variables. Maximize the weighted output to weighted input ratio while ensuring that equivalent ratios for each DMU are less than or equal to one in order to determine each DMUs efficiency. For the sake of computation ease, it could be converted to a linear programming problem.

Formulation of DEA and applications in various fields.

Determination of weights can be attained using the modals, constant return to scale (CRS), and variable return to scale (VRS) methods. Adoption of the model depends on proportional tuning of inputs and outputs. CRS model is used if the escalation of outputs proportion is based on escalating the same proportion of inputs, otherwise VRS model can be applied [45]. Formulation of CRS model is shown below for the selected kth DMU (DMU_k).

Model-1: max $h_k = \frac{\sum_{r=1}^{s} u_r y_{rk}}{\sum_{i=1}^{m} v_i x_{ik}}$

$$\text{Subjected to} \frac{\sum_{r=1}^{s} u_r y_{rj}}{\sum_{i=1}^{m} v_i x_{ij}} \leq 1 j = 1, 2, 3 \ldots n$$

$$u_r \geq 0, v_i \geq 0$$

where v_i represents weights to be determined for ith input, u_r represents the weights for determined of rth output, h_k represents relative efficiency of kth DMU, and m, s, and n refer to total of inputs, outputs, and DMUs, respectively.

DMU_k is taken as efficient if its relative efficiency is $h_k = 1$ and is considered as relatively inefficient if $h_k < 1$. The given fractional linear programming problem is converted into a linear programming (LP) problem by bringing each of the input to 1 and is represented as given below.

Model-2: max $h_k = \sum_{r=1}^{s} u_r y_{rk}$

$$\text{Subjected to} \sum_{i=1}^{m} v_i x_{ik} = 1$$

$$\sum_{r=1}^{s} u_r y_{rj} \leq \sum_{i=1}^{m} v_i x_{ij}$$

The presented LP problem can also be reformulated as dual task problem and shown in a matrix form as given below.

Model-3: $\min_{\theta,\lambda} \theta_k$

$$\text{Subjected to } Y_\lambda \geq Y_k$$

$$X_\lambda \leq \theta_k X_k$$

$\lambda \geq 0$; θ_k free

where $Y = \left(y_{rj}\right)_{n\times s}$ and $X = \left(x_{ij}\right)_{n\times m}$ represent the outputs and inputs matrices, respectively.

This model-3 is known to be input-oriented CRS model and the VRS model can be achieved by adding a constraint $\sum_{j=1}^{n} \lambda_j = 1$ in the respective duality of CRS model. Same efficiency results will be observed in CRS-input and CRS-output models for each DMUs. However, the efficiency results may vary in VRS models.

Numerous bibliographies and extensions of data envelopment analysis are available in the literature since its inception. A review article on DEA stated that there are around 10,300 DEA-related journal articles until the end of 2016. The DEA methodology is being widely applied in the areas of banking, agriculture, transportation, supply chain, hospitality, and public policy [46].

3.2 Integration of Shannon's Entropy with DEA

Steps for combined Shannon's entropy and DEA:

Step-1: Select the DMUs and its identified input and output variables.

Step-2: Generate all the possible input and output subset combinations from the original dataset, $L = (2^m - 1)(2^s - 1)$.

Step-3: Evaluate efficiency scores by using DEA of any traditional models (VRS-OP) and denote relative efficiency scores of each DMU as E_{jl}.

Step-4: Compute the set $e_{jl} = \frac{E_{jl}}{\sum_{j=1}^{n} E_{jl}}$ from the matrix $\left(E_{jl}\right)_{n \times L}$.

Step-5: Set of degree of diversification $d_l = 1 - f_l$ to be calculated where, $f_l = \frac{-1}{\ln(n)} \sum\limits_{j=1}^{n} e_{jl} \ln\left(e_{jl}\right)$.

Step-6: Estimate the CES as $\theta_j = \sum_{l=1}^{L} w_l E_{jl}$ where $w_l = \frac{d_l}{\sum_{l=1}^{L} d_l}$ such that,

$$\sum_{l=1}^{L} w_l = 1.$$

All the above steps are depicted in Fig. 1 flowchart. Combined DEA Shannon's entropy methodology has been used in engineering stream to choose experimental parameters [47]. Shannon's entropy is integrated with SBM method to rank all the DMUs of Chinese provinces, and the study was used to analyze allocation of financial resources from 2008 to 2018 [48].

3.3 Data Collection for Performance Measurement

This study considered 8 of the major ports of India including Kolkata Port Trust (KPT), Paradip Port Trust (PTP), Chennai Port Trust (CPT), V.O. Chidambaramnar Port Trust (VOC), Cochin Port Trust (CoPT), Mumbai Port Trust (MPT), Jawaharlal Nehru Port Trust (JNPT), and Deendayal Port Trust (DDPT). Operational performance parameters include 4 input, i.e., length of berth(s), average draft, number of cranes, and storage area and 2 output variables, namely TEUs handled and average turnaround time ATT) were considered. Data pertaining to the selected major ports is collected from the publications of Indian Ports Association and Ministry of Shipping covering 7 years from 2013–14 to 2019–20 is considered for the study.

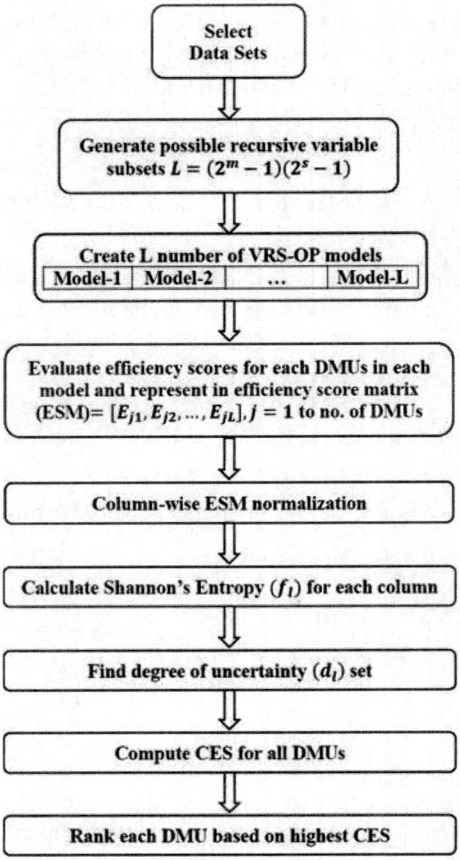

Fig. 1 Combined DEA Shannon's entropy methodology flowchart

4 Analysis of Results

Results derived from DEA variable returns to scale (VRS) analysis and combination of DEA (VRS) with Shannon entropy are presented as under.

Table 1 given above details about the efficiency levels at the selected ports. The results show efficiency level of 1.000 representing almost all the ports to be efficient. This is in spite of considering 4 input and 2 output indicators which satisfies the constraint of DEA modal on variables for consideration.

Table 2 gives results obtained by combining DEA technique with Shannon entropy. It may be observed that as per the results obtained, none of the port had achieved an efficiency level of 1 that generally is expected to be the optimal level under DEA model. However, as per the combined DEA and Shannon entropy technique ports that register the highest value stand efficient vis-à-vis the other ports considered for comparison. Table 3 gives the ranks of all the 8 ports as per their

Table 1 Efficiency scores using DEA VRS-OP method

Port	2013–14	2014–15	2015–16	2016–17	2017–18	2018–19	2019–20
KPT	1.000	1.000	1.000	1.000	1.000	1.000	1.000
VPT	1.000	1.000	1.000	1.000	1.000	1.000	1.000
CPT	1.000	1.000	1.000	1.000	1.000	1.000	1.000
VOC	1.000	0.846	1.000	1.000	1.000	1.000	1.000
CoPT	0.525	0.581	1.000	1.000	1.000	1.000	1.000
MPT	1.000	1.000	1.000	1.000	1.000	1.000	0.552
JNPT	1.000	1.000	1.000	1.000	1.000	1.000	1.000
DDPT	1.000	0.653	0.706	0.186	0.244	0.811	1.000

efficiency levels. From the table, it is evident that JNPT had successfully maintained highest level of efficiency throughout the 7 years. This is due to the existence of high level of physical infrastructure including berths, cranes, and storage area which allows it to handle greater throughput. There exists a close competition among Chennai Port Trust (CPT), Cochin Port Trust (CoPT), and Visakhapatnam Port Trust (VPT) for the second, third, and fourth positions. These ports traditionally had been handling multiple cargos which even today form a major cargo composition for them and containers and container ships are still not a major portion of their cargo composition. Among CPT, CoPT, and VPT, CPT by virtue of its investments in infrastructure has for some years recorded second position among all the 8 ports.

The results show lower efficiency at ports like Mumbai Port Trust (MPT) and Kolkata Port Trust (KPT). This is due to infrastructure constraints at these ports.

From the results, it is evident that investments in physical infrastructure like cranes and berths result in technical efficiency improvements.

Table 2 CES using DEA VRS-OP and Shannon entropy method

Port	2013–14	2014–15	2015–16	2016–17	2017–18	2018–19	2019–20
KPT	0.5879	0.4919	0.5249	0.3518	0.3152	0.3135	0.5660
VPT	0.6062	0.6157	0.6227	0.5471	0.4562	0.7167	0.6828
CPT	0.7841	0.8018	0.8754	0.4596	0.5355	0.4240	0.7256
VOC	0.6563	0.6559	0.7635	0.3660	0.3415	0.6016	0.5825
CoPT	0.3014	0.3341	0.6553	0.5733	0.7648	0.5812	0.7054
MPT	0.5960	0.6568	0.6028	0.7719	0.4019	0.3608	0.2588
JNPT	0.8856	0.9555	0.8977	0.8705	0.8574	0.7929	0.8157
DDPT	0.4912	0.2862	0.2177	0.1385	0.1007	0.3702	0.6448

Table 3 Ranking of Port under DEA VRS-OP and Shannon entropy method

Port	2013–14	2014–15	2015–16	2016–17	2017–18	2018–19	2019–20
KPT	6	6	7	7	7	8	7
VPT	4	5	5	4	4	2	4
CPT	2	2	2	5	3	5	2
VOC	3	4	3	6	6	3	6
CoPT	8	7	4	3	2	4	3
MPT	5	3	6	2	5	8	8
JNPT	1	1	1	1	1	1	1
DDPT	7	8	8	8	8	7	5

5 Findings, Conclusions, and Scope for Further Research

The study measured efficiency levels at container terminals of selected major ports of India using two techniques, namely DEA and DEA in combination with Shannon entropy. The results obtained by DEA combined Shannon entropy technique had given a better chance to check the efficiency levels at these ports. It is also proven that investments in physical infrastructure at ports to achieve better efficiency levels ports lead to improvement in their efficiency levels. It is also proven that DEA when combined with Shannon entropy can give better and comparable results of efficiency. It may be noted that the current research had considered only major ports which account for around 56% of the overall maritime cargo handled in India. Further studies covering both major and non-major ports would help in assessing the efficiency trends at all ports of India.

References

1. Government of India - Ministry of Shipping (2011) Maritime agenda: 2010–2020
2. Delhi N (2015) Annual report. NITI Aayog
3. Kessides IN (2005) Infrastructure privatization and regulation: promises and perils. World Bank Res Obs 20:81–108. https://doi.org/10.1093/wbro/lki003
4. Xie Q, Dai Q, Li Y, Jiang A (2014) Increasing the discriminatory power of DEA using Shannon's entropy. Entropy 16:1571–1585. https://doi.org/10.3390/e16031571bpscorrected.pdf
5. Dooms M (2014) Port industry performance management. Port Technol Int Fev 16–17
6. Blonigen BA, Wilson WW (2008) Port Efficiency and International Trade. Rev Int Econ 16:21–36
7. Brooks MR, Schellinck T, Pallis AA (2011) A systematic approach for evaluating port effectiveness. Marit Policy Manag 38:315–334. https://doi.org/10.1080/03088839.2011.572702
8. Jiang B, Li J (2009) DEA-based performance measurement of seaports in northeast Asia: radial and non-radial approach. Asian J Shipp Logist 25:219–236. https://doi.org/10.1016/S2092-5212(09)80003-5
9. Langen P de, Nijdam M, Horst M van der (2007) New indicators to measure port performance. J Marit Res IV:23–36

10. Pallis A, Vitsounis TK (2008) Towards an alternative measurement of port performance : externally generated information and users' satisfaction towards an alternative measurement of port performance: externally generated information and users'. Int Forum Shipp Ports Airports 1–25
11. Park RK, De Prabir P (2004) An alternative approach to efficiency measurement of seaports. Marit Econ Logist 6:53–69. https://doi.org/10.1057/palgrave.mel.9100094
12. Bichou K, Gray R (2004) A logistics and supply chain management approach to port performance measurement. Marit Policy Manag 31:47–67. https://doi.org/10.1080/0308883032000174454
13. Barros CP, Felício JA, Fernandes RL (2012) Productivity analysis of Brazilian seaports. Marit Policy Manag 39:503–523. https://doi.org/10.1080/03088839.2012.705033
14. Barros CP, Managi S (2008) Productivity Drivers In Japanese Seaports. 33
15. Bergantino AS, Musso E (2011) The role of external factors versus managerial ability in determining seaports' relative efficiency: an input-by-input analysis through a multi-step approach on a panel of Southern European ports. Marit Econ Logist 13:121–141. https://doi.org/10.1057/mel.2011.1
16. Estache A, de la Fé BT, Trujillo L (2004) Sources of efficiency gains in port reform: a DEA decomposition of a Malmquist TFP index for Mexico. Util Policy 12:221–230. https://doi.org/10.1016/j.jup.2004.04.013
17. Haddad EA, Hewings GJD, Perobelli FS, Santos RAC (2007) Port Efficiency in Brazil 1 and regional economic development . The literature provides a number of alternative approaches (1985) and the emerging perspectives associated with the new economic geography, to the work, pp 1–49
18. Navarro-Chávez CL, Zamora-Torres AI (2014) Economic efficiency of the international port system: an analysis through data envelopment. Int Bus Res 7:108–117. https://doi.org/10.5539/ibr.v7n11p108
19. Ng AKY (2012) Port development in east Asia: from efficiency enhancement to regional competitiveness. port technol int 8–9
20. Padilla MJ, Eguia RE (2010) Relative efficiency of seaports in Mindanao
21. Park R-K (2008) A verification of Korean containerport efficiency using the bootstrap approach. J Korea Trade 12:1–30
22. Pjevčević D, Radonjić A, Hrle Z, Čolić V (2012) DEA Window analysis for measuring port efficiencies in Serbia. PROMET Traffic &Transp 24:63–72. https://doi.org/10.7307/ptt.v24i1.269
23. Tongzon J, Heng W (2005) Port privatization, efficiency and competitiveness: Some empirical evidence from container ports (terminals). Transp Res Part A Policy Pract 39:405–424. https://doi.org/10.1016/j.tra.2005.02.001
24. Wang T-F, Cullinane K (2006) The efficiency of European container terminals and implications for supply chain management. Marit Econ; Logist 8:82–99. https://doi.org/10.1057/palgrave.mel.9100151
25. Wanke PF, Barbastefano RG, Hijjar MF (2011) Determinants of efficiency at major Brazilian port terminals. Transp Rev 31:653–677. https://doi.org/10.1080/01441647.2010.547635
26. Cheon SH, Dowall DE, Song DW (2010) Evaluating impacts of institutional reforms on port efficiency changes: Ownership, corporate structure, and total factor productivity changes of world container ports. Transp Res Part E Logist Transp Rev 46:546–561. https://doi.org/10.1016/j.tre.2009.04.001
27. Liu Z (1995) The COMPARATIVE PERFORMANCE OF PUBLIC AND PRIVATE ENTERPRISES: THE CASE OF BRITISH PORts. J Transp Econ Policy 29:263–274
28. Pallis AA, Syriopoulos T (2007) Port governance models: financial evaluation of Greek port restructuring. Transp Policy 14:232–246. https://doi.org/10.1016/j.tranpol.2007.03.002
29. Cullinane K, Song DW, Gray R (2002) A stochastic frontier model of the efficiency of major container terminals in Asia: assessing the influence of administrative and ownership structures. Transp Res Part A Policy Pract 36:743–762. https://doi.org/10.1016/S0965-8564(01)00035-0

30. Tongzon J (2001) Efficiency measurement of selected Australian and other international ports using data envelopment analysis. Transp Res Part A Policy Pract 35:107–122. https://doi.org/10.1016/S0965-8564(99)00049-X
31. Kaisar E, Pathomsiri S, Haghani A, Kourkounakis P, Policy NYuwRC for T, Management, Pb, Maher T, The Port Authority of New Y, New J, Raritan Central R et al (2006) Developing measures of us ports productivity and performance: using DEA and FDH approaches. 47th Annu Transp Res Forum 1:269–280
32. Weille, J de; Ray A (1974) Volume_V111_No_3_244-259.pdf. 244–259
33. Gaur P, Pundir S, Sharma T (2011) Ports face inadequate capacity, efficiency and competitiveness in a developing country: case of India. Marit Policy Manag 38:293–314. https://doi.org/10.1080/03088839.2011.572700
34. Jim WuY, Lin C (2008) National port competitiveness: implications for India. Manag Decis 46:1482–1507. https://doi.org/10.1108/00251740810920001
35. Mokhtar K, Shah M (2013) Efficiency of operations in container terminals: a frontier method. Eur J Bus Manag 5:91–107
36. Rajasekar T, Deo M (2012) International journal of advances in management and economics the size effect of Indian Major ports on its efficiency using Dea-additive models. Int J Adv Manag Econ 1:12–18
37. Rajasekar T, Deo M, Ke R (2014) An exploration in to the causal relationship between performance inputs and traffic of major ports in India: A Panel Data Analysis. 2:72–81. https://doi.org/10.12691/ijefm-2-2-3
38. Charnes A, Cooper WW, Rhodes E (1978) Measuring the efficiency of decision making units. Eur J Oper Res 2:429–444. https://doi.org/10.1016/0377-2217(78)90138-8
39. Banker, R.D.; Chanes, A; Cooper WW (1984) Some models for estimating technical and scale inefficiencies-Banker-Charnes e Cooper.pdf. 1078–1092
40. Wang CN, Lin HS, Hsu HP, Le VT, Lin TF (2016) Applying data envelopment analysis and grey model for the productivity evaluation of Vietnamese agroforestry industry. Sustain 8. https://doi.org/10.3390/su8111139
41. (2009) Incorporating ratios in DEA—Applications to Real Data written by Sanaz Sigaroudi Thesis advisor : Dr . Joseph C . Paradi A thesis submitted in conformity with the requirements for the degree of Master of Applied Science Graduate Department of Mechanical
42. Hosseinzadeh Lotfi F, Toloie Eshlaghy A, Shafiee M (2012) Providers ranking using data envelopment analysis model, cross efficiency and Shannon entropy. Appl Math Sci 6:153–161
43. Qi XG, Guo B (2014) Determining common weights in data envelopment analysis with Shannon's entropy. Entropy 16:6394–6414. https://doi.org/10.3390/e16126394
44. Soleimani-damaneh M, Zarepisheh M (2009) Shannon's entropy for combining the efficiency results of different DEA models: Method and application. Expert Syst Appl 36:5146–5150. https://doi.org/10.1016/j.eswa.2008.06.031
45. Martić M, Novaković M, Baggia A (2009) Data envelopment analysis—basic models and their utilization. Organizacija 42:37–43. https://doi.org/10.2478/v10051-009-0001-6
46. Emrouznejad A, Yang GL (2018) A survey and analysis of the first 40 years of scholarly literature in DEA: 1978–2016. Socioecon Plann Sci 61:4–8. https://doi.org/10.1016/j.seps.2017.01.008
47. Kodavaty J, Pannala RK, Wasson S, Mittal M, Irshad A (2022) A novel method to choose the experimental parameters in large amplitude oscillatory shear rheology. Mater Science Forum 54–64
48. Ma F, Li J, Ma H, Sun Y (2022) Evaluation of the regional financial efficiency based on SBM-Shannon entropy model. Procedia Comput Sci 199:954–961

Ranking of Efficient DMUs Using Super-Efficiency Inverse DEA Model

Swati Goyal, Manpreet Singh Talwar, Shivi Agarwal, and Trilok Mathur

Abstract The overall potential for improving the relative efficiency of decision-making units (DMUs) is revealed by applying the data envelopment analysis (DEA) model. This study proposes a ranking system for ordering efficient DMUs with a super-efficiency inverse DEA (IDEA) model under a constant return to scale (CRS) assumption. IDEA is applied to evaluate the expected output or input variation level while keeping the efficiency value unchanged. For a numerical illustration of the proposed model in real-life problems, firstly, this study calculated the efficiency score of all 52 bus depots of Rajasthan State Road Transport Corporation (RSRTC) for the year 2018–19, applying the DEA model under the CRS assumption. The results revealed that 7 bus depots are efficient. Secondly, these 7 efficient depots have been ranked using the proposed super-efficiency IDEA model.

Keywords Data envelopment analysis · Inverse DEA · Efficiency · Ranking · Public transport sector

1 Introduction

The public transport sector is widely recognized as an accelerator of economic growth and a significant factor in social development. It contributes to the nation's economy by facilitating the movement of passengers and goods from one place to another. Furthermore, a public road transportation system boosts productivity and growth by reducing transportation costs and congestion, the costs of roads, parking facilities,

S. Goyal · M. S. Talwar · S. Agarwal (✉) · T. Mathur
Department of Mathematics, BITS Pilani, Pilani Campus, India
e-mail: shivi@pilani.bits-pilani.ac.in

S. Goyal
e-mail: swati.goyal@pilani.bits-pilani.ac.in

M. S. Talwar
e-mail: manpreetsingh8172@gmail.com

T. Mathur
e-mail: tmathur@pilani.bits-pilani.ac.in

M. Thakur et al. (eds.), *Soft Computing for Problem Solving*, Lecture Notes in Networks and Systems 547, https://doi.org/10.1007/978-981-19-6525-8_47

vehicle operating costs, accidents and environmental effects. Moreover, this sector generates employment as it requires conductors, drivers, booking agents, cleaners and laborers for loading and unloading operations and wayside facilities [1].

India is the world's fifth-largest economy and one of the world's fastest-growing among major economics. In India, more than 300 million passengers travel through public transport in a single day, which provides livelihood to about 40 million people [2]. Due to urbanization, this sector faces formidable pressure to keep pace with the increasing demand and unprecedented challenges in various aspects, including operation and service quality, availability, comfort, affordability and safety [3, 4]. Deterioration of service levels may impact the number of users, which consequently affects the GDP contribution.

In terms of area, Rajasthan is India's largest state, located between the north latitudes $23^{\circ}\ 30'-30^{\circ}\ 11'$ and $69^{\circ}\ 29'-78^{\circ}\ 17'$ and east longitudes in the western part of the country. In 2021, Rajasthan's population is around 82.4 million [5]. Rajasthan State Road Transport Corporation (RSRTC) is a sustainable public road transport service provider of inter-city, with 52 bus depots over 33 districts. The Road Transport Corporation Act of 1950 established RSRTC on October 1, 1964. Buses are also available in the categories of express, ordinary, deluxe, AC sleeper, Volvo Mercedes and Volvo LCD. About 0.91 million passengers were carried per day in the year 2018–19 by RSRTC. Despite the rise in population, RSRTC passenger traffic declined by 0.157 million per year from 2007 to 2019 [6].

Due to subsidies on ticket fares and several inefficient performances of bus depots, RSRTC bus depots are suffering from huge losses. The overall function of RSRTC arises to be uncertain and it would not be inaccurate to suggest that it is on the verge of a debt crisis as its liabilities will keep growing. Because of this, all bus depots reveal inefficient performance and continuously give lousy quality of service. As a result, there is an urgent need to overhaul the existing system and aggressive actions must be taken to resolve public grievances.

DEA is one of the widely used nonparametric linear programming-based techniques developed by Charnes et al. [7] in 1978. The unique ability of DEA to assess the relative efficiency of a group of homogeneous entities known as DMUs with many outputs and inputs has garnered the notice of a significant number of researchers. A vital component of this technique is that it chooses the most optimal weights for inputs and outputs to compute the efficiency for each DMU. Moreover, DEA has become progressively popular for analyzing the relative efficiency in the various areas of management, education, sport, hospital, transportation, etc. [8]. DEA calculates the production possibility set (PPS) depending on the given dataset. The modification in input–output can cause a shift in the relative efficiency and PPS. An appealing topic is how to preserve the efficiency score of an examined DMU if its fundamental structure changes marginally in the short term.

Inverse DEA (IDEA) is a powerful tool for determining the optimal solution to achieve a decision-maker's objective with a given amount of efficiency. Multiple-objective programming (MOP) IDEA model is initiated by Wei et al. [9] to solve the problems of resource allocation and investment analysis. IDEA estimates how much output (input) should be increased corresponding to the modified input (output) amount, while the DMU keeps its current efficiency value same for the other DMUs. Yan et al. [10] proposed a modified IDEA method by adding a cone constraint and it is used to solve the resource allocation problem. More precisely, Hadi-Vencheh and Forouughi [11] employed the IDEA model with MOLP to simultaneously increase one output (input) value and decrease another output (input) value. Jahanshahloo et al. [12] proposed the IDEA model with non-radial enhanced Russell model. Jahanshahloo et al. [13] presented IDEA model as per an inter-temporal dependence assumption and developed a MOLP to solve IDEA problem for weak Pareto solution. Hadi-vencheh et al. [14] improved the IDEA models for interval data instead of crisp values in measuring the efficiency score. Through the IDEA model, a new way for forecasting whether a merger in a market will lead in a major or minor consolidation was suggested by Amin et al. [1]. Recently, Amin et al. [15] formulated a novel method by integrated the goal programming (GP) and inverse DEA for target setting of a merger (decision-makers) in banking industry. Numerical results on instances presented by Le et al. [16] assessed the efficiency and effectiveness of education system using an inverse frontier-based benchmarking DEA model with a mixed integer linear program. Soleimani et al. [17] introduced a single-objective linear programming (SOLP) IDEA model to rank the efficient DMUs by removing one DMU in the PPS.

One of the main issues discussed in DEA is its inability to efficiently rank all the efficient DMUs. In practical applications, decision-makers are interested in ranking all DMUs for resource allocation and performance review. This issue has been thoroughly investigated by DEA experts, giving different approaches such as cross-efficiency, super efficiency, benchmarking ranking, virtual frontier DEA (VFDEA) for ranking efficient DMUs. Sexton et al. [18] suggested a cross-efficiency technique for ranking DMUs that is based on the idea that units are evaluated both self and peer. The weights for each DMU are explicitly determined using the fundamental DEA model, bringing about n sets of weights. Then, each DMU is assessed using the n sets of weights acquiring n efficiency scores. The cross-efficiency of each DMU is the average of the n efficiency scores. Wu et al. [19] utilized the Shannon entropy to calculate the weights for ultimate cross-efficiency scores instead of the average assumption. Jahanshahloo et al. [20] used a method to choose symmetric weights. Andersen and Petersen [21] developed the super-efficiency model by deleting the DMUs under review and building a new frontier with the remaining DMUs. The efficiency score for DMU under review can be equal to 1 or above. Mehrabian et al. [22] offered a comprehensive ranking of efficiency units after addressing issues of infeasibility and stability. Tone [23] presented super efficiency on the SBM model, which was always feasible and stable, as well as having the benefits of non-radial models. Jahanshahloo et al. [24] showed how to rank efficient DMUs using leave one out and L_1 norm. Sueyoshi et al. [25] proposed a "benchmark approach" with the use of a slack-adjusted DEA model and offensive earned-run average (OERA) to overcome the shortcoming of multiple efficient units identified through the DEA model. Jahanshahloo et al. [26] proposed a new model with the idea that efficient units can be the target DMU for inefficient DMUs. Bian and Xu [27] proposed virtual frontier DEA and was further developed by Li et al. [28]. In this model, a new optimal frontier is constructed called the virtual frontier. To differentiate between efficient DMUs in VFDEA models, two different sets, namely the reference and the evaluated DMU sets, are produced using the traditional DEA model. Cui and Li [29] proposed that the reference DMU set remains unchanged during the evaluation process. Wanke and Barros [30] used virtual frontier dynamic range adjusted model data envelopment analysis (VDRAM-DEA) to evaluate the efficiency of Latin American airlines. In addition, Barros et al. [31] applied VDRAM-DEA to assess Angolan hydroelectric power plants, resulting in improved efficiency score discrimination. All ranking methods evaluate units from a particular perspective and each of them has its advantages and hindrances compared to others. Soleimani et al. [17] introduced the single-objective linear programming (SOLP) IDEA model to rank the efficient DMUs by removing one DMU in the PPS.

The inverse super-efficiency DEA method, also known as the growth potential method, is used which is futuristic approach for ranking the DMUs because of its capacity to estimate the inputs changes based on arbitrary increments of outputs. Ranking from this method is more reliable because of the growth potential, wasted inputs (slack variable values) are minimal and outputs variation is manageable. Hence, this suggest that decision-maker can select different increments in output to obtain better understanding of the input variables, giving them more flexibility. In other words, efficient DMUs have no scope for improvement in terms of efficiency, but ranking is vital for preserving service quality.

The literature review has shown that there are limited studies on public transport sector to rank the efficient bus depots. This research proposed a ranking for efficient bus depots of RSRTC by using the super-efficiency IDEA model. This study ranked the efficient bus depots based on the highest DEA efficiency score which have efficiency scores greater than unity. The aim of this study is to assist decision-makers in developing appropriate policies that will improve the overall financial health and productivity of RSRTC bus depots. This research will help them build a new viewpoint on how to provide a satisfactory outcome for travelers.

The remaining part of the paper is written as follows. Section 2 presented the research methodology framework. The description of data and empirical results for 52 RSRTC bus depots are described in Sect. 3. In last section, concluding remarks are mentioned.

2 Research Methodology

In this section, CCR, inverse CCR and super-efficiency inverse DEA models with required definition for the study are discussed.

2.1 CCR Model

The first DEA model is known as Charnes, Cooper and Rhodes which has shortened as CCR [7] model. This model calculates the relative efficiency under CRS assumption technology. For measuring the efficiency of n DMU$_s$, whereas each DMU$_k \in \{1, \ldots.k..., n\}$, considering m inputs to produce s outputs represent by x_{ik} $(i = 1, ..., m)$ and $y_{rk}(r = 1, ..., s)$, respectively, the input-oriented CCR DEA model for measuring the efficiency of kth DMU is defined as follows:

$$\begin{aligned} &\text{Min } \theta_k \\ &\text{Subject to} \\ &\sum_{j=1}^{n} y_{rj}\lambda_{jk} \geq y_{rk}, \qquad r = 1, \ldots, s \\ &\sum_{j=1}^{n} x_{ij}\lambda_{jk} \leq \theta_k x_{ik}, \quad i = 1, \ldots, m \\ &\lambda_{jk} \geq 0, \qquad j = 1, \ldots, n \end{aligned} \tag{1}$$

Considering that θ_k^* represents the optimum efficiency score of CCR DEA model, if the score of $\theta_k^* < 1$ DMU$_k$ is inefficient otherwise DMU$_k$ is efficient. The inverse CCR, super-efficiency and super-efficiency inverse DEA models have been described in the next subsections.

2.2 Inverse DEA Model

Assume the following scenario: if the efficiency score remains same, then how much inputs should decrease when the outputs increase? As a result, evaluate the increase in outputs and inputs to answer this question considering y_k to $y_k + \Delta y_k$ and x_k to $x_k + \Delta x_k$, $y_k, x_k > 0$ and $\Delta y_k \in R_+^{s \times n}$, $\Delta x_k \in R_+^{m \times n}$, respectively. The updated inputs $x_k + \Delta x_k$ are represented by α_k and outputs $y_k + \Delta y_k$ are denoted by β_k for DMU$_k$. The following MOLP model is used to calculate the α_k [19];

$$
\begin{aligned}
&\min \alpha_k = (\alpha_{1k}, \alpha_{2k}, \ldots, \alpha_{mk}) \\
&\text{Subject to} \\
&\sum_{j=1}^{n} y_{rj}\lambda_{jk} \geq \beta_{rk}, r = 1, \ldots., s \\
&\sum_{j=1}^{n} x_{ij}\lambda_{jk} \leq \theta_k \alpha_{ik}, i = 1, \ldots., m \\
&\alpha_{ik} \geq x_{ik}, i = 1, \ldots., m \\
&\lambda_{jk} \geq 0, j = 1, \ldots., n
\end{aligned}
\tag{2}
$$

2.3 Super-Efficiency DEA Model

In CCR DEA model, two types of DMUs exist: efficient and inefficient. When DMUs are on the production frontier at the same time, the standard CCR model has poor discriminating, making further evaluation and comparison between these DMUs problematic. To address the lack of CCR model to evaluate the DMU's efficiency effectively and accurately, Andersen and Petersen [21] suggested super-efficiency DEA model, a type of complete efficiency measure. Super-efficiency DEA model is given as follows:

$$
\begin{aligned}
&\text{Min} \quad \theta_k \\
&\text{Subject to} \\
&\sum_{\substack{j=1 \\ j\neq k}}^{n} y_{rj}\lambda_j \geq y_{rk}, \qquad j = 1, \ldots, n, r = 1, \ldots, s \\
&\sum_{\substack{j=1 \\ j\neq k}}^{n} x_{ij}\lambda_j \leq \theta_k x_{ik}, \qquad j = 1, \ldots, n, i = 1, \ldots, m \\
&\lambda_{jk} \geq 0, \qquad j = 1, \ldots., n
\end{aligned}
\tag{3}
$$

2.4 Super-Efficiency Inverse DEA Model

The fundamental IDEA models produce a new DMU. The production possibility set (PPS) in these models is built by prior DMUs. It is built by n DMUs (DMU_1, ..., DMU_k, ..., DMU_n), while among them changed the DMU_k with $DMU_k^{'}$

by creating a new technology set that is (DMU_1, ..., $DMU_k^{;}$, ..., DMU_n). Hadi-Vencheh et al. [14] proposed a new approach by removing DMU_k that new PPS constructed as (DMU_1, ..., DMU_{k-1}, DMU_{k+1}, ..., DMU_n). Generally, PPS is built by replacing the "perturbed DMU" with a new unit that includes the updated inputs and outputs. Super-efficiency IDEA model [17] is a valuable tool for decision-makers when determining the actual ranking levels of efficient DMUs with changing the PPS. After eliminating kth DMU (DMU_k), , the PPS is not going to change (DMU_1, DMU_2, ….., DMU_{k-1}, $DMU_k^{;}$, DMU_{k+1}…. , DMU_n). The model is used as follows:

$$\begin{aligned}
&\text{Min } \alpha_k = (\alpha_{1k}, \alpha_{2k}, \ldots, \alpha_{mk}) \\
&\text{Subject to} \\
&\sum_{\substack{j=1 \\ j \neq k}}^{n} y_{rj}\lambda_j \geq \beta_{rk}, \; j = 1, \ldots, n, r = 1, \ldots, s \\
&\sum_{\substack{j=1 \\ j \neq k}}^{n} x_{ij}\lambda_j \leq \theta\alpha_{ik}, \quad j = 1, \ldots, n, i = 1, \ldots, m \\
&\alpha_{ik} \geq x_{ik}, \quad i = 1, \ldots, m \\
&\lambda_{jk} \geq 0, \quad j = 1, \ldots., n
\end{aligned} \tag{4}$$

2.5 *Single-Objective IDEA Model*

The main problem in the super-efficiency IDEA model is that it is MOLP, which implies that it will not deliver an optimum solution of MOLP still maintaining a non-exclusive Pareto solution. To overcome this limitation, this study employed an inverse DEA-based single-objective IDEA model. The objective function is constructed by increasing the inputs. Accordingly, in order to produce a single-objective function, for all inputs, the input increment should be the same. As a result, all DMU outputs have the same increase.

Sometimes, the increasing quantity of a unit of them is not always the same. For example, in the road transport sector, one input is fleet size (number of buses) and another input is fuel consumption (1000 K.L.). Therefore, all the variables are normalized by using the described equation,

$$\tilde{x}_{ij} = \frac{x_{ij}}{\max_i x_{ij}}, \; \tilde{y}_{ij} = \frac{y_{rj}}{\max_r y_{rj}} \tag{5}$$

Hence, the increment of inputs will be the same for all inputs to achieve a single-objective function. Thus, outputs are increased by the same amount as all DMUs. All outputs of efficient DMU are increased by b percent ($Y + b\%$) and the growing percentage of inputs (a) is calculated. As a result, the single-objective LP IDEA model for efficient DMUs is given as follows:

$$
\begin{aligned}
&\alpha_k = \text{Min} a \\
&\text{Subject to} \\
&\sum_{\substack{j=1 \\ j \neq k}}^{n} \tilde{y}_{rj}\lambda_j \geq \tilde{y}_{rk} + b, \qquad j = 1, \ldots, n, r = 1, \ldots, s \\
&\sum_{\substack{j=1 \\ j \neq k}}^{n} \tilde{x}_{ij}\lambda_j \leq \tilde{x}_{ik} + a, \qquad j = 1, \ldots, n, i = 1, \ldots, m \\
&\tilde{x}_{ik} + a \geq \tilde{x}_{ik}, \qquad i = 1, \ldots, m \\
&\lambda_{jk} \geq 0, \qquad j = 1, \ldots, n
\end{aligned}
\tag{6}
$$

All DMU_k outputs are enlarged by a comparable amount (b) in the model (6). Hence, all inputs and outputs should be dimensionless. As a result, all data are normalized through model (5).

Definition 1 Suppose DMU_j and DMU_k are estimated by the model (6). The rank of DMU_j is better than DMU_k, when $\alpha_j^* \geq \alpha_k^*$.

3 Numerical Illustration

Passenger transportation is a "service business" and measuring the efficiency of a service business is a complicated matter. This study is exposed to overcome the complex nature of the organizational operation of passenger transportation for 52 RSRTC bus depots. In the present work, DEA is utilized to sweep the changes in efficiency. Step 1 calculates the efficiency value of bus depots using model (1), and in step 2, we apply the proposed super-efficiency IDEA model (6) for ranking the efficient bus depots.

3.1 Data and Parameters Collection:

This study used secondary dataset of 52 bus depots which is provided by the annual reports of RSRTC for year 2018–19 [6]. Various variable affects the transport system; we have selected six significant parameters, in which four inputs and two output

Table 1 Descriptive statistics of inputs, outputs and efficiency value of 52 RSRTC bus depots

	Variables	Max	Min	Average	Stdev
Inputs	Fleet size	140	14	70.08	24.64
	Labor	665	54	271.08	116.88
	Fuel consumption	45.75	4.12	20.91	8.26
	Routes distance	16,769	3496	9064.31	2966.59
Outputs	Passenger K.M. Occupied	9.02	0.77	3.9	1.5
	Vehicle utilization	586	311	391.87	46.39
	Efficiency	1	0.8	0.91	0.05

parameters are used to evaluate the efficiency of 52 bus depots. The descriptive statistics of inputs and outputs are given in Table 1.

Input parameters

Fleet Size: The total number of buses on road in a depot which indicates capital input.

Total Staff: The total number of employees who worked in a depot indicates labor input.

Fuel Consumption: The fuel consumption which indicates energy input and is calculated as,

$$\text{Total Fuel Consumption(1000K.L.)} = (\text{Description of K.M.})/(\text{Average diesel consumption})$$

Route Distance: It is described as the sum of individual distances taken along the route.

Output parameters

Passenger kilometers Occupied: It is the cumulative distance traveled by each passenger, which is defined below,

$$\text{Passenger K.M. occupied (in lakhs)} = \text{Average no. of Buses} \times \text{Description of K.M.} \times \text{Load Factor}$$

Vehicle Utilization: It is the total kilometers traveled by bus/day/hour.

3.2 *Empirical Results*

The result of the model (6) with the constant return to scale (CRS) assumption is given in Table 2. Initially, model (5) is used to normalize the data to feed the model

Table 2 Ranking of efficient bus depots

	Beawar	Jaisalmer	Karauli	Matsya Nagar	Shapur	Tijara	Vidhyadha Nagar
$\text{Min}(\alpha_k^*)$	0.234	0.132	0.156	0.187	0.147	0.136	0.157
Rank	1	7	4	2	5	6	3

(6) to rank the efficient bus depots. Secondly, model (1) recognizes the efficient depots, which shows 7 CCR efficient bus depots: Beawar, Jaisalmer, Karauli, Matsya Nagar, Shapur, Tijara and Vidhyadhar Nagar. Dungarpur has the lowest efficiency score, which is 0.801 of all the depots and 7 bus depots (Beawar, Jaisalmer, Karauli, Matsya Nagar, Shapur, Tijara and Vidhyadhar Nagar) have the highest efficiency score, which is 1. Statistics of efficiency scores are presented in the last column of Table 1.

Furthermore, the ranking of 7 efficient bus depots was evaluated by the super-efficiency IDEA model (6) in Table 2. It is solved by using the software MATLAB2021b. This study applied model (6) to measure the input growth rate by increasing the amount of output of efficient bus depots by 0.15. The bus depot with the greater objective function is ranked higher by the given definition (1).

Table 2 represents the results of the model (6) which concludes Beawar has the rank one with largest value of (α_k^*) and Jaisalmer has the rank 7 with the smallest value of $\left(\alpha_k^*\right)$.

The rank of efficient bus depots of super-efficiency IDEA and super-efficiency methods is given in Table 3. In most cases, the outcomes obtained using this method and the super-efficiency method are comparable. Noted that the results in this section are more reliable due to the fact that (i) this method ranks DMUs based on their growth potential, (ii) wasted inputs (slack variables value) are minimal and (iii) outputs variation is manageable. This means the decision-maker has more flexibility in selecting alternative increments to obtain a good understanding of the input variables.

Table 3 Ranking of efficient bus depots using super-efficiency IDEA and super-efficiency model

Bus depots	Super-efficiency IDEA method	Super-efficiency method
Beawar	1	2
Jaisalmer	7	6
Karauli	4	1
Matsya Nagar	2	3
Shapur	5	4
Tijara	6	7
Vidhyadha Nagar	3	5

4 Conclusion

DEA models evaluated the relative efficiency value of DMUs. Nonetheless, inverse DEA (IDEA) approach has been a popular method in the DEA literature for re-measuring the efficiency of DMUs when the input (output) values of a given DMU are changed. The basic DEA models are good at finding inefficient units but poor at distinguishing between efficient units. To overcome this problem, this study pointed out the ranking of efficient bus depots using super-efficiency IDEA method with a single-objective LP model. This study assessed the relative efficiency of 52 RSRTC bus depots during the period of 2018–19. The proposed model identified that, 7 RSRTC bus depots are efficient and 45 bus depots are inefficient. Now, we ranked efficient bus depots based on super-efficiency IDEA model. This application also gives a novel analytical framework for efficient bus depots-related quality service decision challenges, such as the best allocation of resources among different bus depots. For future suggestion of improving the efficiency of public transport sector, the proposed method in DEA can be suggested for further research to be developed with cross-efficiency inverse DEA method.

References

1. Amin GR, Emrouznejad A, Gattoufi S (2017) Minor and major consolidations in inverse DEA: definition and determination. Comput Ind Eng 103:193–200
2. https://www.financialexpress, August2020. Accessed on 2 Apr 2021
3. Cafiso S, Di Graziano A, Pappalardo G (2013) Road safety issues for bus transport management. Accid Anal Prev 60:324–333
4. De Oña J, De Oña R, Eboli L, Mazzulla G (2013) Perceived service quality in bus transit service: a structural equation approach. Transp Policy 29:219–226
5. https://bit.ly/2WKHOSf. Accessed on 12 May 2020
6. https://transport.rajasthan.gov.in/content/transportportal/en/RSRTC/public-relation/Annual Report.html. Accessed on 23 Sept 2020
7. Charnes A, Cooper WW, Rhodes E (1978) Measuring the efficiency of decision making units. Eur J Oper Res 2(6):429–444
8. Cavaignac L, Petiot R (2017) A quarter century of data envelopment analysis applied to the transport sector: a bibliometric analysis. Socioecon Plann Sci 57:84–96
9. Wei Q, Zhang J, Zhang X (2000) An inverse DEA model for inputs/outputs estimate. Eur J Oper Res 121(1):151–163
10. Yan H, Wei Q, Hao G (2002) DEA models for resource reallocation and production input/output estimation 136(1):19–31
11. Hadi-Vencheh A, Foroughi AA (2006) A generalized DEA model for inputs/outputs estimation. Math Comput Modell 43(5-6):447–457
12. Jahanshahloo GR, Hosseinzadeh Lotfi F, Rostamy-Malkhalifeh M, Ghobadi S (2014) Using enhanced Russell model to solve inverse data envelopment analysis problems. Sci World J
13. Jahanshahloo GR, Soleimani-Damaneh M, Ghobadi S (2015) Inverse DEA under inter-temporal dependence using multiple-objective programming. Eur J Oper Res 240(2):447–456
14. Hadi-Vencheh A, Hatami-Marbini A, Ghelej Beigi Z, Gholami K (2015) An inverse optimization model for imprecise data envelopment analysis Optimization 64(11):2441–2454
15. Amin GR, Al-Muharrami S, Toloo M (2019) A combined goal programming and inverse DEA method for target setting in mergers. Expert Syst Appl 115:412–417

16. Le MH, Afsharian M, Ahn H (2021) Inverse frontier-based benchmarking for investigating the efficiency and achieving the targets in the vietnamese education system. Omega 103:102427
17. Soleimani-Chamkhorami K, Hosseinzadeh Lotfi F, Jahanshahloo G, Rostamy-Malkhalifeh M (2020) A ranking system based on inverse data envelopment analysis. IMA J Manag Math 31(3):367–385
18. Sexton TR, Silkman RH, Hogan AJ (1986) Data envelopment analysis: critique and extensions. New Direct Prog Eval 1986(32):73–105
19. Wu J, Sun J, Liang L, Zha Y (2011) Determination of weights for ultimate cross efficiency using Shannon entropy. Expert Syst Appl 38(5):5162–5165
20. Jahanshahloo GR, Lotfi FH, Jafari Y, Maddahi R (2011) Selecting symmetric weights as a secondary goal in DEA cross-efficiency evaluation. Appl Math Model 35(1):544–549
21. Andersen P, Petersen NC (1993) A procedure for ranking efficient units in data envelopment analysis. Manage Sci 39(10):1261–1264
22. Mehrabian S, Alirezaee MR, Jahanshahloo GR (1999) A complete efficiency ranking of decision making units in data envelopment analysis. Comput Optim Appl 14(2):261–266
23. Tone K (2002) A slacks-based measure of super-efficiency in data envelopment analysis. Eur J Oper Res 143:32–41
24. Jahanshahloo GR, Lotfi FH, Shoja N, Tohidi G, Razavyan S (2004) Ranking using l1-norm in data envelopment analysis. Appl Math Comput 153(1):215–224
25. Sueyoshi T, Ohnishi K, Kinase Y (1999) A benchmark approach for baseball evaluation. Eur J Oper Res 115(3):429–448
26. Jahanshahloo GR, Junior HV, Lotfi FH, Akbarian D (2007) A new DEA ranking system based on changing the reference set. Eur J Oper Res 181(1):331–337
27. Bian YW, Xu H (2013) DEA ranking method based upon virtual envelopment frontier and TOPSIS. Syst Eng Theo Pract 33(2):482–488
28. Li Y, Wang YZ, Cui Q (2015) Evaluating airline efficiency: an application of virtual frontier network SBM. Transp Res Part E: Logist Transp Rev 81:1–17
29. Cui Q, Li Y (2014) The evaluation of transportation energy efficiency: an application of three-stage virtual frontier DEA. Transp Res Part D: Transp Environ 29:1–11
30. Wanke P, Barros CP (2016) Efficiency in Latin American airlines: a two-stage approach combining virtual frontier dynamic DEA and simplex regression. J Air Transp Manag 54:93–103
31. Barros CP, Wanke P, Dumbo S, Manso JP (2017) Efficiency in angolan hydro-electric power station: a two-stage virtual frontier dynamic DEA and simplex regression approach. Renew Sustain Energy Rev 78:588–596
32. Aneja R, Sehrawat N (2020) Depot-wise efficiency of haryana roadways: a data envelopment analysis. Arthaniti: J Econom Theo Pract 0976747920954973

Data Encryption in Fog Computing Using Hybrid Cryptography with Integrity Check

Samson Ejim, Abdulsalam Ya'u Gital, Haruna Chiroma, Mustapha Abdulrahman Lawal, Mustapha Yusuf Abubakar, and Ganaka Musa Kubi

Abstract Over the years, research carried on data encryption in fog computing emphasized confidentiality, authentication, and access control. However, some limitations were encountered and it could not address the issues of inherent cloud computing security challenges such as Man-in-The-Middle and ciphertext attacks. This work incidentally developed an enhanced model that tackled the security challenges found in fog computing using hashing technique by using operational modeling to develop a hybrid cryptography scheme comprising of Advanced Encryption Standard 256 (AES 256), Elliptic Curve Diffie–Helman (ECDH), and Secure Hashing Algorithm 256 (SHA-256). Upon simulation, data obtained was used to analyze the performance and the results gotten showed that the proposed system performed fairly well in terms of execution time and throughput for decryption with a slight compromise in the execution time and throughput for encryption. The essence of this work has proven that it offers a two-way role that guarantees that the data transferred is secured and also validates its integrity. More so, it will provide the basic groundwork for creating a faultless and fitted data encryption mechanism in fog computing.

Keywords Cryptographic hash function · Data security · Encryption · Fog computing · Hybrid cryptography

S. Ejim (✉) · A. Y. Gital · M. A. Lawal
Abubakar Tafawa Balewa University, Bauchi 740102, Nigeria
e-mail: samsonejim@gmail.com; sejim.pg@atbu.edu.ng

H. Chiroma
University of Hafr Al-Batin, Hafr Al-Batin 39524, Saudi Arabia

M. Y. Abubakar
Kano State Polytechnic, Kano 23410, Nigeria

G. M. Kubi
The Federal Polytechnic Nasarawa, Nasarawa 962101, Nigeria

M. Thakur et al. (eds.), *Soft Computing for Problem Solving*, Lecture Notes in Networks and Systems 547, https://doi.org/10.1007/978-981-19-6525-8_48

1 Introduction

Fog computing closes the space between the cloud and the endpoint (Internet of Things (IoT) nodes) by enabling computing, storage, networking, and data management on network nodes within the close vicinity of IoT devices [1, 2]. Keep in mind that fog computing enhances, not substitutes for cloud computing. Because at the network edge, fog permits for short-run analytics, whereas the cloud performs resource-intensive longer-run analytics. Although, data is generated and compiled at locations that house edge devices and sensors, machine learning tasks and complex analytics cannot be executed because of low computing and storage resources [3].

According to a white paper published by Cisco Systems [4], moving all data from the edge of the network to the data center for processing in a traditional cloud computing setup will increase latency. Traffic from thousands of devices will soon exceed the bandwidth capacity. Industry regulations and privacy issues prohibit certain types of data from being stored offsite. In addition, cloud servers communicate only through Internet Protocol (IP) and cannot communicate through many other protocols that IoT devices use, which reduces reliability. The optimal location to evaluate most IoT data is next to the device that generates and processes that data. This is the essence of fog computing.

As an instance of a cloud closer to the edge devices, devices found in fog could face security issues as they are fixed in the fog network [5] or positioned at locations that are not strictly monitored and protected [6]. Traditional attacks become available to compromise the system of fog devices, to realize malicious aims such as eavesdropping and data hijacking. Some of the problems attributed to fog computing were inherited from cloud computing. These inherent challenges of cloud computing include the security and privacy of user data stored in the cloud, intrusion detection, platform reliability, and security threats from multiple virtual machines [7]. To alleviate security and confidentiality issues, encryption technology has been used for many years to ensure the privacy and integrity of data [8].

1.1 Data Security Issues in Fog Computing

To provide real-time services, multiple fog nodes have to handle a large workload. However, the fog network is composed of highly distributed dynamic fog nodes that are susceptible to security threats from malicious devices masquerading as legitimate fog nodes [9]. Consequently, any fog node or IoT device can claim to be legitimate and persuade others to interact with it [10].

Data protection in fog computing mainly focuses on securing network communication and data storage. As IoT devices send data to the closest fog for processing and storage, there is a high chance that the data can be affected by unauthorized changes.

Besides providing services to IoT users, fog nodes also collect sensitive information about users that could violate their privacy. Here, the nearest fog node senses the identity and location of the IoT users and tracks their habits. Therefore, it is easy for an attacker to expose the privacy of the IoT user if they successfully identify the user [11], for example, by intercepting and monitoring the data of a smart meter in a house, an attacker can identify times when a house is empty.

The main contribution of this paper is:

- The design of a data encryption system that incorporates data integrity check, to provide a good encryption effect. This system was developed based on the recommendation made by Alzoubi et al. [12] that data integrity mechanism is one of the key components for securing data processing in fog environment.

The organization of this paper is as follows: An overview of previously related works on encryption of data in fog computing is presented in Sect. 2. The proposed hybrid cryptosystem is introduced in Sect. 3. Section 4 presents the implementation of our model and the comparative analysis of simulated results. Finally, conclusions are drawn in Sect. 5.

2 Related Works

According to [13], the safest approach to guarantee security in fog computing is by using the AES algorithm. However, deploying this system in a fog network consisting of heterogeneous devices could lead to high risk of key exposure; hence, there is a need of a trusted distribution of symmetric keys to establishing secure communication.

The encryption method proposed in [14] focuses on modifying ElGamal Algorithm by using a different prime number and public key for each user. The drawback of this system is that there was no integrity check of the extracted secret data.

In a literature by Saif et al. [15], the AES algorithm was used to provide a security layer against Man-in-the-Middle (MITM) attacks in fog computing. The algorithm was upgraded to a 512-bit key size and implemented on a data block size of 128 bits. Analysis of the algorithm shows that it takes more power and design area as compared to AES 256.

Enhancing the security of user data in the cloud using fog computing, a model developed by Johney et al. [16] used the AES algorithm to encrypt data, Reed-Solomon codes for error detection, and Message Digest algorithm 5 (MD5) to ensure data integrity. However, there are known vulnerabilities in the MD5 algorithm; hence, other hash functions should have been considered in this work.

A dynamic scalable model for securing data at rest in fog computing using Rivest Shamir and Adelman (RSA) algorithm, generalized RSA, and AES with Elliptic Curve Diffie–Hellman (ECDH) as a secure key exchange mechanism was proposed in [17]. Although the model was able to incorporated known cryptosystems, it lacks the presence of a data integrity mechanism that could prevent malicious attacks such as MITM.

A hybrid strategy was presented in [18], which combines steganography and cryptography in the fog environment. Despite the simplicity and low-time penalties attributed to emoticon technique, it is less resistant to common attacks that can take place in a fog environment such as brute force, MITM, and occlusion attacks. Furthermore, this system cannot be used for all types of multimedia, for example, audio.

A fusion of lightweight symmetric and asymmetric encryption algorithms was proposed by Khashan [19] to establish secure communications between fog devices and end-users. In other words, the scheme used Corrected Block Tiny Encryption Algorithm (XXTEA) to encrypt text data, and the symmetric key is encrypted using ElGamal-ECC. However, since XXTEA has a fixed key size (128 bits) and the number of rounds depends on the number of words, this makes the model vulnerable to chosen-plaintext attacks. More so, the generated keys and system parameters are sent to the devices that will perform the encryption in plain form.

3 Proposed System

In our work, we made use of the hybrid cryptographic technique (HCT), a technique that take advantage of the strengths of two or more encryption systems by combining them into a single platform. The rationale behind the usage of HCT is to utilize the strength of one encryption scheme to address the weakness of another type.

Our hybrid cryptosystem consists of AES 256, ECDH, and SHA-256. In our system, we decided to hash the encrypted file instead of hashing the plain file. This is done to prevent access to the plain file until the integrity of the file is checked.

The cryptosystem consists of two architectures: sender's and receiver's architecture.

3.1 Sender's Architecture (Encryption)

The program flow of this architecture as shown in Fig. 1 is described as follows:

1. a. Generate a shared secret key using ECDH.
   b. Take any file as input.
2. Produce a ciphertext by encrypting the selected file using AES 256 algorithm with the shared secret key.
3. Hash the ciphertext by applying SHA-256.
4. Combine the hashed ciphertext and the encrypted file as a payload and send it to the receiver.

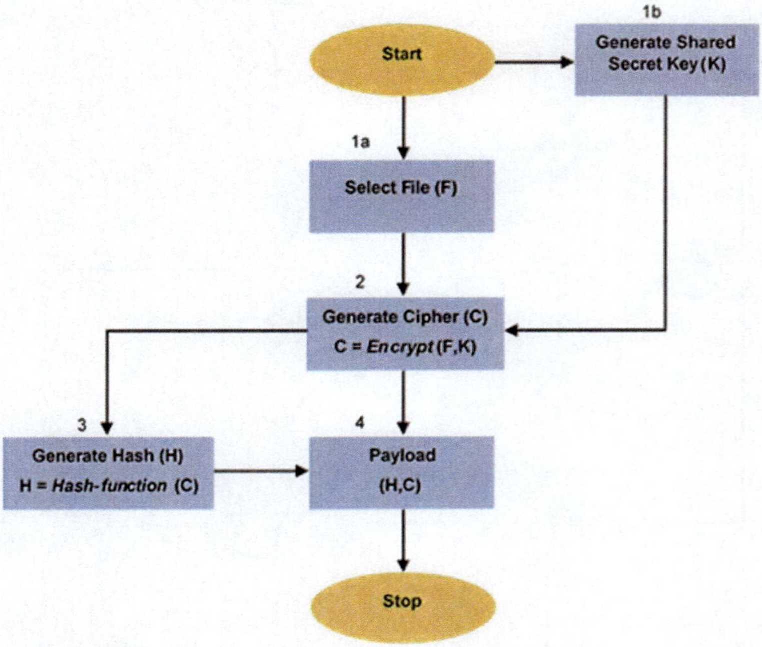

Fig. 1 Encryption process

3.2 Receiver's Architecture (Decryption)

The decryption phase converts the encrypted file into its original form, so that the receiver can access or read the file contents. Figure 2 shows the process follow of this architecture and it is described as follows:

1. Unbundled the payload received from the sender. The payload comprises two parts:
   a. The first part is the hashed ciphertext created by the sender.
   b. The other part is the ciphertext (encrypted file).
2. Hash the ciphertext by applying SHA-256.
3. Compare the receiver hashed ciphertext with the sender hashed ciphertext:
   a. If the comparison finds both hashes to be exact then: go to step 4.
   b. Otherwise: abort the process and discard the data.
4. Generate the shared secret key using ECDH.
5. Using the shared secret key and AES 256 algorithm, decrypt the ciphertext to obtain the plain file.

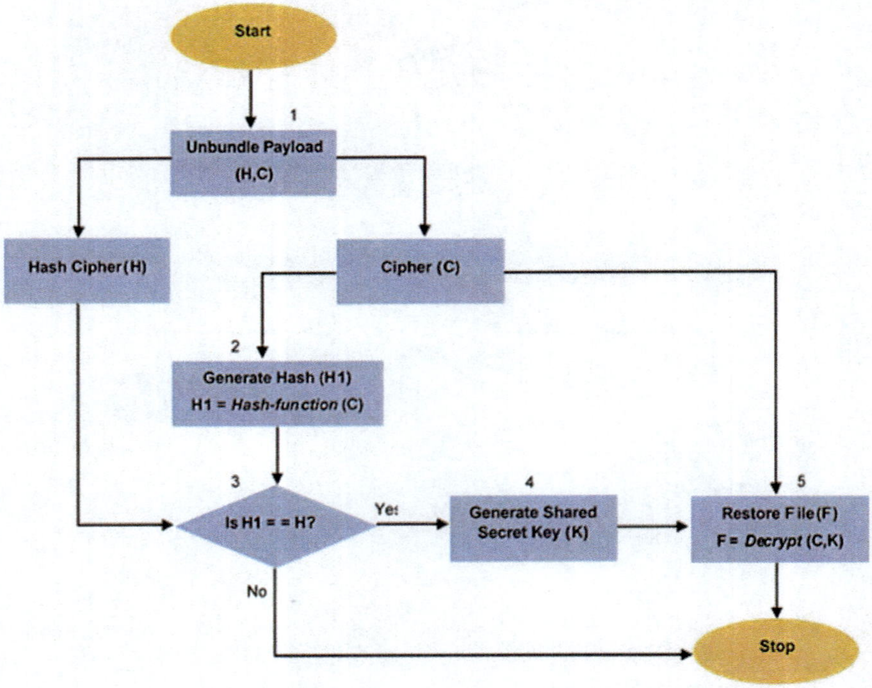

Fig. 2 Decryption process

3.3 Simulation Settings

The algorithms that make up our proposed model were enforced using inbuilt Java container classes inherited from packages named "*java.crypto*" and "*java.security.*" These packages are made available in Java Cryptography Architecture and Extension, for implementing unmanaged operations. The simulation environment was developed using Java Development Kit version 11 (JDK 11) and compiled using NetBeans IDE 12.3 with no changes made to the default settings. The key settings of the algorithms used are given in Table 1:

Table 1 Algorithms key settings

S/No	Algorithm	Key size (bits)	Block size (bits)
1	AES 256	256	128
2	ECDH	256	64
3	SHA-256	256	64

3.4 Data and Performance Metric

In the implementation of our model, the data which were considered are the petroleum datasets (version 1.0, 1.1, and 1.2), containing information of all known oils and gas deposits throughout the world obtained from [20, 21]. More so, we analyzed the performance of our model by measuring the execution time for encryption and decryption and their throughputs. The performance metrics are described as follows:

1. Execution Time (t): This is the time taken to convert plaintext to a ciphertext (i.e., encryption) or vice versa (i.e., decryption). It is measured in milliseconds (ms). For optimum performance of a fog computing security system, a lesser execution time is desired.
2. Throughput (Th) is defined as input file size (Fs) per unit execution time (t). For encryption, the throughput is calculated as the file size (i.e., plaintext) divided by the encryption execution time, whereas for decryption, it is the ciphertext divided by the decryption execution time. For optimum performance of a fog computing security system, a larger throughput is desired.

4 Results and Analysis

Testing of various files was done using the AES 256-ECDH algorithm and the proposed hybrid algorithm (i.e., AES 256-ECDH-SHA256 model). The performance of the models was recorded with the increasing size of files. The files size used in the test varies from 475KB to 4.70 MB. Furthermore, the experiments were conducted several times to ensure that the results are consistent and valid for different file sizes.

4.1 Results Based on Encryption and Throughput-Encryption

Encryption of file sizes ranging from 475KB, 704KB, 1.15MB, 3.20MB, and 4.70 MB was done using our model and the AES 256-ECDH model. The encryption execution time of these files is given in Table 2.

Furthermore, the throughput results for encryption are given in Table 3.

4.2 Results Based on Decryption and Throughput-Decryption

The decryption of file sizes ranging from 475KB, 704KB, 1.15MB, 3.20MB, and 4.70 MB was done using our model and the AES 256-ECDH model. The decryption execution time of these files is given in Table 4.

Table 2 Results of encryption execution time (ms)

File (MB)	Ciphertext (MB)	Proposed model	AES 256-ECDH
0.4745	0.9490	61	27
0.7030	1.4065	126	38
1.1835	2.3670	131	55
3.2864	6.5728	136	112
4.6994	9.3989	140	136

Table 3 Throughput results for encryption (MB/ms)

File (MB)	Proposed model	AES 256-ECDH
0.4745	0.0078	0.0176
0.7030	0.0056	0.0185
1.1835	0.0090	0.0215
3.2864	0.0242	0.0293
4.6994	0.0336	0.0346

Table 4 Results of decryption execution time

File (MB)	Ciphertext (MB)	Proposed model	AES 256-ECDH
0.4745	0.9490	20	16
0.7030	1.4065	44	41
1.1835	2.3670	63	53
3.2864	6.5728	124	102
4.6994	9.3989	163	132

Furthermore, the throughput results for decryption are given in Table 5.

Table 5 Throughput results for decryption (MB/ms)

Ciphertext (MB)	Proposed model	AES 256-ECDH
0.9490	0.0475	0.0593
1.4065	0.0320	0.0343
2.3670	0.0376	0.0447
6.5728	0.0530	0.0644
9.3989	0.0577	0.0712

Table 6 T-test analysis of encryption execution time (ms)

Model	Mean
Proposed model	118.80
AES 256-ECDH	73.60

4.3 Comparative Analysis of Results Obtained

Table 6 presents the result of the t-test analysis of encryption execution time (ms) for the proposed model and the AES 256-ECDH model. The results of the analysis indicated that the proposed model had an average encryption execution time of 118.8 (ms) with a difference of 45.20 (ms) higher than the AES 256-ECDH model. This is attributed to the presence of the message digest generation (i.e., hash). The bar chart presented in Fig. 3 revealed that our model performed fairly well in the last two results.

From Table 7, it can be observed that the proposed model had less throughput-encryption than the AES 256-ECDH model. The average throughput (MB/ms) was 0.0160 and 0.0243, respectively. The bar chart shown in Fig. 4 revealed that the throughput-encryption tends to be exact in both models in the last two results.

Table 8 presents the result of the t-test analysis of decryption execution time (ms) for the proposed model and the AES 256-ECDH model. The results of the analysis indicated that the proposed model had an average decryption execution time of 82.80

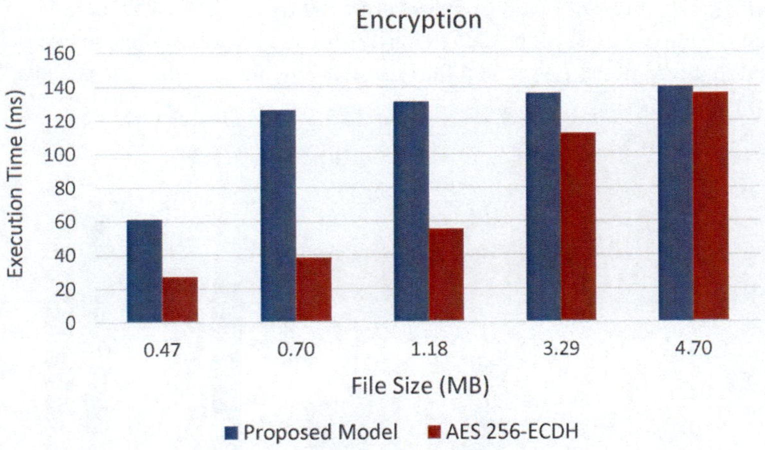

Fig. 3 Bar chart showing encryption execution time

Table 7 T-test analysis of throughput-encryption (MB/ms)

Model	Mean
Proposed model	0.0160
AES 256-ECDH	0.0243

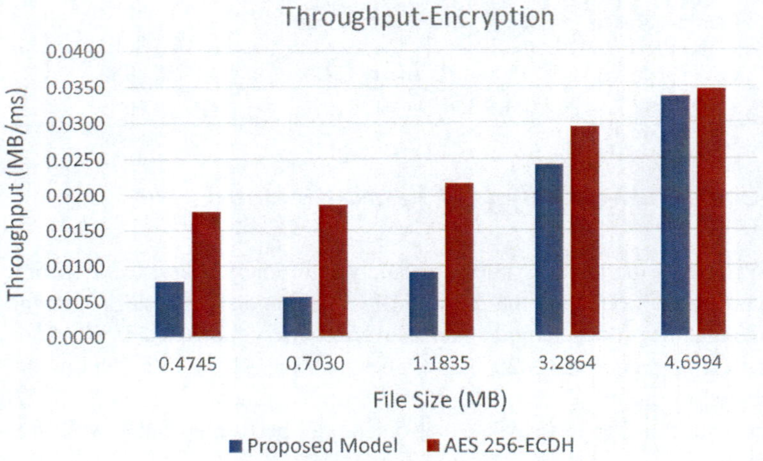

Fig. 4 Bar chart showing throughput-encryption (MB/ms)

Table 8 T-test analysis of decryption execution time (ms)

Model	Mean
Proposed model	82.80
AES 256-ECDH	68.80

(ms) with a difference of 14.00 (ms) higher than the AES 256-ECDH model. This is attributed to the presence of the data integrity check. The bar chart presented in Fig. 5 revealed that our model performed fairly well across the five data sizes considered.

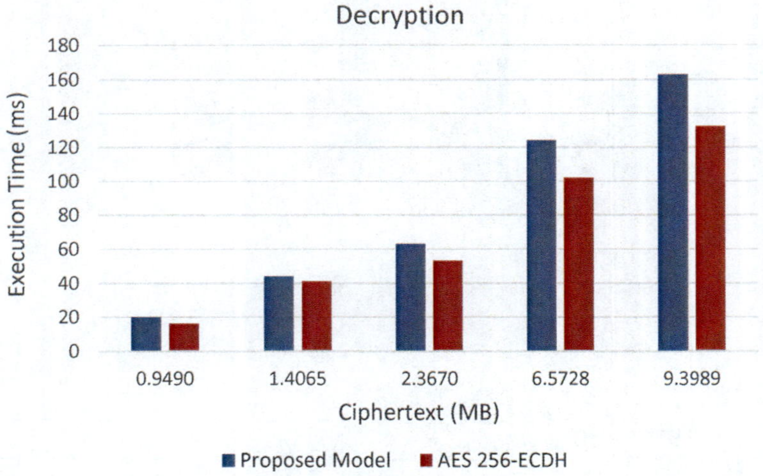

Fig. 5 Bar chart showing decryption execution time

Table 9 T-test analysis of throughput-decryption (MB/ms)

Model	Mean
Proposed model	0.0455
AES 256-ECDH	0.0548

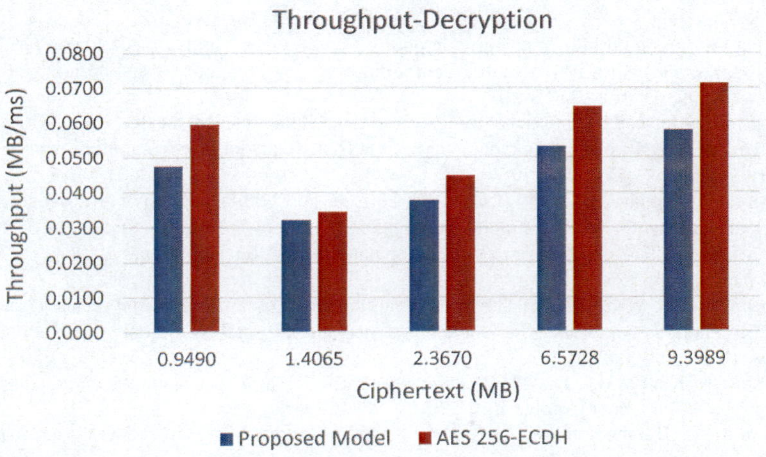

Fig. 6 Bar chart showing throughput-decryption (MB/ms)

From Table 9, it can be observed that the proposed model had less throughput-decryption than the AES 256-ECDH model. The average throughput-decryption (MB/ms) was 0.0455 and 0.0548, respectively. The bar chart shown in Fig. 6 revealed our model performed fairly good across the five data sizes considered.

5 Conclusion

This work aims to enhance the existing system developed using AES 256 and ECDH algorithms and evaluate its performance against the proposed system that incorporated data integrity check (SHA-256) without a recovery technique. The analysis of results shows that the proposed system performed fairly well as compared to the existing model in terms of execution time and throughput for decryption as it guarantees that data sent across the fog network would not be access if its integrity is compromised. However, as seen in Table 2, the size of the ciphertext produced by both models during encryption is two times the size of the original file. Given the characteristics of fog nodes as having limited memory capacity, in the future, we plan to incorporate compression algorithms to help reduce the size of the ciphertext or use other algorithms like Blowfish or Twofish.

References

1. Yousefpour A, Fung C, Nguyen T, Kadiyala K, Jalali F, Niakanlahiji A, Kong J, Jue JP (2019) All one needs to know about fog computing and related edge computing paradigms. J Syst Architect 98:289–330
2. OpenFog Reference Architecture for Fog Computing, https://www.iiconsortium.org/pdf/OpenFog_Reference_Architecture_2_09_17.pdf. Last accessed 10 Dec 2021
3. What is fog computing? Why fog computing trending now? https://medium.com/yeello-digital-marketing-platform/what-is-fog-computing-why-fog-computing-trending-now-7a6bdfd73ef. Last accessed 12 Mar 2022
4. Fog computing and the internet of things: Extend the cloud to where the things are. White Paper. 2016. https://www.cisco.com/c/dam/en_us/solutions/trends/iot/docs/computing-overview.pdf. Last accessed 11 Mar 2022
5. Usman MJ, Ismail AS, Chizari H, Gital AY, Chiroma H, Abdullahi M, Aliyu A (2019) Integrated resource allocation model for cloud and fog computing: toward energy-efficient Infrastructure as a Service (IaaS). In: Advances on computational intelligence in energy. Springer Cham, Switzerland, pp 125–145
6. Stojmenovic I, Wen S (2014) The fog computing paradigm: Scenarios and security issues. In: 2014 Federated conference on computer science and information systems. IEEE, Poland, pp 1–8
7. Khan AN, Kiah MM, Khan SU, Madani SA (2013) Towards secure mobile cloud computing: a survey. Futur Gener Comput Syst 29(5):1278–1299
8. Akomolafe OP, Abodunrin MO (2017) A hybrid cryptographic model for data storage in mobile cloud computing. Int J Comput Netw Inf Secur 9(6):53–60
9. Ashi Z, Al-Fawa'reh M, Al-Fayoumi M (2020) Fog computing: security challenges and countermeasures. Int J Comput Appl 175(15):30–36
10. Ni J, Zhang K, Lin X, Shen X (2017) Securing fog computing for internet of things applications: Challenges and solutions. IEEE Commun Surv Tutor 20(1):601–628
11. Mukherjee M, Matam R, Shu L, Maglaras L, Ferrag MA, Choudhury N, Kumar V (2017) Security and privacy in fog computing: challenges. IEEE Access 5:19293–19304
12. Alzoubi YI, Osmanaj VH, Jaradat A, Al-Ahmad A (2020) Fog computing security and privacy for the internet of thing applications: State-of-the-art. Secur Privacy 4(2):e145
13. Vishwanath A, Peruri R, He J (2016) Security in fog computing through encryption. Int J Inf Technol Comput Sci 8(5):28–36
14. Sowjanya V (2017) Security framework for sharing data in fog computing. Int J Adv Res Comput Commun Eng 6(6):422–427
15. Saif M, Shafique A, Mahfooz S (2019) Providing a security layer for Man-in-the-Middle attack in fog computing. Int J Comput Sci Inf Secur 17(6):19–25
16. Johney RK, Shelry E, Babu KR (2019) Enhanced security through cloud-fog integration. In: 2019 international conference on communication and electronics systems (ICCES). IEEE, India, pp 1530–1535
17. Husein T, Khalique A, Alam MA, Alankar B (2020) A dynamic scalable security model for data at rest in fog computing environment. Int J Innovat Technol Expl Eng 9(11):413–418
18. Kumar H, Shinde S, Talele P (2017) Secure fog computing system using emoticon technique. Int J Recent Innovat Trends Comput Commun 5(7):808–811
19. Khashan OA (2020) Hybrid lightweight proxy re-encryption scheme for secure fog-to-things environment. IEEE Access 8:66878–66887
20. Lujala P, Ketil Rod J, Thieme N (2007) Fighting over oil: Introducing a new dataset. Confl Manag Peace Sci 24(3):239–256
21. Peace Research Institute Oslo (PRIO) Petroleum Dataset, https://www.prio.org/Data/Geographical-and-Resource-Datasets/Petroleum-Dataset/. Last accessed 26 Feb 2022

Reducing Grid Dependency and Operating Cost of Micro Grids with Effective Coordination of Renewable and Electric Vehicle's Storage

Abhishek Kumbhar, Nikita Patil, M. Narule, S. M. Nadaf, and C. H. Hussaian Basha

Abstract The growth of smart grids has led to the number of challenges to maintain the power quality and reliability. On the other hand, the advanced technologies are helping to address the issues that arise due to the heterogeneous entities such as Intermittent Renewable Energy Resources (IRER) and Electric Vehicles (EV). The EV's storage can be utilized to support the energy management in the micro grids and other flexible areas like industries and educational organizations. In order to use EVs as virtual storage units, one has to ensure all the constraints and limitations that are involved. Above all, EV usage for grid energy management should not put EV owner in a chaotic state. There has been numerous EV control strategies proposed for Vehicle to Grid (V2G) and Grid to Vehicle (G2V) operations since last decade. This work mainly focuses on maximization of EV storage usage with consideration of battery degradation. A prioritization based EV strategy is proposed using Adaptive Neuro-Fuzzy Inference System (ANFIS) with four decision constraints.

Keywords EV battery degradation · EV · EV prioritization · Grid dependency · G2V · Optimal energy distribution · V2G

1 Introduction

At present, different EV strategies are proposed for frequency regulation, voltage regulation and other demand response programs at charging fleet. However, there is a gap between the academic research and the existing real-time scenario. Moreover, EVs are the additional loads which cause extra burden to the existing grid [1]. EV's storage can be utilized to support the grid frequency and voltage. Also, it can be

A. Kumbhar · N. Patil · M. Narule · S. M. Nadaf
SCRC, Nanasaheb Mahadik College of Engineering, Walwa, India

C. H. H. Basha (✉)
NITTE Meenakshi Institute of Technology, 6429, NITTE Meenakshi College Rd, BSF Campus, Yelahanka, Bangalore 560064, Karnataka, India
e-mail: hussaian.basha@nmit.ac.in

M. Thakur et al. (eds.), *Soft Computing for Problem Solving*, Lecture Notes in Networks and Systems 547, https://doi.org/10.1007/978-981-19-6525-8_49

exploited as the additional storage to support the stochastic nonconventional energy sources, which are classified as solar and wind. The EV charging fleet can be considered as a huge virtual storage unit that can be used to support the medium and high scale Renewable Energy Sources (RES). It can also be exploited in reducing the grid dependency and total energy cost. Authors have developed the queuing-based charging model in that mean time is determined by focusing on the charge storage points, and starting timings. Also, the charging request of the EVs waiting is also considered [2].

Fan et al. [3] have discussed the impact of the target State of Charge (SoC) of different electric vehicle storage at an instant plus there are introduced a quick charging model. In this model, the maximization of fleet revenue has been done by considering the fleet multiple operators and those are decide the limit and targets of SoC. The effect of Direct Current Fast Charging (DCFC) stations on Low Voltage (LV) and Medium Voltage (MV) distribution transformer's burden has been studied using stochastic EV mobility model [4]. Yang et al. [5] developed optimal strategy to choose the direct current quick charging stations. Here, the electric vehicle energy storage load is estimated through the process of randomizing EV mobility probabilities and a line up method is proposed plus an optimization technology is works based on the facility and distribution network technical confines.

Farkas, and Telek [6] developed an estimation strategy using the Markov chain model for EV entrance procedure for stochastic modelling of the EV charging stations. Roteringand and Ilichas introduced a technique to improve the economical profits of the present electric vehicle owners by considering and evaluating the electrical energy prices [7]. He et al. [8] proposed optimal charging strategy, where the charging rates are minimized for reducing the electrical vehicle charging cost along with the V2G plus G2V functionalities. Xiong et al. [9] illustrated the multiple types charging technologies by considering the major parameters are driver's interest, mobility pattern and electricity pricing [10]. Ota et al. [11] explained the EVs are working as a circulated revolving backup, where everyone performs independently at a basic position of mutual coupling without losing frequency synchronization. The load variation and spinning backups are selected subsidiary provision which is supplied by electric vehicles [12].

Lin et al. [13] introduced a decentralized scheduled technique, here, the electric vehicles are reimbursed to compensate the stochastic nature of production and demand, plus level the power deviations from the expected demand. An extensive amount of literature has been published on optimal ways of exploiting Electric Vehicle storage for supporting grid subsidiary facilities such as frequency variation and voltage [14]. However, the impact of SoC and EV degradation has not been considered fully while EVs are being operated in V2G and G2V modes.

This article's main focus is to minimize the grid dependency with the help of solar PV power and EV's storage. Initially, the 8-h spam is sliced into 96 intervals and regions of energy essential for load destruction are estimated along with energy need in each interval. A prioritization-based model is developed in order to take decision on choosing EVs to participate at a given time interval. A Markova chain model has been developed to estimate the usage probability of EV for V2G and G2V during upcoming intervals which help to prioritize the EVs in such a way that the battery degradation is minimized. Here, in Sect. 2, the EV mobility modeling is explained, and electric vehicle usages probability is given in Sect. 3. Section 4 gives the electric vehicle prioritization, and results and analysis is explained in Sect. 5. Finally, Sect. 6 gives the conclusion of the article.

2 EV Mobility Modeling

It is important thing to determine the basic availability of vehicle sat fleet condition for program them for V2G and G2V modes of operation. Fleet operator should ensure the trip requirements (target SoC at the time of departure) for grid provision. EV mobility technology gives the data about the EVs free time for energy provision and their inclination with essential SoC plus slackness (EV free time at the fleet).

2.1 Electric Vehicle Mobility Data

In this work, employee biometric information has been used to record the vehicle arrival and departure timings. Past one-year data are used to estimate the probability of the vehicle's availability at the institution. The travel distance from residence to institution is taken for each individual vehicle. The technical specifications of the EVs are taken into consideration in order to estimate the SoC droop due to travel. It is considered that each and every EV starts charge up to 0.8 SoC at home. The EV movement statistic is neglecting at the time of holidays by the way of leads to imprecise chances of entrance and exit. Here, the working time table of each employee expected to change from semester to semester and hence, the Probability Distribution Functions (PDFs) are obtained from the respective semester's mobility data. Let, the average travelling distance from home to organization of individual EV is l_i. The distance plus the battery range are considered to be similar for all the EVs. Mileage is taken as15 kWh/100 km, and battery capacity is taken as 40 kWh.

From the data, for each and every day, the arrival plus departure time PDFs of all the EVs are extracted. At starting day of the week, the *i*th electric vehicle running probability in the *j*th time period is considered as $p_{i,j}^{a}$ plus the exit is considered as $p_{i,j}^{d}$. The *i*th vehicle accumulative prospects at *j*th time period is represented as $p_{i,j}^{a,c}$ and $p_{i,j}^{d,c}$ respectively. By utilizing the both entering plus exit probabilities of electric vehicle, the availability probability of every vehicle in every time second determined by utilizing Eq. (1). The trajectory that represents the individual EV travel lengths is represented by L and it is given in Eq. (2). It is expected that all EVs starts charging to 0.8 SoC at the home before their departure. The SoC vector is determined at the time of arrival which is given Eq. (3). The G2V and V2G charge storage capabilities are determined by using Eqs. (4) and (5).

$$P_i^j = P_{i,j}^{a,c} - P_{i,j}^{d,c} \quad (1)$$

$$L = l_1\, l_2\, l_3\, l_4 \cdots l_N \quad (2)$$

$$\mathrm{SoC}_{\mathrm{ev}}^{\mathrm{arr}} = \frac{M}{C} * \left[l_1\, l_2\, l_3\, l_4\, l_5 \cdots l_N \right] \quad (3)$$

$$e_{i,j}^{g2v} = P_i^j * C * (1 - SoC_i^j - 0.2) \quad (4)$$

$$e_{i,j}^{v2g} = P_i^j * C * (1 - \mathrm{SoC}_i^j - 0.2) \quad (5)$$

2.2 Electric Vehicle Laxity

The laxity of vehicle is determined by using Eq. (6). The laxity level is very high then the EV tractability is maintained high for making the grid support [15]. The last term in the Eq. (6) gives the required charge time period of electrical vehicle to which is helpful for charging the EV up to the stage of SoC. The absence of vehicle for going to support the grid is represented as following Eq. (6).

$$L_{ev}^{i,t} = t_d^i - t - \frac{(\mathrm{SoC}_{ev}^{i,t=t_d} - \mathrm{SoC}_{ev}^{i,t})E_{ev}^{i,cap}}{\eta P_{rate}^{i,t}} \quad (6)$$

Let, $x + y + z + N$ then the terms N, x, y, and z are indicated as all vehicles availability, charging started vehicles for going to departure, number of vehicles which are included in the support of grid, and finally, vehicles are not available for any vehicle to power supply grid, and supply grid to vehicle.

2.3 Zones of Energy Need

The amount of power need is estimated in each interval of time using Eq. (7). Here, the grid specified power is the maximum amount of power authenticated to absorb from the grid. The 8 h day span is sliced 96 intervals (each of 5 min). The aggregate energy required from the power supply grid at each and every time period is determined by using Eq. (8). Equation (9) gives the total aggregated energy needed from the electric vehicle storage to supply the constant power to the load. Finally, Eq. (10) gives the total necessity energy required for every zone of charging.

$$P_{\mathrm{need}}^{t,\mathrm{grid}} = \left\{ - P_{\mathrm{solar}}^t + P_{\mathrm{load}}^t + P_{\mathrm{ev}}^{t,\mathrm{ch}} \right\} \quad (7)$$

$$E_{\text{need}}^{t,\text{grid}} = \sum_{i=t}^{t+15} P_{\text{need}}^{i,\text{grid}} \Delta t \tag{8}$$

$$E_{\text{need}}^{t,\text{ev}} = E_{\text{need}}^{t} - E_{\text{spec}}^{\text{grid}} \tag{9}$$

$$E_{\text{need}}^{z} = \sum_{t=1}^{M} E_{\text{need}}^{\text{ev},t} \tag{10}$$

2.4 Energy Distribution

The optimal scheduling of EV range for V2G and G2V is indicated. The scheduling of EV is illustrated in two phases, first phase is at zone level, plus the second is level of interval indication. In each and every zone, the amount energy capacity available from EV backup plus the power essential for the load request is functional for the best supply of vehicle storage including the time intermissions in all zone. Water Filling Algorithm (WFA) is applied for this accomplishment and its applied to the all inter communication systems for the required power supply to the all of the channels depending on the rate of noise range and accessibility of power.

3 Electric Vehicle Usages Probability

The prioritization mainly based on the motto to maximize the EV storage utilization. However, the battery degradation is also considered while doing the prioritization through minimizing the number of transitions from charging mode to discharge mode and vice-versa. In order to minimize this changeover's while EV is being utilized for V2G or G2V, expected EV usage rate for next two consecutive intervals is taken into account. Usage rate is taken as the probability of usage (prt) of the EV for the upcoming intervals i.e. $t + 1$ and $t + 2$. The chances of zones transmission is given in Fig. 1.

Fig.1 Chances of zone transition

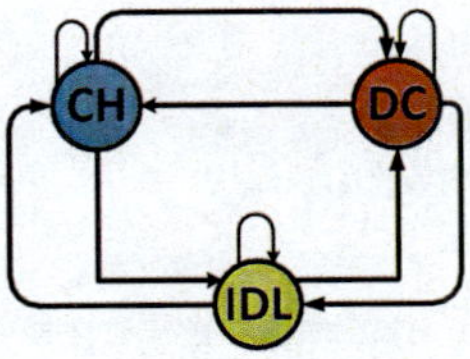

CH: Charging state
DC: Discharging state
IDL: Idle state

3.1 Transition Probability 'CS → CS', p_{11}^{t}

The utilization chance of vehicle at the time of '$t + 1$' is mainly depending on the previous stored electrical energy at the time duration of 't' that can be calculated by applying Eq. (11). Based on Eq. (11), the battery state of charge rate is equal to 0.8 then only the electric vehicle starts taking energy from the battery. At this condition, the probability of electrical vehicle condition is given in Eq. (12).

$$p_{j,u,1}^{t+1} = \frac{p_{j,\mathrm{SoC}}^{ch,t+1} * p_{j,\mathrm{avl}}^{t+1}}{p_{\mathrm{avl}}^{t}} = \frac{\{1 - \{\mathrm{SoC}_j^t + \Delta\,\mathrm{SoC}_j\}\} * p_{\mathrm{avl}}^{t+1}}{p_{\mathrm{avl}}^{t}} \tag{11}$$

$$p_{11}^{t+1} = \frac{\{1 - \mathrm{SoC}_j^t - \Delta\,\mathrm{SoC}_j\} * p_{\mathrm{avl}}^{t+1}}{p_{\mathrm{avl}}^{t}} * \{1 - p_z^{t+1}\} \tag{12}$$

where,

$$\Delta\,\mathrm{SoC}_j = \frac{c_j^{\mathrm{rate}} * n * \Delta t}{E_j}$$

3.2 Transition Probability 'CH → DC', P_{12}^{t}

The usage probability helps in finding the probability of transition from charging mode to discharging. Just as in Eq. (11), the below equation (Eq. (13)) gives the usage probability, which is calculated in the same way as the conditional probability. Now, the transition probability is estimated using the Eq. (14) with the help of zone transition probability.

$$p_{u,2}^{t+1} = \frac{p_{\mathrm{soc}}^{\mathrm{ch},t} * p_{\mathrm{avl}}^{t+1}}{p_{\mathrm{avl}}^{t+1}} = \frac{\{\mathrm{SoC}_j^t + \Delta\,\mathrm{SoC}_j\} * p_{\mathrm{avl}}^{t+1}}{p_{\mathrm{avl}}^{t}} \tag{13}$$

$$p_{1,2}^{t+1} = \frac{\{1 - \mathrm{SoC}_j^t + \Delta\,\mathrm{SoC}_j\}}{p_{\mathrm{avl}}^{t}} * p_z^{t+1} \tag{14}$$

3.3 Transition Probability 'CH → IDL', P_{13}^{t}

Equation (15) helps in determining the probability of the vehicle getting into idling state. Equation (16) gives the probability of mode switchover, i.e. from charging mode to idling mode.

$$p_{j,3}^{t+1,idl} = \{\mathrm{SoC}_j^t + \Delta\,\mathrm{SoC}_j\} \tag{15}$$

$$p_{j,13}^{t+1,idl} = \{\mathrm{SoC}_j^t + \Delta\,\mathrm{SoC}_j\} * \{1 - p_z^{t+1}\} \tag{16}$$

That is the reason behind the framing of the below equation (Eq. (17)) which includes the usage of the term $z(z_{12}^{t+1})$ which is nothing but the zone Transition's mode.

$$p_{j,13}^{t+1,idl} = \{\{\mathrm{SoC}_j^t + \Delta\,\mathrm{SoC}_j\} - Z_{12}^{t+1}\} * \left\{1 - \{1 - Z_{12}^{t+1}\} * p_z^{t+1}\right\} \tag{17}$$

3.4 Transition Probability 'DC → CH', p_{21}^t

Here in this case, it contradicts the charging to discharging transmission. In other words, it is the reversal of the CH → DC transition. The available probability and the usage probability for charging are the factors which are used in the estimation of the probability of transition from discharging mode to charging mode. Equation (17) shows the conditional probability which is being framed from the usage probability, and we use the same strategy in the Eqs. (18) and (19).

$$p_{u,1}^{t+1} = \frac{p_{\mathrm{soc}}^{\mathrm{dc},t} * p_{\mathrm{avl}}^{t+1}}{p_{\mathrm{avl}}^t} = \frac{\left\{1 - \left\{\mathrm{soc}_j^t - \Delta\,\mathrm{soc}_j\right\}\right\}}{p_{\mathrm{avl}}^t} * p_{\mathrm{avl}}^{t+1} \tag{18}$$

$$p_{21}^{t+1} = \frac{\left\{1 - \left\{\mathrm{soc}_j^t - \Delta\,\mathrm{soc}_j\right\}\right\} * p_{\mathrm{avl}}^{t+1}}{p_{\mathrm{avl}}^t} * p_z^{t+1} \tag{19}$$

3.5 Transition Probability 'DC → DC', p_{22}^t

For the electric vehicle's discharging, there are two conditions which need to be satisfied those are the state of charge of the battery should be greater than 0.2 and the electric vehicle should be present at the fleet. Equations (20), and (21) gives out this conditional probability.

$$p_{j,u,2}^{t+1} = \frac{p_{j,\mathrm{soc}}^{\mathrm{dc},t+1} * p_{j,\mathrm{avl}}^{t+1}}{p_{\mathrm{avl}}^t} = \frac{\left\{\mathrm{soc}_j^t - \Delta\,\mathrm{soc}_j\right\}}{p_{\mathrm{avl}}^t} * p_{\mathrm{avl}}^{t+1} \tag{20}$$

$$p_{22}^{t+1} = \frac{\left\{\text{soc}_j^t - \Delta\,\text{soc}_j\right\} * p_{\text{avl}}^{t+1}}{p_{\text{avl}}^t} * \left\{1 - p_z^{t+1}\right\} \tag{21}$$

3.6 *Transition Probability 'DC → IDL', $p_{23}^t p_{23}^t$*

At any interval, the probability of the successive interval can be found out through Eq. (22). In Eq. (23), we use the state of charge to find the probability of the vehicle going into idle state in charging mode Eq. (24) gives the transition probability of CH → IDL.

$$p_{j,3}^{t+1,idl} = \left\{1 - \left\{\text{soc}_j^t - \Delta\,\text{soc}_j\right\}\right\} \tag{22}$$

$$p_{j,23}^{t+1,idl} = \left\{1 - \left\{\text{soc}_j^t - \Delta\,\text{soc}_j\right\}\right\} * \left\{1 - p_z^{t+1}\right\} \tag{23}$$

$$p_{j,23}^{t+1,idl} = \left\{1 - \left\{\text{soc}_j^t - \Delta\,\text{soc}_j\right\} + z_{12}^{t+1}\right\} * \left\{1 - \left\{1 - z_{12}^{t1}\right\} * p_z^{t+1}\right\} \tag{24}$$

4 Electric Vehicle Prioritization

For obtaining the optimum battery storage of the vehicle and to make this practically feasible in the market, wise decision is mandatory regarding the choice of the EV, where state of charge and laxity are the major influencing factors. Needless to say that we must also consider the battery's life span, which is prone to degradation because of the G2V and V2G functionalities. Thus, in this regard, during G2V and V2G operations, it is suggested to maintain a low depth of discharge and moderate state of charge levels as told in Sect. 3.

4.1 *ANFIS Prioritization Procedure*

Jyh-Shing and Roger Jang developed the software ANFIS in the early 1990s. The ANFIS is a hybrid of fuzzy logic controller and artificial neural networks. For the ease of comparison, the input parameters and output parameters are standardized into per unit variables, and then in accordance to that each membership function has been designed. During the training of this hybrid methodology, the error tolerance was chosen to be 0.05 with 100 epochs.

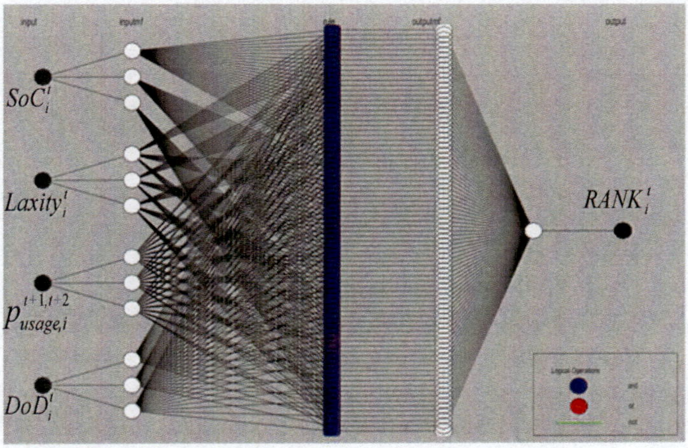

Fig.2 ANFIS structure with four inputs

4.2 *ANFIS Training Data*

For the preparation of the training data, we used to map the input to output ratio (rank). The EV have been sorted in the chronological order according to their rank, and priority was given depending on their rank. Lower the rank, higher will be the priority, and vice versa. That is, the EV with higher rank were preferred the least and those with lower rank were preferred the most. The depth of discharge and usage probability are the factors that influence the rank of the EV. Since there is inter dependency between these four variables (see in Fig. 2), it is hard to map them in a single go.

The preparation of ANFIS training data has been divided into two stages. In the first stage, the utility perspective inputs are considered. After that the customer perspective ones are considered in the second stage. In these stages, the rank of the particular EV is decided. For a given EV, at a particular time interval, the mean of these ranks gives the actual rank of that EV. Due to this, both the perspectives (i.e. utility perspective and customer perspective) are given equal weightage during the utilization of EV for grid support. Although, the entire training data are not shown here, we have presented a few sample cases (Table.1) (charging case).

5 Results and Analysis

The key objectives in this work are: reducing grid dependency and minimizing battery degradation. However, reducing grid dependency is depicted in terms of load flattening (grid intake power should match with authenticated power (200 kW)). From Fig. 3, it can be observed that the SoC is always maintained at moderate levels with

Table 1 Mapping of inputs to output (charging zone)

S. No.	SoC	Laxity	DoD	P_usgae	Rank
1	0.2	0.2	0.2	0.1	0.2
2	0.4	0.4	0.4	0.1	0.4
3	0.6	0.6	0.6	0.1	0.6
4	0.8	0.8	0.8	0.1	0.8
5	1.0	1.0	1.0	0.1	1

the proposed control strategy. This is mainly due to the consideration of SoC and P_usage_t + 1 in the prioritization. Due to consideration of P_usage_t + 1, the top SoC at its peak is ensured not to stay at the same level for more than 2 or 3 intervals of time. It directly impacts on battery degradation. So, with the prioritization parameters i.e. SoC and P_usage_t + 1, the battery SoC is maintained at moderate levels and peak SoC durations are reduced/avoided. Figure 3 shows the comparison of the cases, with and without the proposed prioritization strategy.

The effectiveness of WFA in load flattening is depicted in Fig. 4 by comparing the cases, with and without WFA (optimal energy usage). Here, the load flattening is represented in terms of grid power deviation from the authenticated amount of power. In order to observe the impact of prioritization, Fig. 4 is shown with two cases: with and without prioritization. The prioritization parameters, SoC and Laxity will have impact on usage of EV storage and hence will have impact on load flattening. The consideration of both SoC and Laxity in prioritization can be examined with the help of Fig. 5. The effectiveness of the prioritization is presented in Fig. 6. In this section, each decision variable's impact on SoC levels is presented.

Figure 7 gives a comprehensive view on how the absence of each decision variable impacts the DoD level. Here, 'load flattening' is an indirect objective that reflects the main objective which is 'minimizing the grid dependency'. By maximizing the 'load flattening' means that the load is matching with the authenticated grid power intake.

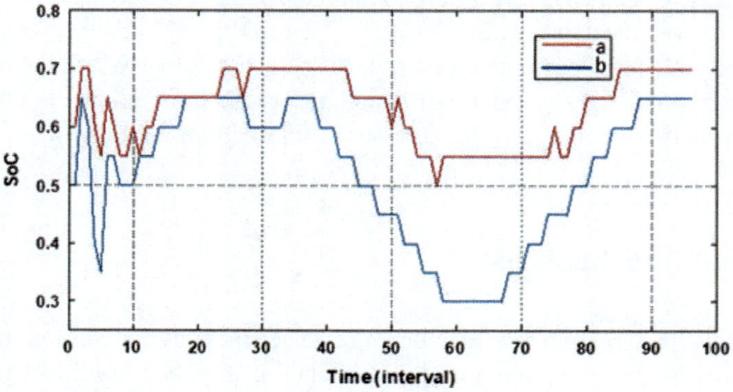

Fig.3 SoC level of EV1: **a** with prioritization and **b** without prioritization

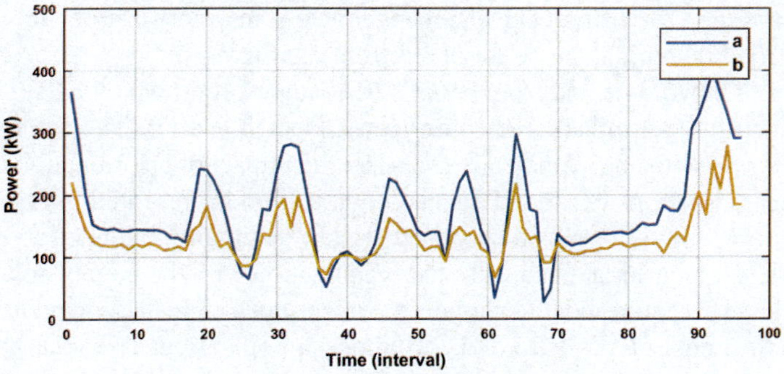

Fig. 4 Load profile: **a** without optimal energy distribution and **b** with optimal energy distribution

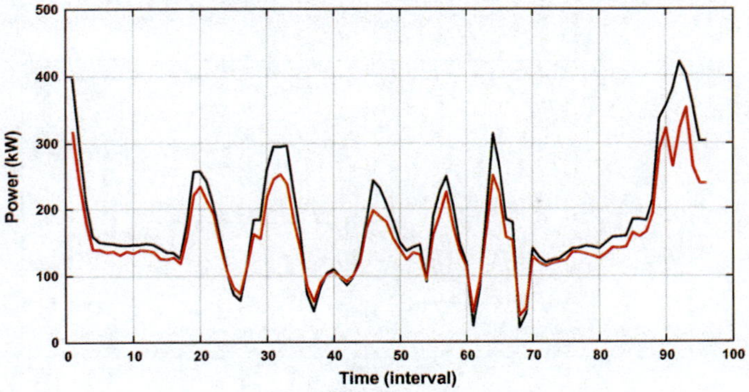

Fig.5 Grid power: **a** with prioritization and **b** with prioritization

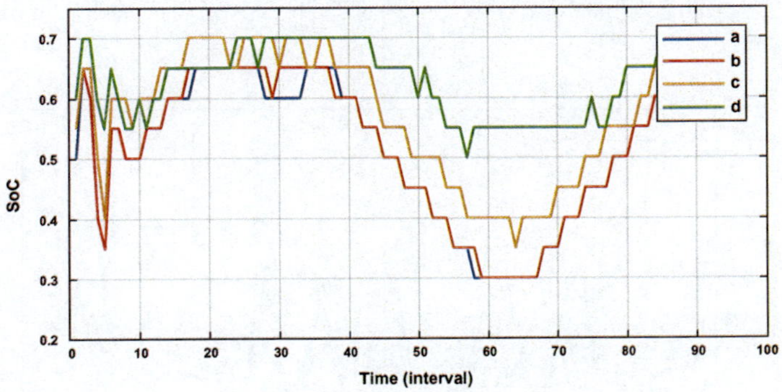

Fig. 6 Impact of prioritization on SoC: **a** Without prioritization **b** If SoC and Laxity are only the decision parameter, **c** If P_usage_t + 1 and Depth of Discharge (DoD) are only the decision parameter, **d** If all parameters are considered

With the absence of different decision parameters in prioritization, the resultant impact on load flattening can be observed from the Fig. 8.

Amount of power injected from EVs (V2G) and power drawn by EVs (G2V) will be influenced by Optimization Electric-load Dispatch (OED). The same is represented in Fig. 9 (with and without OED). Also, the impact of prioritization on aggregate power transactions between EVs and organization is presented in Fig. 10 (with and without prioritization). The proposed strategy is mainly intended to minimize the dependency on grid in an educational organization with the help of solar PV and EVs. The battery requirement to support the uncertainty in load demand is compensated with the help of EV's V2G and G2V modes of operation. The solar power also plotted against EV power transactions (V2G and G2V) in Fig. 11. It is plotted for the day time, it can be seen that both solar power and load are reaching their maximums as the reach midday. It can also be observed that the solar power is exceeding the

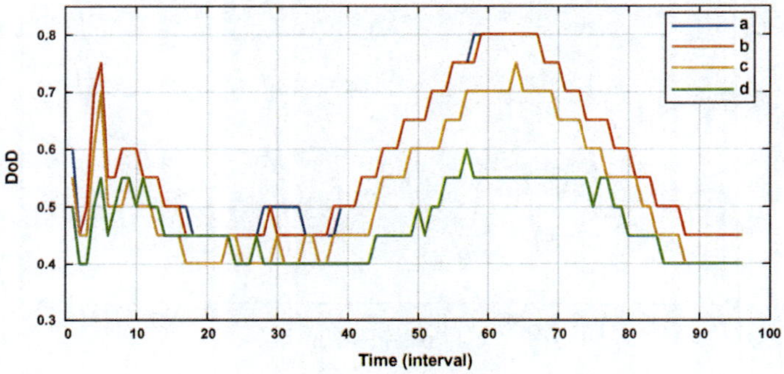

Fig. 7 Impact of prioritization on DoD: **a** Without prioritization **b** If SoC and Laxity are only the decision parameter, **c** If P_usage_t + 1 and DoD are only the decision parameter, **d** If all parameters are considered

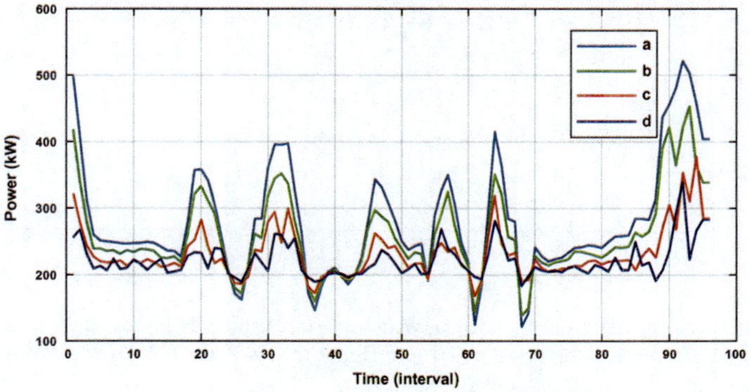

Fig. 8 Impact of prioritization on load flattening: **a** Without prioritization, **b** If P_usage_t + 1 and DoD are only the decision parameter, **c** If SoC and Laxity are only the decision parameter, **d** If all parameters are considered

load demand where we need to utilize EV storage (G2V) in order to store the excess power. Here the –ve power in indicates G2V mode of operation where the EVs will get charge from excess solar power.

The amount of power transactions through EVs (V2G and G2V power) is compared with the actual load in Fig. 12. Due to the excess amount of solar power, EVs will act in G2V mode in order to store the excess power from solar. The effectiveness of the prioritization can also be seen in terms of available aggregate power from all the EVs that are liable to use for energy support (V2G/G2V). Due to consideration of SoC and laxity during EV prioritization, EVs can be chosen in such a way that more number of EVs are available at each interval of time. However, in order to create a win–win situation, there are another two decision parameters which work in favor to battery life time improvement. Hence, the consideration of decision variables leads to variations in amount of power available at a given time interval.

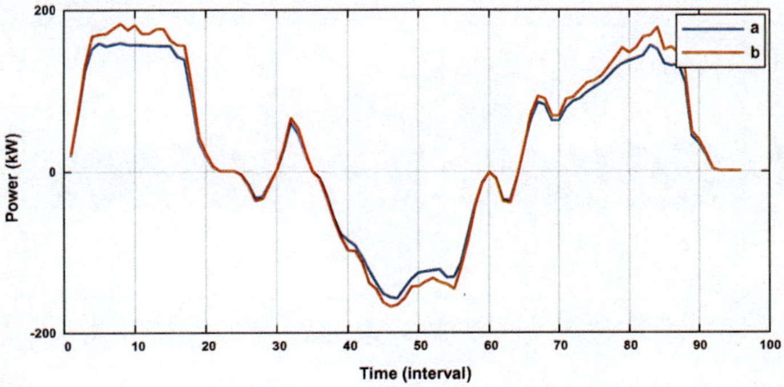

Fig. 9 Amount of V2G and G2V power transaction: **a** without optimal energy distribution and **b** with optimal energy distribution

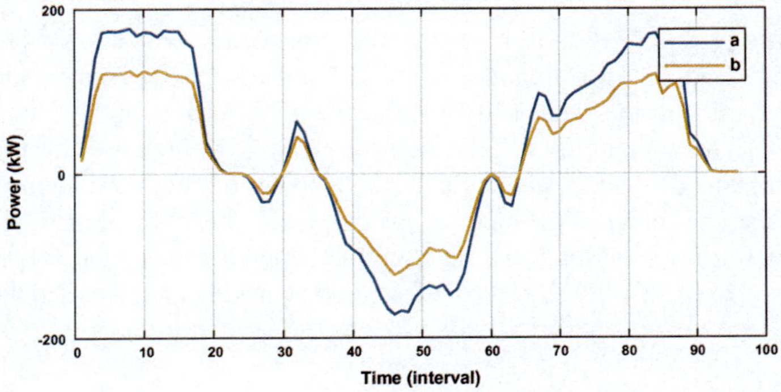

Fig. 10 Amount of V2G and G2V power transaction: **a** with prioritization and **a** without prioritization

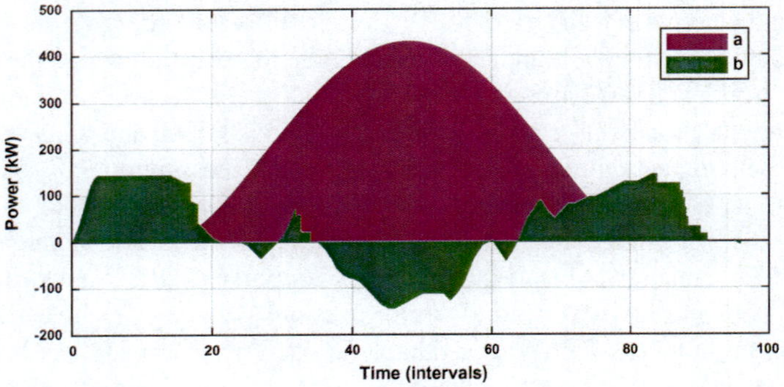

Fig. 11 Solar PV power versus power from EVs

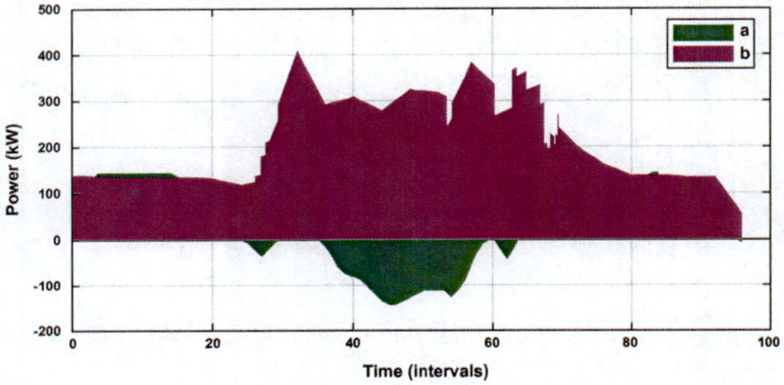

Fig. 12 Load power demand versus Power from EVs

6 Conclusion

The purpose of this work is to minimize the dependency on the grid, through the usage of G2V and V2G functionalities in order to increase the life span of the battery. For the prediction of the availability and the amount of charge in the battery during G2V and V2G functionalities in the successive intervals, we have used the probability analysis. It helps the user in making decision whether or not to use the battery, and also regarding the choice of the mode at any particular interval. In the provided zone, in any interval, for even load sharing from the battery, we used the WFA. For the prioritization of the EV in V2G and G2V support at any interval, we used the ANFIS package.

References

1. Mohseni P, Stevie RG (2009) Electric vehicles: holy grail or fool's gold. In: Power energy society general meeting, 2009, Jul.26–30, 2009, pp.1 Calgary, AB: Institute of Electrical and Electronics Engineer
2. Zenginis I, Vardakas JS, Zorba N, Verikoukis CV (2016) Analysis and quality of service evaluation of a fast charging station for electric vehicles. Energy 112:669–678
3. Fan P, Sainbayar B, Ren S (2015) Operation analysis of fast charging stations with energy demand control of electric vehicles. IEEE Trans Smart Grid 6(4):1819–1826
4. Yunus K, De La Parra HZ, Reza M (2011) Distribution grid impact of plug-in electric vehicles charging at fast charging stations using stochastic charging model. In: Proceedings of the 2011 14th European conference on power electronics and applications. IEEE, pp 1–11
5. Yang Q, Sun S, Deng S, Zhao Q, Zhou M (2018) Optimal sizing of PEV fast charging stations with Markovian demand characterization. IEEE Trans Smart Grid 10(4):4457–4466
6. Farkas C, Telek M (2018) Capacity planning of electric car charging station based on discrete time observations and MAP (2)/G/c queue. Periodica Polytechnica Elect Eng Comput Sci 62(3):82–89
7. Ramalingeswar JT, Subramanian K (2021) A novel energy management strategy to reduce gird dependency using electric vehicles storage in coordination with solar power. J Int Fuzzy Syst (Preprint), 1–17
8. He Y, Venkatesh B, Guan L (2012) Optimal scheduling for charging and discharging of electric vehicles. IEEE Trans Smart Grid 3(3):1095–1105
9. Xiong Y, Gan J, An B, Miao C, Soh YC (2016) Optimal pricing for efficient electric vehicle charging station management. In: Proceedings of the 2016 international conference on autonomous agents and multiagent systems, pp 749–757
10. Liu C, Chau KT, Wu D, Gao S (2013) Opportunities and challenges of vehicle-to-home, vehicle-to-vehicle, and vehicle-to-grid technologies. Proc IEEE 101(11):2409–2427
11. Ota Y, Taniguchi H, Nakajima T, Liyanage KM, Baba J, Yokoyama A (2012) Autonomous distributed V2G (vehicle-to-grid) satisfying scheduled charging. IEEE Trans Smart Grid 3(1):559–564. https://doi.org/10.1109/TSG.2011.2167993
12. Sortomme E, El-Sharkawi MA (2011) Optimal scheduling of vehicle-to-grid energy and ancillary services. IEEE Trans Smart Grid 3(1):351–359
13. Lin J, Leung KC, Li VO (2014) Optimal scheduling with vehicle-to-grid regulation service. IEEE Internet Things J 1(6):556–569
14. Xu B, Dvorkin Y, Kirschen DS, Silva-Monroy CA, Watson JP (2016) A comparison of policies on the participation of storage in US frequency regulation markets. In: 2016 IEEE power and energy society general meeting (PESGM). IEEE, pp 1–5
15. Le Floch C, Kara EC, Moura S (2016) PDE modeling and control of electric vehicle fleets for ancillary services: a discrete charging case. IEEE Trans Smart Grid 9(2):573–581

A Review Survey of the Algorithms Used for the Blockchain Technology

Anjana Rani and Monika Saxena

Abstract A new emerging technology, i.e., blockchain technology, is a distributed, encrypted database model and Peer-to-Peer (P2P) network transaction system that lay outs secure execution of operations that are implemented on a decentralized ledger database shared among multiple nodes which are useful in building the blockchain network; for this, the third party is not needed and it can be done with the help of consensus algorithms. Blockchain depends on cryptography and consensus mechanisms accompanying other algorithms for establishing strong security. The most widely used algorithms are the hashing and the consensus algorithms. But only these two algorithms cannot serve the requirements of every application. So, this paper presents a comprehensive overview of blockchain technology and also focuses on all the different types of algorithms used in blockchain technology. This paper also illuminates the future study directions.

Keywords Blockchain · Digital signature · Hashing · Peer-to-Peer network · Zero-Knowledge Proofs · Consensus algorithms

1 Introduction

At the start, a time-stamp structure similar to the blockchain was introduced by Haber and stornetta in 1991 [1]. However, at the moment, the most exciting tech trend is the blockchain technology, which was first conceptualized in 2008, and its first open-source implementation was installed in 2009 as an essential component of Bitcoin. Both were proposed by an entity called Satoshi Nakamoto [2]. Blockchain technology is a distributed, encrypted database model and Peer-to-Peer (P2P) network transaction system that lay outs the secure implementation of operations that are carried out on a decentralized ledger database that is shared among many nodes which are useful in building the blockchain network. It is a significant security system that is formed on cryptography, consensus algorithms, and also on communication technology [3].

A. Rani (✉) · M. Saxena
Banasthali Vidyapith, Tonk, India
e-mail: anjanarani2814@gmail.com

M. Thakur et al. (eds.), *Soft Computing for Problem Solving*, Lecture Notes in Networks and Systems 547, https://doi.org/10.1007/978-981-19-6525-8_50

However, its possible uses are expansive, encircling computerized "smart" contracts, logistics and supply chain provenance, and security as well as defense in opposition to the stealing of the identity [2]. Cryptocurrencies such as Ethereum and Bitcoin are held up by blockchain technology. In other words, the integrity along with the security in any system can be enhanced by blockchain technology.

By the end of 2022, the market cap of global cryptocurrency reached $3 trillion—higher than before. A worldwide prediction was made that the yearly income of blockchain technology-based enterprise applications will reach $19.9 billion by 2025 [4]. According to Deloitte's global blockchain survey 2021, industries will see the new income streams from the digital assets, blockchain, and cryptocurrency solutions. And global spending on blockchain technology is on the rise as one investigating firm predicts that it will increase from US$5.3 billion (2021) to US$34 billion (2026) [5].

This paper is further divided into 6 sections, where Sect. 2 talks about the literature review. Section 3 introduces blockchain technology. Algorithms used in blockchain are explained in Sect. 4. Section 5 talks about some future trends, and Sect. 6 concludes this paper.

2 Literature Review

Jahan et al. [6], The algorithm, i.e., satellite chain formation algorithm, is used in this paper and also SHA-256 hash algorithm is implemented so that a new blockchain architecture, i.e., parallel blockchain, can be designed, which will be further very helpful in easily, securely, and quickly storing the land-based documents.

Ahmed et al. [7], This paper is based on the smart parking solutions which use the blockchain technology so that the integration of the infrastructure can be easily done. An innovative and intelligent scheme for parking has been set up so that the security of the data of all the stakeholders is maintained accurately. In short, this paper aimed to enhance security as well as privacy via blockchain technology.

White et al. [8], Picture hashing and blockchain technology both are used in this paper to create a system for the verification of an image. So, this paper shows whether authentication of the images can be done via blockchain technology or not.

Parmar et al. [9], The importance of the hash functions is shown in this paper, and it also gives the comparison between the cryptographic techniques such as SHA-1, SHA-2, SHA-3, MD4, and MD5 by using the blockchain technology based on the speed and security. This paper also focuses on the need for using the secured hash algorithm.

Uddin et al. [10], CBCIOT, i.e., combination of the blockchain and IoT, is a consensus algorithm that is designed in this study for the IoT applications which are based on the blockchain. With this algorithm, scalability concerning the validation and also the verification rate can be improved and are compatible with the IoT devices where little delay can be tolerated.

Idrees et al. [11], A thorough analysis of the technology, i.e., blockchain technology, is presented in this paper and it also provides detailed solutions so that the blockchain security for industrial applications can be enhanced. This article also explores the ability of blockchain technology to revolutionize the various industrial applications like health care, IoT, supply chain industry, etc.

Xuan et al. [12], The idea of prestige is introduced in this paper, so that the prestige-based ECBCM (Edge Computing Blockchain Security Consensus Model) can be presented, and also the mechanism of node replacement is also introduced so that the fault tolerance can be ensured. This study provides solutions for the problems of security, efficiency, and the flexibility of the ECBCM.

Sideris et al. [13], This paper provides two architectures of SHA-3 algorithm, first without custom instruction and the second one with the hardware of floating point, in the Keccak hash function by using the NIOS-II processor. Thus, with the help of these architectures, it will help in providing a secure SHA-3 algorithm.

3 Blockchain

A distributed database technology, i.e., blockchain technology, provides the ledger records which are resistant to change. It was first introduced in 2009 by Satoshi Nakamoto for the design as well as the development of the cryptocurrency, i.e., Bitcoin [2]. So, we can say that a blockchain is a shared, immutable, distributed database or ledger in which all the transactions are registered in concern with all the participating forks. According to the Blockchain Revolution of the Don and Alex Tapscott, "Blockchain technology is a trustworthy digital ledger of economic transactions that can be organized to document not only the monetary negotiations but also all the digitally valuable things." It permits the storage of all the transactions into unchangeable records, and every record is distributed among all the participants' forks. In this technology, security is assured because it is using strong public keys, strong cryptographic hash, and also complete decentralization [2, 5].

3.1 Blockchain Structure

A blockchain is a sequence of blocks containing the list of the transactions which are validated by all the participating nodes. So, the data is organized by the blockchain into segments known as blocks. Figure 1 shows the structure of the blockchain, which

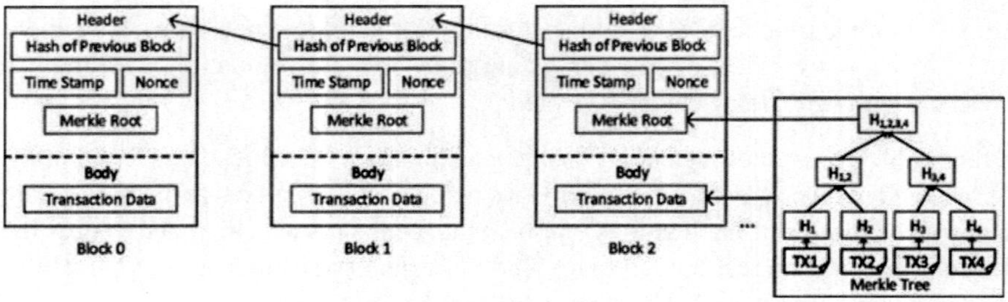

Fig. 1 Structure of blockchain [16]

shows that the blocks are further divided into Block Header and Block Body. Block Body consists of the transactions, and the Block Header consists of the [14–16]:

1. **Hash of the previous block**: The hash value can be obtained after passing the previous block's header to a hash function. And in this way, the blockchain grows by linking new blocks to the hash of the earlier block which is stored in the present block.
2. **A timestamp**: It helps in recording the time at which the block is created.
3. **Nonce**: It is useful in creating as well as validating the block.
4. **Merkle root**: It makes the verification of the transactions effortless as it contains the hash value of all the transactions, so a small change in the transaction produces a different Merkel root, so while verifying the transaction there is no need to verify all the transactions in a block as it can simply be done by comparing the Merkel roots.

3.2 Working of Blockchain

Blockchain is the technology that lies in the dominion of the distributed ledger technology which allows the transactions without using a central authority so that it can function in a decentralized manner [17].

Figure 2 shows the working of the blockchain technology in which a transaction will take place between A and B, and this transaction will be verified by the other nodes. The transaction is not acknowledged if the transaction fails or is not validated, and if all the nodes will verify the transaction, then that transaction will be added to the database or ledger. So, we can say that blockchain is made up of chaining the information of the transactions in the blocks and also storing them in sequential order [18].

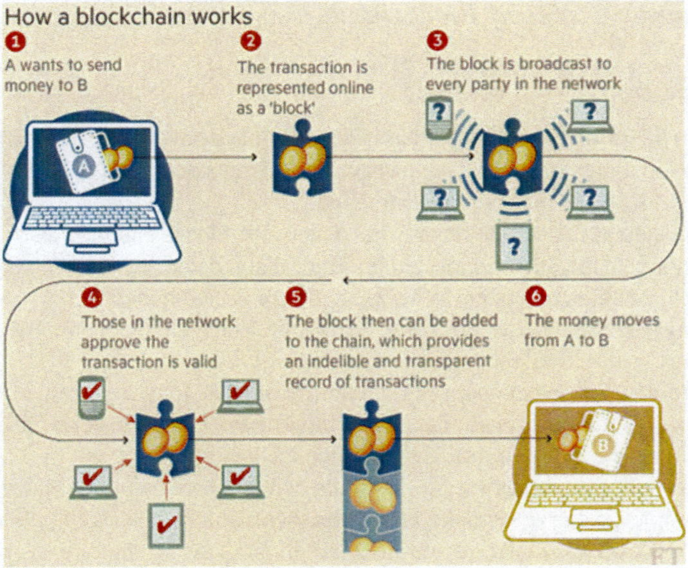

Fig. 2 Working on blockchain [16]

3.3 Types of Blockchain

Earlier, blockchain technology started as a public permissionless technology. After that, other types were also created as a combination of public/private and permissionless/permissioned. Broadly, blockchain can be classified into three categories [14, 19]: 1. public blockchain, 2. private blockchain, and 3. consortium/hybrid blockchain. These three types are explained below:

1. **Public blockchains (permissionless)**: These are permissionless, non-restrictive, and completely independent of organizations, and anyone can sign in to a blockchain network to become an authorized node with the only needed thing, i.e., the Internet.
2. **Private blockchains (permissioned)**: A blockchain network working in a restrictive environment just like a closed network or under the control of a single entity is known as the private blockchain.
3. **Consortium blockchains**: It is also known as the federated blockchain, which has both the features of public as well as private blockchain networks. Essentially, it is a private blockchain with limited access to a particular group; by this, it eliminates the risks that come up with just one entity controlling the network on a private blockchain. It tends to be more secure, scalable, and efficient than the public blockchain network as it offers access controls.

3.4 Characteristics of Blockchain

Following are the key features [4] that are consistent with the technology:

- **Decentralization**: This means that it does not have any administrative authority or a person to look after the framework. Instead, an association of nodes manages the network which makes it decentralized.
- **Transparency**: Each node records the data of the blocks and then distributed the data among all the other connected nodes so that it can be available to each node; by this, it established openness between the connected nodes.
- **Autonomy**: As the nodes are connected, any change can happen only when the change is accepted by the nodes in the majority.
- **Immutable**: For blockchain, immutability means that the data after being combined to the blockchain cannot be modified or changed unless at the same time someone can take control of more than 51% of nodes.
- **Security**: The systems of the blockchain technology are naturally secure because the asymmetrical cryptography is used by them which consists of public key set which are visible to anyone and also the private key set that are visible to the owner only. These keys are very useful to assure the control of the transactions and also the un-tamperability of the transaction.
- **Anonymity**: This feature of the blockchain supports privacy which is achieved by authenticating the transactions but it does not reveal any confidential data of the parties that are intricated in the transaction.

4 Algorithms Used in Blockchain

The network of blockchain abides across a system of nodes. Whenever a new transaction takes place, the blockchain is determined across this system before these new transactions can be integrated as the blocks in the chain. Thus, a coalition of nodes is needed to put that block in the blockchain system. The platform of the blockchain is trustable as an exact copy of the series in which every transaction is shown to the whole network. If there is anyone who tries to cheat, they can be identified easily. But sometimes, a few things can go wrong in other areas of blockchain-based applications which results in compromise. For this reason, blockchain uses several algorithms to increase security and to provide users a safe environment to carry out their transactions.

4.1 Cryptography Algorithms

It is one of the essential requirements in the blockchain as the network grows enormously; it will be difficult to confirm that all the information on the blockchain is

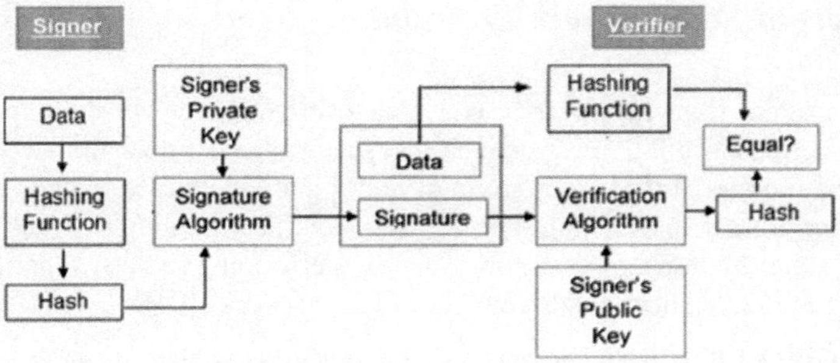

Fig. 3 Block diagram of digital signature [22]

secure from unfavorable threats [20]. The main objective of the cryptography algorithms is to prevent the third party from eavesdropping on private transactions over a blockchain network [21]. Digital signatures and hashing are the two most common types of algorithms used for blockchain security.

Digital Signatures

Trust in the blockchain is established by the cryptographic proof systems, i.e., digital signatures. It is a technique in which a person is bound to the signatures of digital data and then the receiver and the third party can independently verify this binding to access the data. It works on the public key cryptography architecture. According to Fig. 3, digital signature usually is made up of two parts: The first one is the signature algorithm (useful in generating the digital signature over the message, this digital signature is further controlled by the signature key (private key), and this private signature key is controlled by the signer of the message), and the second one is the verification algorithm (which is used to verify the digital signature, and this algorithm is further controlled by the verification key (public key), and this key is kept public so that it can be easily verified by the user) [22].

Hashing

Hashing is another cryptographic method on which blockchain is dependent. It is a method that converts any type of data into a string of characters and provides security to the system through encryption. In simple terms, hashing means taking an input of any length and giving out an output of a length that is fixed, and the fixed-length output is known as an equivalent hash [23].

SHA-256, evolved by the National Security Agency (NSA) in 2001, is the most famous of all the cryptographic hash functions as it is used extensively in blockchain technology. It is a hashing algorithm that is used to convert any length into a fixed-size string of 256 bits (32 bytes). In this, a message is inputted, and then, a hash function, i.e., SHA-256, gives an output known as "hash" or "message digest" [22].

4.2 Peer-To-Peer Network Protocol

The Peer-to-Peer term itself was used by Satoshi Nakamoto in his paper where Bitcoin is defined as a P2P electronic cash system [2]. It is a decentralized network communication model that has a group of nodes that together store and share data where every node acts as a single peer, there is no central authority, and thus, all nodes are equal by having equal power and by performing the tasks that are same for every node [22]. There are three types of P2P networks [24]:

- **Unstructured P2P networks**: The nodes that randomly form the connections to each other are useful in making the unstructured P2P network. Examples are Gnutella, Gossip, and kazoo.
- **Structured P2P networks**: In this, the nodes are organized and every node can effectively search the network for desired data.
- **Hybrid P2P network:** In this, the P2P and client–server models are combined; this network tends to present improved performance than the structured and unstructured networks.

The distributed P2P network and consensus algorithms, when combined, provide blockchain a high degree of resistance to the threatful activities. From file-sharing networks to the platforms of energy trading, P2P can serve many other distributed computing applications. It is the core of the blockchain that makes cryptocurrencies feasible as its architecture provides decentralization and security and also removes dependencies on the third party.

4.3 Zero-Knowledge Proofs

Another example of a blockchain algorithm used for security on blockchain networks is known as Zero-Knowledge Proofs (ZKPS). It serves as a consensual decision-making process in which it allows one party to confirm the authenticity of information to another party. ZKP's efficiency focuses on the fact that the prover does not have to disclose any information about the transaction. Thus, in this way it can safeguard the decentralized nature of the blockchain [25]. So, while making the transaction, it maintains the privacy of the user's sensitive information.

The advantages of ZKP are as follows:

- ZKP does not involve any complex encryption method.
- It is secure by the fact that it does not require anyone to reveal any kind of information.
- It shortens the transactions on blockchain; by this, users would not have to worry about the information storage.

By using Zero-Knowledge Proof in the blockchain world, we will be able to build an end-to-end trust without disclosing any extra information. It can also help avoid data leakage as it builds a secure channel for the users so that they can engage their information without revealing it [26]. Another important use case of ZKP is in the field of storage utility as it comes with a convention that not only prevents the storage unit but also the information within it, and when it is integrated wisely, it helps in making it unfeasible to intercept the private blockchain transactions.

4.4 Consensus Algorithms

It is an approach through which every peer of the blockchain network arrives at a common assent about the available state of the distributed ledger [3]. In this way, the reliability is achieved in the blockchain network by the consensus algorithms and initiates the trust among the unknown peers in a distributed computing domain. The consensus protocol makes sure that a new block that is added to the blockchain is the only accuracy at which all the nodes in the blockchain are agreed upon. To accomplish the following, a consensus mechanism is required:

- To give all participants/nodes an equal opportunity to generate new blocks, i.e., to keep the network decentralized.
- To protect the network from hacking attacks and malicious behavior.
- Based on the predetermined rules reward the block creators.

PoW (Proof of Work)

Satoshi Nakamoto comes up with the original consensus algorithm to control the transactions in the network of Bitcoin [2]. It is used to authenticate the transaction and generate a new block in the chain. According to this algorithm, an association of people, i.e., minors, enters into a competition with each other so that they can complete the transaction easily on the network. Mining is the process in which these minors compete with each other. The minor who successfully generates a valid block gets the reward. It can be executed in the blockchain via the Hashcash proof-of-work system.

This algorithm is specifically used to solve the "double spending problem" as the user double-spends their coins which will inflate the overall supply and degrades other coins, and this makes the currency uncertain and valueless. Bitcoin, Ethereum, Bitcoin Cash, Litecoin, and Monero are the most accepted cryptocurrencies that use the proof-of-work algorithm [27].

PoS (Proof of Stake)

The second most welcomed consensus algorithm is the PoS, which came out as a substitute to PoW, first introduced in 2011 on Bitcoin Talk. PoS comes up with several upgradations to the PoW system:

- No need to use many energy mining blocks, i.e., energy efficiency is better than PoW.

- No need to pick the hardware to create new blocks.
- Provides stronger support to the fragmented chains.

The general concept of PoS is that the nodes that wanted to become validators are needed to lock up a minimum number of tokens that act as a guarantee. Then, the validators have to agree on the transactions that must be added to the following block, just like a guessing game. If their block is chosen, then they will get rewarded via transaction fees and also with the new tokens about their stake.

PoS is resistant to 51% attack because of the fines forced on the validators for any erroneous verification procedure as well as attackers also need to grasp sufficient coins for a long time before the attack on the network; thus, this will increase the difficulty level of the attack [3].

DPoS (Delegated Proof of Stake)

It is a consensus algorithm developed by Daniel Larimer in 2014 [28] as an evolution of the PoS concept. In this, the users of the network vote and elect a representative to validate the next block. So, it is sustained by an election system, by which it chooses the nodes which verify the blocks. These nodes are known as "block producers" and "witnesses" [29].

1. **Voting**: In this, voting is done either directly or users can give their voting power to another user, who can vote in place of that user. Now, the chosen witness is in charge of creating the blocks after verifying the transactions. If the verification is done and all the transactions in the block are signed, a reward is given which is generally shared among those who have voted for the witness. But if the witness failed in his work, i.e., failing to verify the transactions in the specified time, then the block is seeming to be missed and all the transactions are left unverified and unsigned, and thus, no reward is given. That reward is appended to the reward of the next witness who will verify the block. The following witness will collect the failed transaction, and such block is known as stolen.
2. **Witnesses**: The number of witnesses in the top level is round off at a definite number which is in the range of 21–101. These are in charge of the validation of the transactions and of creating the blocks and will be rewarded associated fees in return. They can put a stop to particular transactions from being included in the block without changing any information in the transaction. Voting is a continual process, and the witness at the top level is at risk of being replaced by a user who gets more votes and is thus considered more trusted.
3. **Delegates**: Governance of the blockchain is overseen by the group of delegates, and users in the system also vote for this group. But they do not play any role in the control of transactions. They can purpose change in the size of the block as well as the amount received by the witness.
4. **Block Validators**: These are the full nodes, verifying that blocks that are created by the witnesses follow the consensus rule.

DPoS came into force so that it can strengthen Bit Shares because this algorithm consumes less energy and is much more efficient than PoS and PoW [3]. It is not possible to attack a blockchain network that is based on the DPoS; as to attack, it needs to remove many active delegates and backups who are trusted worldwide.

PBFT (Practical Byzantine Fault Tolerance)

It is a consensus mechanism that was introduced by Barbara Liskov and Miguel Castro in the late 90s. The objective of this algorithm is to alleviate the Byzantine fault, which occurs because of the failure of the consensus and was designed to work efficiently in a system which asynchronous, to be high performance, and also required low overhead time. Distributed computing and blockchain are its main application areas.

It attempts to give a practical Byzantine state machine replication, which can even work when some hostile nodes are in the system [30]. In PBFT, the order of the nodes is sequential, so that one node is the primary, i.e., the leader node, and others are the secondary, i.e., the backup nodes. Now, the main intention is that all the honest nodes will help in reaching the consensus using the majority rule regarding the system state. A Byzantine fault-tolerant system functions on the condition that the maximum number of hostile nodes must not be greater than or equal to 1/3rd of all the nodes in the system. The system becomes more secure with the increment in the number of nodes [30].

RAFT

Diego Ongaro and John Ousterhout developed RAFT in 2014 [31], and it was designed for a preferable understanding of the fact that how consensus can be attained as its predecessor; i.e., Paxos algorithm is very difficult to understand as well implement. This algorithm was put forward to manage the replicated logs that impose strong authority in the system where the process of ledger election is based on the uncertain timer [3]. The security level was improved by this algorithm as there is a constant change in the membership of the server. But RAFT was not able to handle the hostile nodes as it achieves 50% (approx.) crash fault tolerance.

5 Future Trends

After a global pandemic, there is an absolute shift in nearly every field from governance to the manufacturing field. People were restricted to their homes, and all the corporate companies had to run their functioning remotely. This digital transformation also procured momentum, thus the blockchain technology adoption.

- Financial organizations are accepting the use of blockchain for banking operations. According to a study, blockchain can help in reducing the costs for financial service providers. We can say that blockchain is the future of finance as well as the banking industry.
- As the potential of blockchain technology is recognized by the world, there is a need for professionals who have the right blockchain skills and knowledge. Thus,

the future of blockchain holds some career opportunities for aspiring professionals [32].

- Right now, the new governance model is the need of the hour. So, these are the blockchain technology's future as they ensure the standardization of the information from different sources.
- According to the international data corporation (IDC), around 35% of the deployments in IoT would enable blockchain technology by 2025. This combination of IoT and blockchain offers a scalable and secure framework to facilitate communication among IoT devices.
- Next, a futuristic possibility of blockchain technology is the smart contract functionalities, where the basic idea of the smart contract is to execute certain tasks after the fulfillment of some conditions.

So, from above it is clear that the future of the blockchain focuses on the different themes. The future possibility of blockchain technology depends heavily on innovation and could provide favorable benefits for enterprises, and enterprises are looking for new ways to strengthen the blockchain in the future.

6 Conclusion

With the considerable development of technology, the blockchain is attracting more attention in several areas. Many blockchain technologies like Bitcoin and Ethereum are creating their place in the market. But there are several security issues in the blockchain P2P network, and for this, many algorithms are used, which are cryptographic algorithms, P2P networks, and consensus algorithms; these algorithms are also considered the foundation of the blockchain technology. An elaborated description of the blockchain technology with a systematic review of the algorithms used in the blockchain is provided in this paper with some of the future trends of the blockchain technology.

References

1. Haber S, Stornetta W (1991) How to time-stamp a digital document. In: Crypto'90, LNCS, p 537
2. Nakamoto S (2008) Bitcoin: a peer-to-peer electronic cash system. Decentralized Business Review, p 21260.
3. Alsunaidi SJ, Alhaidari FA (2019) A survey of consensus algorithms for blockchain technology. In: 2019 International conference on computer and information sciences (ICCIS). IEEE, Apr 2019, pp 1–6
4. Bhutta MNM, Khwaja AA, Nadeem A, Ahmad HF, Khan MK, Hanif MA, Song H, Alshamari M, Cao Y (2021) A survey on blockchain technology: evolution, architecture and security. IEEE Access 9:61048–61073

5. Rawat DB, Chaudhary V, Doku R (2020) Blockchain technology: Emerging applications and use cases for secure and trustworthy smart systems. J Cybersecurity Privacy 1(1):4–18
6. Jahan F, Mostafa M, Chowdhury S (2020) SHA-256 in parallel blockchain technology: storing land related documents. Int J Comput Appl 975:8887
7. Ahmed S, Rahman MS, Rahaman MS (2019) A blockchain-based architecture for integrated smart parking systems. In: 2019 IEEE international conference on pervasive computing and communications workshops (PerCom workshops). IEEE, Mar 2019, pp 177–182
8. White C, Paul M, Chakraborty S (2020) A practical blockchain framework using image hashing for image authentication. arXiv Prepr. arXiv:2004.06860
9. Parmar M, Kaur HJ (2021) Comparative analysis of secured hash algorithms for blockchain technology and internet of things
10. Uddin M, Muzammal M, Hameed MK, Javed IT, Alamri B, Crespi N (2021) CBCIoT: a consensus algorithm for blockchain-based IoT applications. Appl Sci 11(22):11011
11. Idrees SM, Nowostawski M, Jameel R, Mourya AK (2021) Security aspects of blockchain technology intended for industrial applications. Electronics 10(8):951
12. Xuan S, Chen Z, Chung I, Tan H, Man D, Du X, Yang W, Guizani M (2021) ECBCM: a prestige-based edge computing blockchain security consensus model. Trans Emerg Telecommun Technol 32(6):e4015
13. Sideris A, Sanida T, Dasygenis M (2020) High throughput implementation of the Keccak hash function using the Nios-II processor. Technologies 8(1):15
14. Zheng Z, Xie S, Dai H, Chen X, Wang H (2017) An overview of blockchain technology: architecture, consensus, and future trends. In: 2017 IEEE international congress on big data (BigData congress). IEEE, June 2017, pp 557–564
15. Alsaqqa S, Almajali S (2020) Blockchain technology consensus algorithms and applications: a survey
16. Liang YC (2020) Blockchain for dynamic spectrum management. In: Dynamic spectrum management. Springer, Singapore, pp 121–146
17. Yli-Huumo J, Ko D, Choi S, Park S, Smolander K (2016) Where is current research on blockchain technology?—a systematic review. PLoS ONE 11(10):e0163477
18. Bhowmik D, Feng T (2017) The multimedia blockchain: a distributed and tamper-proof media transaction framework. In: 2017 22nd International conference on digital signal processing (DSP). IEEE, Aug 2017, pp 1–5
19. Velliangiri S. Karthikeyan P (2020) Blockchain technology: challenges and security issues in consensus algorithm. In: 2020 International conference on computer communication and informatics (ICCCI). IEEE, Jan 2020, pp 1–8
20. Yu Z, Liu X, Wang G (2018) A survey of consensus and incentive mechanism in blockchain derived from P2P. In: 2018 IEEE 24th international conference on parallel and distributed systems (ICPADS). IEEE, Dec 2018, pp 1010–1015
21. Saxena M, Jha D (2019) A new pattern mining algorithm for analytics of real time internet of things data. https://doi.org/10.35940/ijitee.A4506.119119
22. Zhai, S., Yang, Y., Li, J., Qiu, C., & Zhao, J. (2019, February). Research on the Application of Cryptography on the Blockchain. In Journal of Physics: Conference Series (Vol. 1168, No. 3, p. 032077). IOP Publishing.
23. Rountree D (2011) Security for Microsoft windows system administrators: introduction to key information security concepts. Elsevier
24. Avendaño JLS, Martín LSM (2019) Communication in microgrids. In: Microgrids design and implementation. Springer, Cham, pp 69–96
25. Pop CD, Antal M, Cioara T, Anghel I, Salomie I (2020) Blockchain and demand response: zero-knowledge proofs for energy transactions privacy. Sensors 20(19):5678
26. Partala J, Nguyen TH, Pirttikangas S (2020) Non-interactive zero-knowledge for blockchain: a survey. IEEE Access 8:227945–227961
27. Bonneau J, Preibusch S, Anderson R (2020) Financial cryptography and data security. Springer International Publishing
28. Larimer D (2014) Delegated proof-of-stake (dpos). Bitshare whitepaper, 81, p 85

29. Yang F, Zhou W, Wu Q, Long R, Xiong NN, Zhou M (2019) Delegated proof of stake with downgrade: a secure and efficient blockchain consensus algorithm with downgrade mechanism. IEEE Access 7:118541–118555
30. Sohail M, Tabet S, Loey M, Khalifa NEM (2022) Using blockchain-based attestation architecture for securing IoT. In: Implementing and leveraging blockchain programming. Springer, Singapore, pp 175–191
31. Goniwada SR (2022) Cloud native architecture and design patterns. In: Cloud native architecture and design. Apress, Berkeley, CA, pp 127–187
32. Kollu PK, Saxena M (2021) Blockchain techniques for secure storage of data in cloud environment. Turk J Comput Math Educ (TURCOMAT). 12:1515–1522

Relay Coordination of OCR and GFR for Wind Connected Transformer Protection in Distribution System Using ETAP

Tarun Nehra, Indubhushan Kumar, Sandeep Gupta, and Moazzam Haidari

Abstract The electrical energy scenario is consistently changing every decade, drifting more toward renewable energies than conventional thermal power generating stations. Current generations are motivated by the availability of high-quality power at a reasonable cost. To maintain the voltage or frequency and reliable supply in distribution of power requires reliable protection equipment and their coordination. The function of the protection system is to identify and segregate the unhealthy sections from the healthy sections of power systems within a stipulated time to avoid large interruptions of power supply as quickly as possible. The power systems are prone to various symmetrical and unsymmetrical faults and short circuits. The protective relays are used for transformer protection using over current relays (OCR) and ground fault relays (GFR). This paper work is focused on the protection of the transformer using the OCR and GFR and analysis of relay coordination connected to the wind energy conversion system of 7.5 MW, 13.8 kV, and Type 2 Generic Induction Generator. The proposed test system is implemented over ETAP and results are evaluated. The calculations of Time Multiplier Setting (TMS) with the plots of TCC are obtained.

Keywords Over current relay · Wind energy conversion system · Ground fault relay · Distribution system

1 Introduction

In any power network, protection should be designed such that protective relays disconnect the defective component of the network as soon as possible, preventing

T. Nehra
Rajasthan Institute of Engineering and Technology, Jaipur, India

I. Kumar · M. Haidari
Saharsa College of Engineering, Saharsa, India

S. Gupta (✉)
Dr. K. N. Modi University, Newai, Rajasthan, India
e-mail: jecsandeep@gmail.com

M. Thakur et al. (eds.), *Soft Computing for Problem Solving*, Lecture Notes in Networks and Systems 547, https://doi.org/10.1007/978-981-19-6525-8_51

equipment damage, minimizing system disturbance, and allowing service to the healthy section of the network to continue. In the event that the primary relays fail, backup relays take over when a sufficient amount of time has passed. Normal, abnormal, and fault circumstances should all be distinguishable by the protective relay [1–4]. The principles of discriminating, selectivity, and backup protection are all included under the phrase relay coordination [5]. Electrical power consumption is expanding at a quicker rate in economically developing countries in the modern period [6–8]. As a result, electrical company networks become extremely complex. It will take several rounds to calculate the TMS of relays so that the required minimum discriminating margin between a relay and all of its backup relays is determined in a big electrical system, and load flow analysis, fault calculations, and listing the primary and backup pairs will be tiresome. This is only possible because of computer programming. ETAP performs numerical computations at breakneck speed, automatically follows industry-accepted standards, and creates simple output reports. ETAP offers a load schedule application that can track up to 10,000,000 load items and report the voltage and short circuit current at each load item's terminals.

Renewable energy (RE) has the potential to significantly reduce global warming and energy costs [9–12]. Every aspect of contemporary civilization relies on energy, which comes from a variety of sources, including nuclear and fossil fuels (coal, oil, and natural gas). The burning of these energy sources produces a potent greenhouse gas (GHGs) such as carbon dioxide (CO_2), sulfur dioxide, and nitrogen dioxide, which are harmful to the environment and public health [13, 14]. Due to their potential environmental benefits, hybrid power systems incorporating RE have recently gained popularity [15].

Although renewable energy sources such as wind and solar offer various advantages, due to their intermittent nature, a solar or wind producer in a stand-alone system cannot provide the demand consistently. Because load needs fluctuate over time, changes in solar or wind energy output may not necessarily correspond to the load distribution over time [16–18]. As a result, extra battery storage or other components are required to provide a continuous power supply to the customer. A hybrid PV/Wind/Battery system has been examined as a dependable source of power [19, 20]. However, a stand-alone system is prohibitively expensive because of the heavy cost of battery energy storage, hence to optimize the advantage, it is critical to identify a good mixture of diverse RE sources.

Ever-increasing load demand has put pressure on power system planners and operators all around the world to ensure that their systems are stable and efficient. Integration of Distributed Generation (DG) in Distribution Systems has recently discovered higher penetration's levels for serving the majority of load demands. The effects of DG units on voltage stability, frequency deviation, and relay synchronization must all be carefully considered. This paper proposes a unique scheme for optimal relay coordination with reliability evaluation in the presence of Wind Turbine Generators (WTGs). The coordination is done using a graphical technique as well as physical changes to the pickup value, plug setting multiplier (PSM), and time dial setting (TDS), as well as further reliability testing with power system faults and short circuit analysis.

2 Power System Faults

Power is a fundamental requirement for every country's economic growth. Throughout history, the availability of electrical power is the most important vehicle for economic growth and social change. Modernization, increased industrial productivity, increased agricultural production, and an improvement in people's standard of living are all dependent on sufficient electrical power supply. The purpose of the power system distribution network is to transmit mass power generated at thermal power stations to power consumers. As a result, it consists of three major components: generation, transmission, and distribution. The power quality is primarily concerned with problems relating to the power distribution. Network of the grid, which distributes energy to customers.

Power system equipment or lines bear natural voltages and currents under operating point, resulting in optimal power distribution. There are the different forms of power system faults. Following the occurrence of a fault, the activity of the power system deviates from normal. In the power system, there are primarily two types of faults: symmetrical faults and unsymmetrical faults.

2.1 Symmetrical Faults

These faults are classified as:

- Phase to Phase to Phase to ground fault (LLLG fault)
- Phase to Phase-to-Phase fault (LLL fault)

2.2 Unsymmetrical Faults

Unsymmetrical faults are those where the system balance is distributed and unbalanced. In the power system, these faults are less serious and more frequent (65–70% faults) [21].

These faults are distinguished as:

(a) Line to ground faults (LG faults):—When insulation breaks down between the one line and ground of the 3ϕ system.
(b) Line to line faults (LL faults):—When insulation break down between the two line of the 3ϕ system.
(c) Double line to ground faults (LLG faults):—When insulation break down between the two line and ground of the 3ϕ system.

2.3 Short Circuit Analysis

The following computations involve the use of short circuit analysis [22, 23]:

i. Three-phase fault level at one or more system nodes (buses). Circuit breakers' short circuit current interruption capacity is determined by the three-phase fault level.
ii. Calculate fault currents for transmission line problems, as well as the contribution from neighboring buses. Overcurrent relays are coordinated using current pairs like these.

In other words, short circuit calculations are primarily used to develop the protective system. Fault currents to ground also generate currents in nearby conductive circuits, such as metallic piping and communications circuits, which can compromise their proper operation and safety.

3 Power System Protection and Relaying

A successful electric power system should ensure that any load connected to it has access to electricity without interruption. Since high voltage transmission lines are exposed, there is a risk of them breaking down due to winds, falling debris, and insulator destruction, among other things. Not only can this cause mechanical disruption, but it can also cause an electrical fault.

When protective relays and relaying devices detect abnormal events, such as electrical circuit faults, they instantly activate switchgear to quickly disconnect problematic equipment from the grid. This minimizes the amount of downtime at the fault site and prevents the fault's impact from spreading throughout the system. The switch gear must be likely to restrict both natural and fault currents. On the other hand, the defensive relay must be able to detect an unexpected situation in the power system and take appropriate action to ensure that normal operations are not affected. The presence of faults is not prevented by relay. Only after the error has happened, it should take action. There are, however, several systems that can predict and avoid major faults. For e.g., a Buchholz relay will detect the gas accumulation caused by an incipient transformer fault.

4 Methodology for Load Flow Analysis

The load flow analysis (LFA) is performed on the proposed system as shown in Figs. 1 and 2, with Adaptive Newton Raphson Iterative method is implemented with maximum iterations of 99 with precision of 0.0001.

In Table 1, CB operation of sequence is shown when a fault was initiated near to transformer and relay coordination was identified. It has been observed that the CB associated with the transformer was tripped with the sequence of CB 19, CB 20, CB 21, CB 2, and CB1 with the increasing time instants. It is plain to see that the CB tripped in the proper sequence by separating the defective section of the system from the healthy portion. Furthermore, Figs. 3 and 4 show the time current characteristic curve for the WTG system shown within describes that the Relay 12 is been considered with the Very Inverse Time Characteristic with OC1-67 as Directional Overcurrent Relay protection for the transformer. The ALSTOM P343 maker MiCOM was implemented at the location. The pickup current, time dial setting, and other settings for the relay 12 are shown within the figure. Figure 3 describes the TCC curve for ground view, i.e., for the ground faults, whereas Fig. 4 describes the TCC Curve for Phase view, i.e., for the phase faults.

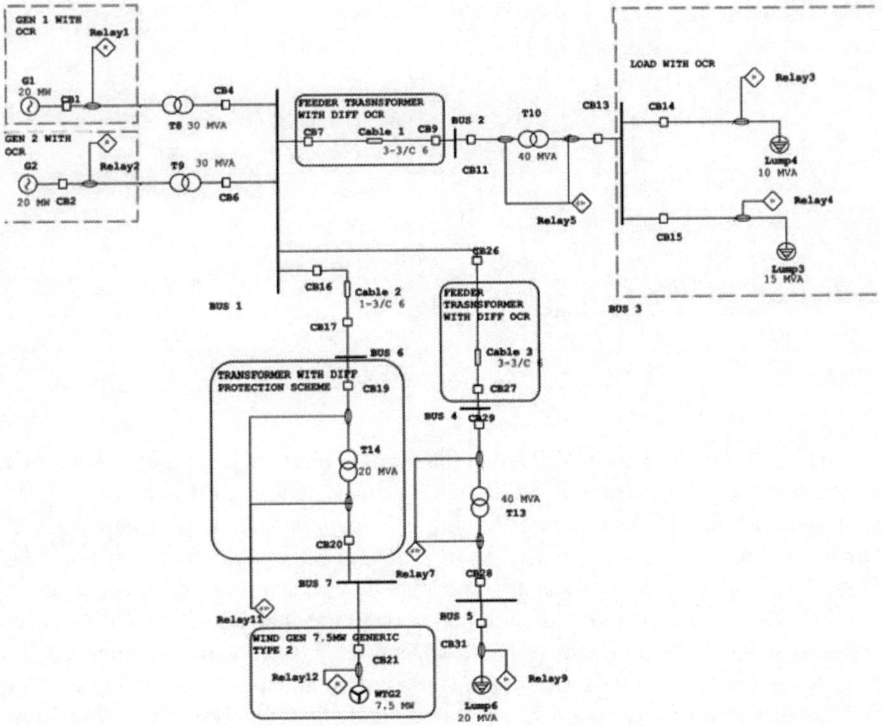

Fig. 1 Complete LFA for the proposed test system

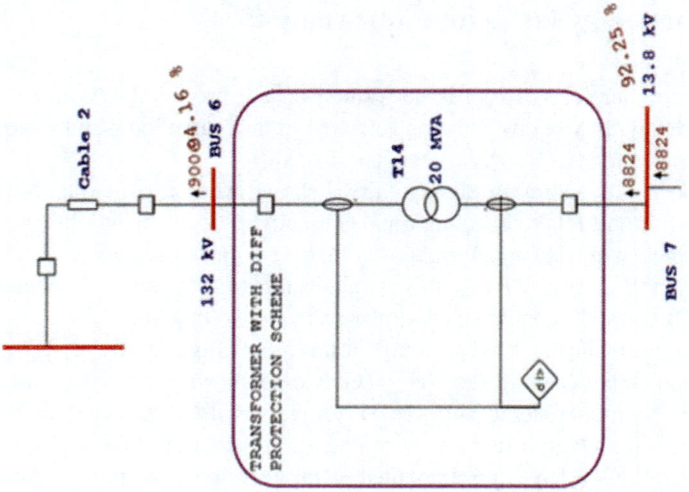

Fig. 2 LFA for the bus 7 for WTG system

Table 1 Sequence of operation of CB and relay for internal fault at transformer at WTG

S No	Time (ms)	ID	T1 (ms)	Condition
1	10	Relay 12	10.0	Phase-OC1-50-Forward
2	20	Relay 11	20.0	Phase 87
3	40	CB19	20.0	Tripped by relay11 phase-87
4	40	CB20	20.0	Tripped by relay11 phase-87
5	70	CB21	60.0	Tripped by relay12 phase-OC1-50-Forward
6	149	Relay 2	149	Phase-OC1-51
7	169	CB2	20.0	Tripped by relay2 phase-OC1-51
8	700	Relay 1	700	Phase-OC1-51
9	760	CB1	60.0	Tripped by relay1 phase-OC1-51
10	3197	Relay 12	3197	Phase-OC1-51-Forward

This has to be noted from Fig. 5 that the thermal limits of generator 1 always be higher than the relay setting for the relay associated with it. If it is relay operates when the fault crosses the thermal limits of the generator than generator will get damage and protection system fails. So, it is always advised and necessary that the relay operating time always must be lower than the thermal limits of the generators.

From the results of the system in terms of load flow analysis (LFA), short circuit analysis (SCA), reliability assessment (RA), and relay coordination settings (RCS), it can be concluded that, the relay coordination not limited to setting of relay settings with the different settings but also need a comprehensive analysis in term of LFA, SCA, RA, and RCS. From the load flow analysis, the nominal power flow and the critical and marginal status of the different components can be estimated which are

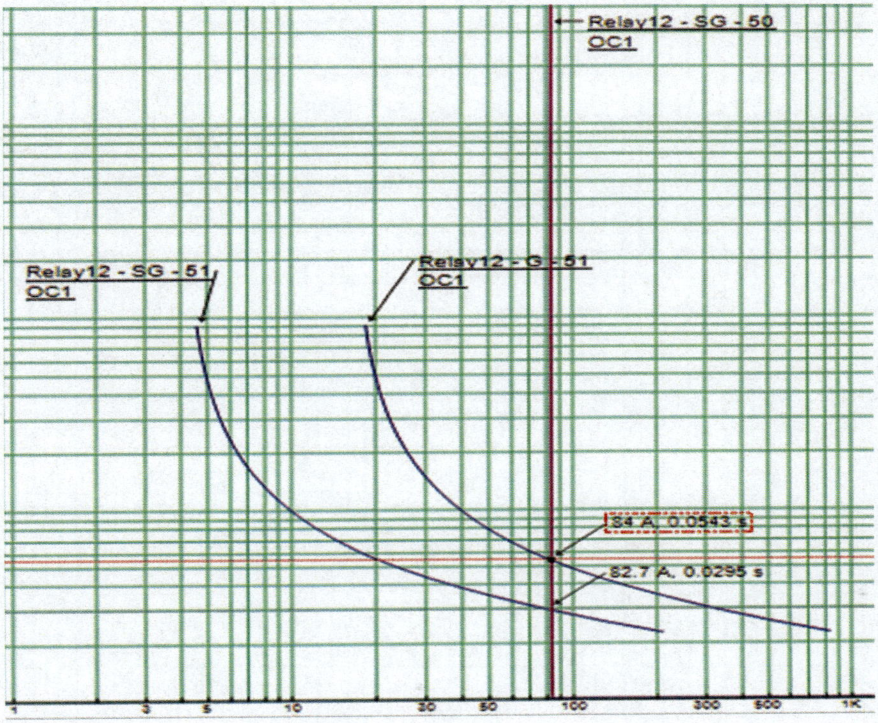

Fig. 3 TCC curve in ground view for WTG system

being used in the RCS along with the SCA values to define and adjust the relay settings, circuit breaker settings, and current transformer settings.

5 Conclusion

When WTG is used in the network under observation, a unique approach is given for successful relay and CB coordination. With the introduction of WTG to the system, it is discovered that the coordination is skewed. As a consequence, the relay's pickup and time dial settings, as well as the low voltage circuit breaker's trip device setting, are all set to their maximum value to improve coordination. The dependability assessment is conducted to ensure network security which is tested with the least possible interruption to consumers. The simulation results are acquired using ETAP. The sequence of operation of various safety mechanisms has been seen when a three-phase or single LG fault develops in any portion of the system. Coordination and reliability evaluation problems for a very large integrated system are difficult and inelegant to solve using a standard technique. As a result of utilizing ETAP software, the amount of labor required and the chance of protective devices failing

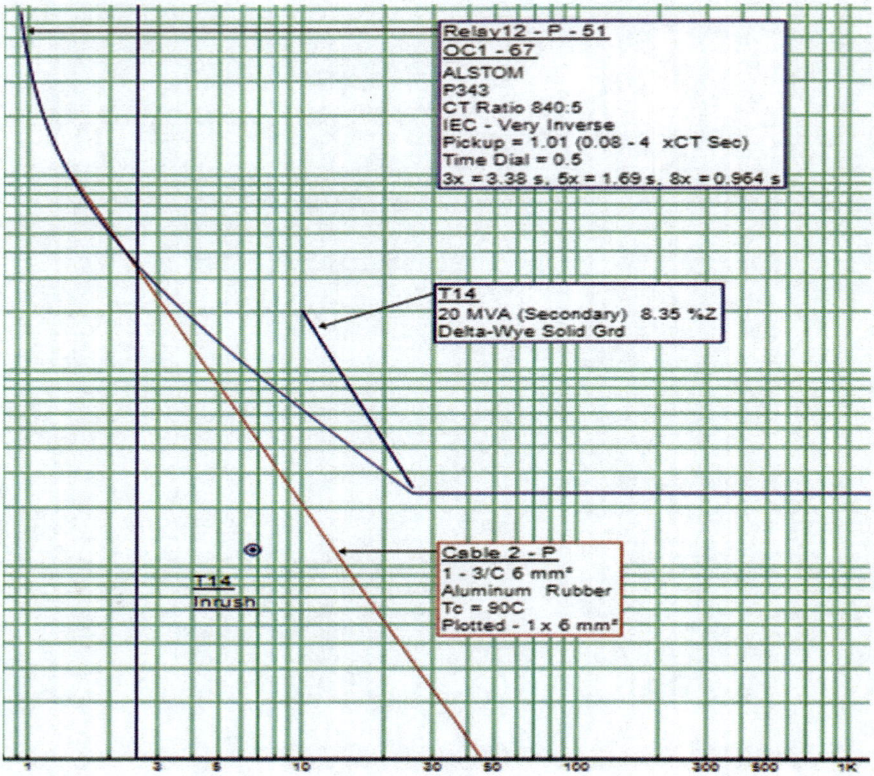

Fig. 4 TCC curve in phase view for WTG system

are reduced, resulting in enhanced speed and accuracy. The system's reliability has increased significantly. This paper results in substantially more optimistic outcomes for enhancing the protection mechanisms of extremely large interconnected power networks.

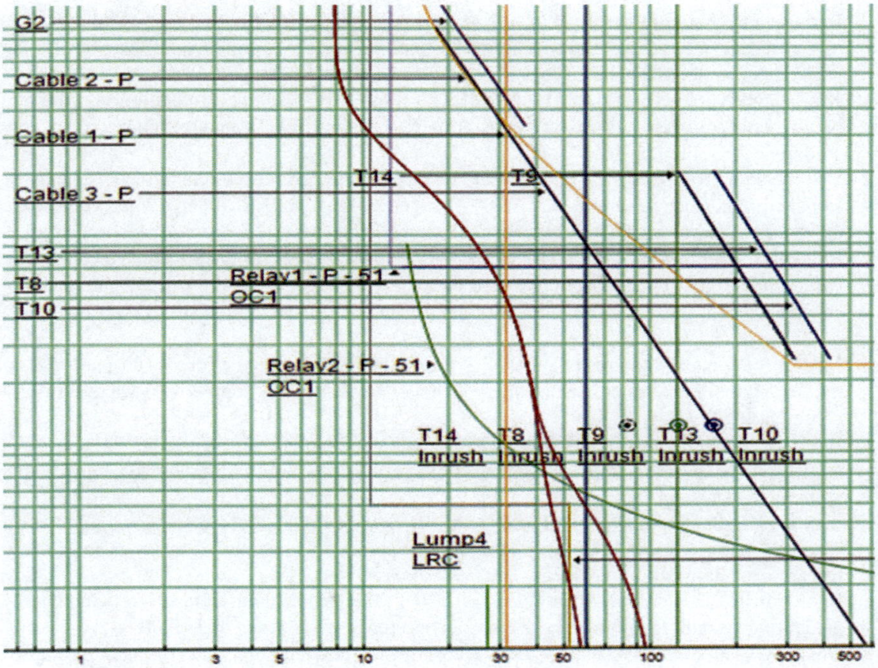

Fig. 5 TCC curves for the complete system

References

1. Lasseter RH (2002) Microgrids, IEEE power engineering society winter meeting. IEEE Conf Proc 1:305–308
2. Montecucco A, Knox AR (2015) Maximum power point tracking converter based on the open-circuit voltage method for thermoelectric generators. IEEE Trans Power Electron 30(2):828–839
3. Gupta S, Kumar TK, Shivaji RM, Nachimuthu K (2020) Wind energy potential, challenges with major technical issues. J Green Eng (JGE) 10(12):12973–12987. ISSN: 1904-4720
4. Chu CC, Chen CL (2009) Robust maximum power point tracking method for photovoltaic cells: a sliding mode control approach. Sol Energy 83:1370–1378
5. Elgendy MA, Zahawi B, Atkinson DJ (2015) Operating characteristics of P&O algorithm at high perturbation frequencies for standalone PV systems. IEEE Trans Energy Convers 30:189–198
6. Elgendy MA, Zahawi B, Atkinson DJ (2012) Assessment of perturb and observe MPPT algorithm implementation techniques for PV pumping applications. IEEE Trans Sustainab Energy 3:21–33
7. Esram T, Chapman PL (2012) Comparison of photovoltaic array maximum power point tracking techniques. IEEE Trans Energy Convers 22:439–449
8. Sangwongwanich A, Blaabjerg F (2019) Mitigation of Inter-harmonics in PV systems with maximum power point tracking modification. IEEE Trans Power Electron 34(9):8279–8282
9. Elbehairy NM, Swief RA, Abdin AM, Abdelsalam TS (2019) Maximum power point tracking for a stand-alone pv system under shading conditions using flower pollination algorithm. In: 21st IEEE international middle east power systems conference (MEPCON), pp. 1–6

10. Babayeva M, Abdullin A, Polyakov N, Konstantin G (2020) Comparative analysis of modeling of a magnus effect-based wind energy system. In: IEEE conference of Russian young researchers in electrical and electronic engineering (EIConRus), pp 602–605
11. Xing C, Xi X, He X, Liu M (2020) Research on the MPPT control simulation of wind and photovoltaic complementary power generation system. In: IEEE sustainable power and energy conference (iSPEC), pp 1058–1063
12. Ji G, Ohyama K (2020) MPPT control of variable speed wind power generation system using switched reluctance generator and AC-AC converter. In: 23rd IEEE international conference on electrical machines and systems (ICEMS), pp 1012–1016
13. Bedoud K, Merabet H, Bahi T, Drici D (2020) Fuzzy observer for MPPT control of variable wind energy conversion system associed to AC-DC converters. In: 20th IEEE international conference on sciences and techniques of automatic control and computer engineering (STA), pp 231–236
14. Mishra J, Pattnaik M, Samanta S (2018) Power management scheme for a wind-photovoltaic hybrid autonomous system with battery storage. In: IEEE 4th southern power electronics conference (SPEC), pp 1–5
15. Syskakis T, Ordonez M (2019) MPPT for small wind turbines: zero-oscillation sensorless strategy. In: IEEE 10th international symposium on power electronics for distributed generation systems (PEDG), pp 1060–1065
16. Jagwani S, Sah GK, Venkatesha L (2018) MPPT based switched reluctance generator control for a grid interactive wind energy system. In: 7th international conference on renewable energy research and applications (ICRERA), pp 998–1003
17. Fu C, Pan T, Liu H, Wu D, Shen Y, Hao Z (2018) MPPT control based fuzzy for wind energy generating system. In: 37th Chinese control conference (CCC), pp 7465–7470
18. Obaid W, Hamid AK, Ghenai C (2019) Hybrid PEM fuel-cell-solar power system design for electric boat with MPPT system and fuzzy energy management. In: International conference on communications, signal processing, and their applications (ICCSPA), pp 1–7
19. Dabboussi M, Hmidet A, Boubaker O (2020) An efficient fuzzy logic mppt control approach for solar PV system: a comparative analysis with the conventional perturb and observe technique. In: 6th IEEE international energy conference (ENERGYCon), pp 366–371
20. Javed MR, Waleed A, Riaz MT, Virk US, Ahmad S, Daniel K, Khan MA (2019) A comparative study of maximum power point tracking techniques for solar systems. In: 22nd international multitopic conference (INMIC), pp 1–6
21. Yuwanda RI, Prasetyono E, Eviningsih RP (2020) Constant power generation using modified MPPT P&O to overcome overvoltage on solar power plants. In: IEEE international seminar on intelligent technology and its applications (ISITIA), pp 392–397
22. Chauhan U, Kumar B, Rani A, Singh V (2019) Optimal perturbation MPPT technique for solar PV system using grey wolf optimization. In: IEEE international conference on computing, power and communication technologies (GUCON), pp 589–592
23. Gupta S, Kumar I (2018) Steady state performance of DFIG with stability assessment in wind system. In: International conference on manufacturing, advance computing, renewable energy and communication (MARC-2018), Lecture Notes in Electrical Engineering (LNEE), Springer, vol 553, pp 11–20

Localized Community-Based Node Anomalies in Complex Networks

Trishita Mukherjee and Rajeev Kumar

Abstract Partitioning complex networks using a community detection method helps in identifying anomalous nodes within the communities in networks. Existing methods on community detection techniques identify anomalies based on network structure as a whole. In this paper, we have proposed an algorithm that detects localized community-based node anomalies present in a network. Our approach uses a widely known non-overlapping community detection method, i.e. the Louvain method to partition a network in different communities. Then, we use two types of centrality measures, namely, Closeness centrality and Katz centrality for extracting anomalous nodes present within communities. We experimented the proposed algorithm on two sets of complex networks and analyse the empirical results. We have illustrated the communities and the localized community-based node anomalies identified based on the principle of these two centrality measures.

Keywords Anomaly detection · Complex networks · Louvain community detection · Node anomalies · Closeness centrality · Katz centrality

1 Introduction

Anomaly detection in complex networks have been an emerging field in the area of social network analysis. It remains an interesting and challenging area due to the heterogeneity and availability of overwhelming amount of network data. Anomaly detection is applied in a multitude of applications such as fraud detection [1], intrusion detection systems [2], online social networks [3], covid-19 datasets [4], etc. Detecting anomalies for large networks are *hard* problems. Conventional algorithmic approaches work conveniently for identifying anomalies in small scale real-world networks. Soft computing approaches, meta-heuristics and statistical measures give a faster convergence for large scale real-world networks, e.g., [5–8].

T. Mukherjee (✉) · R. Kumar
Data to Knowledge (D2K) Lab, School of Computer & Systems Sciences, Jawaharlal Nehru University, New Delhi 110067, India
e-mail: trishitamukherjee77@gmail.com

M. Thakur et al. (eds.), *Soft Computing for Problem Solving*, Lecture Notes in Networks and Systems 547, https://doi.org/10.1007/978-981-19-6525-8_52

In recent years, community detection techniques have been useful in detecting node-based anomalies within networks [9]. Communities in complex networks are a group of nodes that are densely interlinked with each other than to the rest of the nodes in the network. Community detection techniques are decent approaches in partitioning the small scale real-world networks [10].

In this paper, we have used a widely known non-overlapping community detection technique known as the Louvain method. The Louvain method is a hierarchical clustering algorithm, that applies a recursive procedure for finding disjoint communities and executes modularity optimization technique on condensed communities [11].

The anomaly scores of nodes within communities are computed based on node centrality measures. Centrality measures determines a node's influence in the network. In this paper, we use two centrality measures for computing anomaly scores of nodes: (i) Closeness centrality [12], and (ii) Katz Centrality [13]. Closeness centrality of a node is computed by taking the average shortest path that is reachable to all nodes. It basically computes the incoming distance of a node for directed graphs. Nodes belonging to smaller size communities receives small closeness value [12]. Whereas Katz centrality calculates the centrality of a node based on it is influence of its corresponding neighbours. It computes the relative influence of a particular node in the network based on it is corresponding number of 1-hop neighbours and the other nodes that are connected through these 1-hop neighbours [13].

The majority of anomaly detection methods focuses on the local perspective of considering the network as a whole but not on the perspective of anomalous nodes that are hidden locally in the respective communities of a network. To solve this problem, we have proposed an algorithm to find the localized community-based node anomalies in the network. This method is found appropriate for small scale real-world static networks.

The rest of the paper is organized as follows: In Sect. 2, we discuss the related works on anomaly detection based on community detection methods in complex networks. Next, Sect. 3 provides an outline of our problem definition and the proposed algorithm with mathematical explanation. In Sect. 4, the results and discussion are described based on our proposed method, and the experiments performed on the two network data sets. Finally, we conclude this work in Sect. 5.

2 Related Work

This section discusses the related work on developing various algorithms that have been proposed by researchers in detecting anomalies in networks. Most researchers have used non-overlapping community detection methods or clustering techniques. Autopart algorithm [14] is a community-based method to detect network anomalies based on lossless compression and information-theoretic principles where the nodes having similar neighbours are connected together by forming clusters, and the edges that do not belong to any particular cluster are flagged as anomalies. Nodes belonging to different clusters are considered not to belong to any particular cluster; thus, they are also flagged as network anomalies. For extracting communities in graph, the algorithm creates an adjacency matrix of the network and iteratively re-organizes the rows and columns of the matrix into homogeneous blocks of low or high density [10]. This method is parameter-free and iterative established on the concept of minimum description length (MDL) principle [15].

Another community-based network anomaly detection method directly focuses on network clustering. Scan algorithm [16] finds network clusters, hubs and outliers by exploiting the neighbourhood of vertices. This method is based on the concept of common neighbours. Vertices are assigned to a cluster depending on how they share neighbours and strongly correlated. It clusters nodes using the structural similarity measure. If a vertex of a cluster shares a similarity in structure with one of its neighbours, their computed structural similarity is found to be large. The proposed method identifies all structure-connected clusters based on a given parameter setting by checking each vertex of the graph. Vertices that are belonging to many clusters are labelled as hubs, whereas those that cannot be assigned any particular cluster are detected as anomalies.

CADA algorithm [17] is a community-aware approach that identifies anomalous nodes in a global perspective by employing two well-known community algorithms. These algorithms are, namely, the Louvain algorithm [11] and the Infomap approach [18], that scales linearly depending upon the number of edges. This algorithm assigns each node to a community using community detection methods. Each node is assigned an anomaly score, depending on which communities they belong. CADA explores two kinds of anomalies in a network: Random anomaly and Replaced anomaly. Random anomalies are nodes infiltrated in the network to see its changing behaviour or patterns based on the power law distribution of the network. Replaced anomalies replaces a certain number of nodes from the network; then randomly certain nodes are selected wherein an anomaly is injected by changing all the edges from the randomly selected nodes to the new anomaly; later getting removed from the network.

Community Neighbour Algorithm (CNA) [19] partitions a network into meaningful communities by applying Markov Cluster (MCL) algorithm [20] to detect community-based anomalies in a network. Anomaly scores within the communities are assigned. This technique is particularly useful to identify community anomalies wherein there are no local anomalies but anomalies exist within the community.

eRiskCom [1] is another method based on an e-commerce risky community detection algorithm for detection of fraudsters belonging in the same group. Users that interact with fraudsters also likely belong to the same community. A connected graph is constructed containing the potential fraudsters and the graph is partitioned using Louvain algorithm. Next, a pruning mechanism is done by applying Kruskal's algorithm to get top k-core members of the community. Finally, a community search is performed and risk scores of potential fraudster nodes generated. The core members having a high risk score of the respective potential communities form a final risky community.

In this paper, we deviate from the existing methods and propose a novel method to detect the localized node anomalies present in respective communities of the networks.

3 Proposed Methodology

3.1 Problem Definition

Our problem is defined as to identify localized community-based node anomalies in a complex network by applying the non-overlapping community detection method, i.e. Louvain method. Each node's identifier in the complex network is assigned an anomaly score based on two types of centrality measures: (i) Closeness centrality, and (ii) Katz centrality. The node identifier having minimal anomaly score is extracted from the respective communities.

3.2 Proposed Algorithm

The proposed algorithm extracts communities and the anomalous node identifiers present from the respective complex network data sets. The algorithmic steps are included in Algorithm 1 for a simplified overview.

Algorithm 1 : Localized Community-Based Node Anomalies Algorithm

Input: Complex network data set.
Output: Localized community-based node anomalies.

1. The complex network data set is visualized from the given network edgelist, i.e. edges represented amongst the linked node Ids.
2. Extract communities (C) using Louvain community detection method and store the number of communities in N.
3. For each community C_i where $i \in (1, 2, \ldots, N)$ extracted from the network:
4. Compute anomaly scores of the node Id's based on the following two sets of centrality measure:
   (a) $\text{AnomNodeId}_{\text{Closeness}}$ is calculated for each node Id's based on the Closeness centrality measure.
   (b) $\text{AnomNodeId}_{\text{Katz}}$ is calculated for each node Id's based on the Katz centrality measure.
5. Compare each anomaly scores of $\text{AnomNodeId}_{\text{Closeness}}$ and $\text{AnomNodeId}_{\text{Katz}}$.
6. The minimal anomaly score is obtained for each set of centrality measures. The minimal anomaly score node Id based on closeness centrality measure is stored in $\text{Min}_{\text{Closeness}}$ and for Katz centrality measure is stored in Min_{Katz}.
7. Repeat steps 3 to 6 until all minimal anomaly scores node Ids are extracted from each community, respectively.
8. Localized node anomalies from communities are identified for each centrality measure.

3.3 Mathematical Explanation of Our Proposed Algorithm

Given a static, undirected graph $G = (V, E)$ where V denotes the set of nodes and E the set of edges, respectively. Firstly, the graph G is partitioned using the Louvain community detection method [11]. Louvain method is based on the heuristic of modularity maximization. This method works on two phases: Phase 1: Assigns the nodes by local optimization to a particular community, and Phase 2: Computes the maximum positive modularity gain of a node by moving it to all of its neighbouring communities, if there is no positive gain the node remains in it is own community. The computation of modularity gain is as follows:-

$$\Delta Q = \frac{S_{j,\text{in}}}{2m} - \gamma \frac{\sum_{\text{tot}} . S_j}{2m^2} \tag{1}$$

In Eq. 1, j is the isolated node moving into a community C. Here, m is the size of the network, $S_{j,\text{in}}$ is the summation of the weights of the edges from node j to other nodes in the particular community C, S_j is the summation of the weights of the edges incident to node j, $\sum_{\text{tot}}$ is the summation of the weights of the edges incident to nodes in the respective community C and γ is the resolution parameter.

The Anomaly scores of the node Id's based on the Closeness [12] and Katz [13] centrality measures is computed as follows:

$$\text{AnomNode}(u)_{\text{Closeness}} = \frac{n-1}{\sum_{v=1}^{n-1} \text{dist}(v, u)} \tag{2}$$

In Eq. 2, $\text{dist}(v, u)$ denotes the shortest path between node v and node u, and n is the number of nodes that takes to reach node u.

$$\text{AnomNode}(u)_{\text{Katz}} = \sum_{k=1}^{\infty} \sum_{v=1}^{n} \alpha^k (A^k)_{vu} \tag{3}$$

In Eq. 3, A represents the adjacency matrix of the graph G having eigenvalues λ. Here, k is the number of degree connections between nodes u and v. The α value is chosen in such a way that it is the reciprocal of the absolute value of largest eigenvalue λ of A.

For each community C_i generated, where $i \in (1, 2, \ldots, N)$ in Eqs. 4 and 5:

$$\text{Min}_{\text{Closeness}} = \text{minimum}_{C_i}(\text{AnomNode}(u)_{\text{Closeness}}) \tag{4}$$

$$\text{Min}_{\text{Katz}} = \text{minimum}_{C_i}(\text{AnomNode}(u)_{\text{Katz}}) \tag{5}$$

In Eqs. 4 and 5 the node Id's having the minimum anomaly score is extracted from the respective communities.

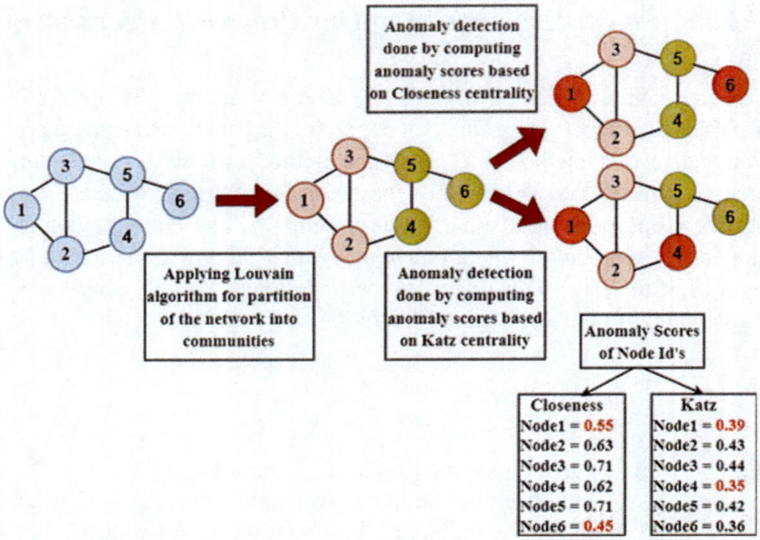

Fig. 1 Visualization of the steps of our proposed method

Figure 1 depicts the visualization of a network on the working mechanism of our proposed algorithm. Firstly, the network is partitioned into two communities after applying Louvain algorithm. The Anomaly scores of node Id's are computed based on Eqs. 2 and 3. The corresponding node Id's having minimum values in their respective communities are computed using Eqs. 4 and 5. Two types of Anomaly detection is done based on the centrality measures. Therefore, Node1 and Node6 are the anomalous nodes detected with respect to the two communities, when Closeness centrality taken as anomaly score whereas, Node1 and Node4 for Katz centrality, respectively.

4 Results and Discussion

4.1 Network Data Statistics

We have selected two different complex networks [21, 22] to qualitatively assess the localized community-based node anomalies present in networks. The network data sets taken are static, unlabelled, and undirected in nature.

Table 1 depicts the features of the two complex networks based on the number of nodes, number of edges, and the average degree.

Table 1 Properties of the complex network data sets

Network data set	Number of nodes	Number of edges	Average degree
Synthetic network	100	200	4.0000
Zachary's Karate Club network	34	78	4.5882

4.2 Results

We have performed experiments based on our proposed algorithm on the two network data sets which are taken for our empirical analysis. Also, we have included figures of the network data sets with respect to the communities and the respective localized community-based node anomalies identified in each of the complex networks.

We have demonstrated the experiments using our proposed algorithm where Figs. 2 and 3 shows the network illustration of the Synthetic network data set which is partitioned into eight communities. Whereas, Figs. 4 and 5 depict the network illustration of the Zachary's Karate Club network data set which is divided into three communities. For each network data set, we have given distinct colours to the communities and coloured the anomalous node Id's as red.

Table 2 summarizes the results of our proposed algorithm based on computation of the number of communities and anomalous node Ids extracted respective to the particular communities on these two network data sets. Anomalous node Id's numbered as 0, 35, and 48 are detected same with respect to three of the communities based on the anomaly scores computed with respect to the two different centrality measures in Synthetic network data set. Similarly, for Zachary's Karate Club network data set, the anomalous node Id's detected same for two of the communities are numbered as 12 and 16. Though, the anomaly scores of the node Id's computed with respect to the two centrality measures generates different values.

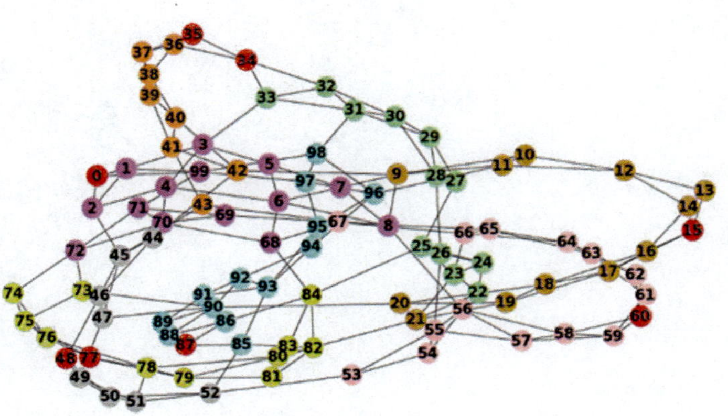

Fig. 2 Synthetic network node anomalies (Closeness centrality)

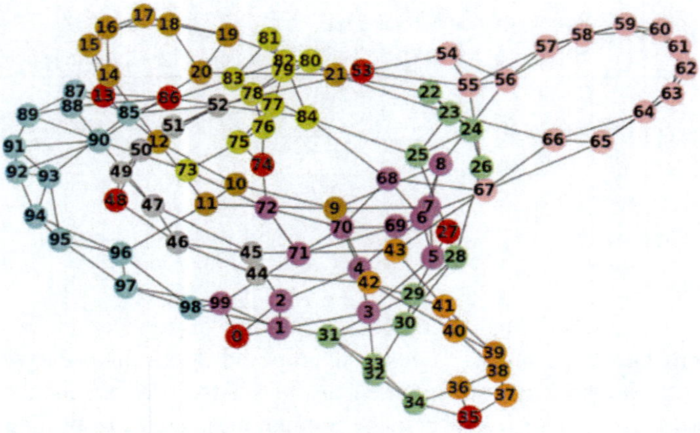

Fig. 3 Synthetic network node anomalies (Katz centrality)

Fig. 4 Zachary's Karate Club network node anomalies (Closeness centrality)

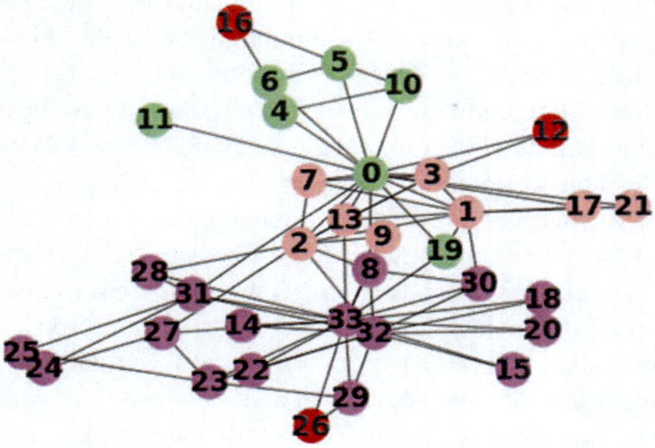

Fig. 5 Zachary's Karate Club network node anomalies (Katz centrality)

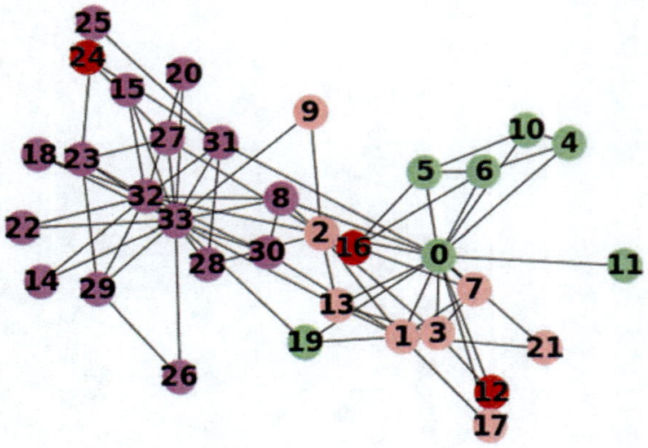

Table 2 Results of the proposed algorithm

Network data set	Closeness centrality		Katz centrality	
	Communities	Anomalous node Id's	Communities	Anomalous node Id's
Synthetic network	8	**0**, 15, 34, **35**, **48**, 60, 77, 87	8	**0**, 13, 27, **35**, **48**, 53, 74, 86
Zachary's Karate Club network	3	**12**, **16**, 26	3	**12**, **16**, 24

Table 3 Synthetic network anomalous node's anomaly scores

$\text{Min}_{\text{Closeness}}$ (anomalous node Id's)	Anomaly score (closeness centrality)	Min_{Katz} (anomalous node Id's)	Anomaly score (Katz centrality)
0	0.2106	0	0.0902
15	0.1755	13	0.0887
34	0.1784	27	0.0891
35	0.1615	35	0.0888
48	0.2071	48	0.0889
60	0.1713	53	0.0890
77	0.1864	74	0.0889
87	0.2143	86	0.0904

Table 4 Zachary's Karate Club anomalous node's anomaly scores

$\text{Min}_{\text{Closeness}}$ (anomalous node Id's)	Anomaly score (closeness centrality)	Min_{Katz} (anomalous node Id's)	Anomaly score (Katz centrality)
12	0.3708	12	0.1161
16	0.2845	16	0.0907
26	0.3626	24	0.1102

Table 3 depicts the anomaly scores of anomalous node Ids based on (i) Closeness, and (ii) Katz centrality for each of the eight communities that is computed using Louvain algorithm for the Synthetic network data set. Similarly, Table 4 shows the anomaly scores of anomalous node Ids for the Zachary's Karate Club network data set. Anomalous node Id's thus selected are based upon the nodes having minimum anomaly scores within the respective communities. The significance of the values depicts the variation of the Anomaly Scores of Node Id's in their respective communities based on Katz centrality and Closeness centrality.

4.3 Discussion

The experiments demonstrated gives an overview on how anomalous nodes within communities are detected in complex networks using our proposed algorithm. Our algorithm works on two phases based on finding communities in a network and anomaly detection within communities. Our findings are suggestive that localized node anomalies that are discovered by our proposed method are the node Id's that exhibits a certain dissimilarity from its other node members present within the particular community and also from previous related works we have seen and proven by our experiments too that Katz centrality is a better measurement to find anomaly scores of the nodes within the communities rather than closeness centrality. Finding anomalous nodes exploits the underlying structure of the communities in a complex networks. Certain anomalous node Ids are found to be similarly identified belonging to their own communities when the two different centrality measures taken as anomaly scores are applied for two different network data sets whilst other anomalous node Ids are differently detected.

5 Conclusion

In this paper, we computed localized community-based node anomalies by partitioning a network into different communities. There is a variation seen in the anomalous node Ids extracted from the communities when two different centrality measures are computed. Our results also establishes the known fact that Katz centrality measure when taken as the Anomaly Score of nodes is likely an effective method as it uses eigenvector centrality approach than the Closeness centrality measure which uses the shortest path approach amongst the node Ids. Katz centrality prunes the network in an effective manner. Our proposed method identifies locally anomalous nodes present within a particular community in a complex network. The proposed algorithm can be applied on various other real-world network data sets by scaling the number of nodes and edges to compute such localized node anomalies within communities. In future, other non-overlapping community detection approaches can be used with these centrality measures. Hybridization approaches, can also be applied for anomaly detection of large real-world networks.

Acknowledgements We thank the anonymous reviewers for their valuable and insightful feedback by which the readability of the paper is improved.

References

1. Liu F, Li Z, Wang B, Wu J Yang J, Huang J, Zhang Y, Wang W, Xue S, Nepal S et al (2022) eRiskCom: an e-commerce risky community detection platform. VLDB J 1–17
2. Chen L, Gao S, Liu B (2022) An improved density peaks clustering algorithm based on grid screening and mutual neighborhood degree for network anomaly detection. Sci Rep 12(1):1–14
3. Mahmood B, Alanezi M (2021) Structural-spectral-based approach for anomaly detection in social networks. Int J Comput Digit Syst 10(1):343–351
4. Francisquini R, Lorena AC, Nascimento MC (2022) Community-based anomaly detection using spectral graph filtering. ArXiv preprint arXiv:2201.09936
5. Raidl GR (2003) A unified view on hybrid metaheuristics. In: Proceedings of international workshop on hybrid metaheuristics. Springer, Berlin, pp 1–12
6. Kumar R, Banerjee N (2011) Multiobjective network topology design. Appl Soft Comput 11(8):5120–5128
7. Saha S, Kumar R, Baboo G (2013) Characterization of graph properties for improved pareto fronts using heuristics and EA for bi-objective graph coloring problem. Appl Soft Comput 13(5):2812–2822
8. Lotf JJ, Azgomi MA, Dishabi MRE (2022) An improved influence maximization method for social networks based on genetic algorithm. Phys A Stat Mech Appl 586:126480
9. Akoglu L, McGlohon M, Faloutsos C (2009) Anomaly detection in large graphs. Technical report, CMU-CS-09-173, CMU
10. Akoglu L, Tong H, Koutra D (2015) Graph based anomaly detection and description: a survey. Data Min Knowl Discov 29(3):626–688
11. Blondel VD, Guillaume JL, Lambiotte R, Lefebvre E (2008) Fast unfolding of communities in large networks. J Stat Mech Theory Exp 2008(10):P10008
12. Freeman LC (1978) Centrality in social networks conceptual clarification. Soc Netw 1(3):215–239
13. Katz L (1953) A new status index derived from sociometric analysis. Psychometrika 18(1):39–43
14. Chakrabarti D (2004) AutoPart: parameter-free graph partitioning and outlier detection. In: Proceedings of European conference on principles of data mining & knowledge discovery, vol 3202. Springer, Berlin, pp 112–124
15. Barron A, Rissanen J, Yu B (1998) The minimum description length principle in coding and modeling. IEEE Trans Inf Theory 44(6):2743–2760
16. Xu X, Yuruk N, Feng Z, Schweiger TA (2007) Scan: a structural clustering algorithm for networks. In: Proceedings of 13th ACM SIGKDD international conference on knowledge discovery & data mining, pp 824–833
17. Helling TJ, Scholtes JC, Takes FW (2018) A community-aware approach for identifying node anomalies in complex networks. In: Proceedings of international conference on complex networks & applications. Springer, Berlin, pp 244–255
18. Rosvall M, Bergstrom CT (2008) Maps of random walks on complex networks reveal community structure. Proc Natl Acad Sci 105(4):1118–1123
19. Vengertsev D, Thakkar H (2015) Anomaly detection in graph: unsupervised learning, graph-based features and deep architecture. Technical report, Department of Computer Science, Stanford University, Stanford, CA, USA
20. van Dongen S (2000) A cluster algorithm for graphs. Rep Inf Syst 10(R 0010)
21. Rossi R, Ahmed N (2015) The network data repository with interactive graph analytics and visualization. In: Proceedings of 29th AAAI conference on artificial intelligence
22. Girvan M, Newman ME (2002) Community structure in social and biological networks. Proc Natl Acad Sci 99(12):7821–7826

Time Series Analysis of National Stock Exchange: A Multivariate Data Science Approach

G. Venkata Manish Reddy, Iswarya, Jitendra Kumar, and Dilip Kumar Choubey

Abstract Stock price prediction attracts individual decisions to invest in share market and may encourage the common people to become active in share trading. Stock trading gives us direct monetary benefit under prevalent uncertainties. Factors like political developments at state, national and international levels, conservative, diverse and complex social conditions, natural disasters, famine, pandemics, economic trade cycle (recession, boom and recursion) and many others have great effects on share prices and stock cost. The present work is based on secondary data sources acquisitioned from various public and private data portals accessed freely. There have been various efforts made in the past to predict the trends of the stock prices on the basis of secondary data. The predicted confidence estimates may enable the common investors to make a profit despite large risk of loss at different point of time under the dynamics of market fluctuations. The present work is based on an inferential methodology, which has been found to be instrumental and particularly suitable to the financial time series analysis. Time series modelling and forecasting have fundamental importance in stock market prediction and analysis. Further, with the application of multivariate methods and regression modelling, we expect better forecast accuracy. The implementation of the proposed approach has been incorporated on real-time data set on daily basis. The software used for various computations is Python, SPSS, MS Excel and MS Solver. The current work is relevant in prediction and analysis of the stock market prices and also seems to be useful to a large number of peoples engaged in day to day trading.

Keywords Forecasting · Economic trade cycle · Time series modelling · Regression analysis · Stock trade · Confidence estimates · Inferential methodology

---

G. Venkata Manish Reddy · Iswarya · J. Kumar (✉)
School of Advanced Sciences, Vellore Institute of Technology, Vellore, Tamil Nadu, India
e-mail: jitendra.kumar@vit.ac.in

D. K. Choubey
Department of Computer Science and Engineering, Indian Institute of Information Technology Bhagalpur, Bhagalpur, Bihar, India
e-mail: dkchoubey.cse@iiitbh.ac.in

M. Thakur et al. (eds.), *Soft Computing for Problem Solving*, Lecture Notes in Networks and Systems 547, https://doi.org/10.1007/978-981-19-6525-8_53

1 Introduction

The stock market is the back bone of any progressive economy like India. Significant capital infusion for various organizations throughout the nation has been made possible through shares sold to the people. So, our national development is directly and strongly related to the performance of our stock markets. Majority of the developing countries depend on their stock exchange for further strengthening of the economy. Considering the general population especially in absence of advanced financial structure, sufficient training programme and timely awareness and methodologies based on stock market prediction play an essential role in bringing more and more people into the market just as to hold the existing investors [1].

Let us consider the role of time series modelling and predictive analysis in stock exchange. It is neither purely exclusive nor implicit in nature. Many financial experts have been attempted to predict stock prices by applying various statistical methodology of exploratory, descriptive and inferential algorithms but achieved the absolute success. This approach suffers from the elements of personal bias, judicious variable selection and model selection suitable to appropriate stock market. The functional and operational mechanism plays a vital role in general. It may happen because in a normal situation, stock market not only depends upon common market factors and prevalent economic arguments rather than highly sensitive towards unforeseen developments in and out. People mindset and their economic adventure defined as risk taking ability and innovative thinking based on proper market research at national and international perspective are highly decisive in nature. Uses of time series and regression modelling for a reasonable forecast can easily overshadow the conventional methodologies. Relatively longer business exposure and sound business knowledge play vital role in conventional approaches of predictive decision-making process, which we cannot afford further, especially in an era of digital technology, machine learning and artificial intelligence based on advanced statistical, stochastic and simulation methods. Stock market prediction has been a research topic for years. There are two hypotheses: efficient market hypothesis and adaptive market hypothesis. The first one proposes that stock prices follow a random walk, and thus, cannot be predicted [2, 3]. The second states that prices of stocks can be predicted precisely [4]. Trading decisions and stock price predictions make use of technical analysis and fundamental analysis. Technical analysis makes use of the corporations/firms historical stock data and applies to suitable statistical methods for analysis, interpretations and prediction to arrive at correct decisions. The fundamental analysis deals with the business performance of a company or a firm.

The rest of the article organized as follows: related work summarized in Sect. 2, methodology and methods has been stated in Sect. 3, results and discussion has been presented and discussed in Sect. 4, conclusion comprehended in Sect. 5 and Sect. 6 consists of future scope.

1.1 Objective

The first objective is to assess and characterize the National Stock Exchange (NSE), using various descriptive statistics, both absolute and relative in nature. The second objective is to develop and assign the consistency ranking to all twenty firms taken under the study and finding among them the top five most consistent companies in respect to the stock closing price. The third objective is to conduct proper regression analysis for fitting the model through estimation and testing of respective parameters involved in the model. In a process, we do explain the possibility of change in share closing value. To conduct, examine and instrument the standard diagnostic measures for validation and testing of the assumptions of the model to claim about the kind of possible relationships among the variables under the study. The fourth objective is to apply ARIMA model to predict the near future. The fifth and the final objective are to assess the quality of prediction of the stock market prices in day to day trading. Summarizing all five together, we can say that our main objective is to predict the share closing value for an effective, accurate, reliable and valid estimation so that risk involved under volatile market conditions should be reduced to avoid possible bankruptcy.

1.2 Background

The suitable stock choice for investment is truly challenging in nature. Accuracy and precision are the foremost among other considerations of model selection criteria. The objective of majority of the investors is to get the most viable returns on their investments. This market can make an individual as a billionaire else a bankrupt overnight. We are interested in studying the fluctuations and volatility in predicting future share prices with better accuracy. That is the reason we use fundamental and technical analysis to foresee stocks. We utilize different prediction methods, and forecasting models to precisely predict stock exchange. The use of the statistical computational methods is an objective way to predict stock prices. Autoregressive integrated moving average (ARIMA) is a form of regression analysis that measures the strength of single dependent variable, Y_i, relative to multiple independent variables, $X_i; i = 1, 2, 3, \ldots, n$. Banerjee [5] "Stock Prediction Model" goal is to predict future stock prices by equities examining the differences between values in the series.

2 Related Work

Nunno [6] states the stock market price prediction using linear and polynomial regression models. He has used the data set of CRSP US stock database and state that "support vector regression (SVR)", is the most effective model used in this project. The model for those stock markets is mainly concerned with the closing price for stocks at the end of the business day. Ariyo et al. [7] present the process of building a stock price predictive model using the autoregressive integrated moving average (ARIMA) model. Stock data obtained from NYSE and NSE used with a stock price predictive model developed and results that the ARIMA model has a strong potential for short-term prediction. Mali et al. [8] present development and implementation of a stock price prediction. The regression modelling and the object-oriented approach have been used for the process. They state that the prediction using multiple regressions based on generalized least square (GLS) yields better results than that of the ordinary least square (OLS) estimates of the parameters. Ashik and Kannan [9] present the time series model for stock price forecasting in India using the data set obtained from stock trade published by the Nigerian Stock Exchange. Olaniyi et al. [10] use data mining tools to reveal the hidden patterns to predict the future estimates. They studied on stock trend prediction using regression model and time series methods to explain the hidden patterns of stock market price and predict the future stock exchange prices of three different banks. Tsai and Wang [11] presented stock price forecasting using hybrid machine learning techniques, and their work conclude that if change in economic environment is fast and the investment objectives varies, investors are unlikely to make a correct decision for stock investment because they lack a detached systemic decision support tool. Therefore, they have combined the two methods, the ANN and the decision tree to obtain a better decision support system and helped the investors to make a correct decisions in stock investments. Sahoo and Charlapally [12] present stock price prediction using regression model and have predicted the future stock price using autoregressive model of order p, AR(p). They have worked on stock data obtained from the New York Stock Exchange. Autoregressive model is implemented using Java programming. They conclude that the comparison between the predicted and actual stock price almost coincides with each other in this model. Nayak et al. [13] have framed two models for daily and monthly prediction using supervised machine learning algorithms. This model has been tested with different stock data available open source. For the selected data set, decision boosted tree (DBT) provides better prediction results than support vector machine (SVM) and logistic regression (LR). Almost seventy per cent of accuracy is noted using supervised machine learning algorithm. Altland [14] has been referred for detail concepts and computational methodology of regression and related estimation of parameters and coefficients. The required data set used for the present work has been acquired and retrieved from various secondary sources and web portals like the web link of the National Stock Exchange, the web link of New York stock exchange, the web link of Yahoo.com and the web link of the Kaggle.com. Kumar [15] presented regression modelling and scan statistics to fit cognizable crime and find out the crime hotspot to predict and monitor the criminal activities in NCT of Delhi. Sukhija et al. [16] presented multiple regression models and computed various measures to fit the spatiotemporal crime data in Haryana. They also identified the crime hotspot for prediction of crimes in Haryana. Bhatia et al. [17] applied machine learning algorithm for drowsiness detection to control road accidents. They also used regression model to fit and predict the road accident cases in Tamil Nadu. Jangir et al. [18] have used functional link convolutional neural network for the classification of diabetes mellitus. Choubey and Paul [19] presented a hybrid intelligent system for diabetes disease diagnosis. Choubey et al. [20] have shown the implementation and analysis of classification algorithms for diabetes. Choubey et al. [21] present comparative analysis of classification methods with PCA and LDA for diabetes. Choubey et al. [22] have presented "performance evaluation of classification methods with PCA and PSO for diabetes". Kumar and Nagpal [23] have done analysis and prediction of crime patterns using big data. Juneja and Rana [24] have presented an improved weighted decision tree approach for breast cancer prediction.

3 Methodology

As mentioned, our work has been evolved through secondary data source obtained from different portals and sources mentioned in the references. Some of the regressor variables like Indian Rupee (INR) to USD conversion rate, gold and crude market values of the New York stock exchange taken from yahoofinance.com. As mentioned above, our objective is to predict the share closing value for better prediction; we believe that a large amount of data should exist. For the same, we have searched for companies that have data at least for ten to twenty years of regular trading. On that criterion, we found twenty companies with sufficient level of relevant information's, namely COAL INDIA, BAJAJ AUTO, GAIL, BPCL, DR. REDDY, CIPLA, HDFC, BHARTI AIRTEL, GRASIM, HDFC BANK, HCL TECH, HERO MOTOR, AXIS BANK, BRITANNIA, ASIAN PAINT, HINDALCO, ADANI PORTS, BAJAJ FINSV, BAJAJ FINANCE and EICHER MOT. **We have used regression and ARIMA model to have inferential conclusions.**

3.1 *Finding and Eliminating Missing Values*

As we were aware that stock exchange has their list of non-working days, so those data were missing, whereas crude and the gold values taken from the New York stock exchange were weekend data which were also missing. So, we have eliminated the complete transactions corresponding to all missing values to avoid the computational difficulties and empirical relevance.

3.2 *Descriptive Analysis*

It summarizes the characteristics of the data. To find out the most consistent and reliable among twenty different companies under the study, we have computed a relative measure known as the coefficient of variation (CV). Furthermore, the National

Stock Exchange considered the share closing value over the period 2001–19. Coefficient of variation (CV) has been used as a relative descriptive statistic to measure the consistency among them and algebraically expressed as following mathematical formulation.

$$\text{CV} = \left[\frac{\sigma}{\mu}\right]100 \tag{1}$$

where σ and μ denote the standard deviation and mean, respectively.

The companies having minimum value of CV will be considered as most consistent and as such assigned rank one and remaining consistency ranking will further be assigned accordingly.

3.3 Multiple Linear Regression (MLR)

The model predicts the relationship between the dependent variable (share value in rupees (INR)) and independent variable (crude oil value, gold value, rupee value (INR)). Multiple linear regression model can develop as follows

$$Y = \beta_0 + \beta_1 X_1 + \beta_2 X_2 + \cdots + \beta_k X_k + \in \tag{2}$$

where Y is a dependent/response variable and $X_1, X_2, X_3, X_4, X_5, \ldots, X_k$ is the independent variables in the model that influences Y's occurrence, i.e. dependent variable and $\in$ is the random or error term that explains the factors influencing the response variable Y, but cannot be explained through the defined model.

3.4 Prediction Analysis

In prediction analysis, we were calculating the prediction accuracy of the model using the root mean square error (RMSE) and the goodness of fit of the model in terms of the values of the coefficient of determination R^2 and the adjusted R^2 obtained from the models for each respective company.

Where

$$\text{Adjusted } R^2 = 1 - \frac{\text{SS}_{\text{res}}/\text{d}f_e}{\text{SS}_{\text{tot}}/\text{d}f_t} = 1 - \frac{(1 - R^2)(n - 1)}{(n - k - 1)} \tag{3}$$

where n is the sample size and k denotes the number of regressors or predictors in the model such that there exists a relationship between p and k such that $p = k + 1$ where p denotes the number of parameters involved in the model.

$$\mathrm{RMSE} = \sqrt{\sum_{i=1}^{n} \frac{\left(\hat{y}_i - y_i\right)^2}{n}} \tag{4}$$

3.5 Rank Correlation

We are finding the relationship between pre-regression ranking computed on basis of the coefficient of variance (CV) and the post regression ranking computed on basis of the mean squared error (MSE).

$$r_S = 1 - \frac{6\sum_{i=1}^{n} D_i^2}{n\left(n^2 - 1\right)} \tag{5}$$

where r_S defined above denotes Spearman rank correlation coefficient and its value always lies between -1 and $+1$.

3.6 Multicollinearity: A Potential Problem

When explanatory variables are correlated tangibly or intangibly among themselves, even partially the problem of multicollinearity said to exist. How serious the problem is depending on the degree of multicollinearity. The presence of a high degree of multicollinearity among the explanatory variables results in following two problems. (a) The standard deviations of the regression coefficients are disproportionately large. (b) The regression coefficient estimates are unstable and as such reliable estimates of the regression coefficients are difficult to obtain.

Pairwise correlations and variance inflation factors (VIFs) are two popular methods to detect multicollinearity in regression modelling. Also, an indication of multicollinearity is relatively larger value of F statistic but smaller value of t-statistic. A serious problem of multicollinearity always results in large standard deviations of the regression coefficients and small t ratios.

The VIF for the variable x_j is

$$\mathrm{VIF}_j = \frac{1}{1 - R_j^2} \tag{6}$$

where R_j^2 denotes the coefficient of determination used to explain the total variation in the response variable on account of the regressors used in the model. Any individual VIF_j larger than 10 indicates that multicollinearity may be influencing the least square estimates of the regression coefficients. Also, if the average value of $\mathrm{VIF}_j > 1$, then serious problem may exist. VIF also need to be evaluated relative to the overall fit of the model. As per concept of statistical sciences, we state that whenever the individual VIFs values are smaller than $\left[\frac{1}{1-R^2}\right]$, the multicollinearity is not strong enough to affect the estimates of the coefficients. In such a case, the independent variables are more strongly related to the y variable than they are among themselves.

3.7 Test of Linearity

Scatterplots of y versus each of the explanatory variables may give an indication of whether the assumption of linearity is preserved, but sometimes, it may not be sufficient. So, residual plots as well exclusive test of linearity are being conducted for the model. A random scatter in the residual plots indicates that linearity and homogeneity (constant variance) of the data have been preserved and indicate absence of violations of some of the assumptions of the model. If some specific patterns in the residual plots persists that indicates violation of assumptions of linearity, normality as well as homogeneity of the model.

3.8 ARIMA

It uses the information in the past values of the time series can alone be used to predict future values.

Where using ARIMA (p, d, q) model, such as

- p Autoregressive,
- d Difference value and
- q Moving average.

We compute our estimates of the parameters using suitable time series methods.

3.9 MLR Versus ARIMA

After the RMSE values obtained from the MLR model and ARIMA, we have compared the results obtained from MLR and ARIMA to compare and assess the accuracy of these two models using root mean squared error values (RMSE). Models having smaller RMSE values will expect to perform better.

4 Results and Discussion

When we compute the aforesaid statistics using suitable methods as explained above on our data set and do estimation, testing and validation for the model implemented, we found the following results that is being discussed to arrive at specific objectives that has been presented and summarized in conclusion section.

4.1 Descriptive Analysis

Table 1 contains the information of Company Name, Mean, Standard Deviation (SD), Coefficient of Variation (CV) and Rank and shows the consistency rankings based on their performances. The five most preferred companies are ranked higher. They are COAL INDIA, BAJAJ AUTO, GAIL, BPCL and DR. REDDY, respectively.

Different values of coefficient of variation for each company have been computed and respective rankings are assigned as shown above.

Table 1 Company's consistency ranking

Company name	Mean	SD	CV	Rank
COAL INDIA	309.5684883	50.9432616	16.45621681	1
BAJAJ AUTO	2084.022555	731.0387139	35.07825346	2
GAIL	305.2927942	124.0990593	40.64919371	3
BPCL	447.6059775	189.6175873	42.36261285	4
DR. REDDY	1666.464225	905.7341936	54.35065331	5
CIPLA	530.7906369	293.7637041	55.34455276	6
HDFC	1227.299297	690.4150292	56.25482154	7
BHARTARTL	369.5067492	210.7036582	57.02295255	8
GRASIM	1815.692757	1161.074656	63.946648	9
HDFC BANK	993.4682138	651.983464	65.62700799	10
HCL TECH	635.8090123	421.7594184	66.33429383	11
HERO MOTOCO	1579.9174	1073.513465	67.94744241	12
AXIS BANK	587.8663351	449.6757847	76.49286204	13
BRITANNIA	1571.766961	1328.266911	84.50787829	14
ASIAN PAINT	1195.125537	1085.546077	90.83113387	15
HINDALCO	365.6407977	364.878733	99.79158107	16
ADANI PORTS	2,415,617.3	2,644,872.9	109.4905596	17
BAJAJ FINSV	2207.547093	2420.030262	109.6253063	18
BAJ FINANCE	1246.052344	1767.358218	141.8365951	19
EICHER MOT	6537.435046	9772.590684	149.48662	20

4.2 Regression Analysis: (MLR)

The regression line to predict the share closing value with the unique change in factor line, company share previous day closing value, current day opening value, current day rupee opening value, last day crude and gold opening value.

Consider and define following set of variables used for developing the MLR mentioned below.

Let us define the following set of variables

Y Companies current day closing value (INR).
X_1 Companies previous day closing value (INR).
X_2 Companies present-day opening value (INR).
X_3 current day Rupee (INR) to USD conversion rate opening value.
X_4 Previous day Crude closing value ($).
X_5 Previous day Gold closing value ($).

After substituting the computed values of coefficients of regression obtained from least square methods and possessing the properties of best linear unbiased estimator (BLUE), as given by the Gauss Markov Theorem of linear estimation computed using the software mentioned above, we obtain the following fitted regression line for top ranked five companies namely the COAL INDIA, BAJAJ AUTO, GAIL, BPCL and DR. REDDY, respectively, as given below.

$$\begin{aligned} Y_{\text{COAL INDIA}} = {} & 6.88716396 + 0.07027468X_1 + 0.9180548X_2 \\ & - 0.02895652X_3 + 0.01398916X_4 - 0.00204746X_5 \end{aligned} \tag{7}$$

$$\begin{aligned} Y_{\text{BAJAJ-AUTO}} = {} & -5.53431251 + 0.07015695X_1 + 0.92520248X_2 \\ & + 0.21176936X_3 - 0.04402971X_4 + 0.00417111X_5 \end{aligned} \tag{8}$$

$$\begin{aligned} Y_{\text{GAIL}} = {} & -0.03905749 + 0.242601253X_1 + 0.753231007X_2 \\ & + 0.015170527X_3 + 0.009368252X_4 - 0.000310685X_5 \end{aligned} \tag{9}$$

$$\begin{aligned} Y_{\text{BPCL}} = {} & 3.00102212 + 0.04998272X_1 + 0.94596782X_2 \\ & - 0.04319829X_3 - 0.0183946X_4 + 0.00195471X_5 \end{aligned} \tag{10}$$

$$\begin{aligned} Y_{\text{DR REDDY}} = {} & -13.42365623 + 0.13954741X_1 + 0.856078126X_2 \\ & + 0.302083353X_3 + 0.058635175X_4 - 0.000122781288X_5 \end{aligned} \tag{11}$$

Figure 1 presents the actual and predicted trend values represented by different colours for the COAL INDIA company obtained from the regression analysis.

Fig. 1 COAL INDIA regression

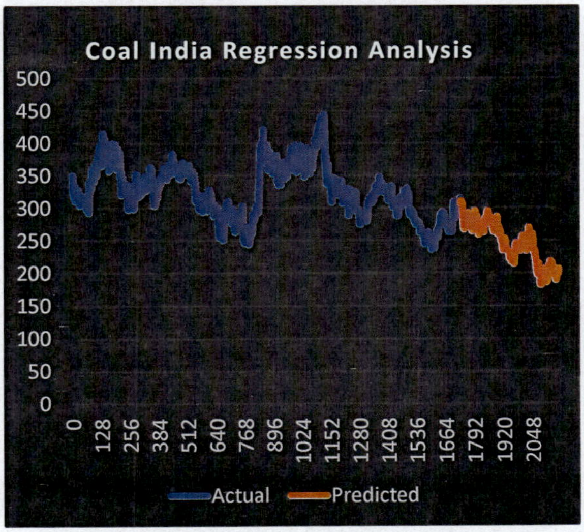

4.3 Prediction Analysis

Table 2 consists of information of the name of the company, R^2 values, adjusted R^2 values, MSE values and RMSE values. Since the sequence of minimum values of RMSE, i.e. 5.5057, 7.6470, 11.4544, 32.9820 and 33.0392 respectively, corresponds to COAL INDIA, GAIL, BPCL, BAJAJ AUTO and DR. REDDY, therefore, we declare and recommend COAL INDIA, GAIL and BPCL as the best three performer companies at a moment. The adjusted R^2 values of these companies are also sufficient to claim the best fit of the model. The R^2 values are approximately 0.99 for all these five companies which are also satisfactory.

Table 2 Summary output of predictive analysis

Name of the company	R^2 value	Adjusted R^2 value	MSE value	RMSE value
COAL INDIA	0.9882	0.9881	30.3136	5.5057
BAJAJ AUTO	0.9978	0.9978	1087.818	32.9820
GAIL	0.9934	0.9934	58.4776	7.6470
BPCL	0.9956	0.9956	131.2042	11.4544
DR. REDDY	0.9988	0.9988	1091.595	33.0392

Table 3 CV and MSE ranking of top five companies

Name of the company	CV ranking (x)	MSE ranking (y)
COAL INDIA	1	1
BAJAJ AUTO	2	4
GAIL	3	2
BPCL	4	3
DR. REDDY	5	5

4.4 Rank Correlation

Table 3 contains the information on Name of the Company, CV Rankings (x), MSE Ranking (y) and shows the respective rankings assigned on basis of two different powerful statistical tools, namely CV and MSE, respectively. We found to have high value of the Pearson rank correlation coefficient $r_s = 0.7$ which shows strong positive relationship between the two ranks. This shows the statistical significance of their ranks.

4.5 Multicollinearity

Table 4 contains information of the constant variable name and shows that the average values of VIFs is greater than ten, that indicate the presence of the effects of multicollinearity on least square estimators and is likely to inflate the value of standard error obtained for the regression model. Though presence of multicollinearity is quite expected and widely common in majority of the financial data set, we may think of reducing it by possible use of variance stabilization transformations like Box Cox method or log method or exponential transformation depending upon the functional relationship between the dependent and independent variables.

Table 4 gives the VIFs values for different companies and variables that enable us to infer that multicollinearity is a problem that need to be addressed separately to ensure better prediction having scope for further extension of this work.

Table 4 Computed values of variance inflation factors (VIFs)

Constant variable name	COAL INDIA	BAJAJ AUTO	GAIL	BPCL	DR. REDDY
Share closing value	273.0	504.91	237.79	175.42	2261.04
Share open values	273.6	505.29	237.30	175.11	2258.27
Rupee value	3.60	3.12	2.33	2.38	4.82
Crude value	2.13	1.88	2.04	2.41	2.00
Gold value	1.96	1.46	3.00	3.28	2.70

Table 5 RMSE value from ARIMA and regression model

Name of the company	ARIMA model	ARIMA RMSE value	Regression analysis RMSE value
COAL INDIA	(0,1,0)	6.112	5.505785
BAJAJ AUTO	(0,1,1)	46.08	32.982090
GAIL	(0,1,7)	8.061	7.647069
BPCL	(0,1,0)	17.559	11.454441
DR. REDDY	(0,1,1)	30.659	33.039299

4.6 ARIMA

Table 5 given below consists of information on name of the company, ARIMA model used, ARIMA RMSE values and regression RMSE values obtained for the data set. The ARIMA RMSE values and the regression RMSE values are different for all five companies. We observe that the least value of ARIMA RMSE is 6.112 belong to COAL INDIA and the highest value 30.659 belong to DR. REDDY. Also, the least value of regression RMSE 5.505785 belongs to the COAL INDIA and the highest value 33.039299 belongs to DR. REDDY. This implies the same findings for both the models under study. **We found the average RMSE value of the regression model (MLR) is 18.1257368 which is significantly smaller than that of the ARIMA model. This further implies that regression model is superior to the ARIMA model in present context.**

Table 6 given below consists of information on results of the model fit. Further Table 6 gives the visual information on prediction accuracy for COAL INDIA Company using the outputs generated by the ARIMA model.

Figure 2 shows the visual description for the predicted values of COAL INDIA company through the ARIMA (0, 1, 0).

4.7 Test of Linearity

Table 7 consists of share closing values versus p-values computed for the COAL INDIA using the model and is shown below. We observe that the share closing value, previous day share and gold closing value do not support the hypothesis of linear relationship (for that p-value > 0.5), whereas current day share, opening value and previous day crude closing value do support the hypothesis of linear relationship (for that p-value < 0.5).

We set the following null and alternative hypothesis

H_0 The relationship is linear.
H_1 The relationship is not linear.

Table 6 Prediction accuracy for COAL INDIA

Model fit											
Fit statistic	Mean	SSE	Minimum	Maximum	Percentile						
					5	10	25	50	75	90	95
Stationary *R*-squared	0.00002639	–	0.00002639	0.00002639	0.00002639	0.00002639	0.00002639	0.00002639	0.00002639	0.00002639	0.00002639
R-squared	0.978	–	0.978	0.978	0.978	0.978	0.978	0.978	0.978	0.978	0.978
RMSE	6.112	–	6.112	6.112	6.112	6.112	6.112	6.112	6.112	6.112	6.112
MAPE	1.336	–	1.336	1.336	1.336	1.336	1.336	1.336	1.336	1.336	1.336
MaxAPE	11.261	–	11.261	11.261	11.261	11.261	11.261	11.261	11.261	11.261	11.261
MAE	4.367	–	4.367	4.367	4.367	4.367	4.367	4.367	4.367	4.367	4.367
MaxAE	43.569	–	43.569	43.569	43.569	43.569	43.569	43.569	43.569	43.569	43.569
Normalized BIC	3629	–	3629	3629	3629	3629	3629	3629	3629	3629	3629

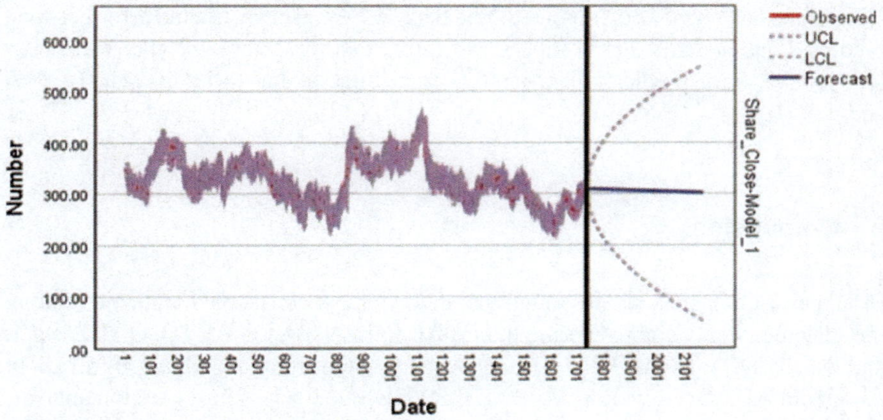

Fig. 2 Visualization of the predicted value for COAL INDIA

Table 7 *p*-values for testing the linearity

Share closing value V_s	*P*-value	*P*-value < or > 0.05	H_0 rejected or accepted
Previous day share closing value	~ 0.000	< 0.05	Reject H_0
Current day share opening value	0.987	> 0.05	Accept H_0
Current day rupee opening value	0.001	< 0.05	Reject H_0
Previous day crude closing value	0.217	> 0.05	Accept H_0
Previous day gold closing value	0.03	< 0.05	Reject H_0

On basis of the result given in Table 6, calculated for the COAL INDIA, we may infer that previous day share closing value, current day rupee opening value and previous day gold closing value follow nonlinearity (relatively difficult to predict), whereas current day share opening value and previous day crude closing value follow the linear relationship (relatively easier to predict).

4.8 Validation

Table 8 consists of dates, predicted value, actual value and the error in rupees terms.

This table represents the predicted share closing values for the company COAL INDIA for two days, which is not the part of the data set, and the selected dates are

Table 8 Error computation

Date	Predicted value	Actual value	Error (INR)
11 May 2020	130.1	128.45	1.65
12 May 2020	128.485	126.85	1.63

six months away from the last date in the data set. We also calculated error between predicted and actual value in rupees and found that the error is around ₹1.63. For later dates, we can predict with error confidence interval that varies between (± ₹1.5 and ± ₹5.0).

5 Conclusion

From descriptive analysis, we found that the five most consistent companies among the selected twenty companies are the COAL INDIA, BAJAJ AUTO, GAIL, BPCL and DR. REDDY. COAL INDIA is the most consistent company obtained from both the model ARIMA as well as MLR. The RMSE value is less in regression analysis than ARIMA for predicting the closing share values of most companies that support regression model as superior to ARIMA model. We found that there is a problem of multicollinearity and nonlinearity for some variables, which results in the high RMSE values. With the current model, we could predict the share closing value of COAL INDIA with an error between ± ₹1.5 and ± ₹5.0. Thus, the present work gives us a complete understanding to achieve all five objectives set forth in the beginning of the research plan.

6 Future Scope

We may set our future objective to predict the share closing value with the least error of [± ₹0.05], i.e. a difference of five paisa. This could be possible by minimizing the effects of multicollinearity in the model that requires more stringent conditions on variable selection and through more precise data set to work with.

Glossary

ARIMA Autoregressive integrated moving average
BLUE Best linear unbiased estimator
CV Coefficient of variation
SD Standard deviation
MSE Mean squared error
RMSE Root mean squared error
. Multicollinearity
. Coefficient of determination (R^2)
. Pearson correlation coefficient

References

1. Gandhmal DP, Kumar K (2019) Systematic analysis and review of stock market prediction techniques. Comput Sci Rev 34:100190
2. Campanella F, Mustilli M, D'Angelo E (2016) Efficient market hypothesis and fundamental analysis: an empirical test in the European securities market. Rev Econ Finan 6:27–42
3. Malkiel BG (2019) A random walk down Wall Street: the time-tested strategy for successful investing. WW Norton & Company
4. Berner R (2019) Adaptive markets: financial evolution at the speed of thought by Andrew Lo
5. Banerjee D (2014, Jan) Forecasting of Indian stock market using time-series ARIMA model. In: 2014 2nd International conference on business and information management (ICBIM), pp 131–135. IEEE
6. Nunno L (2014) Stock market price prediction using linear and polynomial regression models. Computer Science Department, University of New Mexico, Albuquerque
7. Ariyo AA, Adewumi AO, Ayo CK (2014, Mar) Stock price prediction using the ARIMA model. In: 2014 UKSim-AMSS 16th International conference on computer modelling and simulation. IEEE, pp 106–112
8. Mali P, Karchalkar H, Jain A, Singh A, Kumar V (2017) Open price prediction of stock market using regression analysis. Int J Adv Res Comput Commun Eng 6(5)
9. Ashik AM, Kannan KS (2019) Time series model for stock price forecasting in India. In: Logistics, supply chain and financial predictive analytics. Springer, Singapore, pp 221–231
10. Olaniyi SAS, Adewole KS, Jimoh RG (2011) Stock trend prediction using regression analysis—a data mining approach. ARPN J Syst Softw 1(4):154–157
11. Tsai CF, Wang SP (2009, Mar) Stock price forecasting by hybrid machine learning techniques. In: Proceedings of the international multiconference of engineers and computer scientists, vol 1(755), p 60
12. Sahoo PK, Charlapally K (2015) Stock price prediction using regression analysis. Int J Sci Eng Res 6(3):1655–1659
13. Nayak A, Pai MM, Pai RM (2016) Prediction models for Indian stock market. Procedia Comput Sci 89:441–449
14. Altland HW (1999) Regression analysis: statistical modeling of a response variable
15. Kumar J (2014) Some contributions to scan statistics and their applications. PhD thesis, Patna University, pp 230–242
16. Sukhija K, Singh SN, Kumar J (2017, May) Spatial visualization approach for detecting criminal hotspots: an analysis of total cognizable crimes in the state of Haryana. In: 2017 2nd IEEE International conference on recent trends in electronics, information & communication technology (RTEICT). IEEE, pp 1060–1066
17. Bhatia U, Tshering, Kumar J, Choubey DK (2021) Drowsiness image detection using computer vision. In: Soft computing: theories and applications. Springer, Singapore, pp 667–683
18. Jangir SK, Joshi N, Kumar M, Choubey DK, Singh S, Verma M (2021) Functional link convolutional neural network for the classification of diabetes mellitus. Int J Numer Methods Biomed Eng 37(8):e3496
19. Choubey DK, Paul S (2016) GA_MLP NN: a hybrid intelligent system for diabetes disease diagnosis. Int J Intell Syst Appl 8(1):49–59
20. Choubey DK, Paul S, Shandilya S, Dhandhania VK (2020) Implementation and analysis of classification algorithms for diabetes. Curr Med Imaging 16(4):340–354
21. Choubey DK, Kumar M, Shukla V, Tripathi S, Dhandhania VK (2020) Comparative analysis of classification methods with PCA and LDA for diabetes. Curr Diabetes Rev 16(8):833–850
22. Choubey DK, Kumar P, Tripathi S, Kumar S (2020) Performance evaluation of classification methods with PCA and PSO for diabetes. Network Model Anal Health Inf Bioinf 9(1):1–30
23. Kumar R, Nagpal B (2019) Analysis and prediction of crime patterns using big data. Int J Inf Technol 11(4):799–805
24. Juneja K, Rana C (2020) An improved weighted decision tree approach for breast cancer prediction. Int J Inf Technol 12(3):797–804

A TOPSIS Method Based on Entropy Measure for q-Rung Orthopair Fuzzy Sets and Its Application in MADM

Rishu Arora, Chirag Dhankhar, A. K. Yadav, and Kamal Kumar

Abstract In this article, we develop a novel TOPSIS approach for solving the multi-attribute decision-making (MADM) problems under the q-rung orthopair fuzzy numbers (q-RONs) environment. For this, we have proposed a new entropy measure (EM) for q-rung orthopair fuzzy set (q-ROFS) to measure the fuzziness of the q-ROFS. Numerous features of the proposed EM of q-ROFS are also illustrated. Afterwards, by utilizing the proposed EM, a TOPSIS approach has been developed for tackling MADM issues under the q-ROFNs context. To exemplify the proposed TOPSIS technique, a real-life MADM example has been studied. Comparative studies are also developed to illustrate the TOPSIS approach's efficiency.

Keywords TOPSIS · q-ROFSs · Entropy measure · MADM

1 Introduction

Multi-attribute decision-making (MADM) problems are a type of decision-making problem in which we use collective information to choose the best one from a group of finite choices. For handling the MADM issues, the most challenging step for the decision-makers (DMKs) is to choose appropriate environment for providing performance assessments of attributes. To tackle such challenges, [14] presented a new extension of the fuzzy set [15] and intuitionistic fuzzy set [1] known as q-rung orthopair fuzzy number (q-ROFN) $\langle \eta, \upsilon \rangle$ such that $0 \leq \eta^q, \upsilon^q \leq l,\ 0 \leq \eta^q + \upsilon^q \leq l$ and $q \geq 1$. Under this environment, the researchers [4, 5, 8, 12, 13] paid more attentions to handle the real-life MADM issues. Liu et al. [7] defined the q-ROFWEBM aggregating operator (AO) and MADM approach for the q-ROFNs. For

R. Arora
Department of Mathematics and Humanities, MM Engineering College, Maharishi Markandeshwar (Deemed to be University), Mullana, Ambala, Haryana, India

C. Dhankhar · A. K. Yadav · K. Kumar (✉)
Department of Mathematics, Amity School of Applied Sciences, Amity University Haryana, Gurugram 122413, Haryana, India
e-mail: kamalkumarrajput92@gmail.com

M. Thakur et al. (eds.), *Soft Computing for Problem Solving*, Lecture Notes in Networks and Systems 547, https://doi.org/10.1007/978-981-19-6525-8_54

q-ROFNs, [6] defined the MADM approach based on the proposed geometric and averaging AOs. The MADM approach based on Einstein AOs in q-ROFNs context was established by Riaz et al. [10, 11] established the MADM approach and Einstein prioritized weighted averaging AOs for q-ROFNs context. The PDM for q-ROFNs was established by Garg et al. [2]. For q-ROFNs, he knowledge measure is proposed by Khan et al. [4]. [9] defined the various EMs for the q-ROFSs. Garg [2] defined the possibility measure for the interval-valued q-ROFSs. Khan and Ali [3] defined a new ranking method for the q-ROFNs. Wang et al. [12] defined the MABAC method for MADM under the q-ROFNs.

During the DM process, uncertainty plays a crucial role, and one of the most important tasks is to measure the data's uncertainty. To measure the uncertainty of the data, the entropy measure (EM) is a very important tool. Various EM for the q-ROFSs have been defined in [9]. However, on reviewing, we found that the existing EMs proposed in [9] have various limitations. They do not measure the uncertainty in a proper manner. Thus, it is required to develop a new EM to measure the fuzziness of q-ROFSs.

In this study, we propose a novel EM for q-ROFS to measure the data uncertainty. We also prove the different alluring properties of the proposed EM of q-ROFS. The proposed EM can also address the limitations of the existing EMs [9]. Moreover, we construct a TOPSIS approach based on the proposed EM of q-ROFS to tackle MADM issues under the q-ROFNs. Afterwards, to illustrate the proposed TOPSIS approach, a real-life MADM issue has been discussed, and the results are compared with current strategies to reveal the proposed approach's preferences. The proposed TOPSIS method provides a very useful approach to deal with MADM issues in q-ROFNs environment.

Therefore, the key goals of this study are summarized as follows based on the above analysis:

1. A numerical example is presented to show shortcomings of existing EM of q-ROFSs.
2. A new EM of q-ROFSs has been constructed that overcomes the flaws of existing EM of q-ROFSs.
3. The different alluring properties of the proposed EM have been presented in details.
4. By using the proposed EM, a TOPSIS approach has been built for tackling the MADM issues under the q-ROFNs.
5. The new TOPSIS technique has been demonstrated using a real-world MADM problem, and the results have been compared to current strategies to reveal the proposed approach's preferences.

To achieve the intended outcomes, this article is organized as follows: The preliminaries of this study are presented in Sect. 2. Section 3 defines the entropy measure of q-ROFSs. In Sect. 4, we build a new TOPSIS method in view of the proposed entropy of q-ROFSs for tackling MADM issues. Finally, Sect. 6 brings the paper to a close.

2 Preliminaries

The following are some preliminary considerations for this study.

Definition 1 [14] A q-ROFS Q in universal set X is defined by

$$Q = \left\{\langle x, \eta_Q(x), \upsilon_Q(x)\rangle \mid x \in X\right\} \tag{1}$$

where $\eta(x)$ and $\upsilon(x)$ indicate membership and non-membership grades of x, such that $\forall\, x, 0 \le \eta_Q(x), \upsilon_Q(x) \le 1, 0 \le \eta_Q^q(x) + \upsilon_Q^q(x) \le 1$. The hesitance degree of x is given by $\pi_Q = (1 - \eta_Q^q(x) - \upsilon_Q^q(x))^{1/q}$ where $q \ge 1$.

The pair $\langle \eta, \upsilon\rangle$ is frequently known to as a q-ROFN.

Definition 2 For a q-ROFS Q, the entropy measure (EM) E :Q→ [0, 1] fulfils the subsequent properties:

1. $0 \le E(Q) \le 1$;
2. $E(Q) = 0$ iff Q is crisp set;
3. $E(Q) = 1$ iff $\eta_Q(x) = \upsilon_Q(x)$;
4. $E(Q) = E(Q^C)$;
5. $E(Q) \le E(Q')$ if Q is less fuzzy than Q', that is $\eta_Q(x) \le \eta'_Q(x) \le \upsilon'_Q(x) \le \upsilon_Q(x)$ or $\upsilon_Q(x) \le \upsilon'_Q(x) \le \eta'_Q(x) \le \eta_Q(x)$.

In the following, [9] defined some EM for a q-ROFS $Q = \left\{\langle x, \eta_Q(x), \upsilon_Q(x)\rangle \mid x \in X\right\}$ shown as follows:

1. $E_1(Q) = 1 - \frac{1}{n}\sum_{t=1}^{n} |\eta_Q^q(x) - \upsilon_Q^q(x)|$
2. $E_2(Q) = \frac{1}{(\sqrt{2}-1\times n)}\sum_{t=1}^{n}(\sin\frac{1+\eta_Q^q(x)-\upsilon_Q^q(x)}{4}\pi + \sin\frac{1-\eta_Q^q(x)+\upsilon_Q^q(x)}{4}\pi - 1)$
3. $E_3(Q) = \frac{\sum_{t=1}^{n} 1-|\eta_Q^q(x)-\upsilon_Q^q(x)|}{\sum_{t=1}^{n} 1+|\eta_Q^q(x)-\upsilon_Q^q(x)|}$

Definition 3 [14] Let $Q = \left\{\langle x_t, \eta_Q(x_t), \upsilon_Q(x_t)\rangle \mid x_t \in X\right\}, (t = 1, 2, \ldots, n)$ be a q-ROFS. For any $q > 0$ and $k > 0$, the q-ROFS Q^k is defined as follows:

$$Q^k = \left\{\langle x_t, [\eta_Q(x_t)]^k, (1 - [1 - \upsilon_Q^q(x_t)]^k)^q\rangle \mid x_t \in X\right\} \tag{2}$$

Example 1 Consider a q-ROFS $Q = \Big\{\langle x_1, 0.3, 0.7\rangle, \langle x_2, 0.7, 0.6\rangle, \langle x_3, 0.3, 0.5\rangle, \langle x_4, 0.67, 0.19\rangle, \langle x_5, 0.2, 0.8\rangle\Big\}$ as "LARGE".

By utilizing Eq. (2), we generate the q-ROFSs $Q^{1/2}$("Less LARGE"), Q^2("Very LARGE"), Q^3("Quit very LARGE") and Q^4("Very very LARGE") as follows:

For q-ROFSs $Q^{1/2}$, Q, Q^2, Q^3 and Q^4, an effective EM must satisfy the following relation

$$E(Q^{1/2}) > E(Q) > E(Q^2) > E(Q^3) > E(Q^4) \tag{3}$$

Table 1 Existing EMs E_1, E_2 and E_3 for q-ROFSs $Q^{1/2}, Q, Q^2, Q^3$ and Q^4

	E_1	E_2	E_3
$Q^{1/2}$	0.7294	0.8805	0.5741
Q	0.7322	0.9028	0.5776
Q^2	0.6185	0.7869	0.4477
Q^3	0.5170	0.6664	0.3487
Q^4	0.4404	0.5740	0.2824

Now we calculate the above existing EMs E_1, E_2 and E_3 for q-ROFSs $Q^{1/2}$, Q, Q^2, Q^3 and Q^4, and the outcomes are described in Table 1. Table 1 displays that $E_1(Q^{1/2}) < E_1(Q)$, $E_2(Q^{1/2}) < E_2(Q)$ and $E_3(Q^{1/2}) < E_3(Q)$. We observe that the EMs E_1, E_1 and E_3 do not satisfy Eq. (3).

The performance of the existing EMs E_1, E_2 and E_3 of q-ROFS is not satisfactory. Therefore, to conquer the shortcomings of the existing EMs E_1, E_2 and E_3 of q-ROFS, we will define a new EM of q-ROFS.

3 A New Constructive Q-ROF Entropy

This section presents a new entropy measure (EM) for q-ROFS.

Definition 4 The proposed EM for the q-ROFS $Q = \{\langle x_t, \eta_Q(x_t), \upsilon_Q(x_t)\rangle | x_t \in X\}$ is defined as follows:

$$E(Q) = \frac{1}{n}\sum_{t=1}^{n} \cot\left(\frac{\pi}{4} + \frac{|\eta_Q^q(x) - \upsilon_Q^q(x)|}{4}\pi\right) \tag{4}$$

where $q \geq 1$.

Theorem 1 *The proposed EM satisfies the properties listed in Definition 2.*

Proof Let $Q = \{\langle x, \eta_Q(x), \upsilon_Q(x)\rangle | x \in X\}$ be a q-ROFS.

1. Obviously, for each x, we have

$$\frac{\pi}{4} \leq \frac{\pi}{4} + \frac{|\eta_Q^q(x) - \upsilon_Q^q(x)|}{4}\pi \leq \frac{\pi}{2},$$

then

$$0 \leq \frac{\pi}{4} + \frac{|\eta_Q^q(x) - \upsilon_Q^q(x)|}{4}\pi \leq 1,$$

thus we have $0 \leq E(Q) \leq 1$.

2. Let for all $x \in X$, $Q = \{\langle x, \eta_Q(x), \upsilon_Q(x)\rangle | x \in X, \eta_Q(x) = 0, \upsilon_Q(x) = 1$ or $\eta_Q(x) = 1, \upsilon_Q(x) = 0\}$ is a crisp set. So, it is obvious that $E(Q) = 0$.
   If $E(Q) = 0$, *i.e.*, $E(Q) = \frac{1}{n}\sum_{t=1}^{n} \cot(\frac{\pi}{4} + \frac{|\eta_Q^q(x) - \upsilon_Q^q(x)|}{4}\pi) = 0$, then for all $x \in X$, we have

$$\cot\left(\frac{\pi}{4} + \frac{|\eta_Q^q(x) - \upsilon_Q^q(x)|}{4}\pi\right) = 0,$$

   thus $|\eta_Q^q(x) - \upsilon_Q^q(x)| = 1$, then we have $\eta_Q(x) = 0, \upsilon_Q(x) = 1$ or $\eta_Q(x) = 1, \upsilon_Q(x) = 0$. Hence, Q is a crisp set.
3. Let $\eta_Q(x) = \upsilon_Q(x)$, then it could be easily concluded that $E(Q) = 1$.
   Now if $E(Q) = 1$, then for all $x \in X$, we have:

$$\cot\left(\frac{\pi}{4} + \frac{|\eta_Q^q(x) - \upsilon_Q^q(x)|}{4}\pi\right) = 1,$$

   so, $|\eta_Q^q(x) - \upsilon_Q^q(x)| = 0$, we concluded that $\eta_Q(x) = \upsilon_Q(x)$ for all $x \in X$.
4. As we know $Q^C = \langle x, \upsilon_Q(x), \eta_Q(x)\rangle$ and by Eq. 4, we have

$$E(Q^C) = \frac{1}{n}\sum_{t=1}^{n} \cot\left(\frac{\pi}{4} + \frac{|\upsilon_Q^q(x) - \eta_Q^q(x)|}{4}\pi\right)$$
$$= \frac{1}{n}\sum_{t=1}^{n} \cot\left(\frac{\pi}{4} + \frac{|\eta_Q^q(x) - \upsilon_Q^q(x)|}{4}\pi\right) = E(Q).$$

5. Consider a function:

$$f(a, b) = \cot\left(\frac{\pi}{4} + \frac{|a^q - b^q|}{4}\pi\right)$$

   where $a, b \in [0, 1]$.
   Now when $a \leq b$, we have $f(a, b) = \cot(\frac{\pi}{4} + \frac{b^q - a^q}{4}\pi)$, we will show that the function $f(a, b)$ grows with a and shrinks with b.
   We have to find partial derivatives of $f(a, b)$ to a and b, respectively:

$$\frac{\partial f(a, b)}{\partial a} = \frac{\pi q a^{q-1}}{4} \csc^2\left(\frac{\pi}{4} + \frac{b^q - a^q}{4}\pi\right)$$

$$\frac{\partial f(a,b)}{\partial b} = -\frac{\pi q b^{q-1}}{4} \csc^2\left(\frac{\pi}{4} + \frac{b^q - a^q}{4}\pi\right)$$

When $a \le b$, we have $\frac{\partial f(a,b)}{\partial a} \ge 0$, $\frac{\partial f(a,b)}{\partial b} \le 0$, then $f(a,b)$ is increasing with a and decreasing with b, so when $\eta_{Q'}(x) \le \upsilon_{Q'}(x)$ and $\eta_Q(x) \le \eta_{Q'}(x)$, $\upsilon_Q(x) \ge \upsilon_{Q'}(x)$ satisfied, we have $f(\eta_Q(x), \upsilon_Q(x)) \le f(\eta_{Q'}(x), \upsilon_{Q'}(x))$.
Similarly, it can be easily proven that when $a \ge b$,$\frac{\partial f(a,b)}{\partial a} \le 0$, $\frac{\partial f(a,b)}{\partial b} \ge 0$, then $f(a,b)$ is decreasing with x and increasing with b, so when $\eta_{Q'}(x) \ge \upsilon_{Q'}(x)$ and $\eta_Q(x) \ge \eta_{Q'}(x)$, $\upsilon_Q(x) \le \upsilon_{Q'}(x)$ satisfied, we have $f(\eta_Q(x), \upsilon_Q(x)) \le f(\eta_{Q'}(x), \upsilon_{Q'}(x))$.
Therefore, if $Q \preceq Q'$,then $\frac{1}{n}\sum f(\eta_Q(x), \upsilon_Q(x)) \le \frac{1}{n}\sum f(\eta_{Q'}(x), \upsilon_{Q'}(x))$, *i.e.* $E(Q) \le E(Q')$.

□

Example 2 Consider the information given in Example 1. Now, we apply the proposed EM E of q-ROFS and obtain the results as $E(Q^{1/2}) = 0.6685$, $E(Q) = 0.6601$, $E(Q^2) = 0.5528$, $E(Q^3) = 0.4646$ and $E(Q^3) = 0.3952$. Hence, the proposed EM of of q-ROFS satisfies the relation $E(Q^{1/2}) > E(Q) > E(Q^2) > E(Q^3) > E(Q^4)$. Therefore, the performance of the proposed EM E of q-ROFS is good over the existing EMs E_1, E_2 and E_3 for q-ROFSs as given in [9].

4 A TOPSIS Approach for MADM Based on the Proposed Entropy Measure of q-ROFNs

This section builds a novel MADM approach based on TOPSIS methodology by using the proposed EM E of q-ROFS. Assume the alternatives $O_1, O_2, \ldots, O_m$ and the attributes $C_1, C_2, \ldots, C_n$ with weights $\omega_1, \omega_2, \ldots, \omega_3$ such that $\omega_t > 0$ and $\sum_{t=1}^{n} \omega_t = 1$. DMk assesses the alternative O_k towards the attributes C_t by using the q-ROFNs $\widetilde{\chi}_{kt} = \langle \widetilde{\eta}_{kt}, \widetilde{\upsilon}_{kt} \rangle$, $k = 1, 2, \ldots, m$ and $t = 1, 2, \ldots, n$ to build the DMx $\widetilde{D} = (\widetilde{\chi}_{kt})_{m\times n}$, shown as follows:

$$\widetilde{D} = \begin{array}{c} \\ O_1 \\ O_2 \\ \vdots \\ O_m \end{array}\begin{array}{c} \begin{array}{cccc} C_1 & C_2 & \cdots & C_n \end{array} \\ \begin{pmatrix} \widetilde{\chi}_{11} & \widetilde{\chi}_{12} & \cdots & \widetilde{\chi}_{1n} \\ \widetilde{\chi}_{21} & \widetilde{\chi}_{22} & \cdots & \widetilde{\chi}_{2n} \\ \vdots & \vdots & \ddots & \vdots \\ \widetilde{\chi}_{m1} & \widetilde{\chi}_{m2} & \cdots & \widetilde{\chi}_{mn} \end{pmatrix} \end{array}$$

The following are the steps of the proposed MADM method:

Step 1 Transform the DMx $\widetilde{D} = (\widetilde{\chi}_{kt})_{m\times n} = (\langle \widetilde{\eta}_{kt}, \widetilde{\upsilon}_{kt} \rangle)_{m\times n}$ to the NDMx $D = (\chi_{kt})_{m\times n} = \langle \eta_{kt}, \upsilon_{kt} \rangle$ as follows :

$$\chi_{kt} = \begin{cases} \langle \widetilde{\eta}_{kt}, \widetilde{\upsilon}_{kt} \rangle & : \text{if } C_t \text{ is benefit kind attribute} \\ \langle \widetilde{\upsilon}_{kt}, \widetilde{\eta}_{kt} \rangle & : \text{if } C_t \text{ is cost kind attribute} \end{cases}$$

Step 2 Calculate the EM E_t for each attribute C_t as follows:

$$E_t = \frac{1}{m} \sum_{k=1}^{m} \cot\left(\frac{\pi}{4} + \frac{|\eta_{kt}^q - \upsilon_{kt}^q|}{4}\pi\right) \tag{5}$$

and where $q > 0$.

Step 3 Calculate the weight ω_t of the attribute C_t as follows:

$$\omega_t = \frac{1 - E_t}{n - \sum_{t=1}^{n} E_t}, \tag{6}$$

Step 4 Calculate the PIS (S_k^+) and NIS (S_k^+) for each alternative O_k as follows:

$$S_k^+ = \sqrt{\sum_{t=1}^{n} \omega_t (E_{kt})^2}, \tag{7}$$

$$S_k^- = \sqrt{\sum_{t=1}^{n} \omega_t (1 - E_{kt})^2}, \tag{8}$$

where $E_{kt} = \cot(\frac{\pi}{4} + \frac{|\eta_{kt}^q - \upsilon_{kt}^q|}{4}\pi)$ and $q > 0$.

Step 5 Compute the closeness coefficient $Z(O_k)$ for each alternative O_k by using PIS (S_k^+) and NIS (S_k^+) as follows:

$$Z(O_k) = \frac{S_k^-}{S_k^+ + S_k^-} \tag{9}$$

Step 6 Calculate the ranking order (RO) of the alternatives $O_1, O_2, \ldots, O_m$ corresponding to the descending value of $Z(O_k)$.

5 Illustrative Example

A real-life example from Liu et al. [7] has been used to exemplify the above-mentioned approach:

Example 3 A bicycle manufacturer of the USA wants to enter the Asian market. For this, four alternative locations O_1, O_2, O_3 and O_4 in various countries are being considered. The five attributes (C_1 : "$market$"), (C_2 : "$investmentcost$"),(C_3 : "$laborcharacteristics$"), (C_4 : "$infrastructure$") and (C_5 : "$possibilityfor furtherexpansion$") are defined by the firm management to determine the ideal site. By using a q-ROFN $\widetilde{\chi}_{kt}$, the DMk assesses the alternative O_k $(k = 1, 2, 3, 4)$ towards attribute C_t $(t = 1, 2, 3, 4, 5)$ to build the DMx $\widetilde{R} = (\chi_{kt})_{4\times 5} = (\langle \widetilde{\eta}_{kt}, \widetilde{\upsilon}_{kt} \rangle)_{4\times 5}$ shown as follows:

$$\widetilde{D} = \begin{array}{c} \\ O_1 \\ O_2 \\ O_3 \\ O_4 \end{array}\!\!\begin{array}{c} \begin{array}{ccccc} C_1 & C_2 & C_3 & C_4 & C_5 \end{array} \\ \begin{pmatrix} \langle 0.8, 0.1\rangle & \langle 0.6, 0.2\rangle & \langle 0.7, 0.4\rangle & \langle 0.7, 0.6\rangle & \langle 0.6, 0.3\rangle \\ \langle 0.7, 0.5\rangle & \langle 0.7, 0.2\rangle & \langle 0.6, 0.4\rangle & \langle 0.6, 0.3\rangle & \langle 0.9, 0.5\rangle \\ \langle 0.7, 0.2\rangle & \langle 0.5, 0.4\rangle & \langle 0.6, 0.2\rangle & \langle 0.5, 0.4\rangle & \langle 0.5, 0.5\rangle \\ \langle 0.6, 0.3\rangle & \langle 0.7, 0.5\rangle & \langle 0.8, 0.3\rangle & \langle 0.8, 0.2\rangle & \langle 0.9, 0.2\rangle \end{pmatrix} \end{array}$$

To tackle this MADM issue, we apply the proposed MADM approach as follows:

Step 1 Since the attribute C_2 is of the cost type and the remaining all attributes are of the benefit type. So, construct the NDMx $D = (\chi_{kt})_{4\times 5}$ using Eq. (5) as follows:

$$D = \begin{array}{c} \\ O_1 \\ O_2 \\ O_3 \\ O_4 \end{array}\!\!\begin{array}{c} \begin{array}{ccccc} C_1 & C_2 & C_3 & C_4 & C_5 \end{array} \\ \begin{pmatrix} \langle 0.8, 0.1\rangle & \langle 0.2, 0.6\rangle & \langle 0.7, 0.4\rangle & \langle 0.7, 0.6\rangle & \langle 0.6, 0.3\rangle \\ \langle 0.7, 0.5\rangle & \langle 0.2, 0.7\rangle & \langle 0.6, 0.4\rangle & \langle 0.6, 0.3\rangle & \langle 0.9, 0.5\rangle \\ \langle 0.7, 0.2\rangle & \langle 0.4, 0.5\rangle & \langle 0.6, 0.2\rangle & \langle 0.5, 0.4\rangle & \langle 0.5, 0.5\rangle \\ \langle 0.6, 0.3\rangle & \langle 0.5, 0.7\rangle & \langle 0.8, 0.3\rangle & \langle 0.8, 0.2\rangle & \langle 0.9, 0.2\rangle \end{pmatrix} \end{array}$$

Step 2 By using Eq. (5), compute the EM E_t for each attribute $C_t$where $q = 3$, $E_1 = 0.6062$, $E_2 = 0.7266$,$E_3 = 0.6416$,$E_4 = 0.7192$ and $E_5 = 0.5710$.

Step 3 By using Eq. (6), compute the weight ω_t of the attribute C_t where $\omega_1 = 0.2269$, $\omega_2 = 0.1576$, $\omega_3 = 0.2055$, $\omega_4 = 0.1618$ and $\omega_5 = 0.2472$.

Step 4 By using Eqs. (7) and (8), calculate the PIS (S_k^+) and NIS (S_k^+), respectively, for each alternative O_k where $S_1^+ = 0.2990$, $S_2^+ = 0.2843$, $S_3^+ = 0.3771$, $S_4^+ = 0.2362$, $S_1^- = 0.1759$, $S_2^- = 0.2020$, $S_3^- = 0.1145$ and $S_4^- = 0.2561$.

Step 5 By using Eq. (9), calculate the closeness coefficient $Z(O_k)$ for each alternative O_k where $Z(O_1) = 0.3704$, $Z(O_2) = 0.4154$, $Z(O_3) = 0.2330$ and $Z(O_4) = 0.5203$.

Step 6 Since $Z(O_4) > Z(O_2) > Z(O_1) > Z(O_3)$, alternative's RO is $O_4 \succ O_2 \succ O_1 \succ O_3$, and hence, O_4 is the best alternative.

The alternative's ROs acquired by various MADM approaches are contrasted in the accompanying section. From Table 2, MADM method provided by Liu et al. [7] gives

Table 2 For Example 5, a comparison of alternative's RO using various approaches

MADM approaches	ROs
MADM method [7]	$O_4 \succ O_1 \succ O_3 \succ O_2$
MADM method [11]	$O_4 \succ O_1 \succ O_2 \succ O_3$
Proposed MADM method	$O_4 \succ O_2 \succ O_1 \succ O_3$

the RO $O_4 \succ O_1 \succ O_3 \succ O_2$; MADM method provided by[11] gives the RO $O_4 \succ O_1 \succ O_2 \succ O_3$, and the proposed MADM system gets the RO $O_4 \succ O_2 \succ O_1 \succ O_3$. Hence, the best alternative location is O_4 for this task with each technique.

6 Conclusion

In this study, to quantify the uncertainty of the q-ROFS, we introduce an entropy measure (EM). It will assist the DMk in determining the attribute weight of the MADM issues and measuring the uncertainty. The importance of attribute weighting in the MADM process is significant, and it has a direct influence on the ROs of alternatives. The superiority of the proposed EM over the existing EM has been presented by taking a numerical example. Some properties of the proposed EM have also been discussed. Afterwards, according to the proposed EM, a TOPSIS method under the q-ROFNs environment has been developed for solving the MADM issues. At last, a real-life case of MADM issue is given to demonstrate the effectiveness of the developed DM approach, and obtained results are compared with the comes about of other existing approaches. From computed results and the comparative study, the suggested TOPSIS framework of MADM gives a straightforward and realistic solution to handle MADM challenges. In future, we will utilize the proposed EM of q-ROFS in group decision-making problems, medical diagnosis problems, pattern recognition problems, etc.

References

1. Atanassov KT (1986) Intuitionistic fuzzy sets. Fuzzy Sets Syst 20(1):87–96
2. Garg H (2021) A new possibility degree measure for interval-valued q-rung orthopair fuzzy sets in decision-making. Int J Intell Syst 36(1):526–557
3. Khan MJ, Ali MI, Kumam P (2021) A new ranking technique for q-rung orthopair fuzzy values. Int J Intell Syst 36(1):558–592
4. Khan MJ, Kumam P, Shutaywi M (2021) Knowledge measure for the q-rung orthopair fuzzy sets. Int J Intell Syst 36(2):628–655
5. Kumar K, Chen SM (2021) Multiattribute decision making based on the improved intuitionistic fuzzy Einstein weighted averaging operator of intuitionistic fuzzy values. Inf Sci 568:369–383
6. Liu P, Wang P (2018) Some q-rung orthopair fuzzy aggregation operators and their applications to multiple-attribute decision making. Int J Intell Syst 33(2):259–280

7. Liu Z, Liu P, Liang X (2018) Multiple attribute decision-making method for dealing with heterogeneous relationship among attributes and unknown attribute weight information under q-rung orthopair fuzzy environment. Int J Intell Syst 33(9):1900–1928
8. Mishra AR, Rani P, Pardasani KR, Mardani A, Stević Pamučar D (2020) A novel entropy and divergence measures with multi-criteria service quality assessment using interval-valued intuitionistic fuzzy TODIM method. Soft Comput 24(15):11641–11661
9. Peng X, Liu L (2019) Information measures for q-rung orthopair fuzzy sets. Int J Intell Syst 34(8):1795–1834
10. Riaz M, Athar Farid HM, Kalsoom H, Pamuvcar D, Chu YM (2020) A robust q-rung orthopair fuzzy Einstein prioritized aggregation operators with application towards MCGDM. Symmetry 12(6):1058
11. Riaz M, Salabun W, Farid HMA, Ali N, Watrbski J (2020) A robust q-rung orthopair fuzzy information aggregation using Einstein operations with application to sustainable energy planning decision management. Energies 13(9):2155
12. Wang J, Wei G, Wei C, Wei Y (2020) MABAC method for multiple attribute group decision making under q-rung orthopair fuzzy environment. Defence Technol 16(1):208–216
13. Wei G, Gao H, Wei Y (2018) Some q-rung orthopair fuzzy heronian mean operators in multiple attribute decision making. Int J Intell Syst 33(7):1426–1458
14. Yager RR (2017) Generalized orthopair fuzzy sets. IEEE Trans Fuzzy Syst 25(5):1222–1230
15. Zadeh LA (1965) Fuzzy sets. Inf Cont 8(3):338–353

A Novel Score Function for Picture Fuzzy Numbers and Its Based Entropy Method to Multiple Attribute Decision-Making

Sandeep Kumar and Reshu Tyagi

Abstract A picture fuzzy set (PFS) is a comprehensive extension of a fuzzy set which permits to express the vagueness inherent in decision-making information. In this paper, we consider a class of multiple attribute decision-making (MADM) problems under the PFS framework, in which attribute weights are completely unknown. In such problems, the attribute values are given in the form of picture fuzzy numbers (PFNs). In the development of a method to solve such MADM problems, firstly, a novel score function is introduced in order to compare the PFNs with a brief study of related properties. This score function defeats the drawbacks of the existing score functions. After that, a new algorithm is proposed to solve a picture fuzzy MADM problem by using the proposed score function, the picture fuzzy weighted geometric (PFWG) operator and the picture fuzzy hybrid geometric (PFHG) operator. An illustrative example is given to show the applicability of the proposed algorithm.

Keywords Picture fuzzy number · Multiple attribute decision-making · PFHG operator

1 Introduction

In many research fields, such as engineering, economics and management, decision-making issues play an important role. Traditionally, the ratings of the alternatives have been expected to be in the form of crisp numbers. But, in realistic situations, the fuzziness and uncertainties presented in the data cannot be expressed with the help of crisp numbers. To address and characterize the fuzziness presented in the agent opinion in terms of establishing the membership degree (*msd*) and non-membership degree (*nmsd*), the intuitionistic fuzzy set (IFS) [1] and its extension [2] provide a more trustworthy tool in comparative of the fuzzy set (FS) [3]. Until now, lots

S. Kumar (✉) · R. Tyagi
Department of Mathematics, Ch. Charan Singh University, Campus Meerut, Uttar Pradesh 250004, India
e-mail: drsandeepmath@gmail.com

M. Thakur et al. (eds.), *Soft Computing for Problem Solving*, Lecture Notes in Networks and Systems 547, https://doi.org/10.1007/978-981-19-6525-8_55

of researchers have used the IFS theory for modeling and solving many research problems [4–15].

The assignment of the *msd* and the *nmsd* to each element is a distinguishing feature of a given IFS. Sometimes, it is not possible in many real life based situations that the IFS theory can effectively and suitably express to the willingness and propensity of the decision maker. For example, a TV channel selects a finite population in order to collect the votes on an issue like imposing the lockdown across the country (India) during the pandemic COVID-19. In such a case, these votes may be categorized as: 'yes'; 'abstain'; 'no'; and 'refusal.' These votes cannot be precisely expressed in IFSs. Neither a FS nor an IFS can handle this problem.

In such a case, the PFS theory [16, 17] provides a new idea for handling such type of vagueness. This theory is defined by three functions that indicate *msd*, the neutral membership degree (*umsd*) and *nmsd*.

Singh [18] presented a correlation coefficient in the PFS theory for clustering analysis. Son [19] developed a picture fuzzy distance measure and used it to find out the solution of a clustering problem. By using the concept of the picture fuzzy (PF) weighted cross-entropy, Wei [20] proposed a solution method to an MADM problem. Garg [21] developed a series of aggregation operators such as picture fuzzy weighted averaging (PFWA); picture fuzzy ordered weighted averaging (PFOWA) and picture fuzzy hybrid averaging (PFHA) operators in the PFS theory for aggregating PFNs. Wang et al. [22] developed several geometric aggregation operators on PFNs and introduced a new algorithm to solve decision-making problems. In terms of choosing suppliers, Thao [23] investigated the entropy measures for PFSs and proposed similarity measures for multi-criteria decision-making (MCDM) problems. Tian et al. [24] gave a weighted PF power Choquet ordered geometric operator and a weighted PF power Shapley Choquet ordered geometric operator for a PF MCDM method.

After studying the above research works, we have faced the following issues:

(i) For expressing the fuzziness in a better way, there is a need to use a flexible extension of a fuzzy set, i.e., a PFS in the modeling of MADM problems.
(ii) In order to measure a PFS information accurately in a decision process, the decision maker requires a suitable score function which satisfies the basic properties of ordering.
(iii) To compute accurate and reasonable attribute weights in a PF MADM problem.

Motivated by the above-explained issues, therefore, this study is developed into two folds. Firstly, we introduced a novel score function for comparing two or more PFNs. This score function has removed the shortcomings presented in the existing score functions. Secondly, based on the proposed score function, we developed an entropy method for solving an MADM problem, in which attribute weights are completely unknown.

The decoration of this research paper is as follows. In Sect. 2, some basic concepts concerned with the PFS theory have been reviewed. In Sect. 3, the shortcomings of the existing score functions for PFNs are discussed. In Sect. 4, a novel score function is introduced with its basic properties. In Sect. 5, we developed an algorithm for an MADM problem under the PF environment. In Sect. 6, an example is given to demonstrate the developed algorithm. The conclusion and the future scope are presented in Sect. 7.

2 Preliminaries

In this section, we provide some basic concepts related to PFSs and PFNs which are relevant to the present study.

Definition 1 [16, 17] Suppose $I = \{x_1, x_2, \ldots, x_n\}$ is a finite arbitrary set. Then, a subset $\wp \subset I$ is known as PFS if it is defined as follows:

$$\wp = \{\langle x, \psi(x), \kappa(x), \nu(x)\rangle, x \in I\},$$

where the functions $\psi(x), \kappa(x)$ and $\nu(x)$ represent the msd, umsd and nmsd, respectively, for a given x in PFS $\wp$. All these functions are defined from set I to the interval $[0, 1]$. Besides, $\psi(x), \kappa(x)$ and $\nu(x)$ satisfy the following condition

$$0 \leq \psi(x) + \kappa(x) + \nu(x) \leq 1, \forall x \in I.$$

It is observed that a PFS can be reduced into an IFS if $\kappa(x) = 0, \forall x \in I$.

Definition 2 [16, 17] Let $\wp = \{\langle x, \psi(x), \kappa(x), \nu(x)\rangle, x \in I\}$ be a PFS, then the degree of refusal membership is denoted by $r(x)$ and it is given as follows:

$$r(x) = 1 - (\psi(x) + \kappa(x) + \nu(x)), \text{ for } x \in I.$$

Definition 3 [16, 17] A number of the form $g = \langle \psi, \kappa, \nu, r\rangle$ is called PFN, where $\psi \in [0, 1], \kappa \in [0, 1], \nu \in [0, 1] and r \in [0, 1]$ with $\psi + \kappa + \nu + r = 1$, for every $x \in I$. For convenience, a PFN g can be represented as $g = \langle \psi, \kappa, \nu\rangle$.

Definition 4 [22, 25] The arithmetic operations for PFNs $g = \langle \psi, \kappa, \nu\rangle$, $g_1 = \langle \psi_1, \kappa_1, \nu_1\rangle$ and $g_2 = \langle \psi_2, \kappa_2, \nu_2\rangle$ are defined as follows:

(i) $g_1.g_2 = \langle (\psi_1 + \kappa_1)(\psi_2 + \kappa_2) - \kappa_1\kappa_2, \kappa_1\kappa_2, 1 - (1 - \nu_1)(1 - \nu_2)\rangle$,
(ii) $g^\gamma = \langle (\psi + \kappa)^\gamma - \kappa^\gamma, \kappa^\gamma, 1 - (1 - \nu)^\gamma\rangle, \gamma > 0$,
(iii) $g^c = \langle \nu, \kappa, \psi\rangle$, g^c is the complement of g.

Definition 5 [21, 22, 26, 27] Let $g = \langle \psi, \kappa, \nu\rangle$ be any PFN. Then, the score and accuracy functions given by different researchers are described as follows:

(i) The score and accuracy functions defined in [22] are given as

$$S_W(g) = \psi - \nu, \quad S_W(g) \in [-1, 1],$$
$$A(g) = \psi + \kappa + \nu, \quad A(g) \in [0, 1].$$

(ii) The score and accuracy functions developed in [26] are given as

$$S_C(g) = \frac{1+\psi-\nu}{2}, \quad S_C(g) \in [0, 1],$$
$$A(g) = \psi + \kappa + \nu, \quad A(g) \in [0, 1].$$

(iii) The score and accuracy functions investigated in [21] are given as

$$S_G(g) = \psi - \kappa - \nu, \quad S_G(g) \in [-1, 1],$$
$$A(g) = \psi + \kappa + \nu, \quad A(g) \in [0, 1].$$

(iv) The score and accuracy functions in [27] are given as

$$S_F(g) = \frac{1}{2}\left(1 + 2\psi - \nu - (\kappa/2)\right),$$
$$A(g) = \psi + \kappa + \nu, \quad A(g) \in [0, 1].$$

Definition 6 [22] Let g_1 and g_2 be two PFNs, then

(i) If $S(g_1) < S(g_2)$, then $g_2 \succ g_1$,
(ii) If $S(g_1) > S(g_2)$, then $g_1 \succ g_2$,
(iii) If $S(g_1) = S(g_2)$, then

(i)′ If $A(g_1) < A(g_2)$, then $g_2 \succ g_1$,
(ii)′ If $A(g_1) > A(g_2)$, then $g_1 \succ g_2$,
(iii)′ If $A(g_1) = A(g_2)$, then g_1 is equivalent to g_2.

Here, the symbol '$\succ$' represents the relation 'prefers to.'

Definition 7 [22] Suppose $g_i = \langle \psi_i, \kappa_i, \nu_i \rangle (i = 1, 2, \ldots, m)$ are m PFNs. In order to aggregate these PFNs, the PFWG operator and the PFHG operator are, respectively, defined as follows:

$$\mathrm{PFWG}_{\Omega}(g_1, g_2, \ldots, g_m) = \prod_{i=1}^{m} g_i^{\Omega_i},$$

where $\Omega = (\Omega_1, \Omega_2, \ldots, \Omega_m)^T$ is the associated weight vector of the PFNs $g_i (i = 1, 2, \ldots, m)$ with $\sum_{i=1}^{m} \Omega_i = 1$, all $\Omega_i \geq 0$. Then, this operator can be written in the following form

$$\mathrm{PFWG}_{\Omega}(g_1, g_2, \ldots, g_m) = \left\langle \prod_{i=1}^{m} (\psi_i + \kappa_i)^{\Omega_i} - \prod_{i=1}^{m} \kappa_i^{\Omega_i}, \prod_{i=1}^{m} \kappa_i^{\Omega_i}, 1 - \prod_{i=1}^{m} (1 - \upsilon_i)^{\Omega_i} \right\rangle$$

and

$$\text{PFHG}_{\Omega}(g_1, g_2, \ldots, g_m) = \prod_{i=1}^{m} \dot{g}_{\rho(i)}^{\Omega_i},$$

where $\Omega = (\Omega_1, \Omega_2, \ldots, \Omega_m)^T$ is the associated weight vector for the given PFNs $g_i (i = 1, 2, \ldots, m)$ with $\sum_{i=1}^{m} \Omega_i = 1$, all $\Omega_i \geq 0$. The number $g_{\rho(i)} = \langle \psi_{\rho(i)}, \kappa_{\rho(i)}, \nu_{\rho(i)} \rangle$ is the largest PFN and the number $\dot{g}_i = \langle \dot{\psi}_i, \dot{\kappa}_i, \dot{\nu}_i \rangle = g_i^{m\Omega_i}$, Ω_i is the weighting for i^{th} picture fuzzy arguments with $\Omega_i \geq 0$, $\sum_{i=1}^{m} \Omega_i = 1$, and m is the balancing coefficient. Then, this operator takes the following form

$$\text{PFHG}_{\Omega}(g_1, g_2, \ldots, g_m) = \left\langle \prod_{i=1}^{m} \left(\dot{\psi}_{\rho(i)} + \dot{\kappa}_{\rho(i)}\right)^{\Omega_i} - \prod_{i=1}^{m} \dot{\kappa}_{\rho(i)}^{\Omega_i}, \prod_{i=1}^{m} \dot{\kappa}_{\rho(i)}^{\Omega_i}, 1 - \prod_{i=1}^{m} \left(1 - \dot{\nu}_{\rho(i)}\right)^{\Omega_i} \right\rangle.$$

3 Shortcomings of the Existing Score Functions Under PFS Environment

The following numerical examples show that the existing score functions is unable to provide either ranking or accurate ranking among alternatives in an MADM problem under PF environment.

Example 1. Suppose the PF information in terms of two PFNs is given as $g_1 = \langle 0.6, 0.1, 0.2 \rangle$ and $g_2 = \langle 0.5, 0.3, 0.1 \rangle$. The values of score and accuracy functions [22] for g_1 and g_2 are given as $S_W(g_1) = S_W(g_2) = 0.4$ and $A(g_1) = A(g_2) = 0.9$. This indicates that these functions are not able to rank PFNs g_1 and g_2. On the other hand, the score values [26] $S_C(g_1) = S_C(g_2) = 0.7$ for PFNs g_1 and g_2. Again, this score function is unable to rank PFNs.

Example 2. In this example, let us consider two PFNs $g_1 = \langle 0.6, 0.1, 0.2 \rangle$ and $g_2 = \langle 0.6, 0.2, 0.1 \rangle$. The values of score and accuracy functions [21] are given by $S_G(g_1) = S_G(g_2) = 0.3$ and $A(g_1) = A(g_2) = 0.9$. These functions are failed to rank PFNs g_1 and g_2.

Example 3. Let $g_1 = \langle 0.6, 0.1, 0.2 \rangle$ and $g_2 = \langle 0.5, 0.1, 0 \rangle$ be two PFNs. The score values [27] for these PFNs are $S_F(g_1) = S_F(g_2) = 0.975$. This function is again unable to provide a reasonable ranking between PFNs g_1 and g_2.

After reviewing the above mentioned numerical examples, we need a further investigation to find a novel score function for PFNs. The objective of this score function is to provide a reasonable and accurate ranking among the PF information in a decision process.

4 A Novel Score Function for PFNs

Let $g = \langle \psi, \kappa, \nu \rangle$ be a PFN. Then, by using the unknown part, the proposed score function for PFN g, in notational form $PS(g)$, is given as follows:

$$PS(g) = \psi - \nu(1 - \psi - \kappa), \quad PS(g) \in [-1, 1].$$

Some properties (Ppys) of the proposed score function are given as
Ppy 1: For any PFN $g = \langle \psi, \kappa, \nu \rangle$, the proposed score function $PS(g) \in [-1, 1]$.
Ppy 2: If PFN $g = \langle 1, 0, 0 \rangle$, then $PS(g) = 1$.
Ppy 3: If PFN $g = \langle 0, 1, 0 \rangle$, then $PS(g) = 0$.
Ppy 4: If PFN $g = \langle 0, 0, 1 \rangle$, then $PS(g) = -1$.
Ppy 5: If $\kappa = 0$ in a PFN $g = \langle \psi, \kappa, \nu \rangle$, then this PFN reduces into an intuitionistic fuzzy number and $PS(g) = \psi - \nu(1 - \psi)$.
Ppy 6: If $\kappa = 0$ and $\nu = 1 - \psi$, then a PFN reduces into a fuzzy number and $PS(g) = \psi - (1 - \psi)^2$.

Let g_1 and g_2 be two PFNs. Now, we define a comparative law on the basis of the proposed score function $PS(.)$ in the following way:

(i) If $PS(g_1) < PS(g_2)$, then $g_2 \succ g_1$,
(ii) If $PS(g_1) > PS(g_2)$, then $g_1 \succ g_2$,
(iii) If $PS(g_1) = PS(g_2)$, then g_1 equivalent to g_2.

By using the proposed score function for PFNs, we shall resolve the examples 1, 2 and 3 in the following manner.

Example 4. For the PFNs g_1 and g_2 given in Example 1, we get $PS(g_1) = 0.54$ and $PS(g_2) = 0.48$. This gives $g_1 \succ g_2$.

Example 5. Using the data provided in Example 2 for PFNs g_1 and g_2, we have $PS(g_1) = 0.54$ and $PS(g_2) = 0.58$. This gives $g_2 \succ g_1$.

Example 6. From Example 3, the numbers g_1 and g_2 are two PFNs. For these numbers, we get $PS(g_1) = 0.54$ and $PS(g_2) = 0.50$. Therefore, $g_1 \succ g_2$.

5 Proposed Algorithm for Solving MADM Problem Under PFS Framework

In this section, we consider an MADM problem in PFS framework, in which attribute weights are not known. Let, two finite sets of alternatives and attributes be $T = \{T_1, T_2, \ldots, T_m\}$ and $R = \{R_1, R_2, \ldots, R_n\}$, respectively. The steps of the proposed algorithm for computing the best alternative are given as follows:
Step 1: Suppose PF decision matrix is given as $D_{m \times n}(g_{ij}) = [\langle \psi_{ij}, \kappa_{ij}, \nu_{ij} \rangle]$, where ψ_{ij}, κ_{ij} and ν_{ij} are the *msd*, *umsd* and *nmsd* of the alternative T_i with respect to

attribute R_j, respectively, $\psi_{ij} \in [0, 1]$, $\kappa_{ij} \in [0, 1]$, $\nu_{ij} \in [0, 1]$ and $0 \leq \psi_{ij} + \kappa_{ij} + \nu_{ij} \leq 1, i = 1, 2, \ldots, m; j = 1, 2, \ldots, n$.

$$D_{m\times n}(g_{ij}) = \begin{bmatrix} \langle\psi_{11}, \kappa_{11}, \nu_{11}\rangle & \langle\psi_{12}, \kappa_{12}, \nu_{12}\rangle & \ldots & \langle\psi_{1n}, \kappa_{1n}, \nu_{1n}\rangle \\ \langle\psi_{21}, \kappa_{21}, \nu_{21}\rangle & \langle\psi_{22}, \kappa_{22}, \nu_{22}\rangle & \ldots & \langle\psi_{2n}, \kappa_{2n}, \nu_{2n}\rangle \\ \vdots & \vdots & \ddots & \vdots \\ \langle\psi_{m1}, \kappa_{m1}, \nu_{m1}\rangle & \langle\psi_{m2}, \kappa_{m2}, \nu_{m2}\rangle & \ldots & \langle\psi_{mn}, \kappa_{mn}, \nu_{mn}\rangle \end{bmatrix}$$

Step 2: Normalize each entry $g_{ij} = \langle\psi_{ij}, \kappa_{ij}, \nu_{ij}\rangle$ of the decision matrix D into a corresponding entry of the matrix $N = [n_{ij}]_{m\times n}$. If there are different types of attribute, namely cost (C) and benefit (B), then the entries of normalized matrix are obtained by using the following formula:

$$n_{ij} = \begin{cases} g_{ij}^c, & j \in B \\ g_{ij}, & j \in C \end{cases},$$

where g_{ij}^c is the complement of g_{ij}.

Step 3: In this step, the score matrix $E(e_{ij}) = [PS(n_{ij})]_{m\times n}$ is obtained by using the proposed score function on the normalized matrix $N = [n_{ij}]_{m\times n}$.

$$E(e_{ij}) = \begin{bmatrix} PS(n_{11}) & PS(n_{12}) & \ldots & PS(n_{1n}) \\ PS(n_{21}) & PS(n_{22}) & \ldots & PS(n_{2n}) \\ \vdots & \vdots & \ddots & \vdots \\ PS(n_{m1}) & PS(n_{m2}) & \ldots & PS(n_{mn}) \end{bmatrix}$$

Step 4: In this step, Shannon entropy method [28] is used to find out the attribute weights $\Omega_j (j = 1, 2, \ldots, n)$ corresponding to alternative T_i by using the formula given as follows:

$$\Omega_j = \frac{|1 - k_j|}{\sum_{j=1}^{n} |1 - k_j|}, \text{ where } k_j = \sum_{i=1}^{m} E\left(e_{ij}\right).$$

Step 5: After determining PF normalized matrix and attribute weights, two aggregation operators (the PFWG and the PFHG operators) are used to obtain the overall aggregated value c_i of the alternative T_i.

Step 6: To calculate the score values $PS(c_i)$ for all i.

Step 7: With the help of the calculated score values $PS(c_i)$ in Step 6, rank all given alternatives. The finest alternative is selected in accordance with descending order of the values $PS(c_i)$ for all i.

6 Numerical Example

The applicability of the present method has been exhibited by the following numerical example.

Assume that a multi-national corporation is developing its financial strategy for the coming year. After a preliminary screening, the four alternatives are determined as: T_1: investment in the Southern Asian markets, T_2: investment in the Eastern Asian markets, T_3: investment in the Northern Asian markets and T_4: investment in the Local markets. The growth analysis (R_1), the risk analysis (R_2), the social political effect analysis (R_3) and the environmental impact analysis (R_4) are the four parts of this examination.

These four possible alternatives T_i $(i = 1, 2, 3, 4)$ are evaluated by using the PF information under the above four attribute. Then, the following decision matrix is given below.

$$D_{4\times4}(g_{ij}) = \begin{bmatrix} \langle 0.2, 0.1, 0.6\rangle & \langle 0.5, 0.3, 0.1\rangle & \langle 0.5, 0.1, 0.3\rangle & \langle 0.4, 0.3, 0.2\rangle \\ \langle 0.1, 0.4, 0.4\rangle & \langle 0.6, 0.3, 0.1\rangle & \langle 0.5, 0.2, 0.2\rangle & \langle 0.2, 0.1, 0.7\rangle \\ \langle 0.3, 0.2, 0.2\rangle & \langle 0.6, 0.2, 0.1\rangle & \langle 0.4, 0.1, 0.3\rangle & \langle 0.3, 0.3, 0.4\rangle \\ \langle 0.3, 0.1, 0.6\rangle & \langle 0.1, 0.2, 0.6\rangle & \langle 0.1, 0.3, 0.5\rangle & \langle 0.2, 0.3, 0.2\rangle \end{bmatrix}.$$

The computed normalized PF decision matrix $N_{4\times4}(n_{ij})$ is

$$N_{4\times4}(n_{ij}) = \begin{bmatrix} \langle 0.6, 0.1, 0.2\rangle & \langle 0.5, 0.3, 0.1\rangle & \langle 0.5, 0.1, 0.3\rangle & \langle 0.2, 0.3, 0.4\rangle \\ \langle 0.4, 0.4, 0.1\rangle & \langle 0.6, 0.3, 0.1\rangle & \langle 0.5, 0.2, 0.2\rangle & \langle 0.7, 0.1, 0.2\rangle \\ \langle 0.2, 0.2, 0.3\rangle & \langle 0.6, 0.2, 0.1\rangle & \langle 0.4, 0.1, 0.3\rangle & \langle 0.4, 0.3, 0.3\rangle \\ \langle 0.6, 0.1, 0.3\rangle & \langle 0.1, 0.2, 0.6\rangle & \langle 0.1, 0.3, 0.5\rangle & \langle 0.2, 0.3, 0.2\rangle \end{bmatrix}.$$

Now, the proposed score function is used to obtain the score matrix $E_{4\times4}(e_{ij}) = [PS(n_{ij})]_{4\times4}$, and the calculated score matrix is given as follows:

$$E_{4\times4}(e_{ij}) = \begin{bmatrix} 0.54 & 0.48 & 0.38 & 0.00 \\ 0.38 & 0.59 & 0.44 & 0.66 \\ 0.02 & 0.58 & 0.25 & 0.31 \\ 0.51 & -0.32 & -0.20 & 0.10 \end{bmatrix}.$$

From this matrix, the degree of divergence for each attribute is given as follows:

$$|1 - k_1| = 0.45, |1 - k_2| = 0.33, |1 - k_3| = 0.13 \text{ and } |1 - k_4| = 0.07.$$

Using *Step 4*, the calculated attribute weights are as follows:

$$\Omega_1 = 0.4592, \Omega_2 = 0.3367, \Omega_3 = 0.1327, \Omega_4 = 0.0714.$$

(a) Ranking of alternatives by using the PFWG operator

With the calculated attribute weights, the PFWG operator is used to find out the following overall aggregated value of alternative T_i corresponding to the different attributes R_j,

$$c_1 = \langle 0.5437, 0.1566, 0.1988\rangle;\ c_2 = \langle 0.5178, 0.3000, 0.1214\rangle;$$
$$c_3 = \langle 0.3538, 0.1878, 0.2381\rangle;\ c_4 = \langle 0.3190, 0.1580, 0.4403\rangle.$$

The score values of the overall values c_i $(i = 1, 2, 3, 4)$ are given as

$$PS(c_1) = 0.4841,\ PS(c_2) = 0.4957,\ PS(c_3) = 0.2447 \text{ and } PS(c_4) = 0.0887.$$

By utilizing the above-calculated score values, the ranking order of the four alternatives is

$$T_2 \succ T_1 \succ T_3 \succ T_4.$$

Hence, the finest alternative is T_2.

(b) Ranking of alternatives by using the PFHG operator

With the calculated attribute weights, the PFHG operator is used to compute the following overall value of alternative T_i corresponding to the different attributes R_j,

$$c_1 = \langle 0.5435, 0.1752, 0.1761\rangle;\ c_2 = \langle 0.5786, 0.2718, 0.1265\rangle;$$
$$c_3 = \langle 0.5020, 0.1895, 0.1724\rangle;\ c_4 = \langle 0.4776, 0.1007, 0.3682\rangle.$$

The score values of the values c_i $(i = 1, 2, 3, 4)$ are given as

$$PS(c_1) = 0.4940,\ PS(c_2) = 0.5597,\ PS(c_3) = 0.4488 \text{ and } PS(c_4) = 0.3223.$$

By utilizing the above calculated score values, the ranking order of the four alternatives is

$$T_2 \succ T_1 \succ T_3 \succ T_4.$$

Hence, the finest alternative out of four alternatives is T_2.

6.1 A Comparative Study with the Existing Methods

(i) Comparison with Garg's first method [21]

In the above numerical example, Garg's first method is used with the PFWA operator. In this method, the weights (for attributes) are assumed as: $\Omega = (0.2575, 0.3316, 0.1292, 0.2817)^T$. From this method, the ranking of alternatives is $T_2 \succ T_3 \succ T_1 \succ T_4$.

However, this ranking order is different from that of obtained by our proposed method. But, the finest alternative is the same in both the ranking order. This shows the reliability of our method.

(ii) Comparison with Garg's second method [21]

In the above numerical example, Garg's second method is used with the PFHA operator. In this method, the weights (for attributes) are assumed as: $\Omega = (0.2575, 0.3316, 0.1292, 0.2817)^T$. From this method, the ranking of alternatives is $T_2 \succ T_1 \succ T_3 \succ T_4$.

This ranking order is similar to the order, which is obtained by the proposed method. However, the computational procedure is completely different to the Garg's second method.

7 Conclusion

In the present research, we have considered an MADM problem with completely unknown attribute weights under the PFS framework. In this work, firstly, a novel score function is developed for PFNs and some properties of this function are also discussed. Over the established score functions, the superiority of the proposed score function is given by a series of the numerical examples. Secondly, an entropy method is used for solving PF MADM problems, in which weights (for attributes) are not known. In this method, two types of aggregation operators, namely the PFWG and PFHG operators, have been utilized in order to fuse the PF information. The implementation process of the proposed method is shown by a numerical example.

In the future, the authors aim to develop the present study into the interval-valued picture fuzzy environment.

Acknowledgements The first author thanks Ch. Charan Singh University, Meerut-250004, Uttar Pradesh, India for funding this research under the University Grant Scheme via Ref. No. DEV/URGS/2022-23/15.

References

1. Atanassov KT (1986) Intuitionistic fuzzy sets. Fuzzy Sets Syst 20(1):87–96
2. Atanassov K, Gargov G (1989) Interval-valued intuitionistic fuzzy sets. Fuzzy Sets Syst 31(3):343–349
3. Zadeh LA (1965) Fuzzy sets. Inf Control 8(3):338–353
4. Pandey D, Kumar S (2010) Modified approach to multi-objective matrix game with vague payoffs. J Int Acad Phys Sci 14(2):149–157
5. Pandey D, Kumar S (2011) Fuzzy optimization of primal-dual pair using piecewise linear membership functions. Yugoslav J Oper Res 21(2):97–106
6. Kumar S (2018) To solve matrix games with fuzzy goals using piecewise linear membership function. Proc Jangjeon Math Soc 21(4):627–636

7. Kumar S (2016) Max-min solution approach for multi-objective matrix game with fuzzy goals. Yugoslav J Oper Res 26(1):51–60
8. Pandey D, Kumar S (2011) Fuzzy multi-objective fractional goal programming using tolerance. Int J Math Sci Eng Appl 5(1):175–187
9. Kumar S (2020) Duality results in fuzzy linear programming problems based on the concept of goal programming. Int J Syst Sci Oper Logist 7(2):206–216
10. Kumar S (2021) Piecewise linear programming approach to solve multi-objective matrix games with I-fuzzy goals. J Control Decis 8(1):1–13
11. Kumar S, Kumar M (2021) A game theoretic approach to solve multiple group decision making problems with interval-valued intuitionistic fuzzy decision matrices. Int J Manag Sci Eng Manag 16(1):34–42
12. Kumar S, Kumar M (2021) A new order function for interval-valued intuitionistic fuzzy numbers and its application in group decision making. Fuzzy Inf Eng 13(1):111–126
13. Kumar S, Kumar M (2021) A new approach to solve group decision making problems with attribute values and attribute weights represented by interval-valued intuitionistic fuzzy numbers. Int J Appl Comput Math 7(4):163
14. Pandey D, Kumar S (2011) Fuzzy programming approach to solve multi-objective transportation problem. In: Proceedings of international conference on soft computing for problem solving. The Institution of Engineers (India) Roorkee Local Center Indian Institute of Technology Roorkee Campus Roorkee-247667, Uttarakhand, India, December 20–22, pp 525–533
15. Kumar S (2017) The relationship between intuitionistic fuzzy programming and goal programming. In: Proceedings of 6th international conference on soft computing for problem solving. School of Mathematics, Thapar University, Patiaila, Punjab (India), December 23–24, pp 220–229
16. Cuong BC (2013) Picture fuzzy sets first results, part 1. In: Seminar neurofuzzy systems with applications. Preprint 03/2013. Inst Math Hanoi
17. Cuong BC (2013) Picture fuzzy sets first results, part 2. In: Seminar neurofuzzy systems with applications. Inst Math, Hanoi
18. Singh P (2015) Correlation coefficients for picture fuzzy sets. J Intell Fuzzy Syst 28(2):591–604
19. Son LH (2016) Generalized picture distance measure and applications to picture fuzzy clustering. Appl Soft Comput 46:284–295
20. Wei GW (2016) Picture fuzzy cross-entropy for multiple attribute decision making problems. J Bus Econ Manag 17(4):491–502
21. Garg H (2017) Some picture fuzzy aggregation operators and their applications to multi-criteria decision making. Arab J Sci Eng 42(8):5275–5290
22. Wang C, Zhou X, Tu H, Tau S (2017) Some geometric operators based on picture fuzzy sets and their application in multiple decision making. Ital J Pure Appl Math 37:477–492
23. Thao NX (2019) Similarity measures of picture fuzzy sets based on entropy and their application in MCDM. Pattern Anal Appl 23:1–11
24. Tian C, Peng J, Zhang S, Zhang W, Wang J (2019) Weighted picture fuzzy aggregation operators and their applications to multi-criteria decision-making problems. Comput Ind Eng 137(1):106037
25. Cuong BC (2014) Picture fuzzy sets. J Comput Sci Cyb 30(4):409–420
26. Chen SM, Tan JM (1994) Handling multi-criteria fuzzy decision making problems based on vague set theory. Fuzzy Sets Syst 67:163–172
27. Gundogdu FK (2020) Picture fuzzy linear assignment method and its application to selection of pest house location. In: Intelligent and fuzzy techniques: smart and innovative solutions. Springer International Publishing, Heidelberg, pp 101–109
28. Shannon CE (1948) A mathematical theory of communication. Bell Syst Tech J 27(3):379–423

Author Index

M. Thakur et al. (eds.), *Soft Computing for Problem Solving*, Lecture Notes in Networks
and Systems 547, https://doi.org/10.1007/978-981-19-6525-8